The Industrial Electronics Handbook

SECOND EDITION

CONTROL AND MECHATRONICS

The Industrial Electronics Handbook
SECOND EDITION

FUNDAMENTALS OF INDUSTRIAL ELECTRONICS

POWER ELECTRONICS AND MOTOR DRIVES

CONTROL AND MECHATRONICS

INDUSTRIAL COMMUNICATION SYSTEMS

INTELLIGENT SYSTEMS

The Electrical Engineering Handbook Series

Series Editor
Richard C. Dorf
University of California, Davis

Titles Included in the Series

The Avionics Handbook, Second Edition, Cary R. Spitzer
The Biomedical Engineering Handbook, Third Edition, Joseph D. Bronzino
The Circuits and Filters Handbook, Third Edition, Wai-Kai Chen
The Communications Handbook, Second Edition, Jerry Gibson
The Computer Engineering Handbook, Vojin G. Oklobdzija
The Control Handbook, Second Edition, William S. Levine
CRC Handbook of Engineering Tables, Richard C. Dorf
Digital Avionics Handbook, Second Edition, Cary R. Spitzer
The Digital Signal Processing Handbook, Vijay K. Madisetti and Douglas Williams
The Electric Power Engineering Handbook, Second Edition, Leonard L. Grigsby
The Electrical Engineering Handbook, Third Edition, Richard C. Dorf
The Electronics Handbook, Second Edition, Jerry C. Whitaker
The Engineering Handbook, Third Edition, Richard C. Dorf
The Handbook of Ad Hoc Wireless Networks, Mohammad Ilyas
The Handbook of Formulas and Tables for Signal Processing, Alexander D. Poularikas
Handbook of Nanoscience, Engineering, and Technology, Second Edition, William A. Goddard, III, Donald W. Brenner, Sergey E. Lyshevski, and Gerald J. Iafrate
The Handbook of Optical Communication Networks, Mohammad Ilyas and Hussein T. Mouftah
The Industrial Electronics Handbook, Second Edition, Bogdan M. Wilamowski and J. David Irwin
The Measurement, Instrumentation, and Sensors Handbook, John G. Webster
The Mechanical Systems Design Handbook, Osita D.I. Nwokah and Yidirim Hurmuzlu
The Mechatronics Handbook, Second Edition, Robert H. Bishop
The Mobile Communications Handbook, Second Edition, Jerry D. Gibson
The Ocean Engineering Handbook, Ferial El-Hawary
The RF and Microwave Handbook, Second Edition, Mike Golio
The Technology Management Handbook, Richard C. Dorf
Transforms and Applications Handbook, Third Edition, Alexander D. Poularikas
The VLSI Handbook, Second Edition, Wai-Kai Chen

The Industrial Electronics Handbook

SECOND EDITION

CONTROL AND MECHATRONICS

Edited by

Bogdan M. Wilamowski

J. David Irwin

CRC Press is an imprint of the
Taylor & Francis Group, an **informa** business

CRC Press
Taylor & Francis Group
6000 Broken Sound Parkway NW, Suite 300
Boca Raton, FL 33487-2742

First issued in paperback 2017

ISBN-13: 978-1-4398-0287-8 (hbk)
ISBN-13: 978-1-138-07359-3 (pbk)

Library of Congress Cataloging-in-Publication Data

Control and mechatronics / editors, Bogdan M. Wilamowski and J. David Irwin.
p. cm.
"A CRC title."
Includes bibliographical references and index.
ISBN 978-1-4398-0287-8 (alk. paper)
1. Mechatronics. 2. Electronic control. 3. Servomechanisms. I. Wilamowski, Bogdan M. II. Irwin, J. David. III. Title.

TJ163.12.C67 2010
629.8'043--dc22 2010020062

Visit the Taylor & Francis Web site at
http://www.taylorandfrancis.com

and the CRC Press Web site at
http://www.crcpress.com

Contents

PART IV Modeling and Control

PART V Mechatronics and Robotics

Preface

The field of industrial electronics covers a plethora of problems that must be solved in industrial practice. Electronic systems control many processes that begin with the control of relatively simple devices like electric motors, through more complicated devices such as robots, to the control of entire fabrication processes. An industrial electronics engineer deals with many physical phenomena as well as the sensors that are used to measure them. Thus, the knowledge required by this type of engineer is not only traditional electronics but also specialized electronics, for example, that required for high-power applications. The importance of electronic circuits extends well beyond their use as a final product in that they are also important building blocks in large systems, and thus the industrial electronics engineer must also possess knowledge of the areas of control and mechatronics. Since most fabrication processes are relatively complex, there is an inherent requirement for the use of communication systems that not only link the various elements of the industrial process but are also tailor-made for the specific industrial environment. Finally, the efficient control and supervision of factories require the application of intelligent systems in a hierarchical structure to address the needs of all components employed in the production process. This need is accomplished through the use of intelligent systems such as neural networks, fuzzy systems, and evolutionary methods. The Industrial Electronics Handbook addresses all these issues and does so in five books outlined as follows:

1. *Fundamentals of Industrial Electronics*
2. *Power Electronics and Motor Drives*
3. *Control and Mechatronics*
4. *Industrial Communication Systems*
5. *Intelligent Systems*

The editors have gone to great lengths to ensure that this handbook is as current and up to date as possible. Thus, this book closely follows the current research and trends in applications that can be found in *IEEE Transactions on Industrial Electronics*. This journal is not only one of the largest engineering publications of its type in the world but also one of the most respected. In all technical categories in which this journal is evaluated, its worldwide ranking is either number 1 or number 2. As a result, we believe that this handbook, which is written by the world's leading researchers in the field, presents the global trends in the ubiquitous area commonly known as industrial electronics.

The successful construction of industrial systems requires an understanding of the various aspects of control theory. This area of engineering, like that of power electronics, is also seldom covered in depth in engineering curricula at the undergraduate level. In addition, the fact that much of the research in control theory focuses more on the mathematical aspects of control than on its practical applications makes matters worse. Therefore, the goal of *Control and Mechatronics* is to present many of the concepts of control theory in a manner that facilitates its understanding by practicing engineers or students who would like to learn about the applications of control systems. *Control and Mechatronics* is divided into several parts. Part I is devoted to control system analysis while Part II deals with control system design.

Various techniques used for the analysis and design of control systems are described and compared in these two parts. Part III deals with estimation, observation, and identification and is dedicated to the identification of the objects to be controlled. The importance of this part stems from the fact that in order to efficiently control a system, it must first be clearly identified. In an industrial environment, it is difficult to experiment with production lines. As a result, it is imperative that good models be developed to represent these systems. This modeling aspect of control is covered in Part IV. Many modern factories have more robots than humans. Therefore, the importance of mechatronics and robotics can never be overemphasized. The various aspects of robotics and mechatronics are described in Part V. In all the material that has been presented, the underlying central theme has been to consciously avoid the typical theorems and proofs and use plain English and examples instead, which can be easily understood by students and practicing engineers alike.

For MATLAB® and Simulink® product information, please contact

The MathWorks, Inc.
3 Apple Hill Drive
Natick, MA, 01760-2098 USA
Tel: 508-647-7000
Fax: 508-647-7001
E-mail: info@mathworks.com
Web: www.mathworks.com

Acknowledgments

The editors wish to express their heartfelt thanks to their wives Barbara Wilamowski and Edie Irwin for their help and support during the execution of this project.

Editorial Board

Editors

Bogdan M. Wilamowski received his MS in computer engineering in 1966, his PhD in neural computing in 1970, and Dr. habil. in integrated circuit design in 1977. He received the title of full professor from the president of Poland in 1987. He was the director of the Institute of Electronics (1979–1981) and the chair of the solid state electronics department (1987–1989) at the Technical University of Gdansk, Poland. He was a professor at the University of Wyoming, Laramie, from 1989 to 2000. From 2000 to 2003, he served as an associate director at the Microelectronics Research and Telecommunication Institute, University of Idaho, Moscow, and as a professor in the electrical and computer engineering department and in the computer science department at the same university. Currently, he is the director of ANMSTC—Alabama Nano/Micro Science and Technology Center, Auburn, and an alumna professor in the electrical and computer engineering department at Auburn University, Alabama. Dr. Wilamowski was with the Communication Institute at Tohoku University, Japan (1968–1970), and spent one year at the Semiconductor Research Institute, Sendai, Japan, as a JSPS fellow (1975–1976). He was also a visiting scholar at Auburn University (1981–1982 and 1995–1996) and a visiting professor at the University of Arizona, Tucson (1982–1984). He is the author of 4 textbooks, more than 300 refereed publications, and has 27 patents. He was the principal professor for about 130 graduate students. His main areas of interest include semiconductor devices and sensors, mixed signal and analog signal processing, and computational intelligence.

Dr. Wilamowski was the vice president of the IEEE Computational Intelligence Society (2000–2004) and the president of the IEEE Industrial Electronics Society (2004–2005). He served as an associate editor of *IEEE Transactions on Neural Networks*, *IEEE Transactions on Education*, *IEEE Transactions on Industrial Electronics*, the *Journal of Intelligent and Fuzzy Systems*, the *Journal of Computing*, and the *International Journal of Circuit Systems and IES Newsletter*. He is currently serving as the editor in chief of *IEEE Transactions on Industrial Electronics*.

Professor Wilamowski is an IEEE fellow and an honorary member of the Hungarian Academy of Science. In 2008, he was awarded the Commander Cross of the Order of Merit of the Republic of Poland for outstanding service in the proliferation of international scientific collaborations and for achievements in the areas of microelectronics and computer science by the president of Poland.

J. David Irwin received his BEE from Auburn University, Alabama, in 1961, and his MS and PhD from the University of Tennessee, Knoxville, in 1962 and 1967, respectively.

In 1967, he joined Bell Telephone Laboratories, Inc., Holmdel, New Jersey, as a member of the technical staff and was made a supervisor in 1968. He then joined Auburn University in 1969 as an assistant professor of electrical engineering. He was made an associate professor in 1972, associate professor and head of department in 1973, and professor and head in 1976. He served as head of the Department of Electrical and Computer Engineering from 1973 to 2009. In 1993, he was named Earle C. Williams Eminent Scholar and Head. From 1982 to 1984, he was also head of the Department of Computer Science and Engineering. He is currently the Earle C. Williams Eminent Scholar in Electrical and Computer Engineering at Auburn.

Dr. Irwin has served the Institute of Electrical and Electronic Engineers, Inc. (IEEE) Computer Society as a member of the Education Committee and as education editor of *Computer*. He has served as chairman of the Southeastern Association of Electrical Engineering Department Heads and the National Association of Electrical Engineering Department Heads and is past president of both the IEEE Industrial Electronics Society and the IEEE Education Society. He is a life member of the IEEE Industrial Electronics Society AdCom and has served as a member of the Oceanic Engineering Society AdCom. He served for two years as editor of *IEEE Transactions on Industrial Electronics*. He has served on the Executive Committee of the Southeastern Center for Electrical Engineering Education, Inc., and was president of the organization in 1983–1984. He has served as an IEEE Adhoc Visitor for ABET Accreditation teams. He has also served as a member of the IEEE Educational Activities Board, and was the accreditation coordinator for IEEE in 1989. He has served as a member of numerous IEEE committees, including the Lamme Medal Award Committee, the Fellow Committee, the Nominations and Appointments Committee, and the Admission and Advancement Committee. He has served as a member of the board of directors of IEEE Press. He has also served as a member of the Secretary of the Army's Advisory Panel for ROTC Affairs, as a nominations chairman for the National Electrical Engineering Department Heads Association, and as a member of the IEEE Education Society's McGraw-Hill/Jacob Millman Award Committee. He has also served as chair of the IEEE Undergraduate and Graduate Teaching Award Committee. He is a member of the board of governors and past president of Eta Kappa Nu, the ECE Honor Society. He has been and continues to be involved in the management of several international conferences sponsored by the IEEE Industrial Electronics Society, and served as general cochair for IECON'05.

Dr. Irwin is the author and coauthor of numerous publications, papers, patent applications, and presentations, including *Basic Engineering Circuit Analysis*, 9th edition, published by John Wiley & Sons, which is one among his 16 textbooks. His textbooks, which span a wide spectrum of engineering subjects, have been published by Macmillan Publishing Company, Prentice Hall Book Company, John Wiley & Sons Book Company, and IEEE Press. He is also the editor in chief of a large handbook published by CRC Press, and is the series editor for Industrial Electronics Handbook for CRC Press.

Dr. Irwin is a fellow of the American Association for the Advancement of Science, the American Society for Engineering Education, and the Institute of Electrical and Electronic Engineers. He received an IEEE Centennial Medal in 1984, and was awarded the Bliss Medal by the Society of American Military Engineers in 1985. He received the IEEE Industrial Electronics Society's Anthony J. Hornfeck Outstanding Service Award in 1986, and was named IEEE Region III (U.S. Southeastern Region) Outstanding Engineering Educator in 1989. In 1991, he received a Meritorious Service Citation from the IEEE Educational Activities Board, the 1991 Eugene Mittelmann Achievement Award from the IEEE Industrial Electronics Society, and the 1991 Achievement Award from the IEEE Education Society. In 1992, he was named a Distinguished Auburn Engineer. In 1993, he received the IEEE Education Society's McGraw-Hill/Jacob Millman Award, and in 1998 he was the recipient of the

IEEE Undergraduate Teaching Award. In 2000, he received an IEEE Third Millennium Medal and the IEEE Richard M. Emberson Award. In 2001, he received the American Society for Engineering Education's (ASEE) ECE Distinguished Educator Award. Dr. Irwin was made an honorary professor, Institute for Semiconductors, Chinese Academy of Science, Beijing, China, in 2004. In 2005, he received the IEEE Education Society's Meritorious Service Award, and in 2006, he received the IEEE Educational Activities Board Vice President's Recognition Award. He received the Diplome of Honor from the University of Patras, Greece, in 2007, and in 2008 he was awarded the IEEE IES Technical Committee on Factory Automation's Lifetime Achievement Award. In 2010, he was awarded the electrical and computer engineering department head's Robert M. Janowiak Outstanding Leadership and Service Award. In addition, he is a member of the following honor societies: Sigma Xi, Phi Kappa Phi, Tau Beta Pi, Eta Kappa Nu, Pi Mu Epsilon, and Omicron Delta Kappa.

Contributors

Marcelo H. Ang Jr.
Department of Mechanical Engineering
National University of Singapore
Singapore

Ramón Barber
Department of System Engineering
and Automation
University Carlos III
Madrid, Spain

Victor M. Becerra
School of Systems Engineering
University of Reading
Reading, United Kingdom

Miguel Bernal
Centro Universitavio de los Valles
University of Guadalajara
Jalisco, Mexico

Seta Bogosyan
Electrical and Computer Engineering
Department
University of Alaska, Fairbanks
Fairbanks, Alaska

Igor M. Boiko
Department of Electrical and Computer
Engineering
University of Calgary
Calgary, Alberta, Canada

Alain Bouscayrol
Laboratoire d'Electrotechnique et d'Electronique
de Puissance de Lille
University of Lille 1
Lille, France

A. John Boye
Department of Electrical Engineering
University of Nebraska
and
Neurintel, LLC
Lincoln, Nebraska

Shan Chai
School of Electrical and Computer Engineering
RMIT University
Melbourne, Victoria, Australia

Timothy N. Chang
Department of Electrical and Computer
Engineering
New Jersey Institute of Technology
Newark, New Jersey

J. Alexis De Abreu Garcia
Department of Electrical and Computer
Engineering
The University of Akron
Akron, Ohio

Christopher Edwards
Department of Engineering
University of Leicester
Leicester, United Kingdom

Yong Feng
School of Electrical and Computer Engineering
RMIT University
Melbourne, Victoria, Australia

Emilia Fridman
Department of Electrical Engineering—Systems
Tel Aviv University
Tel Aviv, Israel

Leonid Fridman
Control and Advanced Robotics Department
National Autonomus University of Mexico
Mexico City, Mexico

Metin Gokasan
Control Engineering Department
Istanbul Technical University
Istanbul, Turkey

Thierry Marie Guerra
Laboratory of Industrial and Human Automation Control
Mechanical Engineering and Computer Science
University of Valenciennes and Hainaut-Cambresis
Valenciennes, France

Fuat Gurleyen
Control Engineering Department
Istanbul Technical University
Istanbul, Turkey

Fumio Harashima
Tokyo Metropolitan University
Tokyo, Japan

Joel David Hewlett
Department of Electrical and Computer Engineering
Auburn University
Auburn, Alabama

Guan-Chyun Hsieh
Department of Electrical Engineering
Chung Yuan Christian University
Chung-Li, Taiwan

James C. Hung
Department of Electrical Engineering and Computer Science
The University of Tennessee, Knoxville
Knoxville, Tennessee

John Y. Hung
Department of Electrical and Computer Engineering
Auburn University
Auburn, Alabama

Makoto Iwasaki
Department of Computer Science and Engineering
Nagoya Institute of Technology
Nagoya, Japan

Raymond Jarvis
Intelligent Robotics Research Centre
Monash University
Melbourne, Victoria, Australia

Okyay Kaynak
Department of Electrical and Electronic Engineering
Bogazici University
Istanbul, Turkey

Lindsay Kleeman
Department of Electrical and Computer Systems Engineering
Monash University
Melbourne, Victoria, Australia

Tong Heng Lee
Department of Electrical and Computer Engineering
National University of Singapore
Singapore

Arie Levant
Applied Mathematics Department
Tel Aviv University
Tel Aviv, Israel

Chin F. Lin
Department of Electrical Engineering
National Chung Cheng University
Chia-Yi, Taiwan

Ren C. Luo
Department of Electrical Engineering
National Taiwan University
Taipei, Taiwan

María Malfaz
Department of System Engineering and Automation
University Carlos III
Madrid, Spain

Jorge Angel Davila Montoya
Aeronautic Engineering Department
National Polytechnic Institute
Mexico City, Mexico

Toshiyuki Murakami
Department of System Design Engineering
Keio University
Yokohama, Japan

István Nagy
Department of Automation and Applied Informatics
Budapest University of Technology and Economics
Budapest, Hungary

Roberto Oboe
Department of Management and Engineering
University of Padova
Vicenza, Italy

Kouhei Ohnishi
Department of System Design Engineering
Keio University
Yokohama, Japan

Nejat Olgac
Department of Mechanical Engineering
University of Connecticut
Storrs, Connecticut

Sanjib Kumar Panda
Department of Electrical and Computer Engineering
National University of Singapore
Singapore

Eric Rogers
School of Electronics and Computer Science
University of Southampton
Southampton, United Kingdom

James R. Rowland
Department of Electrical Engineering and Computer Science
University of Kansas
Lawrence, Kansas

Asif Šabanović
Faculty of Engineering and Natural Sciences
Sabanci University
Istanbul, Turkey

Nadira Šabanović-Behlilović
Faculty of Engineering and Natural Sciences
Sabanci University
Istanbul, Turkey

Sanjib Kumar Sahoo
Department of Electrical and Computer Engineering
National University of Singapore
Singapore

Miguel A. Salichs
Systems Engineering and Automation Department
University Carlos III
Madrid, Spain

Naresh K. Sinha
Department of Electrical and Computer Engineering
McMaster University
Hamilton, Ontario, Canada

Rifat Sipahi
Department of Mechanical and Industrial Engineering
Northeastern University
Boston, Massachusetts

Zoltán Sütő
Department of Automation and Applied Informatics
Budapest University of Technology and Economics
Budapest, Hungary

Satoshi Suzuki
Department of Robotics and Mechatronics
School of Science and Technology for Future Life
Tokyo Denki University
Tokyo, Japan

Kok Kiong Tan
Department of Electrical and Computer Engineering
National University of Singapore
Singapore

Chee Pin Tan
School of Engineering
Monash University
Malaysia

Kok Zuea Tang
Department of Electrical and Computer Engineering
National University of Singapore
Singapore, Singapore

Robert J. Veillette
Department of Electrical and Computer Engineering
The University of Akron
Akron, Ohio

Liuping Wang
School of Electrical and Computer Engineering
RMIT University
Melbourne, Victoria, Australia

Changyun Wen
School of Electrical and Electronic Engineering
Nanyang Technological University
Singapore

Bogdan M. Wilamowski
Department of Electrical and Computer Engineering
Auburn University
Auburn, Alabama

Tiantian Xie
Department of Electrical and Computer Engineering
Auburn University
Auburn, Alabama

Jian-Xin Xu
Department of Electrical and Computer Engineering
National University of Singapore
Singapore

Choon-Seng Yee
Department of Mechanical Engineering
National University of Singapore
Singapore

Xinghuo Yu
School of Electrical and Computer Engineering
RMIT University
Melbourne, Victoria, Australia

Jing Zhou
Petroleum Department
International Research Institute of Stavanger
Bergen, Norway

I

Control System Analysis

1

Nonlinear Dynamics

István Nagy
Budapest University of Technology and Economics

Zoltán Sütő
Budapest University of Technology and Economics

1.1 Introduction

A new class of phenomena has recently been discovered three centuries after the publication of *Newton's Principia* (1687) in nonlinear dynamics. New concepts and terms have entered the vocabulary to replace time functions and frequency spectra in describing their behavior, e.g., chaos, bifurcation, fractal, Lyapunov exponent, period doubling, Poincaré map, and strange attractor.

Until recently, chaos and order have been viewed as mutually exclusive. Maxwell's equations govern the electromagnetic phenomena; Newton's laws describe the processes in classical mechanics, etc.

They represent the world of order, which is predictable. Processes were called chaotic when they failed to obey laws and they were unpredictable. Although chaos and order have been believed to be quite distinct faces of our world, there were tricky questions to be answered. For example, knowing all the laws governing our global weather, we are unable to predict it, or a fluid system can turn easily from order to chaos, from laminar flow into turbulent flow.

It came as an unexpected discovery that deterministic systems obeying simple laws belonging undoubtedly to the world of order and believed to be completely predictable can turn chaotic. In mathematics, the study of the quadratic iterator (logistic equation or population growth model) [$x_{n+1} = ax_n(1 - x_n)$, $n = 0, 1, 2, \ldots$] revealed the close link between chaos and order [5]. Another very early example came from the atmospheric science in 1963; Lorenz's three differential equations derived from the Navier–Stokes equations of fluid mechanics describing the thermally induced fluid convection in the atmosphere, Peitgen et al. [9]. They can be viewed as the two principal paradigms of the theory of chaos. One of the first chaotic processes discovered in electronics can be shown in diode resonator consisting of a series connection of a *p–n* junction diode and a 10–100 mH inductor driven by a sine wave generator of 50–100 kHz.

The chaos theory, although admittedly still young, has spread like wild fire into all branches of science. In physics, it has overturned the classic view held since Newton and Laplace, stating that our universe is predictable, governed by simple laws. This illusion has been fueled by the breathtaking advances in computers, promising ever-increasing computing power in information processing. Instead, just the opposite has happened. Researchers on the frontier of natural science have recently proclaimed that this hope is unjustified because a large number of phenomena in nature governed by known simple laws are or can be chaotic. One of their principle properties is their sensitive dependence on initial conditions. Although the most precise measurement indicates that two paths have been launched from the same initial condition, there are always some tiny, impossible-to-measure discrepancies that shift the paths along very different trajectories. The uncertainty in the initial measurements will be amplified and become overwhelming after a short time. Therefore, our ability to predict accurately future developments is unreasonable. The irony of fate is that without the aid of computers, the modern theory of chaos and its geometry, the fractals, could have never been developed.

The theory of nonlinear dynamics is strongly associated with the bifurcation theory. Modifying the parameters of a nonlinear system, the location and the number of equilibrium points can change. The study of these problems is the subject of bifurcation theory.

The existence of well-defined routes leading from order to chaos was the second great discovery and again a big surprise like the first one showing that a deterministic system can be chaotic.

The overview of nonlinear dynamics here has two parts. The main objective in the first part is to summarize the state of the art in the advanced theory of nonlinear dynamical systems. Within the overview, five basic states or scenario of nonlinear systems are treated: equilibrium point, limit cycle, quasi-periodic (frequency-locked) state, routes to chaos, and chaotic state. There will be some words about the connection between the chaotic state and fractal geometry.

In the second part, the application of the theory is illustrated in five examples from the field of power electronics. They are as follows: high-frequency time-sharing inverter, voltage control of a dual-channel resonant DC–DC converter, and three different control methods of the three-phase full bridge voltage source DC–AC/AC–DC converter, a sophisticated hysteresis current control, a discrete-time current control equipped with space vector modulation (SVM) and the direct torque control (DTC) applied widely in AC drives.

1.2 Basics

1.2.1 Classification

The nonlinear dynamical systems have two broad classes: (1) *autonomous systems* and (2) *non-autonomous systems*. Both are described by a set of first-order nonlinear differential equations and can be represented in state (phase) space. The number of differential equations equals the *degree of freedom* (or dimension) of the system, which is the number of independent state variables needed to determine uniquely the dynamical state of the system.

1.2.1.1 Autonomous Systems

There are no external input or forcing functions applied to the system. The set of nonlinear differential equations describing the system is

$$\frac{d\boldsymbol{x}}{dt} = \boldsymbol{v} = \boldsymbol{f}(\boldsymbol{x}, \boldsymbol{\mu}) \tag{1.1}$$

where

$\boldsymbol{x}^T = [x_1, x_2, \ldots, x_N]$ is the state vector
$\boldsymbol{v}^T = [v_1, v_2, \ldots v_N]$ is the velocity vector
$\boldsymbol{f}^T = [f_1, f_2, \ldots f_N]$ is the nonlinear vector function
$\boldsymbol{\mu}^T = [\mu_1, \mu_2, \ldots \mu_N]$ is the parameter vector
T denotes the transpose of a vector
t is the time
N is the dimension of the system

The time t does not appear explicitly.

1.2.1.2 Non-Autonomous Systems

Time-dependent external inputs or forcing functions $\boldsymbol{u}(t)$ are applied to the system. It is a set of nonlinear differential equations:

$$\frac{d\boldsymbol{x}}{dt} = \boldsymbol{v} = \boldsymbol{f}(\boldsymbol{x}, \boldsymbol{u}(t), \boldsymbol{\mu}) \tag{1.2}$$

Time t explicitly appears in $\boldsymbol{u}(t)$. (1.1) and (1.2) can be solved analytically or numerically for a given initial condition $\boldsymbol{x}_0$ and parameter vector $\boldsymbol{\mu}$. The solution describes the state of the system as a function of time. The solution can be visualized in a reference frame where the state variables are the coordinates. It is called the *state space* or *phase space*. At any instant, a point in the state space represents the state of the system. As the time evolves, the state point is moving along a path called *trajectory* or *orbit* starting from the initial condition.

1.2.2 Restrictions

The non-autonomous system can always be transformed to autonomous systems by introducing a new state variable $x_{N+1} = t$. Now the last differential equation in (1.1) is

$$\frac{dx_{N+1}}{dt} = \frac{dt}{dt} = 1 \tag{1.3}$$

The number of dimensions of the state space was enlarged by one by including the time as a state variable. From now on, we confine our consideration to autonomous systems unless it is stated otherwise. By this restriction, there is no loss of generality.

The discussion is confined to real, dissipative systems. As time evolves, the state variables will head for some final point, curve, area, or whatever geometric object in the state space. They are called the *attractor* for the particular system since certain dedicated trajectories will be attracted to these objects. We focus our considerations on the long-term behavior of the system rather than analyzing the start-up and transient processes.

The trajectories are assumed to be bounded as most physical systems are. They cannot go to infinity.

1.2.3 Mathematical Description

Basically, two different concepts applying different approaches are used. The first concept considers all the state variables as continuous quantities applying continuous-time model (CTM). As time evolves, the system behavior is described by a moving point in state space resulting in a trajectory (or flow) obtained by the solution of the set of differential equations (1.1) [or (1.2)]. Figure 1.1 shows the continuous trajectory of function $\phi(\boldsymbol{x}_0, t, \boldsymbol{\mu})$ for three-dimensional system, where $\boldsymbol{x}_0$ is the initial point. The second concept takes samples from the continuously changing state variables and describes the system behavior by discrete vector function applying the *Poincaré concept*. Figure 1.1 shows the way how the samples are taken for a three-dimensional autonomous system. A so-called *Poincaré plane*, in general Poincaré section, is chosen and the intersection points cut by the trajectory are recorded as samples. The selection of the Poincaré plane is not crucial as long as the trajectory cuts the surface transversely. The relation between $\boldsymbol{x}_n$ and $\boldsymbol{x}_{n+1}$, i.e., between subsequent intersection points generated always from the same direction are described by the so-called Poincaré map function (PMF)

$$\boldsymbol{x}_{n+1} = \boldsymbol{P}(\boldsymbol{x}_n) \tag{1.4}$$

Pay attention, the subscript of $\boldsymbol{x}$ denotes the time instant, not a component of vector $\boldsymbol{x}$. The Poincaré section is a hyperplane for systems with dimension higher than three, while it is a point and a straight line for a one- and a two-dimensional system, respectively.

For a non-autonomous system having a periodic forcing function, the samples are taken at a definite phase of the forcing function, e.g., at the beginning of the period. It is a stroboscopic sampling, the state variables for a mechanical system are recorded with a flash lamp fired once in every period of the forcing function [8]. Again, the PMF describes the relation between sampled values of the state variables.

Knowing that the trajectories are the solution of differential equation system, which are unique and deterministic, it implies the existence of a mathematical relation between $\boldsymbol{x}_n$ and $\boldsymbol{x}_{n+1}$, i.e., the existence

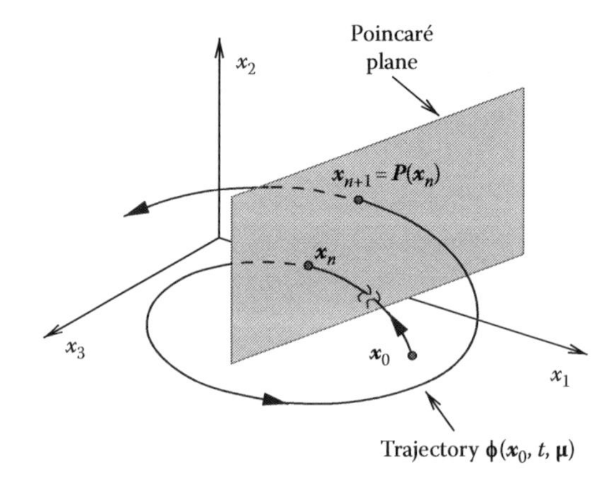

FIGURE 1.1 Trajectory described by $\phi(\boldsymbol{x}_0, t, \boldsymbol{\mu})$. $\boldsymbol{x}_n$, $\boldsymbol{x}_{n+1}$ are the intersection points of the trajectory with the Poincaré plane.

of PMF. However, the discrete-time equation (1.4) can be solved analytically or numerically independent of the differential equation. In the second concept, the discrete-time model (DTM) is used.

The Poincaré section reduces the dimensionality of the system by one and describes it by an iterative, finite-size time step function rather than a differential, infinitesimal time step. On the other hand, PMF retains the essential information of the system dynamics.

Even though the state variables are changing continuously in many systems like in power electronics, they can advantageously be modeled by discrete-iteration function (1.4). In some other cases, the system is inherently discrete, their state variables are not changing continuously like in digital systems or models describing the evolution of population of species.

1.3 Equilibrium Points

1.3.1 Introduction

The nonlinear world is much more colorful than the linear one. The nonlinear systems can be in various states, one of them is the *equilibrium point* (EP). It is a point in the state space approached by the trajectory of a continuous, nonlinear dynamical system as its transients decay. The velocity of state variables $\boldsymbol{v} = \dot{\boldsymbol{x}}$ is zero in the EP:

$$\frac{d\boldsymbol{x}}{dt} = \boldsymbol{v} = \boldsymbol{f}(\boldsymbol{x},\boldsymbol{\mu}) = 0 \tag{1.5}$$

The solution of the nonlinear algebraic function (1.5) can result in more than one EPs. They are $\boldsymbol{x}_1^*, \boldsymbol{x}_2^*, \ldots, \boldsymbol{x}_k^*, \ldots, \boldsymbol{x}_n^*$. The stable EPs are attractors.

1.3.2 Basin of Attraction

The natural consequence of the existence of multiple attractors is the partition of state space into different regions called *basins of attractions*. Any of the initial conditions within a basin of attraction launches a trajectory that is finally attracted by the particular EP belonging to the basin of attraction (Figure 1.2). The border between two neighboring basins of attraction is called *separatrix*. They organize the state space in the sense that a trajectory born in a basin of attraction will never leave it.

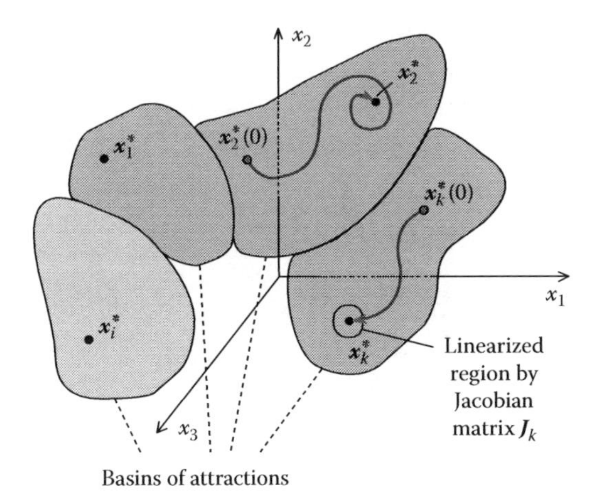

FIGURE 1.2 Basins of attraction and their EP ($N = 3$). x_1, x_2, and x_3 are coordinates of the state space vector $\boldsymbol{x}$. $\boldsymbol{x}_k^*$ is a particular value of state space vector $\boldsymbol{x}$, $\boldsymbol{x}_k^*$ denotes an EP, and $\boldsymbol{x}_k(0)$ is an initial condition in the basin of attraction of $\boldsymbol{x}_k^*$.

1.3.3 Linearizing around the EP

Introducing the small perturbation $\Delta \boldsymbol{x} = \boldsymbol{x} - \boldsymbol{x}_k^*$, (1.1) can be linearized in the close neighborhood of the EP $\boldsymbol{x}_k^*$. Now $\boldsymbol{f}(\boldsymbol{x}_k^* + \Delta \boldsymbol{x}, \boldsymbol{\mu}) = \boldsymbol{f}(\boldsymbol{x}_k^*, \boldsymbol{\mu}) + \boldsymbol{J}_k \Delta \boldsymbol{x} + \cdots$. Neglecting the terms of higher order than $\Delta \boldsymbol{x}$ and substituting it back to (1.1):

$$\frac{d\Delta \boldsymbol{x}}{dt} = \Delta \boldsymbol{v} \cong \boldsymbol{J}_k \Delta \boldsymbol{x} \tag{1.6}$$

where

$$\boldsymbol{J}_k = \begin{bmatrix} \frac{\partial f_1}{\partial x_1} & \frac{\partial f_1}{\partial x_2} & \cdots & \frac{\partial f_1}{\partial x_N} \\ \frac{\partial f_2}{\partial x_1} & \frac{\partial f_2}{\partial x_2} & \cdots & \frac{\partial f_2}{\partial x_N} \\ \vdots & & \ddots & \\ \frac{\partial f_N}{\partial x_1} & \frac{\partial f_N}{\partial x_2} & \cdots & \frac{\partial f_N}{\partial x_N} \end{bmatrix} \tag{1.7}$$

is the Jacobian matrix. The partial derivatives have to be evaluated at $\boldsymbol{x}_k^*$. $\boldsymbol{f}(\boldsymbol{x}^*, \boldsymbol{\mu}) = 0$ was observed in (1.6). $\boldsymbol{J}_k$ is a real, time-independent $N \times N$ matrix. Seeking the solution of (1.7) in the form

$$\Delta \boldsymbol{x} = \boldsymbol{e}_r e^{\lambda t} \tag{1.8}$$

and substituting it back to (1.6):

$$\boldsymbol{J}_k \boldsymbol{e}_r = \lambda \boldsymbol{e}_r \tag{1.9}$$

Its nontrivial solution for λ must satisfy the Nth order polynomial equation

$$\det(\boldsymbol{J}_k - \lambda \boldsymbol{I}) = 0 \tag{1.10}$$

where $\boldsymbol{I}$ is the $N \times N$ identity matrix. From (1.6), (1.8), and (1.9),

$$\Delta \boldsymbol{v} = \boldsymbol{J}_k \boldsymbol{e}_r e^{\lambda t} = \lambda \boldsymbol{e}_r e^{\lambda t} \tag{1.11}$$

Selecting the direction of vector $\Delta \boldsymbol{x}$ in the special way given by (1.8), i.e., its change in time depends only on one constant λ, it has two important consequences:

1. The product $\boldsymbol{J}_k \boldsymbol{e}_r$ ($= \lambda \boldsymbol{e}_r$) only results in the expansion or contraction of $\boldsymbol{e}_r$ by λ. The direction of $\boldsymbol{e}_r$ is not changed.
2. The direction of the perturbation of the velocity vector $\Delta \boldsymbol{v}$ will be the same as that of vector $\boldsymbol{e}_r$.

The direction of $\boldsymbol{e}_r$ is called characteristic direction and $\boldsymbol{e}_r$ is the right-hand side eigenvector of $\boldsymbol{J}_k$ as $\boldsymbol{J}_k$ is multiplied from the right by $\boldsymbol{e}_r$. λ is the eigenvalue (or characteristic exponent) of $\boldsymbol{J}_k$.

We confine our consideration of N distinct roots of (1.10) (multiple roots are excluded). They are $\lambda_1, \lambda_2, \lambda_m, \ldots, \lambda_N$. Correspondingly, we have N distinct eigenvectors as well. The general solution of (1.6):

$$\Delta \boldsymbol{x}(t) = \sum_{m=1}^{N} \boldsymbol{e}_m e^{\lambda_m t} = \sum_{m=1}^{N} \boldsymbol{\delta}_m(t) \tag{1.12}$$

Here we assumed that the initial condition was $\Delta \boldsymbol{x}(t=0)=\sum_{m=1}^{N} \boldsymbol{e}_m$. The roots of (1.10) are either real or complex conjugate ones since the coefficients are real in (1.10).

When we have complex conjugate pairs of eigenvalues $\lambda_m = \hat{\lambda}_{m+1} = \sigma_m - j\omega_m$, the corresponding eigenvectors are $\boldsymbol{e}_m = \hat{\boldsymbol{e}}_{m+1} = \boldsymbol{e}_{m,R} + j\boldsymbol{e}_{m,I}$, where the ^ denotes complex conjugate and where σ_m, ω_m, and $\boldsymbol{e}_{m,R}$, $\boldsymbol{e}_{m,I}$ are all real and real-valued vectors, respectively. From the two complex solutions $\delta_m(t) = \boldsymbol{e}_m \exp(\lambda_m t)$ and $\delta_{m+1}(t) = \boldsymbol{e}_{m+1} \exp(\lambda_{m+1} t)$, two linearly independent real solutions $\boldsymbol{s}_m(t)$ and $\boldsymbol{s}_{m+1}(t)$ can be composed:

$$\boldsymbol{s}_m(t)=\frac{1}{2}\left[\delta_m(t)+\delta_{m+1}(t)\right]=e^{\sigma_m t}\left[\boldsymbol{e}_{m,R}\cos\omega_m t+\boldsymbol{e}_{m,I}\sin\omega_m t\right] \tag{1.13}$$

$$\boldsymbol{s}_{m+1}(t)=\frac{1}{2j}\left[\delta_m(t)-\delta_{m+1}(t)\right]=e^{\sigma_m t}\left[\boldsymbol{e}_{m,I}\cos\omega_m t-\boldsymbol{e}_{m,R}\sin\omega_m t\right] \tag{1.14}$$

When the eigenvalue is real $\lambda_m = \sigma_m$ and the solution belonging to λ_m,

$$\boldsymbol{s}_m(t)=\delta_m(t)=\boldsymbol{e}_m e^{\sigma_m t} \tag{1.15}$$

Note that the time function belonging to an eigenvalue or a pair of eigenvalues is the same for all state variables.

1.3.4 Stability

The EP is stable if and only if the real part σ_m of all eigenvalues belonging to the EP is negative. Otherwise, one or more solutions $\boldsymbol{s}_m(t)$ goes to infinity. $\sigma_m = 0$ is considered as unstable case. When $\sigma_m < 0$, all small perturbation around EP dies eventually, and the system settles back to EP.

Figure 1.3a shows one EP and one of its eigenvectors $\boldsymbol{e}_m$ having real eigenvalue $\lambda_m = \sigma_m$. The orbit is a straight line along $\boldsymbol{e}_m$. Figure 1.3b shows the EP having complex conjugate eigenvalues and the real $\boldsymbol{e}_{m,R}$ and imaginary $\boldsymbol{e}_{m,I}$ component of the complex eigenvector $\boldsymbol{e}_m$. The orbit of the solution $\boldsymbol{s}_m(t)$ is a spiral in the plane spanned by $\boldsymbol{e}_{m,R}$ and $\boldsymbol{e}_{m,I}$ around EP. Both cases, the operation point P is attracted ($\sigma_m < 0$) or repelled ($\sigma_m > 0$).

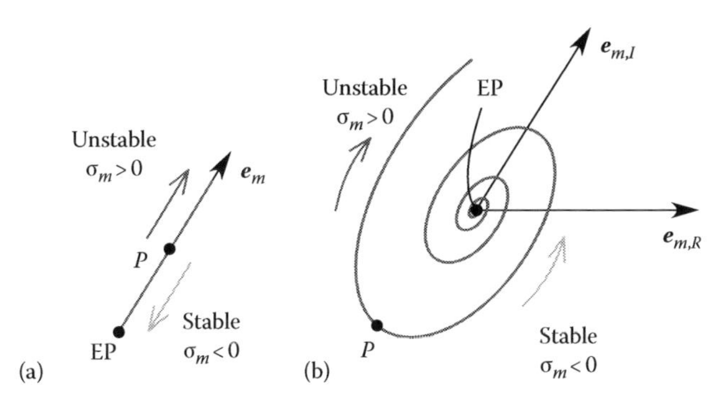

FIGURE 1.3 Stable ($\sigma_m < 0$) and unstable ($\sigma_m > 0$) orbits. (a) Real eigenvalues, and (b) a pair of complex eigenvalues belongs to the eigenvector $\boldsymbol{e}_m$ of EP.

1.3.5 Classification of EPs, Three-Dimensional State Space (N = 3)

Depending on the location of the three eigenvalues in the complex plane, eight types of the EPs are distinguished (Figure 1.4). Three eigenvectors are used for reference frame. The origin is the EP. Eigenvectors $\boldsymbol{e}_1$, $\boldsymbol{e}_2$, and $\boldsymbol{e}_3$ are used when all eigenvalues are real (Figure 1.4a,c,e,g) and $\boldsymbol{e}_1$, $\boldsymbol{e}_{2,R}$, and $\boldsymbol{e}_{2,I}$ are used when we have a pair of conjugate complex eigenvalues (Figure 1.4b,d,f,h). The orbits (trajectories) are moving exponentially in time along the eigenvectors $\boldsymbol{e}_1$, $\boldsymbol{e}_2$, or $\boldsymbol{e}_3$ when the initial condition is on them. The orbits are spiraling in the plane spanned by $\boldsymbol{e}_{2,R}$ and $\boldsymbol{e}_{2,I}$ with initial condition in the plane. The operation points are attracted (repelled) by stable (unstable) EP.

All three eigenvalues are on the left-hand side of the complex plane for *node* and *spiral node* (Figure 1.4a and b). The spiral node is also called *attracting focus*. All three eigenvalues are on the right-hand side for *repeller* and *spiral repeller* (Figure 1.4c and d). The spiral repeller is also called *repelling focus*. For *saddle points*, either one (Index 1) (Figure 1.4e and f) or two (Index 2) (Figure 1.4g and h) eigenvalues are on the right-hand side.

Saddle points play very important role in organizing the trajectories in state space. A trajectory associated to an eigenvector or a pair of eigenvectors can be stable when its eigenvalue(s) is (are) on the left-hand side of the complex plane (Figure 1.4a and b) or unstable when they are on the right-hand side (Figure 1.4c through f). Trajectories heading directly to and directly away from a saddle point are called

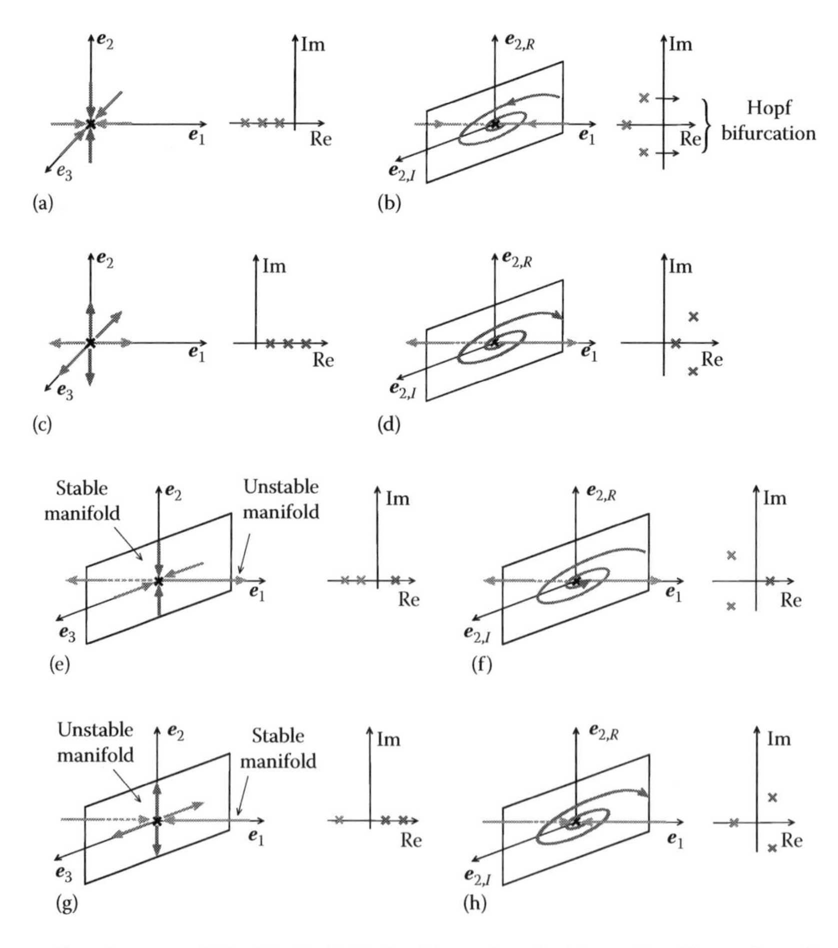

FIGURE 1.4 Classifications of EPs (N=3). (a) Node, (b) spiral node, (c) repeller, (d) spiral repeller, (e) saddle point—index 1, (f) spiral saddle point—index 1, (g) saddle point—2, and (h) spiral saddle point—index 2.

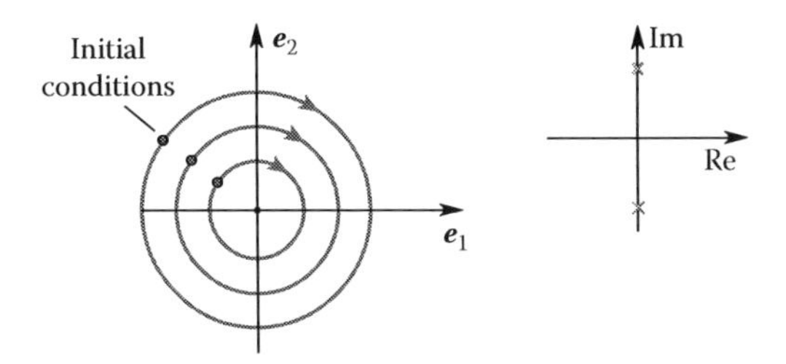

FIGURE 1.5 Center. Eigenvalues are complex conjugate, $N = 2$.

stable and *unstable invariant manifold* or shortly *manifold*. Sometimes, the stable (unstable) manifolds are called *insets* (*outsets*). The operation point either on the stable or on the unstable manifold cannot leave the manifold. The manifolds of a saddle point in its neighborhood in a two-dimensional state space divide it into four regions. A trajectory born in a region is confined to the region. The manifolds are part of the separatrices separating the basins of attractions. In this sense, the manifolds organize the state space.

Finally, when the real part σ_m is zero in the pair of the conjugate complex eigenvalue and the dimension is $N = 2$, the EP is called *center*. The trajectories in the reference frame $\boldsymbol{e}_R$–$\boldsymbol{e}_I$ are circles (Figure 1.5). Their radius is determined by the initial condition.

1.3.6 No-Intersection Theorem

Trajectories in state space cannot intersect each other. The theorem is the direct consequence of the deterministic system. The state of the system is unambiguously determined by the location of its operation point in the state space. As the system is determined by (1.1), all of the derivatives are determined by the instantaneous values of the state variables. Consequently, there is only one possible direction for a trajectory to continue its journey.

1.4 Limit Cycle

1.4.1 Introduction

Two- or higher-dimensional nonlinear systems can exhibit periodic (cyclic) motion without external periodical excitation. This behavior is represented by closed-loop trajectory called *Limit Cycle* (LC) in the state space. There are stable (attracting) and unstable (repelling) LC. The basic difference between the stable LC and the center (see Figure 1.5) having closed trajectory is that the trajectories starting from nearby initial points are attracted by stable limit cycle and sooner or later they end up in the LC, while the trajectories starting from different initial conditions will stay forever in different tracks determined by the initial conditions in the case of center.

Figure 1.6 shows a stable LC together with the Poincaré plane. Here, the dimension is 3. The LC intersects the Poincaré plane at point P_k called *Fixed Point* (FP). It plays a crucial role in nonlinear dynamics. Instead of investigating the behavior of the LC, the FP is studied.

1.4.2 Poincaré Map Function (PMF)

After moving the trajectory from thc LC by a small deviation, the discrete Poinearé Map Function (PMF) relates the coordinates of intersection point P_n to those of the previous point P_{n-1}. All points are the intersection points in the Poincaré plane generated by the trajectory. The PMF in a three-dimensional state space is (Figure 1.6)

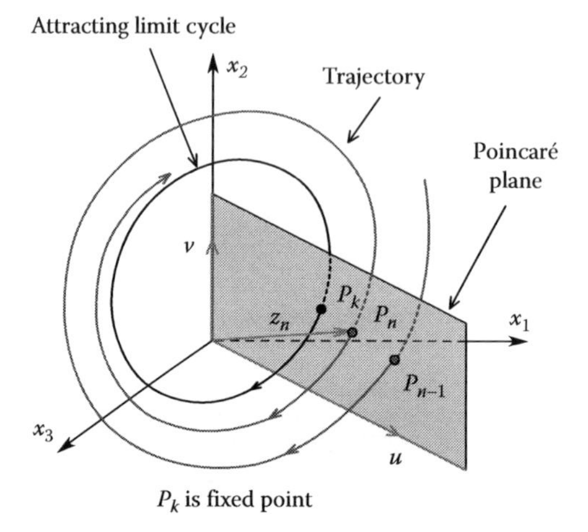

FIGURE 1.6 Stable limit cycle and the Poincaré plane ($N = 3$).

$$u_n = P_1(u_{n-1}, v_{n-1})$$
$$v_n = P_2(u_{n-1}, v_{n-1}) \tag{1.16}$$

where
P_1 and P_2 are the PMF
u_n and v_n are the coordinates of the intersection point P_n in the Poincaré plane

Introducing vector z_n by

$$z_n^T = [u_n, v_n] \tag{1.17}$$

where z_n points to intersection P_n from the origin, the PMF is

$$z_n = \boldsymbol{P}(z_{n-1}) \tag{1.18}$$

At fixed point P_k

$$z_k = \boldsymbol{P}(z_k) \tag{1.19}$$

The first benefit of applying the PMF is the reduction of the dimension by one as the stability of FP P_k is studied now in the two-dimensional state space rather than studying the stability of the LC in the three-dimensional state space, and the second one is the substitution of the differential equation by difference equation.

1.4.3 Stability

The stability can be investigated on the basis of the PMF (1.18). First, the nonlinear function $\boldsymbol{P}(z_{n-1})$ has to be linearized by its Jacobian matrix $\boldsymbol{J}_k$ evaluated at its FP z_k. Knowing $\boldsymbol{J}_k$, (1.18) can be rewritten for small perturbation around the FP z_k as

$$\Delta z_n = J_k \Delta z_{n-1} = J_k^n \Delta z_0 \tag{1.20}$$

where Δz_0 is the initial small deviation from FP P_k.

Substituting J_k by its eigenvalues λ_k and right e_{mr} and left e_{ml} eigenvectors,

$$\Delta z_n = \left[\sum_{m=1}^{N-1} \lambda_m^n e_{mr} e_{ml}^T\right] \Delta z_0 \tag{1.21}$$

Due to (1.21), the LC is stable if and only if all eigenvalues are within the circle with unit radius in the complex plane.

In the stability analysis, both in continuous-time model (CTM) and in discrete-time model (DTM), the Jacobian matrix is operated on the small perturbation of the state vector [see (1.6) and (1.20)]. The essential difference is that it determines the velocity vector for CTM and the next iterate for DTM, respectively.

Assume that the very first point at the beginning of iteration is placed on eigenvector e_m of the Jacobian matrix. Figure 1.7 shows four different iteration processes corresponding to the particular value of eigenvalue λ_m associated to e_m. In Figure 1.7a and b, λ_m is real, but its value is $0 < \lambda_m < 1$ in Figure 1.7a and $\lambda_m > 1$ in Figure 1.7b. The consecutive iteration points along vector e_m are approaching the FP, it is node (Figure 1.7a) and they are repelled from FP, it is repeller (Figure 1.7b). In Figure 1.7c, both λ_m and λ_{m+1} are real but $0 < \lambda_m < 1$ and $\lambda_{m+1} > 1$. The subsequent iterates are converging onto FP along eigenvector e_m while they are repelled along e_{m+1}. The FP is a saddle. In Figure 1.7d, the eigenvalues are complex conjugates. The consecutive iterates move along a spiral path around FP [see (1.13) and (1.14)]. When $|\lambda_m| < 1$, the iterates are converging onto FP, now it is spiral node. When $|\lambda_m| > 1$, the iteration diverges, the FP is spiral repeller.

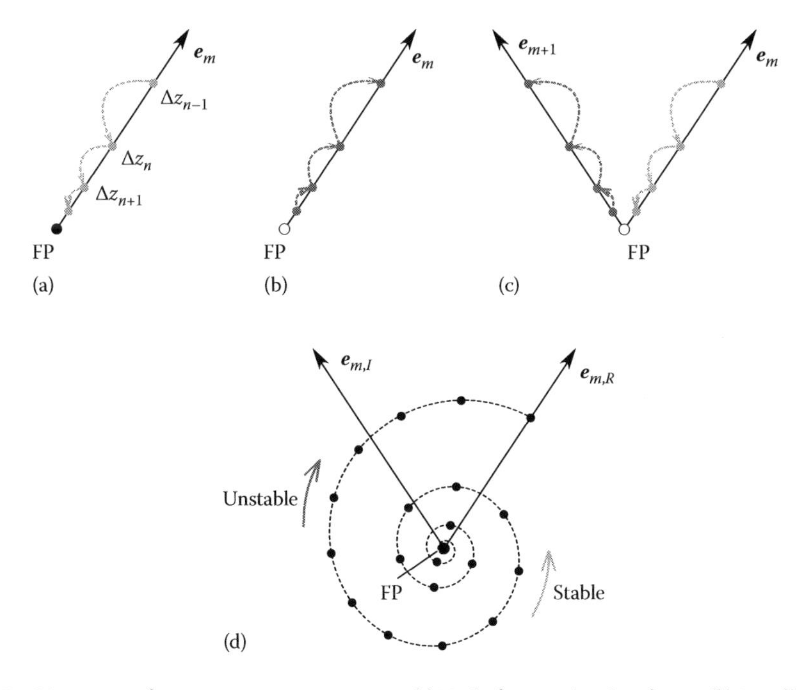

FIGURE 1.7 Movement of consecutive iteration points. (a) Node, λ_m is real and $0<\lambda_m<1$. (b) Repeller, is real and $\lambda_m>1$. (c) Saddle, both λ_m and λ_{m+1} are real, $0<\lambda_m<1$ and $\lambda_{m+1}>1$. (d) λ_m and λ_{m+1} are complex conjugate. Spiral node if $|\lambda_m|<1$ and spiral repeller if $|\lambda_m|>1$.

1.5 Quasi-Periodic and Frequency-Locked State

1.5.1 Introduction

Beside the EP and LC, another possible state or motion is the *quasi-periodic* (Qu-P) motion in CTM. In Qu-P state, the motion—in theory—never exactly repeats itself. Other terms used in literature are *conditionally periodic* or *almost periodic*. Qu-P state is possible in inherently discrete systems as well. The frequency-locked (F-L) state is a special case of Qu-P state. Qu-P state is not possible in $N = 1$ or 2 dimensional systems, i.e., N must be $N \geq 3$. Similarly to chaos, Qu-P state is aperiodic. In chaotic state, two points in state space, which are arbitrarily close, will diverge. In other words, the chaotic system is extremely sensitive to initial conditions and to changes in control parameters. In contrast to chaotic state, two points that are initially close will remain close over time in Qu-P state.

The Qu-P motion is a mixture of periodic motions of several different angular frequencies ω_1, $\omega_2, \ldots, \omega_m$. Depending on the value of their linear combination L,

$$L = k_1\omega_1 + k_2\omega_2 + \cdots + k_m\omega_m \tag{1.22}$$

the motion can be Qu-P, i.e., aperiodic when the sum $L \neq 0$ or it can be F-L, i.e., periodic state when the sum $L = 0$. Here $k_1, k_2, \ldots, k_m$ are any positive (or negative) integer ($k_1 = k_2 = \cdots = k_m = 0$ is excluded).

The EP, LC, Qu-P, and F-L states are *regular attractors* while in chaotic state, the system has *strange attractor* (see later in Section 1.9). The Qu-P motion plays central role in Hamiltonian systems, e.g., in mechanical systems modeled without friction, which are non-dissipative ones. They do not have attractors.

1.5.2 Nonlinear Systems with Two Frequencies

Qu-P and F-L motions occur frequently in practice in systems having a natural oscillation frequency and a different external forcing frequency or two different natural oscillation frequencies. Because of nonlinearity, the superposition of the independent frequencies is not valid.

Starting from (1.22), assume that

$$\frac{\omega_1}{\omega_2} = \frac{T_2}{T_1} = \frac{p}{q} \tag{1.23}$$

where $T_1 = 2\pi/\omega_1$ and $T_2 = 2\pi/\omega_2$ are the periods of respective harmonic oscillations and p and q are positive integers. Here we assume that any common factors in the frequency ratio have been removed, e.g., if $f_1/f_2 = 2/6$, the common factor of 2 will be removed and $f_1/f_2 = p/q = 1/3$ can be written. When the frequency ratio is the ratio of two integers, then the ratio is called *rational* in mathematical sense, i.e., the two frequencies are *commensurate*, the behavior of the system is periodic. It is in F-L state.

On the other hand, when the frequency ratio is irrational, the two frequencies are incommensurate, and the behavior is Qu-P. The last two statements can easily be understood in the geometrical interpretation of the system trajectory.

1.5.3 Geometrical Interpretation

A two-frequency Qu-P trajectory on a toroidal surface in the three-dimensional state space is shown in Figure 1.8. Introducing two angles $\Theta_r = \omega_r t = \omega_1 t$ and $\Theta_R = \omega_R t = \omega_2 t$, they determine point P of the trajectory on the surface of the torus provided that the initial condition point P_0 belonging to $t = 0$ is known. The center of the torus is at the origin. R is the large radius of the torus whose cross-sectional radius is r. The two angles Θ_R and Θ_r are increasing as time evolves and therefore point P' is moving on

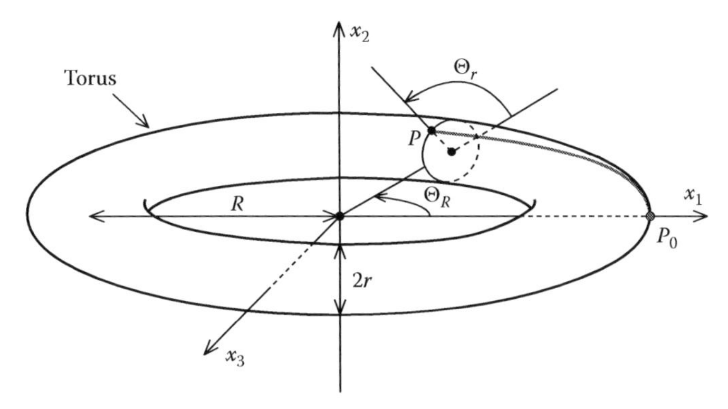

FIGURE 1.8 Two-frequency trajectory on a toroidal surface in state space ($N = 3$).

the surface of the torus tracking the trajectory of state vector $\boldsymbol{x}^T = [x_1, x_2, x_3]$. The three components of the state vector $\boldsymbol{x}$ are given by the equations as follows:

$$\left.\begin{aligned} x_1 &= (R + r\cos\Theta_r)\cos\Theta_R \\ x_2 &= r\sin\Theta_r \\ x_3 &= -(R + r\cos\Theta_r)\sin\Theta_R \end{aligned}\right\} \tag{1.24}$$

The trajectory is winding on the torus around the cross section with minor radius r, making $\omega_R/2\pi$ rotations per unit time. As $\omega_r/\omega_R = p/q$ [see (1.23)] and assuming that p and q are integers, the number of rotations around circle r and circle R in per unit is p and q, respectively. For example, if $p = 1$ and $q = 3$, point P makes three rotations around circle R as long as it makes only one rotation around circle r. Figure 1.9 shows the torus and the Poincaré plane intersecting the torus together with the trajectory on the surface of the torus. The Poincaré plane illustrates its intersection points with the trajectory.

When the frequency ratio $p/q = 1/3$, it is rational. As long as point P rotates once around circle R, it rotates 120° around circle r. Starting from point 0 on the Poincaré plane (Figure 1.9a) after 1–2–3 rotations around circle R, point P intersects the Ponicaré plane successively at point 1–2–3. Point 3

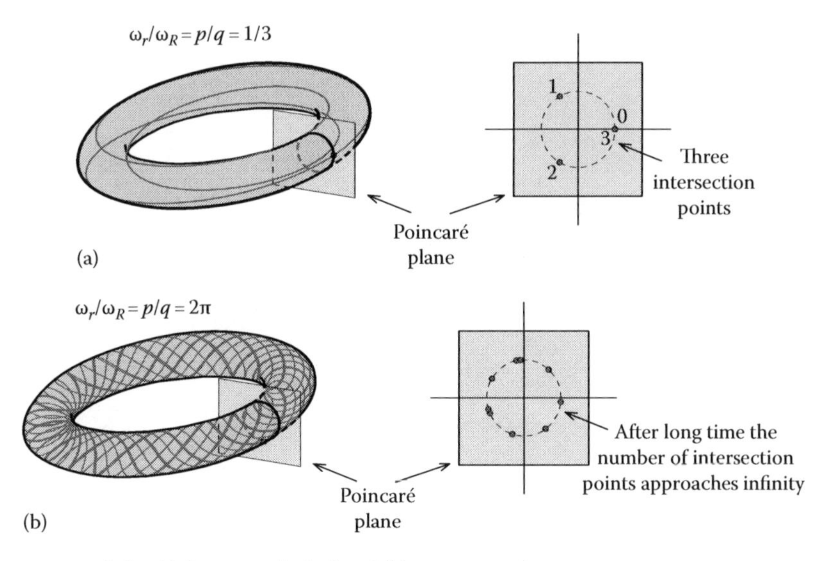

FIGURE 1.9 Example for (a) frequency-locked and (b) quasi-periodic state.

coincides with the starting point 0 as three times 120° is 360°. The process is periodic, the trajectory closes on itself, it is the F-L state.

On the other hand, when the frequency ratio is irrational, e.g., $p/q = 2\pi$ as long as the point P makes one rotation around circle R, it completes 2π rotations around circle r. The phase shift of the first intersection point on the Poincaré plane from the initial point P_0 is given by an irrational angle = 360° × 2π(mod 1) where (mod 1) is the modulus operator that takes the fraction of a number (e.g., 6.28(mod 1) = 0.28). After any further full rotations around circle R, the phase shifts of the intersection points from P_0 on the Poincaré plane remain irrational; therefore, they will never coincide with P_0, and the trajectory will never close on itself. All intersection points will be different. As $t \to \infty$, the number of intersection points will be infinite and a circle of radius r will be visible on the Poincaré plane consisting of infinite number of distinct points (Figure 1.9b). The system is in Qu-P state.

1.5.4 *N*-Frequency Quasi-Periodicity

We have treated up to now the two-frequency quasi-periodicity a little bit in detail. It has to be stressed that in mathematical sense, the N-frequency quasi-periodicity is essentially the same. The N frequencies define N angles $\Theta_1 = \omega_1 t$, $\Theta_2 = \omega_2 t$,... $\Theta_N = \omega_N t$ determining uniquely the position and movement of the operation point P on the surface of the N-dimensional torus. Now again, the trajectory fills up the surface of the N-dimensional torus in the state space.

1.6 Dynamical Systems Described by Discrete-Time Variables: Maps

1.6.1 Introduction

The dynamical systems can be described by difference equation systems with discrete-time variables. The relation in vector form is

$$\boldsymbol{x}_{n+1} = \boldsymbol{f}(\boldsymbol{x}_n) \tag{1.25}$$

where $\boldsymbol{x}_n$ is K-dimensional state variable $\boldsymbol{x}_n^T = [x_n^{(1)}, x_n^{(2)}, \dots x_n^{(K)}]$, $\boldsymbol{f}$ is nonlinear vector function. State vector $\boldsymbol{x}_n$ is obtained at discrete time $n = 1$ by $\boldsymbol{x}_1 = \boldsymbol{f}(\boldsymbol{x}_0)$, where $\boldsymbol{x}_0$ is the initial condition. From $\boldsymbol{x}_1$, the value $\boldsymbol{x}_2 = \boldsymbol{f}(\boldsymbol{x}_1)$ can be calculated, etc. Knowing $\boldsymbol{x}_0$, the orbit of discrete-time system $\boldsymbol{x}_0, \boldsymbol{x}_1, \boldsymbol{x}_2, \dots$ is generated.

We can consider that vector function $\boldsymbol{f}$ maps $\boldsymbol{x}_n$ into $\boldsymbol{x}_{n+1}$. In this sense, $\boldsymbol{f}$ is a *map function*. The number of state variables determines the dimension of the map.

Examples:

One-dimensional map ($K = 1$):

$$\text{Logistic map:} \quad x_{n+1} = a x_n (1 - x_n)$$

$$\text{Tent map:} \quad x_{n+1} = \begin{cases} a x_n & \text{if } x_n \le 0.5 \\ a(1 - x_n) & \text{if } x_n > 0.5 \end{cases}$$

where a is constant.

Two-dimensional map ($K = 2$):

$$\text{Hénon map:} \quad \begin{aligned} x_{n+1}^{(1)} &= 1 + a x_n^{(2)} - b x_n^{(1)^2} \\ x_{n+1}^{(2)} &= c x_n^{(1)} \end{aligned}$$

where a, b, and c are constants.

K-dimensional map:

Poincaré map of an $N = K + 1$ dimensional state space.

Maps can give useful insight for the behavior of complex dynamic systems.

The map function (1.25) can be *invertible* or *non-invertible*. It is invertible when the discrete function

$$\boldsymbol{x}_n = \boldsymbol{f}^{-1}(\boldsymbol{x}_{n+1}) \tag{1.26}$$

can be solved uniquely for $\boldsymbol{x}_n$. $\boldsymbol{f}^{-1}$ is the inverse of $\boldsymbol{f}$. Two examples are given next. First, the invertible Henon map and after, the non-invertible quadratic (logistic) map are discussed. *Henon map* is frequently cited example in nonlinear systems. It has two dimensions and maps the point with coordinate $x_n^{(1)}$ and $x_n^{(2)}$ in the plane to a new point $x_{n+1}^{(1)}$ and $x_{n+1}^{(2)}$. It is invertible, because $x_{n+1}^{(1)}$ and $x_{n+1}^{(2)}$ uniquely determine the value $x_n^{(1)}$ and $x_n^{(2)}$, since

$$x_n^{(1)} = \frac{x_{n+1}^{(2)}}{c}$$

$$x_n^{(2)} = \frac{x_{n+1}^{(1)} - 1 + b x_{n+1}^{(2)\,2}}{a}$$

Here, $a \neq 0$ and $c \neq 0$ must hold. (For some values of a, b, and c, the Henon map can exhibit chaotic behavior.)

Turning now to the non-invertible maps, the *quadratic or logistic map* is taken as example. It was developed originally as a demography model. It is a very simple system, but its response can be surprisingly colorful. It is non-invertible, as Figure 1.10 shows. We cannot uniquely determine $\boldsymbol{x}_n$ from $\boldsymbol{x}_{n+1}$. As we see later, an invertible map can be chaotic only if its dimension is two or more (Henon map). On the other hand, the non-invertible map can be chaotic even in one-dimensional cases, e.g., the logistic map.

1.6.2 Fixed Points

The concept of FP was introduced earlier (see Figure 1.6). The system stays in steady state at FP where $\boldsymbol{x}_{n+1} = \boldsymbol{x}_n = \boldsymbol{x}^*$. When the discrete function (1.25) is nonlinear, it can have more than one FP.

1.6.2.1 One-Dimensional Iterated Maps

For the sake of simplicity, the one-dimensional maps are discussed from now on. The one-dimensional iterated maps can describe the dynamics of large number of systems of higher dimension. To throw light

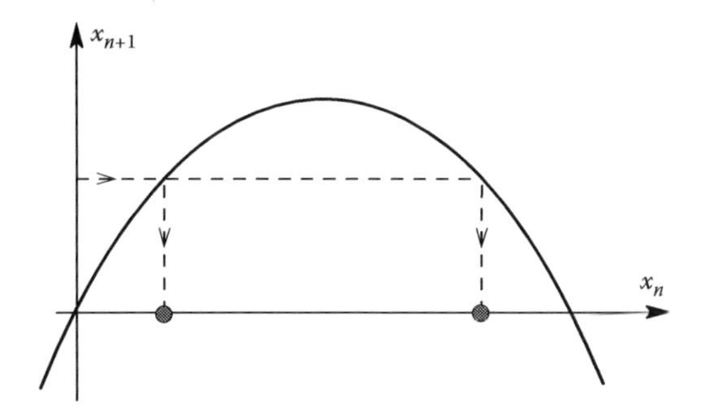

FIGURE 1.10 The quadratic or logistic iterated map.

to the statement, consider a three-dimensional dynamical system with PMF $z_{n+1}^T = [u_{n+1}, v_{n+1}] = \mathbf{P}_n(u_n, v_n)$ [see (1.18)]. In certain cases, we have a relation between the two coordinates u_n and v_n: $v_n = F_v(u_n)$. Substituting it into the map function $u_{n+1} = P_u(u_n, v_n)$, we end up with

$$u_{n+1} = P_u(u_n, F_v(u_n)) = f(u_n) \tag{1.27}$$

one-dimensional map function.

More arguments can be found in the literature (see chapter 5.2 in Ref. [4] and page 66 of Ref. [5]) on the wide scope of applications on the one-dimensional discrete map functions. If the dissipation in the system is high enough, then even systems with dimension more than three can be analyzed by one-dimensional map.

1.6.2.2 Return Map or Cobweb

The iteration in the one-dimensional discrete map function or difference equation

$$x_{n+1} = f(x_n) \tag{1.28}$$

can be done with numerical or graphical method. The return map or cobweb is a graphical method. To illustrate the return map method, we take as example the equation

$$x_{n+1} = \frac{ax_n}{1 + (bx_n)^c} \tag{1.29}$$

where a, b, and c are constants. The iteration has to be performed as follows (Figure 1.11):

1. Plot the function $x_{n+1}(x_n)$ in plane x_{n+1} versus x_n.
2. Select x_0 as initial condition.
3. Draw a straight line starting from origin with slope 1 called *mirror line* or *diagonal.*
4. Read the value x_1 from the graph and draw a line in parallel with the horizontal axis from x_1 to the mirror line (line 1′–1″).
5. Read the value x_2 by vertical line 1″–2′.
6. Repeat the graphical process by drawing the horizontal line 2′–2″ and finally determining x_3.

In order to find the FP x^*, we have to repeat the graphical process.

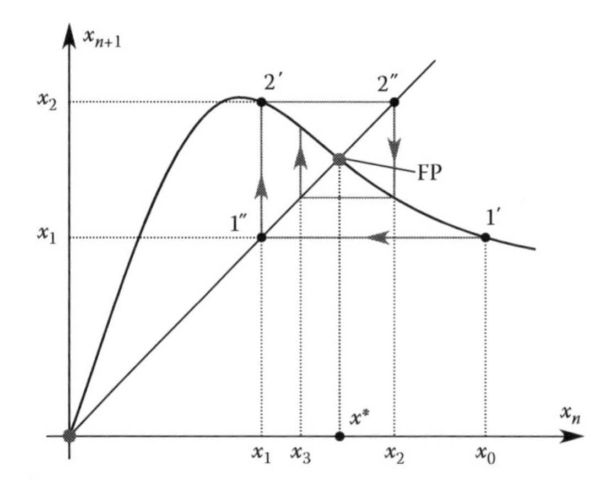

FIGURE 1.11 Iteration in return map.

1.6.2.3 *k*th Return Map

Periodic steady state with period T is represented by a single FP x^* in the mapping, i.e., $x^* = f(x^*)$. kth-order subharmonic solutions with period kT correspond to FPs $\{x_1^* \dots x_k^*\}$, where $x_2^* = f(x_1^*) \dots x_{n+1}^* = f(x_n^*) \dots x_1^* = f(x_k^*)$. The kth iterate of $f(x)$ is defined as the function that results from applying f k times, its notation is $f^{(k)}(x) = f(f(\dots f(x)))$, and its mapping is called kth return map and the process is called period-k.

1.6.2.4 Stability of FP in One-Dimensional Map

The FP is *locally stable* if subsequent iterates starting from a sufficiently near initial point to the FP are eventually getting closer and closer the FP. The expressions attracting FP or asymptotically stable FP are also used. On the other hand, if the subsequent iterates move away from x^*, then the name unstable or repelling FP is used.

1.6.3 Mathematical Approach

By knowing the nonlinear function $f(x)$ and one of its FPs x^*, we can express the first iterate x_1 by applying a Taylor series expansion near x^*:

$$x_1 = f(x_0) = f(x^*) + \left.\frac{df}{dx}\right|_{x^*} \Delta x_0 + \dots \cong f(x^*) + \left.\frac{df}{dx}\right|_{x^*} \Delta x_0 \tag{1.30}$$

where $\Delta x_0 = x_0 - x^*$ and the initial condition x_0 is sufficiently near x^*. The derivative $\lambda = df/dx$ has to be evaluated at x^*. λ is the eigenvalue of $f(x)$ at x^*. The nth iterate is

$$\Delta x_n = \lambda^n (x_0 - x^*) = \lambda^n \Delta x_0 = \left[\left.\frac{df}{dx}\right|_{x^*}\right]^n \Delta x_0 \tag{1.31}$$

It is obvious that x^* is stable FP if $|df/dx|_{x^*} < 1$ and x^* is unstable FP if $|df/dx|_{x^*} > 1$. In general, when the dimension is more than one, λ must be within the unit circle drawn around the origin of the complex plane for stable FP.

The nonlinear $f(x)$ has multiple FPs. The initial conditions leading to a particular x^* constitute the *basin of attraction* of x^*. As there are more basins of attraction, none of FPs can be *globally stable*. They can be only locally stable.

1.6.4 Graphical Approach

Graphical approach is explained in Figure 1.12. Figure 1.12a,c,e, and g presents the return map with FP x^* and with the initial condition (IC). The thick straight line with slope λ at x^* approximates the function $f(x)$ at x^*. Figure 1.12b,d,f, and h depicts the discrete-time evolution $x_n(n)$. $|\lambda| < 1$ in Figure 1.12a,b,c, and d. The subsequent iterates approach x^*, FP is stable. In Figure 1.12e,f,g, and h, $|\lambda| > 1$. The subsequent iterates explode, FP is unstable. Note that the cobweb and time evolution is oscillating when $\lambda < 0$.

1.6.5 Study of Logistic Map

The relation of logistic map is

$$x_{n+1} = a x_n (1 - x_n); \quad a = \text{const.} \tag{1.32}$$

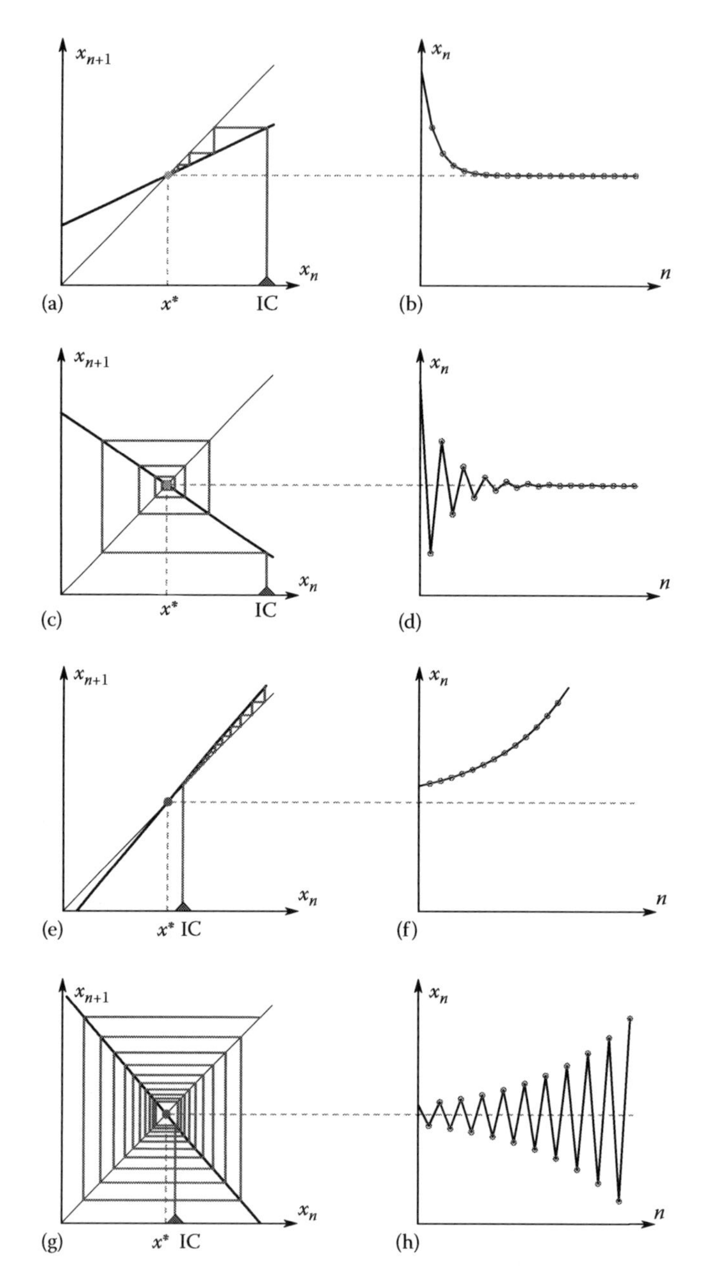

FIGURE 1.12 Iterates in return map and their time evolution.

In general, it has two FPs: $x_1^* = 1 - 1/a$ and $x_2^* = 0$. The respective eigenvalues are $\lambda_1 = 2 - a$ and $\lambda_2 = a$. Figure 1.13 shows the return map (left) and the time evolution (right) at different value a changing in range $0 < a < 4$. When $0 < a < 1$, then λ_1 is outside and λ_2 is inside of the unit circle, i.e., x_1^* is unstable and x_2^* is stable (Figure 1.13a and b). When $1 < a < 3$, then λ_1 is inside and λ_2 is outside of the unit circle, x_1^* is stable and x_2^* is unstable (Figure 1.13c,d,e, and f). There is no oscillation in Figure 1.13c and d $\lambda_1 > 0$ and we have oscillation in Figure 1.13e and f as $\lambda_1 < 0$. Increasing a over $a = 3$, none of the two eigenvalues λ_1 and λ_2 is inside the unit circle, both x_1^* and x_2^* are unstable. However, stable cycle of period-2 (Figure 1.13g and h and the period-4 Figure 1.13i and j) develops at $a = 3.2$ and $a = 3.5$, respectively (see later, the stability of cycles). Increasing a over $a = 3.570$, we enter the chaotic range (Figure 1.13k and l), the iteration is a periodic with narrow ranges of a producing periodic solutions.

As a is increased, first we have period-1 in steady state, later period-2, then period-4 emerge, etc. The scenario is called *period doubling cascade.*

1.6.6 Stability of Cycles

We have already introduced the notation $f^{(k)}(x) = f(f(\ldots f(x))$ for the kth iterate that results from applying f k-times. If we start at x_1^* and after applying f k-times we end up with $x_k^* = x_1^*$, then we say we have period or cycle-k with k separate FPs: $x_1^*, x_2^* = f(x_1^*), \ldots, x_k^* = f(x_{k-1}^*)$.

In the simplest case of period-2, the two FPs are: $x_2^* = f(x_1^*)$ and $x_1^* = f(x_2^*) = f(f(x_1^*))$. Referring to (1.30), we know that the stability of FP x_1^* depends on the value of the derivative

$$\lambda^{(2)} = \left.\frac{df(f(x))}{dx}\right|_{x_1^*} \tag{1.33}$$

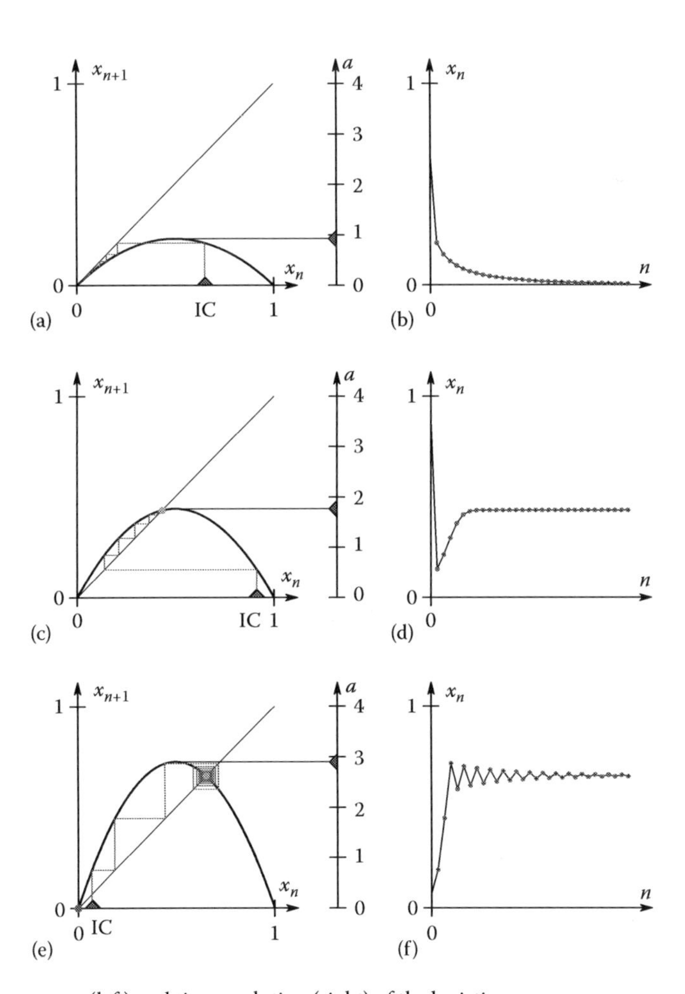

FIGURE 1.13 Return map (left) and time evolution (right) of the logistic map.

(continued)

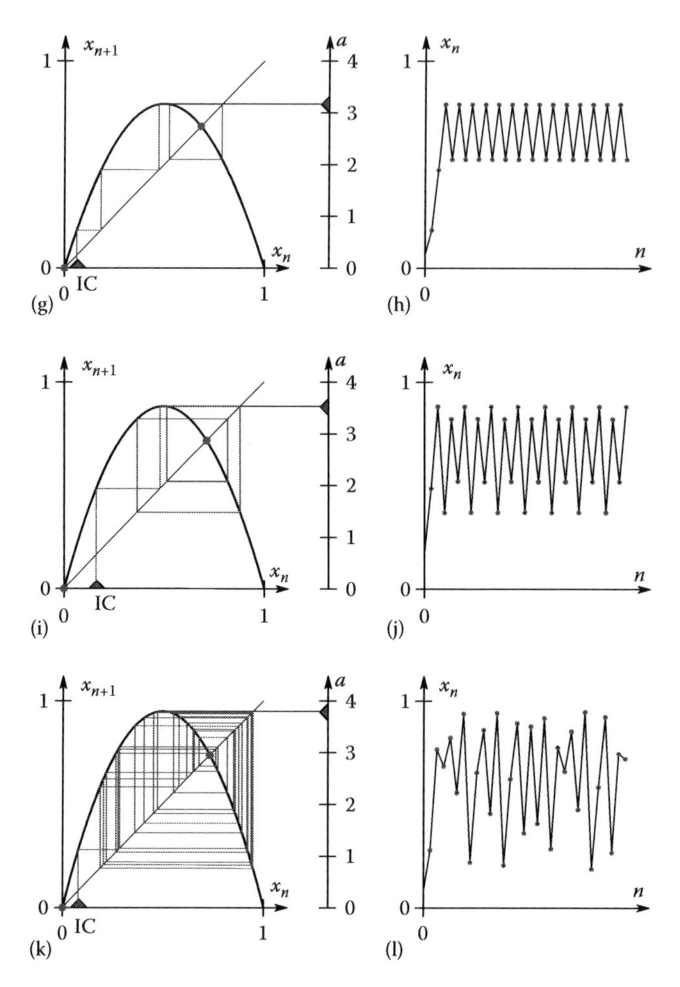

FIGURE 1.13 (continued)

Using the chain rule for derivatives,

$$\left.\frac{df^{(2)}(x)}{dx}\right|_{x_1^*} = \left.\frac{df(f(x))}{dx}\right|_{x_1^*} = \left.\frac{df}{dx}\right|_{f(x_1^*)} \cdot \left.\frac{df}{dx}\right|_{x_1^*} = \left.\frac{df}{dx}\right|_{x_2^*} \cdot \left.\frac{df}{dx}\right|_{x_1^*} \tag{1.34}$$

Consequently,

$$\left.\frac{df^{(2)}(x)}{dx}\right|_{x_1^*} = \left.\frac{df^{(2)}(x)}{dx}\right|_{x_2^*} \tag{1.35}$$

(1.35) states that the derivatives or eigenvalues of the second iterate of $f(x)$ are the same at both FPs belonging to period-2.

As an example, Figure 1.14 shows the return map for the first (Figure 1.14a) and for the second (Figure 1.14b) iterate when $a = 3.2$. Both FPs in the first iterate map are unstable as we have just discussed. We have four FPs in the second iterate map. Two of them, x_1^* and x_2^*, are stable FPs (point S_1 and S_2) and the other two (zero and x^*) are unstable (point U_1 and U_2). The FP of the first iterate must be the FP of the second iterate as well: $x^* = f(x^*)$ and $x^* = f(f(x^*))$.

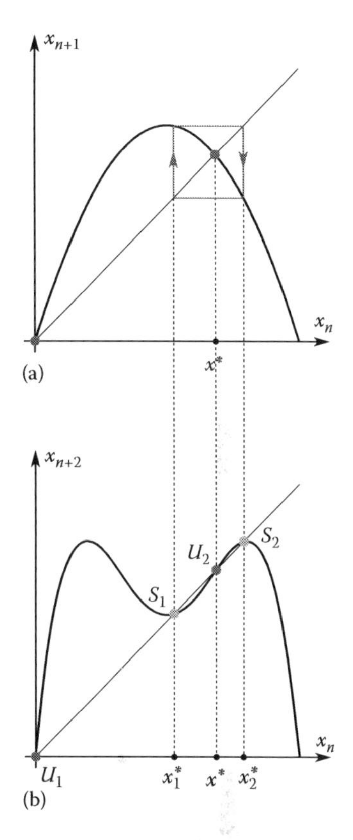

FIGURE 1.14 Cycle of period-2 in the logistic map for $a = 3.3$. (a) Return map for first iterate. (b) Return map for second iterate x_{n+2} versus x_n.

1.7 Invariant Manifolds: Homoclinic and Heteroclinic Orbits

1.7.1 Introduction

To obtain complete understanding of the global dynamics of nonlinear systems, the knowledge of invariant manifolds is absolutely crucial. The invariant manifolds or briefly the manifolds are borders in state space separating regions. A trajectory born in one region must remain in the same region as time evolves. The manifolds organize the state space. There are stable and unstable manifolds. They originate from saddle points. If the initial condition is on the manifold or subspace, the trajectory stays on the manifold. Homoclinic orbit is established when stable and unstable manifolds of a saddle point intersect. Heteroclinic orbit is established when stable and unstable manifolds from different saddle points intersect.

1.7.2 Invariant Manifolds, CTM

The CTM is applied for describing the system. Invariant manifold is a curve (trajectory) in plane ($\boldsymbol{N} = 2$) (Figure 1.15), or curve or surface in space ($\boldsymbol{N} = 3$) (Figure 1.16), or in general a subspace (hypersurface) of the state space ($\boldsymbol{N} > 3$). The manifolds are always associated with saddle point denoted here by $\boldsymbol{x}^*$. Any initial condition in the manifold results in movement of the operation point in the manifold under the action of the relevant differential equations. There are two kinds

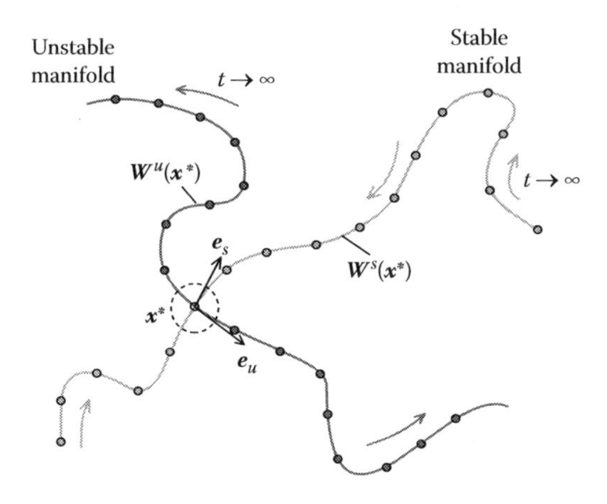

FIGURE 1.15 Stable $\boldsymbol{W}^s$ and unstable $\boldsymbol{W}^u$ manifold ($N = 2$). The CTM is used. $\boldsymbol{e}_s$ and $\boldsymbol{e}_u$ are eigenvectors at saddle point $\boldsymbol{x}^*$ belonging to $\boldsymbol{W}^s$ and $\boldsymbol{W}^u$, respectively.

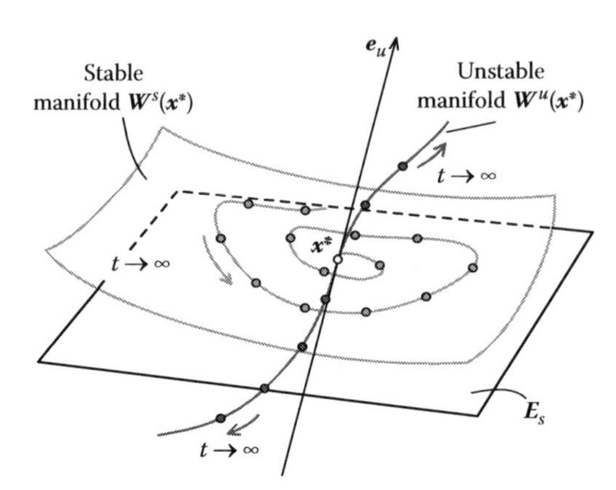

FIGURE 1.16 Stable W^s and unstable W^u manifold ($N = 3$). The CTM is used. $E_s = \text{span}[e_{s1}, e_{s2}]$ stable subspace is tangent of $W^s(x^*)$ at x^*. e_u is an unstable eigenvector at x^*. x^* is saddle point.

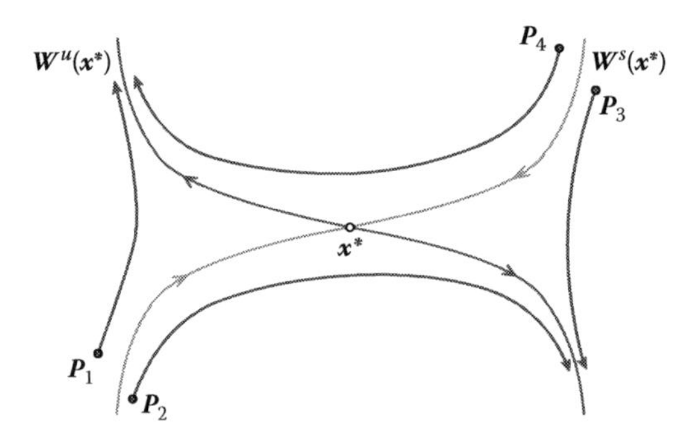

FIGURE 1.17 Initial conditions (P_1, P_2, P_3, P_4) are not placed on any of the invariant manifolds. The trajectories are repelled from $W^s(x^*)$ and attracted by $W^u(x^*)$. Any of the trajectories (orbits) must remain in the region where it was born. $W^u(x^*)$ (or $W^s(x^*)$) are boundaries.

of manifolds: stable manifold denoted by W^s and unstable manifold denoted by W^u. If the initial points are on W^s or on W^u, the operation points remain on W^s or W^u forever, but the points on W^s are attracted by x^* and the points on W^u are repelled from x^*. By considering $t \to -\infty$, every movement along the manifolds is reversed.

If the initial conditions (points P_1, P_2, P_3, P_4 in Figure 1.17) are not on the manifolds, their trajectories will not cross any of the manifolds, they remain in the space bounded by the manifolds. The trajectories are repelled from $W^s(x^*)$ and attracted by $W^u(x^*)$. Any of these trajectories (orbits) must remain in the space where it was born. $W^u(x^*)$ (or $W^s(x^*)$) are boundaries. Consequently, the invariant manifolds organize the state space.

1.7.3 Invariant Manifolds, DTM

Applying DTM, i.e., difference equations describe the system, then mostly the PMF is used. The fixed point x^* must be a saddle point of PMF to have manifolds. Figure 1.18 presents the Ponicaré surface or plane with the stable $W^s(x^*)$ and unstable $W^u(x^*)$ manifold. s_0 and u_0 is the intersection point of the trajectory with the Poincaré surface, respectively. The next intersection point of the same

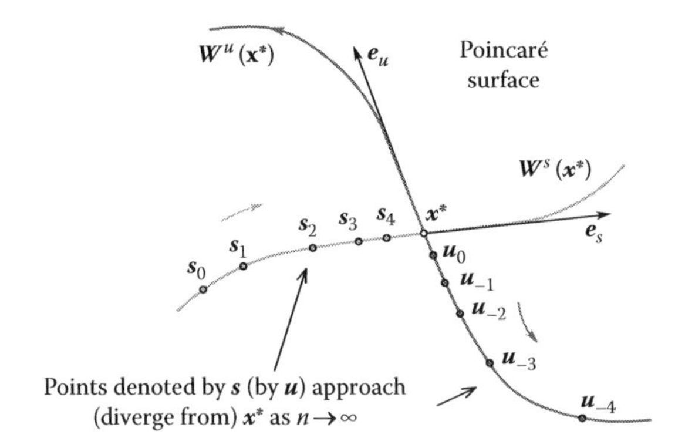

FIGURE 1.18 Stable W^s and unstable W^u manifold. DTM is used. e_s and e_u are eigenvectors at saddle point x^* belonging to W^s and W^u, respectively.

trajectory with the Poincaré surface is s_1 and u_{-1} the following s_2 and u_{-2}, etc. If W^s, W^u are manifolds and s_0, u_0 are on the manifolds, all subsequent intersection point will be on the respective manifold. Starting from infinitely large number of initial point $s_0(u_0)$ on W^s (or on W^u), infinitely large number of intersection points are obtained along W^s (or W^u). Curve $W^s(W^u)$ is determined by using infinitely large number of intersection points.

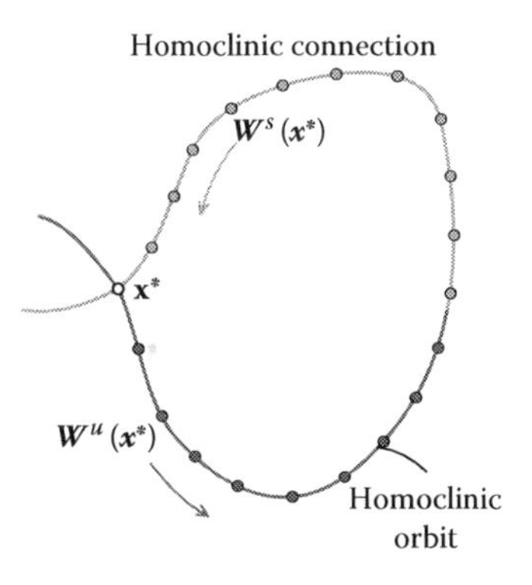

FIGURE 1.19 Homoclinic connection and orbit (CTM).

1.7.4 Homoclinic and Heteroclinic Orbits, CTM

In *homoclinic connection*, the stable W^s and unstable W^u manifold of the same saddle point x^* intersect each other (Figure 1.19). The two manifolds, W^s and W^u, constitute a homolinic orbit. The operation point on the homoclinic orbits approach x^* both in forward and in backward time under the action of the relevant differential equation. In *heteroclinic connection*, the stable manifold $W^s(x_1^*)$ of saddle point x_1^* is connected to the unstable manifold $W^u(x_2^*)$ of saddle point x_2^*, and vice versa (Figure 1.20). The two manifolds $W^s(x_1^*)$ and $W^u(x_2^*)$ [similarly $W^s(x_2^*)$ and $W^u(x_1^*)$] constitute a heteroclinic orbit.

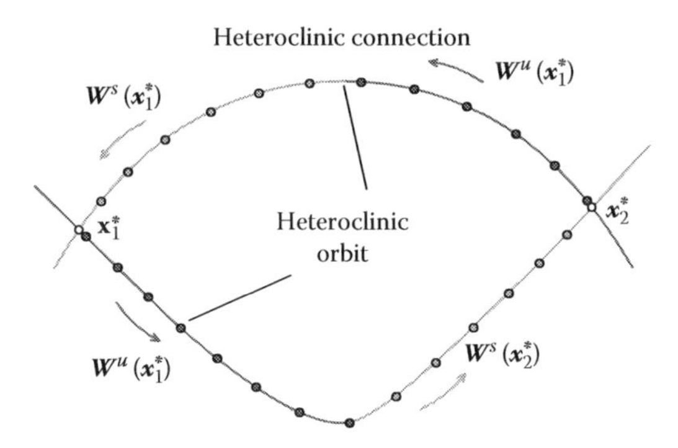

FIGURE 1.20 Heteroclinic connection and orbit (CTM).

1.8 Transitions to Chaos

1.8.1 Introduction

One of the great achievements of the theory of chaos is the discovery of several typical routes from regular states to chaos. Quite different systems in their physical appearance exhibit the same route. The main thing is the universality. There are two broad classes of transitions to chaos: *the local and global bifurcations*. In the first case, for example, one EP or one LC loses its stability as a system parameter is changed. The local bifurcation has three subclasses: period doubling, quasi-periodicity, and intermittency. The most frequent route is the period doubling.

In the second case, the global bifurcation involves larger scale behavior in state space, more EPs, and/or more LCs lose their stability. It has two subclasses, the chaotic transient and the crisis. In this section, only the local bifurcations and the period-doubling route is treated. Only one short comment is made both on the quasi-periodic route and on the intermittency.

In *quasi-periodic route*, as a result of alteration in a parameter, the system state changes first from EP to LC through bifurcation. Later, in addition, another frequency develops by a new bifurcation and the system exhibits quasi-periodic state. In other words, there are two complex conjugate eigenvalues within the unit circle in this state. By changing further the parameter, eventually the chaotic state is reached from the quasi-periodic one.

In the *intermittency route* to chaos, apparently periodic and chaotic states alternately develop. The system state seems to be periodic in certain intervals and suddenly it turns into a burst of chaotic state. The irregular motion calms down and everything starts again. Changing the system parameter further, the length of chaotic states becomes longer and finally the “periodic” states are not restored.

1.8.2 Period-Doubling Bifurcation

Considering now the period-doubling route, let us assume a LC as starting state in a three-dimensional system (Figure 1.21a). The trajectory crosses the Poincaré plane at point P. As a result of changing one system parameter, the eigenvalue λ_1 of the Jacobian matrix of PMF belonging to FP P is moving along the negative real axis within the unit circle toward point -1 and crosses it as λ_1 becomes -1. The eigenvalue λ_1 moves outside the unit circle. FP P belonging to the first iterate loses its stability. Simultaneously, two new stable FPs P_1 and P_2 are born in the second iterate process having two eigenvalues $\lambda_2 = \lambda_3$ (Figure 1.21b). Their value is equal $+1$ at the bifurcation point and it is getting smaller than one as the system parameter is changing further in the same direction. $\lambda_2 = \lambda_3$ are moving along the positive real axis toward the origin.

Continuing the parameter change, new bifurcation occurs following the same pattern just described and the trajectory will cross the Poincaré plane four times (2×2) in one period instead of twice from the new bifurcation point. The period-doubling process keeps going on but the difference between two consecutive parameter values belonging to bifurcations is getting smaller and smaller.

1.8.3 Period-Doubling Scenario in General

The period-doubling scenario is shown in Figure 1.22 where μ is the system parameter, the so-called *bifurcation parameter*, x is one of the system state variables belonging to the FP, or more generally to the intersection points of the trajectory with the Poincaré plane. The intersection points belonging to the transient process are excluded. For this reason, the diagram is called sometime *final state diagram*. The name “bifurcation diagram” refers to the bifurcation points shown in the diagram and it is more generally used.

Feigenbaum [7] has shown that the ratio of the distances between successive bifurcation points measured along the parameter axis μ approaches to a constant number as the order of bifurcation, labeled by k, approaches infinity:

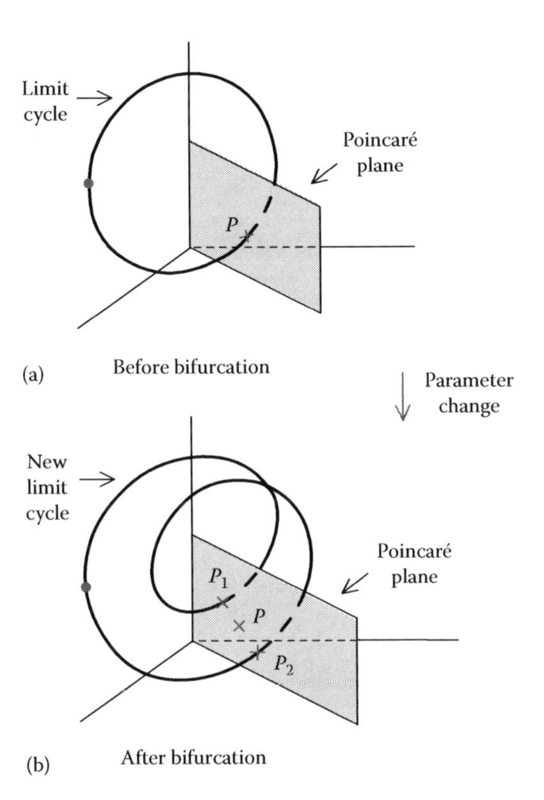

FIGURE 1.21 Period-doubling bifurcation. (a) Period-1 state before, and (b) period-2 state after the bifurcation.

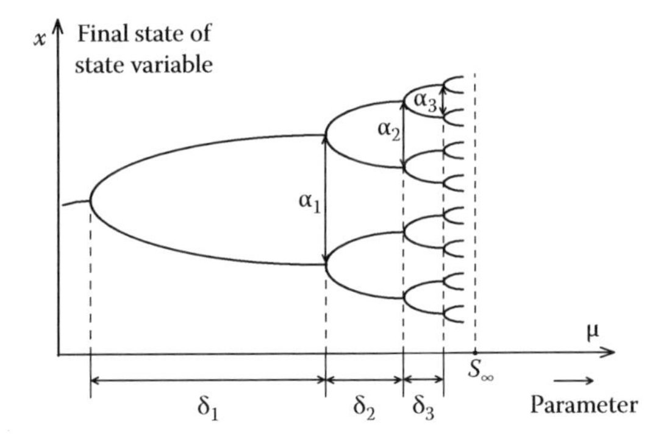

FIGURE 1.22 Final state or bifurcation diagram. S_∞ = Feigenbaum point.

$$\delta = \lim_{k\to\infty} \frac{\delta_k}{\delta_{k+1}} = 4.6693\ldots \tag{1.36}$$

where δ is the so-called *Feigenbaum constant.* It is found that the ratio of the distances measured in the axis x at the bifurcation points approaches another constant number, the so-called *Feigenbaum* α

$$\alpha = \lim_{k\to\infty} \frac{\alpha_k}{\alpha_{k+1}} = 2.5029\ldots \tag{1.37}$$

δ is universal constant in the theory of chaos like other fundamental numbers, for example, $e = 2.718$, π, and the golden mean ratio $(\sqrt{5}-1)/2$.

1.9 Chaotic State

1.9.1 Introduction

The recently discovered new state is the chaotic state. It can evolve only in nonlinear systems. The conditions required for chaotic state are as follows: at least three or higher dimensions in autonomous systems and with at least two or higher dimensions in non-autonomous systems provided that they are described by CTM. On the other hand, when DTM is used, two or more dimensions suffice for invertible iteration functions and only one dimension (e.g., logistic map) or more for non-invertible iteration functions are needed for developing chaotic state. In addition to that, some other universal qualitative features common to nonlinear chaotic systems are summarized as follows:

- The systems are deterministic, the equations describing them are completely known.
- They have extreme sensitive dependence on initial conditions.
- Exponential divergence of nearby trajectories is one of the signatures of chaos.
- Even though the systems are deterministic, their behavior is unpredictable on the long run.
- The trajectories in chaotic state are non-periodic, bounded, cannot be reproduced, and do not intersect each other.
- The motion of the trajectories is random-like with underlying order and structure.

As it was mentioned, the trajectories setting off from initial conditions approach after the transient process either FPs or LC or Qu-P curves in dissipative systems. All of them are called *attractors* or *classical attractors* since the system is attracted to one of the above three states. When a chaotic state evolves in a system, its trajectory approaches and sooner or later reaches an attractor too, the so-called *strange attractor.*

In three dimensions, the classical attractors are associated with some geometric form, the stationary state with point, the LC with a closed curve, and the quasi-periodic state with surface. The strange attractor is associated with a new kind of geometric object. It is called a *fractal structure*.

1.9.2 Lyapunov Exponent

Conceptually, the Lyapunov exponent is a quantitative test of the sensitive dependence on initial conditions of the system. It was stated earlier that one of the properties of chaotic systems is the exponential divergence of nearby trajectories. The calculation methods usually apply this property to determine the Lyapunov exponent λ.

After the transient process, the trajectories always find their attractor belonging to the special initial condition in dissipative systems. In general, the attractor as reference trajectory is used to calculate λ. Starting two trajectories from two nearby initial points placed from each other by small distance d_0, the distance d between the trajectories is given by

$$d(t) = d_0 e^{\lambda t} \tag{1.38}$$

where λ is the Lyapunov exponent. One of the initial points is on the attractor. The equation can hold true only locally because the chaotic systems are bounded, i.e., $d(t)$ cannot increase to infinity. The value λ may depend on the initial point on the attractor. To characterize the attractor by a Lyapunov exponent, the calculation described above has to be repeated for a large number of n of initial points distributed along the attractor. Eventually, the average Lyapunov exponent $\bar{\lambda}$ calculated from the individual λ_n values will characterize the state of the system. The criterion for chaos is $\bar{\lambda} > 0$. When $\bar{\lambda} \leq 0$, the system is in regular state (Figure 1.23).

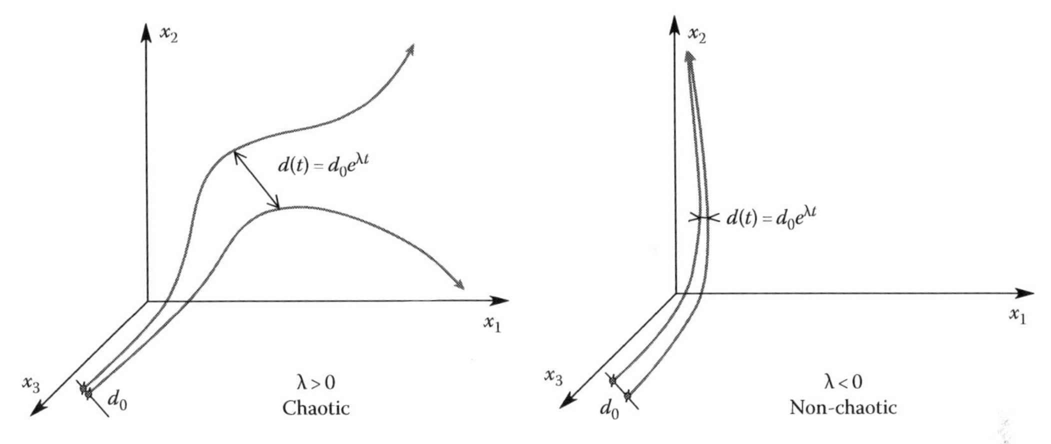

FIGURE 1.23 Positive and negative Lyapunov exponents. The sensitive dependence on initial conditions.

1.10 Examples from Power Electronics

1.10.1 Introduction

Power electronic systems have wide applications both in industry and at home. The structure of these systems keeps changing duc to consecutive turn-on and turn-off processes of electronic switches. They are *variable structure*, piece-wise linear, nonlinear dynamic controlled systems. The primary source of nonlinearity in these systems is that the switching instants depend on the state variables [1]. Furthermore, the frequent nonlinearities can also be found in the systems. In order to show the applications of the theory of nonlinear dynamics summarized in the previous sections, five examples have been selected from the field of power electronics [15,16]. Various system states and bifurcations will be presented without the claim of completeness in this section. Our main objective is to direct the attention of the readers, engineers to the possible strange phenomenon that can be encountered in power electronics and explained by the theory of nonlinear dynamics.

1.10.2 High-Frequency Time-Sharing Inverter

1.10.2.1 The System

It is applied for induction heating where high-frequency power supply is required with frequency of several thousand hertz Figure 1.24 shows the inverter configuration in its simplest form. I_+ and I_-, the so-called positive and negative sub-inverters, are encircled by dotted lines.

The parallel oscillatory circuit L_p–C_p–R_p represents the load. The supply is provided by a center-tapped DC voltage source. In order to explain the mode of action of the inverter, ideal components are assumed. The basic operation of the inverter can be understood by the time functions of Figure 1.25. In Figure 1.25a, the "high" frequency approximately sinusoidal output voltage v_0, the inverter output current pulses i_0, and the condenser voltage v_c can be seen. Thyristors T1 and T2 are alternately fired at instants located at every sixth zero crossings on the positive and on the negative slope of the output voltage, respectively. After firing a thyristor, an output current pulse i_0 is flowing into the load, which changes the polarity of the series condenser voltage v_c, for instance, from $-V_{cm}$ to V_{cm}. The thyristor turn-off time can be a little bit longer than two and half cycles of the output voltage (Figure 1.25b). Figure 1.26 presents a configuration with three positive and three negative sub-inverters. Here, the respective input and output terminals of the sub-inverters are paralleled. The numbering of the sub-inverters corresponds to the firing order. Each sub-inverter pair works in the same way as it was previously described.

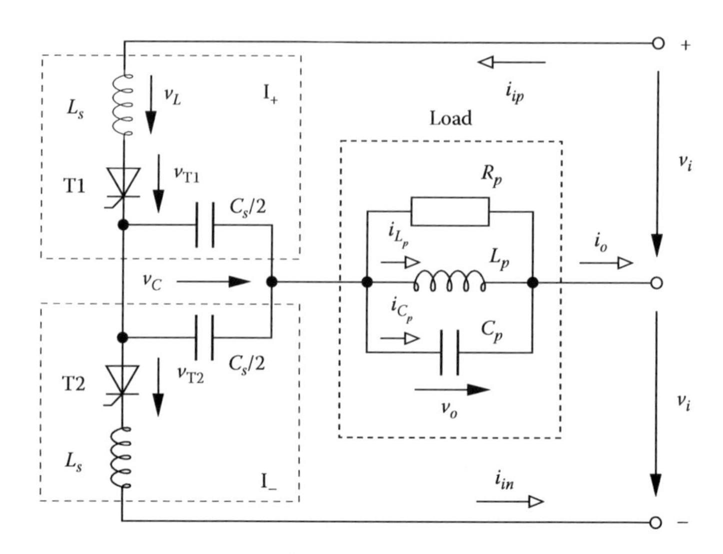

FIGURE 1.24 The basic configuration of the inverter.

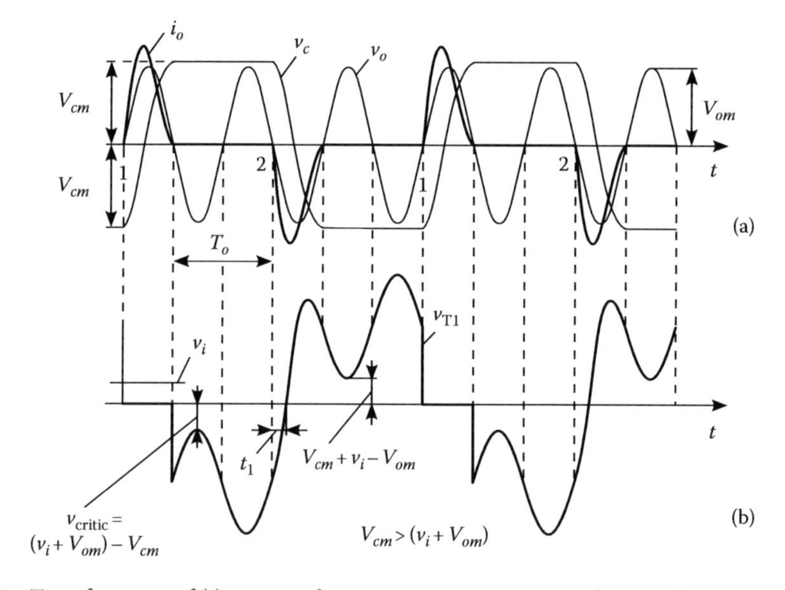

FIGURE 1.25 Time functions of (a) output voltage v_o, output current i_o, condenser voltage v_c, and (b) thyristor voltage v_{T1} in basic operation.

1.10.2.2 Results

One of the most interesting results is that by using an approximate model assuming sinusoidal output voltage, no steady-state solution can be found for the current pulse and other variables in certain operation region and its theoretical explanation can be found in Ref. [6]. The laboratory tests verified this theoretical conclusion. To describe the phenomena in the region in quantitative form, a more accurate model was used. The independent energy storage elements are six L_s series inductances, three C_s series capacitances, one L_p inductance, and one C_p capacitance, altogether 11 elements, with 11 state variables. The accurate analysis was performed by simulation in MATLAB® environment for open- and for closed-loop control.

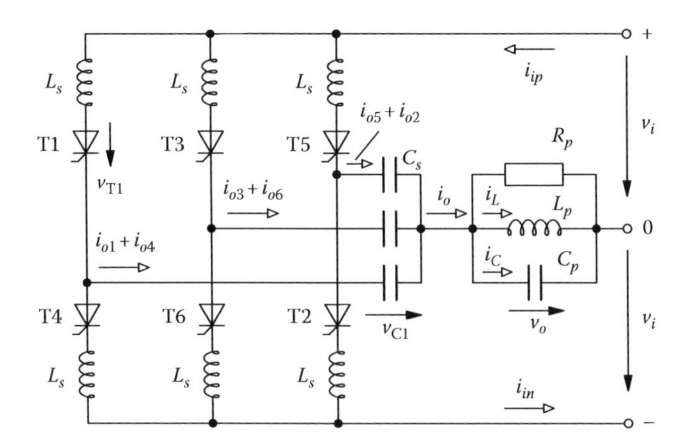

FIGURE 1.26 The power circuit of the time-sharing inverter. (Reprinted from Bajnok, M. et al., Surprises stemming from using linear models for nonlinear systems: Review. In *Proceedings of the 29th Annual Conference on Industrial Electronics, Control and Instrumentation (IECON'03)*, Roanoke, VA, vol. I, pp. 961–971, November 2–6, 2003. CD Rom ISBN:0-7803-7907-1. © [2003] IEEE. With permission.)

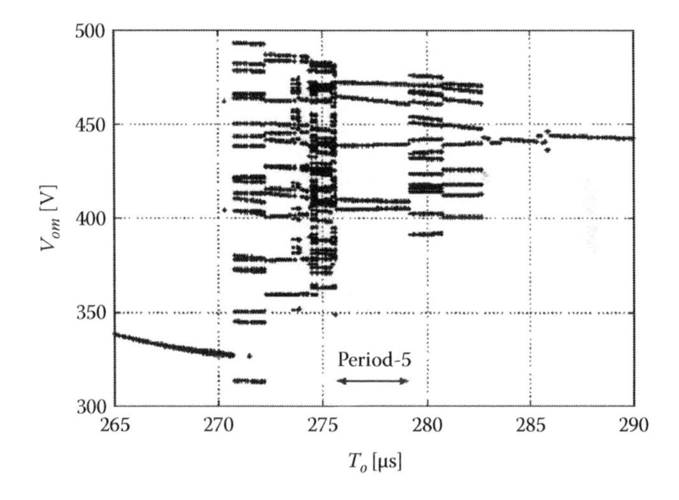

FIGURE 1.27 Bifurcation diagram of inverter with open-loop control.

1.10.2.3 Open-Loop Control

Bifurcation diagram was generated here, the peak values of the output voltage V_{om} were sampled, stored, and plotted as a function of the control parameter T_o, where T_o is the period between two consecutive firing pulses in the positive or in the negative sub-inverters (Figure 1.27). Having just one single value V_{om} for a given T_o, the output voltage v_o repeats itself in each period T_o. This state is called period-1. Similarly, period-5 state develops, for example, at firing period $T_o = 277$ μs. Now, there are five consecutive distinct V_{om} values. v_o is still periodic, it repeats itself after $5T_o$ has elapsed (Figure 1.28).

1.10.2.4 Closed-Loop Control

A self-control structure is obtained by applying a feedback control loop. Now, the approximately sinusoidal output voltage v_o is compared with a DC control voltage V_{DC} and the thyristors are alternatively fired at the crossing points of the two curves. The study is concerned with the effect of the variation of the DC control voltage level V_{DC} on the behavior of the feedback-controlled inverter. Again, the bifurcation diagram is used for the presentation of the results (Figure 1.29). The peak values of the output

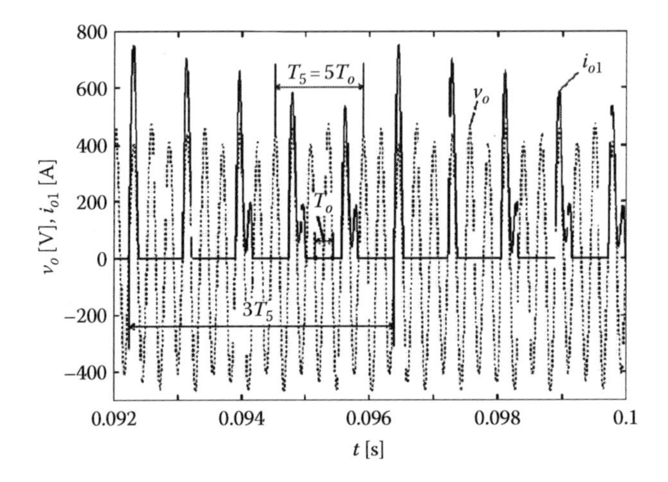

FIGURE 1.28 $v_o(t)$ and $i_{o1}(t)$. $T_o = 277\,\mu s$. Period-5 state. (Reprinted from Bajnok, M. et al., Surprises stemming from using linear models for nonlinear systems: Review. In *Proceedings of the 29th Annual Conference on Industrial Electronics, Control and Instrumentation (IECON'03)*, Roanoke, VA, vol. I, pp. 961–971, November 2–6, 2003. CD Rom ISBN:0-7803-7907-1. © [2003] IEEE. With permission.)

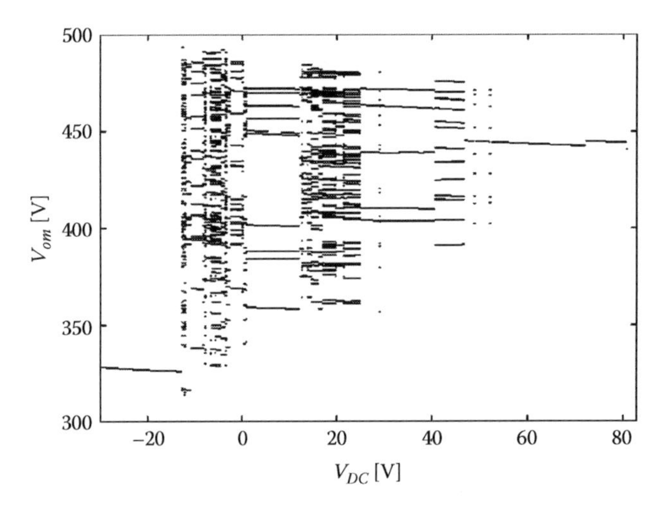

FIGURE 1.29 Bifurcation diagram of inverter with feedback control loop. (Reprinted from Bajnok, M. et al., Surprises stemming from using linear models for nonlinear systems: Review. In *Proceedings of the 29th Annual Conference on Industrial Electronics, Control and Instrumentation (IECON'03)*, Roanoke, VA, vol. I, pp. 961–971, November 2–6, 2003. CD Rom ISBN:0-7803-7907-1. © [2003] IEEE. With permission.)

voltage V_{om} were sampled, stored, and plotted as a function of the control parameter V_{DC}. The results are basically similar to those obtained for open-loop control. As in the previous study, the feedback-controlled inverter generates subharmonics as the DC voltage level is varied.

1.10.3 Dual Channel Resonant DC–DC Converter

1.10.3.1 The System

The dual channel resonant DC–DC converter family was introduced earlier [3]. The family has 12 members. A common feature of the different entities is that they transmit power from input to output through two channels, the so-called positive and negative ones, coupled by a resonating capacitor. The converter can operate both in symmetrical and in asymmetrical mode. The respective variables in the two channels

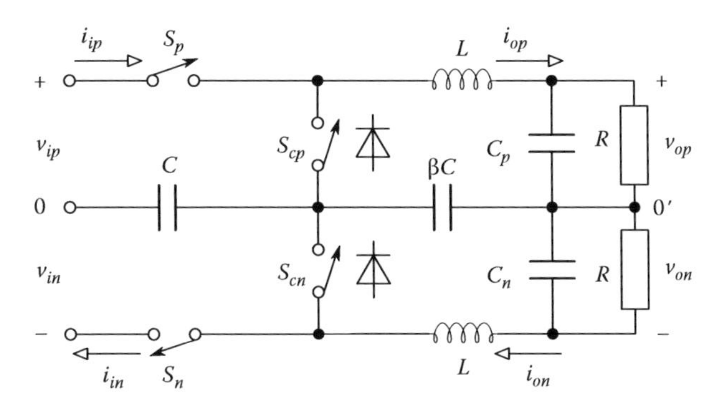

FIGURE 1.30 Resonant buck converter with double load.

vary symmetrically or asymmetrically in the two modes. The energy change between the two channels is accomplished by a capacitor in asymmetrical operation. There is no energy exchange between the positive and the negative channels in the symmetrical case. Our description is restricted to the buck configuration (Figure 1.30) in symmetrical operation. Suffix i and o refer to input and output while suffix p and n refer to positive and negative channel, respectively. Two different converter versions can be derived from the configuration. The first version contains diodes in place of clamping switches S_{cp} and S_{cn}. The second one applies controlled switches conducting current in the direction of arrow. To simplify, the configuration capacitance βC is replaced by short circuit and S_{cp}, S_{cn} are applied. The controlled switches within one channel are always in complementary states (i.e., when S_p is on, S_{cp} is off, and vice versa). By turning on switch S_p, a sinusoidal current pulse i_{ip} is developed from $\omega t = 0$ to α_p ($\omega = 1/\sqrt{LC}$ in circuit S_p, L, v_{op}, C, and v_{ip} (Figure 1.31)). The currents are $i_{ip} = i_{op} = i_{cp}$ in interval $0 < \omega t < \alpha_p$. The capacitor voltage v_c swings from V_{cn} to V_{cp} ($V_{cn} < 0$). By turning on switch S_{cp} at α_p, the choke current commutates from S_p to S_{cp}. The energy stored in the choke at $\omega t = \alpha_p$ is depleted in the interval $\alpha_p < \omega t < \omega T_s$ where $T_s = 1/f_s$ is the switching period. At discontinuous current-conduction mode (DCM), the stored energy is entirely depleted in interval $\alpha < \omega t < \alpha_{ep}$ where α_{ep} denotes the extinction angle of the inductor

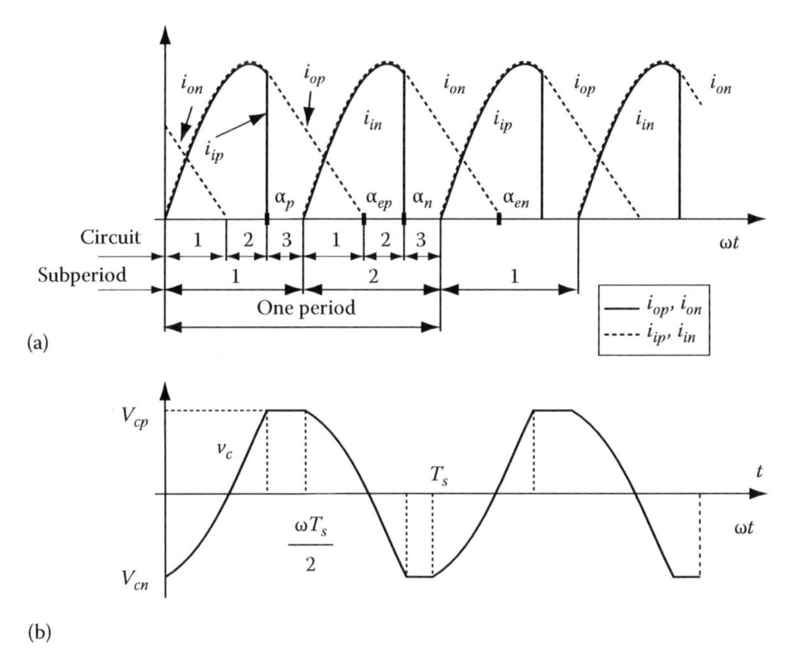

FIGURE 1.31 Time functions of (a) input and output currents and (b) condenser voltage.

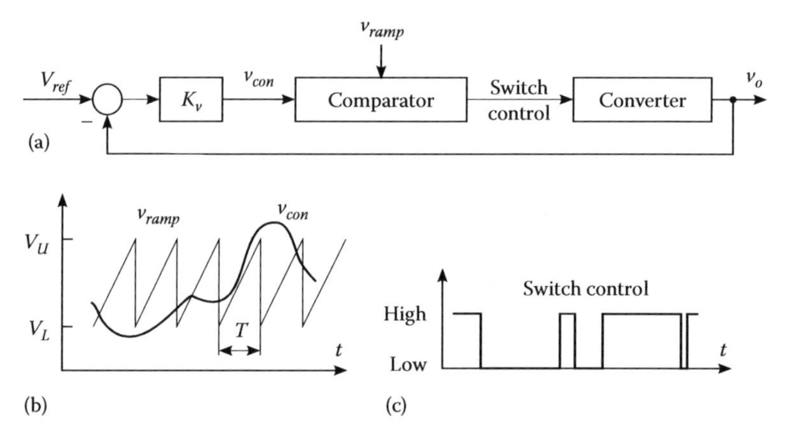

FIGURE 1.32 PWM control: (a) Block diagram, (b) input and (c) output signals of the comparator.

current. In DCM, the current is zero between α_{ep} and ωT_s. In the continuous current-conduction mode (CCM) of operation, the inductor current flows continuously ($i_{op} > 0$). The inductor current i_{op} decreases in both cases in a linear fashion. After turning on S_{cp}, the capacitor voltage v_c stops changing. It keeps its value V_{cp}.

The same process takes place at the negative side, resulting in a negative condenser current pulse and voltage swing after turning on S_n.

1.10.3.2 The PWM Control

For controlling the output voltage $v_o = v_{op} + v_{on}$ by PWM switching, a feedback control loop is applied (Figure 1.32a). The control signal v_{con} is compared to the repetitive sawtooth waveform (Figure 1.32b). The control voltage signal v_{con} is obtained by amplifying the error, the difference between the actual output voltage v_o, and its desired value V_{ref}. When the amplified error signal v_{con} is greater than the sawtooth waveform, the switch control signal (Figure 1.32c) becomes high and the selected switch turns on. Otherwise, the switch is off. The controlled switches are S_p and S_n (the switches within one channel, e.g., S_p and S_{cp} are in complementary states), and they are controlled alternatively, i.e., the switch control signal is generated for the switch in one channel in one period of the sawtooth wave and in the next period, the signal is generated for the switch in the other channel. The switching frequency of these switches is half of the frequency of the sawtooth wave.

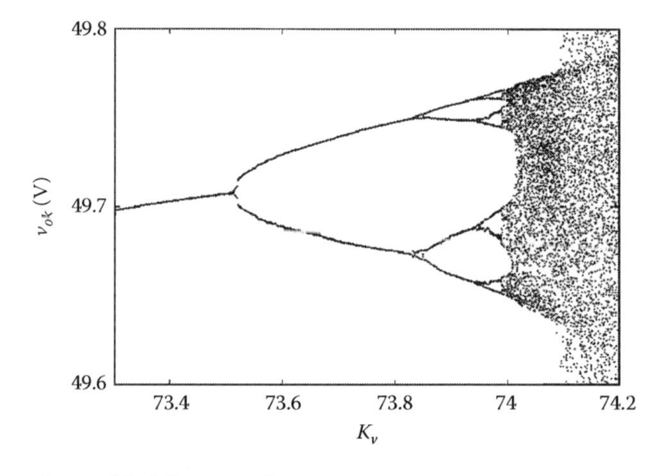

FIGURE 1.33 Enlarged part of the bifurcation diagram.

1.10.3.3 Results

TABLE 1.1 Bifurcation Points in Figure 1.33

k	μ_k	δ_k
0	73.501	
1	73.841	4.0476
2	73.925	4.2000
3	73.945	

The objective is the calculation of the controlled variable v_o in steady state in order to discover the various possible states of this nonlinear variable structure dynamical system. The analysis was performed by simulation in MATLAB® environment. The effect of variation K_v was studied. A representative bifurcation diagram is shown in Figure 1.33. To generate this kind of diagram, the output voltage v_o was sampled and stored, in steady state, at the start of every switching cycle ($v_{ok} = v_o(kT_s)$). With sufficient number of sets of steady-state data, the bifurcation diagram can be obtained by plotting vertically the sampled output voltage whereas the voltage gain K_v is drawn horizontally as a bifurcation parameter. Beautiful period-doubling scenario can be seen. Denoting the first bifurcation point by μ_0 where $\mu_0 = K_v = 73.502$, the next one by μ_1, etc., the values of the first several bifurcation points are in Table 1.1. Table 1.1 contains δ_1 and δ_2 where $\delta_k = (\mu_k - \mu_{k-1})/(\mu_{k+1} - \mu_k)$. Even the first two δ_k values are not far from the Feigenbaum constant [see (1.36)].

The time function of condenser voltage v_c is plotted in Figure 1.34 at $K_v = 73.2$. The period-9 subharmonic state is clearly visible. The period of v_c, together with the period of the other state variables, is $9 \cdot 2T = 9 \cdot T_s$ in period-9 state.

Figure 1.35 shows the Poincaré map for chaotic behavior at $K_v = 74.2$ in plane of output voltage $v_{ok} = v_o(kT_s)$ vs. inductor current $i_{opk} = i_{op}(kT_s)$. This map reveals a highly organized structure. Note also that it lies in a bounded region of state space.

1.10.4 Hysteresis Current-Controlled Three-Phase VSC

1.10.4.1 The System

The objective is to highlight the complex behavior of a three-phase DC–AC voltage source vonverter (VSC) controlled by a closed-loop hysteresis AC current controller (HCC) [10]. The block diagram of the system is shown in Figure 1.36. It consists of a three-phase full-bridge VSC with six bidirectional switches, a simple circuit modeling the AC side, a reference frame transformation, a comparator, and the HCC. All voltage and current vectors used are three-phase space vectors. Space vector is the complex representation of three-phase quantities. The three-phase AC terminals of the VSC are tied through series L–R components to voltage space vector $\bar{e}_s$ rotating with angular speed ω. This simple circuit can approximately model either the AC mains or an induction motor. Space vectors with suffix s are written in stationary reference frame (SRF). There are some benefits by using rotating reference frame (RRF) fixed to voltage space vector $\bar{e}_s$ over SRF. Space vectors without suffix s are in RRF.

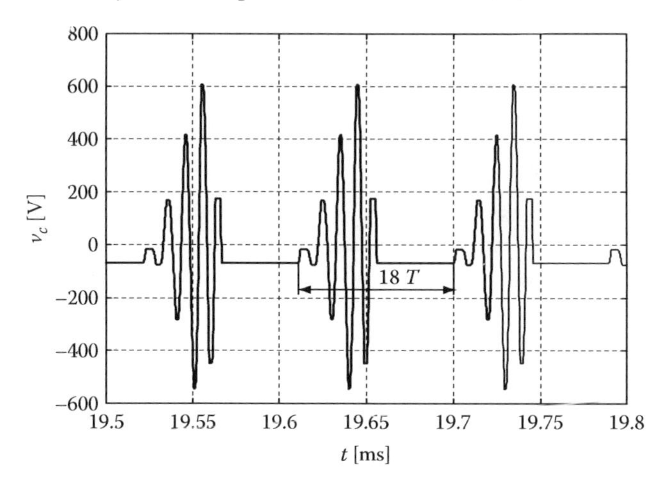

FIGURE 1.34 Condenser voltage in period-9 operation.

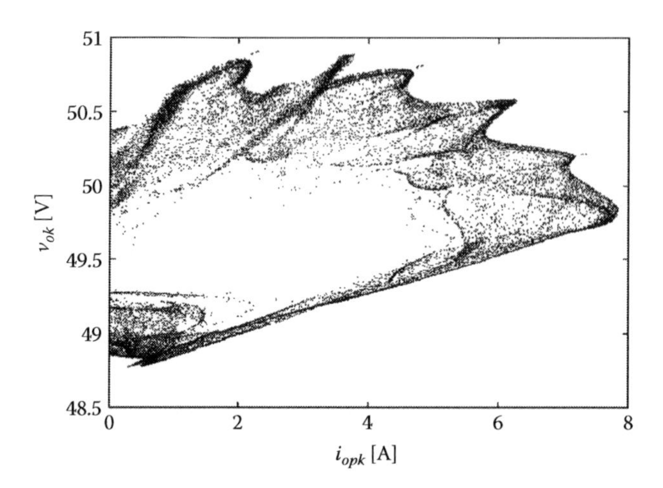

FIGURE 1.35 Poincaré map for chaotic operation ($K_v = 74.2$).

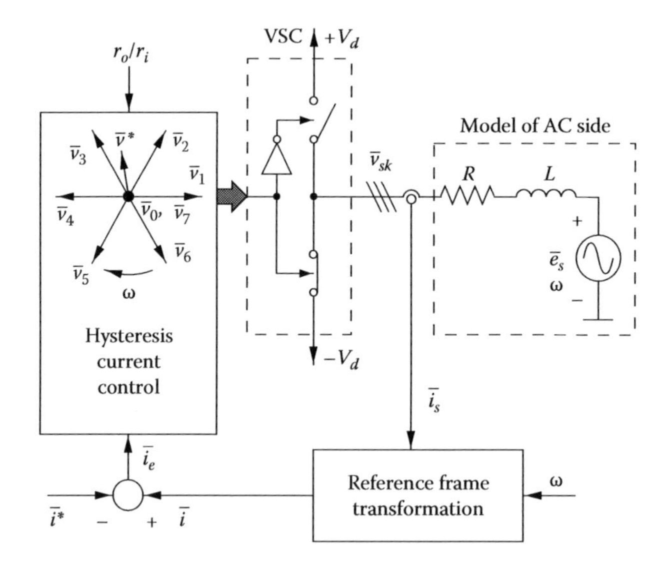

FIGURE 1.36 Voltage source converter with its hysteresis AC current controller. (Reprinted from Sütő, Z. and Nagy, I., *Applied Electromagnetics and Computational Technology II*, Eds H. Tsuboi and I. Vajda, IOS Press, Amsterdam, the Netherlands, pp. 233–244, 2000. With permission.)

1.10.4.2 Hysteresis Control

The controller is a somewhat sophisticated hysteresis AC current control loop. At first, the AC current $\overline{i}_s$ is transformed into RRF revolving with angular frequency ω. In RRF, the reference current $\overline{i}^*$ is a stationary space vector. The shape of the tolerance band (TB) centered around the end point of $\overline{i}^*$ is assumed to be a circle. The HCC selects the switching state of the VSC to keep the space vector of error current $\overline{i}_e$ within the TB. Whenever $\overline{i}_e$ reaches the periphery of TB, a new converter voltage $\overline{v}_k$ is produced by the new switching state of VSC. The new $\overline{v}_k$ forces to rebound $\overline{i}_e$ from the periphery of TB to the interior of TB. To avoid certain shortcomings, two concentric circles as TBs are applied with radius r_i and r_o ($r_o > r_i$).

Figure 1.37a shows a small part of the bifurcation diagram of the system. The sample points of the real component of the error current vector $\overline{i}_e$ are on the vertical axis. T_s is the sampling time, which

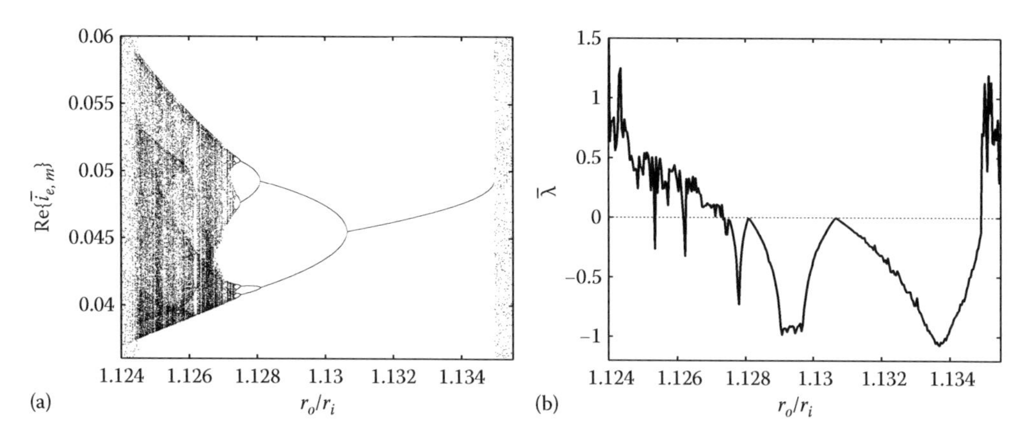

FIGURE 1.37 (a) Period-doubling cascade. (b) Lyapunov exponent. (Reprinted from Sütő, Z. and Nagy, I., *Applied Electromagnetics and Computational Technology II*, Eds H. Tsuboi and I. Vajda, IOS Press, Amsterdam, the Netherlands, pp. 233–244, 2000. With permission.)

is the sixth of the period of vector $\bar{e}_s$, $m = 1, 2\ldots$ The bifurcation parameter is the ratio r_o/r_i. A beautiful period-doubling cascade can be seen. Proceeding from right to left, after a short chaotic range, suddenly the system response becomes periodic from $r_o/r_i \simeq 1.1349$ to 1.1307. Starting from the latter value down to $r_o/r_i \cong 1.1281$, the period is doubled. The response remains periodic with second-order subharmonic oscillation. Period doubling occurs again from $r_o/r_i \cong 1.1281$ and fourth-order subharmonic appears. After a series of period doubling at successively closer border values of r_o/r_i, the system trajectory finally becomes chaotic. Figure 1.37b presents the Lyapunov exponent for the region of the period-doubling cascade showing that at the bifurcation points, the Lyapunov exponent becomes zero. The Poincaré section of the system is shown in Figure 1.38 for a chaotic state at $r_o/r_i = 1.125$. The trajectory is restricted to a smaller region of state space at this r_o/r_i value. The Poincaré section is plotted by the real and imaginary components of the sampled error current space vector. The structure and order of the diagram can be considered as an outcome of the deterministic background of the chaotic systems.

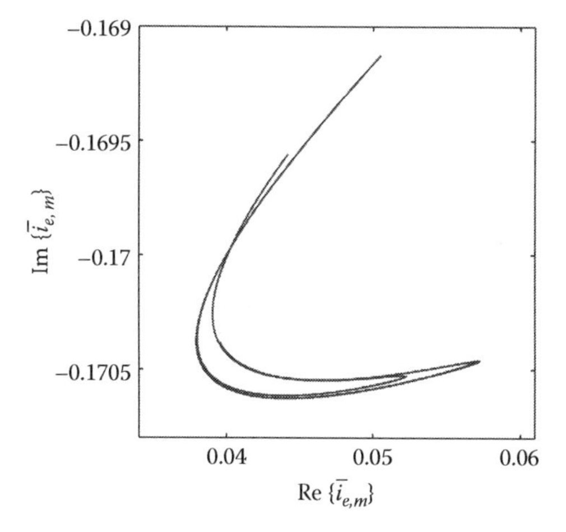

FIGURE 1.38 Poincaré section at $r_o/r_i = 1.125$. The trajectory is restricted to a smaller region of state space. (Reprinted from Sütő, Z. and Nagy, I., *Applied Electromagnetics and Computational Technology II*, Eds H. Tsuboi and I. Vajda, IOS Press, Amsterdam, the Netherlands, pp. 233–244, 2000. With permission.)

1.10.5 Space Vector Modulated VSC with Discrete-Time Current Control

1.10.5.1 The System [11]

In this example, the system studied is a VSC connected to a symmetrical three-phase configuration on the AC side consisting of series *R*–*L* circuits in each phase and a three-phase symmetrical voltage source $\bar{e}_s$ where $\bar{e}_s$ is space vector, the complex representation of three-phase quantities (Figure 1.39). This simple circuit can model either the AC mains or an induction or synchronous machine in steady-state operation. The VSC is a simple three-phase bridge connection with six controlled lossless bidirectional switches. On the DC side, a constant and ideal voltage source with voltage V_{DC} is assumed. This power part of the system is the same as the one in our previous example. The three-phase AC variables are written in space vector form. The space vectors can be expressed in stationary α-β reference frame (SRF) with suffix *s* or in rotating *d*–*q* reference frame (RRF) revolving with ω, the angular frequency of space vector $\bar{e}_s$ and denoted without suffix *s*. The real and imaginary components of space vectors in RRF are distinguished by means of suffix *d* and *q*, respectively.

1.10.5.2 The Control and the Modulation

To track the reference current $\bar{i}^*$, which is usually the output signal of an other controller, by the actual AC current $\bar{i}$, a digital current controller (DCC) is applied. The DCC consists of a Sample&Hold (S&H) unit, a reference frame transformation transforming the three-phase current $\bar{i}_s$ from SRF to RRF with *d*–*q* coordinates, a proportional-integral (PI) digital controller with complex parameters $\bar{K}_0$ and $\bar{K}_1$ including a saturation function $\bar{s}(\bar{y})$ to avoid overmodulation in VSC, and a SVM algorithm. Suffix *m* denotes the discrete-time index, and the z^{-1} blocks delay their input signals with one sampling period T_s. The SVM has to generate a switching sequence with proper timing of state changes in VSC over a fixed modulation period that equals to T_s.

1.10.5.3 Analysis and Results

Omitting here all the mathematics, only the final results are presented. Interested readers can turn to paper [11]. One crucial point in dynamical systems is the stability. It was studied by the behavior of the FPs in the Poincaré plane. Only one FP X_I can exist in the linear region when $\bar{u}_m = \bar{y}_m$, but there are three FPs, X_I, X_{II+}, X_{II-} in the saturation region. It can be shown that the complex control parameters $\bar{K}_0$ and $\bar{K}_1$ can be calculated from the following equations:

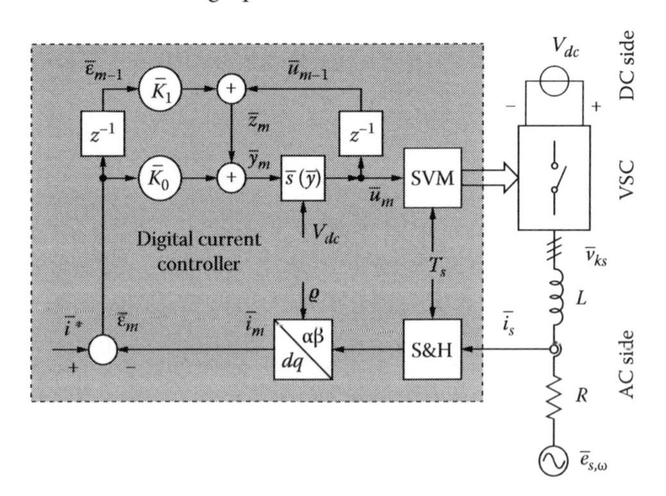

FIGURE 1.39 Block diagram of current-controlled voltage source converter with space vector modulation. (Reprinted from Sütő, Z. and Nagy, I., *IEEE Trans. Cir. Syst: Fund. Theor. Appl.*, 50(8), 1064, 2003. © [2003] IEEE. With permission.)

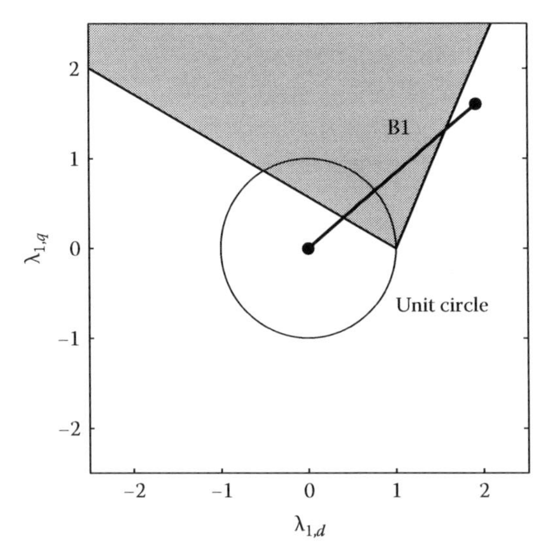

FIGURE 1.40 Bifurcation direction B1 selected to enter and cross the existence region of fixed points $X_{II\pm}$. (Reprinted from Bajnok, M. et al., Surprises stemming from using linear models for nonlinear systems: Review. In *Proceedings of the 29th Annual Conference on Industrial Electronics, Control and Instrumentation (IECON'03)*, Roanoke, VA, vol. I, pp. 961–971, November 2–6, 2003. CD Rom ISBN: 0-7803-7907-1. © [2003] IEEE. With permission.)

$$\bar{K}_0 = \frac{1 + \bar{F} - \bar{\lambda}_0 - \bar{\lambda}_1}{\bar{G}} \tag{1.39}$$

$$\bar{K}_1 = \frac{\bar{\lambda}_0 \bar{\lambda}_1 - \bar{F}}{\bar{G}} \tag{1.40}$$

where

$\bar{\lambda}_0$ and $\bar{\lambda}_1$ are the eigenvalues of the Jacobian matrix calculated at the FP in the linear region
$\bar{F}$ and $\bar{G}$ are complex constants

Selecting $\bar{\lambda}_0$ and $\bar{\lambda}_1$ on the basis of the dynamic requirements, $\bar{K}_0$ and $\bar{K}_1$ are determined from the last two equations. The eigenvalues $\bar{\lambda}_0$ and $\bar{\lambda}_1$ must be inside the unit circle to ensure stable operation. It can be shown that all three FPs can develop in the shaded area in Figure 1.40 even within the unit circle. $\bar{\lambda}_0$ was kept constant in Figure 1.40 at value $\bar{\lambda}_0 = 0.5e^{j45°}$ and $\bar{\lambda}_1$ was changed along line B1, by modifying $\bar{\lambda}_1 = \mu e^{j40°}$ where μ is the bifurcation parameter.

Bifurcation diagram was calculated and it is shown in Figure 1.41a. In the analysis $\bar{x} = \bar{Z}\bar{i} + \bar{e}$ was introduced in place of current $\bar{i}$, where $\bar{Z} = R + j\omega L$ and $\bar{x} = x_d + jx_q$. Since $\bar{\lambda}_1$ crosses the existence region of FPs $X_{II\pm}$, at the critical value $\mu_A = 0.5301$, the FPs $X_{II\pm}$ develop in the state space. Reaching the other critical value $\mu_B = 2.053$, they collide and disappear (Figure 1.41a). There are three more bifurcation values of μ between μ_A and μ_B.

The magnitudes of eigenvalues belonging to FPs $X_{II\pm}$ are also presented in Figure 1.41b. Eigenvalues show that a saddle X_{II-} and a node X_{II+} are born at the left side at μ_A. This bifurcation can be classified as fold bifurcation. By increasing μ very near to μ_A, a spiral node develops from the node due to the creation of a pair of conjugate complex eigenvalues from the merge of two real eigenvalues at X_{II+}.

Looking at the eigenvalues, another critical value μ_3 at 1.292 can be seen. Here, spiral node X_{II+} becomes unstable, the pair of conjugate complex eigenvalues leaves the unit circle. The FP X_{II+} becomes a spiral

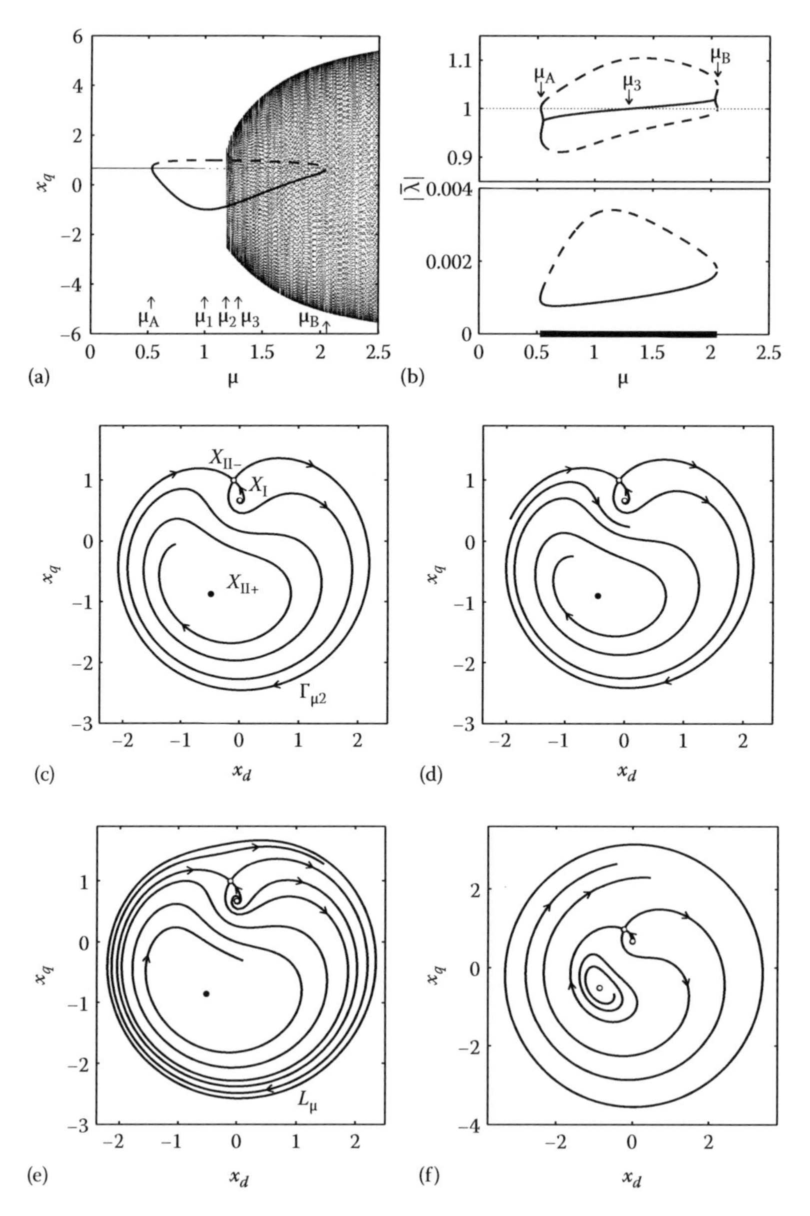

FIGURE 1.41 (a) Bifurcation diagram along B1. (b) Magnitudes of eigenvalues belonging to $X_{II\pm}$ (continuous curves to X_{II+} while dashed curves to X_{II-}). (c) Global phase portrait at the homoclinic connection, $\mu=\mu_2$. (d) Global phase portrait before the homoclinic connection, $\mu = 1.17 < \mu_2$. (e) Global phase portrait after the homoclinic connection. Limit cycle developed. $\mu = 1.2 > \mu_2$. (f) Global phase portrait after fixed point X_{II+} became unstable. $\mu=1.4 > \mu_3$. (Reprinted from Bajnok, M. et al., Surprises stemming from using linear models for nonlinear systems: Review. In *Proceedings of the 29th Annual Conference on Industrial Electronics, Control and Instrumentation (IECON'03)*, Roanoke, VA, vol. I, pp. 961–971, November 2–6, 2003. CD Rom ISBN: 0-7803-7907-1. © [2003] IEEE. With permission.)

saddle at μ_3. Just before μ_B, the curve of the complex eigenvalue splits into two branches indicating that two real eigenvalues develop, and the spiral saddle becomes a saddle. At μ_B, the two saddles disappear. In non-smooth nonlinear system, as in our case, when a pair of FPs appears at some bifurcation value and one FP is unstable and the other one is either stable or unstable, the bifurcation is also called *border collision pair bifurcation*.

When a pair of conjugate complex eigenvalues of a FP leaves the unit circle, the so-called *Neimark-Sacker bifurcation* occurs, which is a kind of local bifurcation. In that case, a closed curve looking like a limit cycle develops in the x_d–x_q plane corresponding to a quasi-periodic "orbit" in the state space. Looking at Figure 1.41b, a pair of conjugate complex eigenvalues of FP X_{II+} leaving the unit circle at the critical value μ_3 can be detected. Indeed, a "limit cycle" develops, as can be seen from the global phase portrait shown in Figure 1.41f. The bifurcation diagram supports this finding but seemingly there is no difference in the bifurcation diagram before and after μ_3. The explanation is that at the critical value μ_2, a global bifurcation, the so-called *homoclinic bifurcation* of the saddle X_{II-} takes place. In Figure 1.41c, the global structure of manifolds belonging to X_{II-} is presented at the critical value μ_2. It can be seen that one branch of unstable manifolds connects to the stable manifold of X_{II-} creating a closed so-called "big" *homoclinic connection* denoted as Γ_{μ_2}. The other unstable and stable branches of manifolds of X_{II-} are directed to the stable X_{II+} and come from the unstable fixed point X_I, respectively. Either increasing or decreasing the bifurcation parameter in respect to μ_2, the homoclinic connection disappears. It is a topologically unstable structure of the state space. At both sides of the homoclinic bifurcation, global phase portraits are calculated (see in Figure 1.41d and e, respectively.) Decreasing μ from μ_2, the unstable manifold comes to the inner side of the stable manifold, and finally is directed to the stable FP X_{II+}. However, when μ is increased over μ_2, the unstable manifold runs to the other side of the stable manifold of X_{II-} and connects to a stable "limit cycle" L_μ that surrounds all three FPs. So, in the region $[\mu_2; \mu_3]$, two attractors, the stable "limit cycle" L_μ and the spiral node X_{II+} exist. When μ_3 is reached, the spiral node X_{II+} becomes a spiral saddle, the only attractor remained is the "limit cycle" L_μ (Figure 1.41f).

The various system states and bifurcations are due to the variations of the complex coefficients $\bar{K}_0$, $\bar{K}_1$ of digital PI controller. Based on the approximate PMF, design conditions were derived to avoid the undesired operations of the system and to select the coefficients $\bar{K}_0$, $\bar{K}_1$ of the controller [11].

1.10.6 Direct Torque Control

1.10.6.1 The System

The control methods applied frequently for induction machine are the field-oriented control (FOC) and the direct torque control (DTC) [2,13]. In DTC, the instantaneous value of the electric torque τ and that of the absolute value ψ of stator flux linkage are estimated from the space vector of stator voltage $\bar{v}$ and stator current $\bar{i}$ and compared to their respective reference value τ^r and ψ^r (Figure 1.42). The error signals $\tau_e = \tau^r - \tau$ and $\psi_e = \psi^r - \psi$ determine the output s_τ and s_ψ of the two hysteresis comparators. The four input signals, the flux orientation $\angle\bar{\psi}$, the digital signal s_ω prescribing the direction of rotation together with s_τ and s_ψ

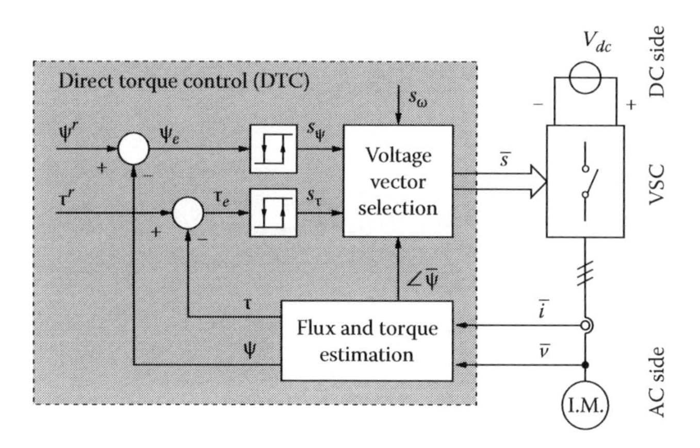

FIGURE 1.42 Direct torque controlled induction machine. (Reprinted from Sütő, Z. and Nagy, I. Bifurcation phenomena of direct torque controlled induction machines due to discontinuities in the operation. In *Proceedings of the IEEE International Conference on Industrial Technology (ICIT2005)*, Hong Kong, December 14–17, 2005. CD Rom ISBN: 0-7803-9484-4. © [2005] IEEE. With permission.)

determine the value of $\bar{s}$, which selects the output voltage vector of the VSC out of the eight possible values [12]. In Ref. [12,14], the DTC system is studied in more detail. A continuous-time model is developed for describing the system between two consecutive switchings under the assumptions that the speed of the rotor and the rotor flux is constant due to the large inertia of the rotating masses and rotor time constant.

1.10.6.2 Mathematical Model

The mathematical model of the system is based on the Poincaré map. The Poincaré map and the PMF is generated and its Jacobian matrix is also calculated for stability analysis [12]. The model is a piece-wise linear one between each switching, but the switching instants depend on the state variables, making the whole model nonlinear. The relations needed for investigating the stability of FPs and the Lyapunov exponents in the function of reference torque are derived.

1.10.6.3 Numerical Results

Numerical values used are given in Ref. [12]. A bifurcation diagram can be seen in Figure 1.43. The bifurcation parameter μ is the reference torque τ^r, and the sample points of the torque error $\tau_{e,m}$ are plotted along the vertical axis. $\tau_{e,m}$ ($m = 1, 2,\ldots$) are taken from the trajectory when the flux space vector is at an appropriate position [12]. The bifurcation parameter spans only a quite narrow region here. In most cases, sample points of $\tau_{e,m}$ covers vertically the full width of the torque hysteresis. (See at the right side of the diagram, where the points suddenly spread vertically.) Various system states and bifurcations can be discovered in the diagram. None of them reveals regular smooth local types of bifurcations such as flip, fold, or Neimark ones. Non-smooth bifurcations like border-collision bifurcations and non-standard bifurcations due to discontinuity of the Poincaré map are present. In order to identify system states, the largest Lyapunov exponent is calculated and plotted in Figure 1.43. Mostly, the Lyapunov exponent is negative, the system is settled down to regular periodic state (one or more FPs appear in the bifurcation diagram). At the right side, there is a wider region from $\mu \approx 0.7645$, where the positive value of $\bar{\lambda}$ implies the presence of chaotic operation. Note that proceeding from left to right, the periodic motion does not become suddenly chaotic at $\mu \approx 0.7626$. Although the points in the bifurcation diagram suddenly messed, the Lyapunov exponent is still clearly negative here and starts to increase more or less linearly. Consequently, the system state is periodic but it must be some higher order subharmonic motion, which becomes chaotic at $\mu \approx 0.7645$. At the left side of Figure 1.43, a sequence of sudden changes (bifurcations) between various orders of subharmonics can be seen. The Lyapunov exponent becomes zero (or positive) at these points, indicating also bifurcations.

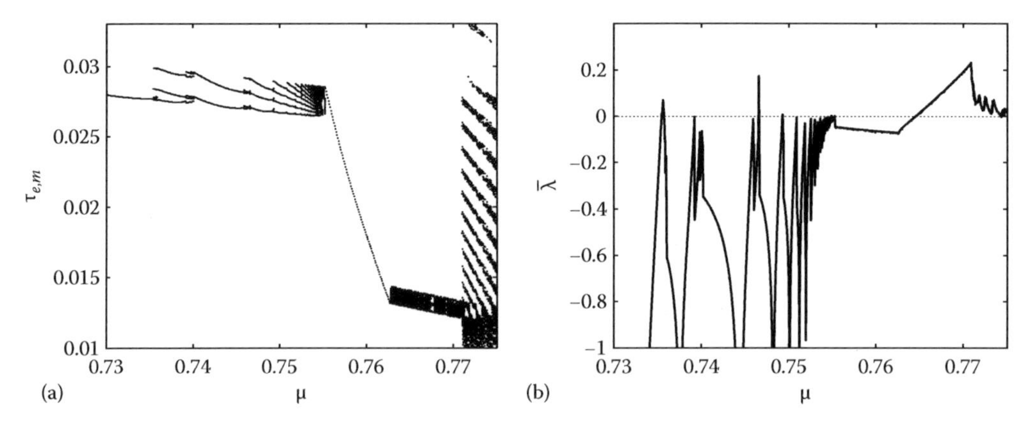

FIGURE 1.43 (a) Bifurcation diagram of the torque error τ_e. Bifurcation parameter μ is the torque reference value τ^r. (b) The largest Lyapunov exponent. (Reprinted from Sütő, Z. and Nagy, I. Bifurcation phenomena of direct torque controlled induction machines due to discontinuities in the operation. In *Proceedings of the IEEE International Conference on Industrial Technology (ICIT2005)*, Hong Kong, December 14–17, 2005. CD Rom ISBN: 0-7803-9484-4. © [2005] IEEE. With permission.)

Acknowledgments

This chapter was supported by the János Bolyai Research Scholarship of the Hungarian Academy of Sciences (HAS), by the Hungarian Research Fund (OTKA K72338), and the Control Research Group of the HAS. The collaboration of B. Buti, O. Dranga, E. Masada, and P. Stumpf is gratefully acknowledged. This work is connected to the scientific program of the "Development of quality-oriented and cooperative R+D+I strategy and functional model at BME" project. This project is supported by the New Hungary Development Plan (Project ID: TÁMOP-4.2.1/B-09/1/KMR-2010-0002).

References

1. S. Banerjee and G. C. Verghese. *Nonlinear Phenomena in Power Electronics: Bifurcations, Chaos, Control and Applications.* IEEE Press, New York, 2001.
2. M. Depenbrok. Direct self-control (DSC) of inverter fed induction machine. *IEEE Transactions on Power Electronics*, PE-3(4):420–429, October 1988.
3. O. Dranga, B. Buti, and I. Nagy. Stability analysis of a feedback-controlled resonant DC–DC converter. *IEEE Transactions on Industrial Electronics*, 50(1):141–152, February 2003.
4. R. C. Hilborn. *Chaos and Nonlinear Dynamics: An Introduction for Scientists and Engineers.* Oxford University Press, New York, 1994.
5. F. C. Moon. *Chaotic and Fractal Dynamics.* John Wiley & Sons, New York, 1992.
6. I. Nagy, O. Dranga, and E. Masada. Study of subharmonic generation in a high frequency time-sharing inverter. *The Transactions of The Institute of Electrical Engineers of Japan*, A Publication of Industry Applications Society, 120-D(4):574–580, April 2000.
7. E. Ott. *Chaos in Dynamical Systems.* Cambridge University Press, New York, 1993.
8. T. S. Parker and L. O. Chua. *Practical Numerical Algorithms for Chaotic Systems.* Springer-Verlag, New York, 1989.
9. H.-O. Peitgen, H. Jürgens, and D. Saupe. *Chaos and Fractals: New Frontiers of Science.* Springer-Verlag, New York, 1993.
10. Z. Sütő and I. Nagy. Study of chaotic and periodic behaviours of a hysteresis current controlled induction motor drive. In H. Tsuboi and I. Vajda, eds., *Applied Electromagnetics and Computational Technology II*, vol. 16 of *Studies of Applied Electromagnetics and Mechanics*, pp. 233–244. IOS Press, Amsterdam, the Netherlands, 2000.
11. Z. Sütő and I. Nagy. Analysis of nonlinear phenomena and design aspects of three-phase space-vector modulated converters. *IEEE Transactions on Circuits and Systems I: Fundamental Theory and Applications—Special Issue on Switching and Systems*, 50(8):1064–1071, August 2003.
12. Z. Sütő and I. Nagy. Bifurcation phenomena of direct torque controlled induction machines due to discontinuities in the operation. In *Proceedings of the IEEE International Conference on Industrial Technology (ICIT2005)*, Hong Kong, December 14–17, 2005. CD Rom ISBN: 0-7803-9484-4.
13. I. Takahashi and T. Noguchi. A new quick-response and high-efficiency control strategy of an induction motor. *IEEE Transactions on Industry Applications*, IA-22(5):820–827, September/October 1986.
14. Z. Sütő, I. Nagy, and E. Masada. Nonlinear dynamics in direct torque controlled induction machines analyzed by recurrence plots. In *Proceedings of the 12th European Conference on Power Electronics and Applications (EPE2007)*, Aalborg, Denmark, September 2–5, 2007. CD Rom ISBN: 9789075815108.
15. I. Nagy, Nonlinear phenomena in power electronics: Review. *Journal for Control, Measurement, Electronics, Computing and Communications (AUTOMATIKA), Zagreb, Croatia*, 42(3–4):117–132, 2001.
16. M. Bajnok, B. Buti, Z. Sütő, and I. Nagy, Surprises stemming from using linear models for nonlinear systems: Review. In *Proceedings of the 29th Annual Conference on Industrial Electronics, Control and Instrumentation (IECON'03)*, Roanoke, VA, vol. I, pp. 961–971, November 2–6, 2003. CD Rom ISBN: 0-7803-7907-1.

2

Basic Feedback Concept

Tong Heng Lee
National University of Singapore

Kok Zuea Tang
National University of Singapore

Kok Kiong Tan
National University of Singapore

2.1 Basic Feedback Concept

Control systems can be broadly classified into two basic categories: open-loop control systems and closed-loop control systems. While both classes of systems have common objectives to provide a desired system response, they differ in physical configurations as well as in performance. An open-loop control system, typically shown in Figure 2.1, utilizes a controller or control actuator in order to obtain the desired response of the controlled variable y. The controlled variable is often a physical variable such as speed, temperature, position, voltage, or pressure associated with a process or a servomechanism device. An input signal or command r is applied to the controller $C(s)$, whose output acts as the actuating signal e, which then actuates the controlled plant $G(s)$ to drive the controlled variable y toward the desired value. The external disturbance, or noise affecting the plant, is represented in the figure by a signal n, appearing at the input of the plant.

An example of an open-loop control system is the speed control of some types of variable-speed electric drills as in Figure 2.2. The user will move the trigger that activates an electronic circuit to regulate the voltage V_i to the motor. The shaft speed w is strongly load dependent, so the operator must observe and counter any speed deviation through the trigger which varies V_i.

In contrast to an open-loop control system, a closed-loop control system utilizes an additional measure and feedback of the controlled variable in order to compare the actual output with the desired output response. A measurement system will, therefore, be a part of a closed-loop system to provide this feedback. A simple closed-loop feedback control system is shown in Figure 2.3. The difference between the controlled variable y of the system $G(s)$ under control and the reference input r is amplified by the controller $C(s)$ and used to control the system so that this difference is continually reduced. Figure 2.3 represents the simplest type of feedback systems and it forms the backbone of many more complex feedback systems such as those involving multiple feedback as in cascade control, Smith predictor control, Internal Model Control, etc. It is interesting to note that many systems in nature also inherently employ a feedback system in them. For example, while driving the car, the driver gets visual feedback on the speed and position of the car by looking at a speedometer and checking through the mirrors and windows.

A simple position control system for a dc motor is shown in Figure 2.4. In the figure, solid lines are drawn to indicate electrical connections, and dashed lines show the mechanical connections. A position

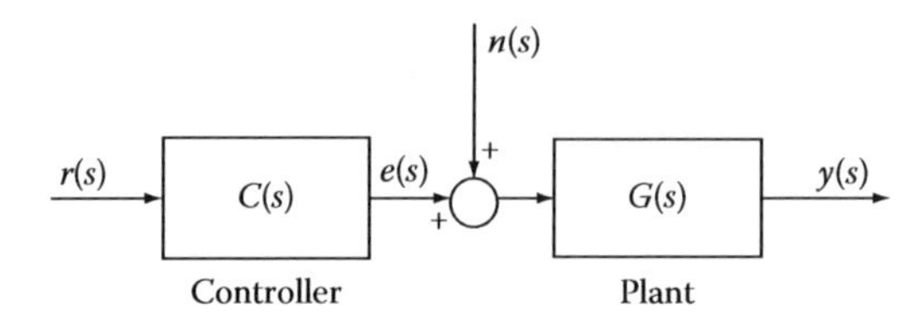

FIGURE 2.1 Block diagram of an open-loop control system.

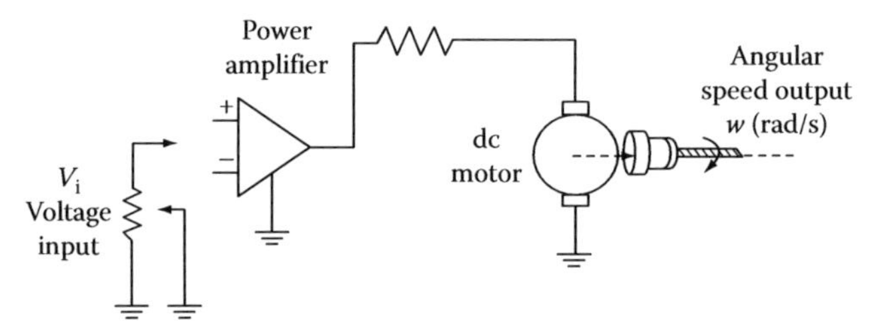

FIGURE 2.2 Example of an open-loop control system—variable-speed electric drill.

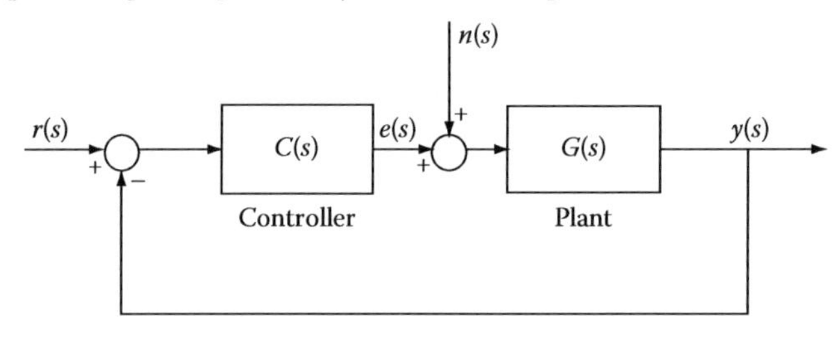

FIGURE 2.3 Block diagram of a closed-loop control system.

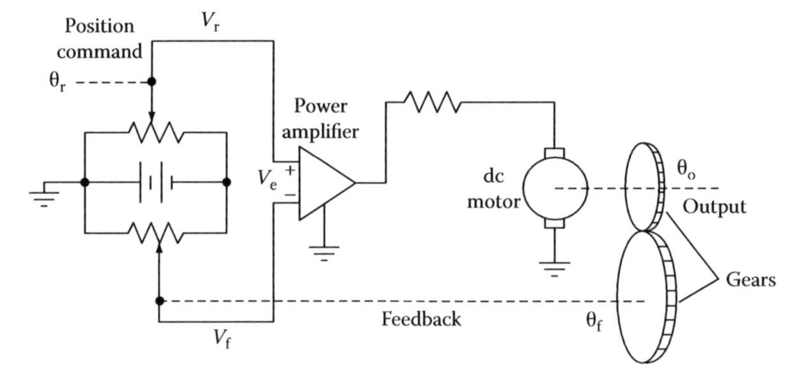

FIGURE 2.4 Example of a closed-loop control system—position control system for a dc motor.

command θ_r is dialed into the system and it is converted into an electrical signal V_r, representative of the command via an input potentiometer. A gear connection on the output shaft transfers the output angle θ_o to a shaft, feeding back the sensed output θ_f into an electric signal V_f, which thus represents the output variable θ_o. The power amplifier amplifies the difference V_e between V_r and V_f and drives the dc motor. The motor turns until the output shaft is at a position such that V_f is equal to V_r. When this situation is reached, the error signal V_e is zero. The output of the amplifier is thus zero, and the motor stops turning.

2.1.1 Effects of Feedback

Compared to open-loop control systems, feedback systems or closed-loop systems are relatively complex and costly as expensive measurement devices such as sensors are often necessary to yield the feedback.

Despite the cost and complexity, they are highly popular because of many beneficial characteristics associated with feedback. The continual self-reduction of system error is but one of the many effects that feedback brings upon a system. We shall now show that feedback also brings about other desirable characteristics such as a reduction of the system sensitivity, improvement of the transient response, reduction of the effects of external disturbance and noise, and improvement of the system stability.

2.1.1.1 Reduction of the System Sensitivity

To construct a suitable open-loop control system, all the components of the open-loop transfer function $G(s)C(s)$ must be very carefully selected, so that they respond accurately to the input signal. In practice, all physical elements will have properties that change with environment and age. For instance, the winding resistance of an electric motor changes as the temperature of the motor rises during operation. The performance of an open-loop control system will thus be extremely sensitive to such changes in the system parameters. In the case of a closed-loop system, the sensitivity to parameter variations is reduced and the tolerance requirements of the components can be less stringent without affecting the performance too much. To illustrate this, consider the open-loop system and the closed-loop system shown in Figures 2.1 and 2.3, respectively. Suppose that, due to parameter variations, $G(s)$ is changed to $G(s) + \Delta G(s)$, where $|G(s)| \gg |\Delta G(s)|$. Then, in the open-loop system shown in Figure 2.1, the output is given by

$$y(s) + \Delta y(s) = [G(s) + \Delta G(s)]C(s)r(s)$$

Hence, the change in the output is given by

$$\Delta y(s) = \Delta G(s)C(s)r(s)$$

In the closed-loop system shown in Figure 2.3,

$$y(s) + \Delta y(s) = \frac{[G(s) + \Delta G(s)]}{1 + [G(s) + \Delta G(s)]C(s)} C(s)r(s)$$

$$\Delta y(s) \approx \frac{\Delta G(s)}{1 + C(s)G(s)} C(s)r(s)$$

The change in the output of the closed-loop system, due to the parameter variations in $G(s)$, is reduced by a factor of $1 + G(s)C(s)$. To illustrate, consider the speed control problem of a simple dc motor as an example. In this case, $G(s)$ represents the transfer function between the speed of the motor and its input voltage, and suppose for the illustration, $G(s) = 1/(10s + 1)$. Further, consider a simple proportional controller given by $C(s) = 10$ and a unit step reference signal, $r(s) = 1/s$. Now suppose there is a 10% drift in the static gain of the plant, leading to $\Delta G(s) = 0.1/(10s + 1)$. With open-loop control as in Figure 2.1, the resultant change in the controlled speed variable would be $\Delta y(s) = 1/(s(10s + 1))$. Under feedback control as in Figure 2.3, the corresponding variation in y is given by $\Delta y(s) = 1/(s(10s + 11))$, clearly reduced as compared to open-loop control. Thus, feedback systems have the desirable advantage of being less sensitive to inevitable changes in the parameters of the components in the control system.

2.1.1.2 Improvement of the Transient Response

One of the most important characteristics of control systems is their transient response. Since the purpose of control systems is to provide a desired response, the transient response of control systems often must be adjusted until it is satisfactory. Feedback provides such a means to adjust the transient response of a control system and thus a flexibility to improve on the system performance. Consider the open-loop control system shown in Figure 2.1 with $G(s) = K/(Ts + 1)$ and $C(s) = a$. Clearly, the time constant of the system is T. Now consider the closed-loop control system shown in Figure 2.3 with the same feedforward transfer function as that for Figure 2.1. It is straightforward to show that the time constant of this system has been reduced to $T/(1 + Ka)$. The reduction in the time constant implies a gain in the system bandwidth and a corresponding increase in the system response speed. Using the speed control problem of a dc motor as an example (i.e., $K = 1$, $T = 10$, $a = 10$), the time constant of the closed-loop system is computed as $T_c = (1/11)T = 10/11$, i.e., the closed-loop system is about 10 times faster compared to the open-loop system.

2.1.1.3 Reduction of the Effects of External Disturbance and Noise

All physical control systems are subjected to some types of extraneous signals and noise during operation. These signals may cause the system to provide an inaccurate output. Examples of these signals are thermal noise voltage in electronic amplifiers and brush or commutator noise in electric motors. In process control, for instance, considering a room temperature control problem, external disturbance often occurs when the doors and windows in the room are opened or when the outside temperature changes. A good control system should be reasonably resilient under these circumstances. In many cases, feedback can reduce the effect of disturbance/noise on the system performance. To elaborate, consider again the open-loop and closed-loop systems of Figures 2.1 and 2.3, respectively, with a nonzero disturbance/noise component n, at the input of the plant $G(s)$. Assume that both systems have been designed to yield a desired response in the absence of the disturbance/noise component, n. In the presence of the disturbance/noise, the change in the output for the open-loop system is given by

$$\Delta y(s) = G(s)n(s)$$

For the closed-loop control system, the change in the output is given by

$$\Delta y(s) = \frac{G(s)}{1 + G(s)C(s)} n(s)$$

Clearly, the effects of the disturbance/noise has been reduced by a factor of $1 + G(s)C(s)$ with feedback. Consider the same speed control problem of a dc motor, but with the motor subject to a step type of load disturbance, say, $n(s) = 0.1/s$. With an open-loop system, the effect of the loading on the motor speed y is given by $\Delta y(s) = 0.1/(s(10s + 1))$. In comparison, under closed-loop control, the variation of the controlled variable is $\Delta y(s) = 0.1/(s(10s + 11))$, clearly reduced as compared to open-loop control.

2.1.1.4 Improvement of the System Stability

Stability is a notion that describes whether the system will be able to follow the input command as it changes. In a non-rigorous manner, a system is said to be unstable if its output is out of control or increases without bound. Here, it suffices to say that feedback could be a double-edged knife. If properly used, it can cause an originally unstable system to be stable. However, if improperly applied, it may possibly destabilize a stable system. Again consider the speed control problem of the dc motor. The pole of the original plant is at $s = -(1/10)$. Under closed-loop control, the pole is shifted deeper into the left-hand half of the complex plane at $s = -(11/10)$. This phenomenon implies that an improved system stability is achieved (more details on stability analysis will be given in a later chapter).

2.1.2 Analysis and Design of Feedback Control Systems

There are numerous analysis and design techniques for feedback control systems. Depending on the complexity of the plant and the control objectives, one might want to employ a linear design method such as the classical root-locus or frequency-response method for the controller, or resort to nonlinear design methods useful for adaptive control, optimal control, system identification, etc. These techniques are fields in their own rights and entire books can be dedicated to each one of them. In later sections, more will be discussed on these topics.

Currently, there are many tools available for computer-based simulation and the design of feedback control systems such as MATLAB®, MATRIXx, VisSim, Workbench, LabVIEW, etc. With these tools available, the analysis and design of feedback systems can be efficiently done. These simulation tools are mainly Windows-based applications which support either text-based or graphical programming. Some of the software tools allow rapid product prototyping and hardware-in-the-loop tests. They provide a seamless path from the design of controllers, selection of control parameters, implementing to actual hardware, and final testing. In many cases, hardware manufacturers of controllers provide library functions and drivers for users to port their controller designs to actual system implementation. Interesting user interfaces can also be created using these design tools. These tools allow real-time signals and control commands to be viewed by the users. These software design and analysis features are valuable for controller debugging and fine-tuning purposes.

2.1.3 Implementation of Feedback Control Systems

Feedback control systems may be implemented via an analog circuit patch up or with a digital computer. An electronic analog controller typically consists of operational amplifiers, resistors, and capacitors. The operational amplifier is often used as the function generator, and the resistors and capacitors are arranged to implement the transfer functions of the desired control combinations. An analog implementation of control system is more suited to fairly simple systems such as those using a three-term PID controller as $C(s)$ in the feedback system of Figure 2.3. As the target control system design becomes more complex, the analog implementation will become increasingly bulky and tedious to handle. In these cases, a digital controller will be more useful. Microprocessor-based digital controllers have now become very popular in industrial control systems. There are many reasons for the popularity of such controllers. The power of the microprocessor provides advanced features such as adaptive self-tuning, multivariable control, and expert systems. The ability of the microprocessor to communicate over a field bus or local area network is another reason for the wide acceptance of the digital controller. A digital controller measures the controlled variable at specific times which are separated by a time interval called the sampling time, Δt. Each sample (or measurement) of the controlled variable is converted to a binary number for input to a digital computer. Based on these sampled information, the digital computer will execute an algorithm to calculate the controller output. More will be dealt with on the subject of digital control in a later section.

2.1.4 Application Examples

Feedback control systems can be readily found in diverse applications and industries. For systems with time-critical requirements, feedback controllers have been implemented on dedicated real-time processors. PC-based controllers provide an economical solution to implement feedback controllers, though they may not fulfill the time-critical demands of the systems. The following are some sample examples of feedback control systems:

2.1.4.1 Process Control

Coupled tanks are commonly encountered in the process control industry. A pilot-scale setup is shown in Figure 2.5. The schematic block diagram of this process is shown in Figure 2.6. The variables to be controlled are the liquid levels in the two tanks which arc interconnected via a small linking pipe. Control valves are used to manipulate the flow of liquid to the two tanks. These valves can be electrically controlled by control signals from the controller via the DAQ card. Level sensors installed in the tanks provide feedback signals of the liquid levels in the tanks.

FIGURE 2.5 Application example—coupled-tanks in process control.

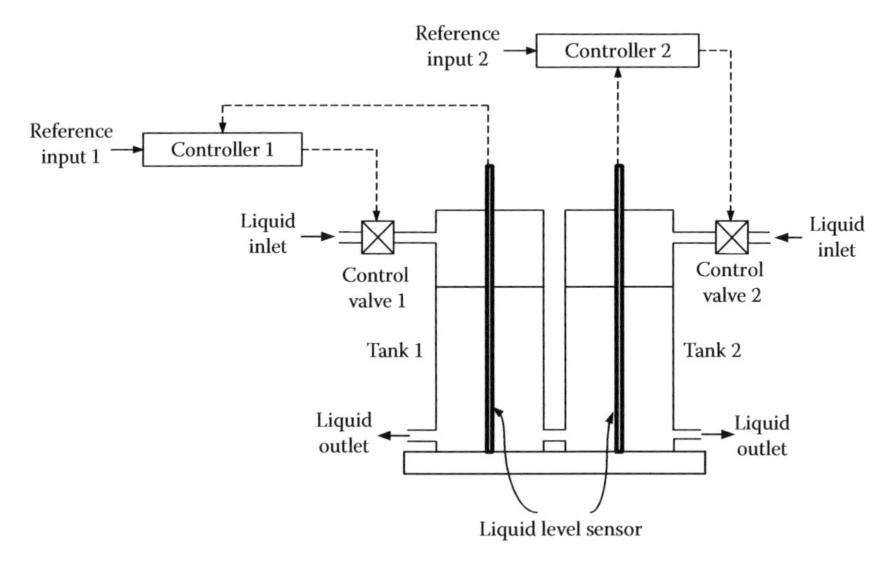

FIGURE 2.6 Schematic of the coupled-tanks.

2.1.4.2 Semiconductor Manufacturing

Lithography is a key process in silicon semiconductor manufacturing, which requires a high-precision control of processing parameters. The rapid transition to smaller microelectronic feature sizes involves the introduction of new lithography technologies, new photoresist materials, and tighter processes specifications. This transition has become increasingly difficult and costly. The application of advanced computational and control methodologies have seen increasing utilization in recent years to improve yields, throughput, and, in some cases, to enable the actual process to manufacture smaller devices.

There are specific applications in the lithography process where feedback control has made an impact. For example, with the introduction of chemically amplified photoresists, the temperature sensitivity of the postexposure thermal process is now significant. Poor nonuniformity directly impacts the line-width distribution and the chip performance. The temperature control system for this process requires careful consideration, including the equipment design and temperature-sensing techniques. A multi-zone programmable thermal-processing system is developed to improve the uniformity of temperature during these thermal processing steps as shown in Figure 2.7. The system has been demonstrated for uniform temperature control and the optimization of post-apply and postexposure thermal-processing

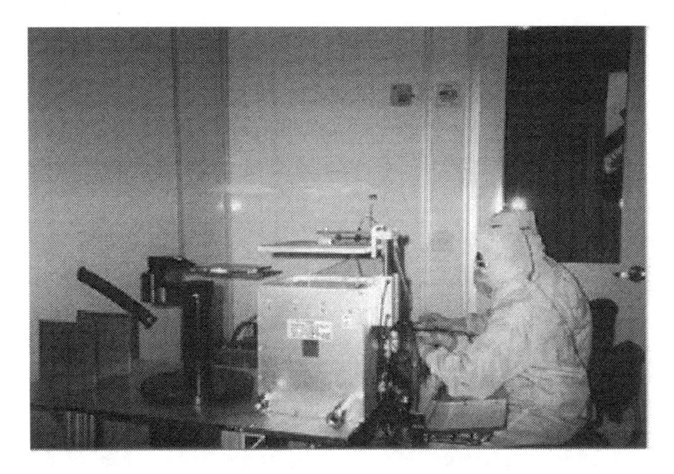

FIGURE 2.7 Application example—multi-zone thermal cycling.

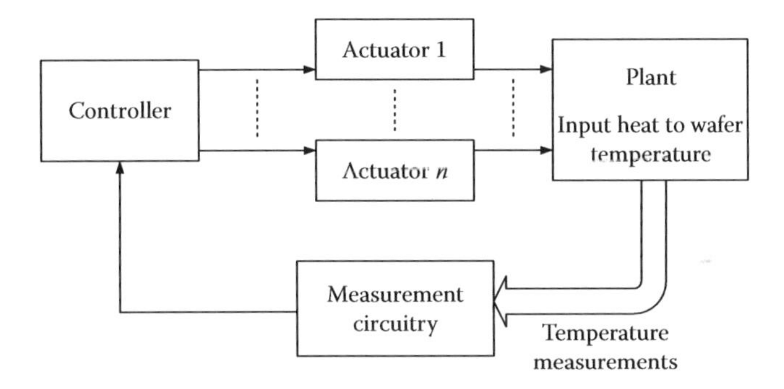

FIGURE 2.8 Schematic diagram of the multi-zone temperature control system.

conditions for chemically amplified photoresists used in the fabrication of quartz photomasks. The schematic block diagram of this system is shown in Figure 2.8, highlighting the concepts of feedback control.

2.1.4.3 Biomedical Systems

Biological injection has been widely applied in transgenic tasks. In spite of the increasing interest in biomanipulation, it is still mainly a time-consuming and laborious work performed by skilled operators, relying only on the visual information through the microscope. The skilled operators require professional training and success rate has not been high. Besides, an improper operation may cause irreversible damage to the tissue of the cell due to the delicate membrane, which can arise directly from errors and a lack of repeatability of human operators. All of these result in low efficiency and low productivity associated with the process. Hence, in order to improve the biological injection process, an automated bio-manipulation system (Figure 2.9) using a piezoelectric motor can be used to efficiently control the whole injection process. In the schematic block diagram of this system (Figure 2.10), the control signal enables the piezoelectric actuator to move precisely while the current position of the actuator is obtained from the encoder attached to the moving part.

2.1.4.4 Precision Manufacturing

Highly precise positioning systems operating at high speed arc enablers behind diverse applications, including metrology systems, micromachining and machine tools, and MEMS/MST. An example of a long-travel position stage used for wafer features' inspection is given in Figure 2.11. In these systems, positioning accuracy to sub-micrometer or even nanometer regimes is often necessary and to achieve

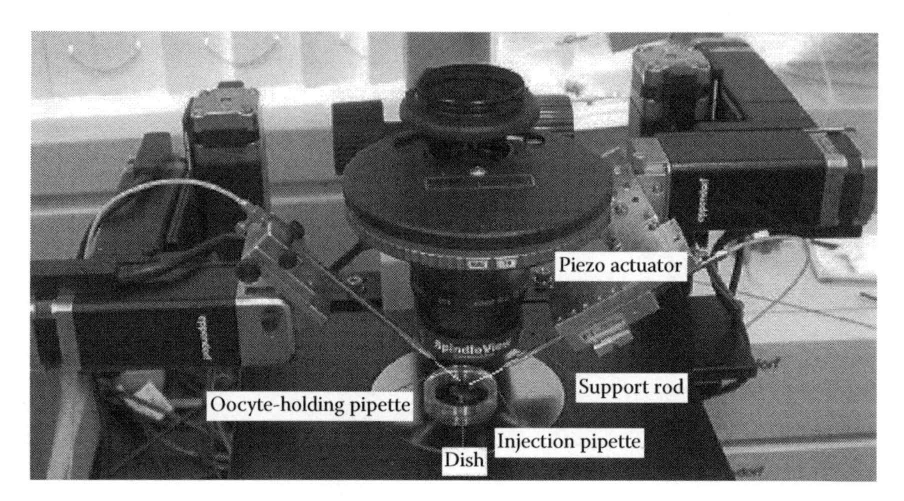

FIGURE 2.9 Application example—Bio-manipulation system.

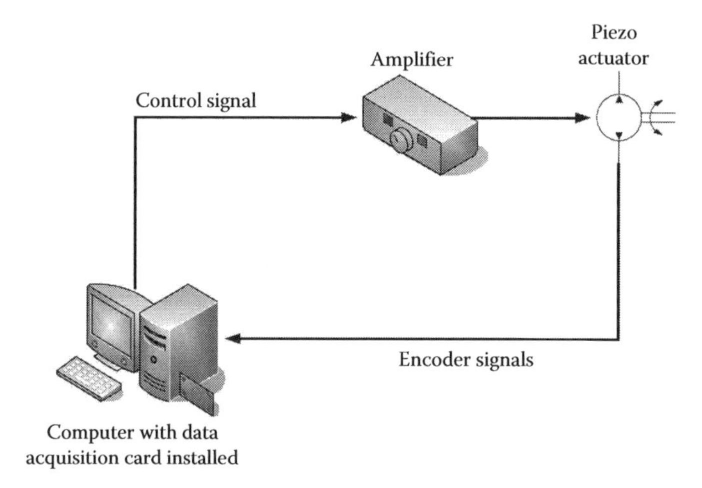

FIGURE 2.10 Schematic of the piezoelectric motor injection system.

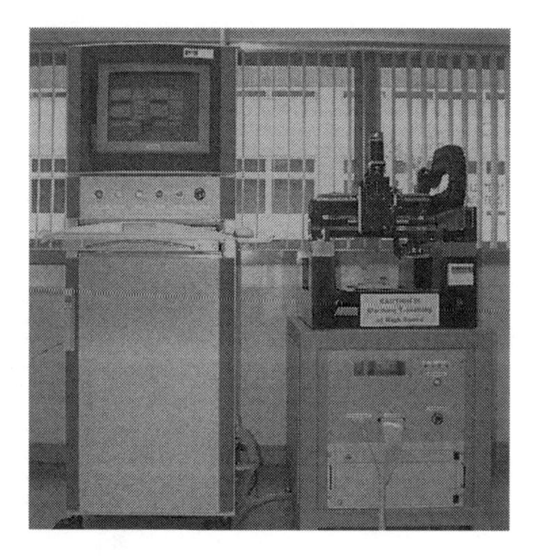

FIGURE 2.11 Application example—Precision manufacturing.

these objectives, the feedback of multiple variables is necessary. In the face of such challenges, concepts of feedback control can be used to design multiple feedback controllers to achieve high precision and high speed control. Apart from the design of parameters for the main positioning controller, other variables as in velocity and acceleration of the moving part will be necessary to feedback to the control system, so that other effects impeding fine control can be compensated, including friction, force ripples, and other disturbances. A schematic of a precision motion control system is given in Figure 2.12.

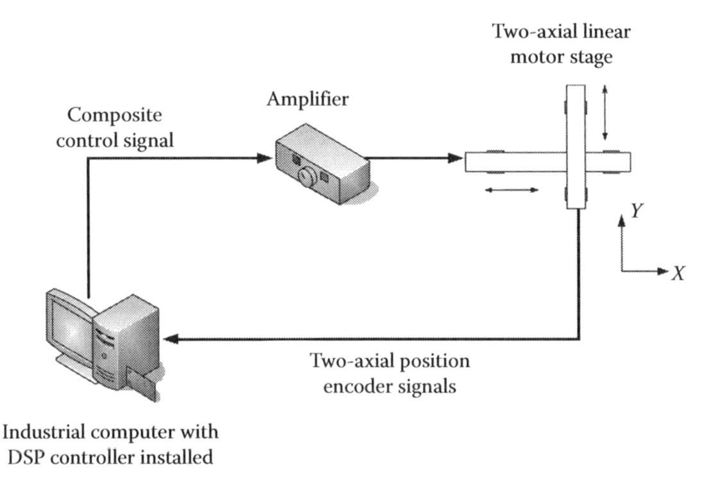

FIGURE 2.12 Schematic of a precision motion control system.

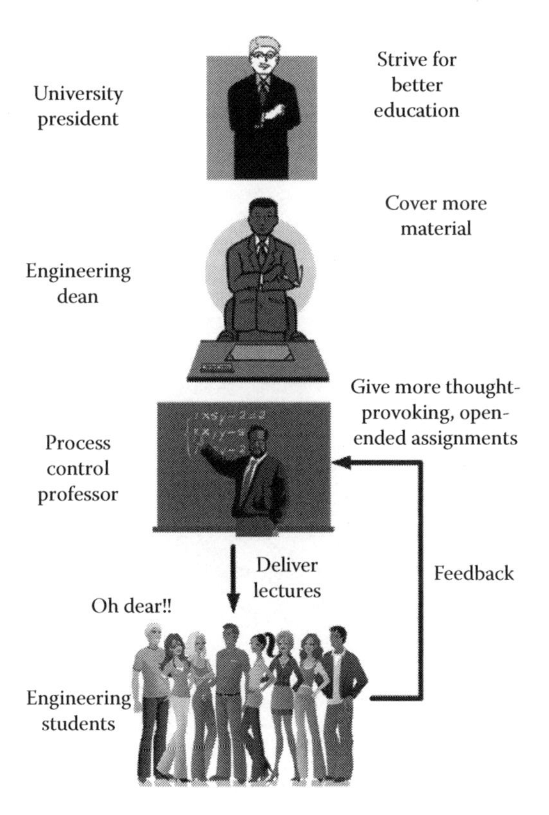

FIGURE 2.13 Application example—Feedback control in the university.

2.4.4.5 Management Systems

Feedback concepts are utilized far beyond engineering applications. Figure 2.13 shows a feedback system in operation at the university to ensure efficient operations at various entities of the university. The president is concerned with high-level university objectives and will use feedback on the performance of the university in the various aspects of education to set goals for the faculties. The engineering dean will translate the high-level objectives into faculty-level programs and curriculum, based on feedback on the position of that faculty measured with respect to the university goals. The control professor will deliver the lectures to the students in a way as to meet the faculty's directions and he will use students' feedback to gauge how well he is functioning at the level.

Such a multiple feedback and cascade system functions better than the single-loop case where the president deals directly with the students. It is able to respond to students' problems faster and provides more frequent and direct feedback.

Bibliography

1. Dorf, R.C. *Modern Control Systems*, 9th edn., Prentice-Hall Inc., Upper Saddle River, NJ, 2000.
2. Miron, D.B. *Design of Feedback Control Systems*, Harcourt Brace Jovanovich, San Diego, CA, 1989.
3. D' Souza, A.F. *Design of Control Systems*, Prentice-Hall Inc., Englewood Cliffs, NJ, 1988.
4. Ogata, K. *Modern Control Engineering*, Prentice-Hall Inc., Upper Saddle River, NJ, 2002.

3

Stability Analysis

Naresh K. Sinha
McMaster University

3.1 Introduction

The basis of automatic control is negative feedback since the actual output is compared with the desired output and the difference (or error) is utilized in such a way that the two (actual output and the desired output) are brought closure. One advantage of this negative feedback is that the system is made less sensitive to small variation in the parameters of the forward path; an important consideration in making the system more economical to manufacture.

As with most good things, this advantage is coupled with the danger that the system may become unstable. This is due to the fact that inherent time lag (phase shift) in the system may change the negative feedback into positive feedback at certain higher frequencies with the possibility that the system may become oscillatory.

In view of this, stability analysis of systems is very important. Since the systems are dynamic in nature, they are modeled mathematically through differential equations. For the general case, these will be nonlinear differential equations of order n, where n is a positive integer. Such a differential equation can be easily converted into a set of n first-order differential equations that can be expressed in the following compact form:

$$\dot{x} = f(x,u) \tag{3.1}$$

$$y = g(x,u) \tag{3.2}$$

where

$x(t)$ is the n-dimensional state vector
$u(t)$ is the m-dimensional input vector
$y(t)$ is the p-dimensional output vector
the vectors f and g are nonlinear functions of x and u, respectively

The elements of the state vector x are said to be the state variables of the system.

The system is said to be *autonomous* if the input $u(t)$ is identically zero. For this case, Equation 3.1 is reduced to

$$\dot{x} = f(x) \tag{3.3}$$

For the special case of linear time-invariant system, Equations 3.1 and 3.2 take the simplified form

$$\dot{x} = Ax + bu \tag{3.4}$$

$$y = cx \tag{3.5}$$

where

A is an $n \times n$ matrix with constant real elements
b is an n-dimensional column vector
c is an n-dimensional row vector (for a single-output system)

These are the equations for a single-input single-output system. For a multivariable system, the vectors b and c are replaced with the matrices B and C of appropriate dimensions, respectively.

Taking the Laplace transform of both sides of Equations 3.4 and 3.5, it is easily shown that the transfer function of the system is given by

$$G(s) = \frac{Y(s)}{U(s)} = c(sI - A)^{-1}b = \frac{c\,adj(sI - A)b}{\det(sI - A)} \tag{3.6}$$

From Equation 3.6, it follows immediately that the poles of the transfer function $G(s)$ and the eigenvalues of the matrix A are identical.

3.2 States of Equilibrium

An autonomous system, as defined in Equation 3.3, will reach a state of equilibrium when the derivative of the state vector is identically zero. It is equivalent to saying that an object is at a standstill when its velocity is zero. Hence, we can state formally that all values of x that satisfy the equation

$$f(x) = 0 \tag{3.7}$$

are states of equilibrium.

It will be seen that Equation 3.7 may have many possible solutions. Therefore, a nonlinear system will normally have several states of equilibria. Therefore, stability analysis of nonlinear systems involves examining each state of equilibrium and determining if it is a stable state.

Since the case of nonlinear systems is rather complicated, we shall first study the stability of linear systems in Section 3.3. Later on, we shall return to nonlinear systems.

For a linear time-invariant autonomous system, Equation 3.4 leads to

$$Ax = 0 \tag{3.8}$$

This equation has a "trivial" solution given by

$$x = 0 \tag{3.9}$$

Nontrivial solutions exist if the matrix A is singular, that is, it has one or more zero eigenvalues.

In Section 3.3, we shall give a definition of stability of linear systems that will rule out zero eigenvalues. Therefore, the existence of nontrivial solutions to Equation 3.9 will also be ruled out for stable linear systems.

3.3 Stability of Linear Time-Invariant Systems

Stability of a system can be defined in many possible ways. For a linear system, the following is a rather simple way to define stability:

Definition: A linear system is stable if and only if its output is bounded* for all possible bounded inputs.

This is also known as bounded-input bounded-output stability.

From this definition, it follows that a linear time-invariant system will be stable if and only if all the eigenvalues of the matrix A (or correspondingly, all the poles of the transfer function of the system) have negative real parts; in other words, all poles must be on the left of the $j\omega$-axis of the s-plane.

To understand this further, consider a transfer function with a pole at the origin (equivalent to the A-matrix having an eigenvalue at zero, i.e., a singular A). In this case, application of a step input to the system will cause the steady-state output to be the unbounded ramp input. Similarly, consider a transfer function with a pair of poles $\pm jb$ on the imaginary axis. In this case, if we apply the input $C \sin bt$, the steady-state output will be an unbounded function of the form $Dt \sin(bt + \phi)$, where C, D, and φ are constants.

Some alternative definitions of stability, which are equivalent to the above, are given in Equations 3.10 and 3.11 in terms of the impulse response of the system:

$$\lim_{t \to \infty} w(t) = 0 \tag{3.10}$$

$$\int_0^\infty w^2(t)dt < \infty \tag{3.11}$$

These can be easily proved due to the fact that the impulse response, $w(t)$, is the inverse Laplace transform of the transfer function, $G(s)$, and will be bounded if and only if all the poles of the transfer function have negative real parts.

It is evident from the above that the most direct approach for investigating the stability of a linear system is to determine the location of the poles of its transfer function, which are also the eigenvalues of the matrix A. Often this is not very convenient as it requires evaluating the roots of a polynomial, the degree of which may be high. An analytical method for evaluating the roots of a polynomial of degree greater than five is not known. In the general case, therefore, one must use numerical search techniques to find the roots. There are good computer programs to find the roots of a polynomial using these techniques. Also, there are robust programs for finding the eigenvalues of a square matrix. These can be used whenever a computer and the required software are available.

Fortunately, it is not necessary to determine the actual location of the poles of the transfer function for investigating the stability of a linear system. We only need to find out if the number of poles in the right half of the s-plane is zero or not. A procedure for determining the number of roots of a polynomial $P(s)$ in the right half of the s-plane without evaluating the roots is discussed in Section 3.3.1. It may be

* A function of time is said to be bounded if it is finite for all values of time. Some examples of bounded functions are (1) a step function, (2) a sinusoid, and (3) a decaying exponential, for example, $4e^{-3t} \cos 4t$. Some examples of unbounded functions are (1) a ramp function, for example, t, (2) an increasing exponential, for example, e^{2t} or $e^{3t} \sin 4t$, and (3) t^n, where $n > 0$.

noted that even this will be unnecessary if any coefficient of the polynomial is zero or negative as that indicates a factor of the form $(s - a)$ for $(s - \alpha \pm j\beta)$, where a, α, and β are positive numbers.

3.3.1 Routh–Hurwitz Criterion

The Routh–Hurwitz criterion was developed independently by A. Hurwitz (1895) in Germany and E.J. Routh (1892) in the United States.

Let the characteristic polynomial (the denominator of the transfer function after cancellation of common factors with the numerator) be given by

$$\Delta(s) = a_0 s^n + a_1 s^{n-1} + \cdots + a_{n-1} s + a_n \tag{3.12}$$

Then the following Routh table is obtained:

$$\begin{array}{lllll} s^{n-1} & a_1 & a_3 & a_5 & \cdots \\ s^{n-2} & b_1 & b_3 & b_5 & \cdots \\ s^{n-3} & c_1 & c_3 & c_5 & \cdots \\ \cdot & & & & \\ \cdot & & & & \\ \cdot & & & & \\ s^0 & h_1 & & & \end{array}$$

where the first two rows are obtained from the coefficients of $\Delta(s)$. The elements of the following rows are obtained as shown below:

$$b_1 = \frac{a_1 a_2 - a_0 a_3}{a_1} \tag{3.13}$$

$$b_3 = \frac{a_1 a_4 - a_0 a_5}{a_1} \tag{3.14}$$

$$c_1 = \frac{b_1 a_3 - a_1 b_3}{b_1} \tag{3.15}$$

and so on.

In preparing the Routh table for a given polynomial as suggested above, some of the elements may not exist. In calculating the entries in the line that follows, these elements are considered to be zero. The procedure will be clearer from the examples that follow.

The Routh–Hurwitz criterion states that the number of roots with positive real parts is equal to the number of changes in sign in the first column of the Routh table.

Consequently, the system is stable if and only if there are no sign changes in the first column of the table.

Example 3.1

$$\Delta(s) = s^4 + 5s^3 + 20s^2 + 30s + 40 \tag{3.16}$$

The Routh table is as follows:

$$\begin{array}{llll} s^4 & 1 & 20 & 40 \\ s^3 & 5 & 30 & \\ s^2 & 14 & 40 & \\ s^1 & \frac{220}{14} & & \\ s^0 & 40 & & \end{array}$$

No sign changes in the first column indicate no root in the right half of the s-plane, and hence a stable system.

Example 3.2

$$\Delta(s) = s^3 + 2s^2 + 4s + 30 \tag{3.17}$$

The following Routh table is obtained

$$\begin{array}{lll} s^3 & 1 & 4 \\ s^2 & 2 & 30 \\ s^1 & -11 & \\ s^0 & 30 & \end{array}$$

Two sign changes in the first column indicate that there are two roots in the right half of the s-plane. Therefore, this is the characteristic polynomial of an unstable system.

Two special difficulties may arise while obtaining the Routh table for a given characteristic polynomial.

Case 3.1: If an element in the first column turns out to be zero, it should be replaced by a small positive number, ε, in order to complete the table. The sign of the elements of the first column is then examined as ε approaches zero. The following example will illustrate this.

Example 3.3

$$\Delta(s) = s^5 + 2s^4 + 3s^3 + 6s^2 + 12s + 18 \tag{3.18}$$

The Routh table is shown below, where zero in the third row is replaced by ε.

$$\begin{array}{llll} s^5 & 1 & 3 & 12 \\ s^4 & 2 & 6 & 18 \\ s^3 & \varepsilon & 3 & \\ s^2 & 6-\frac{6}{\varepsilon} & 15 & \\ s^1 & \frac{18\varepsilon - 18 - 18\varepsilon^2}{6\varepsilon - 6} & & \\ s^0 & 15 & & \end{array}$$

As ε approaches zero, the first column of the table may be simplified to obtain

$$\begin{array}{ll} s^5 & 1 \\ s^4 & 2 \\ s^3 & \varepsilon \\ s^2 & \dfrac{-6}{\varepsilon} \\ s^1 & 3 \\ s^0 & 18 \end{array}$$

Two sign changes in this column indicate two roots in the right half of the s-plane.

Case 3.2: Sometimes we may find that an entire row of the Routh table is zero. This indicates the presence of some roots that are negative of each other. In such cases, we should form an auxiliary polynomial from the row preceding the zero row. This auxiliary polynomial has only alternate powers of s, starting with the highest power indicated by the power of the leftmost column of the row, and is a factor of the characteristic polynomial. The number of roots of the characteristic polynomial in the right half of the s-plane will be the sum of the number of right-half plane roots of the auxiliary polynomial and the number determined from the Routh table of the lower order polynomial obtained by dividing the characteristic polynomial by the auxiliary polynomial. The following example will elucidate the procedure.

Example 3.4

$$\Delta(s) = s^5 + 6s^4 + 10s^3 + 35s^2 + 24s + 44 \tag{3.19}$$

The Routh table is as follows:

$$\begin{array}{llll} s^5 & 1 & 10 & 24 \\ s^4 & 6 & 35 & 44 \\ s^3 & \dfrac{25}{6} & \dfrac{50}{3} & \\ s^2 & 11 & 44 & \\ s^1 & 0 & & \quad \omega \text{ zero row} \end{array}$$

The auxiliary row gives the equation

$$11s^2 + 44 = 0 \tag{3.20}$$

Hence, $s^2 + 4$ is a factor of the characteristic polynomial, and we get

$$q(s) = \frac{\Delta(s)}{s^2 + 4} = s^3 + 6s^2 + 11s + 6 \tag{3.21}$$

The Routh table for $q(s)$ is as follows:

$$\begin{array}{lll} s^3 & 1 & 11 \\ s^2 & 6 & 6 \\ s^1 & 10 & \\ s^0 & 6 & \end{array}$$

Since this shows no sign changes in the first column, it follows that all roots of $\Delta(s)$ are in the left half of the s-plane, with one pair of roots $s = \pm j2$ on the imaginary axis. Consequently, $\Delta(s)$ is the characteristic polynomial of an oscillatory system. According to our definition, this is an unstable system.

Example 3.5

$$\Delta(s) = s^6 + 2s^5 + 9s^4 + 12s^3 + 43s^2 + 50s + 75 \tag{3.22}$$

The Routh table is as follows:

$$\begin{array}{lcccc} s^6 & 1 & 9 & 43 & 75 \\ s^5 & 2 & 12 & 50 & \\ s^4 & 3 & 18 & 75 & \\ s^3 & 0 & 0 & & \end{array}$$

The auxiliary polynomial is

$$p(s) = 3s^2 + 18s^2 + 75 = 3(s^2 + 2s + 3)(s^2 - 2s + 3) \tag{3.23}$$

which has two roots in the right half of the s-plane. Also,

$$q(s) = \frac{\Delta(s)}{p(s)} = s^2 + 2s + 3 \tag{3.24}$$

The Routh table for $q(s)$ is as follows:

$$\begin{array}{lcc} s^2 & 1 & 3 \\ s^1 & 2 & \\ s^0 & 3 & \end{array}$$

Therefore, $\Delta(s)$ has two roots in the right half of the s-plane and is the characteristic polynomial of an unstable system.

3.3.2 Relative Stability

Application of the Routh criterion, as discussed above, only tells us whether the system is stable or not. In many cases, we need more information. For example, if the Routh test shows that a system is stable, we often like to know how close it is to instability, that is, how far from the $j\omega$-axis is the pole closest to it. This information can be obtained from the Routh criterion by shifting the vertical axis in the s-plane to obtain the p-plane, as shown in Figure 3.1.

Hence, in the polynomial $\Delta(s)$, if we replace s by $p - \alpha$, we get a new polynomial $\Delta(p)$. Applying the Routh test to this polynomial will tell us how many roots $\Delta(p)$ has in the right half of the p-plane. This is also the number of roots of $\Delta(s)$ located toward the right of the line $s = -\alpha$ in the s-plane.

Example 3.6

Consider the polynomial

$$\Delta(s) = s^3 + 10.2s^2 + 21s + 2 \tag{3.25}$$

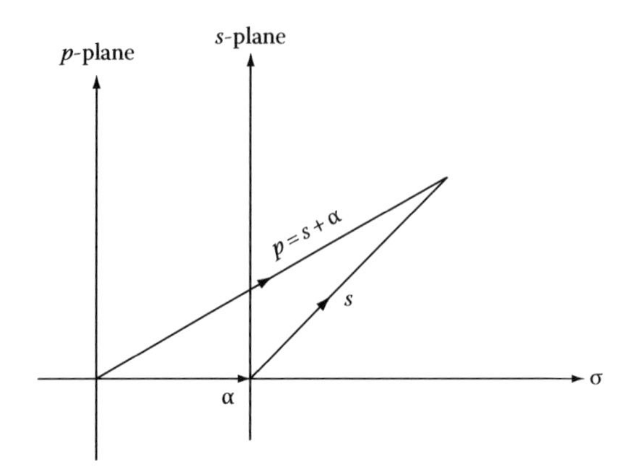

FIGURE 3.1 Shift of the axis to the left by α.

The Routh table shown below tells us that the system is stable.

s^3	1	21
s^2	10.2	2
s^1	20.804	
s^0	2	

Now we shall shift the axis to the left by 0.2, that is,

$$p = s + 0.2 \tag{3.26}$$

$$\text{or} \quad s = p - 0.2 \tag{3.27}$$

Substituting Equation 3.27 into Equation 3.25, we get

$$\Delta(p) = (p-0.2)^2 + 10.2(p-2)^2 + 21(p-2) + 2 = p^3 + 9.6p^2 + 170.4p - 1.80 \tag{3.28}$$

The Routh table shows that $\Delta(p)$ shows that it has one root in the right half of the p-plane. This implies one root between 0 and −0.2 in the s-plane.

p^3	1	17.8
p^2	9.6	−1.80
p^1	17.27	
p^0	−1.804	

It follows that shifting by a smaller amount would enable us to get a better idea of the location of this root.

We see from this example that although $\Delta(s)$ is the characteristic polynomial of a stable system, it has a root between 0 and −0.2 in the s-plane. This indicates a large time constant of the order of 5 s.

3.3.3 Stability under Parameter Uncertainty

In our discussion so far, we have assumed that the characteristic polynomial $\Delta(s)$ for the system is known precisely. In practice, for a product to be economical, some tolerance must be allowed in the values of the components. As a result, for most systems, the coefficients a_i of the characteristic polynomial

$$\Delta(s) = a_0 s^n + a_1 s^{n-1} + \cdots + a_{n-1}s + a_n \tag{3.29}$$

are unknown except for the bounds

$$\alpha_i \le a_i \le \beta_i, \quad i = 0, 1 \ldots, n \tag{3.30}$$

Consequently, the system will be stable if the roots of $\Delta(s)$ will be in the left half of the s-plane for each set of values of the coefficients within the ranges defined by Equation 3.30. This is called *robust parametric stability*.

According to an interesting result proved by Kharitonov (1978), it is necessary to investigate the stability for only four polynomials formed from Equations 3.29 and 3.30 [K78]. Now, we discuss how one can obtain these four *Kharitonov polynomials*. These are defined in the form of the polynomial

$$\Delta(s) = c_0 s^n + c_1 s^{n-1} + \cdots + c_{n-1}s + c_n \tag{3.31}$$

where the coefficients c_i are defined in a pairwise fashion, c_{2k}, c_{2k+1}, $k = 0, 1, \ldots, m$, where

$$m = \begin{cases} \dfrac{n}{2} & \text{if } n \text{ is even} \\ \dfrac{n-1}{2} & \text{if } n \text{ is odd} \end{cases} \tag{3.32}$$

Based on the pairwise assignment of the coefficients, the four polynomials are assigned a mnemonic {k = even, k = odd} name as shown below, for $(2k + 1) < n$:

$$\Delta_1(s)[\max,\max;\min,\min] \qquad c_{2k} = \begin{cases} \beta_{2k} & \text{if } k \text{ is even} \\ \alpha_{2k} & \text{if } k \text{ is odd} \end{cases}$$

$$c_{2k+1} = \begin{cases} \beta_{2k+1} & \text{if } k \text{ is even} \\ \alpha_{2k+1} & \text{if } k \text{ is odd} \end{cases}$$

$$\Delta_2(s)[\min,\min;\max,\max] \qquad c_{2k} = \begin{cases} \alpha_{2k} & \text{if } k \text{ is even} \\ \beta_{2k} & \text{if } k \text{ is odd} \end{cases}$$

$$c_{2k+1} = \begin{cases} \alpha_{2k+1} & \text{if } k \text{ is even} \\ \beta_{2k+1} & \text{if } k \text{ is odd} \end{cases}$$

$$\Delta_2(s)[\min,\max;\max,\min] \qquad c_{2k} = \begin{cases} \alpha_{2k} & \text{if } k \text{ is even} \\ \beta_{2k} & \text{if } k \text{ is odd} \end{cases}$$

$$c_{2k+1} = \begin{cases} \beta_{2k+1} & \text{if } k \text{ is even} \\ \alpha_{2k+1} & \text{if } k \text{ is odd} \end{cases}$$

$$\Delta_1(s)[\max,\min;\min,\min] \quad c_{2k} = \begin{cases} \beta_{2k} & \text{if } k \text{ is even} \\ \alpha_{2k} & \text{if } k \text{ is odd} \end{cases}$$

$$c_{2k+1} = \begin{cases} \alpha_{2k+1} & \text{if } k \text{ is even} \\ \beta_{2k+1} & \text{if } k \text{ is odd} \end{cases}$$

Proofs of this remarkable result can be found in references [A82,B85,G63]. We illustrate its usefulness by a simple example.

Example 3.7

Consider a feedback system shown in Figure 3.2 with the forward path transfer function

$$G(s) = \frac{K}{s(s+a)(s+b)} \tag{3.33}$$

and

$$H(s) = 1 \tag{3.34}$$

where, the uncertainty ranges of the parameters K, a, and b are as given below:

$$K = 10 \pm 2, \quad a = 3 \pm 0.5, \quad \text{and} \quad b = 4 \pm 0.2 \tag{3.35}$$

We determine the stability of the closed-loop system using Kharitonov's four polynomials. First we determine the characteristic polynomial

$$\Delta(s) = a_0 s^3 + a_1 s^2 + a_2 s + a_3 = s^3 + (a+b)s^2 + abs + K \tag{3.36}$$

From the given ranges on the parameters, the coefficients of the characteristic polynomial are

$$\begin{aligned} &a_0 = 1 \\ &6.3 \le a_1 \le 7.7 \\ &9.5 \le a_2 \le 14.7 \\ &8 \le a_3 \le 12 \end{aligned} \tag{3.37}$$

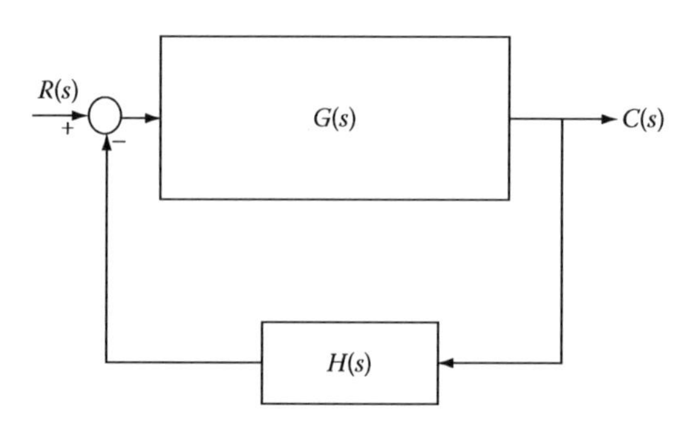

FIGURE 3.2 A closed-loop system.

We now obtain the four polynomials as

$$\begin{aligned}
\Delta_1(s) &= s^3 + 7.7s^2 + 9.5s + 8 \\
\Delta_2(s) &= s^3 + 6.3s^2 + 14.7s + 12 \\
\Delta_3(s) &= s^3 + 7.7s^2 + 14.7s + 8 \\
\Delta_4(s) &= s^3 + 6.3s^2 + 9.5s + 12
\end{aligned} \tag{3.38}$$

Application of the Routh–Hurwitz criterion to each of these four polynomials shows that the first column of Routh table is positive for all of them. This shows that the system in this example is stable for all possible sets of parameters within the ranges given by Equation 3.37.

3.3.4 Stability from Frequency Response

The method described above assumes that the transfer function of the closed-loop system is known and is in the form of a rational function of the complex frequency variable s, that is the ratio of two polynomials of finite degree. In many practical situations, the transfer function may not be known, or it is not a rational function, as will be the case when the forward path of a closed-loop system contains an ideal delay. In such cases, one can use the Nyquist criterion to investigate the stability of the closed-loop system from the frequency response of the open-loop transfer function. The main advantage here is that the frequency response can be obtained experimentally if the transfer function is not known.

Consider the closed-loop system shown in Figure 3.2. The main reason for possible instability of this system is that although it is designed as negative feedback, it may turn out to be positive feedback if at some frequency a phase lag of 180° occurs in the loop. If the feedback at this frequency is sufficient to sustain oscillations, the system will act as an oscillator. The frequency response of the open-loop system can therefore be utilized for investigating the stability of the closed-loop system.

From superficial considerations, it will appear that the closed-loop system shown in Figure 3.2 will be stable, provided that the open-loop frequency response $GH(j\omega)$ does not have gain of 1 or more at the frequency where the phase shift is 180°. However, this is not always true. For example, consider the polar plot of the frequency response shown in Figure 3.3.

It is seen that two values of ω, the phase shift is 180° and the gain is more than one (at points A and B), and yet this system will be stable when the loop is closed. A thorough understanding of the problem is possible by applying the criterion of stability developed by *H.* Nyquist in 1932. It is based on a theorem

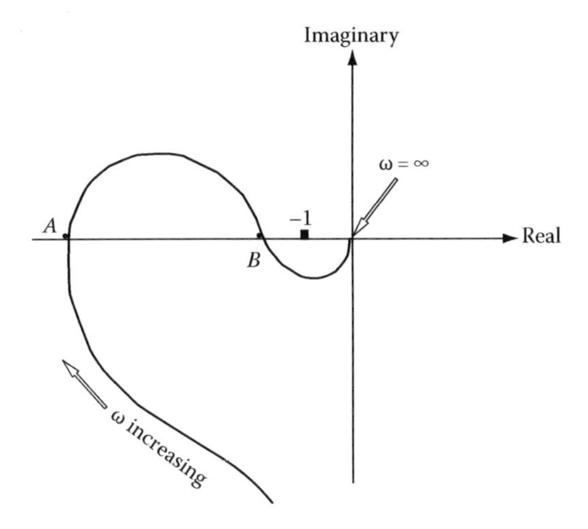

FIGURE 3.3 Polar plot of the frequency response of the forward path of a conditionally stable system.

in complex variable theory, due to Cauchy, called the principle of the argument, which is related to mapping of a closed path (or contour) in the s-plane for a function $F(s)$.

For example, consider the polar plot of the frequency response shown in Figure 3.3. For two values of ω, the phase shift is 180° and the gain is more than one (at points A and B, yet this system will be stable when the loop is closed).

A thorough understanding of the problem is possible by applying the criterion of stability developed by *H.* Nyquist in 1932. It is based on a theorem in complex variable theory, due to Cauchy, called the principle of the argument, and related to mapping of a closed path (or contour) in the s-plane for a function.

The overall transfer function of the system shown in Figure 3.2 is given by

$$T(s) = \frac{G(s)}{1 + G(s)H(s)} \tag{3.39}$$

To find out if the closed-loop system will be stable, we must determine whether $T(s)$ has any pole in the right half of the s-plane (including the $j\omega$-axis), that is, whether $F(s) = 1 + G(s)H(s)$ has any root in this part of the s-plane. For this purpose, we must take a contour in s-plane that encloses the entire right half plane as shown in Figure 3.4a.

If $F(s)$ has a pole or zero at the origin of the s-plane or at some points on the $j\omega$-axis, we must make a detour along an infinitesimal semicircle, as shown in Figure 3.4b. This is called the Nyquist contour, and our object is to determine the number of zeros of $F(s)$ inside this contour.

Let $F(s)$ have P poles and Z zeros within the Nyquist contour. Note that the poles of $F(s)$ are also the poles of $GH(s)$, but the zeros of $F(s)$ are different from those of $GH(s)$, and not known. For the system to be stable, we must have $Z = 0$, that is, the characteristic polynomial must not have any root within the Nyquist contour.

From the principle of the argument, a map of the Nyquist contour in the F-plane will encircle the origin of the F-plane N times in the clockwise direction, where

$$N = Z - P \tag{3.40}$$

Thus, the system is stable if and only if $N = -P$, so that Z will be zero. Further, note that the origin of the F-plane is the point $-1 + j0$ in the GH-plane. Hence, we get the following criterion in terms of the loop transfer function $GH(s)$.

> A feedback system will be stable if and only if the number of counterclockwise encirclements of the point $-1 + j0$ by the map of the Nyquist contour in the GH-plane is equal to the number of poles of $GH(s)$ inside the Nyquist contour in the s-plane.

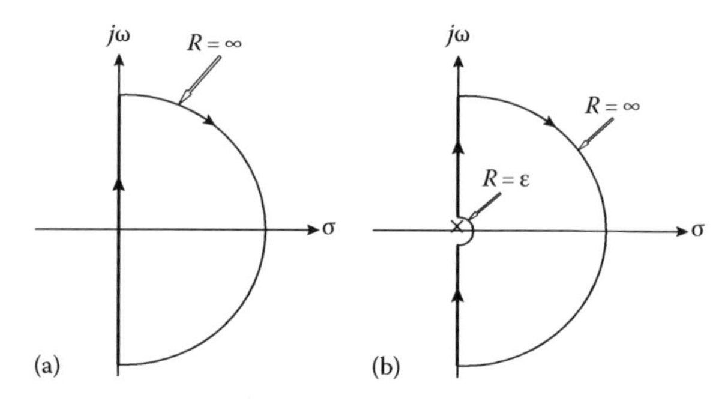

FIGURE 3.4 The Nyquist contour.

The map of the Nyquist contour in the *GH*-plane is called the Nyquist plot of *GH*(*s*). The polar plot of the frequency response of *GH*(*s*), which is the map of the positive part of the $j\omega$-axis, is an important part of the Nyquist plot and can be obtained experimentally or by computation if the transfer function is known. The procedure of completing the rest of the plot will be illustrated by a number of examples.

Example 3.8

Consider the transfer function

$$GH(s) = \frac{60}{(s+1)(s+2)(s+5)} \tag{3.41}$$

The frequency response is readily calculated and is given in Table 3.1. Sketches of the Nyquist contour and the Nyquist plot are shown in Figure 3.5. The various steps of the procedure for completing the Nyquist plot are given below:

1. Part *AB* in the *s*-plane is the $j\omega$-axis from $\omega = 0$ to ω, and maps into the polar plot *A′B′* in the *GH*-plane.
2. The infinite semicircle *BCD* maps into the origin of the *GH*-plane.
3. The part *DA*, which is the negative part of the $j\omega$-axis maps into the curve *D′A′*. It may be noted that *D′A′* is the mirror image of *A′B′* about the real axis. This follows from the fact that $GH(-j\omega)$ is the complex conjugate for $GH(j\omega)$. Consequently, it is easily sketched from the polar plot of the frequency response.

TABLE 3.1 Frequency Response of the Transfer Function Given by Equation 3.41

ω	Gain	Phase Shift	ω	Gain	Phase Shift
0	0	0°	3.0	0.90	−158.8°
0.25	5.77	−24.0°	4.123	0.476	−180°
0.6	5.18	−46.3°	5.0	0.309	−191.9°
1.0	3.72	−82.9°	10.0	0.052	−226.4°
2.0	1.76	−130.2°	20.0	0.007	−247.4°

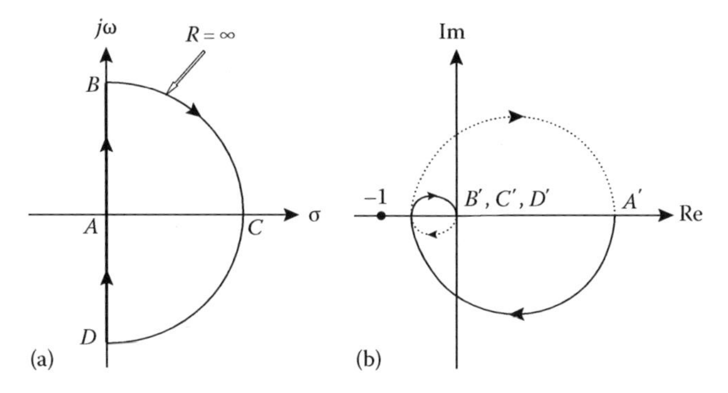

FIGURE 3.5 Nyquist contour (a) and plot for the transfer function (b) given by Equation 3.41.

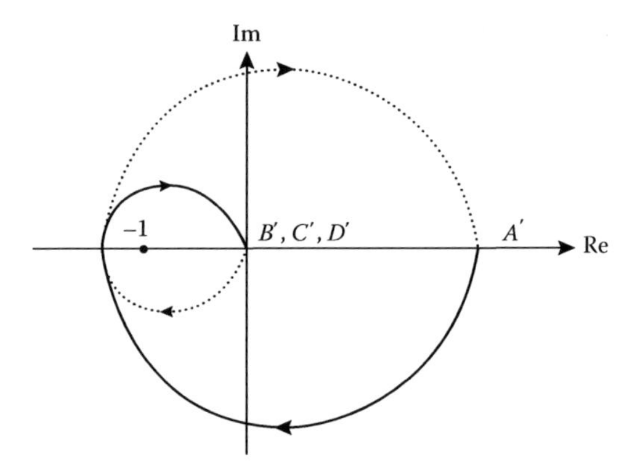

FIGURE 3.6 Nyquist plot for the transfer function given by Equation 3.41 if the open-loop gain is increased by a factor of more than 2.15.

The Nyquist plot does not encircle the point $-1 + j0$ in the GH-plane. Hence we have $N = 0$. Also, the function $GH(s)$ does not have any pole inside the Nyquist contour. This makes the system stable since $Z = N + P = 0$.

If we increase the open-loop gain of this system by a factor of 2.15 or more, the resulting Nyquist plot will encircle the point $-1 + j0$ twice in the GH-plane, as shown in Figure 3.6. Consequently, for this case, $N = 2$. Since P is still zero, the closed-loop system is unstable, with two poles in the right half of the s-plane.

Example 3.9

Consider the open-loop transfer function

$$G(s) = \frac{K}{s(s+2)(s+6)} \tag{3.42}$$

Since this time we have a pole at the origin of the s-plane, the Nyquist contour must take a small detour around this pole while still attempting to enclose the entire right half of the s-plane, including the $j\omega$-axis. Hence, we get the semicircle EFA of infinitesimal radius, ε. The resulting Nyquist contour and plot are shown in Figure 3.7. The various steps in obtaining the Nyquist plot are described below:

1. Part AB of in the s-plane is the positive part of the $j\omega$-axis from $j\varepsilon$ to $j\infty$ and maps into the polar plot of the frequency response $A'B'$ in the GH-plane (shown as a solid curve).
2. The infinite semicircle BCD in the s-plane maps into the origin of the GH-plane.
3. The part DE (negative $j\omega$-axis) maps into the image $D'E'$ of the frequency response $A'B'$.
4. The infinitesimal semicircle EFA is the plot of the equation $s = \varepsilon e^{j\theta}$, with θ increasing from −90° to 90°. It maps into the infinite semicircle, $E'F'A'$, since $GH(s)$ can be approximated as $K/12\varepsilon e^{j\theta}$ for small s. Note that $s = \varepsilon e^{j\theta}$ in the denominator will cause the infinitesimal semicircle to map into an infinite semicircle with traversal in the opposite direction due to the change in the sign of θ as it is brought into the numerator from the denominator. For the plot shown, $N = 2$ and $P = 0$; consequently, the system will be unstable, with $Z = 2$. If the gain is sufficiently reduced, the point $-1 + j0$ will not be encircled, with the result that the system will be stable, since $N = 0$ and $Z = 0$.

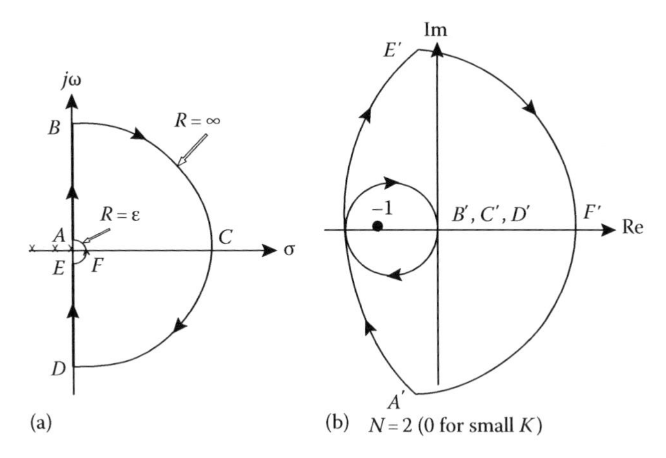

FIGURE 3.7 Nyquist contour and plot for the transfer function given by Equation 3.42: (a) s-plane and (b) GH-plane.

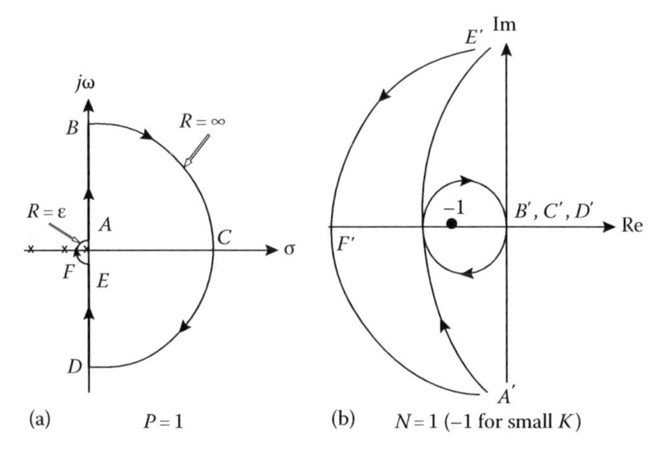

FIGURE 3.8 Nyquist plot for the transfer function give by Equation 3.42 with the Nyquist contour enclosing the pole at the origin: (a) s-plane and (b) GH-plane.

Example 3.10

We shall reconsider the transfer function of the preceding example, but this time we shall take a different Nyquist contour. The detour around the pole at the origin will be taken from its left, as shown in Figure 3.8a. The resulting Nyquist plot is shown in Figure 3.8b.

The essential difference with Figure 3.7b is in the map of the infinitesimal semicircle in the Nyquist contour, which is an infinite semicircle in the counterclockwise direction in this case. Consequently, we have $P = 1$. Here, the Nyquist plot encircles the point $-1 + j0$ once in the clockwise direction, giving us $N = 1$, so that $Z = N + P = 2$, and the system will be unstable. Also, reducing the gain sufficiently will cause the Nyquist plot to encircle the point $-1 + j0$ once in the counterclockwise direction, making $N = -1$, with the result that now we shall have $Z = N + P = 0$. Thus, the system will be stable for small values of K. Both of these results agree with those obtained in the previous example, as expected.

3.4 Stability of Linear Discrete-Time Systems

In order to use a digital computer as part of control systems, the input and the output are sampled at $t = kT$, where k is an integer and T is the sampling interval. This has the effect of converting the continuous-time input and output functions into discrete-time functions defined only at the sampling instants.

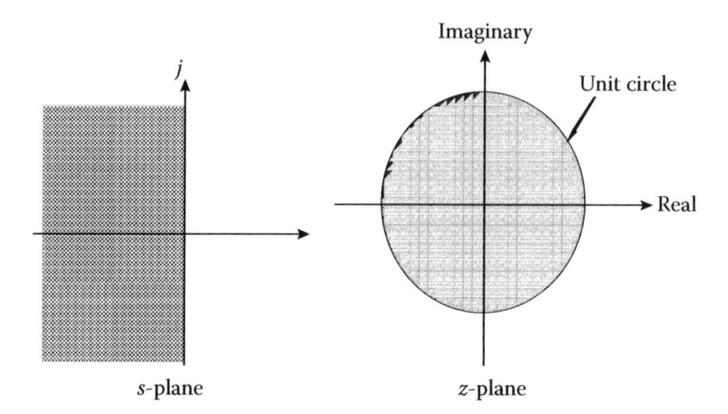

FIGURE 3.9 Mapping from s-plane to z-plane.

These are called sequences and are related by difference equations instead of differential equations as in the continuous-time case.

Just as continuous-time systems are represented either by differential equations or by transfer functions obtained by taking the Laplace transform of the differential equation, discrete-time systems can be represented either through difference equations or by their z-transfer function where

$$z = e^{sT}$$

The mapping between the s-plane and the z-plane is shown in Figure 3.9.

Using the state-variable formulation, a difference equation of order n can be written as

$$x(kT + T) = Fx(T) + Gu(T) \tag{3.43}$$

The necessary and sufficient condition for the stability of a discrete-time system is that all the poles of its z-transfer function lie inside the unit circle of the z-plane. This follows from the transformation given by Equation 3.43. Alternatively, the system will be stable if and only if the magnitude of all eigenvalues of F is less than one.

Keeping this in mind, the two basic techniques studied for continuous-time systems—(a) the Routh–Hurwitz criterion and (b) the Nyquist criterion—can be used with slight modifications for discrete-time systems.

3.4.1 Routh–Hurwitz Criterion

The Routh–Hurwitz criterion, discussed in Section 3.1, is an attractive method for investigating system stability without requiring evaluation of the roots of the characteristic polynomial. Since it enables us to determine the number of roots with positive real parts, it cannot be used directly for discrete-time systems, where we need to find the number of roots outside the unit circle. It is possible to use the Routh–Hurwitz criterion to determine if a polynomial $Q(z)$ has roots outside the unit circle by using the *bilinear transformation*

$$w = u + fv = \frac{z+1}{z-1} \quad \text{or} \quad z = \frac{w+1}{w-1} = \frac{u+1+jv}{u-1+jv} \tag{3.44}$$

(also called the Möbius transformation) that maps the unit circle of the z-plane into the imaginary axis of the w-plane and the interior of the unit circle into the left half of the w-plane. This is shown in Figure 3.10. We can now apply the Routh criterion to $Q(w)$ as in the s-plane.

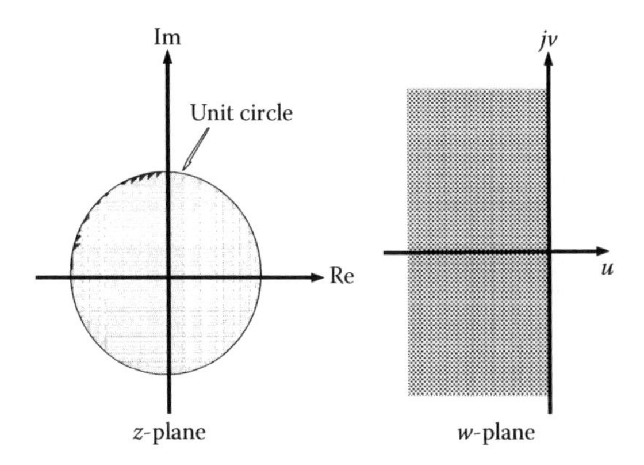

FIGURE 3.10 Mapping in bilinear transformation.

Example 3.11

The characteristic polynomial of a discrete-time system is given by

$$Q(z) = z^3 - 2z^2 + 1.5z - 0.4 = 0 \tag{3.45}$$

Then

$$Q(w) = \left(\frac{w+1}{w-1}\right)^2 - 2\left(\frac{w+1}{w-1}\right)^2 + 1.5\left(\frac{w+1}{w-1}\right) - 0.4 = 0$$

$$\text{or} \quad (w+1)^3 - 2(w-1)(w+1)^2 + 1.5(w+1)(w-1)^2 - 0.4(w-1)^2 = 0$$

This can be further simplified to obtain the polynomial

$$0.1\,w^3 + 0.7\,w^2 + 2.3\,w + 4.9 = 0 \tag{3.46}$$

The Routh table is shown below:

w^3	0.1	2.3
w^2	0.7	4.9
w^1	1.6	
w^0	4.9	

No change in sign in the first column of the Routh table indicates that all roots of $Q(w)$ are in the left half of the w-plane, and correspondingly, all roots of $Q(z)$ are inside the unit circle. Hence, $Q(z)$ is the characteristic polynomial of a stable system.

It will be seen that this method requires a lot of algebra. In Section 3.4.2, we discuss another test for stability that can be performed directly in the z-plane, in a manner reminiscent of the conventional Routh test in the s-plane, although requiring more terms and more computation.

3.4.2 Jury Stability Test

This is similar to the Routh test and can be used directly in the z-plane. The procedure is shown below.

Let the characteristic polynomial be given by

$$Q(z) = a_0 z^n + a_1 z^{n-1} + \cdots + a_{n-1} z + a_n, \quad a_0 > 0 \tag{3.47}$$

Then we form the array shown in the following table.

a_0	a_1	a_2	$\cdots$	a_k	$\cdots$	a_{n-1}	a_n	
a_n	a_{n-1}	a_{n-2}	$\cdots$	a_{n-k}	$\cdots$	a_1	a_0	$\alpha_n = a_n/a_0$
b_0	b_1	b_2	$\cdots$	b_{n-k}	$\cdots$	b_{n-1}		
b_{n-1}	b_{n-2}	b_{n-3}	$\cdots$	b_{k-1}	$\cdots$	b_0		$\alpha_{n-1} = b_{n-1}/b_0$
c_0	c_1	c_2	$\cdots$	c_{n-k}	$\cdots$			
c_{n-2}	c_{n-3}	c_{n-4}	$\cdots$	c_{k-2}	$\cdots$			$\alpha_{n-2} = c_{n-2}/c_0$
	.	.	$\cdots$	.	$\cdots$			

The elements of the first and second rows are the coefficients of $Q(z)$, in the forward and the reverse order, respectively. The third row is obtained by multiplying the second row by $\alpha_n = a_n/a_0$ and subtracting this from the first row. The fourth row is the third row written in the reverse order. The fifth row is obtained in a similar manner, that is, by multiplying the fourth row by b_{n-1}/b_0 and subtracting this from the third row. The sixth row is the fifth row written in the reverse order. The process is continued until there are $2n + 1$ rows. The last row has only one element.

Jury's stability test states that if $a_0 > 0$, then all roots of $Q(z)$ are inside the unit circle if and only if b_0, c_0, d_0, etc. (i.e., the first element in each odd-numbered row) are all positive. Furthermore, the number of negative elements in this set is equal to the number of roots of $Q(z)$ outside the unit circle and a zero element implies a root on the unit circle.

Example 3.12

Consider the characteristic polynomial, $Q(z)$, given in Example 3.11. The Jury's table is shown below:

1	−2	1.5	−0.4	
−0.4	1.5	−2	1	$\alpha_3 = -0.4$
0.84	−1.4	0.7		
0.7	−1.4	0.84		$\alpha_2 = 5/6$
0.2567	−0.2333			
−0.2333	0.2567			$\alpha_1 = -0.909$
0.0445				

Since the coefficients a_0, b_0, c_0, and d_0 are all positive, it follows that all roots of $Q(z)$ are inside the unit circle and it is the characteristic polynomial of a stable system.

3.4.3 Nyquist Criterion

It was seen in Section 3.3.3 that one can use the frequency response of the open-loop to determine stability of a closed-loop continuous-time system using the Nyquist criterion. The main objective there was to determine the number of zeros of $1 + GH(s)$ in the right half of the s-plane from the number of encirclements of the point $-1 + j0$ in the GH-plane.

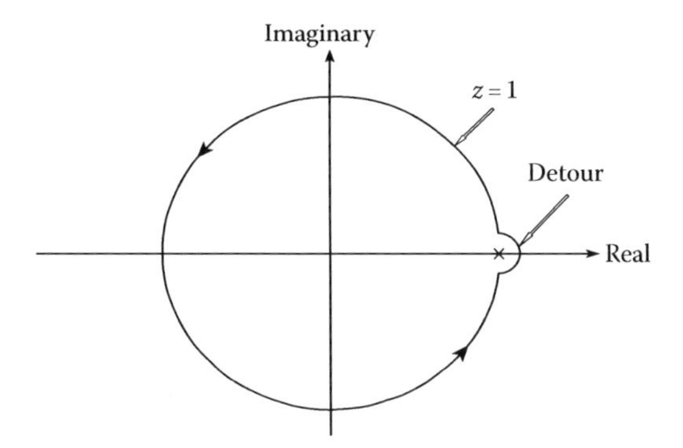

FIGURE 3.11 Nyquist contour in the z-plane.

To apply the Nyquist criterion to determine stability of discrete-time systems, we need to determine the number of roots of $1 + GH(z)$ in the region outside the unit circle of the z-plane. The Nyquist contour in this case is the unit circle of the z-plane, with detours around poles on the path, traversed in the counterclockwise direction, as shown in Figure 3.11.

Note that in the s-plane, the Nyquist contour was taken in the clockwise direction. This can be understood by appreciating that the right half of the s-plane maps outside the unit circle in the z-plane, and we enclose this region if it is to our right while we traverse along the Nyquist contour.

Let N be the number of counterclockwise encirclements of the Nyquist plot of the point $-1 + j0$ in the $GH(z)$-plane. The Nyquist criterion states that for a stable system, we must have $N = -P$, where P is the number of poles of $GH(z)$ inside the Nyquist contour (i.e., in the region to our right while we traverse the contour in the z-plane).

Example 3.13

The Nyquist criterion will be illustrated using the unity-feedback system, where the transfer function of the forward path is

$$G(z) = \frac{0.6(z+0.4)}{(z-1)(z-0.4)} \tag{3.48}$$

since the transfer function has a pole at $z = 1$, the Nyquist contour will be as shown in Figure 3.11, with a detour around that point. On this detour

$$z = 1 + \varepsilon e^{j\theta}, \quad \varepsilon \ll 1 \tag{3.49}$$

and

$$G(z)\Big|_{z=1+ze^{j\theta}} = \frac{0.6(1.4)}{\varepsilon e^{j\theta}(0.6)} = \frac{1.4}{\varepsilon} e^{-j\theta} \tag{3.50}$$

This leads to an arc of infinite radius on the Nyquist plot. For z on the unit circle, we have

$$G(z)\Big|_{z=e^{jwT}} = \frac{0.6(e^{jwT}+0.4)}{(e^{jwT}-1)(e^{jwT}-0.4)} \tag{3.51}$$

TABLE 3.2 Frequency Response of the Transfer Function Given by Equation 3.49

ωT	M	φ	ωT	M	φ
0.1	13.91	−98.3	1.2	0.69	−166.0
0.2	6.83	−106.4	1.4	0.52	−173.3
0.3	4.43	−114.3	1.6	0.41	−179.4
0.4	3.20	−121.8	1.8	0.33	−184.4
0.5	2.45	−128.9	2.0	0.26	−188.2
0.6	1.94	−135.5	2.2	0.22	−190.6
0.7	1.57	−141.6	2.4	0.18	−191.5
0.8	1.30	−147.3	2.6	0.16	−190.6
0.9	1.09	−152.5	2.8	0.14	−187.9
1.0	0.93	−157.4	π	0.13	−180.0

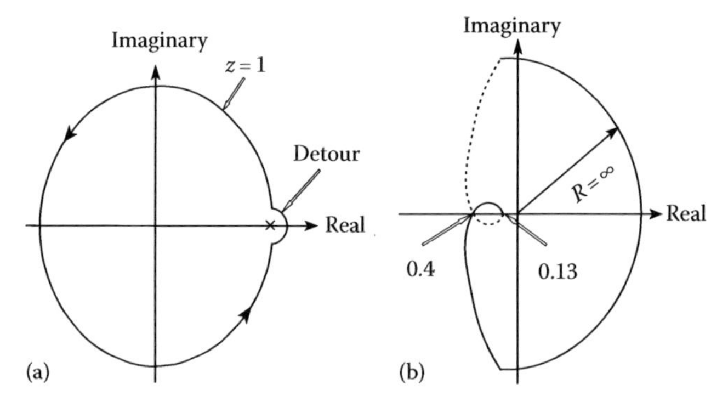

FIGURE 3.12 Nyquist contour and plot for Example 3.13: (a) Nyquist contour and (b) Nyquist plot.

In Equation 3.50, $0 < \omega T < 2\pi$. Since $G(e^{j\omega T})$ is the complex conjugate of $G(e^{-j\omega T})$, it is necessary to calculate the frequency response only for $0 < \omega < \pi$. Some values are shown in Table 3.2.

The polar plot of the frequency response is the map of the upper half of the unit circle of the z-plane. The lower half maps as the mirror image of this response about the real axis. The plot is completed from the map of the detour, which is a semicircle of infinite radius, obtained from Equation 3.49. Figure 3.12 shows the Nyquist plot for this system, with the Nyquist contour as in Figure 3.11. From the Nyquist plot, it is evident that the system is stable. It can, however, be forced into instability if the gain is increased by a factor of 2.5, so that the point $-1 + j0$ is enclosed within the Nyquist plot.

As an alternative to using the z-plane, one may apply the bilinear transformation to the w-plane making it similar to the plot for continuous-time systems. A minor modification of the bilinear transformation due to Tustin is often used and is described in Equation 3.44.

$$w = \frac{2}{T}\frac{z-1}{z+1}$$
$$z = \frac{(2/T)+w}{(2/T)-w} \tag{3.52}$$

where T is the sampling interval.

Tustin had originally used this transformation for approximate design of digital filters.

To understand the implications of this transformation, let us consider the stability boundary in the w-plane by writing

$$w = u + jv = \frac{2}{T}\frac{e^{sT}-1}{e^{sT}+1} = \frac{2}{T}\tanh\frac{sT}{2} \tag{3.53}$$

For $s = j\omega$, we get $Z = e^{jwT}$, which is the unit circle in the z-plane, and

$$w = 0 + jv = j\frac{2}{T}\tan\frac{\omega t}{2} \tag{3.54}$$

Thus, the w-plane frequency, v, is related to the s-plane frequency ω through the relationship

$$v = \frac{2}{T}\tan\frac{\omega t}{2} \tag{3.55}$$

Note that v is approximately equal to ω for small values of ωT. The approximation is valid as long as $\tan \omega t/2$. The error in this approximation is less than 4% for

$$\frac{\omega t}{2} \le \frac{\pi}{10} \quad \text{or} \quad \omega \le \frac{2\pi}{10T} = \frac{\omega_s}{10} \tag{3.56}$$

Thus, if the sampling frequency is chosen so that Equation 3.56 is satisfied over the bandwidth of the system, then it is possible to use the w-plane frequency response of a discrete-time system exactly like the s-plane frequency of a continuous-time system, for the study of system stability as well as for the design of compensators based on frequency response.

3.5 Stability of Nonlinear Systems

Unlike linear systems, the stability of nonlinear systems depends not only on the physical properties of the system but also on the magnitude and nature of the input as well as the initial conditions. Hence, the study of stability of nonlinear systems is more complicated than for linear systems.

Several definitions of stability have been used in the literature for nonlinear systems. We shall consider here mainly the case of an autonomous or unforced system. Since, in genera, nonlinear systems may have more than one state of equilibrium, each of these states must be examined for stability.

As stated in Section 3.2, all values of the n-dimensional state vector that satisfy the equation

$$f(x) = 0 \tag{3.57}$$

are states of equilibrium. Each of these states, visualized as a particular point in an n-dimensional space, must be investigated for stability.

Some of these may be points of stable equilibrium, while others may be points of unstable equilibrium. A good example is the bistable multivibrator, an electronic circuit with three states of equilibrium, two of which are stable and one unstable.

For any nonlinear system with many states of equilibrium, it is necessary to investigate stability at each of these points.

A system is said to have *local stability* at an equilibrium state if, after a small perturbation, it eventually returns to that state,

A system is said to have *global stability* at an equilibrium state if, for any perturbation (small or large), it eventually returns to that state.

We can investigate local stability of nonlinear systems by examining the effect of small perturbations at each point of equilibrium. This can be done by obtaining an approximate linear model at each of these points and testing it for stability. This is described below.

3.5.1 Linearization

Linearization is based on the Taylor series expansion of a nonlinear function about an operating point. For example, consider a nonlinear function, $f(x)$. It can be written as

$$f(x) = f(x_0) + \left.\frac{df}{dx}\right|_{x=x_0}(x - x_0) + \left.\frac{d^2 f}{d^2 x}\right|_{x=x_0}\frac{(x - x_0)^2}{2!} + \cdots \tag{3.58}$$

We get a linear approximation of Equation 3.58 if we ignore all terms except the first two. Clearly, this will be a good approximation if either $(x - x_0)$ is very small or the higher order derivatives of f are very small. This is the main idea behind the incremental linear models used for the analysis of electronic circuits.

We shall now generalize this to the case of the state equations for nonlinear systems. Assuming that the dimension of x is n, we rewrite Equation 3.1 as

$$\dot{x} = f(x,u) = \begin{bmatrix} f_1(x,u) \\ f_2(x,u) \\ f_1(x,u) \\ \cdot \\ \cdot \\ \cdot \\ f_n(x,u) \end{bmatrix} \tag{3.59}$$

Ignoring the higher order terms in the Taylor series expansion of this vector, differential equation leads to the linearized model (assuming $x_0 = 0$, i.e., the coordinates have been transformed to make the origin the point of equilibrium, and $u_0 = 0$)

$$\dot{x} = Ax + Bu \tag{3.60}$$

where

$$A = \begin{bmatrix} \frac{\partial f_1}{\partial x_1} & \frac{\partial f_1}{\partial x_2} & \cdots & \cdots & \frac{\partial f_1}{\partial x_n} \\ \frac{\partial f_2}{\partial x_1} & \frac{\partial f_2}{\partial x_2} & \cdots & \cdots & \frac{\partial f_2}{\partial x_n} \\ & & \vdots & & \\ \frac{\partial f_n}{\partial x_1} & \frac{\partial f_n}{\partial x_2} & \cdots & \cdots & \frac{\partial f_n}{\partial x_n} \end{bmatrix} \tag{3.61}$$

and

$$B = \begin{bmatrix} \frac{\partial f_1}{\partial u_1} & \frac{\partial f_1}{\partial u_2} & \cdots & \cdots & \frac{\partial f_1}{\partial u_m} \\ \frac{\partial f_2}{\partial u_1} & \frac{\partial f_2}{\partial u_2} & \cdots & \cdots & \frac{\partial f_2}{\partial u_m} \\ & & \vdots & & \\ \frac{\partial f_n}{\partial u_1} & \frac{\partial f_n}{\partial u_2} & \cdots & \cdots & \frac{\partial f_n}{\partial u_m} \end{bmatrix} \tag{3.62}$$

A and B are said to be Jacobian matrices. Again, this linear model will be valid only for small deviations around the point of equilibrium. Nevertheless, it can be used for investigating local stability around the point of equilibrium (for the autonomous) case by simply applying the Routh criterion to the characteristic polynomial of A. The following example illustrates the main idea behind this approach.

Note that in Equation 3.52, we have used x instead of the deviation $x - x_0$. This is a common practice, even if x_0 is not the origin of the state space. It is to be understood that x represents the variation of the state from the point of equilibrium or "set-point" in the terminology of process control.

Example 3.14

Consider the following second-order nonlinear differential equation:

$$\frac{d^2x}{d^2t} + 2x\frac{dx}{dt} + 2x^2 - 4x \quad (3.63)$$

a. Determine the points of equilibrium.
b. Investigate the stability of the system near each point of equilibrium.

Solution

We shall first derive the state equations for the given differential equation. Let

$$\left.\begin{aligned} x_1 &= x \\ x_2 &= \dot{x} \end{aligned}\right\} \quad (3.64)$$

Then we obtain the following state equations

$$f_1 = \dot{x}_1 = x_2 \quad (3.65)$$

The points of equilibrium are obtained by setting the two derivatives in Equation 3.57 to zero and are readily found as (0,0) and (2,0).

The resulting Jacobian matrix is shown in Equation 3.58

$$\begin{bmatrix} \dfrac{\partial f_1}{\partial x_1} & \dfrac{\partial f_1}{\partial x_2} \\ \dfrac{\partial f_2}{\partial x_1} & \dfrac{\partial f_2}{\partial x_2} \end{bmatrix} = \begin{bmatrix} 0 & 1 \\ -4x_1 - x_2 + 4 & -2x_1 \end{bmatrix} \quad (3.66)$$

For the equilibrium point given by $x_1 = x_2 = 0$,

$$A = \begin{bmatrix} 0 & 1 \\ 4 & 0 \end{bmatrix} \quad (3.67)$$

The characteristic polynomial for this case is

$$\det(sI - A) = \begin{bmatrix} s & 4 \\ 1 & s \end{bmatrix} = s^2 - 4 \quad (3.68)$$

indicating an unstable system with a pole in the right half of the s-plane.

For the equilibrium point given by $x_1 = x_2 = 0$, we have

$$A = \begin{bmatrix} 0 & 1 \\ -4 & -4 \end{bmatrix} \tag{3.69}$$

The characteristic polynomial for this case is

$$\det(sI - A) = \begin{bmatrix} s & -1 \\ 4 & s+4 \end{bmatrix} = s^2 + 4s + 4 \tag{3.70}$$

which has roots in only left half of the **s**-plane. This indicates that the system will be stable in the neighborhood of this point.

3.5.2 Lyapunov's Method

The simple stability criteria developed for linear systems are not applicable to nonlinear systems since the concept of the roots of a characteristic polynomial is no longer valid. As stated earlier, many different classes of stability have been defined for nonlinear systems. In this section, we discuss the stability in the sense of Lyapunov.*

For any given state of equilibrium, it is common practice to transform coordinates in the state space so that the origin becomes the point of equilibrium. This is convenient for examining *local stability* and can be done for each point of equilibrium.

Let us now consider a hypersphere of finite radius surrounding the origin of the state space (the point of equilibrium), that is, the set of points described by the equation

$$x_1^2 + x_2^2 + \cdots x_n^2 = R^2 \tag{3.71}$$

in the n-dimensional state space. Let this region be denoted by $S(R)$.

The system is said to be stable in the sense of Lyapunov if there exists a region $S(\varepsilon)$ such that a trajectory starting from any point $x(0)$ in this region does not go outside the region $S(R)$. This is illustrated in Figure 3.13 for the two-dimensional case, $n = 2$.

Note that with this definition, it is not necessary for the trajectory to approach the point of equilibrium. It is only required that the trajectory be within the region ε. This permits the existence of oscillations of limited amplitude, like limit cycles.

The system is said to be *asymptotically stable* if there exists a $\delta > 0$ such that the trajectory starting from any point $x(0)$ within $S(\delta)$ does not leave $S(R)$ at any time and finally returns to the origin. The trajectory in Figure 3.14a shows asymptotic stability.

The system is said to be *monotonically stable* if it is asymptotically stable and the distance of the state from the origin decreases monotonically with time. The trajectory in Figure 3.14b shows monotonic stability.

A system is said to be *globally stable* if the regions $S(\delta)$ and $S(R)$ extend to infinity.

A system is said to be *locally stable* if the region $S(\delta)$ is small and when subjected to small perturbations, the state remains within the small specified region $S(R)$.

We can investigate local stability of nonlinear systems by examining the effect of small perturbations at each point of equilibrium. This can be done by obtaining an approximate linear model at each of these points and testing it for stability. This is described below.

Consider a region ε in the state space enclosing an equilibrium point x_0. Then this is a point of stable equilibrium provided that there is a region $\delta(\varepsilon)$ contained within ε such that any trajectory starting in the region δ does not leave the region ε.

* After A.M. Lyapunov, a Russian mathematician who did pioneering work in this area during the late nineteenth century.

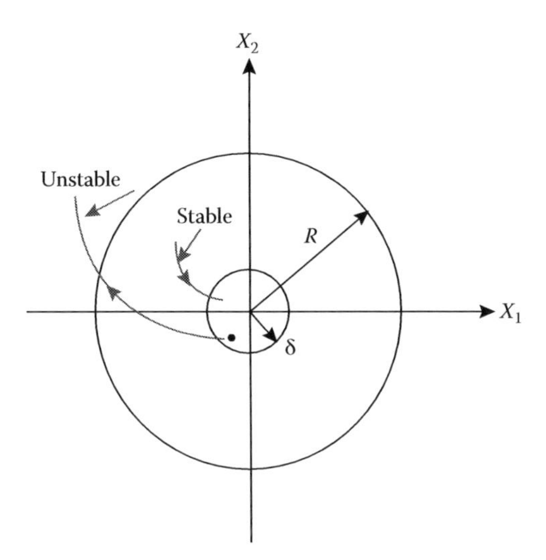

FIGURE 3.13 Stability in the sense of Lyapunov.

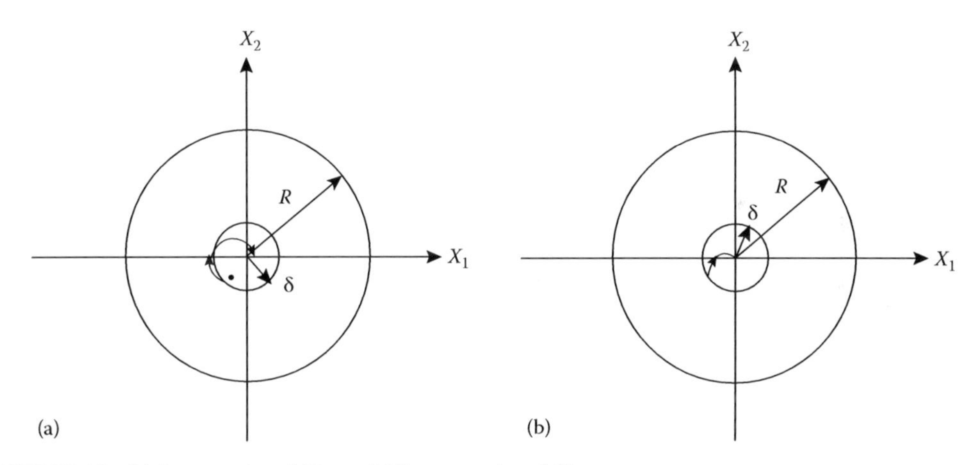

FIGURE 3.14 (a) Asymptotic stability and (b) monotonic stability.

Lyapunov's direct method provides a means for determining the stability of a system without actually solving for the trajectories in the state space. It is based on the simple concept that the energy stored in a stable system cannot increase with time. Given a set of nonlinear state equations, one first defines a scalar function $V(x)$ that has properties similar to energy and then examines its derivative with respect to time.

Theorem 3.1

A system described by $\dot{x} = f(x)$ is asymptotically stable in the vicinity of the point of equilibrium at the origin of the state space if there exists a scalar function V such that

1. $V(x)$ is continuous and has continuous first partial derivatives at the origin
2. $V(x) > 0$ for $x \neq 0$ and $V(0) = 0$
3. $V(\dot{x}) < 0$ for all $x \neq 0$

Note that these conditions are *sufficient but not necessary* for stability. $V(x)$ is often called a Lyapunov function.

Theorem 3.2

A system described by $\dot{x} = f(x)$ is unstable in a region Ω about the equilibrium at the origin of the state space if there exists a scalar function V such that

1. $V(x)$ is continuous and has continuous first partial derivatives in Ω
2. $V(x) \geq 0$ for $x \neq 0$ and $V(0) = 0$
3. $V(\dot{x}) > 0$ for all $x \neq 0$

Again, it should be noted that these conditions are *sufficient but not necessary.*

Example 3.15

Consider the system described by the equations

$$\left.\begin{aligned} \dot{x}_1 &= x_2 \\ \dot{x}_2 &= -x_1 - x_2^3 \end{aligned}\right\} \tag{3.72}$$

If we make

$$V = x_1^2 + x_2^2 \tag{3.73}$$

which satisfies conditions 1 and 2, then we get

$$\dot{V} = 2x_1\dot{x}_1 + 2x_2\dot{x}_2 = 2x_1x_2 + 2x_2(-x_1 - x_2^3) = -2x_2^4 \tag{3.74}$$

It will be seen that $\dot{V} < 0$ for all nonzero values of x_2, and hence the system is asymptotically stable.

Example 3.16

Consider the system described by

$$\dot{x}_1 = -x_1(1 - 2x_1x_2) \tag{3.75}$$

Let

$$V = \frac{1}{2}x_1^2 + x_2^2 \tag{3.76}$$

which satisfies conditions 1 and 2. Then

$$\dot{V} = x_1\dot{x}_1 + 2x_2\dot{x}_2 = -x_1^2(1 - 2x_1x_2) - 2x_2^2 \tag{3.77}$$

Although it is not possible to make a general statement regarding global stability for this case, it is clear that $\dot{V}$ is negative if $1 - 2x_1x_2$. This defines a region of stability in the state space, bounded by all points for which $x_1x_2 < 0.5$.

Evidently, the main problem with this approach is the selection of a suitable function $V(x)$ such that its derivative is either positive definite or negative definite. Unfortunately, there is no general method that will work for every nonlinear system. Gibson (1963) has described the variable gradient method for generating Lyapunov functions [G63]. It is not described here, but the interested reader may refer to Section 8.1 of the Gibson text. Atherton (1982) provides further information on Lyapunov functions [A82].

References

[A82] Atherton, D. P. 1982. *Nonlinear Control Engineering,* Van Nostrand-Reinhold, London, U.K.

[B85] Bose, N. K. 1985. A systematic approach to stability of sets of polynomials, *Contemporary Mathematics,* 47:25–34.

[G63] Gibson, J. E. 1963. *Nonlinear Automatic Control,* McGraw-Hill, New York.

[K78] Kharitonov, V. L. 1978. Asymptotic stability of an equilibrium position of a family of systems of linear differential equations, *Differential'llye UraFclliya,* 14:1483–1485.

[MAD89] Minnichelli, R. J., Anagnost, J. J., and Desoer, C. A. 1989. An elementary proof of Kharitonov's stability theorem with extensions, *IEEE Transactions on Automatic Control,* AC-34:995–998.

[S94] Sinha, N. K. 1994. *Control Systems,* 2nd edn., Wiley Eastern, New Delhi, India.

4

Frequency-Domain Analysis of Relay Feedback Systems

Igor M. Boiko
University of Calgary

4.1 Relay Feedback Systems

4.1.1 Introduction

Relay feedback systems are one of the most important types of nonlinear systems. The term “relay” came from electrical applications where on–off control has been used for a long time. To describe the nonlinear phenomenon typical of electrical relays, the nonlinear function also received the name “relay” comprising a number of discontinuous nonlinearities. Thus, when applied to the type of the control systems, the term “relay” is now associated not with applications but with the kind of nonlinearities that are found in the system models.

The applications of the relay feedback principle have evolved from vibrational voltage regulators and missile thruster servomechanisms of the 1940s [3] to numerous on–off process parameter closed-loop control systems, sigma–delta modulators, process identification and automatic tuning of PID controllers techniques [9,10], DC motors, hydraulic and pneumatic servo systems, etc. It is enough to mention an enormous number of residential temperature control systems available throughout the world to understand how popular the relay control systems are. A number of industrial examples of relay systems were given in the classical book on relay systems [7]. Furthermore, many existing sliding mode algorithms [8] can be considered and analyzed via the relay control principle.

The use of relay control provides a number of advantages over the use of linear control: simplicity of design, cheaper components, and the ability to adapt the open-loop gain in the relay feedback system as parameters of the system change [1,4,6]. As a rule, they also provide a higher open-loop gain and a higher performance [5] than the linear systems. In some applications, smoothing of the Coulomb friction and of other plant nonlinearities can also be achieved.

Relay control systems theory has been consistently receiving a lot of attention since the 1940s from the worldwide research community. Traditionally, the scope of research is comprised of the following problems: existence and parameters of periodic motions, stability of limit cycles, and input–output problem (set point tracking or external signal propagation through the system), which also includes the disturbance attenuation problem. The theory of relay feedback systems is presented in a number of classical and recent publications, some of which are given in the list of references.

The method of analysis presented in this chapter is similar from the methodological point of view to the describing function method. Some concepts (like the notion of the *equivalent gain*) are the same. However, the presented method is exact and the notions that are traditionally used within the describing function method are redefined, so that they now describe the system properties in the exact sense. For the reason of this obvious connection between the presented method and the describing function method, the latter is presented as a logical extension of the former. And some introductory material of the chapter is devoted to the describing function method. Section 4.2 presents the locus of a perturbed relay system (LPRS) method, and Section 4.3 gives the methodology of compensator design in relay feedback system, based on the LPRS method.

4.1.2 Relay Feedback Systems

It is a well-known fact that relay feedback systems exhibit self-excited oscillations as their inherent mode of operation. Analysis of the frequency and the amplitude of these oscillations is an objective of analysis of relay systems. However, in application of relay feedback systems, the *autonomous* mode, when no external signals are applied to the system, does not normally occur. An external signal always exists either in the form of an exogenous disturbance or a set point. In the first case, the system is supposed to respond to this disturbance in such a way as to provide its compensation. In the second case, the system is supposed to respond to the input, so that the output can be brought in alignment with this external input. In both cases, the problem of analysis of the effect of the external signals on the system characteristics is an important part of system performance analysis.

One can see that the problem of analysis of external signal propagation applies to both types of systems considered above. Moreover, if we consider a model of the system, the difference between those two signals would only be in the point of application, and from the point of the methodology of analysis, they are not different. Since we are going to deal with models in this chapter, we can consider only one signal applied to the system and consider it being a disturbance or a reference input signal depending on the system task. Naturally, the problem of propagation of external signals cannot be solved without the autonomous mode analysis having been carried out first. Therefore, a complex analysis, which is supposed to include the analysis of the autonomous mode and the analysis of external signal propagation, needs to be carried out in practical applications.

In a general case (which includes time-delay linear plant), relay feedback systems can be described by the following equations:

$$\begin{aligned} \dot{\mathbf{x}}(t) &= \mathbf{A}\mathbf{x}(t) + \mathbf{B}u(t-\tau) \\ y(t) &= \mathbf{C}\mathbf{x}(t) \end{aligned} \tag{4.1}$$

where

$\mathbf{A} \in R^{n \times n}$, $\mathbf{B} \in R^{n \times 1}$, $\mathbf{C} \in R^{1 \times n}$ are matrices
$\mathbf{A}$ is nonsingular
$\mathbf{x} \in R^{n \times 1}$ is the state vector
$y \in R^1$ is the system output

τ is the time delay (which can be set to zero if no time delay is present)
$u \in R^1$ is the control defined as follows:

$$u(t) = \begin{cases} +c & \text{if } \sigma(t) = f(t) - y(t) \geq b \\ & \text{or } \sigma(t) > -b, \quad u(t-) = c \\ -c & \text{if } \sigma(t) = f(t) - y(t) \leq -b \\ & \text{or } \sigma(t) < b, \quad u(t-) = -c \end{cases} \tag{4.2}$$

where

$f(t)$ is an external relatively slow input signal to the system
σ is the error signal
c is the relay amplitude
$2b$ is the hysteresis value of the relay
$u(t-) = \lim_{\varepsilon \to 0, \varepsilon > 0} u(t - \varepsilon)$ is the control at time instant immediately preceding time t

We shall consider that time $t = 0$ corresponds to the time of the error signal becoming equal to the positive half-hysteresis value (subject to $\dot{\sigma} > 0$): $\sigma(0) = b$ and call this time the *time of switch initiation.*

We represent the relay feedback system as a block diagram (Figure 4.1). In Figure 4.1, $W_l(s)$ is the transfer function of the linear part (of the plant in the simplest case), which can be obtained from the matrix-vector description (4.1) as

$$W_l(s) = e^{-\tau s}\mathbf{C}(\mathbf{I}s - \mathrm{A})^{-1}\mathbf{B} \tag{4.3}$$

We shall assume that the linear part is strictly proper, i.e., the relative degree of $W_l(s)$ is one or higher.

4.1.3 From Describing Function Analysis to LPRS Analysis

4.1.3.1 Symmetric Oscillations in Relay Feedback Systems

A mode that may occur in a relay feedback system is a self-excited (nonvanishing) oscillation, which is also referred to as a self-excited periodic motion or a *limit cycle*. If the system does not have asymmetric nonlinearities, this periodic motion is symmetric in the autonomous mode (no external input applied). However, if a constant external input (disturbance) is applied to the system that has a periodic motion, the self-excited oscillations become biased or asymmetric. The key approach to the analysis of the relay feedback systems is a method of analysis of self-excited oscillations (symmetric and asymmetric). The following sections give a general methodology of such an analysis based on the frequency-domain concepts.

Consider the autonomous mode of system operation ($f(t) \equiv 0$). Due to the character of the nonlinearity, which results in the control having only two possible values, the system in Figure 4.1 cannot have

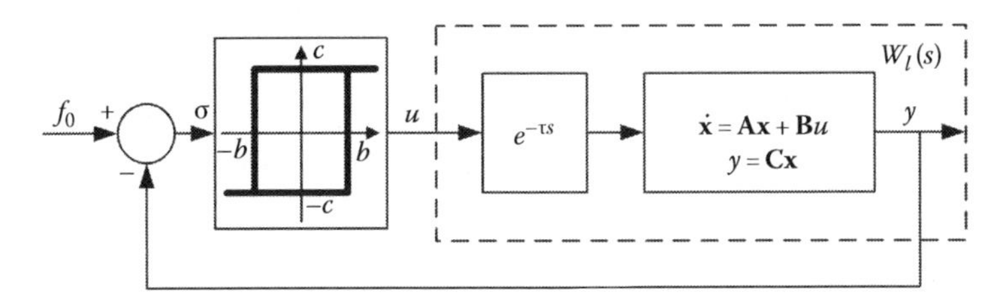

FIGURE 4.1 Relay feedback system.

an equilibrium point. We can assume that a symmetric periodic process of unknown frequency Ω and amplitude a of the input to the relay occurs in the system. Finding the values of the frequency and amplitude is one of the main objectives of the analysis of relay feedback systems.

Very often, an approximate method known as the describing function (DF) method is used in engineering practice. The main concepts of this method were developed in the 1940s and 1950s [6]. A comprehensive coverage of the DF method is given in Refs. [1,4]. The DF method provides a simple and efficient but not exact solution of the periodic problem. For this method to provide an acceptable accuracy, the linear part of the system must be a low-pass filter to be able to filter out higher harmonics, so that the input to the relay could approximately be considered a sinusoid.

In accordance with the DF method concepts, an assumption is made that the input to the nonlinearity is a harmonic signal, and the so-called DF of the nonlinearity is found as a function of the amplitude and frequency as per the following definition:

$$N(a,\omega) = \frac{\omega}{\pi a}\int_0^{2\pi/\omega} u(t)\sin\omega t\,dt + j\frac{\omega}{\pi a}\int_0^{2\pi/\omega} u(t)\cos\omega t\,dt \tag{4.4}$$

DF given by formula (4.4) is essentially a complex gain of the transformation of the harmonic input by the nonlinearity into the control signal with respect to the first harmonic in the control signal. For the hysteretic relay nonlinearity, the formula of the DF can be obtained analytically. It is given as follows [1]:

$$N(a) = \frac{4c}{\pi a}\sqrt{1-\left(\frac{b}{a}\right)^2} - j\frac{4cb}{\pi a^2}, \quad (a \ge b) \tag{4.5}$$

For the hysteretic relay, the DF is a function of the amplitude only and does not depend on the frequency. The periodic solution in the relay feedback system can be found from the equation of the *harmonic balance* [1],

$$W_l(j\Omega) = -\frac{1}{N(a)} \tag{4.6}$$

which is a complex equation with two unknown values: frequency Ω and amplitude a. The expression in the left-hand side of the Equation 4.6 is the frequency response (Nyquist plot) of the linear part of the system. The value in the right-hand side is the negative reciprocal of the DF of the hysteretic relay. Let us obtain it from formula (4.5):

$$-N^{-1}(a) = -\frac{\pi a}{4c}\sqrt{1-\left(\frac{b}{a}\right)^2} - j\frac{\pi b}{4c}, \quad (a \ge b) \tag{4.7}$$

One can see from Equation 4.7 that the imaginary part does not depend on the amplitude a, which results in a simple solution of Equation 4.6. Graphically, the periodic solution of the equations of the relay feedback system would correspond to the point of intersection of the Nyquist plot of the linear part (being a function of the frequency) and of the negative reciprocal of the DF of the hysteretic relay (being a function of the amplitude) given by formula (4.7), in the complex plane. This periodic solution is not exact, which is a result of the approximate nature of the DF method itself that is based upon the assumption about the harmonic shape of the input signal to the relay. However, if the linear part of the system has the property of the low-pass filter, so that the higher harmonics of the control signal are attenuated well enough, the DF method may give a relatively precise result.

4.1.3.2 Asymmetric Oscillations in Relay Feedback Systems and Propagation of External Constant Inputs

Now we turn to the analysis of asymmetric oscillations in the relay feedback system, which is the key step to the analysis of the system response to constant and slow varying disturbances and reference input signals.

Assume that the input to the system is a constant signal f_0: $f(t) \equiv f_0$. Then, an asymmetric periodic motion occurs in the system (Figure 4.2), so that each signal now has a periodic and a constant term: $u(t) = u_0 + u_p(t)$, $y(t) = y_0 + y_p(t)$, $\sigma(t) = \sigma_0 + \sigma_p(t)$, where subscript "0" refers to the constant term in the Fourier series, and subscript p refers to the periodic term of the function (the sum of periodic terms of the Fourier series).

The constant term is the mean or averaged value of the signal on the period. Now let us imagine that we quasi-statically (slowly) slew the input from a certain negative value to a positive value, so that at each value of the input, the system would exhibit a stable oscillation and measure the values of the constant term of the control (mean control, which can quantitatively be represented by the shaded area in Figure 4.2, with account of the signs) versus the constant term of the error signal (mean error). By doing this, we can determine the constant term of the control signal as a function of the constant term of the error signal, which would be not a discontinuous but a smooth function $u_0 = u_0(\sigma_0)$. This function will be further referred to as the *bias function*. A typical bias function is depicted in Figure 4.3. The described

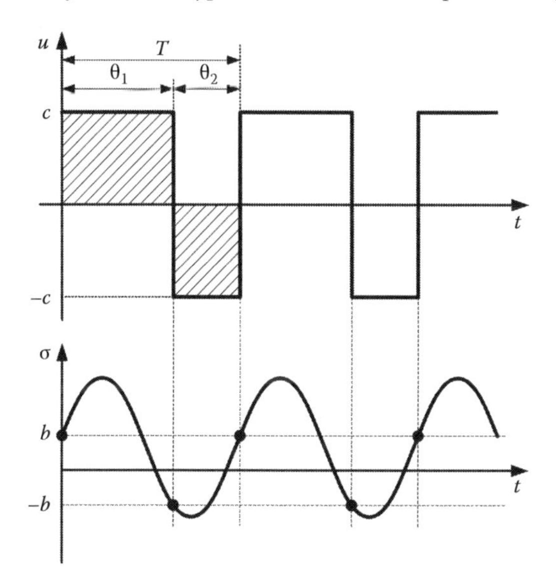

FIGURE 4.2 Asymmetric oscillations at unequally spaced switches.

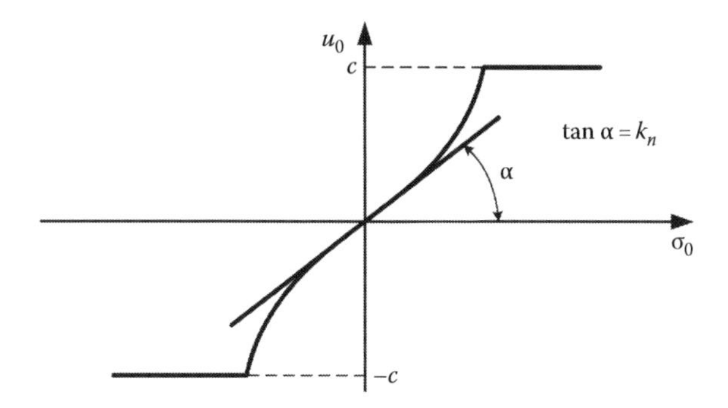

FIGURE 4.3 Bias function and equivalent gain.

smoothing effect is known as the *chatter smoothing* phenomenon, which is described in [5]. The derivative of the mean control with respect to the mean error taken in the point of zero mean error $\sigma_0 = 0$ (corresponding to zero constant input) provides the *equivalent gain* of the relay k_n, which conceptually is similar to the so-called the *incremental gain* [1,4] of the DF method. The *equivalent gain* of the relay is used as a local approximation of the bias function $k_n = du_0/d\sigma_0\big|_{\sigma_0=0} = \lim_{f_0\to 0}(u_0/\sigma_0)$.

Now we carry out the analysis of asymmetric oscillations in the system Figure 4.1 caused by a nonzero constant input $f(t) \equiv f_0 \neq 0$. The DF of the hysteresis relay with a biased sine input is represented by the following well-known formula [1]:

$$N(a,\sigma_0) = \frac{2c}{\pi a}\left[\sqrt{1-\left(\frac{b+\sigma_0}{a}\right)^2} + \sqrt{1-\left(\frac{b-\sigma_0}{a}\right)^2}\right] - j\frac{4cb}{\pi a^2}, \quad \left(a \geq b + |\sigma_0|\right) \tag{4.8}$$

where a is the amplitude of the oscillations. The mean control as a function of a and σ_0 is given by the following formula:

$$u_0(a,\sigma_0) = \frac{c}{\pi}\left(\arcsin\frac{b+\sigma_0}{a} - \arcsin\frac{b-\sigma_0}{a}\right) \tag{4.9}$$

From (4.8) and (4.9), we can obtain the DF of the relay and the derivative of the mean control with respect to the mean error for the symmetric sine input

$$k_{n(DF)} = \frac{\partial u_0}{\partial \sigma_0}\bigg|_{\sigma_0=0} = \frac{2c}{\pi a}\frac{1}{\sqrt{1-\left(\frac{b}{a}\right)^2}} \tag{4.10}$$

which is the *equivalent gain* of the relay (the subscript *DF* refers to the DF method that was used for finding this characteristic).

Since at the slow inputs, the relay feedback system behaves similarly to a linear system with respect to the response to those input signals, finding the *equivalent gain* value is the main point of the input–output analysis. Once it is found, all subsequent analysis of propagation of the slow input signals can be carried out exactly like for a linear system with the relay replaced with the *equivalent gain*. The model obtained via the replacement of the relay with the *equivalent gain* would represent the model of the averaged (on the period of the oscillations) motions in the system.

4.2 Locus of a Perturbed Relay System Theory

4.2.1 Introduction to the LPRS

As we considered above, the motions in relay feedback systems are normally analyzed as motions in two separate dynamic subsystems: the "slow" subsystem and the "fast" subsystem. The "fast" subsystem pertains to the self-excited oscillations or periodic motions. The "slow" subsystem deals with the forced motions caused by an input signal or by a disturbance and the component of the motion due to nonzero initial conditions. It usually pertains to the averaged (on the period of the self-excited oscillation) motion. The two dynamic subsystems interact with each other via a set of parameters: the results of the solution of the "fast" subsystem are used by the "slow" subsystem. This decomposition of the dynamics is possible if the external input is much slower than the self-excited oscillations, which is normally the case. Exactly like within the DF method, we shall proceed from the assumption that the external signals applied to the system are slow in comparison with the oscillations.

Consider again the harmonic balance equation (4.6). With the use of formulas of the negative reciprocal of the DF (4.7) and the equivalent gain of the relay (4.10), we can rewrite formula (4.6) as follows:

$$W_l(j\Omega) = -\frac{1}{2}\frac{1}{k_{n(DF)}} + j\frac{\pi}{4c}y_{(DF)}(0) \tag{4.11}$$

In the imaginary part of Equation 4.11, we consider that the condition of the switch of the relay from minus to plus (defined as zero time) is the equality of the system output to the negative half hysteresis ($-b$): $y_{(DF)}(t = 0) = -b$. It follows from (4.8), (4.10), (4.11) that the frequency of the oscillations and the equivalent gain in the system can be varied by changing the hysteresis value $2b$ of the relay. Therefore, the following two mappings can be considered: $M_1{:}b \to \Omega$, $M_2{:}b \to k_n$. Assume that M_1 has an inverse mapping (it follows from (4.8), (4.10), (4.11) for the DF analysis and is proved below via deriving an analytical formula of that mapping), $M_1^{-1}: \Omega \to b$. Applying the chain rule, consider the mapping $M_2\left(M_1^{-1}\right): \Omega \to b \to k_n$. Now let us define a certain function J exactly as the expression in the right-hand side of formula (4.11) but require from this function that the values of the equivalent gain and the output at zero time should be exact values. Applying mapping $M_2\left(M_1^{-1}\right): \omega \to b \to k_n$, $\omega \in \left[0;\infty\right)$, in which we treat frequency ω as an independent parameter, we write the following definition of this function:

$$J(\omega) = -\frac{1}{2}\frac{1}{k_n} + j\frac{\pi}{4c}y(t)\bigg|_{t=0} \tag{4.12}$$

where

$k_n = M_2\left(M_1^{-1}(\omega)\right)$

$y(t)\big|_{t=0} = M_1^{-1}(\omega)$

$t = 0$ is the time of the switch of the relay from "$-c$" to "$+c$"

Thus, $J(\omega)$ comprises the two mappings and is defined as a characteristic of the response of the linear part to the unequally spaced pulse input $u(t)$ subject to $f_0 \to 0$ as the frequency ω is varied. The real part of $J(\omega)$ contains information about the gain k_n, and the imaginary part of $J(\omega)$ comprises the condition of the switching of the relay and, consequently, contains information about the frequency of the oscillations. If we derive the function that satisfies the above requirements, we will be able to obtain the exact values of the frequency of the oscillations and of the *equivalent gain*.

We will refer to the function $J(\omega)$ defined above and to its plot in the complex plane (with the frequency ω varied) as the LPRS. With LPRS of a given system computed, we are able to determine the frequency of the oscillations (as well as the amplitude) and the equivalent gain k_n (Figure 4.4): the point of intersection of the LPRS and of the straight line, which lies at the distance $\pi b/(4c)$ below (if $b > 0$) or above (if $b < 0$) the horizontal axis and parallel to it (line "$-\pi b/4c$"), offers the computation of the frequency of the oscillations and of the equivalent gain k_n of the relay. According to (4.12), the frequency Ω of the oscillations can be computed via solving the equation

$$\operatorname{Im} J(\Omega) = -\frac{\pi b}{4c} \tag{4.13}$$

(i.e., $y(0) = -b$ is the condition of the relay switch) and the gain k_n can be computed as

$$k_n = -\frac{1}{2\operatorname{Re} J(\Omega)} \tag{4.14}$$

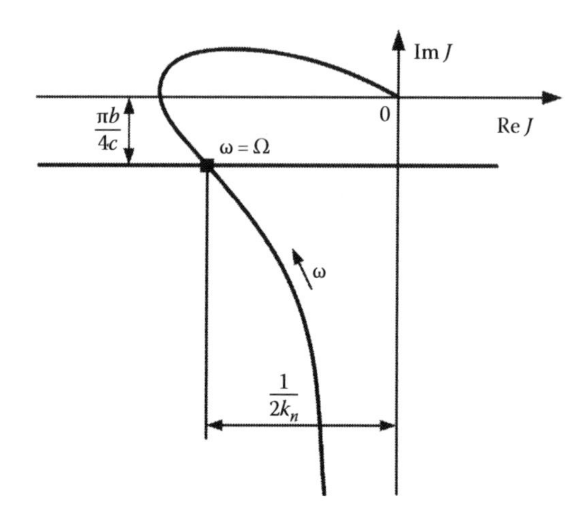

FIGURE 4.4 LPRS and analysis of relay feedback system.

Formula (4.13) provides a periodic solution and is, therefore, a *necessary condition of the existence of a periodic motion* in the system.* Formula (4.12) is only a definition and not intended for the purpose of computing of the LPRS $J(\omega)$. It is shown below that although $J(\omega)$ is defined via the parameters of the oscillations in a closed-loop system, it can be easily derived from the parameters of the linear part without employing the variables of formula (4.12).

4.2.2 Computation of the LPRS from Differential Equations

4.2.2.1 Matrix State-Space Description Approach

We now derive the LPRS formula for a time-delay plant in accordance with its definition (4.12), from the system description given by formulas (4.1) and (4.2). Derivation of the LPRS formula necessitates finding a periodic solution of (4.1) and (4.2), which for an arbitrary nonlinearity is a fundamental problem in both mathematics and control theory. Its solution can be traced back to the works of Poincaré who introduced a now widely used geometric interpretation of this problem as finding a closed orbit in the state space. The approach of Poincaré involves finding a certain map that relates subsequent intersection by the state trajectory of certain hyperplane. A fixed point of this map would give a periodic solution in the dynamic system. We can define the hyperplane for our analysis by the following equation: $f_0 - \mathbf{Cx} = b$, which corresponds to the initiation of the relay switch from $-c$ to $+c$ (subject to the time derivative of the error signal being positive). Consider the solution for the constant control $u = \pm 1$:

$$\mathbf{x}(t) = e^{\mathbf{A}(t-\tau)}\mathbf{x}(\tau) \pm \mathbf{A}^{-1}\left(e^{\mathbf{A}(t-\tau)} - \mathbf{I}\right)\mathbf{B}, \quad t > \tau$$

and also an expression for the state vector at time $t = \tau$: $\mathbf{x}(\tau) = e^{\mathbf{A}\tau}\mathbf{x}(0) - \mathbf{A}^{-1}(e^{\mathbf{A}\tau} - \mathbf{I})\mathbf{B}$. A fixed point of the Poincaré return map for asymmetric periodic motion with positive pulse length θ_1 and negative pulse length θ_2, and the unity amplitude (the LPRS does not depend on the control amplitude, and it can be considered $c = 1$ when deriving the LPRS formula) is determined as follows:

$$\eta_p = e^{\mathbf{A}\theta_1}\rho_p + \mathbf{A}^{-1}\left(e^{\mathbf{A}\theta_1} - \mathbf{I}\right)\mathbf{B} \tag{4.15}$$

* The actual existence of a periodic motion depends on a number of other factors too, including orbital stability of the obtained periodic solution and initial conditions.

$$\rho_p = e^{\mathbf{A}\theta_2}\eta_p - \mathbf{A}^{-1}\left(e^{\mathbf{A}\theta_2} - \mathbf{I}\right)\mathbf{B} \tag{4.16}$$

where

$\rho_p = \mathbf{x}(\tau) = \mathbf{x}(T + \tau)$

$\eta_p = \mathbf{x}(\theta_1 + \tau)$

T is the period of the oscillations

We suppose that θ_1 and θ_2 are known. Then (4.15) and (4.16) can be solved for ρ_p and η_p as follows. Substitute (4.16) for ρ_p in (4.15):

$$\eta_p = e^{\mathbf{A}\theta_1}\left[e^{\mathbf{A}\theta_2}\eta_p - \mathbf{A}^{-1}\left(e^{\mathbf{A}\theta_2} - \mathbf{I}\right)\mathbf{B}\right] + \mathbf{A}^{-1}\left(e^{\mathbf{A}\theta_1} - \mathbf{I}\right)\mathbf{B}$$

which leads to $\eta_p = e^{\mathbf{A}(\theta_1+\theta_2)}\eta_p - e^{\mathbf{A}\theta_1}\mathbf{A}^{-1}\left(e^{\mathbf{A}\theta_2} - \mathbf{I}\right)\mathbf{B} + \mathbf{A}^{-1}\left(e^{\mathbf{A}\theta_1} - \mathbf{I}\right)\mathbf{B}$. Regroup the above equation as follows:

$$\left(\mathbf{I} - e^{\mathbf{A}(\theta_1+\theta_2)}\right)\eta_p = \left[-e^{\mathbf{A}\theta_1}\mathbf{A}^{-1}\left(e^{\mathbf{A}\theta_2} - \mathbf{I}\right) + \mathbf{A}^{-1}\left(e^{\mathbf{A}\theta_1} - \mathbf{I}\right)\right]\mathbf{B}$$

Express η_p from the above formula:

$$\eta_p = \left(\mathbf{I} - e^{\mathbf{A}(\theta_1+\theta_2)}\right)^{-1}\left[-e^{\mathbf{A}\theta_1}\mathbf{A}^{-1}\left(e^{\mathbf{A}\theta_2} - \mathbf{I}\right) + \mathbf{A}^{-1}\left(e^{\mathbf{A}\theta_1} - \mathbf{I}\right)\right]\mathbf{B}$$

Considering that $\mathbf{A}e^{\mathbf{A}t}\mathbf{A}^{-1} = e^{\mathbf{A}t}$, we obtain the following formula:

$$\eta_p = \left(\mathbf{I} - e^{\mathbf{A}(\theta_1+\theta_2)}\right)^{-1}\mathbf{A}^{-1}\left[2e^{\mathbf{A}\theta_1} - e^{\mathbf{A}(\theta_1+\theta_2)} - \mathbf{I}\right]\mathbf{B}$$

Similarly, substitute (4.15) into (4.16):

$$\rho_p = e^{\mathbf{A}(\theta_1+\theta_2)}\rho_p + e^{\mathbf{A}\theta_2}\mathbf{A}^{-1}\left(e^{\mathbf{A}\theta_1} - \mathbf{I}\right)\mathbf{B} - \mathbf{A}^{-1}\left(e^{\mathbf{A}\theta_2} - \mathbf{I}\right)\mathbf{B}$$

or $(\mathbf{I} - e^{\mathbf{A}(\theta_1+\theta_2)})\rho_p = e^{\mathbf{A}\theta_2}\mathbf{A}^{-1}(e^{\mathbf{A}\theta_1} - \mathbf{I})\mathbf{B} - \mathbf{A}^{-1}(e^{\mathbf{A}\theta_2} - \mathbf{I})\mathbf{B}$, which gives

$$\rho_p = \left(\mathbf{I} - e^{\mathbf{A}(\theta_1+\theta_2)}\right)^{-1}\mathbf{A}^{-1}\left[e^{\mathbf{A}(\theta_1+\theta_2)} - 2e^{\mathbf{A}\theta_2} + \mathbf{I}\right]\mathbf{B}$$

Considering $\theta_1 + \theta_2 = T$, solution of (4.5), (4.6) results in

$$\eta_p = \left(\mathbf{I} - e^{\mathbf{A}T}\right)^{-1}\mathbf{A}^{-1}\left[2e^{\mathbf{A}\theta_1} - e^{\mathbf{A}T} - \mathbf{I}\right]\mathbf{B} \tag{4.17}$$

$$\rho_p = \left(\mathbf{I} - e^{\mathbf{A}T}\right)^{-1}\mathbf{A}^{-1}\left[e^{\mathbf{A}T} - 2e^{\mathbf{A}\theta_2} + \mathbf{I}\right]\mathbf{B} \tag{4.18}$$

Now, considering that

$$\rho_p = \mathbf{x}(\tau) = e^{\mathbf{A}\tau}\mathbf{x}(0) - \mathbf{A}^{-1}(e^{\mathbf{A}\tau} - \mathbf{I})\mathbf{B}$$

and

$$\eta_p = \mathbf{x}(\theta_1 + \tau) = e^{\mathbf{A}\tau}\mathbf{x}(\theta_1) + \mathbf{A}^{-1}(e^{\mathbf{A}\tau} - \mathbf{I})\mathbf{B}$$

we find $\mathbf{x}(0)$ as follows:

$$\mathbf{x}(0) = e^{-\mathbf{A}\tau}\left[\left(\mathbf{I} - e^{\mathbf{A}T}\right)^{-1}\mathbf{A}^{-1}\left[e^{\mathbf{A}T} - 2e^{\mathbf{A}\theta_2} + \mathbf{I}\right]\mathbf{B} + \mathbf{A}^{-1}\left(e^{\mathbf{A}\tau} - \mathbf{I}\right)\mathbf{B}\right] \tag{4.19}$$

Reasoning along similar lines, find the formula for $\mathbf{x}(\theta_1)$.

$$\mathbf{x}(\theta_1) = e^{-\mathbf{A}\tau}\left[\left(\mathbf{I} - e^{\mathbf{A}T}\right)^{-1}\mathbf{A}^{-1}\left[2e^{\mathbf{A}\theta_1} - e^{\mathbf{A}T} - \mathbf{I}\right]\mathbf{B} - \mathbf{A}^{-1}\left(e^{\mathbf{A}\tau} - \mathbf{I}\right)\mathbf{B}\right] \tag{4.20}$$

Consider now the symmetric motion as a limit of (4.19) at $\theta_1; \theta_2 \to \theta = T/2$:

$$\lim_{\theta_1;\theta_2 \to \theta = T/2} \mathbf{x}(0) = \left(\mathbf{I} + e^{\mathbf{A}T/2}\right)^{-1}\mathbf{A}^{-1}\left[\mathbf{I} + e^{\mathbf{A}T/2} - 2e^{\mathbf{A}(T/2-\tau)}\right]\mathbf{B} \tag{4.21}$$

The imaginary part of the LPRS can be obtained from (4.21) in accordance with its definition as follows:

$$\begin{aligned}\operatorname{Im} J(\omega) &= \frac{\pi}{4}\mathbf{C}\lim_{\theta_1;\theta_2 \to \theta = T/2} \mathbf{x}(0) \\ &= \frac{\pi}{4}\mathbf{C}\left(\mathbf{I} + e^{\mathbf{A}\pi/\omega}\right)^{-1}\left(\mathbf{I} + e^{\mathbf{A}\pi/\omega} - 2e^{\mathbf{A}(\pi/\omega - \tau)}\right)\mathbf{A}^{-1}\mathbf{B}\end{aligned} \tag{4.22}$$

For deriving the expression of the real part of the LPRS, consider the periodic solution (4.17) and (4.18) as a result of the feedback action:

$$\begin{cases} f_0 - y(0) = b \\ f_0 - y(\theta_1) = -b \end{cases} \tag{4.23}$$

Having solved the set of equations (4.23) for f_0, we can obtain $f_0 = (y(0) + y(\theta_1))/2$. Hence, the constant term of the error signal $\sigma(t)$ is $\sigma_0 = f_0 - y_0 = ((y(0) + y(\theta_1))/2) - y_0$. The real part of the LPRS definition formula can be transformed into

$$\operatorname{Re} J(\omega) = -\frac{1}{2}\lim_{\gamma \to \frac{1}{2}} \frac{0.5\left[y(0) + y(\theta_1)\right] - y_0}{u_0} \tag{4.24}$$

where $\gamma = \theta_1/(\theta_1 + \theta_2) = \theta_1/T$. Then $\theta_1 = \gamma T$, $\theta_2 = (1 - \gamma)T$, $u_0 = 2\gamma - 1$, and (4.24) can be rewritten as

$$\operatorname{Re} J(\omega) = -\frac{1}{2}\lim_{\gamma \to \frac{1}{2}} \frac{0.5\mathbf{C}\left[\mathbf{x}(0) + \mathbf{x}(\theta_1)\right] - y_0}{2\gamma - 1} \tag{4.25}$$

where $\mathbf{x}(0)$ and $\mathbf{x}(\theta_1)$ are given by (4.19) and (4.20), respectively. Now, at first, considering the limit

$$\lim_{\gamma\to\frac{1}{2}}\frac{e^{\mathbf{A}\gamma T}-e^{-\mathbf{A}\gamma T}e^{\mathbf{A}T}}{2\gamma-1}=\mathbf{A}Te^{\mathbf{A}T/2} \tag{4.26}$$

find the following limit:

$$\begin{aligned}\lim_{u_0\to0(\theta_1+\theta_2=T=const)}\frac{\mathbf{x}(0)+\mathbf{x}(\theta_1)}{u_0}&=\lim_{\gamma\to1/2}\frac{\mathbf{x}(0)+\mathbf{x}(\theta_1)}{2\gamma-1}\\&=2e^{-\mathbf{A}\tau}T\left(\mathbf{I}-e^{\mathbf{A}T}\right)^{-1}e^{\mathbf{A}T/2}\mathbf{B}\end{aligned} \tag{4.27}$$

To find the limit $\lim_{u_0\to0}(y_0/u_0)$, consider the equations for the constant terms of the variables (averaged variables), which are obtained from the original equations of the plant via equating the derivatives to zero:

$$\begin{cases}\mathbf{0}=\mathbf{A}\mathbf{x}_0+\mathbf{B}u_0\\y_0=\mathbf{C}\mathbf{x}_0\end{cases}$$

From those equations, obtain $\mathbf{x}_0=-\mathbf{A}^{-1}\mathbf{B}u_0$ and $y_0=-\mathbf{C}\mathbf{A}^{-1}\mathbf{B}u_0$. Therefore,

$$\lim_{u_0\to0}\frac{y_0}{u_0}=-\mathbf{C}\mathbf{A}^{-1}\mathbf{B}, \tag{4.28}$$

which is the gain of the plant. The real part of the LPRS is obtained by substituting (4.27) and (4.28) for respective limits in (4.25):

$$\begin{aligned}\operatorname{Re}J(\omega)&=-\frac{1}{2}\lim_{u_0\to0}\frac{\frac{1}{2}\mathbf{C}\left[\mathbf{x}(0)+\mathbf{x}(\theta_1)\right]-y_0}{u_0}\\&=-\frac{1}{2}T\mathbf{C}\left(\mathbf{I}-e^{\mathbf{A}T}\right)^{-1}e^{\mathbf{A}(T/2-\tau)}\mathbf{B}-\frac{1}{2}\mathbf{C}\mathbf{A}^{-1}\mathbf{B}\end{aligned} \tag{4.29}$$

Finally, the state-space description based formula of the LPRS can be obtained by combining formulas (4.21) and (2.29) as follows:

$$\begin{aligned}J(\omega)=&-0.5\mathbf{C}\left[\mathbf{A}^{-1}+\frac{2\pi}{\omega}\left(\mathbf{I}-e^{\frac{2\pi}{\omega}\mathbf{A}}\right)^{-1}e^{\left(\frac{\pi}{\omega}-\tau\right)\mathbf{A}}\right]\mathbf{B}\\&+j\frac{\pi}{4}\mathbf{C}\left(\mathbf{I}+e^{\frac{\pi}{\omega}\mathbf{A}}\right)^{-1}\left(\mathbf{I}+e^{\frac{\pi}{\omega}\mathbf{A}}-2e^{\left(\frac{\pi}{\omega}-\tau\right)\mathbf{A}}\right)\mathbf{A}^{-1}\mathbf{B}\end{aligned} \tag{4.30}$$

The real part of the LPRS was derived under the assumption of the limits at $u_0\to0$ and at $\gamma\to1/2$ being equal. This can only be true if the derivative dT/du_0 in the point $u_0=0$ is zero, which follows from the symmetry of the function $T=T(u_0)$. A rigorous proof of this is given in Ref. [2].

4.2.2.2 Orbital Asymptotic Stability

The stability of periodic orbits (limit cycles) is usually referred to as the orbital stability. The notion of orbital stability is different from the notion of the stability of an equilibrium point, as for an orbitally stable motion, the difference between the perturbed and unperturbed motions does not necessarily vanish. What is important is that, if orbitally stable, the motion in the perturbed system converges to the orbit of the unperturbed system. In relay feedback systems, analysis of orbital stability can be reduced to the analysis of certain equivalent discrete-time system with time instants corresponding to the switches of the relay, which can be obtained from the original system by considering the Poincaré map of the motion having an initial perturbation. If we assume that the initial state is $\mathbf{x}(0) = \rho + \delta\rho$, where $\delta\rho$ is the initial perturbation, and find the mapping $\delta\rho \rightarrow \delta\eta$, we will be able to make a conclusion about the orbital stability of the system from consideration of the Jacobian matrix of this mapping. The local orbital stability criterion can be formulated as the following statement.

Local orbital stability. The solution of system (4.1), (4.2) provided by condition (4.13) is locally orbitally asymptotically stable if all eigenvalues of matrix

$$\Phi_0 = \left[\mathbf{I} - \frac{\mathbf{v}\left(\tau + \frac{T}{2} -\right)\mathbf{C}}{\mathbf{C}\mathbf{v}\left(\tau + \frac{T}{2} -\right)} \right] e^{\mathbf{A}\frac{T}{2}} \tag{4.31}$$

where the velocity vector is given by

$$\mathbf{v}\left(\tau + \frac{T}{2} -\right) = \dot{\mathbf{x}}\left(\tau + \frac{T}{2} -\right) = 2\left(\mathbf{I} + e^{\mathbf{A}T/2}\right)^{-1} e^{\mathbf{A}(T/2-\tau)}\mathbf{B}$$

have magnitudes less than 1.

Proof. Consider the process with time $t \in (-\infty; \infty)$ and assume that by the time $t = 0$, where time $t = 0$ is the switch initiation time, a periodic motion has been established, with the fixed point of the Poincaré map given by formulas (4.17) and (4.18). Also, assume that there are no switches of the relay within the interval $(0;\tau)$. At the switch time $t = \tau$, the state vector receives a perturbation (deviation from the value in a periodic motion) $\mathbf{x}(\tau) = \rho = \rho_p + \delta\rho$. Find the mapping of this perturbation into the perturbation of the state vector at the time of the next switch and analyze if the initial perturbation vanishes. For the time interval $t \in [\tau; t^*]$, where t^* is the time of the switch of the relay from "+1" to "−1," the state vector (while the control is $u = 1$) is given as follows:

$$\begin{aligned} \mathbf{x}(t) &= e^{\mathbf{A}(t-\tau)}(\rho_p + \delta\rho) + \mathbf{A}^{-1}(e^{\mathbf{A}(t-\tau)} - \mathbf{I})\mathbf{B} \\ &= e^{\mathbf{A}(t-\tau)}\rho_p + \mathbf{A}^{-1}(e^{\mathbf{A}(t-\tau)} - \mathbf{I})\mathbf{B} + e^{\mathbf{A}(t-\tau)}\delta\rho \end{aligned} \tag{4.32}$$

where the first two addends give the unperturbed motion, and the third addend gives the motion due to the initial perturbation $\delta\rho$. Let us denote

$$\mathbf{x}(t^*) = \eta = \eta_p + \delta\eta \tag{4.33}$$

It should be noted that η_p is determined not at time $t = t^*$ but at time $t = \tau + \theta_1$. The main task of our analysis is to find the mapping $\delta\rho \rightarrow \delta\eta$. Therefore, it follows from (4.33) that

$$\delta\eta = \mathbf{x}(t^*) - \eta_p \tag{4.34}$$

Considering that all perturbations are small and times t^* and $\tau + \theta_1$ are close, evaluate $\mathbf{x}(t^*)$ via linear approximation of $\mathbf{x}(t)$ in the point $t = \tau + \theta_1$:

$$\mathbf{x}(t^*) - \mathbf{x}(\tau+\theta_1) = \dot{\mathbf{x}}(\tau+\theta_1-)(t^* - \tau - \theta_1)$$

where $\dot{\mathbf{x}}(\tau+\theta_1-)$ is the value of the derivative at time immediately preceding time $t = \tau + \theta_1$. Express $\mathbf{x}(t^*)$ from the last equation as follows:

$$\mathbf{x}(t^*) = \mathbf{x}(\tau+\theta_1) + \dot{\mathbf{x}}(\tau+\theta_1-)(t^* - \tau - \theta_1) \tag{4.35}$$

Now, evaluate $\mathbf{x}(\tau + \theta_1)$ using (4.32):

$$\mathbf{x}(\tau+\theta_1) = \eta_p + e^{\mathbf{A}\theta_1}\delta\rho \tag{4.36}$$

Substitute (4.35) and (4.36) into (4.34):

$$\begin{aligned}\delta\eta &= \mathbf{x}(\tau+\theta_1) + \dot{\mathbf{x}}(\tau+\theta_1-)(t^* - \tau - \theta_1) - \eta_p \\ &= \mathbf{v}(\tau+\theta_1-)(t^* - \tau - \theta_1) + e^{\mathbf{A}\theta_1}\delta\rho \end{aligned} \tag{4.37}$$

where $\mathbf{v}(t) = \dot{\mathbf{x}}(t)$. Now, find $(t^* - \tau - \theta_1)$ from the switch condition

$$\begin{aligned}(t^* - \tau - \theta_1)\dot{y}(\tau+\theta_1-) &= -\delta y(\tau+\theta_1) \\ &= -\mathbf{C}\delta\mathbf{x}(\tau+\theta_1-) = -\mathbf{C}e^{\mathbf{A}\theta_1}\delta\rho \end{aligned} \tag{4.38}$$

From (4.38), we obtain

$$t^* - \tau - \theta_1 = -\frac{\delta y(\tau+\theta_1)}{\dot{y}(\tau+\theta_1-)} = -\frac{\mathbf{C}e^{\mathbf{A}\theta_1}}{\dot{y}(\tau+\theta_1-)}\delta\rho$$

Now, substitute the expression for $t^* - \tau - \theta_1$ into (4.37):

$$\begin{aligned}\delta\eta &= -\mathbf{v}(\tau+\theta_1-)\frac{\mathbf{C}e^{\mathbf{A}\theta_1}}{\dot{y}(\tau+\theta_1-)}\delta\rho + e^{\mathbf{A}\theta_1}\delta\rho \\ &= \left[\mathbf{I} - \frac{\mathbf{v}(\tau+\theta_1-)\mathbf{C}}{\mathbf{C}\mathbf{v}(\tau+\theta_1-)}\right]e^{\mathbf{A}\theta_1}\delta\rho \end{aligned} \tag{4.39}$$

Denote the Jacobian matrix of the mapping $\delta\rho \to \delta\eta$ as follows:

$$\Phi_1 = \left[\mathbf{I} - \frac{\mathbf{v}(\tau+\theta_1-)\mathbf{C}}{\mathbf{C}\mathbf{v}(\tau+\theta_1-)}\right]e^{\mathbf{A}\theta_1} \tag{4.40}$$

To be able to assess local orbital stability of the periodic solution, we need to find the Jacobian matrix of the Poincaré return map, which will be the chain-rule application of two mappings: $\delta\rho \to \delta\eta$ and

$\delta\rho \to \delta\eta$. However, for the symmetric motion, checking only half a period of the motion is sufficient, which gives formula (4.31).

In addition to the stability analysis, it can be recommended that the direction of the relay switch should be verified too. This condition is formulated as the satisfaction of the following inequality:

$$\dot{y}\left(\frac{T}{2}-\right)=\mathbf{C}\mathbf{v}\left(\frac{T}{2}-\right)>0$$

where $\mathbf{v}\left(\frac{T}{2}-\right)$ is given above.

4.2.3 Computation of the LPRS from Plant Transfer Function

4.2.3.1 Infinite Series Approach

Another formula of $J(\omega)$ can now be derived for the case of the linear part given by a transfer function. Suppose the linear part does not have integrators. We write the Fourier series expansion of the signal $u(t)$ (Figure 4.2):

$$u(t)=u_0+\frac{4c}{\pi}\sum_{k=1}^{\infty}\sin\frac{(\pi k\theta_1/(\theta_1+\theta_2))}{k}\times\left\{\cos\left(\frac{k\omega\theta_1}{2}\right)\cos(k\omega t)+\sin\left(\frac{k\omega\theta_1}{2}\right)\sin(k\omega t)\right\}$$

where

$u_0 = c(\theta_1-\theta_2)/(\theta_1+\theta_2)$

$\omega = 2\pi/(\theta_1+\theta_2)$

Therefore, $y(t)$ as a response of the linear part with the transfer function $W_l(s)$ can be written as

$$y(t)=y_0+\frac{4c}{\pi}\sum_{k=1}^{\infty}\frac{\sin(\pi k\theta_1/(\theta_1+\theta_2))}{k}\times\left\{\cos\left(\frac{k\omega\theta_1}{2}\right)\cos\left[k\omega t+\varphi_l(k\omega)\right]+\sin\left(\frac{k\omega\theta_1}{2}\right)\sin\left[k\omega t+\varphi_l(k\omega)\right]\right\}A_l(k\omega) \tag{4.41}$$

where

$\varphi_l(k\omega) = \arg W_l(jk\omega)$

$A_l(k\omega) = |W_l(jk\omega)|$

$y_0 = u_0|W_l(j0)|$

The conditions of the switches of the relay have the form of Equation (4.23) where $y(0)$ and $y(\theta_1)$ can be obtained from (4.41) if we set $t = 0$ and $t = \theta_1$, respectively:

$$y(0)=y_0+\frac{4c}{\pi}\sum_{k=1}^{\infty}\left[\frac{0.5\sin\left(2\pi k\theta_1/(\theta_1+\theta_2)\right)\operatorname{Re}W_l(jk\omega)+\sin^2\left(\pi k\theta_1/(\theta_1+\theta_2)\right)\operatorname{Im}W_l(jk\omega)}{k}\right] \tag{4.42}$$

$$y(\theta_1) = y_0 + \frac{4c}{\pi}\sum_{k=1}^{\infty}\frac{\left[0.5\sin(2\pi k\theta_1/(\theta_1+\theta_2))\mathrm{Re}\,W_l(jk\omega) - \sin^2(\pi k\theta_1/(\theta_1+\theta_2))\mathrm{Im}\,W_l(jk\omega)\right]}{k} \tag{4.43}$$

Differentiating (4.23) with respect to f_0 (and taking into account (4.42) and (4.43)), we obtain the formulas containing the derivatives in the point $\theta_1 = \theta_2 = \theta = \pi/\omega$. Having solved those equations for $d(\theta_1 - \theta_2)/df_0$ and $d(\theta_1 + \theta_2)/df_0$, we shall obtain $d(\theta_1 + \theta_2)/df_0\,|_{f_0=0} = 0$, which corresponds to the derivative of the frequency of the oscillations, and

$$\left.\frac{d(\theta_1 - \theta_2)}{df_0}\right|_{f_0=0} = \frac{2\theta}{c\left(|W_l(0)| + 2\sum_{k=1}^{\infty}\cos(\pi k)\mathrm{Re}\,W_l(\omega k)\right)} \tag{4.44}$$

Considering the formula of the closed-loop system transfer function we can write

$$\left.\frac{d(\theta_1 - \theta_2)}{df_0}\right|_{f_0=0} = \frac{2\theta k_n}{c\left(1 + k_n\,|\,A_l(0)\,|\right)} \tag{4.45}$$

Having solved together Equations 4.44 and 4.45 for k_n, we obtain the following expression:

$$k_n = \frac{0.5}{\sum_{k=1}^{\infty}(-1)^k\mathrm{Re}\,W_l(k\pi/\theta)} \tag{4.46}$$

Taking into account formula (4.46), the identity $\omega = \pi/\theta$, and the definition of the LPRS (4.12), we obtain the final form of expression for Re$J(\omega)$. Similarly, having solved the set of equations (4.23) where $\theta_1 = \theta_2 = \theta$, and $y(0)$ and $y(\theta_1)$ have the form (4.42) and (4.43), respectively, we obtain the final formula of Im$J(\omega)$. Having put the real and the imaginary parts together, we obtain the final formula of the LPRS $J(\omega)$ for relay feedback systems:

$$J(\omega) = \sum_{k=1}^{\infty}(-1)^{k+1}\mathrm{Re}\,W_l(k\omega) + j\sum_{k=1}^{\infty}\frac{1}{2k-1}\mathrm{Im}\,W_l\left[(2k-1)\omega\right] \tag{4.47}$$

4.2.3.2 Partial Fraction Expansion Technique

It follows from (4.47) that LPRS possesses the property of *additivity*, which can be formulated as follows.

Additivity property. If the transfer function $W_l(s)$ of the linear part is a sum of n transfer functions, $W_l(s) = W_1(s) + W_2(s) + \cdots + W_n(s)$, then the LPRS $J(\omega)$ can be calculated as a sum of n LPRS: $J(\omega) = J_1(\omega) + J_2(\omega) + \cdots + J_n(\omega)$, where $J_i(\omega)$ $(i = 1,\ldots,n)$ is the LPRS of the relay system with the transfer function of the linear part being $W_i(s)$.

The considered property offers a technique of LPRS calculation based on the expansion of the plant transfer function into partial fractions. Indeed, if $W_l(s)$ is expanded into the sum of first- and second-order dynamics, LPRS $J(\omega)$ can be calculated via summation of the component LPRS $J_i(\omega)$ corresponding to each of addends in the transfer function expansion, subject to available analytical formulas for LPRS of first- and second-order dynamics. Formulas for $J(\omega)$ of first- and second-order dynamics are presented in Table 4.1. LPRS of some of these dynamics are considered below.

TABLE 4.1 Formulas of the LPRS $J(\omega)$

Transfer Function $W(s)$	LPRS $J(\omega)$
$\frac{K}{s}$	$0 - j\frac{\pi^2 K}{8\omega}$
$\frac{K}{Ts+1}$	$\frac{K}{2}(1-\alpha\,\mathrm{csch}\alpha) - j\frac{\pi K}{4}\tanh(\alpha/2)$, $\alpha = \pi/(T\omega)$
$\frac{Ke^{-\tau s}}{Ts+1}$	$\frac{K}{2}(1-\alpha e^{\gamma}\mathrm{csch}\alpha) + j\frac{\pi K}{4}\left(\frac{2e^{-\alpha}e^{\gamma}}{1+e^{-\alpha}} - 1\right)$, $\alpha = \frac{\pi}{T\omega},\quad \gamma = \frac{\tau}{T}$
$\frac{K}{(T_1 s+1)(T_2 s+1)}$	$\frac{K}{2}[1 - T_1/(T_1 - T_2)\alpha_1\mathrm{csch}\alpha_1 - T_2/(T_2 - T_1)\alpha_2\mathrm{csch}\alpha_2)]$ $-j\frac{\pi K}{4}/(T_1 - T_2)[T_1\tanh(\alpha_1/2) - T_2\tanh(\alpha_2/2)]$, $\alpha_1 = \pi/(T_1\omega),\quad \alpha_2 = \pi/(T_2\omega)$
$\frac{K}{s^2+2\xi s+1}$	$\frac{K}{2}[(1-(B+\gamma C)/(\sin^2\beta + \sinh^2\alpha)]$ $-j\frac{\pi K}{4}(\sinh\alpha - \gamma\sin\beta)/(\cosh\alpha + \cos\beta)$, $\alpha = \pi\xi/\omega,\quad \beta = \pi(1-\xi^2)^{1/2}/\omega,\quad \gamma = \alpha/\beta$ $B = \alpha\cos\beta\sinh\alpha + \beta\sin\beta\cosh\alpha$, $C = \alpha\sin\beta\cosh\alpha - \beta\cos\beta\sinh\alpha$
$\frac{Ks}{s^2+2\xi s+1}$	$\frac{K}{2}[\xi(B+\gamma C) - \pi/\omega\cos\beta\sinh\alpha]/(\sin^2\beta + \sinh^2\alpha)]$ $-j\frac{\pi K}{4}(1-\xi^2)^{-1/2}\sin\beta/(\cosh\alpha + \cos\beta)$, $\alpha = \pi\xi/\omega,\quad \beta = \pi(1-\xi^2)^{1/2}/\omega,\quad \gamma = \alpha/\beta$ $B = \alpha\cos\beta\sinh\alpha + \beta\sin\beta\cosh\alpha$, $C = \alpha\sin\beta\cosh\alpha - \beta\cos\beta\sinh\alpha$
$\frac{Ks}{(s+1)^2}$	$\frac{K}{2}[\alpha(-\sinh\alpha + \alpha\cosh\alpha)/\sinh^2\alpha - j0.25\pi\alpha/(1+\cosh\alpha)]$, $\alpha = \pi/\omega$
$\frac{Ks}{(T_1 s+1)(T_2 s+1)}$	$\frac{K}{2}/(T_2 - T_1)[\alpha_2\mathrm{csch}\,\alpha_2 - \alpha_1\mathrm{csch}\,\alpha_1]$ $-j\frac{\pi K}{4}/(T_2 - T_1)[\tanh(\alpha_1/2) - \tanh(\alpha_2/2)]$, $\alpha_1 = \pi/(T_1\omega),\quad \alpha_2 = \pi/(T_2\omega)$

4.2.4 LPRS of Low-Order Dynamics

4.2.4.1 LPRS of First-Order Dynamics

As it was mentioned above, one of possible techniques of LPRS computing is to represent the transfer function as partial fractions, compute the LPRS of the component transfer functions (partial fractions), and add those partial LPRS together in accordance with the additivity property. To be able to realize this methodology, we have to know the formulas of the LPRS for first- and second-order dynamics. It is of similar meaning and importance as the knowledge of the characteristics of first- and second-order dynamics in linear system analysis. The knowledge of the LPRS of the low-order dynamics is important for other reasons too because some features of the LPRS of low-order dynamics can be extended to higher order systems. Those features are considered in the following section.

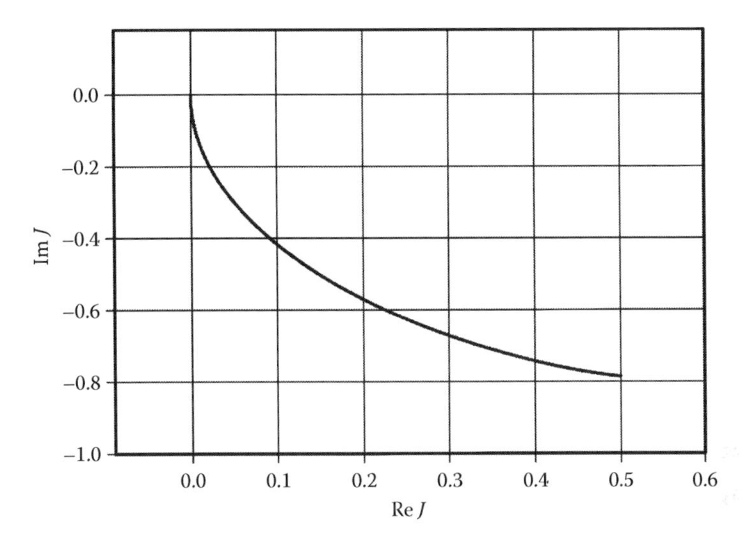

FIGURE 4.5 LPRS of first-order dynamics.

The LPRS formula for the first-order dynamics given by the transfer function $W(s) = K/(Ts + 1)$ is given by the following formula [2]:

$$J(\omega) = \frac{K}{2}\left(1 - \frac{\pi}{T\omega}\operatorname{csch}\frac{\pi}{T\omega}\right) - j\frac{\pi K}{4}\tanh\frac{\pi}{2\omega T} \quad (4.48)$$

where csch(x) and tanh(x) are hyperbolic cosecant and tangent, respectively.

The plot of the LPRS for $K = 1$, $T = 1$ is given in Figure 4.5. The point $(0.5K; -j\frac{\pi}{4}K)$ corresponds to the frequency $\omega = 0$ and the point $(0; j0)$ corresponds to the frequency $\omega = \infty$. The high-frequency segment of the LPRS has an asymptote being the imaginary axis. With the formula for the LPRS available, we can easily find the frequency of periodic motions in the relay feedback system with the linear part being the first-order dynamics. The LPRS is a continuous function of the frequency and for every hysteresis value from the range $b \in [0;cK]$, there exists a periodic solution of the frequency that can be determined from (4.13) and (4.48), which is $\Omega = \pi/2T\tanh^{-1}(b/cK)$. It is easy to show that when the hysteresis value b tends to zero, then the frequency of the periodic solution tends to infinity: $\lim_{b\to 0}\Omega = \infty$, and when the hysteresis value b tends to cK, then the frequency of the periodic solution tends to zero: $\lim_{b\to cK}\Omega = 0$. From (4.48), we can also see that the imaginary part of the LPRS is a monotone function of the frequency. Therefore, the condition of the existence of a finite frequency periodic solution holds for any nonzero hysteresis value from the specified range and the limit for $b \to 0$ exists and corresponds to infinite frequency.

It is easy to show that the oscillations are always orbitally stable. The stability of a periodic solution can be verified via finding eigenvalues of the Jacobian of the corresponding Poincaré map. For the first-order system, the only eigenvalue of this Jacobian will always be zero, as there is only one system variable, which also determines the condition of the switch of the relay.

4.2.4.2 LPRS of Second-Order Dynamics

Now we shall carry out a similar analysis for the second-order dynamics. Let the matrix **A** of (4.1) be $\mathbf{A} = [0\ 1; -a_1\ -a_2]$ and the delay be $\tau = 0$. Here, consider a few cases, all with $a_1 > 0$, $a_2 > 0$.

A. Let $a_2^2 - 4a_1 < 0$. Then plant transfer function can be written as $W(s) = K/(T^2s^2 + 2\xi Ts + 1)$. The LPRS formula can be found, for example, via expanding the above transfer function into partial fractions and applying to it formula (4.48) obtained for the first-order dynamics. However, the coefficients of those partial fractions will be complex numbers and this circumstance must be considered. The formula of the LPRS for the second-order dynamics can be written as follows:

$$J(\omega) = \frac{K}{2}\left(1 - \frac{g + \gamma h}{\sin^2\beta + \sinh^2\alpha}\right) - j\frac{\pi K}{4}\frac{\sinh\alpha - \gamma\sin\beta}{\cosh\alpha + \cos\beta} \tag{4.49}$$

where

$\alpha = \pi\xi/\omega T$

$\beta = \pi\sqrt{1-\xi^2}/\omega T$

$\gamma = \alpha/\beta$

$g = \alpha\cos\beta\sinh\alpha + \beta\sin\beta\cosh\alpha$

$h = \alpha\sin\beta\cosh\alpha + \beta\cos\beta\sinh\alpha$

The plots of the LPRS for $K = 1$, $T = 1$ and different values of damping ξ are given in Figure 4.6: (#1 – $\xi = 1$, #2 – $\xi = 0.85$, #3 – $\xi = 0.7$, #4 – $\xi = 0.55$, #5 – $\xi = 0.4$). The high-frequency segment of the LPRS of the second-order plant approaches the real axis.

It can be easily shown that $\lim_{\omega\to\infty} J(\omega) = (0; j0)$, $\lim_{\omega\to 0} J(\omega) = (0.5K; -j\pi/4K)$. These limits give the two boundary points of the LPRS corresponding to zero frequency and infinite frequency. Analysis of function (4.49) shows that it does not have intersections with the real axis except the origin. Since $J(\omega)$ is a continuous function of the frequency ω (that follows from formula (4.49)), a solution of equation (4.3) exists for any $b \in (0; cK)$. Therefore, a periodic solution of finite frequency exists for the considered second-order system for every value of b within the specified range, and there is a periodic solution of infinite frequency for $b = 0$.

B. Consider the case when $a_2^2 - 4a_1 = 0$. To obtain the LPRS formula, we can use formula (4.49) and find the limit for $\xi \to 1$. The LPRS for this case is given in Figure 4.6 (#1). All subsequent analysis and conclusions are the same as in case A.

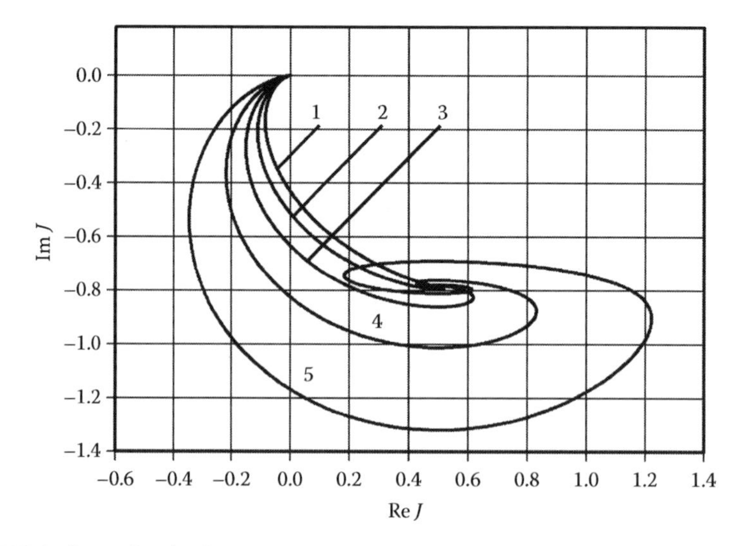

FIGURE 4.6 LPRS of second-order dynamics.

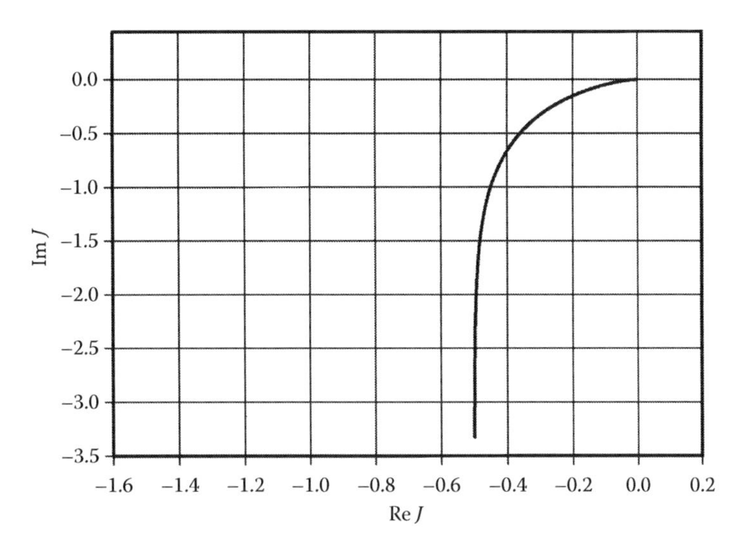

FIGURE 4.7 LPRS of integrating second-order dynamics.

C. Assume that $a_2^2 - 4a_1 > 0$. The transfer function can be expanded into two partial fractions, and then according to the *additivity property*, the LPRS can be computed as a sum of the two components. The subsequent analysis is similar to the previous one.
D. For $a_1 = 0$, the transfer function is $W(s) = K/[s(Ts + 1)]$. For this plant, the LPRS is given by the following formula, which can be obtained via partial fraction expansion technique:

$$J(\omega) = \frac{K}{2}\left(\frac{\pi}{T\omega}\operatorname{csch}\frac{\pi}{T\omega} - 1\right) + j\frac{\pi K}{4}\left(\tanh\frac{\pi}{2\omega T} + \frac{\pi}{2\omega}\right) \tag{4.50}$$

The plot of the LPRS for $K = 1$, $T = 1$ is given in Figure 4.7. The whole plot is totally located in the third quadrant. The point $(0.5K; -j\infty)$ corresponds to the frequency $\omega = 0$ and the point $(0; j0)$ corresponds to the frequency $\omega = \infty$. The high-frequency segment of the LPRS has an asymptote being the real axis.

4.2.4.3 LPRS of First-Order Plus Dead-Time Dynamics

Many industrial processes can be adequately approximated by the first-order plus time-delay transfer function $W(s) = Ke^{-\tau s}/(Ts + 1)$, where K is a gain, T is a time constant, τ is a time delay (dead time). This factor results in the particular importance of analysis of these dynamics. The formula of the LPRS for the first-order plus dead-time transfer function can be derived as follows [2]:

$$J(\omega) = \frac{K}{2}\left(1 - \alpha e^{\gamma}\operatorname{csch}\alpha\right) + j\frac{\pi}{4}K\left(\frac{2e^{-\alpha}e^{\gamma}}{1+e^{-\alpha}} - 1\right) \tag{4.51}$$

where $\gamma = \tau/T$. The plots of the LPRS for $\gamma = 0$ (#1), $\gamma = 0.2$ (#2), $\gamma = 0.5$ (#3), $\gamma = 1.0$ (#4), and $\gamma = 1.5$ (#5) are depicted in Figure 4.8. All the plots begin in the point $(0.5, -j\pi/4)$ that corresponds to the frequency $\omega = 0$. Plot number 1 (that corresponds to zero dead time) comes to the origin that corresponds to infinite frequency. Other plots are defined only for the frequencies that are less than the frequency corresponding to half of the period.

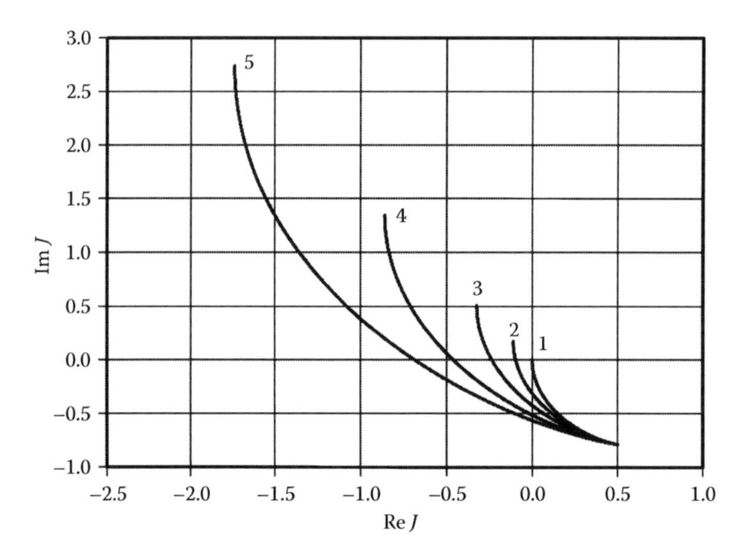

FIGURE 4.8 LPRS of first-order plus dead-time dynamics.

4.2.5 Some Properties of the LPRS

The knowledge of some properties of the LPRS is helpful for its computing, checking the results of computing, and especially for the design of linear compensators with the use of the LPRS method. One of the most important properties is the additivity property. A few other properties that relate to the boundary points corresponding to zero frequency and infinite frequency are considered below. We now find the coordinates of the initial point of the LPRS (which corresponds to zero frequency) considering the LPRS formula (4.30). For that purpose, we find the limit of function $J(\omega)$ for ω tending to zero. At first, evaluate the following two limits: $\lim_{\omega\to 0}\left[\frac{2\pi}{\omega}\left(\mathbf{I}-e^{\frac{2\pi}{\omega}\mathbf{A}}\right)^{-1}e^{\frac{\pi}{\omega}\mathbf{A}}\right]=0$; $\lim_{\omega\to 0}\left[\left(\mathbf{I}+e^{\frac{\pi}{\omega}\mathbf{A}}\right)^{-1}\left(\mathbf{I}-e^{\frac{\pi}{\omega}\mathbf{A}}\right)\right]=\mathbf{I}$. With those two limits available, we can write the limit for the LPRS as follows:

$$\lim_{\omega\to 0} J(\omega)=\left[-0.5+j\frac{\pi}{4}\right]\mathbf{CA}^{-1}\mathbf{B} \tag{4.52}$$

The product of matrices $\mathbf{CA}^{-1}\mathbf{B}$ in (4.52) is the negative value of the gain of the plant transfer function. Therefore, the following statement has been proven. For a nonintegrating linear part of the relay feedback system, the initial point of the corresponding LPRS is $(0.5K; -j\pi K/4)$, where K is the static gain of the linear part. This totally agrees with the above analysis of the LPRS of the first- and second-order dynamics; see, for example, Figures 4.5 and 4.6.

To find the limit of $J(\omega)$ for ω tending to infinity, consider the following two limits of the expansion into power series of the exponential function:

$$\lim_{\omega\to\infty}\exp\left(\tfrac{\pi}{\omega}\mathbf{A}\right)=\lim_{\omega\to\infty}\sum_{n=0}^{\infty}\frac{(\pi/\omega)^n}{n!}\mathbf{A}^n=\mathbf{I}$$

and

$$\lim_{\omega\to\infty}\left\{\tfrac{2\pi}{\omega}\left[\mathbf{I}-\exp\left(\tfrac{2\pi}{\omega}\mathbf{A}\right)\right]^{-1}\right\}=\lim_{\lambda=\frac{2\pi}{\omega}=\to 0}\{\lambda[\mathbf{I}-\exp(\lambda\mathbf{A})]^{-1}\}=-\mathbf{A}^{-1}$$

Finally, taking account of the above two limits, we prove that the end point of the LPRS for the non-integrating linear part without time delay is the origin

$$\lim_{\omega\to\infty,\tau=0} J(\omega) = 0 + j0 \tag{4.53}$$

For time-delay plants, the end point of the LPRS is also the origin, which can be shown using the infinite-series formula (4.47). However, the type of approach of the LPRS to the origin at $\omega \to \infty$ is not asymptotic.

4.3 Design of Compensating Filters in Relay Feedback Systems

4.3.1 Analysis of Performance of Relay Feedback Systems and LPRS Shaping

With the presented methodology of analysis of the effect of the constant input (via the use of the *equivalent gain* and LPRS), we can now consider analysis of the slow signals propagation through a relay feedback system. We assume that signals f_0, σ_0, y_0 that previously were considered constant are now slow-changing signals (slow in comparison with the periodic motions). We shall call them the slow components of the motion. By comparatively slow, we shall understand the signals that can be considered constant over the period of the self-excited oscillation without significant loss of accuracy of the oscillations estimation. Although this is not a rigorous definition, it outlines a framework for the following analysis.

Due to the feedback action, the system always tries to decrease the value of the error signal σ. This is also true with respect to the averaged value (or slow varying component) of the error signal σ_0. As a result, the averaged value of the error signal normally stays within the "linear zone" of the bias function (Figure 4.3). In that case, the equivalent gain value found for the infinitesimally small constant input will be a good estimation of the ratio between the averaged control and the averaged error signal even for large amplitudes of the input. The only difference from the analysis of the response to the constant inputs would be the effect of the dynamics of the linear part, which must be accounted for. The dynamics of the averaged motions can be represented by the block diagram in Figure 4.9. The system in Figure 4.9 is linear, and the dynamics of the averaged motions in the relay feedback system are governed by linear equations—due to the chatter smoothing of the relay nonlinearity.

Relay control involves modes with the frequency of self-excited oscillations (switching frequency) typically much higher than the closed-loop system bandwidth (that corresponds to the frequency range of the external input signals). Therefore, the equivalent gain of the relay in the closed-loop system and input–output properties of the plant are defined by different frequency ranges of the LPRS. The input–output properties of the relay (the equivalent gain) are defined by the shape of the LPRS at the frequencies near the switching frequency while the input–output properties of the plant depend on the characteristics of the plant at the frequencies within the system bandwidth. In order to have a larger equivalent gain, one would want that the point of intersection with the real axis should be closer to the origin. At the same time, the LPRS shape in the frequency range of the system bandwidth should be preserved. Therefore, the compensating filters design can be based on the change of the shape of the LPRS at the frequencies near the frequency of the self-excited oscillations without affecting the LPRS values at the frequencies of the input signals, which would result in the increase of the equivalent gain of the relay and enhancement of the closed-loop performance of the system (Figure 4.10).

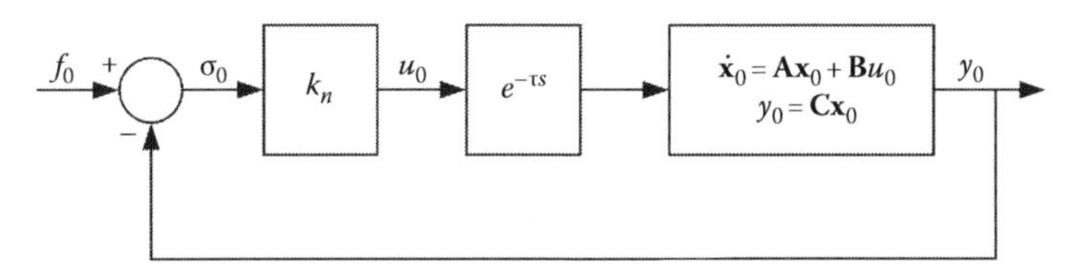

FIGURE 4.9 Dynamics of the averaged motions.

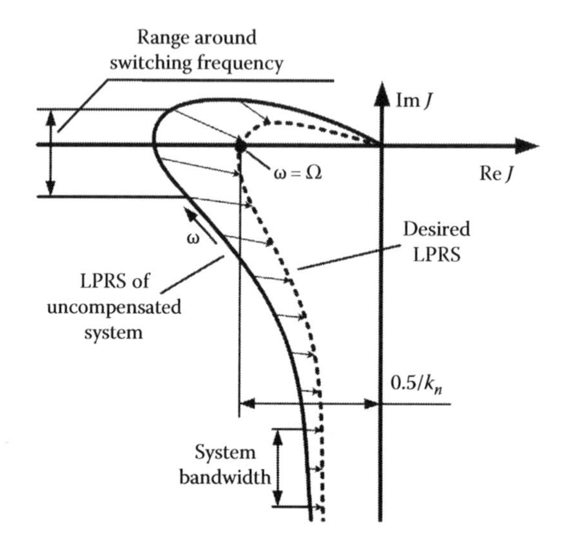

FIGURE 4.10 Desired LPRS shape.

We consider compensation of relay feedback systems assuming that the frequency of the self-excited oscillations (switching frequency) of the uncompensated system meets the specifications and does not need to be significantly modified by the introduction of compensating filters (insignificant changes are still possible), so that the purpose of the compensation is not the design of the switching frequency but the enhancement of the input–output dynamic characteristics of the system.* The compensation objective in terms of LPRS can be defined as selection of such linear filters and their connection within the system, which enforce the LPRS of the system to have a desired location at the frequencies near the switching frequency and do not change the open-loop characteristics of the system at the frequencies within the specified bandwidth.

The desired shape of the LPRS can be evaluated from the specification to the system closed-loop performance in the specified bandwidth (provided that the performance enhancement is going to be achieved only via the increase of the *equivalent gain* k_n, so that the frequency properties of the open-loop system in the specified bandwidth will not change). The methodology of the design of the system can be described as follows. At first, an uncompensated system consisting of the plant and the relay is analyzed. At this step, the frequency characteristics and the LPRS of the plant, the frequency of self-excited oscillations, and the frequency response of the closed-loop system are computed.

At the second step, the requirements to the desired LPRS can be calculated on the basis of the results of the uncompensated system analysis. Namely, a desired location of the point of intersection of the LPRS with the straight line "$-\pi b/(4c)$" and the corresponding frequency of self-excited oscillation can be determined. This can be done with the use of formulas (4.13) and (4.14) and the results of the analysis of the uncompensated system.

Depending on the application of the system, its performance can be specified in a few different ways. Therefore, there can be various ways of the desired LPRS generation. Assume that the compensator is not supposed to change the frequency response of the open-loop system in the bandwidth. Then, the desired LPRS location can be determined by

$$\begin{cases} \mathrm{Re}J_d(\Omega_d) > -0.5/k_{nd} \\ \Omega_1 < \Omega_d < \Omega_2 \\ \mathrm{Im}J_d(\Omega_d) = -\pi b/(4c) \end{cases} \tag{4.54}$$

* Design of the switching frequency can also be done with the use of LPRS; however, it is specific for a particular type of system.

where

d refers to desired

k_{nd} is the desired value of gain k_n of the relay

$[\Omega_1; \Omega_2]$ is the specified range for the switching frequency

b is the specified value of the hysteresis (usually small value)

c is the specified output level of the relay (control)

4.3.2 Compensator Design in the Relay Feedback Systems

Considering the implementation of the idea of changing the LPRS shape (location at higher frequencies), we should assess what kinds of filters and signals are available. We might consider the use of low-pass, high-pass, phase-lead, phase-lag, lead–lag, lag–lead, band-pass, and band-rejecting filters and two possible points of connection: the output signal (error signal), which would provide the cascade connection of the filter (in series with the plant), and the output of the relay, so that the output of the filter is summed with the system input, which would be a parallel connection of the filter (parallel with the plant). Among possible connections and filters, consider two types of compensators performing the described function of the LPRS transformation: (1) the cascade connection with the use of the phase-lag filter and (2) the parallel connection with the use of the band-pass filter.

The Cascade Compensation. The output of the system is always available. Therefore, the cascade compensation can be implemented in most cases. The circuit connected in series with the plant must enforce the LPRS to have a desired location. This can be done with a filter having the transfer function $W(s) = (T_1 s + 1)/(T_2 s + 1)$, where $\omega_{max} < 1/T_2 < 1/T_1 < \Omega$, ω_{max} is the upper boundary of the specified bandwidth. This compensator does not strongly affect the frequency response of the open-loop system at the frequencies within the bandwidth. It changes the shape of the LPRS at higher frequencies only (beginning from a certain frequency, the LPRS of the compensated system can be approximately calculated as the product of the plant LPRS and the coefficient $T_1/T_2 < 1$). Other dynamic filters of higher order with similar amplitude-frequency response can be used for such compensation too. One of the features of such cascade compensation is that this compensation causes the frequency of the self-excited oscillations to *decrease*.

The Parallel Compensation. The output of the relay is always available and can be used for the parallel compensation. The circuit connected in parallel with the plant must enforce the LPRS to have the desired location (Figure 4.10). The transfer function of the open-loop system is calculated as the sum of the transfer functions of the plant and of the compensator. As a result, in the case of the parallel compensation, the LPRS of the linear part is calculated as the sum of the plant LPRS and the compensator LPRS (see the *additivity property*). Yet, addition of the compensator must not change the frequency response of the open-loop system at the frequencies within the bandwidth.

Such properties are typical of the band-pass filters with the bandwidth encompassing the frequency of the self-excited oscillations of the system. Formulas of the LPRS of some band-pass filters are presented in Table 4.1. The LPRS of the second-order band-pass filters are depicted in Figure 4.11. One of the features of the parallel compensation is that it can cause the frequency of the self-excited oscillations to *increase*. Note that Table 4.1 and Figure 4.11 show the normalized LPRS $J_n(\omega)$ (computed for unity gain and time constants). To obtain the LPRS for the transfer function, $W(s) = (KTs)/(T^2 s^2 + 2\xi Ts + 1)$ (including the case of the general form with $\xi \to 1$), one has to recalculate the LPRS according to

$$J(\omega) = K\,J_n(T\omega) \tag{4.55}$$

where $J_n(\omega)$ is the normalized LPRS of the band-pass compensator.

The main idea of the design technique is to match the point corresponding to the maximal amplitude of the LPRS Figure 4.11 with the desired point of intersection of the LPRS and of the line "$-\pi b/(4c)$," and to find the gain of the normalized LPRS that would ensure this transformation. After that, the filter gain

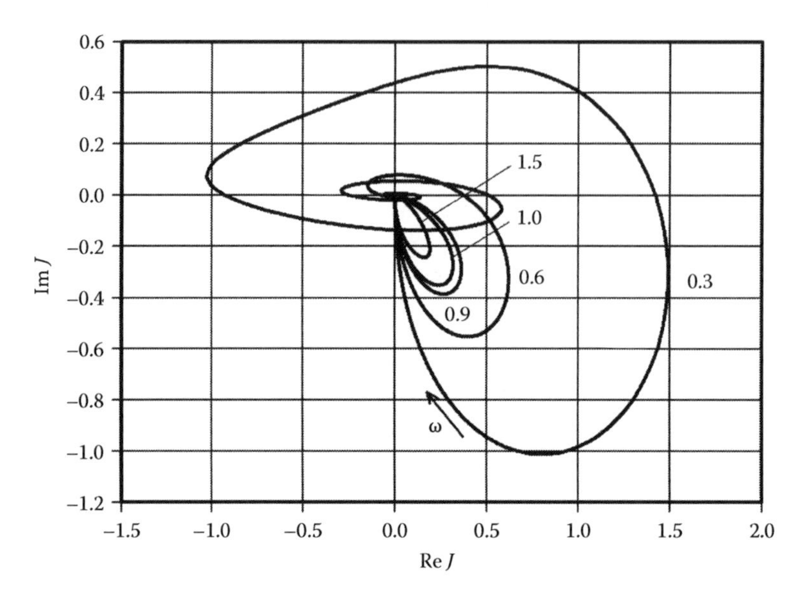

FIGURE 4.11 LPRS $J_n(\omega)$ of band-pass filters $W(s) = s/(s^2 + 2\xi s + 1)$ for $\xi = 0.3$–1.5.

and time constant can be found from (4.55). Design of compensating filters is specific for the particular type of relay feedback system. An example of design is given in Ref. [2].

4.3.3 Conclusions

The LPRS method is presented in the chapter beginning with basic ideas and progressing to computing formulas and algorithms of system analysis and compensator design. It is shown that the use of the LPRS method not only allows for a precise analysis of the complex dynamics of the relay feedback system but also provides a convenient tool for compensator selection and design. Because of such LPRS properties as exactness, simplicity, and convenience of use, the problems of analysis and design of many relay feedback systems can be efficiently solved.

References

1. Atherton D.P. (1975). *Nonlinear Control Engineering—Describing Function Analysis and Design*. Van Nostrand Company Limited, Workingham, Berks, U.K.
2. Boiko I. (2009). *Discontinuous Control Systems: Frequency-Domain Analysis and Design*. Birkhauser, Boston, MA.
3. Flugge-Lotz I. (1953). *Discontinuous Automatic Control*. Princeton University Press, Princeton, NJ.
4. Gelb A. and Vander Velde W.E. (1968). *Multiple-Input Describing Functions and Nonlinear System Design*. McGraw-Hill, New York.
5. Hsu J.C. and Meyer A.U. (1968). *Modern Control Principles and Applications*. McGraw Hill, New York.
6. MacColl L.A. (1945). *Fundamental Theory of Servomechanisms*. D. Van Nostrand Co., New York.
7. Tsypkin Ya.Z. (1984). *Relay Control Systems*. Cambridge University Press, Cambridge, U.K.
8. Utkin V. (1992). *Sliding Modes in Control and Optimization*. Springer-Verlag, Berlin, Germany.
9. Wang Q.-G., Lee T.H., and Lin C. (2003). *Relay Feedback: Analysis, Identification and Control*. Springer, London, U.K.
10. Yu C.-C. (1999) *Automatic Tuning of PID Controllers: Relay Feedback Approach*. Springer-Verlag, New York.

5

Linear Matrix Inequalities in Automatic Control

Miguel Bernal
University of Guadalajara

Thierry Marie Guerra
University of Valenciennes and Hainaut-Cambresis

5.1 What Are LMIs?

5.1.1 Preliminaries

This chapter is a short introduction to a broad and active field, that of linear matrix inequalities (LMIs) in automatic control; therefore, as expected, many subjects are left out or been just outlined. Nevertheless, details are available in [1,2,9,12].

In broad sense, an LMI is a set of mathematical expressions whose variables are linearly related matrices. This ample definition highlights the fact that an LMI can adopt nonobviously linear or matrix expressions. The following formal definition is normally used for an LMI [1]:

$$F(x) = F_0 + \sum_{i=1}^{n} x_i F_i > 0, \tag{5.1}$$

where

$x \in \mathbb{R}^m$ is the *decision variable vector*

$F_i = F_i^T \in \mathbb{R}^{n \times n}$, $i = 0,\ldots,n$ are given constant symmetric matrices

In this context, (>) stands for positive-definiteness (strict LMI); a *non-strict LMI* has (≥) instead of (>) in (5.1). Note also that variables in this definition are not matrices. Should an LMI be expressed with matrices as variables, it can be transformed to (5.1) by decomposing any matrix variable in a base of symmetric matrices.

The solution set of LMI (5.1), denoted by $S = \{x \in \mathbb{R}^m, F(x) > 0\}$, is called its *feasibility set*, which is a convex subset of $\mathbb{R}^m$. More generally, given a set $S = \{x \in \mathbb{R}^m, g(x) \le 0, h(x) = 0\}$ with $g\colon \mathbb{R}^m \to \mathbb{R}^n$ and $h\colon \mathbb{R}^m \to \mathbb{R}^l$, a *convex optimization problem* consists in minimizing a convex function $f(x)$ provided that $g(x)$ is also convex and $h(x)$ is affine. Finding a solution x to (5.1), if any exists, is a convex optimization problem, which implies that it is amenable for computer solution in a polynomial time. Moreover, convexity guarantees no local minima to be found and finite feasibility tests. If no solution exists, the corresponding problem is called *infeasible*.

The following well-known convex or quasi-convex optimization problems are relevant for analysis and synthesis of control systems [1,2]:

1. Finding a solution x to the LMI system (5.1) or determining that there is no solution is called the *feasibility problem* (FP). This problem is equivalent to minimizing the convex function $f: x \to \lambda_{min}(F(x))$ and then deciding whether the solution is positive (strictly feasible solution), zero (feasible solution), or negative (unfeasible case).
2. Minimizing a linear combination of the decision variables $c^T x$ subject to (5.1) is called an *eigenvalue problem* (EVP), also known as an *LMI optimization*. This problem is a generalization of linear programming for the cone of positive semidefinite matrices; thus, it belongs to semidefinite programming.
3. Minimizing eigenvalues of a pair of matrices, which depend affinely on a variable, subject to a set of LMI constraints or determining that the problem is infeasible is called a *generalized eigenvalue problem* (GEVP), i.e.,

$$\text{minimize } \lambda \quad \text{subject to } \lambda B(x) - A(x) > 0, \quad B(x) > 0, \quad C(x) > 0$$

where $A(x)$, $B(x)$, and $C(x)$ are symmetric and affine with respect to x. The problem can also be rewritten as

$$\text{minimize } \lambda_{max}(A(x), B(x)) \quad \text{subject to } B(x) > 0, \quad C(x) > 0$$

where $\lambda_{max}(X, Y)$ denotes the largest generalized eigenvalue of $\lambda Y - X$ with $Y > 0$.

Several methods have been developed for solving the aforementioned problems. Historically, the first attempts to solve LMIs were *analytical* [3,4], followed by some *graphical techniques* [5] and algebraic Ricatti equations [6]. Since the early 1980s, LMIs have been solved *numerically*, first through convex programming [7], then through interior-point algorithms [8]. The interested reader in the latter methods is referred to Sections 5.2.3 and 5.2.4 in [1], where comprehensive description is provided.

A word about the actual LMI solvers proceeds since the algorithms above have been implemented under several toolboxes and software programs. Most of them are MATLAB® related, such as LMI Toolbox (LMILAB) [9], SeDuMi, SDPT3, VSDP, or LMIRank. MAXDET [1] and CSDP do not run under MATLAB related. LMILAB exploits projective methods and linear algebra, while SeDuMi is a semidefinite programming solver.

5.1.2 Some Properties

As stated in the previous section, an LMI can appear in seemingly nonlinear or non-matrix expressions. Moreover, some solutions of nonlinear inequalities can be subsumed as the solution of an LMI problem. To recast matrix inequalities as LMI expressions, some properties and equivalences follow:

Property 1: A set of LMIs $F_1(x) > 0, \ldots, F_k(x) > 0$ is equivalent to the single LMI $F(x) = \text{diag}[F_1(x), \ldots, F_k(x)] > 0$ where $\text{diag}[F_1(x), \ldots, F_k(x)]$ denotes the block-diagonal matrix with $F_1(x), \ldots, F_k(x)$ on its main diagonal.

Property 2: (Schur Complement) Given a matrix $Q(x) \in \mathbb{R}^{n \times m}$, $Q(x) > 0$, a full rank-by-row matrix $S(x) \in \mathbb{R}^{n \times m}$, and a matrix $R(x) \in \mathbb{R}^{n \times n}$, all of them depending affinely on x, the following inequalities are equivalent:

$$\begin{bmatrix} R(x) & S(x) \\ S(x)^T & Q(x) \end{bmatrix} > 0 \tag{5.2}$$

$$R(x) - S(x)Q(x)^{-1}S(x)^T > 0 \tag{5.3}$$

Property 3: (Slack variables I) Given matrices A, G, L, P, and Q with the proper sizes, the following inequalities are equivalent [10,11]:

$$A^T PA - Q < 0, \quad P > 0 \tag{5.4}$$

$$\begin{bmatrix} -Q & A^T P \\ PA & -P \end{bmatrix} < 0 \tag{5.5}$$

$$\exists G \begin{bmatrix} -Q & A^T G \\ G^T A & -G - G^T + P \end{bmatrix} < 0, \quad P > 0 \tag{5.6}$$

$$\exists G,L \begin{bmatrix} -Q + A^T L + LA & -L + A^T G \\ -L^T + G^T A & -G - G^T + P \end{bmatrix} < 0, \quad P > 0 \tag{5.7}$$

Property 4: (Slack variables II) Given matrices A, G, L, $P = P^T$ and Q with the proper sizes, the following inequalities are equivalent [11]:

$$A^T P + PA + Q < 0 \tag{5.8}$$

$$\exists G,L \begin{bmatrix} A^T L + LA + Q & P - L + A^T G \\ P - L^T + G^T A & -G - G^T + P \end{bmatrix} < 0 \tag{5.9}$$

Property 5: (S-procedure) Given matrices $F_i = F_i^T$, $i = 0,\ldots,p$ and quadratic functions with inequalities conditions:

$$x^T F_0 x > 0, \quad \forall x \neq 0 : x^T F_i x \geq 0, \quad \forall i \in \{1,\ldots,p\} \tag{5.10}$$

A sufficient condition for these conditions to hold is:

$$\exists \tau_1,\ldots,\tau_p \geq 0 : F_0 - \sum_{i=1}^{p} \tau_i F_i > 0 \tag{5.11}$$

Property 6: (Finsler's Lemma) Given a vector $x \in \mathbb{R}^n$ and matrices $Q = Q^T \in \mathbb{R}^{n\times n}, R \in \mathbb{R}^{m\times n}$, and $S \in \mathbb{R}^{m\times n}$ such that rank(R) < n, rank(S) < n, the following inequalities are equivalent:

$$x^T Qx < 0, \quad \forall x \neq 0 : Rx = 0, \quad Sx = 0 \tag{5.12}$$

$$R_\perp^T QR_\perp < 0, \quad S_\perp^T QS_\perp < 0 \tag{5.13}$$

$$\exists \sigma \in \mathbb{R} : Q - \sigma R^T R < 0, \quad Q - \sigma S^T S < 0 \tag{5.14}$$

$$\exists X \in \mathbb{R}^{n\times m} : Q + S^T XR + R^T X^T S < 0 \tag{5.15}$$

Getting expression (5.13) from (5.15) is usually referred as the *elimination lemma* for the obvious reason that expression (5.13) is equivalent to (5.15) without X.

Property 7: (Congruence I) Given a matrix $P = P^T$ and a full rank-by-column matrix Q, $P > 0 \Rightarrow QPQ^T > 0$.

Property 8: (Congruence II) Given two square matrices P and full rank Q, $P > 0 \Leftrightarrow QPQ^T > 0$.

Property 9: (Completion of squares) Given two matrices X, Y of proper size, $\forall Q = Q^T > 0$, $XY^T + YX^T \leq XQX^T + YQ^{-1}Y^T$.

5.2 What Are LMIs Good For?

5.2.1 Model Analysis

A number of classical stability issues can be recast as LMI problems and therefore solved, for linear time-invariant (LTI) models as well as for linear parameter-varying (LPV) ones. In the sequel, some of them are briefly presented and exemplified.

LTI model stability: Consider a continuous-time LTI model $\dot{x}(t) = Ax(t)$ and a Lyapunov function candidate $V = x^T(t)Px(t)$, $P > 0$, The time derivative of the Lyapunov function is given by $\dot{V} = \dot{x}^T(t)Px(t) + x^T(t)P\dot{x}(t) = x^T(t)(A^TP + PA)x(t)$, which guarantees the origin to be globally asymptotically stable if $\exists P > 0{:}A^TP + PA < 0$. The latter is indeed an LMI FP with P as a matrix decision variable. Applying Property 1, this problem is summarized as solving the following LMI:

$$\begin{bmatrix} P & 0 \\ 0 & -A^TP - PA \end{bmatrix} > 0 \tag{5.16}$$

Following a similar procedure and using Schur complement (property 2), for a discrete-time LTI model $x(t + 1) = Ax(t)$, the following LMI problem arises:

$$\begin{bmatrix} P & A^TP \\ PA & P \end{bmatrix} > 0 \tag{5.17}$$

LPV model stability: Consider a continuous-time LPV model $\dot{x}(t) = A(\delta)x(t)$ with $A(\delta) \in \Gamma$, Γ, being a closed convex set. Quadratic stability of this model implies robust stability, this is

$$\exists P > 0{:}\ A(\delta)^T P + PA(\delta) < 0, \quad \forall \delta \in \Delta \Rightarrow \text{ LPV model stable } \forall\ A \in \Gamma.$$

To translate the previous statement in LMI tests, two cases should be distinguished according to the type of uncertainties involved: structured and unstructured. LPV models correspond to the former case.

As an example of *structured uncertainties*, consider a *polytopic set* $\Gamma = co[A_1,\ldots,A_r] = \left\{A(\delta) : A(\delta) = \sum_{i=1}^{r} \delta_i A_i,\ \forall \delta \in \Delta\right\}$, provided that $\Delta = \left\{\delta = [\delta_1,\ldots,\delta_r],\ \delta_i \geq 0,\ \sum_{i=1}^{r} \delta_i = 1\right\}$. Quadratic stability of continuous-time LPV model $\dot{x}(t) = A(\delta)x(t)$ with $A(\delta) \in \Gamma = co[A_1,\ldots,A_r]$ is then guaranteed if

$$\exists P > 0 : A_i^T P + PA_i < 0, \quad \forall i \in \{1,\ldots,r\} \tag{5.18}$$

since $\forall A(\delta) \in \Gamma$, $\exists \delta \in \Delta$, $\delta_i \geq 0 : A(\delta) = \sum_{i=1}^{r} \delta_i A_i$ which means that $\delta_i > 0$, $A_i^T P + PA_i < 0 \Rightarrow (\delta_i A_i)^T P + P(\delta_i A_i) < 0$. Taking into account that $P > 0$, it follows that $\sum_{i=1}^{r} (\delta_i A_i)^T P + P \sum_{i=1}^{r} (\delta_i A_i) = A^T(\delta)P + PA(\delta) < 0$.

Similarly, quadratic stability of a discrete-time LPV model $x(t+1) = A(\delta)x(t)$ with $A(\delta) \in \Gamma = co[A_1, \ldots, A_r]$ is guaranteed if

$$\begin{bmatrix} P & A_i^T P \\ PA_i & P \end{bmatrix} > 0, \quad \forall i \in \{1, \ldots, r\} \tag{5.19}$$

To illustrate the results above, consider the following discrete-time LPV model with structured uncertainties:

$$x(t+1) = A(\delta)x(t), \quad A(\delta) \in co\left[A_1, \ldots, A_2\right] \tag{5.20}$$

with $A_1 = \begin{bmatrix} 0.1a - 0.1 & 0 \\ 0.3 & -0.1 - 0.1a \end{bmatrix}$, $A_2 = \begin{bmatrix} 0.1b - 0.2 & 0.5 \\ 0 & -0.1b - 0.2 \end{bmatrix}$.

Stability of the previous model has been investigated for parameters a and b in $-10 \le a \le 10$, $-10 \le b \le 10$. For each pair of integers in this grid, feasibility of LMI conditions in (5.19) has been tested via MATLAB LMI toolbox. The results are shown in Figure 5.1, where a cross mark indicates that the stability problem is feasible for the corresponding parameter pair (a,b), i.e., model (5.20) is guaranteed to be quadratically stable for those parameter pairs.

Uncertain LTI models: A typical example of *unstructured uncertainties* is given by the norm-bounded ones. In this case, $\Gamma = \{A(\Delta) = A + B\Delta C, \Delta \in \Theta\}$ where $\Theta = \left\{\Delta \in \mathbb{R}^{m \times q}, \|\Delta\| = \sqrt{\lambda_{\max}(\Delta^T \Delta)} \le \gamma\right\}$ for the uncertain LTI model

$$\begin{aligned} \dot{x}(t) &= Ax(t) + Bw(t) \\ z(t) &= Cx(t) \end{aligned} \tag{5.21}$$

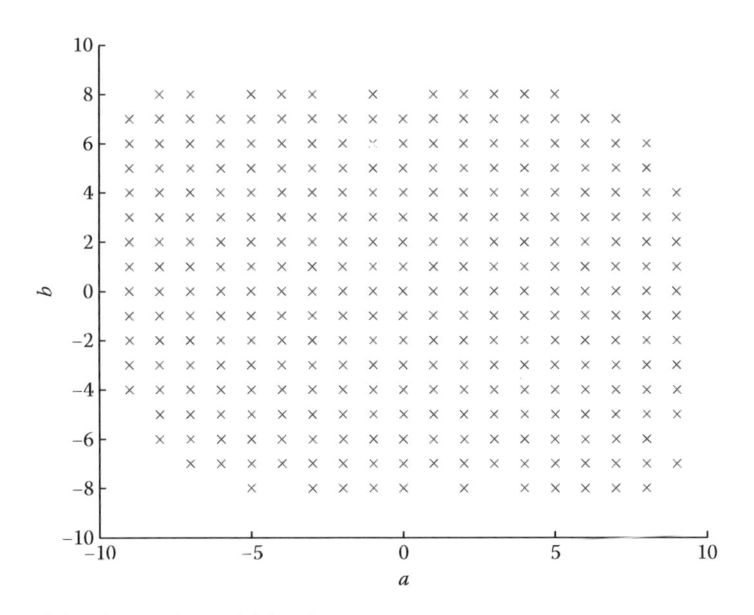

FIGURE 5.1 Feasibility domain for model (5.20).

with $w(t) = \Delta z(t)$. The well-known *bounded real lemma* states that the model (5.21) is quadratically stable if and only if $\exists P > 0$, such that

$$\begin{bmatrix} A^T P + PA + C^T C & PB \\ B^T P & -\gamma^2 I \end{bmatrix} < 0 \tag{5.22}$$

which is obviously an LMI FP with P as a single decision variable. Moreover, since γ^2 can be considered a single decision variable, (5.22) admits an EVP formulation to find the minimum $\gamma > 0$ that renders it feasible.

This LMI arises naturally from two developments: the first one concerning the time derivative of the Lyapunov function candidate $V = x^T(t)Px(t)$,

$$\dot{V}(x) = (Ax + Bw)^T Px + x^T P(Ax + Bw) = x^T(A^T P + PA)x + 2x^T PBw < 0 \tag{5.23}$$

$$\Leftrightarrow \dot{V} = \begin{bmatrix} x \\ w \end{bmatrix}^T \begin{bmatrix} A^T P + PA & PB \\ B^T P & 0 \end{bmatrix} \begin{bmatrix} x \\ w \end{bmatrix} < 0 \tag{5.24}$$

and the second one concerning the fact that $w^T w = z^T \Delta^T \Delta z \le \gamma^2 z^T z$ because $\Delta^T \Delta \le \gamma^2 I$:

$$w^T w - \gamma^2 z^T z \le 0 \quad \Leftrightarrow \quad \begin{bmatrix} x \\ w \end{bmatrix}^T \begin{bmatrix} -C^T C & 0 \\ 0 & \gamma^2 I \end{bmatrix} \begin{bmatrix} x \\ w \end{bmatrix} \le 0 \tag{5.25}$$

Combining both expressions via property 5 (S-procedure), the result in (5.22) arises.

Another example of unstructured uncertainties is given by the *sector-bounded* ones, i.e., $f: \mathbb{R}^q \to : \mathbb{R}^q : f(0) = 0, f \in \{z^T f(z) \ge 0\}$ in a square feedback interconnection as

$$\begin{aligned} \dot{x}(t) &= Ax(t) + Bw(t) \\ z(t) &= Cx(t) + Dw(t) \\ w(t) &= f(z(t)) \end{aligned} \tag{5.26}$$

Following the same outline of the previous results, this problem can also be cast as an LMI problem, leading to the well-known *positive real lemma*, which states that model (5.26) is quadratically stable if

$$\begin{bmatrix} A^T P + PA & PB - C^T \\ B^T P - C & -D^T - D \end{bmatrix} < 0 \tag{5.27}$$

Notice that the Nyquist plot of (5.26) provides a graphical condition for the feasibility of LMI (5.27).

As expected, the previous results can be straightforwardly extended to discrete-time uncertain LTI models. The interested reader is referred to [1] for details.

LMI regions: It is well known that exponential stability of an LTI model can be influenced by placing the model modes in different regions. Given matrices $Q = Q^T$, $R = R^T$, and S, the set of complex numbers

$$R_{LMI} = \left\{ z \in C, \begin{bmatrix} I \\ zI \end{bmatrix}^* \begin{bmatrix} Q & S \\ S^T & R \end{bmatrix} \begin{bmatrix} I \\ zI \end{bmatrix} \right\} \tag{5.28}$$

is called an *LMI region*. Depending on the structure of matrices Q, R, and S, different regions can be defined; for instance, consider the region $\mathbb{R}(z) < -\alpha, \alpha > 0$, corresponding to a damping factor α. In terms of definition (5.28), it can be described as $Q = 2\alpha$, $S = 1$, and $R = 0$, which imply $z \in \mathbb{C}$, $\bar{z} + z < -2\alpha$. Circles, strips, and cones can be likewise described and added to any LTI stability criteria since LMI constraints can be summed up to be simultaneously held.

Moreover, all the eigenvalues of a matrix A lie in the LMI region (5.28) if and only if there exists a matrix $P > 0$ such that the following LMI is feasible:

$$\begin{bmatrix} I \\ A \otimes I \end{bmatrix}^{*} \begin{bmatrix} P \otimes Q & P \otimes S \\ P \otimes S^T & P \otimes R \end{bmatrix} \begin{bmatrix} I \\ A \otimes I \end{bmatrix} < 0 \tag{5.29}$$

H_2- and H_∞-norm upper bounds: Consider a globally asymptotically stable continuous-time LTI model (5.21) with $x(0) = 0$, transfer function $F(s) = C(sI - A)^{-1}B$ and (A,B,C) controllable and observable. In the sequel, the controllability and observability grammians of $F(s)$ will be denoted as P_c and P_o, respectively, which correspond to the unique positive solutions to $AP_c + P_cA^T + BB^T = 0$ and $A^TP_o + P_oA + C^TC = 0$. Then, guaranteeing an H_2-norm upper bound of $F(s)$ turns out to be an LMI problem, since

$$\|F\|_2 \le \gamma \Leftrightarrow \mathrm{Tr}\left(CP_cC^T\right) < \gamma^2 \Leftrightarrow \exists X > 0, \quad AX + XA^T + BB^T < 0, \quad \mathrm{Tr}\left(CXC^T\right) < \gamma^2$$

$$\Leftrightarrow \exists P = X^{-1} > 0, \quad A^TP + PA + PBB^TP < 0, \quad \mathrm{Tr}\left(CP^{-1}C^T\right) < \gamma^2$$

$$\Leftrightarrow \exists P > 0, \quad Z : A^TP + PA + PBB^TP < 0, \quad CP^{-1}C^T < Z, \quad \mathrm{Tr}(Z) \le \gamma^2$$

$$\Leftrightarrow \exists P > 0, \quad Z : \begin{bmatrix} A^TP + PA & PB \\ B^TP & -\gamma I \end{bmatrix} < 0, \quad \begin{bmatrix} P & C^T \\ C & Z \end{bmatrix} > 0, \quad \mathrm{Tr}(Z) < \gamma \tag{5.30}$$

Note that several properties have been employed in the previous development, such as congruence and Schur complement.

Similarly, guaranteeing $\|F(s)\|_\infty < \gamma$ is equivalent to the feasibility of the following LMI problem with P as a decision variable:

$$\begin{bmatrix} A^TP + PA & PB & C^T \\ B^TP & -\gamma I & D^T \\ C & D & -\gamma I \end{bmatrix} < 0 \tag{5.31}$$

Note that since γ is not multiplied by any decision variable, searching its minimum value is also possible via an LMI optimization problem. Moreover, both results above have their equivalent LMI expressions for discrete-time LTI models.

5.2.2 Controller Design

Some controller design issues for LTI models follow. It is important to keep in mind that LMI applications extend well beyond this survey to LPV and quasi-LPV models (Takagi-Sugeno, for instance) [13], both continuous [15] and discrete-time [14], uncertain [16], and delayed [17]. Nonetheless, the elementary results in the sequel as well as the proofs behind them give an outline to the aforementioned extensions.

State-feedback controller design: Consider an open-loop continuous-time LTI model $\dot{x}(t) = Ax(t) + Bu(t)$ with state-feedback control law $u(t) = -Lx(t)$. Quadratic stability of the closed-loop LTI model $x(t) = (A - BL)x(t)$ can be studied by replacing A by $A-BL$ in (5.16), though the result is not an LMI since it asks for $P = P^T > 0$, L such that

$$(A - BL)^T P + P(A - BL) < 0 \tag{5.32}$$

Defining $X = P^{-1} > 0$ and multiplying left and right by it (property 8), expression (5.32) yields the LMI expression

$$XA^T - M^T B^T + AX - BM < 0 \tag{5.33}$$

with $M = LP^{-1} = LX$.

Consider a feedback control law $u = -Lx(t)$ for a polytopic LPV model $\dot{x}(t) = A(\delta)x(t) + B(\delta)u(t)$, $(A,B) \in co[(A_1,B_1),\ldots,(A_r,B_r)]$. Using the result in (5.18) and the aforementioned transformations, it can be found that this model is quadratically stabilizable if $\exists X > 0$ and M such that $XA_i^T + A_iX - B_iM - M^T B_i^T < 0$. As before, the control gain is given by $L = MX^{-1}$.

To exemplify the LMI-based controller design described above as well as the pole-placement properties of LMI regions in (5.28), consider the continuous-time LPV model

$$\dot{x} = A(\delta)x(t) + B(\delta)u(t), \quad A(\delta) \in co\left[A_1, A_2\right], \quad B(\delta) \in co\left[B_1, \ldots, B_2\right] \tag{5.34}$$

with $A_1 = \begin{bmatrix} -1 & 0 \\ 2 & -1 \end{bmatrix}$, $A_2 = \begin{bmatrix} 2 & 1 \\ 0 & -3 \end{bmatrix}$, $B_1 = \begin{bmatrix} -1 \\ -3 \end{bmatrix}$, $B_2 = \begin{bmatrix} -1 \\ -2 \end{bmatrix}$.

LMIs (5.33) allow a feedback controller $u = -Lx$ to be designed if a feasible solution is found. Consider, in addition, that the poles of the nominal models are desired to lie in the convex D-region $D = \{z \in \mathbb{C}: Re(z) < -1, Re(z) < -|Im(z)|, |z| < 10\}$, illustrated in Figure 5.2. This region is characterized by (5.29) with three different sets of matrices Q, S, and R [2]:

$$Q = 2, \quad S = 1, \quad R = 0 \tag{5.35}$$

$$Q = 0_{2\times 2}, \quad S = \begin{bmatrix} \sin\left(\frac{\pi}{4}\right) & \cos\left(\frac{\pi}{4}\right) \\ -\cos\left(\frac{\pi}{4}\right) & \sin\left(\frac{\pi}{4}\right) \end{bmatrix}, \quad R = 0_{2\times 2} \tag{5.36}$$

$$Q = -100, \quad S = 0, \quad R = 1 \tag{5.37}$$

and $A = A_i - B_iL$, $i = 1,2$ since these are the nominal models whose poles are to be placed.

LMIs in (5.33) and (5.29) under definitions (5.35), (5.36), and (5.37) for $i = 1,2$ were found feasible with

$$P = \begin{bmatrix} 0.8502 & 0.2070 \\ 0.2070 & 0.2166 \end{bmatrix}, \quad L = \begin{bmatrix} -3.6303 & -1.1689 \end{bmatrix} \tag{5.38}$$

As shown in Figure 5.2 with cross marks, all the poles of the closed-loop nominal models lie in the desired region D.

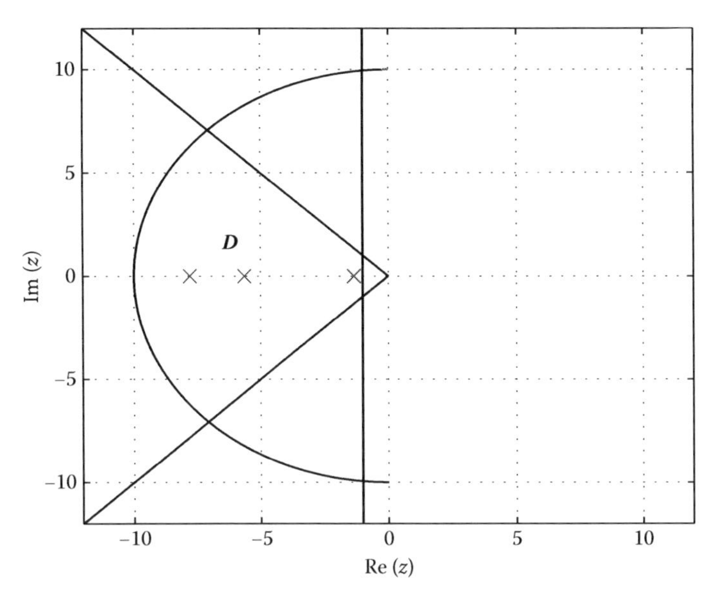

FIGURE 5.2 An LMI region for stabilization of (5.34).

H_2- and H_∞-norm bounded controller design: Consider the following model

$$\dot{x}(t) = Ax(t) + B_w w(t) + B_u u(t)$$
$$z(t) = Cx(t) + D_w w(t) + D_u u(t) \tag{5.39}$$

with $x(t) \in \mathbb{R}^n$ the state-space vector, $u(t) \in \mathbb{R}^m$ the control input vector, $z(t) \in \mathbb{R}^q$ the controlled outputs, and $w(t) \in \mathbb{R}^l$ the generalized disturbances.

Under a state-feedback control law $u(t) = -Lx(t)$, conditions in (5.30) can be used to guarantee the H_2 norm of the closed-loop transfer function $F_{cl}(s)$ of model (5.39) to be bounded by $\gamma > 0$. Nonetheless, additional modifications are needed to render them LMI, as shown in the following development:

$$\exists L \in \mathbb{R}^{m\times n}, \quad \|F_{cl}(s)\|_2^2 \le \gamma$$
$$\Leftrightarrow \exists L \in \mathbb{R}^{m\times n}, \quad X = X^T > 0 : X(A - B_u L)^T + (A - B_u L)X + B_w B_w^T < 0$$
$$\mathrm{Tr}((C - D_u L)X(C - D_u L)^T) < \gamma, \quad D_w = 0 \tag{5.40}$$

Defining $M = LX$ and introducing an auxiliary matrix $W : (C - D_u L)X(C - D_U L)^T < W$, $\mathrm{Tr}(W) < \gamma$, we have that the previous conditions can be rewritten as LMIs on decision variables X, M, and W:

$$X > 0, \quad XA^T + AX - M^T B_u^T - B_u M + B_w B_w^T < 0, \quad \mathrm{Tr}(W) < \gamma$$
$$\begin{bmatrix} W & CX - D_u M \\ XC^T - M^T D_u^T & X \end{bmatrix} > 0 \tag{5.41}$$

Consider now the problem of guaranteeing the H_∞ norm of the closed-loop transfer function $F_{cl}(s)$ of model (5.39) to be bounded by $\gamma > 0$. As with the procedure above for H_2-norm bounded design, the H_∞

stability conditions in (5.31) will be invoked to be adapted for the closed-loop case, substituting any A by $(A - B_uL)$, then applying congruence with diag([X I I]), $X = P^{-1}$, and finally replacing any LX by M to get the following LMI expression:

$$\begin{bmatrix} XA^T - M^T B_u^T + AX - B_u M & B_w & XC^T - M^T D_u^T \\ B_w^T & -\gamma I & D_w^T \\ CX - D_u M & D_w & -\gamma I \end{bmatrix} < 0 \qquad (5.42)$$

A number of results on delayed systems, descriptor forms, observer design, and output feedback can be found in the literature, though they exceed the scope of this introductory chapter. The interested reader is referred to the references below.

References

1. S. Boyd, L.E. Ghaoui, E. Feron, and V. Balakrishnan, *Linear Matrix Inequalities in System and Control Theory* (SIAM: Studies In Applied Mathematics International), Society for Industrial Mathematics, Philadelphia, PA, 1994.
2. C. Scherer and S. Weiland, Linear matrix inequalities in control, Delft University, Delft, the Netherlands, 2005.
3. A.M. Lyapunov, Problème général de la stabilité du mouvement, *Annals of Mathematics Studies*, 17, Princeton University Press, Princeton, NJ, 1947.
4. A.I. Lu're, Some nonlinear problems in the theory of automatic control, H.M. Stationery Off., London, U.K., 1957.
5. V.A. Yakubovich, The method of matrix inequalities in the stability theory of nonlinear control systems I, II, III, *Automation and Remote Control*, 25–26(4); 905–917, 577–592, 753–763, 1967.
6. J.C. Willems, Least squares stationary optimal control and the algebraic Ricatti equation, AC-16(6): 621–634, 1971.
7. N. Karmarkar, A new polynomial-time algorithm for linear programming, *Combinatorica*, 4(4): 373–395, 1984.
8. Yu. Nesterov and A. Nemirovsky, A general approach to polynomial-time algorithms design for convex programming. Technical report, Centr. Econ. and Math. Inst., USSR, Acad. Sci., Moscow, USSSR, 1988.
9. P. Gahinet, A. Nemirovski, A.J. Laub, and M. Chilali, *LMI Control Toolbox*, The Mathworks Inc., Natick, MA, 1995.
10. M.C. Oliveira and R.E. Skelton, Stability tests for constrained linear systems, *Perspectives in Robust Control*, Lecture Notes in Control and Information Sciences 268, Springer, 2001.
11. D. Peaucelle, D. Arzelier, O. Bachelier, and J. Bernussou, A new robust D-stability condition for real convex polytopic uncertainty, *Systems and Control Letters*, 40, 21–30, 2000.
12. D. Henrion, Course in LMI optimization with applications in control (Lecture notes), 2003.
13. K. Tanaka and H.O. Wang, *Fuzzy Control Systems Design and Analysis. A Linear Matrix Inequality Approach*, John Wiley & Sons, New York, 2001.
14. T.M. Guerra, A. Kruszewski, and M. Bernal, Control law proposition for the stabilization of discrete Takagi-Sugeno models, *IEEE Transactions on Fuzzy Systems*, 17(3), 724–731, 2009.
15. M. Johansson, A. Rantzer, and K. Arzen, Piecewise quadratic stability of fuzzy systems, *IEEE Transactions on Fuzzy Systems*, (7):713–722, 1999.
16. S. Xu and T. Chen, Robust H_∞ control for uncertain discrete-time systems with time-varying delays via exponential output feedback controllers, *Systems and Control Letters*, 3–4(51):171–18, 2004.
17. E. Fridman, A. Seuret, and J.P. Richard, Robust sampled-data stabilization of linear systems: An input delay approach, *Automatica*, 40(8):1441–1446, 2004.

6

Motion Control Issues

Roberto Oboe
University of Padova

Makoto Iwasaki
Nagoya Institute of Technology

Toshiyuki Murakami
Keio University

Seta Bogosyan
University of Alaska, Fairbanks

6.1 Introduction

Motion control (MC) is concerned with all the issues arising in the control of the movement of a physical device (Ohnishi et al. 1996). This means that MC studies not only the use of proper devices for sensing and actuation, in addition to suitable control algorithms, but also all possible interactions of the controlled device with the environment. This is the most significant challenge of MC, as the control of the motion rarely occurs in a closed, well-defined environment. Instead, such environments can be populated by humans or unknown objects, remotely located, or characterized by an uncertain or approximate knowledge of the operating conditions.

In each of the possible environments, rules and specifications are rather different. Traditionally, MC has been dealing with the control of the movement in macro-world scenarios, where the performances are usually stated in terms of speed and accuracy of the movement. To achieve the desired level of performance, powerful tools have been developed through MC research, to be blended into the design of control algorithms for different applications, such as point-to-point motion, servo-positioning, control of moving flexible structures, etc. Those tools can only be partially used when the MC problem scales down to micro and nano level, where the laws governing the motion and the interactions (e.g., tribological issues) are rather different from those used in macroscaled problems, and their knowledge less consolidated when compared with that of the macro-world. In this scenario, robust solutions must be utilized to cope with the high degree of uncertainty in the description of the device to be controlled and its environment.

The above scenario drastically changes, when humans come into play, i.e., when the operators may share the operating space with a moving device (Katsura et al. 2007). In this case, the target of the MC system is to provide a motion that will not be harmful to humans. Additionally, humans and devices may cooperate by touching each other (Katsura et al. 2006), so the motion of the devices should be controlled in order to provide a useful cooperation with the human's task. As an example, haptic devices can exert a force on users and exploit the proprioceptive system of humans to convey information to operators and, in turn, modify their behavior. In all the possible interactions with humans, an effective sensing process must be implemented, using all the possible technologies available.

Finally, the availability of a fast and reliable communication infrastructure now opens devices located in remote and/or harsh environments to new applications of MC, with all the additional concerns arising from variable communication delays and data losses.

Summarizing all the above, the design of an MC system is a multifold task. Devices, control strategies, analysis and control of the interaction with the environment, accuracy improvement, and performance optimization are all concurring to the realization of a modern MC system. For this reason, this chapter discusses the most relevant issues in the different areas of interest for this discipline.

Section 6.2 describes issues relevant to high-accuracy MC systems in detail. Section 6.3 addresses the problem of the MC in human–machine interaction. Finally, Section 6.4 presents the most recent results in the field of remote MCs.

6.2 High-Accuracy Motion Control

Fast-response and high-precision MC, on the levels of force, torque, velocity, and/or position, is one of the indispensable requirements in a wide variety of high-performance mechatronic systems, including micro-/nanoscale motion, such as data-storage devices, machine tools, manufacturing tools for electronics components, and industrial robots, from the standpoints of high productivity, high quality of products, and total cost reduction (Ohishi 2008). In those applications, the required specifications in the motion performance, e.g., response/settling time and trajectory/settling accuracy, should be sufficiently achieved. In addition, robustness against disturbances and/or uncertainties, mechanical vibration suppression, and adaptation capability against variations in mechanisms should be essential properties to be provided in the performance.

As well known, a simple approach to achieve the fast and precise motion performance is to expand the control bandwidth of feedback control system: a higher bandwidth allows the system to be more robust against disturbances and uncertainties. However, mechanical vibration modes, dead-time/delay components, and the presence of nonlinearities generally do not allow widening the control bandwidth without impairing the overall system stability.

One of the most effective and practical approaches to improve the motion performance is the control design in a 2-degrees-of-freedom (2DOF) control framework (Umeno and Hori 1991), as shown in Figure 6.1,

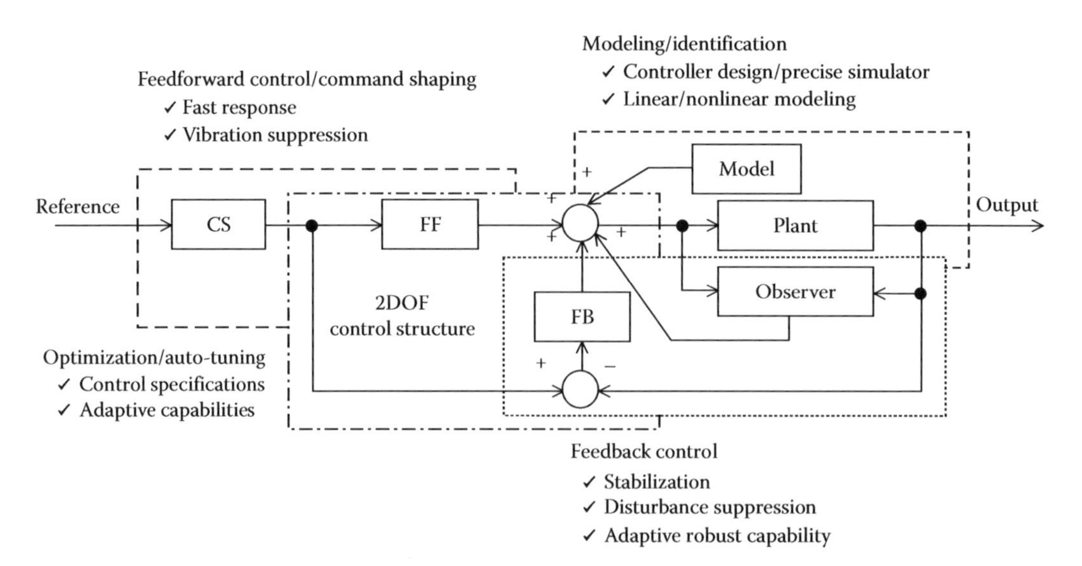

FIGURE 6.1 Block diagram of MC system.

with feedback and feedforward compensators designed in order to achieve the desired performance, while limiting the control bandwidth demand.

Indeed, the design of both feedforward and feedback controllers calls for a precise modeling of the plant to be controlled and (possibly) of the disturbances acting on it. However, such accurate knowledge is sometimes hard or very expensive to achieve, especially when the plant operates in an off-laboratory environment, where operating conditions may drastically change with time. For this reason, it is vital to develop model identification procedures that may accurately track the plant and environment characteristics during operations and, in turn, drive a change in the control strategy, if needed.

All of the above explains why the design of a modern high-accuracy MC system can be roughly divided into four tightly related parts: (1) feedback control system, (2) feedforward control system and command shaping, (3) system modeling and identification, and (4) optimization and auto-tuning of the MC system.

Each component will be analyzed in the following.

6.2.1 Feedback Control System

In precision MC applications, inherent requirements for the feedback system are a robust performance on system stability and/or controllability, disturbance suppression, and possibly adaptation capability against system variations.

Robust design can be achieved by using standard tools, but it is often desirable to use a set of simple components, in order to achieve the desired loop transfer function, with the prescribed stability margins. In this sense, novel approaches to achieve the robust stability have been proposed by applying simple filters, such as a resonant filter (Atsumi et al. 2006) and complex lead/lag filters (Messner 2008). With such techniques, the gain and phase of the plant to be controlled are shaped by repeatedly using an element with a customizable response in order to modify the original plant response (e.g., by depleting the resonant peak or by locally modifying the phase response), in order to allow for a larger closed-loop bandwidth, while keeping large stability margin. To clarify with an example, let us consider a simple plant, characterized by two low-frequency real poles and two high-frequency complex poles (Figure 6.2).

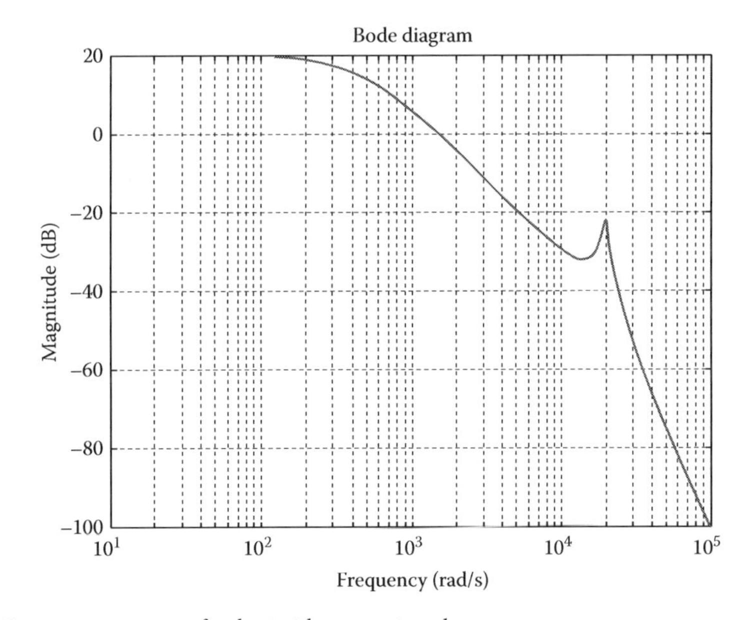

FIGURE 6.2 Frequency response of a plant with resonant modes.

A simple approach for the feedback controller design is to use a standard proportional–integral–derivative (PID) controller, tuned to achieve a certain open-loop crossing frequency and a desired phase margin at that frequency. Enlarging the closed-loop bandwidth by increasing the 0 dB crossing frequency while maintaining the phase margin can lead to multiple crossings of the loop transfer function, due to the presence of lightly damped resonant modes, as shown in Figure 6.3.

Cascading a notch filter with the PID controller can eliminate the multiple 0 dB crossings, but this notch reduces disturbance rejection and causes phase warping that adversely affects the overall servo performance. An alternative approach to dealing with the effects of mechanical resonances employs the so-called secondary phase margin, which is defined when the open loop has multiple 0 dB crossings. Kobayashi et al. (2003) formally define the secondary phase margin, and the concept is easily understood by examining the Nyquist plot of Figure 6.4. The damping of the resonance in closed-loop depends on the secondary phase margin.

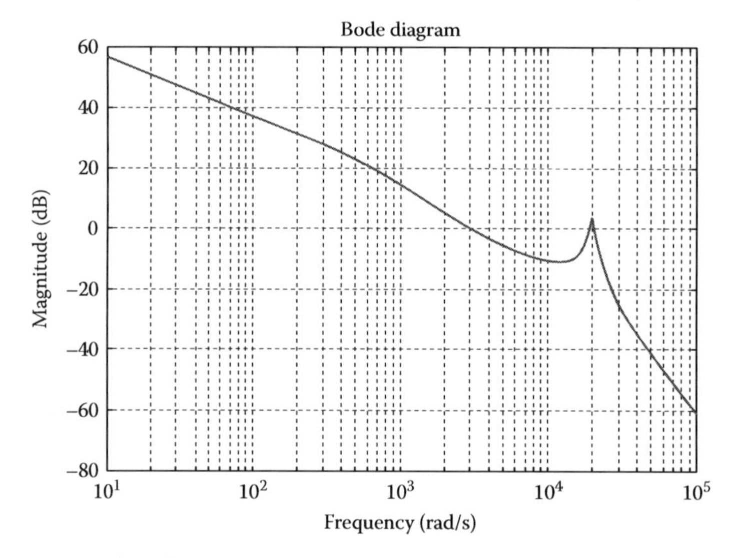

FIGURE 6.3 Resonant mode spillover.

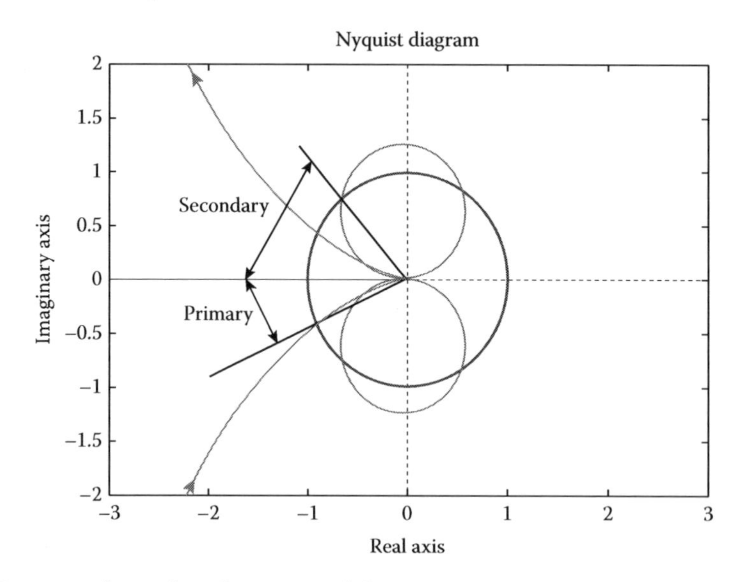

FIGURE 6.4 Primary and secondary phase margin definition.

Phase stabilized design explicitly addresses this limitation, by using a phase-shaping network to increase the secondary phase margin, while maintaining a high primary crossing frequency and, in turn, a large closed-loop bandwidth. Typical phase-shaping networks are properly tuned lag compensators, which move the secondary crossing point clockwise in the Nyquist plot by reducing both the phase and the gain of the loop transfer function after the first crossing. A limitation of this approach is that the gain and phase characteristics of standard lag–lead compensators also alter both the primary crossing frequency and the primary phase margin. If the lag compensator is realized with real pole–zero pairs, the servo designer has little freedom in the overall design since the maximum phase lag is a function of the ratio of the pole to the zero (Franklin et al. 1994). In order to overcome this limitation, Messner and Oboe (2004) proposes the use of complex phase lag networks, with poles and zeros as in Figure 6.5.

In such type of networks, both the center frequency and the phase variations can be tuned independently and, as shown in Figure 6.6, the achieved modifications are more localized when compared with standard lag networks with real poles and zeros.

By exploiting this feature, the desired phase profile for the loop transfer function can be easily achieved by locally modifying the original plant transfer function, using a set of properly placed complex lead/lag networks. A typical result is shown in Figure 6.7, where it can be seen that the same primary and secondary phase margins can be achieved with real complex networks, but only with the complex ones the gain margin is kept large, as well as the overall robustness.

As for the design of the feedback control for the suppression of the disturbance effects, it is well-known that a disturbance observer is one of the powerful tools to enhance the disturbance suppression performance. Some concerns on its use in such applications, which are characterized by variations in plant parameters, have been explicitly addressed and solved in Kobayashi et al. (2007), where it has been shown how the proper design of the observer leads to a user-definable phase lead or lag introduced by a mismatched disturbance observer.

However, a single controller, even when robust, often fails in achieving the desired performance, due to the fact that the plant may drastically change its characteristics, especially in the presence of nonlinear phenomena. A relevant example for this problem can be found in high-precision MC for data storage. In this case, the mechanical system in charge of moving the read/write equipment must

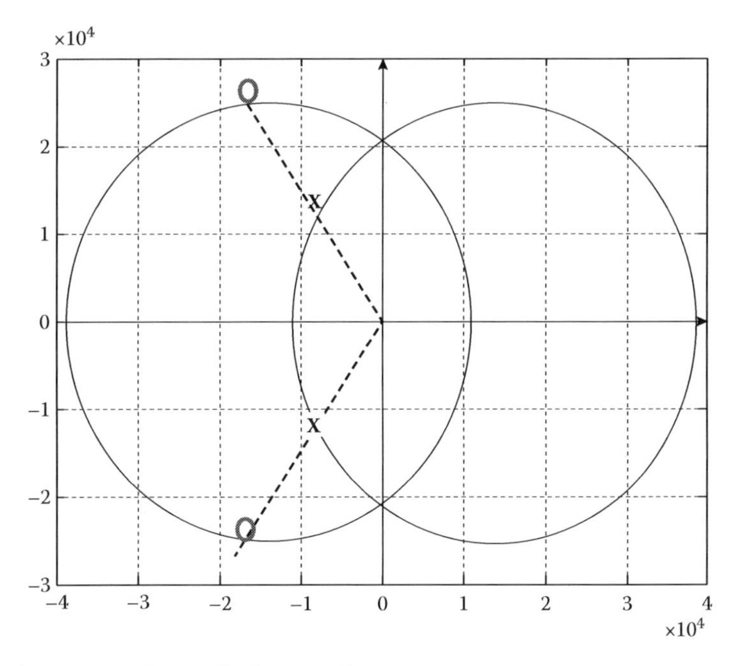

FIGURE 6.5 Pole–zero map of a complex lag network.

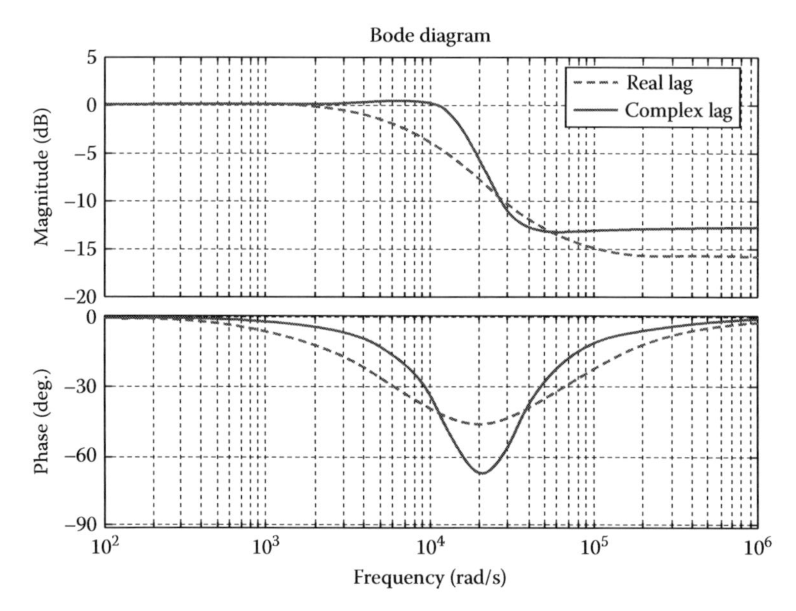

FIGURE 6.6 Comparison of frequency responses: real lag vs. complex lag.

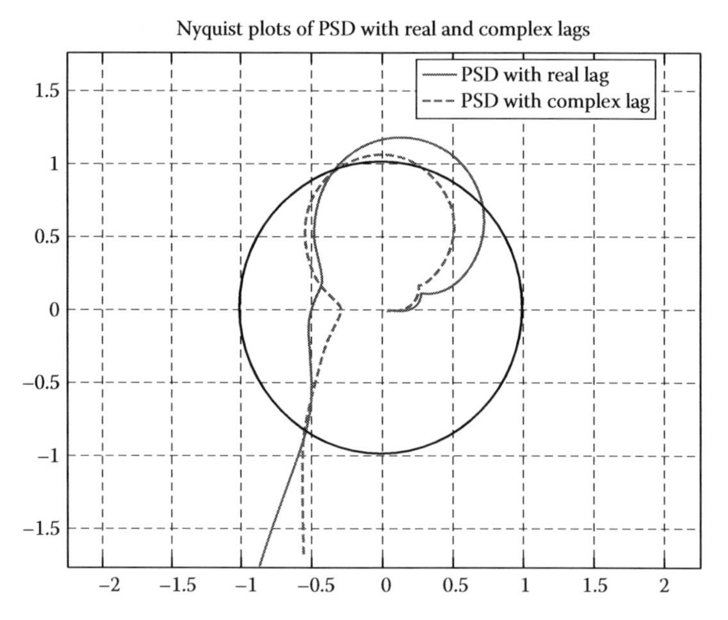

FIGURE 6.7 Phase stabilized design: real lag vs. complex lag.

be controlled in order to make rapid and wide movements from one data area to another. Then, once close to a target position, the controller must provide a ultrahigh accuracy and an excellent disturbance rejection. During fast motion, the accuracy is traded with the speed and the friction of the mechanical system, which is mainly viscous. On the other hand, when the speed goes close to zero, the control must cope with nonlinear friction and possible slip-stick motion around the target position. For this reason, it is a common practice to use the so-called mode switching control strategy, which selects the controller that best fits the current operating conditions among a set of predefined controllers. With this type of strategy, the controller design is greatly simplified, but a side effect that arises is the handling of possible unwanted transients induced by switching. A solution to this problem is the initial value compensation

(IVC), which initializes the states of the controllers in order to guarantee high degree of smoothness during transitions. Varieties of approaches on the IVC have been reported according to the application characteristics (Oboe and Marcassa 2002, Yang and Tomizuka 2006, Hirose et al. 2008).

Adaptive robust control framework has also been applied in order to cope with plant changes in MC systems. For example, an adaptive transient performance against parametric uncertainties (Taghirad and Jamei 2008) and a multirate perfect tracking control for plant with parametric uncertainty (Fujimoto and Yao 2005) could be achieved.

Other approaches including the computational intelligence have been introduced to improve the feedback control performance: repetitive learning control provides fast response and precise table positioning (Ding and Wu 2007), fuzzy control gives the maximum sensitivity for servo systems (Precup et al. 2007), and neuron-based PI fuzzy controller suppresses the mechanical vibration (Li and Hori 2007).

6.2.2 Feedforward Control System and Command Shaping

Feedforward compensation can improve the motion performance in the 2DOF control framework, e.g., fast response, high accuracy, and vibration suppression, independent of the characteristics of a feedback system (system stability, disturbance suppression, etc.). The feedforward compensator can be systematically designed by, e.g., a coprime factorization (Hioki et al. 2004), with consideration of mechanical vibration suppression. Perfect tracking performance can also be achieved by a multirate feedforward control approach (Saiki et al. 2008).

Command shaping, on the other hand, should be an additional control freedom to the feedforward compensation. This is particularly effective in some MC applications, where the system to be controlled exhibits some flexibility. This characteristic may lead to unwanted oscillations, especially in point-to-point motion applications. One well-known solution used to address this issue is the "input shaping," i.e., the offline design of an input (usually a sequence of delayed impulses) that brings the system from the initial position to the final one, with zero residual energy at the end of the motion. This approach, initially proposed by Singer and Seering (1990), has been improved by adding robustness features (Singhose et al. 1994) and applied in different areas on MC, like robotics, crane control, etc. All classical techniques, however, rely on the assumption that the system to be moved can be considered as a second-order system. This limitation has been partially solved in (Hyde 1991), where higher order systems have been addressed, with the main drawbacks arising during long operation, in addition to increased complexity in the derivation of the solution. Moreover, none of the approaches is conceived to directly address the variation of the process characteristics during the motion. In this area, the most innovative results (e.g., Van den Broeck et al. 2008) have been obtained by exploiting high-performance CPUs, in order to numerically solve the problem of the cancellation of residual vibrations on higher order flexible systems, while satisfying several constraints (like maximum available actuation power, robustness, etc.). On the other hand, variations of system characteristics have been recently addressed (De Caigny 2008a,b) for linear parameter varying (LPV) systems, with a solution in which the LPV system is embedded into a spline optimizer, which computes the optimal trajectories for the LPV system, achieving the desired degree of smoothness, while satisfying possible constraints on inputs and outputs.

Of course, shaping is not limited to the reference trajectory of an MC system. This technique can also be used to change the characteristics of a system subject to oscillatory disturbances (e.g., data-storage servo-positioners), by properly "shaping" the command to the process. The most common technique to address this problem is the repetitive controller, which can be seen as a filtering (i.e., a shaping of spectral content) on the command of the servo-controller, aimed at increasing the gain of the loop at the frequencies where the disturbances can be found. Similarly, other techniques (like adaptive feedforward compensation [AFC]) add repetitive compensating signals to the controller output. The success of such techniques in data storage comes from the relative simplicity of the design, which relies on the strong assumption that the sampling frequency is an integer multiple of the frequency of the disturbance. In addition, with repetitive control, information on the spectrum of the periodic disturbance cannot be used in the design (all harmonics are treated in the same way). Recent results (Steinbuch 2002, Steinbuch et al. 2007) have proved that it is possible to design repetitive controllers that are robust against variations in the period of the disturbances. Moreover, in Pipeleers et al. (2008), it has been shown that it is possible to realize repetitive controllers, which require less memory that traditional ones, while showing some robustness against variations on the disturbance period and accounting for spectral information on disturbances.

6.2.3 System Modeling and Identification

When designing an MC system, one of the most important issues is to have an accurate model of the process to be controlled at hand, in order to achieve the most effective model-based feedforward compensation and/or more progressive design of feedback controllers. Such a model can be nonparametric (e.g., a frequency response) or parametric (e.g., an ARMA model). This, however, is not always enough to properly design the controller for the system. In fact, the final performance is determined by the disturbance acting on the system and the strategies used in compensating for their effects. In this context, starting from the seminal work of many researchers (Van Den Hof and Sharma 1993, Karimi and Landau 1998), new results have been obtained in the so-called simultaneous identification, i.e., the use of experimental data obtained from the system (operating in closed-loop fashion), to get a discrete-time model of both the system and the spectral description of the disturbances acting on it (Zeng and de Callafon 2004), i.e., Po(s) and Ho(s) in the block diagram of Figure 6.8.

A relevant application of this approach is in MC systems where it is impossible to operate the identification in open loop (e.g., in data storage; Antonello et al. 2007), without incurring in failures or major errors in the identified models. It is worth noticing that technological limitations usually set an upper limit in the sampling rate of the measured data and, in turn, to the maximum frequency of the plant modes that can be identified. In order to overcome this limitation, linear identification and control of plant dynamics above the system Nyquist frequency has been introduced in Atsumi et al. (2007) and Kawafuku et al. (2007).

This and many other identification techniques found in literature rely on the assumption that the dynamic system to be identified is linear or weakly nonlinear. In almost all MC systems, however, there are always some nonlinear phenomena, which should be carefully identified by analyzing their complicated behaviors and their effects on the MC system. In cases of nanoscale positioning, such as for ultraprecision stage control in semiconductor and/or LCD manufacturing, nonlinear components directly deteriorate the motion performance. Nonlinear friction in mechanisms

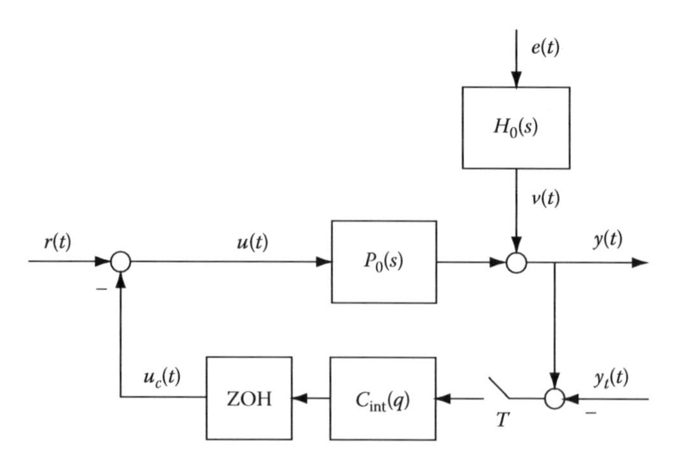

FIGURE 6.8 Block diagram of closed-loop identification.

interferes with motion performance as a disturbance; hence, friction may cause a tracking error or deviation, from the trajectory reference generated by the feedforward compensator in the 2DOF control system, deteriorating the motion performance, i.e., in terms of accuracy and vibration suppression.

In Jamaludin et al. (2008), the nonlinear friction affecting a linear motion system is described by using generalized Maxwell-slip model, which is experimentally tuned and used for the quadrant glitch compensation. According to the span of the motion, the friction model can be tailored in order to better capture the motion-relevant characteristics. This is the case of micro- and nanosized systems, where the friction between the elements in relative motion can be properly described by a dynamic system, which implements a state machine that determines whether there is an actual motion or not, i.e., if there is a "pre-sliding" or "sliding" friction between the parts (Kawafuku et al. 2008). Another interesting application of the nonlinear modeling in MC can be found in Ruderman et al. (2008), where the nonlinear behavior of a robotic transmission is used in order to compensate for the joint deflection due to the payload and, in turn, to improve the overall accuracy.

6.2.4 Optimization and Auto-Tuning of Motion Control System

From a practical viewpoint, it is quite important to optimize the whole MC system (from the system modeling to the compensator design and tuning) and to flexibly adapt it to variations in various motion environments. Soft computing techniques and/or computational intelligence can be an attractive alternative to conventional approaches to possess autonomous functions in the controller design.

As for the optimization, it can be performed in order to achieve various targets, like the minimization of the energy consumption or the maximization of some performance index, e.g., the comfort or the safety in a vehicle. In robotics, e.g., recent results have been reached in the area of the time–energy optimal path tracking. In Verscheure et al. (2008) and Lombai Szederkenyi (2008), the computation of the optimal trajectories, which satisfy several constraints (collision avoidance, torque limitation, motion smoothness, power consumption, minimum time, etc.), has been carried out by exploiting the availability of powerful hardware and tailored numerical routines, in order to achieve an "almost" real-time performance. It is worth noticing that in robotics, the optimal solution (i.e., the motion that satisfies some performance index) can be usually computed offline. On the other hand, unexpected obstacles, changes in dynamics, etc., call for online optimization schemes, capable to replan the reference trajectory in order to accommodate for any possible change in operating conditions. Some early approaches (Fiorini and Shiller 1997) for the optimal trajectory planning with moving obstacles were aimed at the collision avoidance in the presence of moving obstacles, all characterized by a rather simplified dynamic behavior. With the advent of more efficient CPUs, it has been possible to realize more refined optimization schemes, capable of multi-objective optimization, in a receding-horizon framework. Among all, a relevant result can be found in Bertolazzi et al. (2007), where the optimal reference maneuver for a vehicle is computed in real time, while considering moving objects, road conditions, safety and comfort constraints, power limits, etc.

As for the use of soft computing techniques in optimization, genetic algorithms have been reported as effective tools in the design of optimized controllers (Itoh et al. 2004, Low and Wong 2007), where multiple objectives for the target system can be handled, e.g., motion speed and accuracy, vibration suppression, and energy minimization.

Optimization, of course, can implicitly drive the adaptation process by modifying one or more blocks of the control system of Figure 6.1, in order to achieve the desired performance. On the other hand, adaptation capability has been introduced to address specific modifications of the plant or the environment. For instance, an adaptive neural network compensation for nonlinearity in HDDs has been proposed in Hong and Du (2008), and fault-tolerant schemes against system failures have been reported in Delgado et al. (2005) and Izumikawa et al. (2005).

6.3 Motion Control and Interaction with the Environment

Recent advances of computer technology makes it possible to realize environment recognition based on visual information, signal processing, and many kinds of sensors. In the conventional sensor feedback system, however, controller is constructed, so that the whole system is insensitive against environmental variations (Kim and Hori 1995, Ohnishi et al. 1996). This brings improvement to the system reliability, but the adaptive control for the environment is not achieved. On the other hand, the fusion of flexible control, based on sensor information and MC, can be considered to represent a human–machine interactive system, though the system construction is not optimal. Then, the recognition of environment including human becomes a key issue for a reliable interaction between the environment, human, and machine from a control viewpoint. The related technologies of intelligent transport systems (ITS) will be given as the typical system (Daily and Bevly 2004). For instance, the stabilization of a vehicle is realized by information-intensive control technology with the use of several kinds of sensors. In such a system, it is necessary to construct an adaptive recognition system for the environmental change and the MC system that acts on the environment adaptively through human manipulation. Hence, it is necessary to achieve the unification of sensor information–based environment recognition with MC technologies while considering an environmental change. For example, Mutoh et al. (2007) proposes safety and stable driving for an electric vehicle by the estimation of braking force on various road surfaces. To improve the stability of yaw motion in an electric vehicle, Ohara and Murakami (2008) implements an active angle control of the front wheel based on the estimation of road cornering stiffness. The adaptation to environment is also considered in the development of the biped robot, which is another type of a mobile system, just like vehicle. Minakata et al. (2008) proposes a flexible shoe system for biped robots to optimize energy consumption in lateral gait motion as an environment adaptation method, and an optimal combination of hardware (shoes) and software (controller) is developed. This combination strategy of software and hardware design is an important aspect of advanced MC.

In human–machine interaction, a typical example is a wheel chair system (Cooper et al. 2002, Seki et al. 2006). Various welfare devices have been developed by the influence of the society's aging population. They are required to support caregivers and cared-for people these days because the burden of the nursing care is increasing. Once again, in such a system, unifying the sensor information–based environment recognition and the MC technologies is of paramount importance. Tashiro and Murakami (2008) proposes a power assist control of step passage motion for an electric wheelchair by the estimation of reaction force from environment. Shimizu and Oda (2007) introduces a virtual power assist control of an electric wheelchair based on visual information. From these examples, by utilizing the wheelchair system, a schematic illustration of the key components of advanced mechatronics for MC system is summarized in Figure 6.9.

To achieve sophisticated and advanced MC for the systems mentioned in the above examples, due to the importance of information processing between the regulated system and the MC system, it is necessary to develop the new and emergent research area of motion and information control technology. Some key issues to this aim can be summarized as

1. Information-intensive control technology using heterogeneous and redundant sensor systems
2. Recognition technology of environmental change in the system, including environmental model
3. Development of a multifunctional MC system based on information-intensive control technology
4. Adaptive MC for the interaction among human, environment, and machines

Considering these key issues, a schematic illustration of system integration in advanced MC is shown in Figure 6.10. As described in this figure, MC systems are connected through the information network, and the physical sensation of human–machine interactive system is transmitted by this information network. This is one of the emergent technologies arising from advanced and networked applications of MC systems, currently also known as cyber-physical systems (CPS). A typical example of such systems is a remote surgery system by the bilateral control, as shown in Figure 6.11. In this system, described

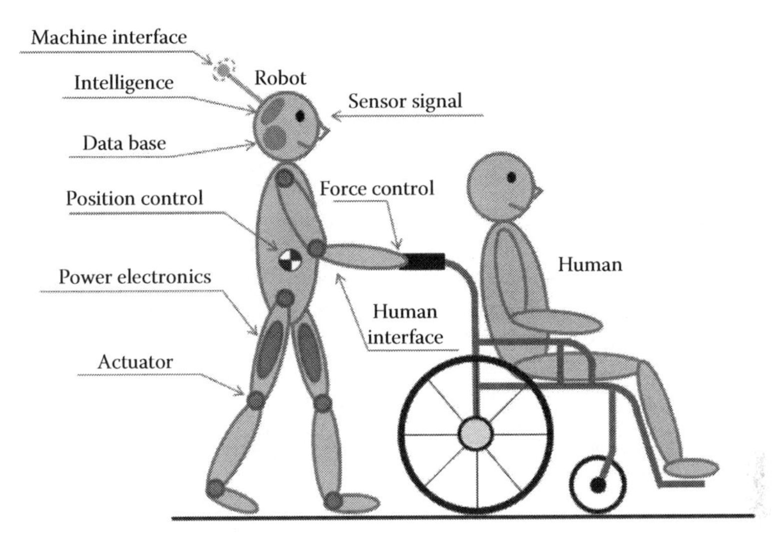

FIGURE 6.9 A schematic illustration of the key components of advanced mechatronics for MC.

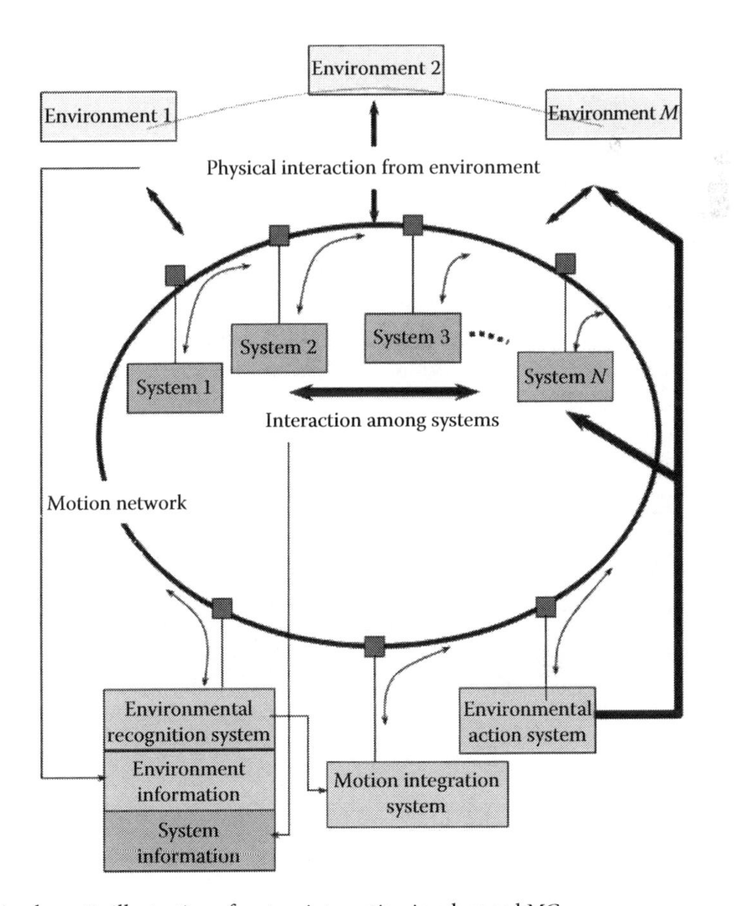

FIGURE 6.10 A schematic illustration of system integration in advanced MC.

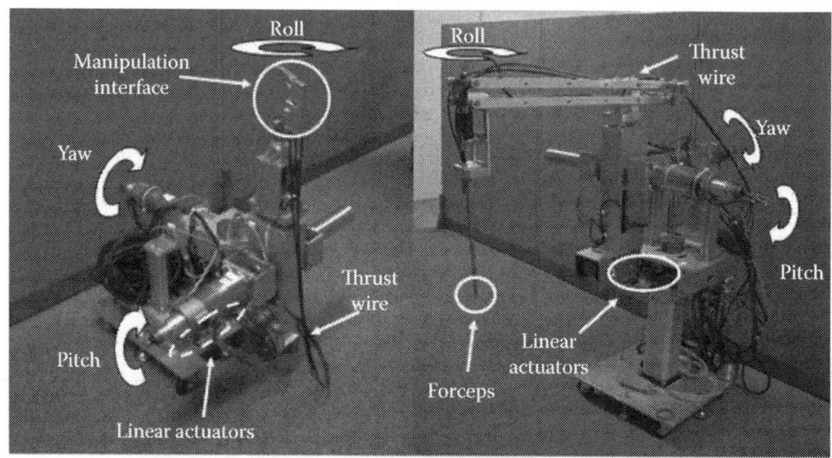

Master robot

Slave robot

Haptic surgical robot

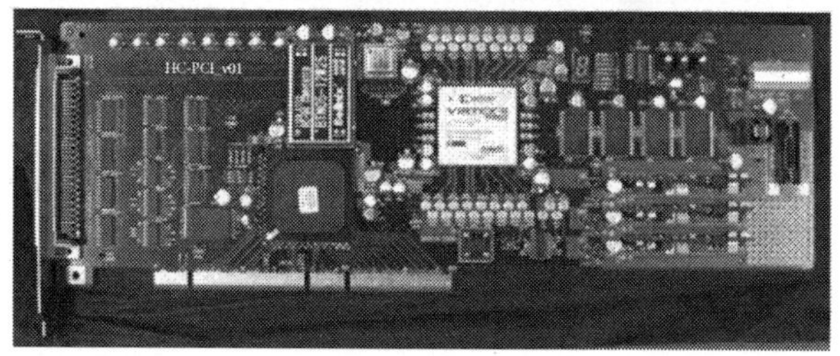

FPGA board

FIGURE 6.11 An example of bilateral surgery system.

in Matsumoto et al. (2007) and Shimono et al. (2007), a novel structure of acceleration-based bilateral system is implemented to achieve high transparency of force sensation (haptics) through LAN-based motion network system. To increase the feasibility of practical remote surgery system, the development of synchronous transmission of not only force sensation but also visual information is required as one of the future works. In the advanced MC with respect to this issue, delay compensation of data transmission will be a key technology (Natori and Ohnishi 2008).

The next section will introduce the most relevant results in this field.

6.4 Remote Motion Control

An emerging area of MC concentrates on the MC of remote systems. In fact, in addition to the traditional applications like telerobotics (Hokayem and Spong 2006) and field bus–based applications (Thomesse 2005), there are new ones, aimed at the MC of multiple objects, allocated remotely or at some distance with respect to one another, and yet, contributing in some way to the accomplishment of a certain "general" task. A typical case is the control of a fleet of robots (or, more generally, vehicles), which must coordinate their motion in order to maintain a relative position or to safely transport a payload that is too big for a single robot (Ercarnacao and Pascoal 2001, Fax and Murray 2002). Control strategies may vary according to the task and the communication resources available, and they may range from master–slave (or leader–follower) schemes to self-arranging fleets, with the global coordination arising from the interaction of each element with immediate neighbors.

In all scenarios, there is an underlying communication infrastructure that allows for the communication among two or more elements of the system and, in this framework, one of the open issues regards the handling of the limitations of the communication systems: packet losses, variable delays, and unpredictable available bandwidth. There are two possible ways to address this limitation. The first one consists of the realization of communication systems with reliable and predictable performance, while the second is the implementation of a control scheme that is inherently robust against the failures or limitations of the communication system.

Considering the first approach, an effective solution seems to be achieved with the introduction of new real-time communication protocols, like real-time Ethernet (RTE) (Thomesse 2005), which overcomes (at least in principle) the aforementioned limitations. These are high-speed networks (100/1000 Mbits/s) based on the well-known, original Ethernet specification and tailored for industrial applications, in order to guarantee very short transmission times as well as tight determinism for periodic traffic and hard real-time operation for acyclic data delivery.

As for the explicit addressing of the uncertainty on data delivery and delays, many of the possible solutions, applied to telerobotics, can be found in Hokayem and Spong (2006). Most of the proposed solutions, however, make some simplifying assumption to make the problem treatable, which are not applicable to an MC application. The first one is that the time is considered as a continuous variable, while the use of digital communication network leads naturally to a discrete-time scenario. Moreover, many of the approaches consider the delay as a variable with a time derivative strictly less than one. As shown in Oboe et al. (2007), this approach cannot be applied to digital communication networks, so the discrete-time issue of the remotely controlled systems must be directly addressed.

Recently, new approaches to prove the stability of a telerobotic equipment in a discrete-time framework, based on passivity, have been proposed in Secchi et al. (2003). Another promising approach is the bilateral generalized predictive control (BGPC) proposed in Slama et al. (2007), which is based on an extension of model predictive control (MPC). MPC is an advanced method for process control that has been used in several process industries such as chemical plants, oil refineries, and in robotics area. The major advantages of MPC are the possibility to handle constraints and the intrinsic ability to compensate large or poorly known time delays. The main idea of MPC is to rely on dynamic models of the process in order to predict the future process behavior on a receding horizon and, accordingly, to select command input with respect to the future reference behavior. Motivated by all the advantages of this method, the MPC was applied to teleoperation systems (Sheng and Spong 2004). The originality of the approach proposed in Slama et al. (2007) lies in an extension of the general MPC, so-called bilateral MPC (BMPC), allowing to take into account the case where the reference trajectory is not a priori known in advance due to the slave force feedback. The bilateral term is employed to specify the use of the signal feedback, which alters the reference system dynamic in the controller.

Another way to explicitly address the variable delay is to measure and then compensate it. This approach, proposed in Seuret et al. (2006), is based on the use of GPS-based synchronization of the nodes, by means of which the random delay due to the communication can be compensated in a deterministic manner.

A more recent approach in bilateral control is the consideration of the communication delay in the control input and feedback loop as a lumped disturbance, which is further estimated via a communication delay observer (CDOB) as in Natori and Ohnishi (2008). In the literature, there are also variations of this approach treating the communication delay as a disturbance, using a sliding mode observer, and by combining the sliding mode observer with an extended Kalman filter, which estimates the load on the slave side (Bogosyan et al. 2009).

6.5 Conclusions

MC is a "broad" term to define all the strategies, tools, and technologies that concur in realizing a system that moves in its environment. Indeed, the controlled motion must satisfy one or more requirements, in terms of span, accuracy, execution time, cost, safety of the humans who may interact with the motion system, etc. Such systems may largely differ in their characteristics, so a general approach to MC is not possible. A simple example can be given to support this point with micro-/nanoscaled motion systems, where phenomena like surface adhesion, usually neglected in the macroscaled systems, have a large influence in the overall behavior. Another example comes from the MC of robotic systems, which can be conceived in different ways, according to the degree of interaction with the surrounding environment or with humans populating its workspace. This chapter presented some relevant results achieved in different areas pertaining to MC, which indeed represent a multidisciplinary area, where many competences concur at the realization of innovative system. Some of these systems have already proven capable, while some demonstrate great potential to improve the overall quality of life.

References

Antonello, R., R. Oboe, and R.A. de Callafon. 2007. An identification experiment for simultaneous estimation of low-order actuator and windage models in a hard disk drive, *IEEE International Symposium on Industrial Electronics, 2007* (*ISIE 2007*), Vigo, Spain, June 4–7, 2007, pp. 3102–3107.

Atsumi, T., A. Okuyama, and M. Kobayashi. 2007. Track-following control using resonant filter in hard disk drives. *IEEE/ASME Transactions on Mechatronics*, 12:472–479.

Atsumi, T., A. Okuyama, and S. Nakagawa. 2006. Vibration control above the Nyquist frequency in hard disk drives, *Proceedings of the 9th International Workshop on Advanced Motion Control*, Istanbul, Turkey, 2006, pp. 103–108.

Bertolazzi, E., F. Biral, and M. Da Lio. 2007. Real-time motion planning for multibody systems. *Multibody System Dynamic*, 17(2–3):119–139.

Bogosyan, S., B. Gadamsetty, M. Gokasan, and A. Sabanovic. 2009. Novel observers for compensation of communication delay in bilateral control systems, *Proceedings of the 35th Annual Conference of IEEE Industrial Electronics Society* (*IECON 2009*), Porto, Portugal, November 3–6, 2009.

Cooper, R.A., T.A. Corfman, S.G. Fitzgerald, M.L. Boninger, D.M. Spaeth, W. Ammer, and J. Arva. 2002. Performance assessment of a Pushrim-activated power-assisted wheelchair control system. *IEEE Transactions on Control Systems Technology*, 10(1):121–126.

Daily, R. and D.M. Bevly. 2004. The use of GPS for vehicle stability control systems. *Transactions on Industrial Electronics*, 51(2):270–277.

De Caigny, J., B. Demeulenaere, J. De Schutter, and J. Swevers. 2008a. Dynamically optimal polynomial splines for flexible servo-systems: Experimental results, *10th IEEE International Workshop on Advanced Motion Control, 2008* (*AMC '08*), Trento, Italy, March 26–28, 2008a, pp. 98–103.

De Caigny, J., B. Demeulenaere, J. De Schutter, and J. Swevers. 2008b. Polynomial spline input design for LPV motion systems, *10th IEEE International Workshop on Advanced Motion Control, 2008* (*AMC '08*), Trento, Italy, March 26–28, 2008b, pp. 86–91.

Delgado, D.U.C., S.M. Martínez, and K. Zhou. 2005. Integrated fault-tolerant scheme for a DC speed drive. *IEEE/ASME Transactions on Mechatronics*, 10(4):419–427.

Ding, H. and J. Wu. 2007. Point-to-point motion control for a high-acceleration positioning table via cascaded learning schemes. *IEEE Transactions on Industrial Electronics*, 54(5):2735–2744.

Encarnacao, P. and A. Pascoal. 2001. Combined trajectory tracking and path following: An application to the coordinated control of marine craft, *IEEE Conference on Decision and Control* (*CDC'2001*), Orlando, FL, 2001.

Fax, A. and R. Murray. 2002. Information flow and cooperative control vehicle formations, *Proceedings of the 2002 IFAC*, Barcelona, Spain, 2002.

Fiorini, P. and Z. Shiller. June 1997. Time optimal trajectory planning in dynamic environments. *Journal of Applied Mathematics and Computer Science*, special issue on *Recent Development in Robotics*, 7(2):101–126.

Franklin, G.F., J.D. Powell, and A. Emani-Naeini. 1994. *Feedback Control of Dynamic Systems*. Addison-Wesley, Reading, MA.

Fujimoto, H. and B. Yao. 2005. Multirate adaptive robust control for discrete-time non-minimum phase systems and application to linear motors. *IEEE/ASME Transactions on Mechatronics*, 10:371–377.

Hioki, T., K. Yubai, and J. Hirai. 2004. Joint design method based on coprime factorization of 2DOF control system, *Proceedings of the 8th IEEE International Workshop on Advanced Motion Control*, Kawasaki, Japan, 2004, pp. 523–527.

Hirose, N., M. Iwasaki, M. Kawafuku, and H. Hirai. 2008. Initial value compensation using additional input for semi-closed control systems, *Proceedings of the 10th IEEE International Workshop on Advanced Motion Control*, Trento, Italy, 2008, pp. 218–223.

Hokayem, P.F. and M.W. Spong. 2006. Bilateral teleoperation: An historical survey. *Automatica*, 42:2035–2057.

Hong F. and C. Du. 2008. An improved adaptive neural network compensation of pivot nonlinearity in hard disk drives, *Proceedings of the 10th International Workshop on Advanced Motion Control*, Trento, Italy, 2008, pp. 440–443.

Hyde, J.M. 1991. Multiple mode vibration suppression in controlled flexible systems. MIT Space Engineering Research Center Report (SERC).

Itoh K., M. Iwasaki, and N. Matsui. 2004. Optimal design of robust vibration suppression controller using genetic algorithms. *IEEE Transaction on Industrial Electronics*, 51(5):947–953.

Izumikawa, Y., K. Yubai, and J. Hirai. 2005. Fault-tolerant control system of flexible arm for sensor fault by using reaction force observer. *IEEE/ASME Transactions on Mechatronics*, 10(4):391–396.

Jamaludin, Z., H.V. Brussel, and J. Swevers. 2008. Quadrant glitch compensation using friction model-based feedforward and an inverse-model-based disturbance observer, *Proceedings of the 10th International Workshop on Advanced Motion Control*, Trento, Italy, 2008, pp. 212–217.

Karimi, A. and I.D. Landau. 1998. Comparison of the closed-loop identification methods in terms of the bias distribution. *Systems & Control Letters*, 32:159–167.

Katsura, S., J. Suzuki, and K. Ohnishi. October 2006. Pushing operation by flexible manipulator taking environmental information into account industrial electronics. *IEEE Transactions on Industrial Electronics*, 53(5):1688–1697.

Katsura, S., T. Suzuyama, and K. Ohishi. December 2007. A realization of multilateral force feedback control for cooperative motion. *IEEE Transactions on Industrial Electronics*, 54(6):3298–3306.

Kawafuku, M., J.H. Eom, M. Iwasaki, and H. Hirai. 2007. Identification method for plant dynamics over Nyquist frequency, *Proceedings of the 33rd Annual Conference of the IEEE Industrial Electronics Society*, Taipei, Taiwan, 2007, pp. 362–367.

Kawafuku, M., A. Ohta, M. Iwasaki, and H. Hirai. 2008. Comparison of rolling friction behavior in HDDs, *10th IEEE International Workshop on Advanced Motion Control, 2008 (AMC '08)*, Trento, Italy, March 26–28, 2008, pp. 272–277.

Kim, K. and Y. Hori. December 1995. Experimental evaluation of adaptive and robust schemes for robot manipulator control. *IEEE Transactions on Industrial Electronics*, 42(6):653–662.

Kobayashi, H., S. Katsura, and K. Ohnishi. 2007. An analysis of parameter variations of disturbance observer for motion control. *IEEE Transactions on Industrial Electronics*, 54:3413–3421.

Kobayashi, M., S. Nakagawa, and S. Nakamura. 2003. A phase-stabilized servo controller for dual-stage actuators in hard disk drives. *IEEE Transactions on Magnetics*, 39:844–850.

Li, W. and Y. Hori. 2007. Vibration suppression using single neuron-based PI fuzzy controller and fractional-order disturbance observer. *IEEE Transactions on Industrial Electronics*, 54(1):117–126.

Lombai, F. and G. Szederkenyi. 2008. Trajectory tracking control of a 6-degree-of-freedom robot arm using nonlinear optimization, *10th IEEE International Workshop on Advanced Motion Control, 2008 (AMC '08)*, Trento, Italy, March 26–28, 2008, pp. 655–660.

Low, K.S. and T.S. Wong. 2007. A multiobjective genetic algorithm for optimizing the performance of hard disk drive motion control system. *IEEE Transactions on Industrial Electronics*, 54(3):1716–1725.

Matsumoto, Y., S. Katsura, and K. Ohnishi. 2007. Dexterous manipulation in constrained bilateral teleoperation using controlled supporting point. *IEEE Transactions on Industrial Electronics*, 54(2):1113–1121.

Messner, W.C. 2008. Classical control revisited: Variations on a theme, *Proceedings of 10th IEEE International Workshop on Advanced Motion Control*, Trento, Italy, 2008, pp. 15–20.

Messner, W.C. and R. Oboe. 2004. Phase stabilized design of a hard disk drive servo using the complex lag compensator, *Proceedings of the IEEE American Control Conference*, Boston, MA, 2004, pp. 1165–1170.

Minakata, H., K. Seki, and S. Tadakuma. 2008. A study of energy-saving shoes for robot considering lateral plane motion. *Transactions on Industrial Electronics*, 55(3):1271–1276.

Mutoh, N., Y. Hayano, H. Yahagi, and K. Takita. 2007. Electric braking control methods for electric vehicles with independently driven front and rear wheels. *Transactions on Industrial Electronics*, 54(2):1168–1176.

Natori, K. and K. Ohnishi. 2008. A design method of communication disturbance observer for time-delay compensation, taking the dynamic property of network disturbance into account. *IEEE Transactions on Industrial Electronics*, 55(5):2152–2168.

Oboe, R. and F. Marcassa. 2002. Initial value compensation applied to disturbance observer-based servo control in HDD, *Proceedings of the 7th International Workshop on Advanced Motion Control*, Maribor, Slovenia, 2002, pp. 34–39.

Oboe, R., T. Slama, and A. Trevisani. 2007. Telerobotics through Internet: Problems, approaches and applications. *Annals of the University of Craiova-Series: Automation, Computers, Electronics and Mechatronics*, 4(31)(2):81–90.

Ohara, H. and T. Murakami. 2008. A stability control by active angle control of front-wheel in vehicle system. *Transactions on Industrial Electronics*, 55(3):1277–1285.

Ohishi, K. 2008. Realization of fine motion control based on disturbance observer, *Proceedings of 10th IEEE International Workshop on Advanced Control*, Trento, Italy, 2008, pp. 1–8.

Ohnishi, K., M. Shibata, and T. Murakami. 1996. Motion control for advanced mechatronics. *IEEE/ASME Transactions on Mechatronics*, 1(1):56–57.

Otsuka, J. and T. Masuda. 1998. The influence of nonlinear spring behavior of rolling Elements on ultraprecision positioning. *Nanotechnology*, 9:85–92.

Pipeleers, G., B. Demeulenaere, J. De Schutter, and J. Swevers. 2008. Generalized repetitive control: Better performance with less memory, *10th IEEE International Workshop on Advanced Motion Control, 2008 (AMC '08)*, Trento, Italy, March 26–28, 2008, pp. 104–109.

Precup, R.E., J. Preitl, and P. Korondi. 2007. Fuzzy controllers with maximum sensitivity for servosystems. *IEEE Transactions on Industrial Electronics*, 54(3):1298–1310.

Ruderman, M., F. Hoffmann, and T. Bertram. 2008. Preisach model of nonlinear transmission at low velocities in robot joints, *10th IEEE International Workshop on Advanced Motion Control, 2008 (AMC '08)*, Trento, Italy, March 26–28, 2008, pp. 721–726.

Saiki, K., A. Hara, K. Sakata, and H. Fujimoto. 2008. A study on high-speed and high-precision tracking control of large-scale stage using perfect tracking control method based on multirate feedforward control, *Proceedings of the 10th International Workshop on Advanced Motion Control*, Trento, Italy, 2008, pp. 206–211.

Secchi, C., S. Stramigioli, and C. Fantuzzi. 2003. Digital passive geometric telemanipulation, *Proceedings of the IEEE International Conference on Robotics and Automation*, Taipei, Taiwan, 2003, Vol. 3, pp. 3290–3295.

Seki, H., T. Iijima, H. Minakata, and S. Tadakuma. 2006. Novel step climbing control for power assisted wheelchair based on driving mode switching, *32nd Annual Conference on IEEE Industrial Electronics*, Paris, France, 2006, pp. 3827–3832.

Seuret, A., M. Termens-Ballester, A. Toguyeni, S.E. Khattabi, and J.-P. Richard. 2006. Implementation of an internet-controlled system under variable delays, *Proceedings of ETFA '06*, Prague, Czech Republic, 2006, pp. 675–680.

Sheng, J. and M.W. Spong. 2004. Model predictive control for bilateral teleoperation systems with time delays, *Canadian Conference on Electrical and Computer Engineering*, Tampa, FL, 2004, Vol. 4, pp. 1877–1880.

Shimizu, H. and N. Oda. 2007. Vision-based force following control for mobile robots, *SICE Annual Conference 2007*, 2007, pp. 2507–2512.

Shimono, T., S. Katsura, and K. Ohnishi. 2007. Abstraction and reproduction of force sensation from real environment by bilateral control. *IEEE Transactions on Industrial Electronics*, 54(2):907–918.

Singer, N.C. and W P. Seering. 1990. Preshaping command inputs to reduce system vibration. *Transactions of the ASME, Journal of Dynamic Systems, Measurement, and Control*, 112:76–82.

Singhose, W.E., W.P. Seering, and N.C. Singer. 1994. Residual vibration reduction using vector diagrams to generate shaped inputs. *Transactions of the ASME, Journal of Dynamic Systems, Measurement, and Control*, 116:654–659.

Slama, T., D. Aubry, R. Oboe, and F. Kratz. 2007. Robust bilateral generalized predictive control for teleoperation systems, *Proceedings of the 15th IEEE Mediterranean Conference on Control and Automotion'07*, Athens, Greece, 2007.

Steinbuch, M. 2002. Repetitive control for systems with uncertain period-time. *Automatica*, 38(12):2103–2109.

Steinbuch, M., S. Weiland, and T. Singh. 2007. Design of noise and period-time robust high order repetitive control, with application to optical storage. Reprint accepted for publication in *Automatica*, 43: 2086–2095.

Taghirad, H.D. and E. Jamei. 2008. Robust performance verification of adaptive robust controller for hard disk drive. *IEEE Transactions on Industrial Electronics*, 55:448–456.

Tashiro, S. and T. Murakami. 2008. Step passage control of a power-assisted wheelchair for a caregiver. *Transactions on Industrial Electronics*, 55(4):1715–1721.

Thomesse, J.P. June 2005. Fieldbus technologies in industrial automation. *Proceedings of the IEEE*, 93(6):1073–1101.

Umeno, T. and Y. Hori. 1991. Robust speed control of DC servomotors using modern two degrees-of-freedom controller design. *IEEE Transactions on Industrial Electronics*, 38:363–368.

Van den Broeck, L., G. Pipeleers, J.D. Caigny, B. Demeulenaere, J. Swevers, and J.D. Schutter, 2008. A linear programming approach to design robust input shaping, *Proceedings of the 10th International Workshop on Advanced Motion Control*, Trento, Italy, March 26–28, 2008, pp. 80–85.

Van Den Hof, P. and R. Schrama. 1993. An indirect method for transfer function estimation from closed loop data. *Automatica*, 29:1523–1527.

Van Den Hof, P. and R. Schrama. 1995. Identification and control-closed loop issues. *Automatica*, 31(12):1751–1770.

Verscheure, D., B. Demeulenaere, J. Swevers, J. De Schutter, and M. Diehl. 2008. Time-energy optimal path tracking for robots: A numerically efficient optimization approach, *10th IEEE International Workshop on Advanced Motion Control, 2008 (AMC '08)*, Trento, Italy, March 26–28, 2008, pp. 727–732.

Yang, L. and M. Tomizuka. 2006. Short seeking by multirate digital controllers for computation saving with initial value adjustment. *IEEE/ASME Transactions Mechatronics*, 11(1):9–16.

Zeng, J. and R.A. de Callafon. 2004. Servo experiments for modeling of actuator and windage dynamics in a hard disk drive, *Proceedings of the IEEE International Conference on Mechatronics (ICM)*, Istanbul, Turkey, June 3–5, 2004, Vol. 1, pp. 218–223.

7

New Methodology for Chatter Stability Analysis in Simultaneous Machining

Nejat Olgac
University of Connecticut

Rifat Sipahi
Northeastern University

7.1 Introduction and a Review of Single Tool Chatter

We present a new methodology on the stability of *simultaneous (or parallel) machining* (SM), where multiple tools operate at different cutting conditions (spindle speeds, depths-of-cut) on the same workpiece. It is obvious that such a parallel processing operation as opposed to conventional single-tool machining (STM) (also known as *serial process*) is more time efficient [1–5]. It is also known that SM dynamics is considerably more complex, especially when metal removal rates are maximized and the dynamic coupling among the cutting tools, the workpiece, and the machine tool(s) becomes very critical. Primary difficulty arises due to the regenerative cutting forces that may ultimately violate the instability boundaries. The occurring dynamics is known as the "unstable regenerative chatter" at the interface between the workpiece and the cutting tool. The aim is to create working conditions that facilitate chatter rejection (i.e., the "stable regenerative chatter"). The stability repercussions of such dynamics are poorly understood at the present, both in the metal-processing field as well as in the mathematics. It is partly due to the modeling difficulty and the large number of parameters involved in the emanating dynamics. Furthermore, there is no evidence of a broadly agreed upon analytical methodology to assess the stability/instability boundaries for a perfectly known dynamic process model, even for the simplest cases of simultaneous machining.

It is well known from numerous investigations that the conventional STM introduces some important stability concerns. Similar problems result in much more complex form for SM. In lack of a solid mathematical methodology to study the SM chatter, the existing practice is likely to be and it really is suboptimal and guided by trial-and-error or ad hoc procedures. There is certainly room for much-needed improvement in the field, and this chapter presents some key steps in this direction. We compose the document with the support of several earlier publications. The segments on STM, orthogonal cutting, and SM follow primarily the directions of [34,35]. Further discussions on the selection of the variable-pitch milling process and the optimal design of multi-flute milling cutters as well as the operating conditions with the primary intent of avoiding the unstable chatter are attributed to a large extent to [36,37].

7.1.1 Regenerative Chatter and Its Impacts on Machining

Optimum machining aims to maximize the material removal rate while maintaining a sufficient disturbance rejection (i.e., the "stability margin in cutting") to assure the surface quality. The machine tool instability primarily relates to "chatter." As accepted in the manufacturing community, there are two groups of machine tool chatter: *regenerative* and *nonregenerative* [6]. Regenerative chatter occurs due to the periodic tool passing over the undulations on the previously cut surface, and nonregenerative chatter has to do with mode coupling among the existing modal oscillations. The former brings a delay argument into the inherent dynamics since part of the cutting force emanates from the tooth position one revolution earlier (thus, delayed by a cutter-passing period). The latter, however, does not have such a term. This text evolves mainly on regenerative type, thus we refer to it simply as "chatter" in the rest of the document.

In order to prevent the onset of chatter, one has to select the operational parameters appropriately, namely, the chip loads and spindle speeds. Existing studies on machine tool stability address conventional STM processes. They are not expandable to SM because of the substantial difference in the underlying mathematics. Leaving the details to the later sections, we can say that SM requires analyzing a more complex mathematical problem stemming from the general class of *parametric quasi-polynomials with multiple-delay terms*. There is no practical methodology known at this point to resolve the complete stability mapping for such constructs. This text presents a novel procedure in that sense.

Machine tool chatter is an undesired engineering phenomenon. Its negative effects on the surface quality, tool life, etc., are well known. Starting with the early works [7–10], many researchers very cleverly addressed the issues of physical modeling, the dynamic progression, structural reasoning, and stability limit aspects of this seemingly straightforward and very common behavior. Further research focused on the particulars of parameter selections in machining to avoid the build up of these undesired oscillations and on the analytical predictions of chatter. Some interesting readings on this are found in [6,10–12]. Most commonly, chatter research has focused on the conventional STM. The aim, again, is to increase the metal removal rate while avoiding the onset of chatter [13–15]. A natural progressive trend is to increase the productivity through SM. This process can be further optimized by determining the best combination of the chip loads and spindle speeds with the constraint of chatter instability. For SM, however, multiple spindle speeds, which cross-influence each other, create governing differential equations with multiple time delay terms. Their characteristic equations are known in mathematics as "quasi-polynomials with multiple time delays." Multiplicity of the delays (which arise from multiplicity of spindle speeds) present enormously more complicated problem compared with the conventional STM chatter governed by a single delay, and especially when they are rationally independent.

7.1.2 Basics of Single-Tool Machining

Let us review the basics of STM chatter dynamics for the clarity of the argument borrowing some parts from [6]. Consider an *orthogonal turning* process for the sake of example (Figure 7.1). The underlying mechanism for regenerative chatter is quite simple to state. The desired (and nominal) chip thickness, h_o,

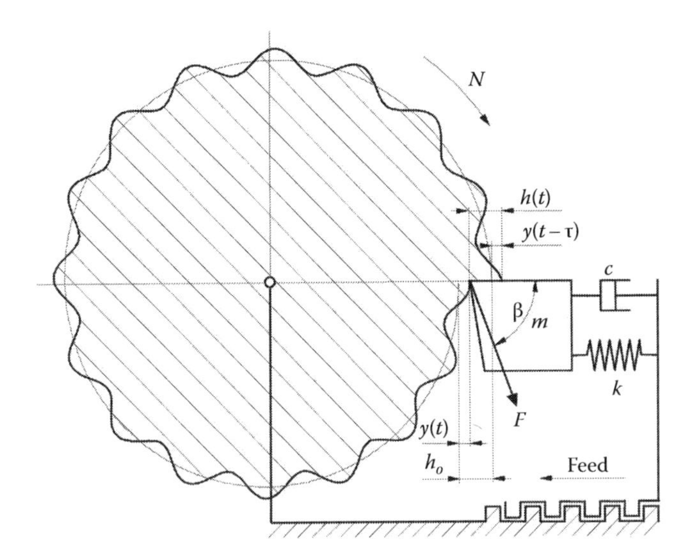

FIGURE 7.1 Orthogonal turning process. (From Olgac, N. and Sipahi, R., *ASME J. Manuf. Sci. Eng.*, 127, 791, 2005. With permission.)

is considered constant. The tool actually cuts the chip from the surface, which is created during the previous pass. The process-generated cutting force, F, is realistically assumed to be proportional to the dynamic chip thickness, $h(t)$. It carries the signature of $y(t) - y(t - \tau)$, where $y(t)$ is the fluctuating part of the chip thickness at time t (so-called offset chip thickness from the nominal value h_o), and τ [s] is the period of successive passages of the tool, which is equal to $60/N$, where N is the spindle speed (rpm).

The block diagram in Figure 7.2 gives a classical causality representation of the dynamics for this orthogonal turning operation. Nominal chip thickness, h_o, is disturbed by the undulating offset chip thickness, y. These undulations create driving forces for the y dynamics τ s later during the next passage, and thus the attribute "regenerative." In other words, what is uncut τ s earlier, returns as additional chip thickness that the cutting tool has to remove. Therefore, the tool–workpiece interface experiences different amplitude of excitation forces. Taking $G(s)$ as the transfer function between the cutting force F and y, the governing dynamics becomes a linear delayed differential equation, for which the characteristic equation is a quasi-polynomial.

For the sake of streamlining the presentation, a single-degree-of-freedom cutting dynamics is taken into account instead of higher degree-of-freedom and more complex models. The workpiece and its rotational axis are considered to be rigidly fixed, and the only tool flexibility is taken in the radial direction. In Figure 7.2, the following causal relations are incorporated:

$$\text{the cutting force: } F(t) = Cbh(t) \tag{7.1}$$

$$\text{actual chip thickness: } h(t) = h_o - y(t) + y(t - \tau) \tag{7.2}$$

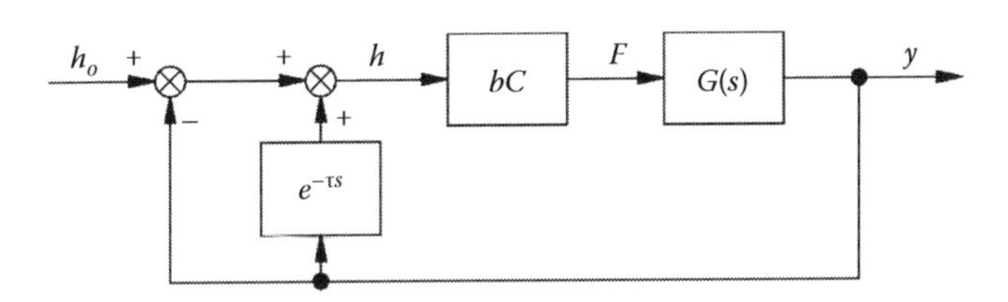

FIGURE 7.2 Functional block diagram of chatter regeneration. (From Olgac, N. and Sipahi, R., *ASME J. Manuf. Sci. Eng.*, 127, 791, 2005. With permission.)

where

b is the chip width, which is user specified and assumed constant

C is the cutting force constant

τ [s] corresponds to the period of one spindle revolution, $\tau = 60/N$, N (rpm), which is also user selected

7.1.3 Single-Tool Machining and Stability

Assuming that the force–displacement relationship is linear, the entire cutting mechanism described by Figure 7.1 becomes linear as well. Cutting is at equilibrium when $y = 0$, which means an ideal cut with no waviness. The cutting force F remains constant, and the tool support structure (i.e., stiffness and damping characteristics) yields a static deflection throughout the cutting operation. This equilibrium is called "stable" or "asymptotically stable," if the loop characteristic equation of the block diagram in Figure 7.2

$$1 + \left(1 - e^{-\tau s}\right) bCG(s) = 0 \tag{7.3}$$

has all its roots on the stable left half-plane [16,17]. This equation is transcendental, and it possesses infinitely many *finite characteristic roots*, all of which have to be taken into account for stability. It is clear that the selection of b and τ influences the stability of the system and the complete stability map of this system in b and τ domain is the well-known "chatter stability lobes," Figure 7.3 (using the parameters from Ref. [34]). There are several such lobes marked on this figure. They represent the (b, τ) plane mapping of the dynamics with dominant characteristic root at $Re(s) = -a$. When $a = 0$ (corresponding to the thick line), these curves show the well-known stability lobes that form the stability constraints for any process optimization. For instance, if we increase the metal removal rate (by increasing the chip width b while keeping the spindle speed and related delay τ constant), aggressive cutting ultimately invites chatter instability, which is obviously not desirable. This transition corresponds to marching along a vertical line in Figure 7.3.

There are two cutting conditions under the control of the machinist, namely τ, which is the inverse of the spindle speed $(60/N)$, and b, which is the chip width. The other parameters, that is, C and $G(s)$, represent the existing cutting characteristics, which are considered to remain unchanged. The open-loop transfer function $G(s)$ typically manifests high-impedance, damped, and stable dynamic behavior.

Certain selections of b and $\tau(= 60/N)$ can introduce marginal stability to the system on the boundaries separating stability from instability in Figure 7.3. At the operating points on these boundaries, the characteristic equation (7.3) possesses a pair of imaginary roots $\pm\omega i$ as dominant roots. As such the

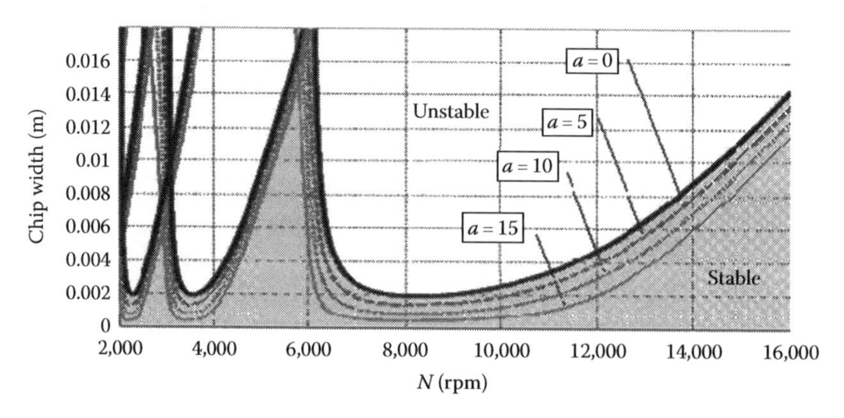

FIGURE 7.3 Typical stability lobes and constant stability margin lobes. (From Olgac, N. and Sipahi, R., *ASME J. Manuf. Sci. Eng.*, 127, 791, 2005. With permission.)

complete system is resonant at ω, i.e., the entire structure mimics a spring-mass resonator (a conservative system) at the respective frequency of marginal stability, also known as "chatter frequency." The desirable operating point should lie in the shaded region, marked as "stable." For the simple mathematical construct as in (7.3), the determination of the stability boundaries (lobes) is relatively simple. Although the problem looks complex, the complete stability maps of single-delay problems represented by (7.3) are obtainable (e.g., for the case of STM). One should notice that in (7.3), the transcendental term appears with only single delay τ and no integer multiples of it (i.e., 2τ, 3τ, etc.). In mathematical terms, such systems are known as "single delay with no commensurate conditions." Even for much more complicated dynamics than what is shown in Figure 7.1 (e.g., including higher order modes of the tool holder), the characteristic equation still has no commensurate terms. For this class of systems, a simple procedure is given in [34].

It is logical to select the cutting parameters (b, N) sufficiently away from the chatter stability bounds. Conventional terminology alluding to this feature is called the "stability margin." It refers to $a = -Re$ (dominant characteristic root). The bigger the value of a, the higher the "chatter rejection speed," therefore, the better the surface quality. A set of operating points are shown where a = 5, 10, and 15, just to give an idea of the distribution of the loci where the chatter rejection speed is constant. Optimum working conditions are reached by increasing b and N up to the physical limitations provided that a desirable stability margin (i.e., a) is guaranteed. Figure 7.3 represents a unique declaration of what is known as "equal stability margin" lobes. The determination of these curves also offers a mathematically challenging problem, which can be found in Ref. [38]. In this text, we will treat a set of more challenging problems as the following chapters are formed.

7.2 Regenerative Chatter in Simultaneous Machining

The functional block diagram in Figure 7.2 expands in dimension by the number of tools involved in the case of SM, as depicted in Figure 7.4a and b for two example operations. The flexures are shown at one point and in 1D sense to symbolize the most complex form of the restoring forces in 3D. Spindle speeds and their directions are also selected symbolically just to describe the types of operations we consider here.

The crucial difference between the STM and SM is the coupling among the individual tool–workpiece interfaces, either through the flexible workpiece (as in Figure 7.4a) or the machine tool compliance characteristics (as in Figure 7.4b), or both. Clearly, the chip load at tool i ($i = 1,\ldots,n$), which carries the signature of $y_i(t) - y_i(t - \tau_i)$, will influence the dynamics at tool j. This is the *cross-regenerative effect*, which implies that the state of the ith tool one revolution earlier (i.e., τ_i s) affects the present dynamics of the jth tool. Consequently, the dynamics of tool j will reflect the combined regenerative effects from all the tools, including itself. We assume these tool-to-tool interfaces are governed under linear relations.

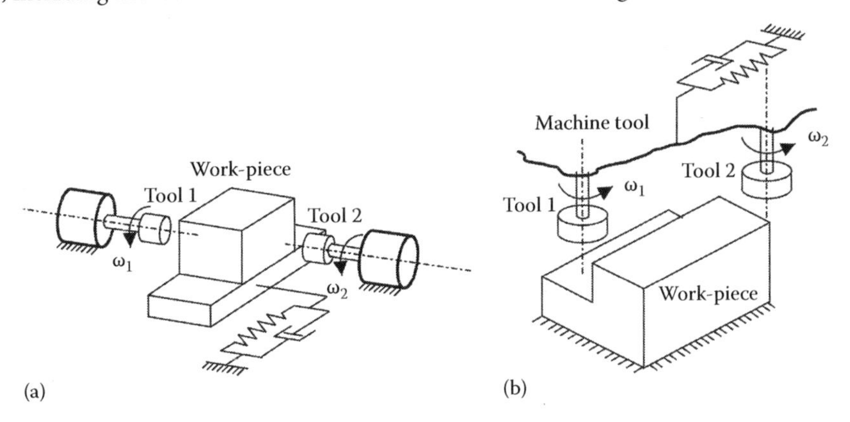

FIGURE 7.4 Conceptual depiction of simultaneous face milling: (a) workpiece coupled and (b) machine tool coupled. (From Olgac, N. and Sipahi, R., *ASME J. Manuf. Sci. Eng.*, 127, 791, 2005. With permission.)

The overall dynamics becomes a truly cross-coupled linear multiple time-delay system. It is known that this dynamics also carries time varying and periodic nature of the cutting force directions; however, as it was pointed in the literature [33], we take the fundamental elements of these forces in their Fourier expansions. As such, the dynamics reduces to linear time-invariant form.

7.2.1 Simultaneous Machining and Multiple Delays

The stability of linear time-invariant multiple-delay systems is poorly known in the mathematics community. There is no simple extension of the conventional STM stability treatment to multiple spindle and consequently to multiple-delay cases of SM, mainly because the stability problem is NP-hard with respect to increased number of delays. The specific objective of this paper is to develop a new mathematical framework to address the STM stability problem. A notational selection here is that bold capital fonts represent the vector or matrix forms of the lower case elements, such as $\{\tau\} = (\tau_1, \tau_2)$.

We present next the multi-spindle cutting tool dynamics in a generic form (see Figure 7.5). Analogous to Figure 7.2, $\mathbf{H}_0$ $(n \times 1)$ and $\mathbf{H}$ $(n \times 1)$ represent the commanded and actual depth-of-cut (d-o-c) vectors, components of which denote the d-o-c at n individual tools. $\mathbf{Y}$ $(n \times 1)$ is the tool displacement fluctuations vector along the d-o-c directions. Similarly, $\text{BC}_{\text{diag}} = \text{diag}(b_i C_i)$, $i = 1,\ldots, n$ represents the influence of b_i (chip width) and C_i (the cutting force constants) vis-à-vis the tool i. $\text{BC}_{\text{diag}} = \text{diag}(b_1\, C_1, b_2\, C_2, \ldots, b_n\, C_n)$ are the same terms as in the single-tool case, except the multiple-tool version of them. The exponential diagonal matrix $\mathbf{e}_{\text{diag}}^{-\tau s} = \text{diag}(e^{-\tau_1 s}, e^{-\tau_2 s}, \ldots, e^{-\tau_n s})$, τ_i $(= 60/N_i)$, $i = 1, \ldots\, n$, represents the delay effects (i.e., the regenerative elements), and N_is are the spindle speeds. Notice that the relation between τ_j and N_j should go through the number of flutes in case of milling. Mathematically, however, this nuance introduces only a scale factor along the delays. As such, it is overlooked at this stage, and this point is revisited in Case Study II. $G(s)$ $(n \times n)$ is the dynamic influence transfer function, which entails all the auto- and cross-coupling effects between the cutting force vector, F $(n \times 1)$, and the tool displacements in the d-o-c direction $\mathbf{Y}$ $(n \times 1)$. We are making the assumption that these relations are all linear, as it is common in most fundamental chatter studies. It is clear that $G(s)$ may appear in much more complicated forms, for instance, in milling, it becomes periodically time-variant matrix [33]. The past investigations suggest the use of Fourier expansion's fundamental term in such cases, to avoid the mathematical complexity in order to extract the underlying regenerative characteristics.

7.2.2 Stability of Simultaneous Machining

The characteristic equation of the loop in Figure 7.5 is

$$CE(s, \boldsymbol{\tau}, \mathbf{B}) = \det\left[\mathbf{I} + \mathbf{G}(s)\mathbf{BC}_{\text{diag}}\left(\mathbf{I} - \mathbf{e}_{\text{diag}}^{-\tau s}\right)\right] = 0, \quad (7.4)$$

which is representative of a dynamics with *multiple time delays* (τ) and *multiple parameters* (B) as opposed to a STM, where there is a single delay τ and a single parameter, b. Equation 7.4 is a parameterized

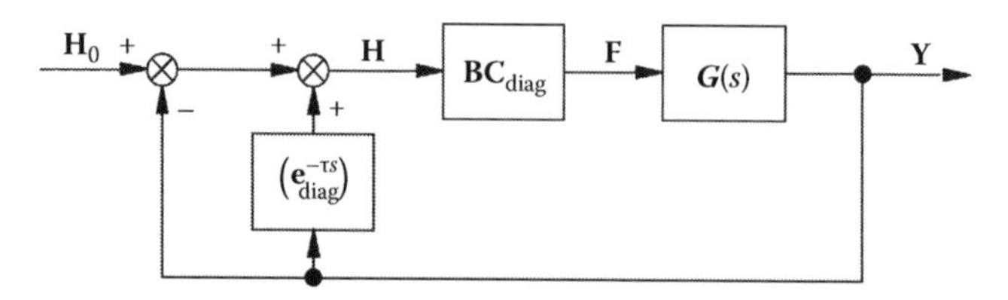

FIGURE 7.5 Block diagram representing the simultaneous machining. (From Olgac, N. and Sipahi, R., *ASME J. Manuf. Sci. Eng.*, 127, 791, 2005. With permission.)

quasi-polynomial in *s* with multiple time delays. For a stable operation, its infinitely many characteristic roots should all be on the left half of the complex plane. The most general form of $CE(s, \boldsymbol{\tau}, \mathbf{B})$ contains terms like $e^{-\sum_{j=1}^{n}\alpha_j\tau_j s}$ with $\alpha_j = 0$ or 1 representing the cross talk among the delay terms (i.e., among the tools). For instance, $e^{-(\tau_1+\tau_3)s}$ term would indicate the cross-coupling between the regenerative effects of spindles 1 and 3.

For the exceptional case where $G(s)$ is also diagonal, Equation 7.4 decouples all the delay effects so that the problem reduces to *n* independent STM chatter problems. Otherwise, the cross-talk terms will appear in Equation 7.4, which add to the complexity of the analysis considerably. A critical point to note is that there is no possible *commensurate delay* formation here, i.e., no terms will appear with $e^{-k\tau_i s}$, $k \geq 2$. Physically, this implies that the regenerative effect of tool *j* acts on itself only once.

One can find substantial literature on the stability of systems with only one single time delay [18–22,38] even for *commensurate* cases. Therefore, STM dynamics is a simpler subclass of what is considered here. The published studies address the question of the "stability margin" in time delay, τ_{max}. They consider all the parameters (**B**) to be fixed and claim that the higher delay values than a certain τ_{max} would invite instability. These methods typically stop once τ_{max} is computed. A recent paradigm by the authors enables the determination of *all* the stable regions of τ completely (including the first stability interval of $0 \leq \tau < \tau_{max}$) [23–27]. This new framework, called "cluster treatment of characteristic roots" (CTCR), yields the complete stability picture for single-delay systems. Multiple time-delay systems (MTDS) are, however, significantly more complex, and their parameterized form (i.e., for varying **B** matrix) bring another order of magnitude to the difficulty. This is the focal highlight of this chapter.

Again, the system depicted in Figure 7.5 is asymptotically stable if and only if all the roots of the transcendental characteristic equation (7.4) are on the left half of the complex *s* plane. Infinitely many such roots will have to be tracked for the stability condition. This is obviously a prohibitively difficult task. There exists no commonly accepted methodology at present to study the stability of such systems. We wish to clarify this statement that many investigations performing time-domain simulations or iterative numerical determinations of stability boundaries are not included simply because they are not considered comparable approaches. First, they are performed point-by-point in (τ_1,τ_2) space and, therefore, the application is computationally overwhelming. Second, they are *numerical* as opposed to *analytical* nature of CTCR. Very few mathematical investigations reported on the subject also declare strict limitations and restrictions to their pursuits [18,28–30]. Reference [28] handles the same problem using simultaneous nonlinear equation solvers but is only able to treat systems in the form of $a + be^{-\tau_1 s} + ce^{-\tau_2 s} = 0$ with *a*, *b*, *c* being constants. This imposes, clearly, a very serious restriction. References [29,30] tackle a limited subclass of (7.4) with $\det[G(s)]$ of degree 2, but without a damping term, and their approach is not expandable to damped cases. We study in this text, the most general form of (7.4) under the conceptual framework of the CTCR [23–26]. This is a new approach in mathematics, and it is uniquely suitable for SM.

7.3 CTCR Methodology

In order to avoid the notational complexity, we present the description of the CTCR methodology in the next section, using a reduced system for $n = 2$, without loss of generality. The most general form of the characteristic equation (7.4) representing the two-tool SM regenerative chatter dynamics becomes

$$CE(s,\tau_1,\tau_2,b_1,b_2) = a_0 + a_1e^{-\tau_1 s} + a_2e^{-\tau_2 s} + a_3e^{-(\tau_1+\tau_2)s} = 0, \tag{7.5}$$

where $a_j(s,b_1,b_2)$, $j = 0, 1, 2, 3$ are polynomials in *s* with parameterized coefficients in b_1 and b_2. The highest degree of *s* in (7.5) resides within $a_0(s)$ and it has no time delay accompanying it. We wish to recover the stability portrait (referred to as the "lobes") in four-dimensional $(\tau_1, \tau_2, b_1, b_2)$ space, where the delay terms τ_i $(= 60/N_i)$ are rationally independent of each other. The dynamics at hand is a *Retarded Multiple Time-Delay System*, as it is known in the mathematics [18,19,21,28–32].

7.3.1 Characteristics of Multiple-Dimensional Stability Maps

Counterpart of the stability lobes given in Figure 7.3 is in four-dimensional parameter space of (τ_1, τ_2, b_1, b_2) for this SM operation. The transition, however, from single to multiple time delays is not trivial even when the parameters (b_1, b_2) are fixed. We revisit a mathematical tool for this operation in this section from [24–26,32].

We start with a simpler stability problem for $\mathbf{B} = (b_1, b_2)^T$ parameters being fixed. Although the dynamics represented by (7.5) possesses infinitely many characteristic roots, the most critical ones are those that are purely imaginary. Obviously, any stability switching (from stable to unstable or vice versa) takes place when the parameters cause such purely imaginary roots. These imaginary roots display some very interesting constructs. Based on our preliminary work, we claim two novel propositions to summarize these peculiar features. These two propositions form the foundations of CTCR paradigm. A nontrivial transition of the CTCR from single delay to multiple delays leads to an *exhaustive stability analysis tool* in the space of the time delays (τ_1, τ_2). In the interest of streamlining the discussions, we provide these propositions without proof and recommend [32] to the interested reader for the theoretical development. We present later some example cases to demonstrate the strengths and the uniqueness of CTCR.

Proposition 7.1. Equation 7.5 can have an imaginary root only along countably infinite number of hypersurfaces $\wp(\tau_1,\tau_2)$, τ_1, and $\tau_2 \in \Re^+$. These hypersurfaces, which are simply "curves" in two dimension, are indeed offspring of a manageably small number of hyperplanes, which we will call the "kernel hypersurfaces," $\wp_0(\tau_1,\tau_2)$. All the hypersurfaces in $\wp(\tau_1,\tau_2)$ are some descendants of $\wp_0(\tau_1,\tau_2)$. ■

Related to this proposition, we present two explanatory remarks, which are also critical to the CTCR framework:

Remark 7.1. Kernel and offspring: If there is an imaginary root at $s = \mp\omega_c i$ (subscript "c" is for crossing) for a given set of time delays $\{\tau_0\} = (\tau_{10}, \tau_{20})$, the same imaginary root will appear at all the countably infinite grid points of

$$\{\tau\} = \left(\tau_{10} + \frac{2\pi}{\omega_c} j, \quad \tau_{20} + \frac{2\pi}{\omega_c} k\right) \quad j = 1,2,\ldots, \quad k = 1,2,\ldots, \tag{7.6}$$

Notice that for a fixed ω_c, the distinct points in (7.6) generate a grid in $\{\tau\} \in \Re^{2+}$ space with equidistant grid size in both dimensions. When ω_c is varied continuously, the respective grid points also display a continuous variation, which ultimately form the hyperplanes $\wp(\tau_1,\tau_2)$. Therefore, instead of generating these grid points and studying their variational properties, one can search only for the critical building block, "the kernel," for $j = k = 0$ and for *all* possible ω_c's, $\omega_c \in \Re^+$. The kernel can alternatively be defined by $\min(\tau_1,\tau_2)_{\omega_c}, (\tau_1,\tau_2) \in \Re^{2+}$ for all possible ω_c s. ■

Remark 7.2. The determination of the kernel. As stated earlier, if there is any stability switching (i.e., from stability to instability or vice versa), it will take place at a point on $\wp(\tau_1,\tau_2)$ curves. Therefore, we need *all* possible $\wp(\tau_1,\tau_2)$ and the representative ω_c s, exhaustively. In other words, we have to determine the kernel $\wp_0(\tau_1,\tau_2)$ along which *all* the imaginary roots $s = \omega_c i$ of (7.5) are found. That means the *description of kernel* has to be exhaustive. The determination of the kernel and its offspring is mathematically very challenging problem. For this, we deploy a unique transformation called the Rekasius substitution [22,32]:

$$e^{-\tau_i s} = \frac{1-T_i s}{1+T_i s}, \quad T_i \in \Re, \quad i=1,2, \tag{7.7}$$

which holds identically for $s = \omega_c i$, $\omega_c \in \Re$. This is an exact substitution for the exponential term, not an approximation, for $s = \omega_c i$, with the obvious mapping condition of

$$\tau_i = \frac{2}{\omega_c}\left[\tan^{-1}(\omega_c T_i) + j\pi\right], \quad j=0,1,\ldots. \tag{7.8}$$

Equation 7.8 describes an asymmetric mapping in which T_i (distinct in general) is mapped into countably infinite τ_i sets, each of which has periodically distributed time delays for a given ω_c with periodicity $2\pi/\omega_c$. Substitution of (7.7) into Equation 7.5 converts it from $CE(s, \tau_1, \tau_2)$ to $\overline{\overline{CE}}(s, T_1, T_2)$. Notice the slight breach of notation, which drops b_1, b_2 parameters from the arguments both in CE and $\overline{\overline{CE}}$. We next create another equation as described below:

$$\overline{CE}(s, T_1, T_2) = \overline{\overline{CE}}(s, T_1, T_2)(1+T_1 s)(1+T_2 s) = \sum_{k=0}^{4} \lambda_k(T_1, T_2) s^k. \tag{7.9}$$

Since the transcendental terms have all disappeared, this equation can now be studied much more efficiently. Notice that *all* the imaginary roots of $CE(s, \tau_1, \tau_2)$ and $\overline{CE}(s, T_1, T_2)$ are identical, i.e., they coincide, while there is no enforced correspondence between the remaining roots of these equations. That is, considering the root topologies

$$\begin{aligned} \Omega_1 &= \left\{ s \middle| CE(s, \tau_1, \tau_2) = 0, \quad (\tau_1, \tau_2) \in \Re^{2+} \right\}, \\ \Omega_2 &= \left\{ s \middle| \overline{CE}(s, T_1, T_2) = 0, \quad (T_1, T_2) \in \Re^{2} \right\}, \end{aligned} \tag{7.10}$$

the imaginary elements of these two topologies are identical. In another notation, one can write $\Omega_1 \cap \mathbf{C}^0 \equiv \Omega_2 \cap \mathbf{C}^0$, where $\mathbf{C}^0$ represents the imaginary axis. It is clear that the exhaustive determination of the $(T_1, T_2) \in \Re^2$ loci (and the corresponding ω_cs) from Equation 7.9 is much easier task than the exhaustive evaluation of the same loci in $(\tau_1, \tau_2) \in \Re^{2+}$ from Equation 7.5. Once these loci in (T_1, T_2) are found, the corresponding *kernel* and the *offspring* in (τ_1, τ_2) can be determined as per (7.8). ■

We give a definition next, before stating the second proposition. The *root sensitivities* of each purely imaginary characteristic root crossing, $\omega_c i$, with respect to one of the time delays is defined as

$$S_{\tau_j}^{s}\Big|_{s=\omega_c i} = \frac{ds}{d\tau_j}\Big|_{s=\omega_c i} = -\frac{\partial CE/\partial \tau_j}{\partial CE/\partial s}\Bigg|_{s=\omega_c i} \quad i=\sqrt{-1}, \quad j=1,2, \tag{7.11}$$

and the corresponding *root tendency* with respect to one of the delays is given as

$$\text{Root tendency} = RT\Big|_{s=\omega_c i}^{\tau_j} = \text{sgn}\left[\text{Re}\left(S_{\tau_j}^{s}\Big|_{s=\omega_c i}\right)\right]. \tag{7.12}$$

This property represents the direction of the characteristic root's crossing when only one of the delays varies.

Proposition 7.2 [following, 32]. Assume that there exists an imaginary root at $i\omega_c$, which is caused by the delay compositions (τ_{10}, τ_{20}) on the kernel and its two-dimensional offspring

$$\left(\tau_1, \tau_2\right)_{\omega_c} = \left(\tau_{1j} = \tau_{10} + \frac{2\pi}{\omega_c} j, \quad \tau_{2k} = \tau_{20} + \frac{2\pi}{\omega_c} k; \quad j = 0, 1, 2, \ldots, k = 0, 1, 2, \ldots\right). \tag{7.13}$$

The root tendency at this point is invariant with respect to j (or k) when k (or j) is fixed. That is, regardless of which offspring (τ_{1j}, τ_{2k}) of the kernel set (τ_{10}, τ_{20}) causes the crossing, $RT\big|_{s=\omega_c i}^{\tau_1}$ $\left(RT\big|_{s=\omega_c i}^{\tau_2}\right)$ are the same for all $j = 1, 2 \ldots$ $(k = 1, 2 \ldots)$, respectively. In other words, the imaginary root always crosses either to the unstable right half-plane ($RT = +1$) or to the stable left half-plane ($RT = -1$), when one of the delays is kept fixed, and the other one is skipping from one grid point to the next, regardless of the actual values of the time delays as long as they are derived from the same kernel (τ_{10}, τ_{20}). ■

Note: There are several other computational tools that can be used to determine the kernel hypercurves. Interested reader is directed to the following texts:

1. Kronecker summation technique [39], which operates on an augmented characteristic equation that is independent of the delays.
2. Building block method [40], which considers a finite dimensional space, the so-called spectral delay space, to simplify the process.
3. Frequency sweeping algorithm [41], which offers a practical numerical procedure to obtain the kernel hypersurfaces, using geometry.

The two propositions of the CTCR and the steps associated with the two remarks stated above ultimately generate a complete stability mapping in (τ_1, τ_2) space. We next present three case studies to show how CTCR applies to STM and SM chatter problems. The first is to demonstrate how one can deploy CTCR to the conventional STM chatter. The second is the analysis of an experimental study, which is detailed in [35]. And the third study presents how CTCR can be deployed for designing six-flute milling cutters.

7.4 Example Case Studies

7.4.1 Case Study 1: Application of CTCR for Single-Tool Machining

In order to gain better understanding of the CTCR procedure, we take the conventional chatter stability problem first with a single cutter ($n = 1$). For this case, a regenerative dynamics with one single time delay appears. The characteristic equation (7.5) reduces to

$$CE(s, \tau, b) = a_0(b, s) + a_1(b, s)e^{-\tau s} = 0, \tag{7.14}$$

where the only delay is $\tau[\mathrm{s}] = 60/N[\mathrm{RPM}]$, N is the only spindle speed, and b is the chip width. There are numerous case studies in the literature on this problem. Just for demonstration purposes, we take the orthogonal turning example given in [34, Equation 7], which starts from the characteristic equation

$$1 + bC\frac{\cos\beta\left(1 - e^{-\tau s}\right)}{ms^2 + cs + k} = 0, \tag{7.15}$$

where C, β, m, c, and k are the constants related to the cutting dynamics. This equation can be rewritten as

$$CE(s, \tau, b) = ms^2 + cs + k + bC\cos\beta\left(1 - e^{-\tau s}\right) = 0, \tag{7.16}$$

which is in the same form as (7.14). The parametric values are taken from practice as follows

$$C = 2\times10^9\ \text{N/m}^2 \quad \beta = 70° \quad m = 50\ \text{kg} \quad c = 2\times10^3\ \text{kg/s} \quad k = 2\times10^7\ \text{N/m}$$

Equation 7.16 with these numerical values becomes

$$s^2 + 40s + 400{,}000 + 1{,}368{,}0805.73b - 1{,}368{,}0805.73be^{-\tau s} = 0. \tag{7.17}$$

We next search for the stability pockets in $\tau \in \Re^+$ space for varying chip widths $b \in \Re^+$. This picture is conventionally known as the "stability lobes" for regenerative chatter. Unlike multiple-delay cases, this problem with a single delay is solvable using a number of different procedures given in the literature [6,21,34,43]. Our aim here is to show step-by-step how the CTCR paradigm applies in this case.

Take b as a fixed parameter.

Use the Rekasius substitution of (7.7) in (7.20) to obtain

$$\begin{aligned}\overline{CE}(s,b,T) &= (1+Ts)(s^2 + 40s + 400{,}000 + 1{,}368{,}0805.73b) - 1{,}368{,}0805.73b\,(1-Ts)\\ &= Ts^3 + (40T+1)s^2 + [40 + (400{,}000 + 2{,}736{,}1611.46b)T]\,s + 400{,}000\\ &= \lambda_3(T,b)s^3 + \lambda_2(T,b)s^2 + \lambda_1(T,b)s + \lambda_0 = 0,\end{aligned} \tag{7.18}$$

where $\lambda_j(T, b), j = 0\ldots3$ are self-evident expressions.

Search for the values of T, which create $s = \omega i$ as a root for (7.18). This is best handled using the Routh's Array, Table 7.1.

Apply the standard rules of Routh's array dictating that the only term in s^1 row to be zero for Equation 7.18 to possess a pair of imaginary roots, i.e.,

$$\lambda_2\lambda_1 - \lambda_3\lambda_0 = 0, \quad \lambda_2 \neq 0. \tag{7.19}$$

Equation 7.19 yields a quadratic equation in T for a given chip width b, which results in at most two real roots for T. If these roots are real (call them T_1 and T_2) we proceed further. It is obvious that for those real T values, the imaginary characteristic roots of (7.18) will be

$$s_j = \omega_j i\sqrt{\frac{\lambda_0}{\lambda_2}}\Bigg|_j = \sqrt{\frac{\lambda_1}{\lambda_3}}\Bigg|_j, \quad \text{with } T = T_j, \quad j = 1, 2, \tag{7.20}$$

if and only if $\lambda_2\,\lambda_0 > 0$. These are the only two imaginary roots (representing the two chatter frequencies) that can exist for a given b value. Obviously, if T_1 and T_2 are complex conjugate numbers for a value of b, it implies that there is no possible imaginary root for (7.18) regardless of $\tau \in \Re^+$. In other words, this d-o-c causes no stability switching for any spindle speed $N > 0$.

Using Equation 7.8, we determine the τ values corresponding to these (T_1, ω_1) and (T_2, ω_2) pairs. There are infinitely many τ_{1j} and τ_{2j}, $j = 0, 1, 2, \ldots$, respectively. As per the definition of *kernel*, following (7.6), the smallest values of these τ_1 and τ_2 form the two-point kernel for this case. For instance, when $b = 0.005$ [m], these values are found as

$$\begin{aligned} T_1 &= -0.006142 & T_2 &= -0.0003032\\ \omega_1 &= 728.21\ \text{rad/s} & \omega_2 &= 636.33\ \text{rad/s}\\ \tau_1 &= 0.0049\ \text{s} & \tau_2 &= 0.0093\ \text{s}\end{aligned} \tag{7.21}$$

TABLE 7.1 The Routh's Array (b and T Arguments Are Suppressed)

S^3	λ_3	λ_1
S^2	λ_2	λ_0
S^1	$\dfrac{\lambda_2\lambda_1 - \lambda_0\lambda_3}{\lambda_2}$	
S^0	λ_0	

The first clustering is already achieved. For $b = 0.005$, the kernel consists of two discrete points (as opposed to a curve in two-delay cases, such as in Case II below) $\tau_1 = 0.0049$ s and $\tau_2 = 0.0093$ s, for which the characteristic equation has two imaginary roots. These are the *only* two imaginary roots that any $\tau \in \Re^+$ can ever produce for the given $b = 0.005$. Each τ_1 (or τ_2) resulting in ω_1 (or ω_2) also describes a set of countably infinite delay sets, called the "offspring" of the original τ_1 (or τ_2). They are given by the single-delay version of Equation 7.8 as

$$\tau_{1j} = \tau_1 + \frac{2\pi}{\omega_1} j \quad \text{with the cluster identifier} \quad \omega_1 = 728.21 \text{ rad/s}, \quad j = 1, 2, \ldots$$

$$\tau_{2j} = \tau_2 + \frac{2\pi}{\omega_2} j \quad \text{with the cluster identifier} \quad \omega_2 = 636.33 \text{ rad/s}, \quad j = 1, 2, \ldots$$

The second clustering feature (Proposition 7.2) is slightly more subtle. The root tendencies associated with the transitions of τ defined by

$$\tau_{1j} - \varepsilon \to \tau_1 \to \tau_{1j} + \varepsilon \quad \text{are all} \quad RT = +1 \text{ (i.e., destabilizing)}$$

$$\tau_{2j} - \varepsilon \to \tau_2 \to \tau_{2j} + \varepsilon \quad \text{are all} \quad RT = -1 \text{ (i.e., stabilizing)}$$

Let us examine this invariance property for $\{\tau_{1j}\}$ cluster, every element of which renders the same $s = \omega_1 i$ characteristic root. The differential form of (7.18) is

$$\begin{aligned} dCE(s,\tau,b) &= \frac{\partial CE(s,\tau,b)}{\partial s} ds + \frac{\partial CE(s,\tau,b)}{\partial \tau} d\tau \\ &= \left(2ms + c + bC\cos\beta\tau e^{-\tau s}\right) ds + bC\cos\beta s e^{-\tau s} d\tau = 0, \end{aligned} \tag{7.22}$$

which results in

$$\frac{ds}{d\tau} = -\frac{bC\cos\beta s e^{-\tau s}}{2ms + c + bC\cos\beta\tau e^{-\tau s}}. \tag{7.23}$$

It is easy to show two features in (7.23): (1) $e^{-\tau s}$ remains unchanged for $s = \omega_{1i}$ and $\tau = \tau_{1j}$, $j = 0, 1, 2, \ldots$, and (2) $RT_1 = \text{Re}(ds/d\tau)|_{s=\omega_1 i}$ is independent of τ despite the varying s term in the denominator, and this root tendency is +1 for all τ_{1j}, $j = 0, 1, 2, \ldots$. The invariance feature for (τ_2, ω_2) pair can be obtained in a similar way, which becomes stabilizing, i.e., $RT_2 = -1$. This demonstrates the second clustering feature as per Proposition 7.2.

We can now declare the stability posture of the system for a given b (say 0.005 m), and the deployment of CTCR is completed for this d-o-c (Figure 7.6). Notice the invariant RT from points $C_1, C_1', C_1'', \ldots$ (all destabilizing) and $C_2, C_2', C_2'', \ldots$ (all stabilizing) as τ increases. As a consequence, the number of unstable roots, NU, can be declared in each region very easily as sparingly shown on the figure. Obviously, when $NU = 0$, the cutting is stable; $0 < \tau < \tau_1$ stable, $\tau_1 < \tau < \tau_2$ unstable, $\tau_2 < \tau < \tau_1 + 2\pi/\omega_1$ stable, etc., stability switchings occur. We marked the stable intervals with thicker line style for ease of recognition. If we sweep $b \in [0 \ldots 30\,\text{mm}]$, we obtain the complete stability chart of Figure 7.6.

Stable cutting appears below the dark curve (also known as the chatter bound). The conventional *chatter stability lobes* (as the machine tools community calls them) in the rpm domain can readily be obtained using $\tau = 60/N$ coordinate conversion.

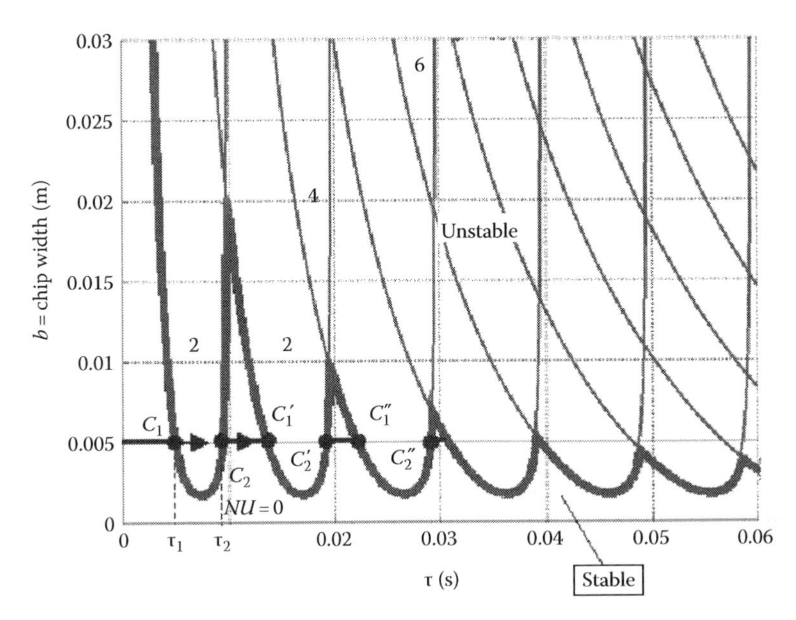

FIGURE 7.6 STM chatter stability chart in delay domain. (From Olgac, N. and Sipahi, R., *ASME J. Manuf. Sci. Eng.*, 127, 791, 2005. With permission.)

7.4.2 Case Study 2: Deployment of CTCR on Variable-Pitch Milling Cutters

We report the following experimental validation of CTCR on an equally interesting high-tech machining process: milling with variable-pitch cutters, following the scientific data provided in Refs. [33,35–37]. The practice of variable-pitch cutters originates from the desire of attenuating the regenerative chatter. Instead of four equidistant flutes located around the cutting tool (4 × 90° as described in Figure 7.7a), variable pitch is used (θ_1, θ_2 as shown in Figure 7.7b). Differently from the conventional uniform-pitch milling (Figure 7.7a), this cutter distributes the cutting energy to different frequencies and, therefore, it would attenuate the onset of undesirable chatter behavior until very high rates of metal removal are attempted. This is a very advantageous property from the increased efficiency standpoint. Many interesting variations of this clever idea have been developed and put to practice over the last 40 years. Due to the nonuniform pitch distribution in this system, multiple regenerative effects appear in the governing equation with multiple time delays, which are proportional to the respective pitch angles. The main problem reduces to the stability analysis of multiple-delay system as in (7.5).

We adopt the core problem from Ref. [33] as follows: A four-fluted uniform-pitch cutter is used in milling Al356 alloy workpiece. The cutter has 19.05 mm diameter, 30° helix, and 10° rake angles. Stability chart indicates that this milling process is unstable for axial d-o-c a = 5 mm and spindle speed N = 5000 rpm. A natural question follows: Which pitch angles should be selected for the best chatter stability

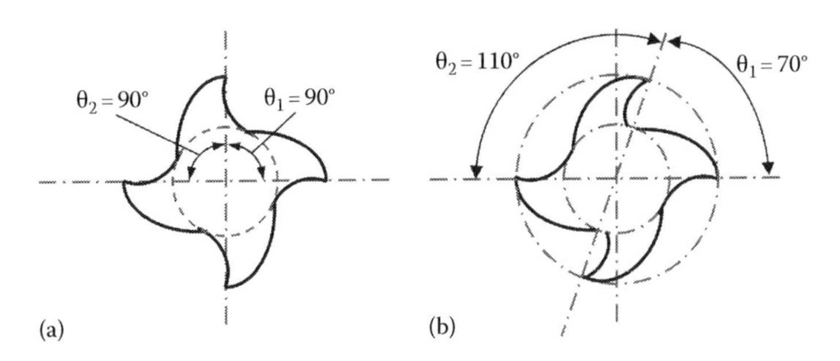

FIGURE 7.7 Cross section of the end mill. (a) uniform pitch and (b) variable pitch.

margins when variable-pitch milling is considered? [34] further develops this case study exactly, except it frees the pitch ratio, therefore introducing a truly multiple time-delayed dynamics. In other words, the selection of the two "pitch angles" is completely relaxed. The steps of this exercise and the numerical results are explained next.

The system characteristic equation is taken as [33, Equation 15]:

$$\det\left[\mathbf{I}-\frac{1}{4\pi}K_t a\left(4-2\left(e^{-\tau_1 s}+e^{-\tau_2 s}\right)\right)\Phi_0(s)\right]=0 \tag{7.24}$$

with $K_t = 697\,\text{MPa}$, a = axial d-o-c [m], $\tau_1[\text{s}]=(\theta_1[\text{deg}]/360)(60/N[\text{RPM}])$ is the delay occurring due to the pitch angle θ_1, $\tau_2=(\theta_2/\theta_1)\,\tau_1$ is the second delay due to θ_2, N is the spindle speed, and s is the Laplace variable. The matrix Φ_0 containing the transfer functions and the mean cutting directions is defined as

$$\Phi_0=\begin{pmatrix}\phi_{xx}\alpha_{xx}+\phi_{yx}\alpha_{xy} & \phi_{xy}\alpha_{xx}+\phi_{yy}\alpha_{xy}\\ \phi_{xx}\alpha_{yx}+\phi_{yx}\alpha_{yy} & \phi_{xy}\alpha_{yx}+\phi_{yy}\alpha_{yy}\end{pmatrix}, \tag{7.25}$$

where

$$\alpha_{xx}=0.423,\qquad \alpha_{xy}=-1.203$$
$$\alpha_{yx}=1.937,\qquad \alpha_{yy}=-1.576$$

are the fundamental Fourier components of the periodically varying directional coefficient matrix. The transfer functions in (7.25) are populated from [33, Table 1], as

$$\phi_{xx}=\frac{0.08989}{s^2+159.4s+0.77\times10^7}+\frac{0.6673}{s^2+395.2s+0.1254\times10^8}+\frac{0.07655}{s^2+577.2s+0.2393\times10^8}$$

$$\phi_{xy}=\phi_{yx}=0$$

$$\phi_{yy}=\frac{0.834}{s^2+162.2s+0.1052\times10^8}$$

Inclusion of these expressions and parameters in the characteristic equation and expanding it into a scalar expression, we reach the starting point of the CTCR paradigm. For $a = 4\,\text{mm}$ d-o-c, the characteristic equation of this dynamics is

$$\begin{aligned}CE(s,\tau_1,\tau_2)={}&625\,s^8+808{,}750\,s^7+3{,}5068.83\times10^6 s^6+3{,}1376.027\times10^9 s^5+0.68610^{18}s^4\\&+0.385\times10^{21}s^3+0.565\times10^{25}s^2+0.150\times10^{28}s+0.166\times10^{32}\\&+(-0.266\times10^9 s^6-0.324\times10^{12}s^5-0.127\times10^{17}s^4-0.924\times10^{19}s^3-0.18\times10^{24}s^2\\&-0.572\times10^{26}s-0.756\times10^{30})(e^{-\tau_1 s}+e^{-\tau_2 s})+(284{,}943.13\times10^9 s^4+0.212\times10^{18}s^3\\&+0.889\times10^{22}s^2+0.253\times10^{25}s+0.538\times10^{29})e^{-(\tau_1+\tau_2)s}+(14247.16\times10^{10}s^4\\&+0.106\times10^{18}s^3+0.445\times10^{22}s^2+0.126\times10^{25}s+0.269\times10^{29})(e^{-2\tau_1 s}+e^{-2\tau_2 s})=0\end{aligned} \tag{7.26}$$

Parametric form of (7.26), that is, $CE(s,\tau_1,\tau_2,a)$ expression, is prohibitive to display due to space limitations, thus the substitution of $a = 4\,\text{mm}$. Please note that all the numerical values above are given in their truncated form to conserve space.

Notice the critical nuance between Equations 7.4 and 7.14. The former represents a truly two-spindle, two-cutter setting, while 7.14 is for single-spindle turning process or a milling process with uniformly distributed flutes in a milling cutter. Mathematical expression for the characteristic equation becomes two-time delayed quasi-polynomial of 7.26 which is, in fact, much more complex than 7.16 due to the commensurate delay formation (i.e., $e^{-2\tau_1 s}$, $e^{-2\tau_2 s}$ terms are present). The stability assessment of this equation is a formidable task. We put CTCR to test and present the results below.

The CTCR takes over from Equation 7.26 and creates the complete stability outlook in (τ_1, τ_2) space as in Figure 7.8a. The *kernel* is marked thicker to discriminate it from its *offspring*. The grids of D_0 (kernel) and D_1, D_2, D_3, ... (offspring sets) are marked for ease of observation. Notice the equidistant grid size $2\pi/\omega$ as per Equation 7.6, $\overline{D_0D_1} = \overline{D_1D_2} = \overline{D_2D_3} = \overline{D_3D_4}$, etc. The (τ_1, τ_2) delays at all of these sibling points impart the same ωi imaginary root for (7.26). Figure 7.8b shows the possible chatter frequencies of this system for all $(\tau_1, \tau_2) \in \Re^+$ for varying pitch ratios $(0, \infty)$ whether they are operationally feasible or not. It is clear that this system can exhibit only a restricted set of imaginary roots (from Figure 7.8b, 3250 rad/s [517 Hz] $< \omega <$ 3616 rad/s [575 Hz]). All of these *chatter frequencies* are created by the kernel and no (τ_1, τ_2) point on the offspring can cause an additional chatter frequency outside the given set.

Figure 7.8a shows the stable (shaded) and unstable regions in (τ_1,τ_2) space for a given axial d-o-c (a = 4 mm). It contains some further information as well. Each point (τ_1,τ_2) in this figure represents a spindle speed. It is obvious that, considering the relation $\tau_1 + \tau_2 = 30/N$, all the constant spindle speed lines are with slope −1, as annotated on the figure. The constant pitch ratio lines pass through the origin, where pitch ratio = τ_2/τ_1.

Notice that the most desirable pitch ratios are close to $\tau_1/\tau_2 = 1$ for effective chip removal purposes. Therefore, very high or very low pitch ratios are not desirable, and $55° < \theta_1 < 90°$ as declared in [33]. The pitch ratios between $\tau_2/\tau_1 \in [1.374, 2.618]$ offer stable operation (marked as points A and B on Figure 7.8a) for 5000 rpm. The corresponding pitch angles are ($\theta_{1A} = 75.8°$, $\theta_{2A} = 104.2°$) and ($\theta_{1B} = 49.8°$, $\theta_{2B} = 130.2°$). They coincide precisely with the results declared in [33, Figure 3]. Figure 7.8a provides a very powerful tool in the hands of the manufacturing engineer. One can select uniform-pitch cutter and 7500 rpm speed (point O_2) as opposed to variable-pitch cutter (pitch ratio 11/7) and 5000 rpm (point O_1), increasing the metal removal rate by 50%. This figure is for a constant d-o-c, a. A 3D stability plot can be produced scanning the values of a in, for instance, a = 1...6 mm, $\tau_1 \in \Re^+$, $\tau_2 \in \Re^+$ domain. A cross section of this 3D plot with constant-N planes is comparable with [33, Figure 3].

We next look at the chatter stability variations for two different settings (a) Uniform pitch cutters ($\theta_1 = \theta_2 = 90°$) where the pitch ratio is unity, (b) Variable pitch cutters ($\theta_1 = 70°$, $\theta_2 = 110°$), where the pitch ratio is 7/11, both with Al356 workpiece. It is clear from Figure 7.8a that the system goes in and out of stable behavior as the spindle speed decreases (i.e., $\tau_1 + \tau_2$ increases). User could select the optimum operating condition within many available regions (such as between the points of O_1 and O_2, as discussed earlier). This optimization procedure is presented in Section 7.5.

7.4.3 Case Study 3: Six-Flute Milling Cutter Design

The core problem in this case study is stated as follows. We wish to design a six-flute milling cutter. Given a fixed tool diameter and tool holding structure, for a desired axial d-o-c, a, find which pitch angles and spindle speeds impart stable cutting. We refer to Figure 7.9 for the assumed distribution of the pitch angles θ_1, $\theta_1 + \varphi$, $\theta_1 + 2\varphi$. The fixed increment φ separates the three pitch angles. The corresponding tooth passage times are given as $\tau_1 = \theta_1/\Omega$, $\tau_2 = \tau_1 + h$, $\tau_3 = \tau_1 + 2h$, where $h = \varphi/\Omega$ with Ω [rad/s] being the spindle speed. The characteristic equation is taken from [36, Equation 17], where the preliminary machining parameters and dynamic characteristics are also described. For a 2 mm d-o-c, it is given by

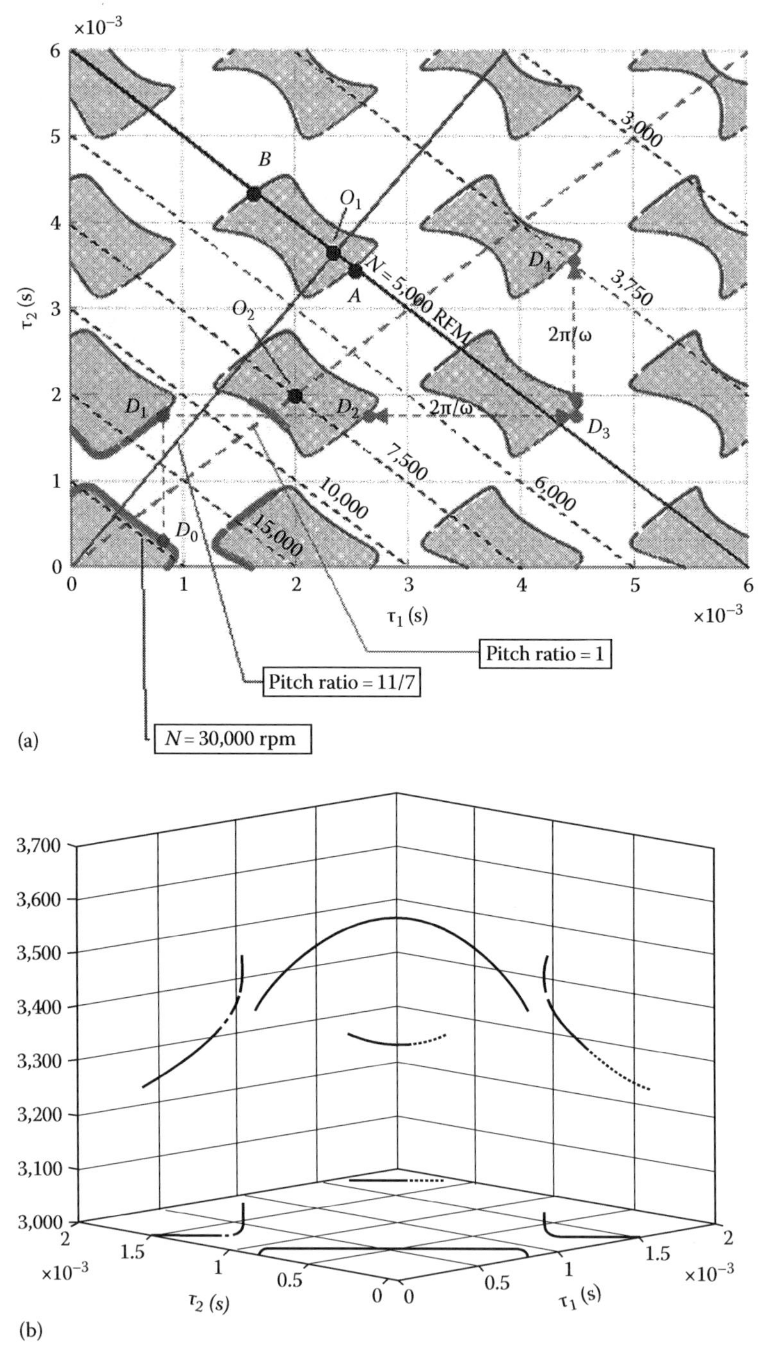

FIGURE 7.8 (a) Stability regions for 4 mm axial d-o-c, with four-flute cutter milling on Al 356 (kernel, thick; offspring, thin; stable zones, shaded). (From Olgac, N. and Sipahi, R., *ASME J. Manuf. Sci. Eng.*, 127, 791, 2005. With permission.) (b) Chatter frequencies (exhaustive) [rad/s] vs. the kernel (τ_1, τ_2). (From Olgac, N. and Sipahi, R., *ASME J. Manuf. Sci. Eng.*, 127, 791, 2005. With permission.)

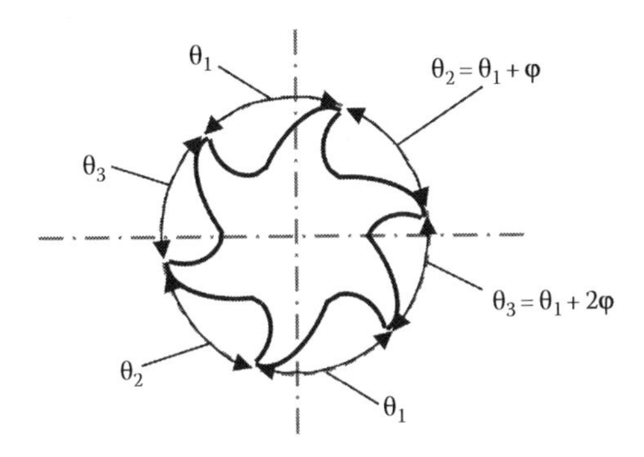

FIGURE 7.9 Cross section of variable-pitch six-flute end mill cutter.

$$\begin{aligned}CE(s,\tau_1,h) &= 125\,s^6 + 89{,}600\,s^5 + 0.3947\times10^{10}s^4 + 0.1823\times10^{13}s^3 + 0.4065\times10^{17}s^2\\ &+ 0.9062\times10^{19}\,s + 0.1357\times10^{24} + (-0.2757\times10^{8}s^4 - 0.1711\times10^{11}s^3 - 0.6117\times10^{15}s^2\\ &- 0.1616\times10^{18}s - 0.3068\times10^{22})e^{-\tau_1 s} + (0.647\times10^{13}s^2 + 0.1212\times10^{16}s + 0.5353\times10^{20})e^{-2\tau_1 s}\\ &+ (-27{,}569{,}203.11\,s^4 - 0.1711\times10^{11}s^3 - 0.6117\times10^{15}s^2 - 0.1616\times10^{18}s\\ &- 0.3068\times10^{22})(e^{-(\tau_1+h)s} + e^{-(\tau_1+2h)s}) + (0.1294\times10^{14}s^2 + 0.2425\times10^{16}s\\ &+ 0.1071\times10^{21})(e^{-(2\tau_1+h)s} + e^{-(2\tau_1+3h)s}) + (0.1941\times10^{14}s^2 + 0.3637\times10^{16}s\\ &+ 0.1606\times10^{21})e^{-2(\tau_1+h)s} + (0.647\times10^{13}s^2 + 0.1212\times10^{16}s + 0.5353\times10^{20})e^{-(2\tau_1+4h)s} = 0.\end{aligned} \tag{7.27}$$

One can see the fundamental similarity between (7.5) and (7.27). Main difficulties in resolving the stable regions arise in several fronts: (1) The rationally independent delays τ_1 and τ_2, both appear in cross-talking forms (e.g., $e^{-(\tau_1+\tau_2)s}$), (2) Their commensurate forms also appear (e.g., $e^{-2\tau_1 s}$ and $e^{-2\tau_2 s}$), (3) the coefficients of the equation are badly ill-conditioned, in that they exhibit a difference of about 25 orders of magnitude.

The primary tool we utilize is, again, the stability assessment method, CTCR. It results in the stability tableaus as shown in Figures 7.10 and 7.11. Stable operating regions in these figures are shaded. The constant spindle speed lines are marked using the relation $\tau_1 + h = 10/N$, N [rpm]. Notice the kernel curve, which is shown in thick lines, and the offspring curves in thin lines. To facilitate the practical usage of Figure 7.10, we marked the spindle speeds $N = 625 \ldots 5000$ rpm. For instance, if the machinist wishes to operate at 3000 rpm with $a = 2$ mm axial d-o-c without changing the metal removal rate (i.e., very critical), an unstable (i.e., chatter prone) uniform-pitch operation at $\tau_1 = 3.33$ ms, $h = 0$, can easily be replaced by stable (i.e., chatter rejecting) operation at $\tau_1 = 20$ ms and $h = 1.33$ ms (variable-pitch format). This is a considerable flexibility in the selection process of the user.

Note that the stability pictures in Figures 7.10 and 7.11 can be efficiently produced for a variety of d-o-c selections, and similar design discussion can be performed. Just to give an idea to the reader, we wish to present some measures of computational complexity. Figures 7.10 and 7.11 take about 40 s CPU time on a computer with 3.2 GHz Intel Centrino chip and 512 MB RAM. Using this mechanism, one can select appropriate cutting tool geometry (i.e., the pitch variations) and determine the chatter-free maximal d-o-c for increased productivity. According to the discussions in this section, it is obvious that Figures 7.10 and 7.11 offer a unique representation for the ultimate goal at hand, and they can be achieved using the stability analysis technique, CTCR.

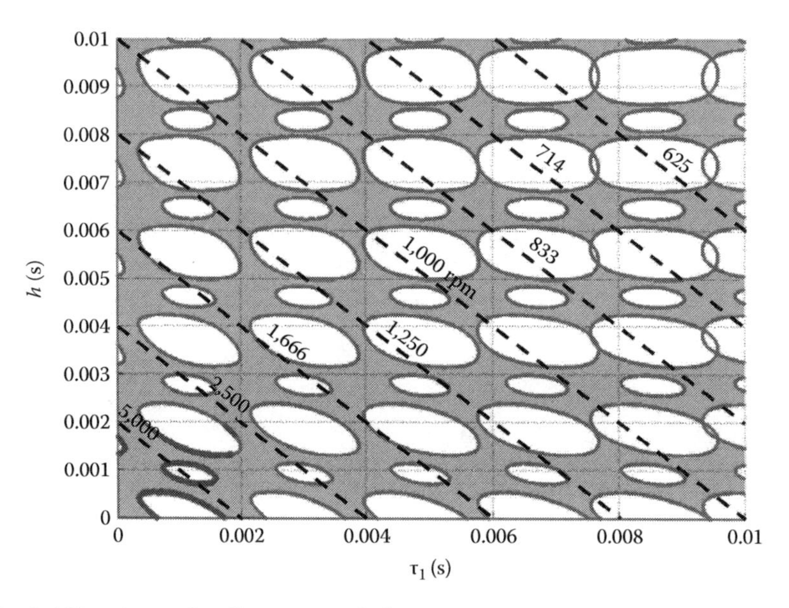

FIGURE 7.10 Stability picture of six-flute cutter with d-o-c = 2 mm.

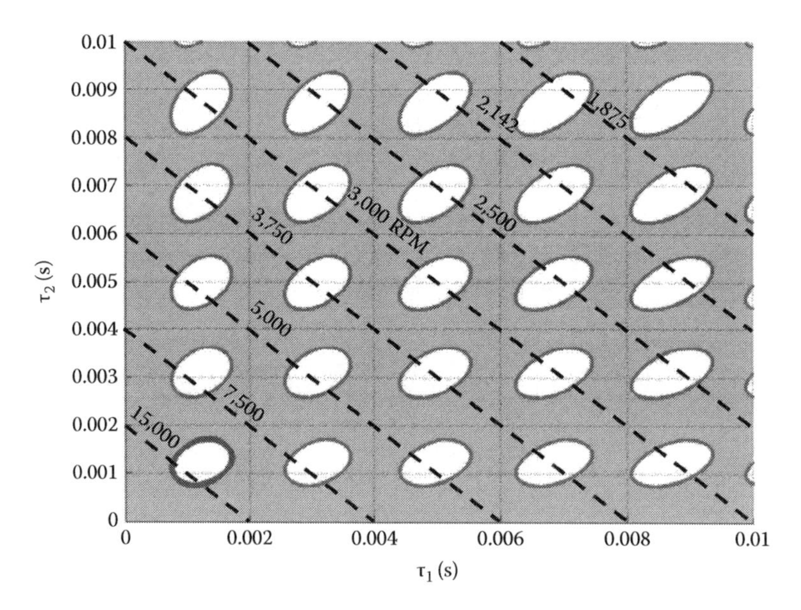

FIGURE 7.11 Stability picture of four-flute cutter with d-o-c = 2 mm.

7.5 Optimization of the Process

In the shaded regions of Figure 7.8a, the chatter has stable nature. Thus, these regions are preferred for operation. We then search for a measure of chatter rejection within these stable zones. It is well known that this property is represented by the real part of the dominant characteristic root of the system characteristic equation. The smaller this real part, the faster the chatter rejection (faster settling after disturbances) is.

7.5.1 Optimization Problem

Two conflicting objectives in metal cutting are

1. *Increased metal removal rate*, which is directly proportional to axial d-o-c, a, and spindle speed $N[\text{rpm}] = 30/(\tau_1 + \tau_2)$, as in the Case Study II above.
2. Generating desirably chatter-free *surface quality* on the workpiece.

In order to serve for both objectives, we introduce the following objective function:

$$J = -\alpha\left\{\Re e\left(s_{dom}\right)\right\} + \beta\left\{\frac{a}{\tau_1 + \tau_2}\right\}, \tag{7.28}$$

in which α and β are constant positive weights that can be selected based on the importance given on the conflicting items.

This is a three-dimensional optimization problem, in which τ_1, τ_2, and a are independent parameters, and $|\Re e(s_{dom})|$ is a measure of chatter rejection, as explained earlier. CTCR paradigm helps us obtain the stability boundaries of this system in the time-delay domain, but obtaining the real part of the rightmost characteristic roots requires that we look inside each stability region. Finding the real part of rightmost characteristic root of a quasi-polynomial is still an open problem in mathematics. One of the few numerical schemes available in the literature to "approximate" the rightmost (dominant) roots for different values of time delays is given in [42] that will be used for this optimization problem. It is clear that the variation of τ_1, τ_2 as well as axial d-o-c, a, is continuous. Figure 7.12 displays an example case study of four-flute milling cutter, taken from [37]. For that figure, a grid-point evaluation of the objective function is performed, which results in the respective chatter rejection abilities. One can seek the most desirable points in the stable operating regions. At any point (τ_1, τ_2), we can characterize the tool geometry (i.e., the pitch angles θ_1 and θ_2) as well as the spindle speed $N = 30/(\tau_1 + \tau_2)$.

Notice that the pitch ratios $\tau_1/\tau_2 = \theta_1/\theta_2$ are varied between [0.5…1] due to physical constraints. Too small pitch angles cause practical difficulty in extracting the metal chips after cutting. The best

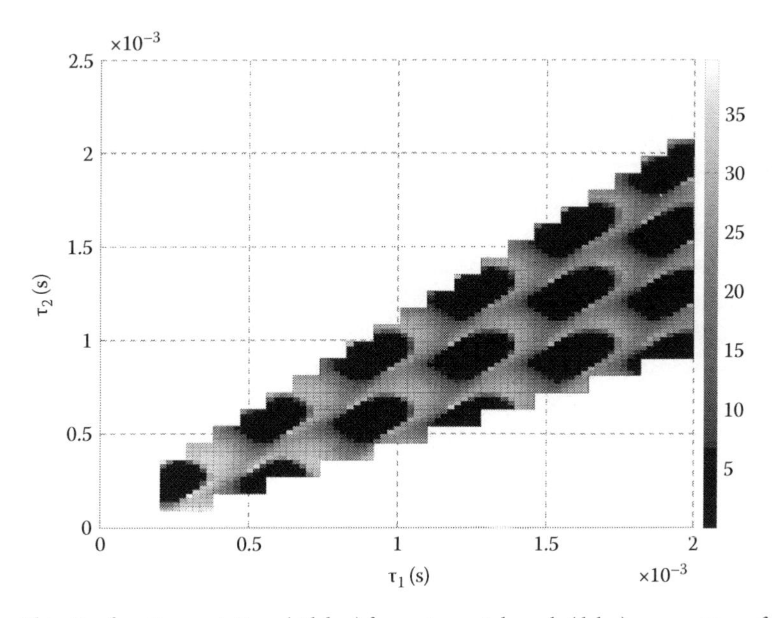

FIGURE 7.12 Objective function variations (sidebar) for various pitch angle (delay) compositions, for 5 mm d-o-c.

TABLE 7.2 Description of the Optimum Points for Different d-o-c

a (mm)	J_{max} ($\times 10^3$)	$\lvert\Re e(s_{dom})\rvert$ (1/s)	τ_1 (ms)	τ_2 (ms)	θ_1 (deg)	θ_2 (deg)	N (rpm)
3	36.6088	97.1846	0.225	0.180	99.5	80.5	74,534
3.5	36.8306	117.9176	1.50	0.90	112.5	67.5	12,500
4	40.2018	100.8796	0.245	0.1575	109.6	70.4	74,534
4.5	38.9324	94.4836	0.2675	0.1575	113.3	66.7	70,588
5	38.6183	91.4836	0.290	0.1575	116.6	63.4	67,039
5.5	39.4497	94.0572	0.20	0.3375	67	113	55,814

Source: Olgac, N. and Sipahi, R., *ASME J. Manuf. Sci. Eng.*, 127, 791, 2005. With permission.

operating point in each stable region is numerically determined for a given d-o-c, a. We then create the matrix of objective function values at various combinations or grid points of (τ_1, τ_2) and collect the corresponding best operating points (maxima of the objective function, J_{max}) for each d-o-c on Table 7.2. Highlighted columns show the best cutter design (θ_1 and θ_2) and the most desirable spindle speed N for different depths-of-cut. The numerial values of these points are shown in Table 7.2 along with the corresponding values of the objective function J_{max}. These maximum values of J represent the best selections of operating conditions that result in maximum metal removal rate, while maximizing the chatter rejection capability in parallel. Accordingly, we can conclude that the best milling cutter should be designed with $\theta_1 = 109.6°$, $\theta_2 = 70.4°$, so that it accommodates the highest productivity. Obviously, this comparison is only valid under the given objective function and its weighthing factors (which are taken as $\alpha = 0.3$ and $\beta = 1$). So, for this particular operation, when performed at the setting of $a = 4$ mm, $N = 74{,}534$ rpm with $\theta_1 = 109.6°$, $\theta_2 = 70.4°$ pitch angles on the cutter, one could produce 8122.1 mm^3/s material removal with desirable surface quality (considering 0.05 mm/tooth feed rate and approximately 6 HP of spindle power).

We wish to draw the attention of the reader to the point that Figure 7.12 is symmetric with respect to τ_1 and τ_2. Departing from a geometry-based intuition, we make a qualified statement that τ_1 and τ_2(θ_1 and θ_2, in essence) are commutable in this process without changing the results, and the mathematical expressions validate this statement.

7.6 Conclusion

The chapter treats the machine tool chatter from an analytical perspective to bridge the mathematical novelties with the physics of the operation. An ideally suiting mathematical procedure, CTCR, is presented for determining the complete stability posture of the regenerative chatter dynamics in SM. The novelty appears in the exhaustive declaration of the stability regions and the complete set of chatter frequencies for the given process. The outcome of CTCR forms a very useful tool to select the operating conditions, such as d-o-c and spindle speed, in SM applications. We also present an example case study addressing the same selection process on a simpler problem: the conventional STM chatter. Another case study is for the chatter dynamics of currently practiced "variable-pitch milling" operation. In this particular study, we deploy CTCR from the design of the milling cutter geometry to the spindle speed and d-o-c selections and further to the optimization of these decisions.

Acknowledgments

This research has been supported in part by the awards from DoE (DE-FG02-04ER25656), NSF CMS-0439980, NSF DMI 0522910, and NSF ECCS 0901442.

References

1. I. Lazoglu, M. Vogler, S. G. Kapoor, and R. E. DeVor, Dynamics of the simultaneous turning process, *Transactions of North American Manufacturing Research Institute of SME*, 26, 135–140, 1998.
2. C. Carland, Use of multi-spindle turning automatics, *International, Industrial and Production Engineering Conference (IPE)*, Toronto, Canada, pp. 64–68, 1983.
3. C. Carland, 2 Spindle are Better than 1, *Cutting Tool Engineering*, 49(2), 186–192, 1997.
4. A. Allcock, Turning towards fixed-head 'Multis', *Machinery and Production Engineering*, 153(3906), 47–54, 1995.
5. C. Koepfer, Can this new CNC multispindle work for your shop? *Modern Machine Shop*, 69, 78–85, 1996.
6. J. Tlusty, *Machine Dynamics, Handbook of High Speed Machining Technology*, Chapman and Hall, New York, 1985.
7. M. E. Merchant, Basic mechanics of the metal-cutting process, *ASME Journal of Applied Mechanics*, 66, A-168, 1944.
8. S. Doi and S. Kato, Chatter vibration of lathe tools, *Transactions of ASME*, 78, 1127, 1956.
9. S. A. Tobias, *Machine Tool Vibration*, Wiley, New York, 1961.
10. J. Tlusty, L. Spacek, M. Polacek, and O. Danek, *Selbsterregte Schwingungen an Werkzeugmaschinen*, VEB Verlag Technik, Berlin, Germany, 1962.
11. H. E. Merritt, Theory of self excited machine tool chatter, *Journal of Engineering for Industry*, 87, 447, 1965.
12. R. L. Kegg, Cutting dynamics in machine tool chatter, *Journal of Engineering for Industry*, 87, 464, 1965.
13. Y. Altintas and E. Budak, Analytical prediction of stability lobes in milling, *Annals of the CIRP*, 44, 357–362, 1995.
14. S. Smith and J. Tlusty, Efficient simulation program for chatter in milling, *Annals of the CIRP*, 42, 463–466, 1993.
15. S. Smith and J. Tlusty, Update on high speed milling dynamics, *Transactions of the ASME, Journal of Engineering for Industry*, 112, 142–149, 1990.
16. B. C. Kuo, *Automatic Control Systems*, Prentice Hall, Englewood Cliffs, NJ, 1987.
17. C. L. Phillips and R. D. Harbor, *Feedback Control Systems*, Prentice-Hall, Englewood Cliffs, NJ, 1988.
18. C. S. Hsu, Application of the tau-decomposition method to dynamical systems subjected to retarded follower forces, *ASME Journal of Applied Mechanics*, 37, 258–266, 1970.
19. C. S. Hsu and K. L. Bhatt, Stability charts for second-order dynamical systems with time lag, *ASME Journal of Applied Mechanics*, 33, 119–124, 1966.
20. J. Chen, G. Gu, and C. N. Nett, A new method for computing delay margins for stability of linear delay systems, *Systems & Control Letters*, 26, 107–117, 1995.
21. K. L. Cooke and P. van den Driessche, On zeros of some transcendental equations, *Funkcialaj Ekvacioj*, 29, 77–90, 1986.
22. Z. V. Rekasius, A stability test for systems with delays, *Proceedings of the Joint Automatic Control Conference*, San Francisco, CA, Paper No. TP9-A, 1980.
23. R. Sipahi and N. Olgac, Active vibration suppression with time delayed feedback, *ASME Journal of Vibration and Acoustics*, 125(3), 384–388, 2003. Best Student Paper Award at ASME-IMECE, New Orleans, LA, 2002.
24. R. Sipahi and N. Olgac, Degenerate cases in using direct method, *Transaction of ASME, Journal of Dynamic Systems, Measurement, and Control*, 125, 194–201, 2003.
25. R. Sipahi and N. Olgac, Active vibration suppression with time delayed feedback, *ASME Journal of Vibration and Acoustics*, 125, 384–388, 2003.

26. N. Olgac and R. Sipahi, An exact method for the stability analysis of time delayed LTI systems, *IEEE Transactions on Automatic Control*, 47, 793–797, 2002.
27. N. Olgac and R. Sipahi, A practical method for analyzing the stability of neutral type LTI-time delayed systems, *Automatica*, 40, 847–853, 2004.
28. J. K. Hale and W. Huang, Global geometry of the stable regions for two delay differential equations, *Journal of Mathematical Analysis and Applications*, 178, 344–362, 1993.
29. S.-I. Niculescu, On delay robustness analysis of a simple control algorithm in high-speed networks, *Automatica*, 38, 885–889, 2002.
30. G. Stepan, *Retarded Dynamical Systems: Stability and Characteristic Function*, Longman Scientific & Technical, New York, co-publisher John Wiley & Sons Inc., New York, 1989.
31. N. MacDonald, An interference effect of independent delays, *IEE Proceedings*, 134, 38–42, 1987.
32. R. Sipahi and N. Olgac, A unique methodology for the stability robustness of multiple time delay systems, *Systems & Control Letters*, 55(10), 819–825, 2006.
33. Y. Altintas, S. Engin, and E. Budak, Analytical stability prediction and design of variable pitch cutters, *ASME Journal of Manufacturing Science and Engineering*, 121, 173–178, 1999.
34. N. Olgac and M. Hosek, A new perspective and analysis for regenerative machine tool chatter, *International Journal of Machine Tools & Manufacture*, 38, 783–798, 1998.
35. N. Olgac and R. Sipahi, A unique methodology for chatter stability mapping in simultaneous machining, *ASME Journal of Manufacturing Science and Engineering*, 127, 791–800, 2005.
36. N. Olgac and R. Sipahi, Dynamics and stability of multi-flute variable-pitch milling, *Journal of Vibration and Control*, 13(7), 1031–1043, 2007.
37. H. Fazelinia and N. Olgac, New perspective in process optimization of variable pitch milling, *International Journal of Materials and Product Technology*, 35(1/2), 47–63, 2009.
38. D. Filipovic and N. Olgac, Delayed resonator with speed feedback, design and performance analysis, *Mechatronics*, 12(3), 393–413, 2002.
39. A. F. Ergenc, N. Olgac, and H. Fazelinia, Extended kronecker summation for cluster treatment of LTI systems with multiple delays, *SIAM Journal on Control and Optimization*, 46(1), 143–155, 2007.
40. H. Fazelinia, R. Sipahi, and N. Olgac, Stability analysis of multiple time delayed systems using 'Building Block' concept, *IEEE Transactions on Automatic Control*, 52(5), 799–810, 2007.
41. R. Sipahi and I. I. Delice, Extraction of 3D stability switching hypersurfaces of a time delay system with multiple fixed delays, *Automatica*, 45, 1449–1454, 2009.
42. D. Breda, S. Maset, and R. Vermiglio, Pseudospectral differencing methods for characteristic roots of delay differential equations, *SIAM Journal on Scientific Computing*, 27, 482–495, 2005.
43. R. Sipahi and N. Olgac, Stability robustness of retarded LTI systems with single delay and exhaustive determination of their imaginary spectra, *SIAM Journal on Control and Optimization*, 45(5), 1680–1696, 2006.

II

Control System Design

8

Internal Model Control

James C. Hung
The University of Tennessee, Knoxville

Internal model control (IMC) is an attractive method for designing a control system if the plant is inherently stable. The method was first formally reported in Garcia and Morari (1982). Since then, additional research results on this subject have been reported in journals and conferences. The method has been formulated using Laplace transform, and therefore is a frequency-domain technique. The concept is summarized here for a single-input–single-output (SISO) system.

8.1 Basic IMC Structures

A single-degree-of-freedom (SDF) IMC structure is shown in Figure 8.1. Notice that the plant model, G_{pm}, is explicitly included in the overall controller. When the actual plant, G_{p}, matches the plant model G_{pm}, there is no feedback signal and the control is open loop.

$$Y(s) = G_{\text{a}}G_{\text{p}}R(s) + (1 - G_{\text{a}}G_{\text{p}})D(s)$$

The design of a cascade compensator G_{a} for a specified response to input r is simple. Clearly, the stability of both the plant and the compensator is necessary and sufficient for the system to be stable. The system is a feedback system for disturbance d, thus attenuating its effect.

It should be pointed out that, corresponding to each IMC controller G_{a} of Figure 8.1, an equivalent controller G_{c} of a conventional control structure, shown in Figure 8.2, exists. In fact,

$$G_{\text{c}} = \frac{G_{\text{a}}}{1 - G_{\text{a}}G_{\text{pm}}}.$$

Independent reference input response and disturbance rejection can be achieved by adopting a two-degree-of-freedom (TDF) IMC structure, as shown in Figure 8.3. In this structure, the controller consists of the plant model G_{pm} and two compensators G_{a} and G_{b}. When $G_{\text{p}} = G_{\text{pm}}$, the system responses to reference input r and to disturbance d are given, respectively, by

$$Y_r(s) = G_{\text{a}}G_{\text{p}}R(s)$$

and

$$Y_{\text{d}}(s) = (1 - G_{\text{a}}G_{\text{b}}G_{\text{p}})D(s),$$

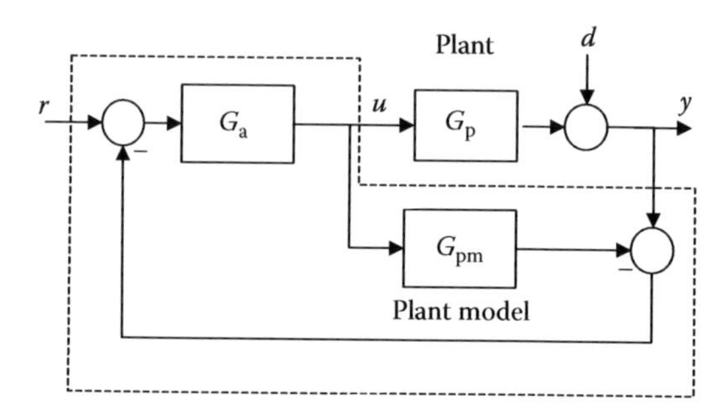

FIGURE 8.1 SDF internal model control.

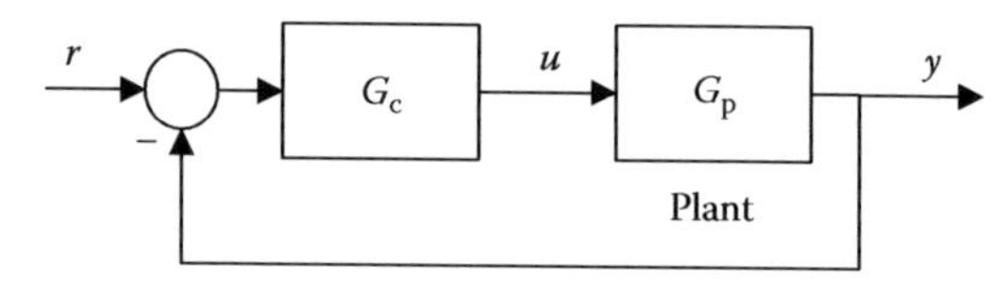

FIGURE 8.2 Conventional control.

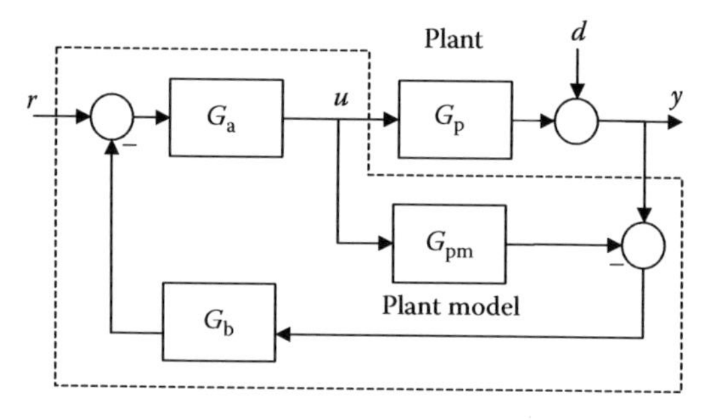

FIGURE 8.3 TDF internal mode control.

where compensators G_a and G_b can be adjusted independently to achieve the desired individual responses. Notice that these responses are linear in G_a and G_b, making compensator design easy.

When there are differences between the plant and the model, a feedback signal exists and can be used for achieving system robustness.

Under this condition,

$$Y_r(s) = \frac{G_a G_p R(s)}{Q}$$

and

$$Y_d(s) = \frac{(1 - G_a G_b G_p) D(s)}{Q}$$

where

$$Q = 1 + G_a G_b (G_p - G_{pm})$$

Q is the characteristic polynomial of the system. The conditions for stability are as follows: all roots of the characteristic equation are on the left-half plane and there is no pole zero cancellation among G_a, G_b, and $(G_p - G_{pm})$ on the right-half plane. Robust stability is achieved by an appropriate choice of G_b.

8.2 IMC Design

Two-degree-of-freedom IMC design usually begins with the design of the forward compensator G_a for a desired input–output response. Then the feedback compensator is designed to achieve a specified system stiffness with respect to disturbances and a specified robustness with respect to model error.

8.3 Discussion

In the IMC approach, the controller is easy to design and the IMC structure provides an easy way to achieve system robustness. For the MIMO case and for an in-depth treatment of IMC, readers are referred to Morari and Zafiriou (1989) and the reference list therein.

One limitation of the IMC method is that it cannot handle plants that are open-loop unstable. As mentioned before, a stable plant using a stable compensator results in a stable IMC system. An equivalent controller can be determined for the conventional control configuration giving the same response characteristics. On the other hand, an unstable plant may be stabilized by a controller using the conventional control configuration; however, the corresponding IMC control system is not stable. As an example, consider a system in the conventional control configuration. The system consists of an unstable plant

$$G_p = G_{pm} = \frac{1}{(s-1)(S+5)}$$

and a cascade controller

$$G_c = 18$$

The closed-loop system is stable having a transfer function

$$G_{CL} = \frac{18}{s^2 + 4s + 13}$$

The equivalent single-degree-of-freedom compensator G_a of the IMC is given by

$$G_a = \frac{G_c}{1 + G_c G_{pm}} = \frac{18(s-1)(s+5)}{s^2 + 4s + 13}$$

but it does not help to stabilize the system. Any noise entering the system at the plant input will drive the plant output to infinity. Here one sees that the stability of a system is, in general, implementation dependent.

References

Garcia, C. E. and Morari, M. 1982. Internal model control. 1. A unifying review and some new results, *Ind. Eng. Chem. Process Des. Dev.*, 21(2):308–323.

Morari, M. and Zafiriou, E. 1989. *Robust Process Control*, Prentice-Hall, Englewood Cliffs, NJ.

9

Dynamic Matrix Control

James C. Hung
The University of Tennessee, Knoxville

Dynamic matrix control (DMC) is a method suitable for the control "of processes." The method was developed by control professionals in the oil industry in the 1970s. It is suitable for control by a computer and, since the concept is intuitively transparent, is attractive to technical people with a limited background in control theory. It is intended for linear processes, including both single-input-single-output (SISO) and multi-input–multi-output (MIMO) cases. The method provides a continuous projection of a system's future output for the time horizon required for the system to reach a steady state. The projected outputs are based on all past changes in the measured input variables. The control effort is then computed to alter the projected output to satisfy a chosen performance specification. The method is intended for the control of a linear process which is open-loop stable and does not have pure integration. Cutler (1982) gives a very lucid description of the concept. The method is based on the step response characteristics of a process, and therefore, is a time domain technique. The concept is summarized here for a SISO system.

9.1 Dynamic Matrix

A key tool of the DMC is the *dynamic matrix* which can be constructed from a process' unit-step response data. Consider a SISO discrete-data process. Its output at the end of the kth time interval is given by the following convolution summation

$$y_k = \sum_{i=1}^{k} h_{k-i+1}\delta u_i$$

where

h_i is the discrete-data step response
δu_i is the incremental step input at the beginning of the ith time interval

For $k = 1$ to n, the following *dynamic matrix equation* can be formed:

$$\begin{bmatrix} y_1 \\ y_2 \\ \vdots \\ \vdots \\ y_n \end{bmatrix} = \begin{bmatrix} h_1 & 0 & 0 & \cdots & 0 \\ h_2 & h_1 & 0 & \cdots & 0 \\ h_2 & h_2 & h_1 & \cdots & 0 \\ \vdots & \vdots & \vdots & \ddots & \vdots \\ h_n & h_{n-1} & \cdots & \cdots & h_{n-m+1} \end{bmatrix} \begin{bmatrix} \delta u_1 \\ \delta u_2 \\ \vdots \\ \delta u_m \end{bmatrix}$$

or

$$y = H\delta u$$

The $n \times m$ H matrix shown is the so-called *dynamic matrix* of the SISO process and its elements are called dynamic coefficients. This matrix is used for projecting the future output. The vector at the left-hand side of the equation is called the *output projection vector.*

9.2 Output Projection

Assume that the control computation is repeated in every time interval. Under this condition, only the most current input change δu_1 is involved in the output projection computation and the H matrix becomes a column vector $\mathbf{h}$. The iterative projection computation steps proceed as follows:

1. Initialize the output projection vector $\mathbf{y}$ by setting all its elements to the currently observed value of the output.
2. At the beginning of a subsequent time interval, shift the projection forward one time interval and change its designation to $\mathbf{y}^*$.
3. Compute the change in the input from the last time interval to the present.
4. Compute the changes in the output vector by using the input change and the dynamic matrix equation.
5. Compute the updated output projection vector $\mathbf{y}$ using the *updating equation*.

$$\mathbf{y} = \mathbf{y}^* + h\delta u_1$$

6. Loop back to Step 2, to the next time interval.

Since the process is assumed to be open-loop stable, errors due to erroneous initialization will diminish with time. In general, the projected output values do not match the observed values due to disturbances. The discrepancy provides the feedback data for the output vector adjustment.

9.3 Control Computation

Let the process setpoint (reference input) be r, the output error at time instant i is given by

$$e_i = r - y_i$$

For i = 1 to n, e_i form an error vector $\mathbf{e}$. The incremental control input vector δu needed to eliminate the error $\mathbf{e}$ can be obtained by one of the variety of methods, depending on the performance criterion chosen. For example, it can be obtained by solving the equation

$$\mathbf{e} = H\delta u$$

Since the output response from the last input must have the time to reach steady state, the H matrix will have more rows than columns. A solution for δu can be obtained by using the classical least-square formula

$$\delta u = (H^{\mathrm{T}}H)^{-1}H^{\mathrm{T}}\mathbf{e}$$

Figure 9.1 shows the block diagram of a DMC system. The DMC block performs output projection and control computation.

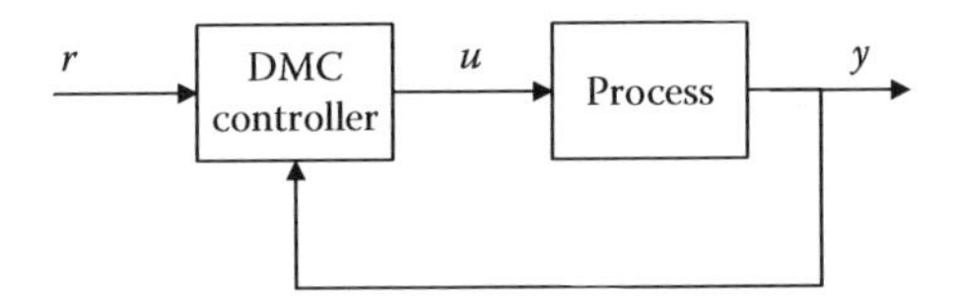

FIGURE 9.1 Dynamic matrix control (DMC).

9.4 Remarks

The extension to MIMO processes is conceptually straightforward. For a two-input–two-output process, the dynamic matrix equation has the form

$$\begin{bmatrix} y_1 \\ y_2 \end{bmatrix} = H \begin{bmatrix} \delta u_1 \\ \delta u_2 \end{bmatrix}$$

where
y_1 and y_2 are the two-output projection vectors
δu_1 and δu_2 are the two incremental step-input vectors
H is the dynamic matrix of the process

Merits of DMC are as follows:

1. It is easy to understand and easy to apply without involving sophisticated mathematics.
2. Coefficients of the dynamic matrix can be obtained by testing, without relying solely on the mathematical model of the process.
3. Since the transport lag characteristics can easily be included in the dynamic matrix, the design of control is independent of whether there is transport lag or not.
4. When disturbances are observable, feed-forward control can conveniently be implemented via the DMC structure.

A demerit of DMC is that it is limited to open-loop bounded-input–bounded-output (BIBO) stable type of processes.

Theoretical work has been done on DMC by control researchers since the early 1980s. The additional results obtained have further revealed the capability of the method and its comparison to other control methods. Prett and Garcia (1988) is a good source for additional information.

References

Cutler, C. R. 1982. Dynamic matrix control of imbalanced systems, *ISA Trans.* 21(1):1–6.

Prett, D. M. and Garcia, C. E. 1988. Chapters 5 and 6, *Fundamental Process Control*, Butterworths, Boston, MA.

10

PID Control

James C. Hung
The University of Tennessee, Knoxville

Joel David Hewlett
Auburn University

10.1 Introduction

PID control was developed in the early 1940s for controlling processes of the first-order-lag-plus-delay (FOLPD) type. The type is modeled by the transfer function

$$G_p(s) = \frac{Ke^{-\tau s}}{1 + Ts} \tag{10.1}$$

where

K is the d-c gain of the process
T is the time constant of the first-order lag
τ is the transport-lag or delay

The controller generates a control signal that is proportional to the system error, its time integral, and its time derivative.

A PID controller is often modeled in one of the following two forms:

$$G_C(s) = K_p + \frac{K_I}{S} + K_D S \tag{10.2}$$

$$G_C(s) = K_p\left(1 + \frac{1}{T_I S} + T_D S\right) \tag{10.3}$$

The effect of each term on the closed-loop system is intuitively clear. The proportional term is to affect system error and system stiffness; the integral term is mainly for eliminating the steady-state error; the derivative term is for damping oscillatory response. Note that having the pure integration term makes the PID controller an active network element. Dropping appropriate terms in these equations gives P, PD, or PI controllers. Determination of controller parameters for a given process is the controller design.

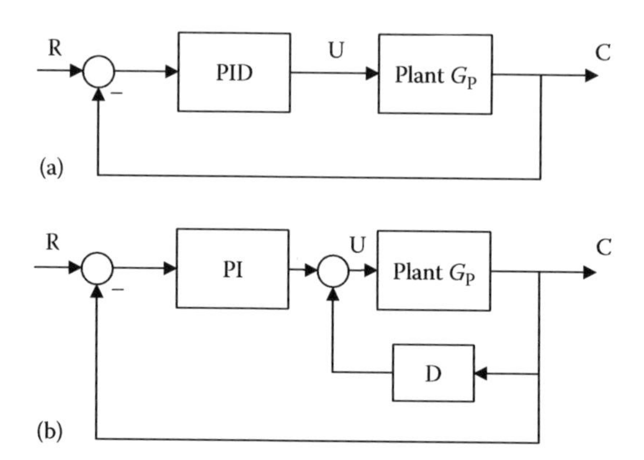

FIGURE 10.1 Two configurations for PID controlled systems: (a) cascade compensation and (b) cascade and feedback compensation.

Two commonly used PID control configurations are shown in Figure 10.1. In Figure 10.1a, the entire PID forms a cascade compensator, while in Figure 10.1b, the derivative part is in the feedback. The latter is sometimes more convenient for practical reasons. For example, in motor control, the motor shaft may already have a tachometer, so the derivative output is already available.

10.2 Pole Placement

For processes which have well-defined low-order models, the closed loop poles of the system can be chosen arbitrarily using analytical methods (Hägglund and Åström, 1996). Take for example the general second-order case in which the transfer function is of the form

$$G_p(s) = \frac{K}{(s\tau_1 + 1)(s\tau_2 + 1)} \quad (10.4)$$

where

K is the gain

τ_1 and τ_2 are time constants associated with the system's two poles

When the transfer function for the PID controller (10.3) is combined with (10.4), the result is a third-order closed-loop system of the form

$$G_{CL}(s) = \frac{Ks/\tau_1\tau_2}{s^2 + s^2\left(\dfrac{\tau_1 + \tau_2 + KK_pT_D}{\tau_1\tau_2}\right) + s\left(\dfrac{1 + K_pK}{\tau_1\tau_2}\right) + \dfrac{K_pK}{\tau_1\tau_2T_I}} \quad (10.5)$$

For a general third-order closed-loop system, the characteristic equation can be expressed as

$$(s + a\omega)(s^2 + 2\xi\omega s + \omega^2) = 0 \quad (10.6)$$

where

ξ is the damping ratio

ω is the natural frequency

Equating the coefficients of (10.5) to (10.6),

$$a\omega = 2\xi\omega = \frac{\tau_1 + \tau_2 + KK_p T_D}{\tau_1 \tau_2}$$

$$\omega^2 + 2\xi a\omega^2 = \frac{1 + K_p K}{\tau_1 \tau_2}$$

$$a\omega^2 = \frac{K_p K}{\tau_1 \tau_2 T_I} \quad (10.7)$$

Finally, choosing ξ, a, and ω to yield the desired closed-loop poles leaves three equations and three unknowns. Through some manipulation of (10.7), the following solution is obtained:

$$K_p = \frac{\tau_1 \tau_2 \omega^2 (1 + 2\xi a) - 1}{K}$$

$$T_I = \frac{\tau_1 \tau_2 \omega^2 (1 + 2\xi a) - 1}{a \tau_1 \tau_2 \omega^3}$$

$$T_D = \frac{\tau_1 \tau_2 \omega (a + 2\xi) - \tau_1 - \tau_2}{\tau_1 \tau_2 (1 + 2\xi a)\omega^2 - 1} \quad (10.8)$$

Assuming the system is in fact FOLPD, ξ and ω may be chosen arbitrarily. However, some care must be taken to ensure that the chosen values are reasonable. For instance, looking at (10.8), it is clear that for

$$\omega = \frac{\tau_1 + \tau_2}{\tau_1 \tau_2 (a + 2\xi)} \quad (10.9)$$

$T_D = 0$, resulting in purely PI control. Furthermore, for

$$\omega < \frac{\tau_1 + \tau_2}{\tau_1 \tau_2 (a + 2\xi)} \quad (10.10)$$

The resulting solution is unobtainable since this forces T_D to take on a negative value. Therefore, (10.9) represents a lower bound for realizable choices of ω.

10.3 Ziegler–Nichols Techniques

Two celebrated Ziegler–Nichols techniques were developed in the early 1940s for tuning PID control of FOLPD systems (Ziegler and Nichols, 1942). Both techniques are very easy to execute, making them very popular.

10.3.1 First Technique

The first technique can be performed online and does not require knowledge of the model parameters. It involves an adjustment procedure, which can be carried out by a technician. The result is often satisfactory but can be crude. Equation 10.2 is the controller model for this technique. The tuning procedure is given as follows.

To begin, set K_p, K_I, and K_D all to zero. Increase the value of K_p until oscillation occurs. At this setting, record $K_{max} = K_p$ and T = oscillation period.

For PID controller, set

$$K_p = 0.6K_{max}$$

$$K_I \le \frac{2K_p}{T}$$

$$K_D \ge 0.125K_pT$$

For PI controller, set

$$K_p = 0.45K_{max}$$

$$K_I \le \frac{1.2K_p}{T}$$

For PD controller, set

$$K_p = 0.65K_{max}$$

$$K_D \ge .125K_pT$$

For P controller, set

$$K_p = 0.5K_{max}$$

10.3.2 Second Technique

The second technique consists of two steps:

1. The determination of the model parameters K, τ, and T, from a step response test.
2. Setting controller parameters.

The steps can be computerized, which have been incorporated into some of the commercial PID controllers (Rovira et al., 1969). The result obtained from the second technique is more refined. Equation 10.3 is the controller model for this technique. The process parameters are determined from its step response. Graphical techniques as well as computer techniques are available for such determination.

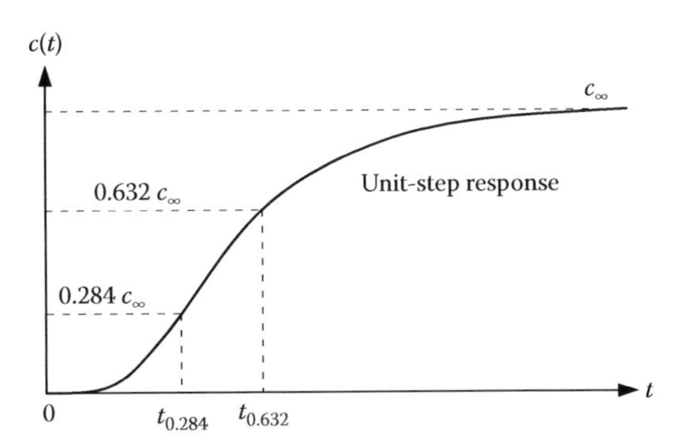

1. $K = c_\infty - c(0)$
2. Locate $t_{0.284}$ and $t_{0.632}$, the time instants when $c(t) = 0.284c_\infty$ and $c(t) = 0.632c_\infty$, respectively.
3. Solve T and τ from the two equations

$$t_{0.284} = \tau + \frac{T}{3} \quad \text{and} \quad t_{0.632} = \tau + T$$

FIGURE 10.2 A graphical FOLPD parameter identification method.

Figure 10.2 shows typical graphical techniques and Figure 10.3 gives a least-square-based computer algorithm. The parameter setting is done by using the formula and the table contained in Figure 10.4 where values in the table were established empirically (Cheng and Hung, 1985).

Both techniques have become classics. Details of PID tuning and some further development can be found in Cheng and Hung (1985) and Franklin et al. (1994).

Unit-step response equation

$$c(t_i) = K\left[1 - e^{-(t_i - \tau)/T}\right]$$

$$K = c_\infty - c(0)$$

Let $x(t_i) = \text{In}\left[1 \quad \frac{c(t_i)}{K}\right] = (\tau \quad t_i) \qquad i = 1, 2, \ldots$

Choose $\tau = \tau_1$, and determine T by least-square using n measurement $x(t_i)$, $i = 1$ to n.

$$\mathbf{x} - \begin{bmatrix} x(t_1) \\ \vdots \\ x(t_n) \end{bmatrix} - \begin{bmatrix} (\tau_1 - t_1) \\ \vdots \\ (\tau_1 - t_n) \end{bmatrix} \frac{1}{T} - \mathbf{h}\frac{1}{T}$$

$$\frac{1}{T} = (\mathbf{h}^{\mathrm{T}}\mathbf{h})^{-1}\mathbf{h}^{\mathrm{T}}\mathbf{x}$$

Compute $y(t) = K[1 - e^{-(t - \tau_1)/T}]$

$$J = \sum_{i=1}^{n} [c(t_i) - y(t_i)]^2$$

Repeat the above process using different τ's until a minimum J is

FIGURE 10.3 A least-square based FOLPD parameter identification algorithm.

$$K_P = \frac{A_P}{K}\left(\frac{\tau}{T}\right)^{E_P}$$

$$T_I = \frac{T}{A_I\left(\frac{\tau}{T}\right)^{B_I}} \text{ for regulator, } \quad T_I = \frac{T}{A_I + B_I\left(\frac{\tau}{T}\right)} \text{ for tracker.}$$

$$T_D = T\, A_D\left(\frac{\tau}{T}\right)^{B_D}$$

R/T	A_P	B_P	A_I	B_I	A_D	B_D
P	0.902/–	−0.985/–	–/–	–/–	–/–	–/–
PI	0.984/0.586	−0.986/−0.916	0.608/1.03	−0.707/−0.165	–/–	–/–
PID	1.435/0.965	−0.921/−0.855	0.878/0.796	−0.749/−0.147	0.482/0.308	1.137/0.929

R/T: (value of regulator)/(value of tracker)

FIGURE 10.4 Formulas and data for tuning PID family of controllers.

Example 10.1

The unit-step response of a process is shown in Figure 10.5. The computer algorithm of Figure 10.3 is used to identify the parameters of the FOLPD model. The unit-step response of the identified model is also shown in Figure 10.5 where values of the identified parameters are also shown. Computer formulas of Figure 10.4 are used to tune the PID controller. For regulator control, $K_p = 0.814$, $T_I = 2.442$, and $T_D = 0.980$; for tracking control, $K_p = 0.543$, $T_I = 3.558$, and $T_D = 0.644$. Comparing the two sets of parameter values, one sees that the tracking control system is more damped than the regulator system. The closed-loop regulator system responding to a disturbance is shown in Figure 10.6; the closed-loop tracking system responding to a unit-step input is shown in Figure 10.7.

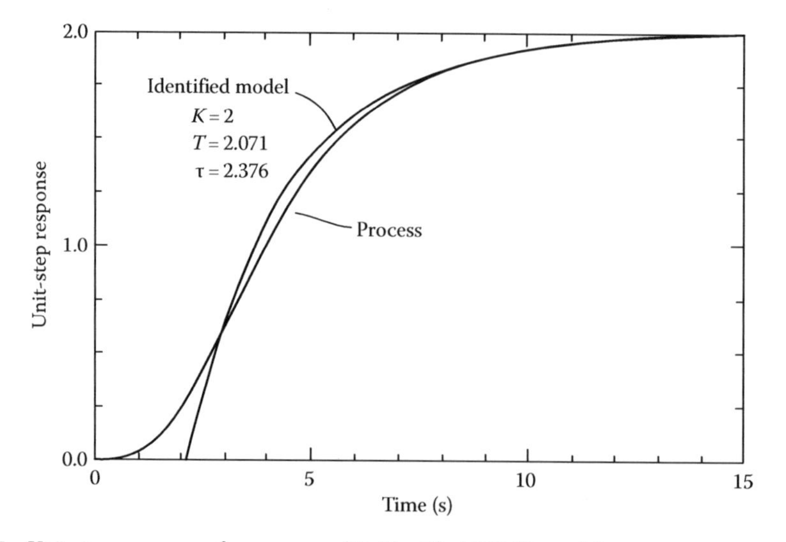

FIGURE 10.5 Unit-step responses of a process and its identified FOLPD model.

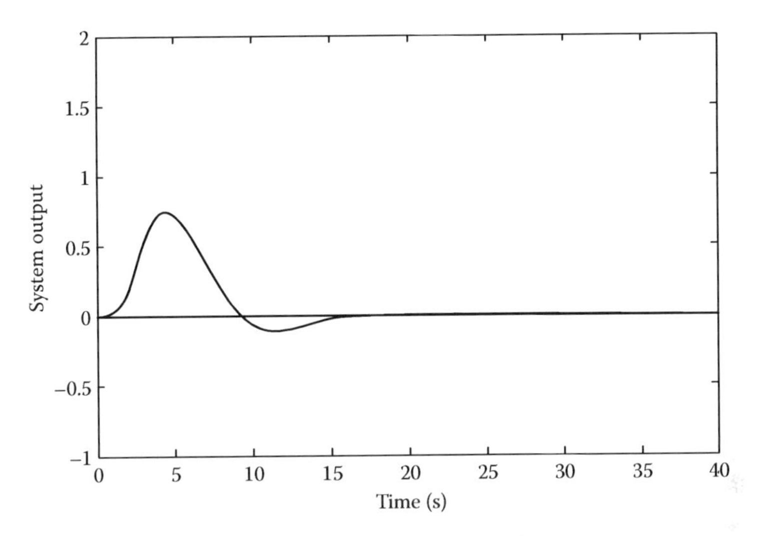

FIGURE 10.6 Unit-step disturbance response of a PID controlled regulator.

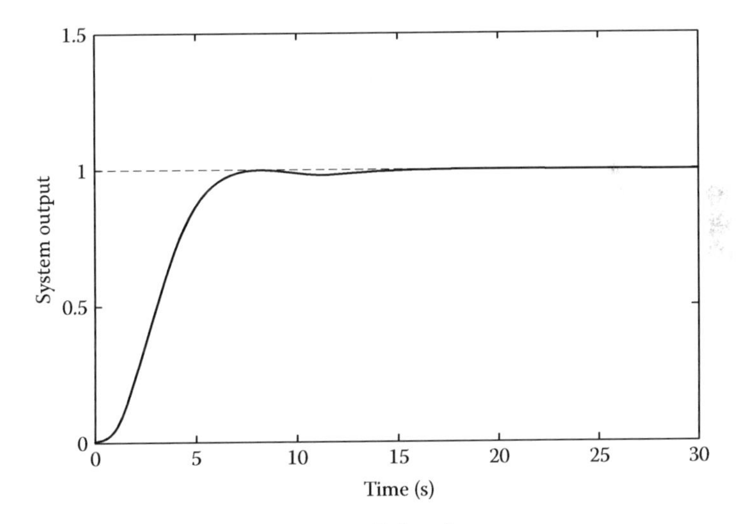

FIGURE 10.7 Unit-step input response of a PID controlled tracking system.

10.4 Remarks

The classical PID techniques have been very useful for process control since many industrial processes can be well approximated by FOLPD models. The techniques are also effective for processes modeled by other forms of transfer functions, as long as their input–output characteristics resemble that of FOLPD. In general, these techniques can be applied to any linear process whose step response does not have overshoot. In fact, the true transfer function of the process in the above example is not a FOLPD one. It is given by

$$G_p(s) = \frac{2}{(s-1)^2(s+2)(s+0.5)}$$

But the step response of this transfer function resembles that of a FOLPD one. So the PID control using the Ziegler–Nichols tuning technique can be applied.

The classical PID control can make a closed-loop system have a satisfactory damping ratio ξ and zero steady-state error in response to a set-point input. However, Ziegler–Nichols tuning does not address other important concerns of control systems, such as risetime, bandwidth, robustness, and system stiffness. Therefore, the approach is for very limited objectives.

It should be pointed out that a PID controller can also be designed using Bode-plot, root-locus, pole-placement, or other techniques to achieve more sophisticated objectives. However, one then needs to know the plant transfer function. Knowing the transfer function, he or she may consider all classical compensation networks (lead, lag, lead-lag networks, etc.) and has much more freedom in designing a controller.

The popularity of PID control is due to the fact that a technician, who has no exposure to control theory, can do the tuning of PID controller without knowledge of the transfer function.

References

Cheng, G. S. and Hung, J. C. (1985). A least-square based selftuning of PID controller, *Proceeding of IEEE Southeastcon*, Raleigh, NC, IEEE Catalog no. 85C3H2161-8, p. 325.

Franklin, G. E., Powell, D., and Emami-Naeini, A. (1994). *Feedback Control of Dynamic Systems*, 3rd edn. Addison-Wesley, Reading, MA.

Hägglund, T. and Åström, K. J. (1996). *The Control Handbook*, CRC Press LLC, Boca Raton, FL.

Rovira, A. A., Murrill, P. W., and Smith, C. L. (1969). Tuning controllers for set-point changes, *Instruments and Control Systems*, 42(12):67.

Ziegler, J. G. and Nichols, N. B. (1942). Optimum setting for automatic controllers, *ASME Transactions*. 64(11):759.

11

Nyquist Criterion

James R. Rowland
University of Kansas

11.1 Introduction and Criterion Examples

The Nyquist criterion is a frequency-based method used to determine the stability or relative stability of feedback control systems [1–4]. Let $GH(s)$ be the transfer function of the open-loop plant and sensor in a single-loop negative-feedback continuous-time system. A contour that includes the entire righthalf of the s-plane is first mapped into the $GH(s)$-plane by using either s-plane vectors or Bode diagrams of magnitude and phase versus frequency ω. This polar plot mapping is often considered to be the most difficult step in the procedure. The number of closed-loop poles in the righthalf s-plane Z is calculated as the sum of P and N, where P is the number of open-loop poles in the righthalf s-plane and N is the number of encirclements of the $-1 + j0$ point in the $GH(s)$-plane. The closed-loop system is stable if and only if $Z = 0$. It is important to specify N as positive if the encirclements are in the same direction as the contour in the s-plane and negative if in the opposite direction. Different ranges of gain K can yield different numbers of encirclements and, hence, different stability results. Moreover, the application of the Nyquist criterion for the stability of a closed-loop system can be presented succinctly in a table for different ranges of K.

Example 11.1

Consider a negative unity feedback system having $G(s) = K/(s + 1)^3$. For the section of the **s**-plane contour along the positive $j\omega$-axis shown in Figure 11.1a, the corresponding $G(s)$-plane contour is the polar plot of $G(j\omega)$, as ω varies from 0 to $+\infty$, shown in Figure 11.1b, for $K = 10$. The frequency at which the phase of $G(j\omega)$ is $-180°$ is designated as the phase crossover frequency ω_{pc}, which for this example is $\sqrt{3} = 1.732$ rad/s. Observe that ω_{pc} could be obtained alternatively by setting the imaginary part of $G(j\omega)$ equal to 0. Next, the infinite arc that encloses the entire righthalf s-plane maps into a point at the origin in the $G(s)$-plane, and the remaining section (the negative $j\omega$-axis) is obtained as the mirror reflection of Figure 11.1b about the real axis, as shown in Figure 11.1c. Finally, the $G(s)$-plane plot for $K < 0$ is the mirror reflection of Figure 11.1c about the $j\omega$-axis, as shown in Figure 11.1d. Table 11.1 shows stability results for different ranges of K.

The three open-loop poles are located at $s = -1$. Therefore, none of the open-loop poles are inside the righthalf s-plane contour and, thus $P = 0$. The real part of $G(j\omega_{pc}) = G(j\sqrt{3}) = -1$ when $K = 8$, which is the upper limit of the range shown in Row 1 of the table. In words, the $G(s)$-plane contour for positive

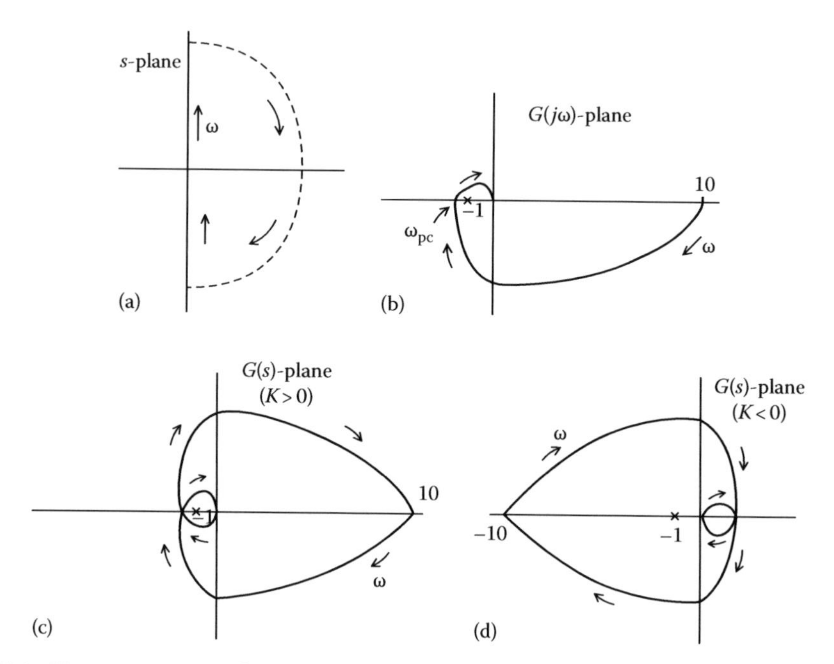

FIGURE 11.1 Nyquist criterion application to Example 11.1.

TABLE 11.1 Nyquist Criterion Results for Example 11.1

Range of K	P	N	Z	Stability
$0 < K < 8$	0	0	0	Yes
$8 < K < \infty$	0	2	2	No
$-1 < K < 0$	0	0	0	Yes
$-\infty < K < -1$	0	1	1	No

K in Figure 11.1c passes through the $-1 + j0$ point for $K = 8$; the $-1 + j0$ point is not encircled ($N = 0$) for $K < 8$; and it is encircled twice ($N = 2$) for $K > 8$. These results are shown in Rows 1 and 2 of the table. For the negative K case, shown in Figure 11.1d, the real part of $G(j0) = -1$ when $K = -1$. Similarly, the results in Rows 3 and 4 are obtained for the ranges of negative K. In brief, the Nyquist criterion has shown (Rows 1 and 3) that the closed-loop system is stable for $-1 < K < 8$.

Example 11.2

Let a negative unity feedback system have a plant transfer function $G(s) = K/[s(s + 2)^2]$. The s-plane contour along the positive $j\omega$-axis from Example 11.1 is modified by inserting an infinitesimal arc of radius ε that excludes the open-loop pole at the origin from inside the contour, as shown in Figure 11.2a. The $G(s)$-plane contour corresponding to this arc is an infinite arc extending from the positive real axis clockwise through the fourth quadrant and connecting with the polar plot of $G(j\omega)$, as ω varies from ε to $+\infty$. The complete $G(s)$-plane contour is shown in Figure 11.2b for $K = 20$, and ω_{pc} is equal to 2 rad/s. Table 11.2 shows stability results for different ranges of K.

The open-loop poles are located at $s = -2, -2$, and the origin. Therefore, again no open-loop poles are inside the righthalf s-plane contour and, thus $P = 0$. The real part of $G(j\omega_{pc}) = G(j1) = -1$ when $K = 16$, which is the upper limit of the range, shown in Row 1 of the table. In words, the $G(s)$-plane contour for positive K in Figure 11.2b passes through the $-1 + j0$ point for $K = 16$; the $-1 + j0$ point is not encircled

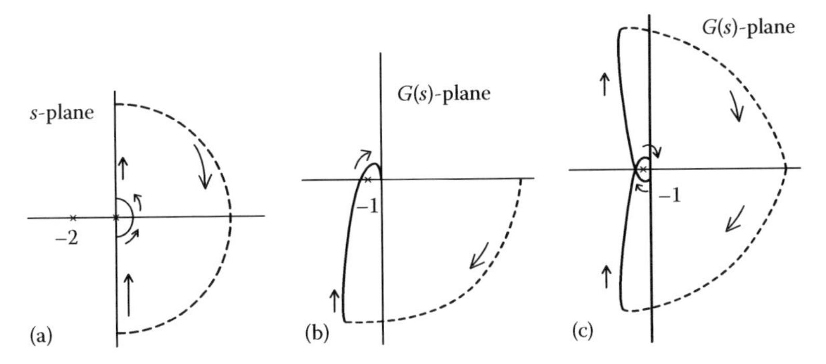

FIGURE 11.2 Nyquist criterion application to Example 11.2.

TABLE 11.2 Nyquist Criterion Results for Example 11.2

Range of K	P	N	Z	Stability
$0 < K < 16$	0	0	0	Yes
$16 < K < \infty$	0	2	2	No
$-\infty < K < 0$	0	1	1	No

($N = 0$) for $K < 16$; and it is encircled twice ($N = 2$) for $K > 16$. These results are shown in Rows 1 and 2 of the table. For the negative K case, there is one encirclement of $-1 + j0$ for all $K < 0$ and consequently $Z = P + N = 0 + 1 = 1$, as shown in Row 3. In summary, the Nyquist criterion has shown (Row 1) that the closed-loop system is stable for $0 < K < 16$.

Example 11.3

To illustrate the Nyquist criterion for a system having an unstable open-loop plant, consider the negative unity feedback system with $G(s) = K(s + 1)/[s(s - 1)]$. The s-plane contour along the positive $j\omega$-axis is again modified with an infinitesimal arc of radius ε. The infinite arc extends from the negative real axis clockwise through the second quadrant and connects with the polar plot of $G(j\omega)$, as ω varies from ε to $+\infty$. The complete $G(s)$-plane contour is shown in Figure 11.3 for $K = 2$, and ω_{pc} is equal to 1 rad/s. Table 11.3 shows stability results for different ranges of K.

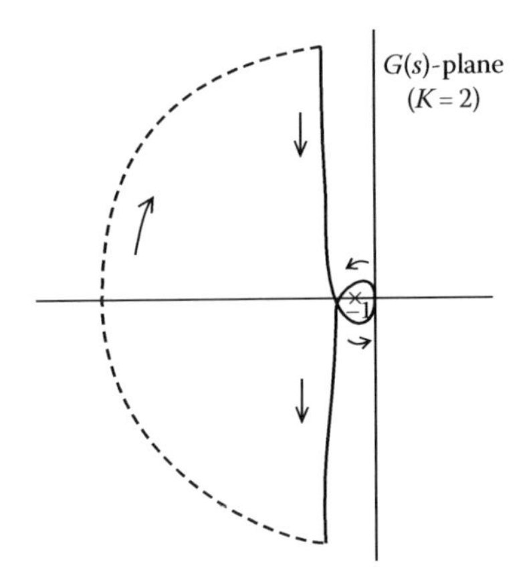

FIGURE 11.3 Nyquist criterion application to Example 11.3.

TABLE 11.3 Nyquist Criterion Results for Example 11.3

Range of K	P	N	Z	Stability
$0 < K < 1$	1	+1	2	No
$1 < K < \infty$	1	−1	0	Yes
$-\infty < K < 0$	1	0	1	No

The open-loop poles are located at $s = +1$ and the origin. Therefore, one open-loop pole ($s = +1$) lies inside the righthalf s-plane contour and, thus $P = 1$. The real part of $G(j\omega_{pc}) = G(j1) = -1$ when $K = 1$, which is the upper limit of the range, shown in Row 1 of the table. In words, the $G(s)$-plane contour for positive K in Figure 11.3 passes through the $-1 + j0$ point for $K = 1$; the $-1 + j0$ point is encircled once in the same direction as the s-plane contour ($N = +1$) for $K < 1$; and it is encircled once in the opposite direction as the s-plane contour ($N = -1$) for $K > 1$. These results are shown in Rows 1 and 2 of the table. For the negative K case, there are no encirclements of $-1 + j0$ for all $K < 0$ and consequently $Z = P + N = 1 + 0 = 1$, as shown in Row 3. In summary, the Nyquist criterion has shown (Row 2) that the closed-loop system is stable for $1 < K < \infty$.

11.2 Phase Margin and Gain Margin

Relative stability can be described by using measures that designate how close the $G(s)$-plane contour comes to an encirclement of the $-1 + j0$ point for open-loop stable systems. In such cases, the value of N is 0 because there is no encirclement. Two measures of relative stability are phase margin (PM) and gain margin (GM). The PM is determined as the absolute difference between −180° and the phase at the gain crossover frequency ω_{gc} defined by the equation $|G(j\omega_{pc})| = 1$. In words, PM is the amount of phase that could be added to an open-loop plant transfer function before the system becomes unstable. A typical desired PM is 40° or greater, which requires that the phase at ω_{gc} must be less negative than −140°. The GM is defined as the reciprocal of $|G(j\omega_{pc})|$ or alternatively as K^*/K, where K is the actual gain and K^* is the gain for which the system becomes unstable. In words, GM is the amount of gain that can be multiplied by the actual plant transfer function gain before the system becomes unstable. A typical desired GM is 5 or greater.

Example 11.4

Let $K = 15$ for a negative unity gain feedback system having a plant transfer function $G(s) = K/s(s + 4)$. Using $|G(j\omega_{gc})| = 1$ yields $\omega_{gc} = 3$ rad/s from which the phase of $G(j\omega_{pc}) = G(j3) = -153.1°$, and the PM is calculated as 36.9°. Figure 11.4 shows the PM marked on the polar plot in the $G(j\omega)$-plane. Since there is no finite ω for which the phase is −180° (only $\omega_{pc} = \infty$), the GM is infinite.

Example 11.5

Consider again the system of Example 11.1 having a plant $G(s) = K/(s + 1)^3$, and let $K = 5$. Using $|G(j\omega_{gc})| = 1$ yields $\omega_{gc} = \sqrt{(5^{2/3} - 1)} = 1.387$ rad/s from which the PM is calculated as 17.37°. Figure 11.5 shows the PM on the polar plot in the $G(j\omega)$-plane. The phase crossover frequency ω_{pc} was calculated in Example 11.1 as $\omega_{pc} = \sqrt{3} = 1.732$ rad/s. Since $K^* = 8$ from Example 11.1, the GM can easily be calculated as GM $= K^*/K = 8/5$ or as $1/|G(j\omega_{gc})| = 1/|Gj\sqrt{3}| = 1.60$. This GM is shown in Figure 11.5.

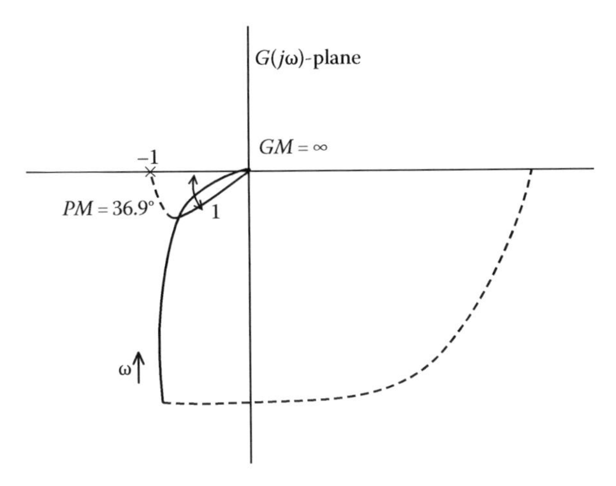

FIGURE 11.4 Phase margin and gain margin for Example 11.4.

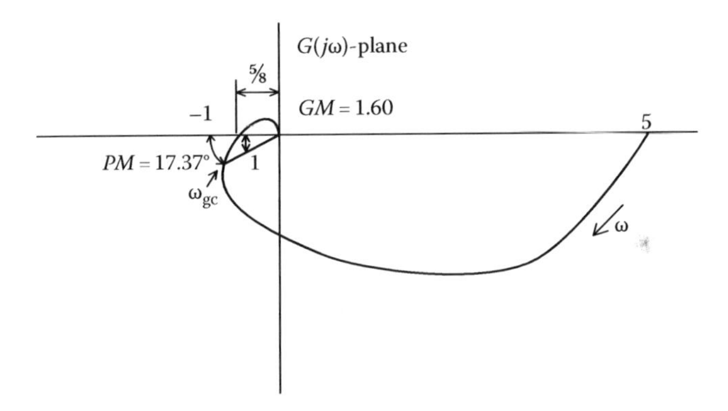

FIGURE 11.5 Phase margin and gain margin for Example 11.5.

11.3 Digital Control Applications

The application of the Nyquist criterion to digital control systems differs from the application to continuous-time systems. Single-loop digital control systems having a single sampler inside the loop yield the characteristic equation $1 + GH(z) = 0$, where $GH(z)$ is the pulsed transfer function expressed in terms of z-transforms. The sampler is often followed by a zero-order hold device for smoothing. In applying the Nyquist criterion, an important difference from continuous-time systems is that the Bode diagrams, which were useful for mapping the $j\omega$-axis into the $G(s)$-plane, cannot be used for digital control systems. Instead, z-plane vectors are used to map the unit circle in the z-plane into a corresponding contour in the $GH(z)$-plane. A second difference between digital control systems and continuous-time systems is the value of Z required for stability. Continuous-time feedback systems are stable if and only if $Z = 0$, i.e., the number of closed-loop poles inside the righthalf s-plane is zero. Digital control systems are stable if and only if Z is equal to the order of the system, i.e., all of the closed-loop poles lie inside the unit circle in the z-plane.

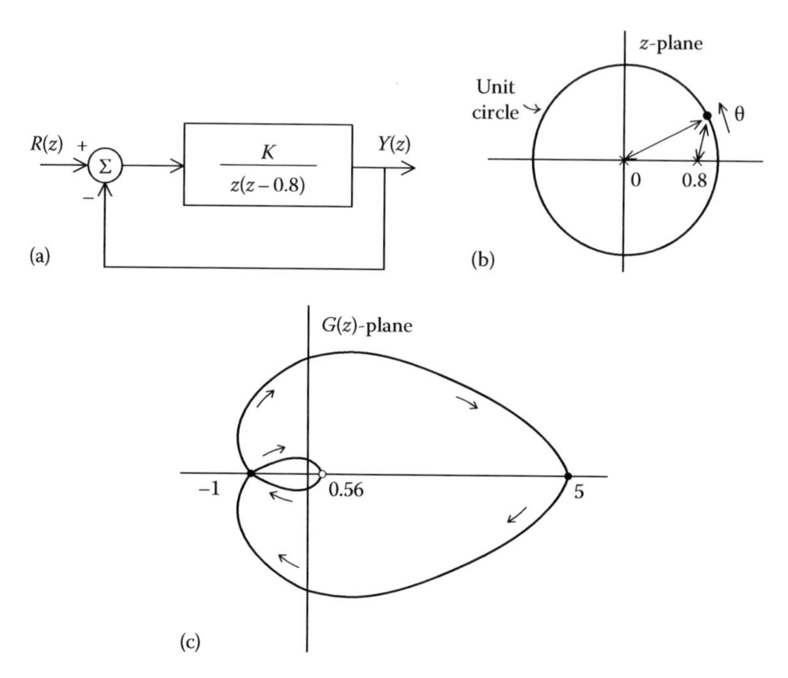

FIGURE 11.6 Nyquist criterion application to the digital control system of Example 11.6.

TABLE 11.4 Nyquist Criterion Results for Example 11.6

Range of K	P	N	Z	Stability
$0 < K < 1$	2	0	2	Yes
$1 < K < \infty$	2	−2	0	No
$-0.2 < K < 0$	2	0	2	Yes
$-1.8 < K < -0.2$	2	−1	1	No
$-\infty < K < -1.8$	2	−2	0	No

Example 11.6

Consider the closed-loop digital control system having an open-loop pulsed transfer function given by $G(z) = K/[z(z-0.8)]$, as shown in Figure 11.6a. Use z-plane vectors as θ varies from 0° to 180° clockwise around the unit circle (Figure 11.6b), resulting in the $G(z)$-plane mapping for $K = 1$, shown in Figure 11.6c. The magnitude of $G(z)$ is unity at the phase crossover angle $\theta_{pc} = \cos^{-1}(0.4)$, and the phase of $G(z)$ is −360° and its magnitude is 0.556 for $\theta = 180°$. A mirror reflection about the real axis in the $G(z)$-plane completes the mapping for θ between 180° and 360°. Table 11.4 shows the Nyquist criterion results for different ranges of K, both positive and negative. In summary, the Nyquist criterion has shown (Rows 1 and 3) that the closed-loop system is stable for $-0.2 < K < 1$.

Example 11.7

As a final example, consider a closed-loop digital control system with negative unity feedback having an open-loop pulsed transfer function $G(z) = K(z+1)/[z(z-1)]$, which places an open-loop pole on the unit circle. Again, use z-plane vectors, as θ varies from 0° to 180° clockwise around the unit circle (Figure 11.7a), including an infinitesimal arc around the open-loop pole at $z = 1$. Thus, all open-loop poles are inside the

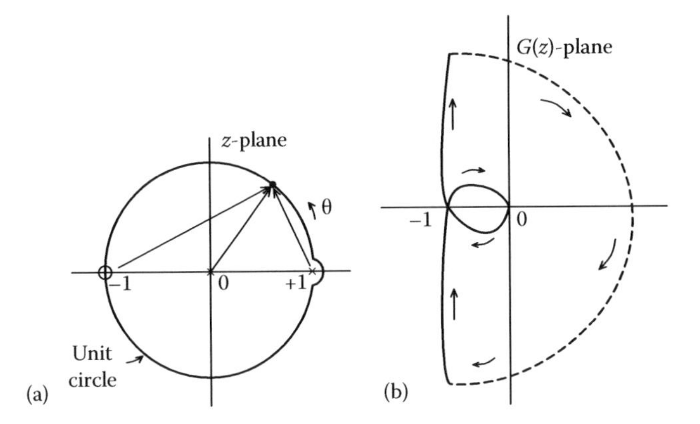

FIGURE 11.7 Nyquist criterion application to the digital control system of Example 11.7.

TABLE 11.5 Nyquist Criterion Results for Example 11.7

Range of K	P	N	Z	Stability
$0 < K < 1$	2	0	2	Yes
$1 < K < \infty$	2	−2	0	No
$-\infty < K < 0$	2	−1	1	No

contour in the z-plane. This infinitesimal arc in the z-plane results in an infinite arc that starts on the positive real axis in the $G(z)$-plane and moves clockwise through the fourth quadrant, connecting with the z-plane vector plot for $\theta = \varepsilon$. The complete $G(z)$-plane plot is shown in Figure 11.7b, and Nyquist criterion results are provided in Table 11.5.

11.4 Comparisons with Root Locus

Root locus, described elsewhere in this book provides a plot of closed-loop poles in the s-plane for continuous-time feedback systems as the gain K varies. Imaginary axis crossings occur for gains K^* at the imaginary axis location $s = j\omega^*$, which can be found by using the Routh table. On the other hand, the Nyquist criterion is based on an s-plane contour that contains the $j\omega$-axis and an infinite arc around the righthalf s-plane. A phase crossover frequency ω_{pc} is determined when applying the Nyquist criterion. The value of ω^* from root locus is the same as the phase crossover frequency ω_{pc} from the Nyquist criterion. Figure 11.8 shows this relationship between root locus and Nyquist criterion procedures.

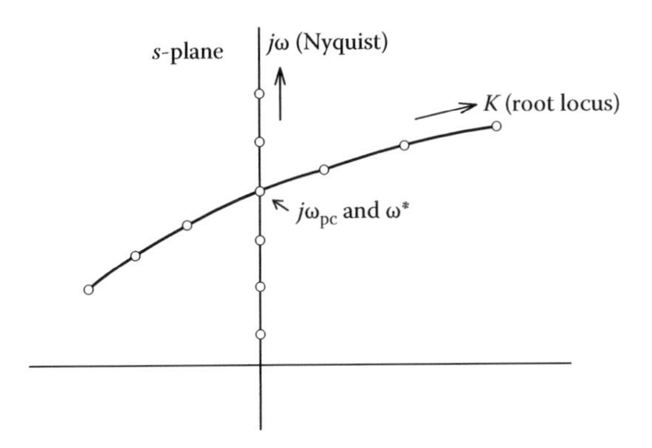

FIGURE 11.8 Relationships between root locus and the Nyquist criterion.

Finally, the reader should consider determining root locus sketches for the Nyquist criterion examples presented here. The same ranges of stability would be obtained. Moreover, GMs in Examples 11.4 and 11.5 can also be obtained by root locus plots. However, the PM calculations are exclusive for frequency-based methods (Nyquist criterion and Bode diagrams).

11.5 Summary

The Nyquist criterion was described for both continuous-time and digital control systems having a single feedback loop. Three examples were given for continuous-time systems and two for digital control systems. The GM and PM calculations were also shown. For continuous-time systems, the right-half s-plane was mapped into the $GH(s)$-plane and the number of encirclements N of the $-1 + j0$ point recorded. The number of closed-loop poles in the righthalf s-plane Z is equal to the sum of N and the number of open-loop poles (P) inside the righthalf s-plane. The closed-loop system is stable if and only if Z equals zero. Ranges of K for different N (and hence Z) are often listed in a table. The same procedure applies for digital control systems, except that the contour in the z-plane is the unit circle and the closed-loop system is stable if and only if Z equals to the order of the system.

References

1. R.C. Dorf and R.H. Bishop, *Modern Control Systems*, 11th edn., Prentice-Hall, Lebanon, IN, 2007.
2. G. Franklin, J.D. Powell, and A. Emami-Naeini, *Feedback Control of Dynamic Systems*, 5th edn., Prentice-Hall, Upper Saddle River, NJ, 2005.
3. B.C. Kuo and F. Golnaraghi, *Automatic Control Systems*, 8th edn., John Wiley & Sons, New York, 2002.
4. N.S. Nise, *Control Systems Engineering*, 5th edn., John Wiley & Sons, New York, 2008.

12

Root Locus Method

Robert J. Veillette
The University of Akron

J. Alexis De Abreu Garcia
The University of Akron, Fairbanks

12.1 Motivation and Background

The form of the natural response of any linear dynamic system is determined by the complex plane locations of the poles and zeros of the system transfer function. Any negative real pole of the s-domain transfer function of a continuous-time system corresponds to an exponential mode of response; the rate of decay of the exponential depends on the magnitude of the pole. Any pair of complex-conjugate poles in the left half-plane corresponds to an exponentially decaying oscillatory mode of response; the real part of the poles determines the rate of decay, while the imaginary part determines the frequency of oscillation. Any pole or poles in the right half of the complex plane correspond to an unbounded (exponentially growing) mode of response; the system is then said to be unstable.

The pole locations of a feedback control system are a crucial consideration in its design. Loosely speaking, if a system is to respond to inputs quickly and without excessive oscillation, it should be designed such that its poles lie far enough into the left half-plane. A more detailed discussion relating the pole and zero locations of a system transfer function to the time-domain system response and control design specifications is presented in Franklin et al. (2009, pp. 108–129).

Root locus analysis was first developed by Evans (1948) as a means of determining the set of positions (loci) of the poles of a closed-loop system transfer function as a scalar parameter of the system varies over the interval from $-\infty$ to ∞. It constitutes a powerful tool in the design of a compensator for a feedback control system. The compensator poles and zeros are chosen to shape the branches of the root locus; then the compensator gain is chosen to place the closed-loop poles at the desired positions along the branches. By displaying the closed-loop poles, the root locus analysis complements the frequency-domain methods, which provide information on the magnitude and phase of the system frequency response. It is especially useful for systems that are unstable or marginally stable in open loop since the frequency-domain methods are more cumbersome for such systems.

12.2 Root Locus Analysis

12.2.1 Problem Definition

Consider the closed-loop feedback control system shown in Figure 12.1. The quantity $KG(s)H(s)$ is referred to as the open-loop transfer function of the system. Usually, the poles and zeros of the open-loop transfer function are known or are easily determined. Moreover, these poles and zeros do not depend on the gain K. If $G(s)$ and $H(s)$ are each expressed as the ratio of polynomials in s as

$$G(s) = \frac{n_G(s)}{d_G(s)}, \quad H(s) = \frac{n_H(s)}{d_H(s)}, \tag{12.1}$$

then the poles of the open-loop transfer function are the solutions of

$$d_G(s)d_H(s) = 0, \tag{12.2}$$

and the zeros of the open-loop transfer function are the solutions of

$$n_G(s)n_H(s) = 0. \tag{12.3}$$

The transfer function of the closed-loop system is

$$T(s) = \frac{KG(s)}{1 + KG(s)H(s)} = \frac{Kn_G(s)d_H(s)}{d_G(s)d_H(s) + Kn_G(s)n_H(s)}. \tag{12.4}$$

The poles of $T(s)$ are the solutions of the characteristic equation

$$1 + KG(s)H(s) = 0, \tag{12.5}$$

or of the equivalent characteristic equation

$$d_G(s)d_H(s) + Kn_G(s)n_H(s) = 0, \tag{12.6}$$

which clearly depend on the gain K. It is of interest to determine how the closed-loop poles move in the s plane as K varies. The root locus method is a means of plotting these poles as K ranges from $-\infty$ to ∞.

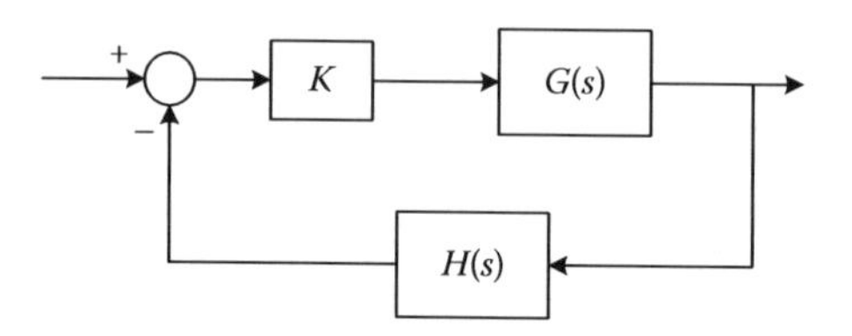

FIGURE 12.1 Closed-loop system.

Definition 12.1 The *root locus* (sometimes called the *complete root locus*) of the system in Figure 12.1 is the set of loci of the poles of the closed-loop system as the gain K ranges from $-\infty$ to ∞.

In order to determine and plot the root locus of the system, write Equation 12.5 as

$$KG(s)H(s) = -1 \tag{12.7}$$

or

$$KG(s)H(s) = 1\angle(2k+1)\pi, \quad k = 0, \pm 1, \pm 2, \ldots, \tag{12.8}$$

which is equivalent to the two simultaneous conditions

$$|KG(s)H(s)| = 1 \tag{12.9}$$

and

$$\angle G(s)H(s) = \begin{cases} (2k+1)\pi, & K > 0 \\ 2k\pi, & K < 0 \end{cases}, \quad k = 0, \pm 1, \pm 2, \ldots. \tag{12.10}$$

Equations 12.9 and 12.10 are sometimes given as an alternative definition of the root locus. Equation 12.9 is known as the *magnitude condition*, and Equation 12.10 is known as the *angle condition*.

Since K can be any real number, any point s automatically satisfies the magnitude condition; therefore, the magnitude condition is mostly useful for determining the value of K that corresponds to a point s already known to lie on the root locus.

It is the angle condition that determines the locus. According to the angle condition, a point s is on the root locus if the angle of $G(s)H(s)$ is any multiple of π. If the angle is an odd multiple of π, then s is said to be on the *standard root locus* (sometimes called simply the *root locus*), which is the set of solutions of Equation 12.5 for $K > 0$; if the angle is an even multiple of π, then s is said to be on the *complementary* (or *negative*) *root locus*, which is the set of solutions of Equation 12.5 for $K < 0$. For the standard root locus, the angle condition reduces to

$$\angle G(s)H(s) = (2k+1)\pi, \quad k = 0, \pm 1, \pm 2, \ldots. \tag{12.11}$$

We concentrate here on the construction of the standard root locus. The development of the *complementary* (or *negative*) *root locus* is similar to that of the standard root locus and is not discussed here. A summary of the guidelines for sketching the complementary root locus is given in Franklin et al. (2009, pp. 266–270). For a complete treatment of the complementary root locus, interested readers are referred to Kuo (1995, 472–509).

12.2.2 Development of Rules for Constructing Root Locus

Considering the availability of special-purpose software for control systems analysis, it can be argued that the most convenient approach to plotting a root locus is to use a computer. Nevertheless, the ability to sketch a root locus by hand is essential. It affords confidence that a computer-generated solution is free of errors. In addition, it improves the intuition necessary for designing a feedback controller—a process of altering a system to obtain the desired root locus.

The intuition and understanding that attend the ability to sketch the root locus by hand and the use of the computer for precise root locus calculations are complementary tools in feedback control design. It is possible, with hardly any calculations, to get a clear idea about the general form of the root locus of even a complicated system. However, except for simple systems, precise calculations of the detailed features of the locus are best done by computer.

The rules for plotting the root locus are now developed. The development is intended to be intuitively convincing and complete, but not too mathematical. Other developments, with varying degrees of mathematical rigor, may be found in Franklin et al. (2009, pp. 228–234), Kuo (1995, pp. 477–505), Nise (2008, pp. 381–395), and D'Azzo and Houpis (1988, pp. 225–235). This presentation starts with the rules that are the simplest and therefore the most useful for sketching the root locus by hand and proceeds to the rules that are more complex and tedious to apply. Usually, a useful approximate sketch can be made without any tedious calculations.

Rule 12.1 The number of branches of the root locus is equal to the number of poles of the open-loop transfer function $KG(s)H(s)$.

As K varies, each of the closed-loop poles moves along a continuous path in the complex plane, called a "branch" of the root locus. Thus, the number of branches is equal to the number of closed-loop poles for any value of K. But the number of closed-loop poles is equal to the number of open-loop poles since the characteristic polynomials in Equations 12.2 and 12.6 have the same degree. (It is assumed that both $G(s)$ and $H(s)$ are proper—i.e., neither $G(s)$ nor $H(s)$ has more zeros than poles.)

Rule 12.2 The root locus is symmetric about the real axis in the s plane.

For any physical system, the coefficients of the open-loop transfer function, and hence of the characteristic, Equation 12.6, are real. As a result, for each K, any complex closed-loop poles occur in conjugate pairs; therefore, so do the root locus branches.

Rule 12.3 The root locus starts (for $K = 0$) at the poles of $G(s)H(s)$ and ends (for $K = \infty$) at the zeros of $G(s)H(s)$, including the zeros at infinity.

It seems reasonable from an intuitive point of view that the branches of the root locus should originate from the open-loop poles. When $K = 0$, there is no feedback to alter the open-loop system dynamics. This is also easy to establish mathematically. Simply observe that setting $K = 0$ makes the closed-loop characteristic equation (12.6) identical with the open-loop characteristic equation (12.2).

Let the open-loop transfer function have n poles and m zeros. Then Rule 12.3 claims that m of the branches of the root locus terminate at the finite zeros of $G(s)H(s)$, while the other $n - m$ branches go off to infinity as $K \to \infty$. This claim also seems reasonable. According to the magnitude condition, as K approaches infinity, $|G(s)H(s)|$ must approach zero. To establish the claim mathematically, Equation 12.6 can be rewritten as

$$\frac{d_G(s)d_H(s)}{K} + n_G(s)n_H(s) = 0. \tag{12.12}$$

It is clear that each of the m open-loop zeros will satisfy Equation 12.12 as $K \to \infty$. For s equal to any of the open-loop zeros, both terms on the left-hand side of Equation 12.12 vanish, the first one since $d_G(s)$ $d_H(s)$ is bounded and $1/K \to 0$, and the second one since the open-loop zeros are defined by Equation 12.3. Therefore, m of the branches of the root locus must terminate at the open-loop zeros. To see that the other $n - m$ branches go to infinity, note that the left-hand side of Equation 12.12 is a polynomial of degree n, whose $n - m$ highest order coefficients decrease as K increases. As these coefficients go to zero, the polynomial approximates $n_G(s)$ $n_H(s)$ in a larger and larger region in the s plane. The additional $n - m$ roots must lie outside this growing region, where the highest order terms of the polynomial still dominate.

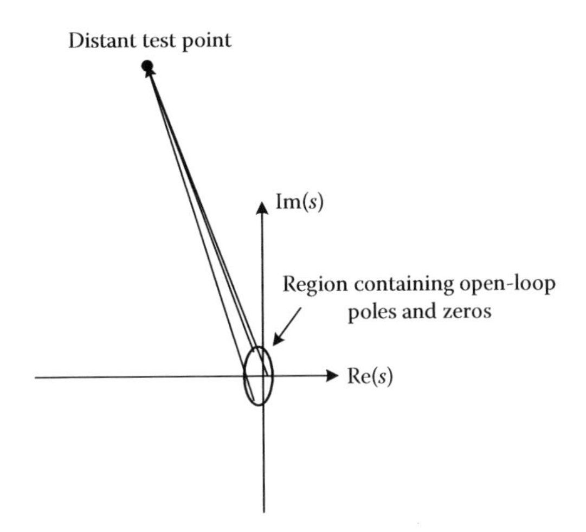

FIGURE 12.4 The angle condition for test points far from the origin.

for some integer k. For large s, Equation 12.18 is approximated by

$$\sum_{i=1}^{m} \angle(s) - \sum_{i=1}^{n} \angle(s) = -(2k+1)\pi \tag{12.19}$$

or

$$m\angle(s) - n\angle(s) = -(2k+1)\pi. \tag{12.20}$$

Solving for $\angle s$ yields the given condition, Equation 12.17.

Rule 12.6 The intersection of the asymptotes lies on the real axis and is given by

$$\sigma = \frac{\sum_{i=1}^{n} p_i - \sum_{i=1}^{m} z_i}{n-m}, \tag{12.21}$$

where p_i and z_i denote the open-loop poles and zeros, respectively.

If $n - m = 0$, then there are no asymptotes, and if $n - m = 1$, then there is a single asymptote at an angle of 180°. In either of these cases, the "intersection of the asymptotes" is irrelevant; therefore, only the situation where $n - m \geq 2$ needs to be considered. The illustrations in Figure 12.3 indicate that, for $n - m \geq 2$, the roots that go to infinity are balanced about some point on the real axis. This point, the "center of gravity" or average of those roots, is the intersection of the asymptotes, denoted as σ.

The center of gravity of the roots that go to infinity may be found by taking advantage of a simple property of polynomials: For any polynomial

$$q(s) = s^n + a_1 s^{n-1} + \cdots + a_n, \tag{12.22}$$

the sum of the roots is simply the negative of the coefficient a_1. (This is easily verified by writing $q(s)$ in factored form.) In the case where $n - m \geq 2$, the second coefficient of the closed-loop characteristic polynomial

$$q_{cl}(s) = d_G(s)d_H(s) + Kn_G(s)n_H(s), \tag{12.23}$$

is the same as that of the open-loop characteristic polynomial $d_G(s)\ d_H(s)$; therefore, the sum of the closed-loop poles does not depend on K. For $K = 0$, the closed-loop poles r_i are at the open-loop pole locations; therefore, the sum of the closed-loop poles is established as

$$\sum_{i=1}^{n} r_i = \sum_{i=1}^{n} p_i, \tag{12.24}$$

for all K. As $K \rightarrow \infty$, m of the roots is found at the open-loop zeros z_i, and the other $n - m$ roots are on the asymptotes. Therefore, the sum of the roots is also given by

$$\sum_{i=1}^{n} r_i = \sum_{i=1}^{m} z_i + (n-m)\sigma. \tag{12.25}$$

Comparison of Equations 12.24 and 12.25 yields

$$\sum_{i=1}^{m} z_i + (n-m)\sigma = \sum_{i=1}^{n} p_i, \tag{12.26}$$

which in turn yields Equation 12.21.

Rule 12.7 The points at which the root locus intersects the $j\omega$ axis, and the corresponding values of K, may be determined from the magnitude and phase plots of the open-loop transfer function.

This method is based on the angle and magnitude conditions for the root locus. An accurate (computer-generated) phase plot of the open-loop transfer function will show the frequency or frequencies ω at which the angle of $G(j\omega)H(j\omega)$ equals an odd multiple of 180°. By the angle condition, these frequencies are the values on the imaginary axis intersected by the root locus. The magnitudes of $G(j\omega)H(j\omega)$ at these frequencies can then be read from an accurate magnitude plot. By the magnitude condition, these magnitudes are the reciprocals of the values of K that correspond to the $j\omega$-axis crossings.

Rule 12.8 The angle of departure of the root locus from a complex pole p_1 is given by

$$\theta_{p_1} = \sum_{i=1}^{m} \angle(p_1 - z_i) - \sum_{i=2}^{n} \angle(p_1 - p_i) + 180°, \tag{12.27}$$

where

$p_1, \ldots, p_n$ are the open-loop poles
$z_1, \ldots, z_m$ are the open-loop zeros

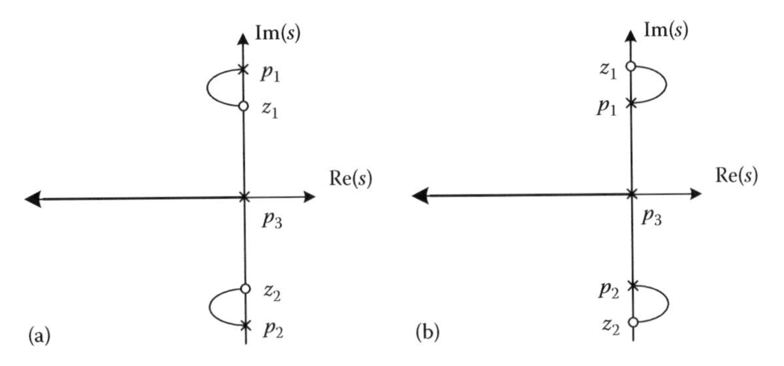

FIGURE 12.5 Root locus plots demonstrating angles of departure and arrival: (a) $\theta_{z_1} = -180°$ and (b) $\theta_{z_1} = 0$.

The angle of arrival of the root locus at a complex zero is given by

$$\theta_{z_1} = \sum_{i=1}^{n} \angle(z_1 - p_i) - \sum_{i=2}^{m} \angle(z_1 - z_i) - 180°. \tag{12.28}$$

Equations 12.27 and 12.28 are found by applying the angle condition to test points near the complex pole or zero. This calculation is especially useful for determining the position of a short branch of the root locus between a complex pole and a complex zero lying close together. As an example, Figure 12.5 shows two pole–zero configurations for which the position of the locus between a pole and a zero is critical to determining the closed-loop system stability. The approximate position is easily determined by computing the angle of departure from the pole or the angle of arrival at the zero. For Figure 12.5a, the angle condition near the zero z_1 is

$$\theta_{z_1} + \angle(z_1 - z_2) - \angle(z_1 - p_1) - \angle(z_1 - p_2) - \angle(z_1 - p_3) = -180° \tag{12.29}$$

or

$$\theta_{z_1} + 90° - (-90°) - (90°) - (90°) = -180°, \tag{12.30}$$

which gives

$$\theta_{z_1} = -180°. \tag{12.31}$$

This calculation implies that the branch approaches the zero from the left. In fact, the branch lies entirely in the left half-plane; the closed-loop system is stable. For Figure 12.5b, the only difference in the calculation is that $\angle(z_1 - p_1) = 90°$, instead of –90°; as a result, Equation 12.29 yields

$$\theta_{z_1} = 0. \tag{12.32}$$

This means that the branch approaches the zero from the right. The branch lies entirely in the right half-plane, and the closed-loop system is unstable. The subtle difference between the two systems in Figure 12.5 results in two opposite conclusions concerning the closed-loop system stability. The computation of the angle of departure or the angle of arrival is helpful in this case for accurately reaching these conclusions.

Rule 12.9 Points where the root locus breaks away from the real axis (breakaway points) or breaks into the real axis (break-in points) are the real values of s on the root locus that satisfy

$$\frac{d[G(s)H(s)]}{ds} = 0. \tag{12.33}$$

At a breakaway point, two branches of the root locus on the real axis meet for a certain value of K, and then leave the axis as a complex-conjugate pair as K increases. Considering only the real axis, the value of K corresponding to the roots reaches a relative maximum at the breakaway point. By the magnitude condition, this implies that $G(s)H(s)$ reaches a relative minimum on the real axis at a breakaway point. Similarly, $G(s)H(s)$ reaches a relative maximum at a break-in point. Hence, a breakaway or break-in point is a point on the real locus for which the derivative of $G(s)H(s)$ is zero. This condition may be checked directly by hand; or the function $G(s)H(s)$ may simply be plotted for real s to find its relative minima and maxima on the real axis.

The same derivative rule presented for finding real breakaway and break-in points may also be used for finding complex breakpoints. A complex breakpoint is a point off the real axis at which two or more branches of the root locus meet, resulting in a repeated root for the corresponding value of K, and split up again.

12.2.3 Steps for Sketching the Root Locus

As a summary, the steps for sketching the root locus of a system are now presented. These steps do not always need to be taken sequentially but may be used as general guidelines for the root locus method of analysis. It is assumed that the open-loop transfer function is proper:

1. Write the open-loop transfer function with numerator and denominator polynomials in factored form.
2. Plot the open-loop poles and zeros in the s plane.
3. Fill in all intervals of the real axis that lie to the left of an odd number of real poles and zeros (Rule 12.4).
4. Sketch the asymptotes of the $n - m$ branches that go to infinity (Rules 12.5 and 12.6).
5. Start a root locus branch at each open-loop pole (Rule 12.1). Keeping in mind that the locus is symmetric with respect to the real axis (Rule 12.2), try to visualize how each branch will arrive at a zero or approach an asymptote as it goes to infinity (Rule 12.3).
6. If the exact frequency or gain corresponding to a $j\omega$-axis crossing is critical, or if uncertain whether a $j\omega$-axis crossing occurs, plot accurate magnitude and phase plots to determine it (Rule 12.7). Note that the range of gains for which the system is stable may also be determined by applying the Routh–Hurwitz test (Nise 2008, 291–302) to the closed-loop characteristic polynomial. This test can be especially useful if the root locus has multiple $j\omega$-axis crossings.
7. If necessary to further determine the positions of certain branches of the locus, compute approximate angles of departure from complex poles or arrival at complex zeros (Rule 12.8).
8. If the exact location (or the existence) of a breakaway or break-in point is critical, use the derivative rule (Rule 12.9) or a computer plot of the locus to determine it.

Remember that the angle condition may be checked at any point in the complex plane to determine whether that point lies on the root locus. If a given point in the vicinity of the open-loop poles and zeros satisfies the angle condition only *approximately* (not exactly), then it is safe to assume that it lies *near* the root locus, at least in the sense that a modest perturbation in the plant could result in a closed-loop pole there.

12.3 Compensator Design by Root Locus Method

An essential aspect of feedback control design is the introduction of a compensator to alter the characteristics of the open-loop transfer function. In a few cases, a satisfactory design may result simply from introducing a gain into the loop. Then, a straightforward root locus analysis is sufficient to determine the value of the gain required to achieve the desired transient response. In most cases, however, a satisfactory design requires the introduction of a dynamic compensator. Then, the overall system transient response depends upon the selection of the pole and zero locations, as well as the gain, of the compensator.

A root locus design method consists of selecting the compensator poles and zeros to position the locus branches and then selecting the compensator gain to place the roots at some desired positions along the branches. The first part of the design is the more complicated one, requiring first of all an understanding of how the root locus is affected by the inclusion of additional poles or zeros in a given loop transfer function.

12.3.1 Effect of Including Additional Poles in the Open-Loop Transfer Function

The addition of poles to the open-loop transfer function increases the number of branches in the root locus and pushes some branches to the right in the s plane. This is a consequence of the angle condition. It is clear from Equation 12.14 that a pole of the open-loop transfer function contributes a negative angle to $G(s)H(s)$ for points s in the upper half-plane. That is, for a given test point in the upper half-plane, an additional open-loop pole will cause the angle of $G(s)H(s)$ to become more negative; therefore, the angle of $G(s)H(s)$ will reach $-180°$ for test points at smaller angles in the s plane. In other words, some branches of the root locus will lie farther to the right.

The effect on the root locus of adding poles to two different open-loop transfer functions is illustrated in Figure 12.6. For each system, the original root locus is shown, followed by the root locus with first one and then two additional poles. For both systems, some branches of the root move to the right as open-loop poles are added.

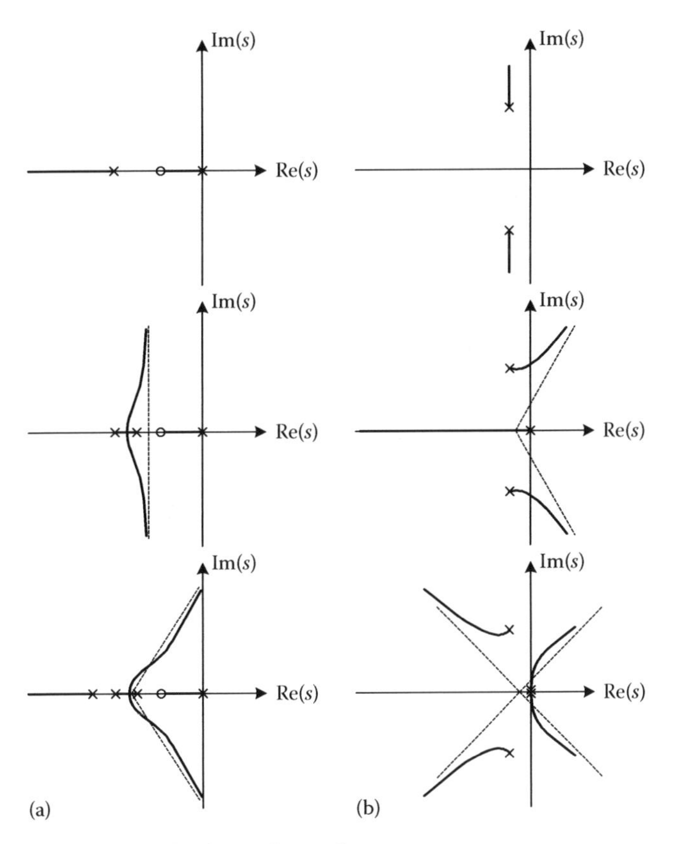

FIGURE 12.6 The effect of additional poles on the root locus.

Adding poles to a system has an undesirable effect on the root locus. In particular, for systems of higher relative degree $n - m$, the feedback tends to be more destabilizing. On the other hand, frequency-domain analysis shows that the addition of a pole or poles near the origin can improve the low-frequency performance of the closed-loop system. In any case, for a system with a relative degree $n - m \geq 2$, care must be taken to keep the feedback gain sufficiently low that all of the closed-loop poles lie in the left half-plane.

12.3.2 Effect of Including Additional Zeros in the Open-Loop Transfer Function

The addition of zeros to the open-loop transfer function pulls the closed-loop poles to the left in the s plane. Also, each zero attracts a branch of the root locus to terminate there for $K = \infty$. In short, the effect of additional zeros is roughly the opposite of that of additional poles.

The effect on the root locus of adding zeros to two different open-loop transfer functions is illustrated in Figure 12.7. For each system, the original root locus is shown, followed by the root locus with first one and then two additional zeros. For both systems, some branches of the root locus move to the left as more open-loop zeros are added.

Adding left half-plane zeros to a system has a desirable effect on the root locus. Systems with low relative degree (i.e., $n - m \leq 1$) can be stable for all values of the feedback gain. Even a system with some open-loop poles in the right half-plane can be stabilized using feedback if there are some zeros in the left half-plane. On the other hand, a zero that has a stabilizing effect on the system may cause a deterioration of the low-frequency performance of the system. Such an undesirable effect can be discerned by the use of frequency-domain analysis techniques.

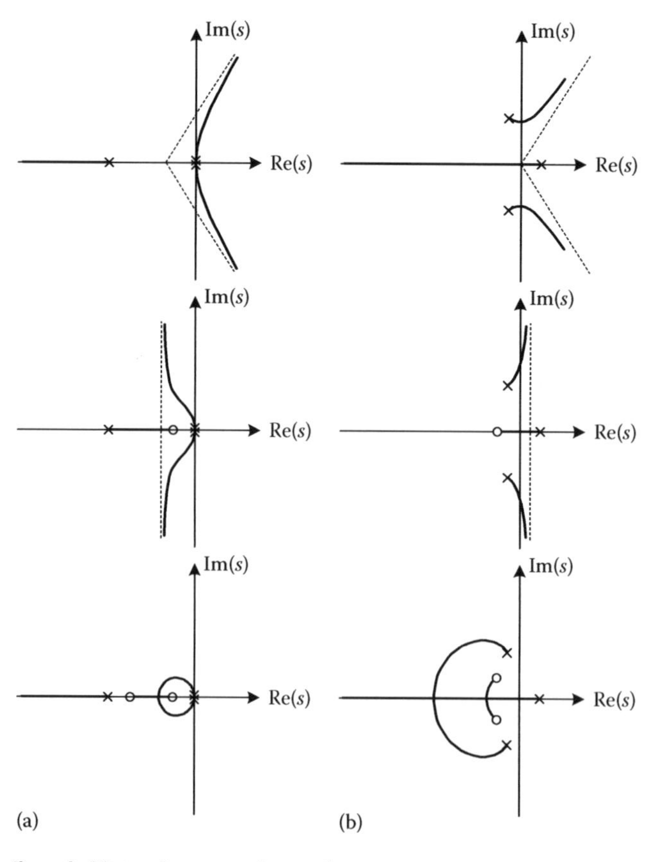

FIGURE 12.7 The effect of additional zeros on the root locus.

12.3.3 Effect of a Lead Compensator

If the branches of the root locus of a given system lie too far to the right, they could be moved to the left if it were possible to introduce an additional zero to the open-loop transfer function. Of course, a proper compensator has at least as many poles as zeros, so adding zeros also requires adding poles. When the effect of a zero is desired, the design calls for a lead compensator. A lead compensator has a transfer function of the form

$$G_c(s) = \frac{K(s+a)}{(s+b)}, \tag{12.34}$$

with $0 < a < b$. It is called a lead compensator because its frequency response $G_c(j\omega)$ has a positive angle (phase lead) for all frequencies $\omega > 0$.

The effect of a lead compensator on the root locus of two different open-loop transfer functions is illustrated in Figure 12.8. (These are the same two systems used to illustrate the effect of additional zeros in Figure 12.7.) For each system, the original root locus is shown, followed by the root locus with the lead compensator. The angles of the asymptotes of the root locus branches remain unchanged; however, the intersection of the asymptotes is moved to the left. In effect, the zero of the compensator, being locally more influential than the pole, initially draws the root locus to the left.

12.3.4 Lead Compensator Design

The goal of the lead compensator design is to improve the transient response of the system over that of the uncompensated system, in order to obtain adequate speed of response and damping. In many cases, the speed and damping of the system response can be correlated to the locations of the dominant closed-loop poles, which are the poles nearest the origin of the s plane, but not too close to a zero of the forward transfer function $G(s)$. Generally speaking, the larger the magnitude of the dominant poles, the faster the system response; the larger the angle of the dominant poles, measured counter-clockwise from the positive $j\omega$ axis, the more heavily damped (i.e., the less oscillatory) the system

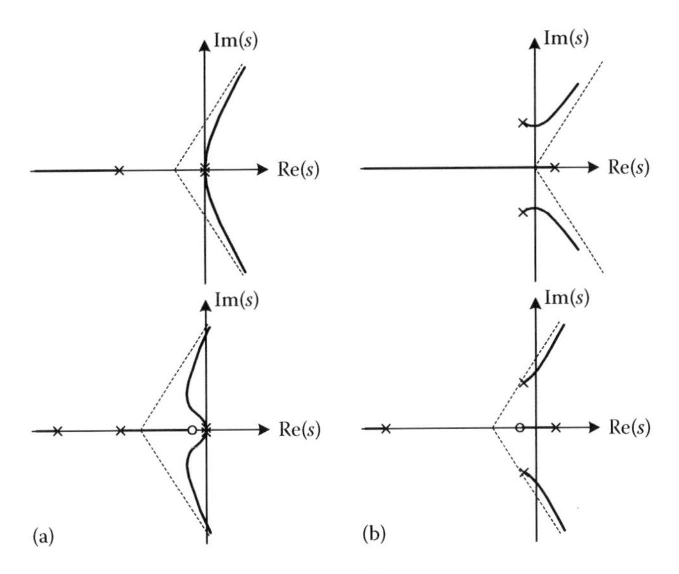

FIGURE 12.8 The effect of a lead compensator on the root locus.

response. Therefore, the goal of the lead compensator design may be restated as the placement of the dominant poles at a desired magnitude and angle.

Given an open-loop transfer function $G_p(s)$, and assuming the transient response of the uncompensated closed-loop system is not satisfactory, one method for the design of a lead compensator may proceed as follows:

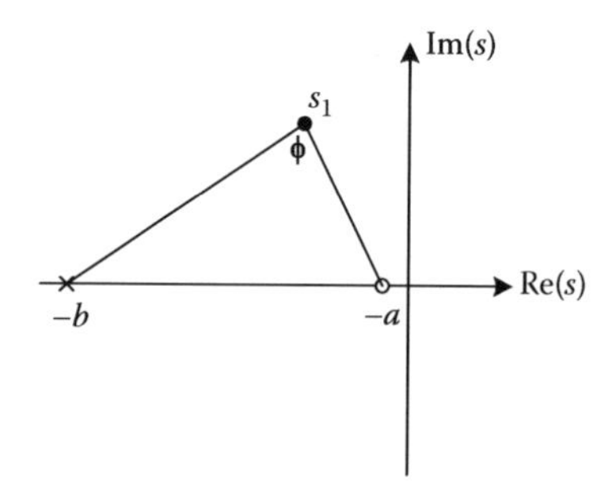

FIGURE 12.9 Graphical determination of the angle of a lead compensator at a point s_1.

1. Choose the desired locations (magnitude and angle) of the dominant poles in the s plane. Assume the dominant poles are a complex-conjugate pair, and call the desired upper half-plane location s_1.
2. Find $\angle G_p(s_1)$.
3. Choose the parameters a and b (i.e., the pole and the zero) of the compensator such that

$$\angle\frac{(s_1+a)}{(s_1+b)}+\angle G_p(s_1)=\pm 180^\circ. \tag{12.35}$$

   Then, according to the angle condition, the point s_1 will lie on the compensated root locus.
4. Choose the parameter K (i.e., the gain parameter) of the compensator such that

$$\left|\frac{K(s_1+a)}{(s_1+b)}G_p(s_1)\right|=1. \tag{12.36}$$

   Then, according to the magnitude condition, the point s_1 will be a pole of the compensated closed-loop system.

There is some design freedom inherent in Step 3 above. The angle of the compensator at s_1 is found graphically as the angle between the line segments connecting s_1 with the compensator pole and zero; see Figure 12.9, where this angle is labeled ϕ. It is clear that the choice of the compensator pole and zero that yield a particular angle is not unique. An additional criterion may be imposed to fix the two parameters a and b. For example, a may be chosen to cancel a real pole of the plant near (not on) the $j\omega$ axis and then b determined to satisfy Equation 12.35. Or a and b may be chosen together to minimize the required value of K in Step 4, while still achieving the desired angle. A graphical method for this approach is given in D'Azzo and Houpis (1988, 370–371).

Note that the steady-state performance is not considered in the design of the lead compensator as presented. The steady-state performance of the system depends not on the closed-loop pole locations but on the open-loop dc gain of the system, which cannot easily be visualized using the root locus technique. If necessary, the steady-state performance of the system may be improved by use of a lag compensator; however, the lag compensator design is best done by frequency-domain techniques, where the open-loop gains can be clearly seen.

12.4 Examples

The use of the root locus method as a design technique is now illustrated by use of two examples. The details of the construction of the loci are omitted for the most part since they are routine and can be filled in by the reader, either by hand or by computer.

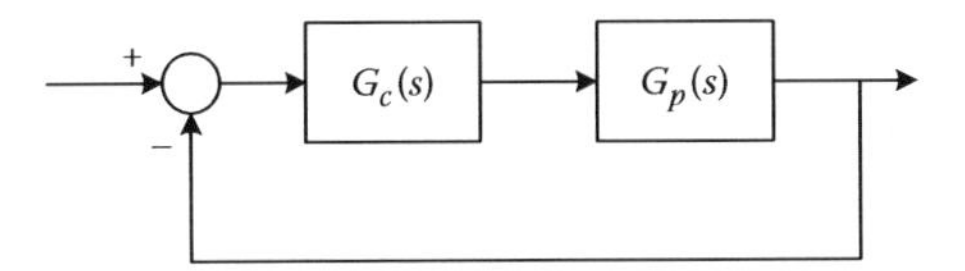

FIGURE 12.10 Form of closed-loop system for examples.

12.4.1 Compensation of an Inertial System

Consider the feedback system shown in Figure 12.10. By comparison with Figure 12.1, $G(s) = G_p(s)G_c(s)$ and $H(s) = 1$. The plant is represented by

$$G_p(s) = \frac{1}{s^2}, \tag{12.37}$$

which describes an inertial system, such as a mass in linear motion with a force input along the direction of motion. The compensator is to have the form

$$G_c(s) = \frac{K(s+1)}{(s+p)}, \tag{12.38}$$

where $K > 0$ and $p > 0$ are design parameters. The design procedure consists of choosing p so that the branches of the root locus assume a desired shape and then choosing K to place the closed-loop poles at the desired positions along the branches. The root locus is given in Figure 12.11 for several values of p. Some of the corresponding step responses are shown in Figures 12.12 and 12.13.

For $p = 1$, the pole and zero of $G_c(s)$ cancel, and the compensator becomes just a gain K. This choice corresponds to the usual first step for a compensator design. The open-loop transfer function is

$$G(s) = G_p(s)G_c(s) = \frac{K}{s^2}, \tag{12.39}$$

which has no zeros and two poles at the origin. The root locus is shown in Figure 12.11a. The relative degree of $G(s)$ is two, so the asymptotes of the loci are at angles of ±90° in the complex plane. As the gain K is increased, the closed-loop poles simply move along the $j\omega$ axis away from their starting points at the origin. Hence, the system is not effectively stabilized for any value of K. The system response will be characterized by undamped oscillations. Increasing gain results in increasing frequency of oscillations.

For $p = 3$, $G_c(s)$ takes the form of the a lead compensator, and the open-loop transfer function is

$$G(s) = \frac{K(s+1)}{s^2(s+3)}. \tag{12.40}$$

The root locus is shown in Figure 12.11b. The relative degree of $G(s)$ is still two, so the root locus asymptotes are still at angles of ±90°; however, the zero of the compensator attracts the locus somewhat, so that the vertical asymptotes intersect the real axis at $\sigma = -1$. All of the closed-loop poles are in the left half-plane for all values of $K > 0$. (Frequency-domain analysis would show that the compensator has added stabilizing phase lead to the open-loop transfer function.)

If the compensator parameters are chosen as $p = 3$ and $K = 7$, then the closed-loop poles lie at −1.47 on the real axis and at $-0.76 \pm j2.04$ along the complex-conjugate branches of the root locus. These complex-conjugate positions have been chosen at the point of greatest damping along the root locus of Figure 12.11b. Nevertheless, the step response of the resulting closed-loop system, shown in Figure 12.12, is oscillatory and shows over 50% overshoot. The frequency of the oscillations can be increased or decreased by varying the gain K, but the damping cannot be improved if the poles lie on the given branches of the root locus.

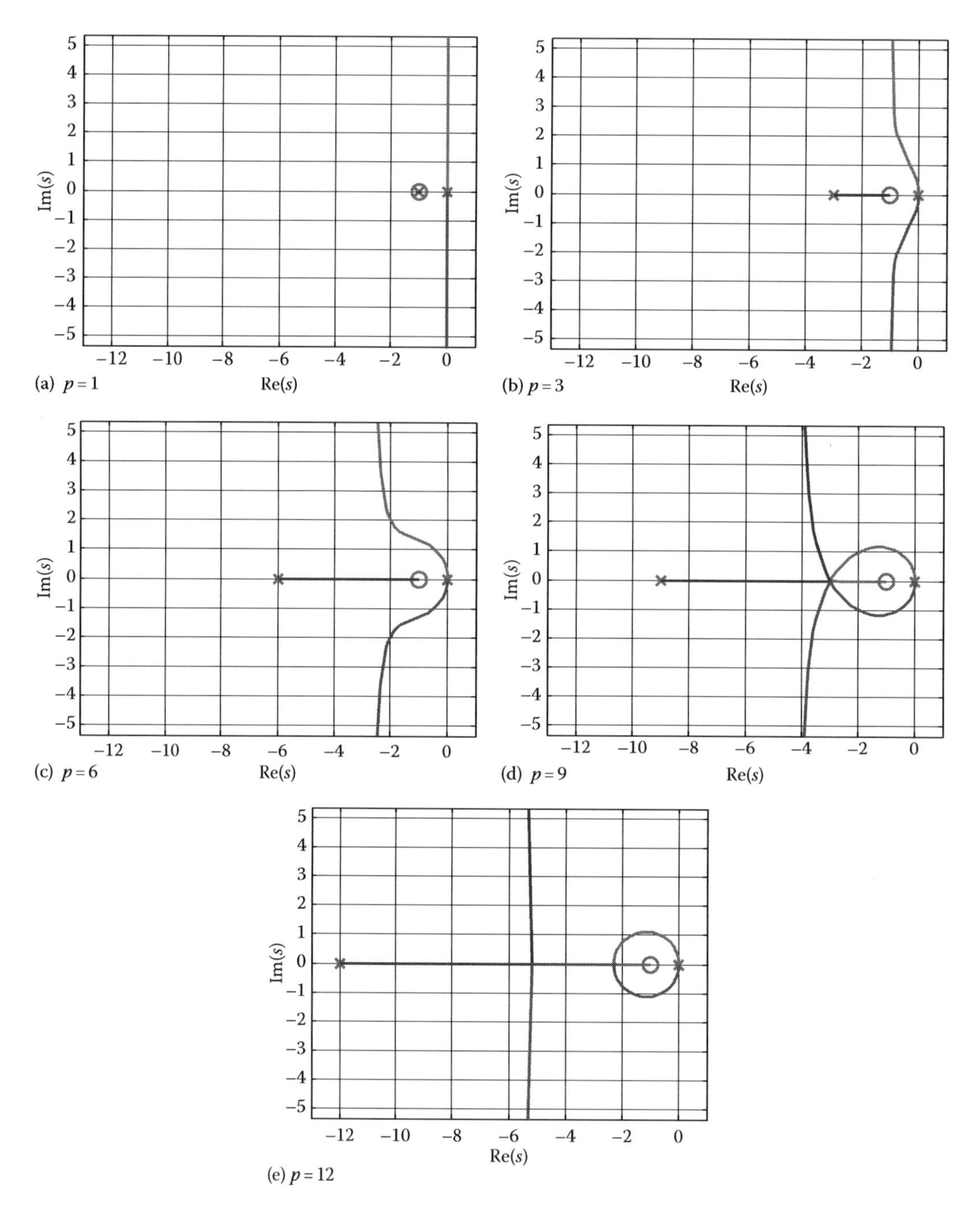

FIGURE 12.11 Root locus plots for inertial plant with lead compensator: (a) $p = 1$, (b) $p = 3$, (c) $p = 6$, (d) $p = 9$, and (e) $p = 12$.

To obtain a better-damped response, the pole of the compensator can be moved farther to the left. With the compensator pole more distant, the compensator zero attracts the locus more strongly to the left. (Frequency-domain analysis would show that the stabilizing phase lead of the compensator is increased.) Figure 12.11c shows the root locus plots for $p = 6$.

If the compensator parameters are chosen as $p = 6$ and $K = 15$, then the closed-loop poles lie at -2.32 on the real axis and at $-1.84 \pm j1.75$ along the complex-conjugate branches of the root locus. Again, these

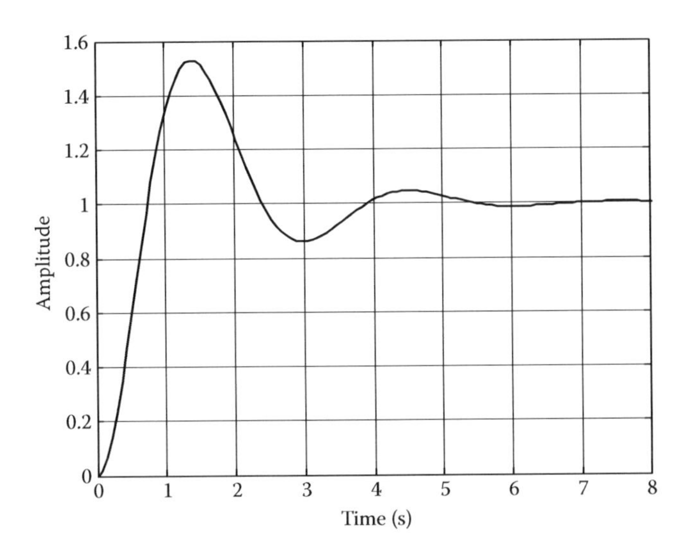

FIGURE 12.12 Step response of inertial plant with lead compensator, $p = 3$, $K = 7$. Closed-loop pole locations are -1.47 and $-0.76 \pm j2.04$.

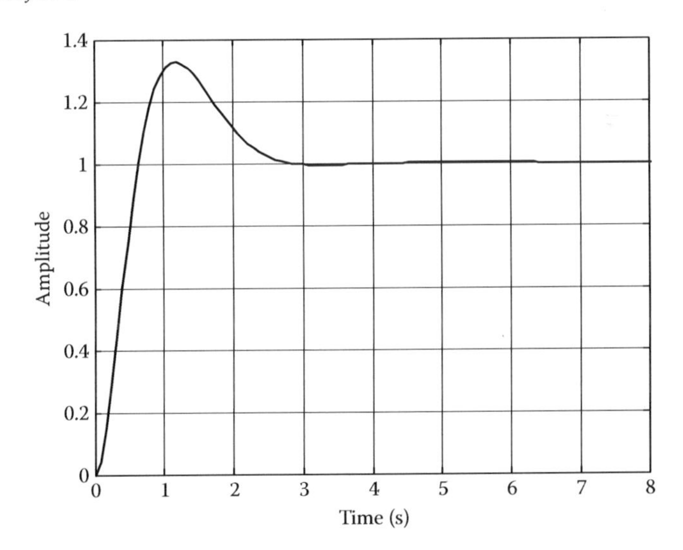

FIGURE 12.13 Step response of inertial plant with lead compensator, $p = 6$, $K = 15$. Closed-loop pole locations are -2.32 and $-1.84 \pm j1.75$.

complex-conjugate positions have been chosen at the point of greatest damping along the root locus of Figure 12.11c. The step response, shown in Figure 12.13, is well damped but still shows considerable overshoot resulting from the zero at $s = -1$.

For $p > 9$, the complex-conjugate branches of the locus are attracted to the left and downward sufficiently that they meet the real axis, resulting in break-in and breakaway points to the left of the compensator zero, as shown in Figure 12.11d through e. It might seem desirable to fix $p > 9$ and to choose the gain K to obtain repeated poles at the break-in point on the negative real axis, and thus achieve critical damping. However, there are at least two disadvantages to such a design. First, the actual closed-loop pole locations will be extremely sensitive to the value of K, so that the poles will not be accurately placed in practice. Second, regardless of the increased damping associated with the pole locations, the system response will still show substantial overshoot because of the zero at $s = -1$.

12.4.2 PID Compensation of an Oscillatory System

Consider again the feedback system shown in Figure 12.10. For this example, let the plant transfer function be given by

$$G_p(s) = \frac{K}{(s^2+1)}, \quad K = 1. \tag{12.41}$$

This transfer function has two poles on the imaginary axis and may represent an undamped spring-mass system with an external force input acting on the mass. The root locus method will be used in the design of a PID (proportional + integral + derivative) controller for this system. The assumed form of the compensator is

$$G_c(s) = K_d s + K_p + \frac{K_i}{s}, \tag{12.42}$$

where K_d, K_p, and K_i are the design parameters.

Since $G_c(s)$ is not a proper transfer function, the PID controller can be realized only approximately in practice. Specifically, the derivative term will have to be implemented as

$$\frac{K_d s}{\varepsilon s + 1} \approx K_d s, \tag{12.43}$$

where ε is a positive parameter much smaller than 1. The parameter ε will be ignored in the presentation of this example.

For the compensator given in Equation 12.42, the open-loop transfer function of Figure 12.10 is

$$G_p(s)G_c(s) = \frac{K_d s^2 + K_p s + K_i}{s(s^2+1)}, \tag{12.44}$$

which gives the closed-loop transfer function $T(s)$ as

$$T(s) = \frac{G_p(s)G_c(s)}{1 + G_p(s)G_c(s)} = \frac{K_d s^2 + K_p s + K_i}{s^3 + K_d s^2 + (1+K_p)s + K_i}. \tag{12.45}$$

The system is third order; therefore, by choice of the three compensator parameters, the three closed-loop poles may be placed arbitrarily. In a typical design procedure, K_p and K_d are chosen to meet transient response specifications and K_i is chosen to satisfy steady-state requirements.

In this example, the design approach will be first to define the ratio

$$\frac{K_i}{K_p} = \alpha, \tag{12.46}$$

so that the closed-loop transfer function may be written as

$$T(s) = \frac{K_d s^2 + K_p(s+\alpha)}{s(s^2 + K_d s + 1) + K_p(s+\alpha)}. \tag{12.47}$$

The poles of the closed-loop system are determined by the characteristic equation

$$s(s^2 + K_d s + 1) + K_p(s + \alpha) = 0. \tag{12.48}$$

If α and K_d are considered as constant parameters, and K_p is allowed to vary from 0 to ∞, then finding the solutions of Equation 12.48 is equivalent to a root locus problem as defined by Equation 12.6. The equivalent numerator polynomial

$$n(s) = (s + \alpha) \tag{12.49}$$

fixes the zero of the equivalent open-loop transfer function at $s = -\alpha$. The equivalent denominator polynomial

$$d(s) = s(s^2 + K_d s + 1) \tag{12.50}$$

sets the poles of the equivalent open-loop transfer function. Two of these poles depend on the parameter K_d. The other one always lies at $s = 0$, corresponding to the integral term in the compensator, which guarantees that the closed-loop system output will track constant inputs with zero steady-state error.

Figure 12.14a through d shows some root locus plots of the closed-loop characteristic equation (12.48). The value of α (i.e., the zero in the root locus plots) is kept fixed at 0.5. Four different values of K_d are used for the four plots, resulting in four different locations for the open-loop complex poles. The branches of each root locus plot represent the closed-loop pole locations as K_p varies from 0 to ∞, with α and K_d held constant. Note that the two zeros of $T(s)$ are not seen directly on the root locus plots but may be found directly as the roots of the polynomial

$$n_T(s) = K_d s^2 + K_p(s + \alpha). \tag{12.51}$$

In this example, the design may be done without much regard for these zeros, as they turn out not to dominate the system response.

For $K_d = 0$ (Figure 12.14a), the controller provides no damping. Two branches of the root locus lie entirely in the right half-plane, meaning that the system is unstable for every $K_p > 0$. As K_d is increased, the damping ratio associated with the open-loop poles increases, and the complex branches of the root loci lie farther to the left. For $K_d = 0.5$ (Figure 12.14b), the root locus lies entirely in the left half-plane, meaning that the system is stable for every $K_p > 0$. Note that, as K_d is increased from 0 to 2 (Figure 12.14a through d), the complex-conjugate pair of open-loop poles follow a semicircular path at a unit distance from the origin. For $K_d = 2$ (Figure 12.14e), the open-loop poles are repeated at $s = -1$.

Let the derivative gain be chosen as $K_d = 2$. Then different values of the proportional gain K_p produce closed-loop poles at different positions along the locus of Figure 12.14e. Figure 12.15a through d shows four different step-response plots for the closed-loop system, corresponding to four different gains K_p. As K_p increases, the speed of response increases as the real pole moves to the left, but the overshoot increases as the damping ratio associated with the complex-conjugate poles decreases.

For $K_p = 1$, the closed-loop poles lie at -0.35 on the real axis and at $-0.82 \pm j0.86$ along the complex-conjugate branches of the root locus. The zeros of $T(s)$ lie at $-0.25 \pm j0.43$. The real-axis pole is associated with a slow exponential decay in the error, which causes a relatively long settling time, as shown in Figure 12.15a.

For $K_p = 2$, the closed-loop poles lie at -0.43 on the real axis and at $-0.78 \pm j1.31$ along the complex-conjugate branches of the root locus. The zeros of $T(s)$ lie at $-0.50 \pm j0.50$. As the real-axis pole has moved to the left, the settling time is reduced, as shown in Figure 12.15b.

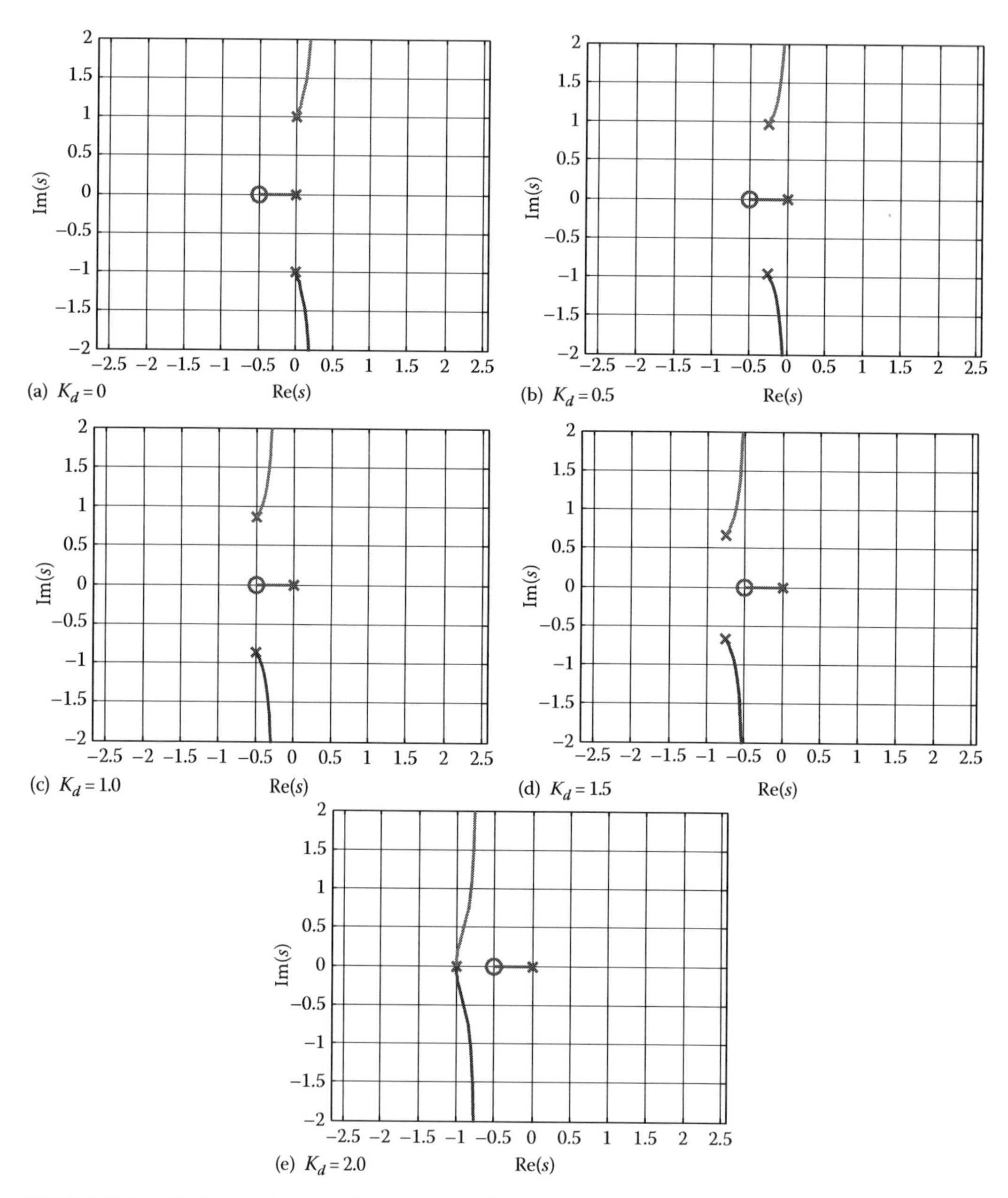

FIGURE 12.14 Root locus plot for oscillatory plant with PID compensator, with $K_i/K_p = 0.5$: (a) $K_d = 0$, (b) $K_d = 0.5$, (c) $K_d = 1.0$, (d) $K_d = 1.5$, and (e) $K_d = 2.0$.

For $K_p = 4$, the closed-loop poles lie at −0.47 on the real axis and at $-0.77 \pm j1.92$ along the complex-conjugate branches of the root locus. The zeros of $T(s)$ are repeated at −1.00. The settling time is further reduced, although the complex-conjugate poles are beginning to produce some oscillation in the response, as shown in Figure 12.15c.

For $K_p = 8$, the closed-loop poles lie at −0.48 on the real axis and at $-0.75 \pm j2.77$ along the complex-conjugate branches of the root locus. The zeros of $T(s)$ lie at −0.59 and −3.41. The complex-conjugate poles now cause more oscillation and hence a slight increase in the settling time, and the zero at −0.59 contributes to the large first peak in the response, as shown in Figure 12.15d.

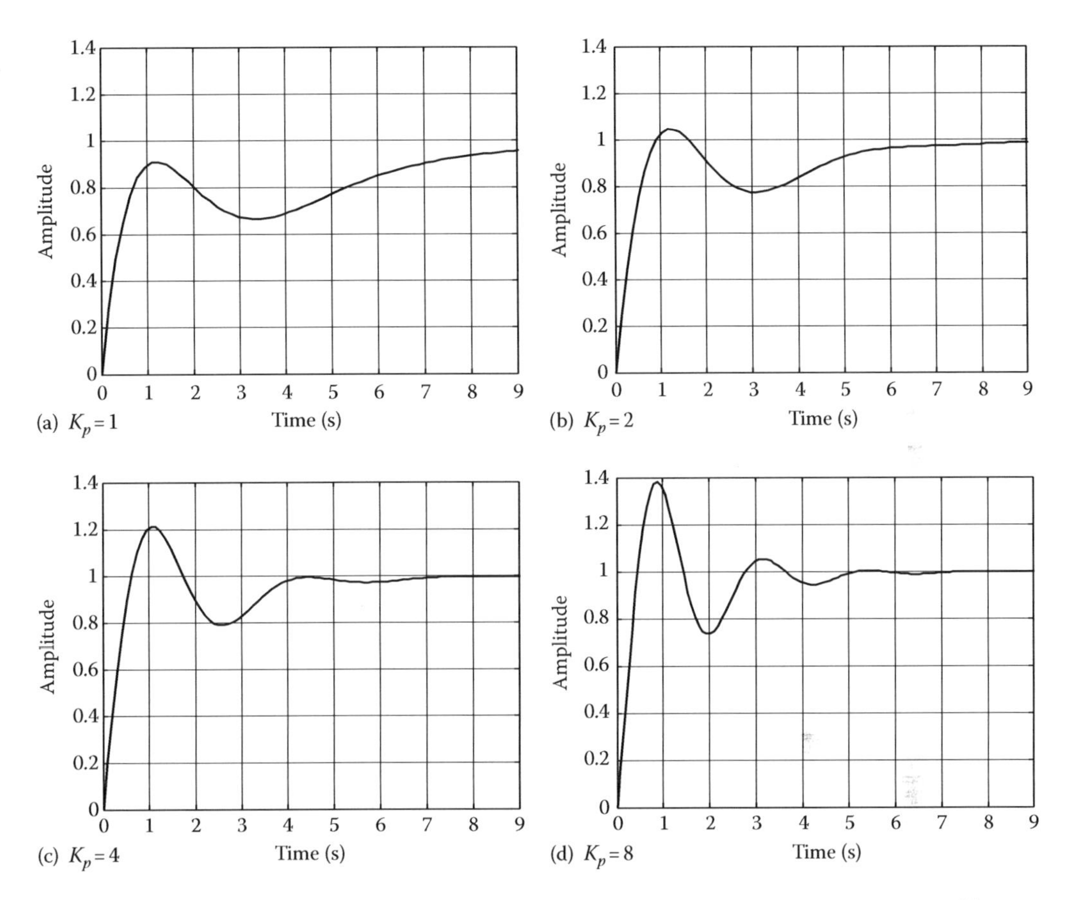

FIGURE 12.15 Step response plots for oscillatory plant with PID compensator, $K_i/K_p = 0.5$, $K_d = 2$: (a) $K_p = 1$, (b) $K_p = 2$, (c) $K_p = 4$, and (d) $K_p = 8$.

The design may be concluded by choosing $\alpha = K_i/K_p = 0.5$, $K_d = 2$, and $K_p = 4$. These parameters give a closed-loop system with the step response shown in Figure 12.15c. The damping of the response may generally be improved by increasing K_d; however, if K_d is increased too much, the zeros of $T(s)$ will become complex and move to the right, possibly causing severe overshoot in the step response. As an alternative, the damping may be improved by decreasing the parameter α, although this will tend to produce a slower decay in the exponential mode of the response.

12.5 Conclusion

The root locus plots reveal the general effects of feedback compensator parameters on the poles of a closed-loop system transfer function. Without tedious calculations, hand-sketched root locus plots yield considerable useful information on the closed-loop pole locations. Computer-generated root locus plots yield more detailed and precise information that may be needed to exactly specify the desired compensator parameters.

The root locus method is by itself a useful tool in feedback control design. Its focus is naturally on the system transient response, which is closely related to the closed-loop pole and zero locations. However, a complete design must also account for the system steady-state response and the stability margins, which may not be easily seen from the root locus plots. Thus, the root locus method is more powerful when used in conjunction with frequency-response plots, which show the loop gain and phase directly.

References

D'Azzo, J. J. and Houpis, C. H. 1988. *Liner Control System Analysis and Design: Conventional and Modern*, 3rd edn., McGraw-Hill, New York.

Evans, W. R. 1948. Graphical analysis of control systems, *Trans. Am. Inst. Electrical Eng.*, 67:547–551.

Franklin, G. F., Powell, J. D., and Emami-Naeini, A. 2009. *Feedback Control of Dynamic Systems*, 6th edn., Pearson Prentice Hall, Upper Saddle River, NJ.

Kuo, B. C. 1995. *Automatic Control Systems*, 7th edn., Prentice Hall, Englewood Cliffs, NJ.

Nise, N. S. 2008. *Control Systems Engineering*, 5th edn., Wiley, New York.

13

Variable Structure Control Techniques

Asif Šabanović
Sabanci University

Nadira Šabanović-Behlilović
Sabanci University

The study of the discontinuous control systems has been maintained on high level in the course of the development of the control system theory. Relay regulators were present in very early days of the feedback control system application [FL53,TS55]. The study of dynamical systems with discontinuous control is a multifaceted problem, which embraces control theory and application aspects. The problem of dynamical plant with discontinuous control has been approached by mathematicians, physicists, and engineers to solve problems that arise in their own field of interest.

The discontinuity of control results in a discontinuity of the right-hand side of the differential equations describing the system motion. If such discontinuities are deliberately introduced on certain manifold in the system state space, the motion in a sliding mode may occur in the system. This type of motion features some attractive properties and has long since been applied in relay systems. In some fields (power electronics and electrical drives), the switching is a "way of life" and the systems that are treated in these fields are discontinuous by themselves without any artificial introduction of switching. Sliding modes are the basic motion in Variable Structure Systems (VSS) [EM70,VI92].

While in the sliding mode the trajectories of the state vector belong to the manifold of lower dimension than of the state space, so the order of differential equations describing motion is reduced. In most of the practical systems, the motion in sliding mode is control independent and is determined by the plant parameters and the equations of the discontinuity surfaces. This allows the initial control problem to be decoupled into problems of lower dimension dealing with enforcement of sliding mode and desired equations of motion in sliding mode. The control is used to restrict the motion of the system on the sliding mode subspace while the desired character of the motion in the intersection of discontinuity surfaces is provided by an appropriate selection of the equations of the discontinuity surfaces. Under certain condition, the system with sliding mode may become invariant to variation of the dynamic properties of the control plant. The unremitting interest to discontinuous control systems enhanced by their effective application to solution of problems diverse in their physical nature and functional purpose is a basic element in favor of the importance of this area.

In this chapter, basic methods of the analysis and design of VSS with sliding modes will be discussed. Our intention is not to give full account of the VSS methods but to introduce and explain those that will be later used in the chapter. The discussion on the VSS application will focus on switching converters and electrical drives—systems that structurally have discontinuous element—and on motion control systems.

13.1 Sliding Mode in Continuous-Time Systems

VSS are originally defined in continuous time for dynamic systems described by ordinary differential equations with discontinuous right-hand side. In such a system, the so-called sliding mode motion can arise. This motion is represented by the state trajectories in the sliding mode manifold and high-frequency switching in the control inputs. The fact that motion belongs to certain manifold in state space with a dimension lower than order of the system results in the motion equations order reduction.

In general, motion of a control system with discontinuous right-hand side may be described by differential equation

$$\begin{aligned}
&\dot{x} = f(x,t,u) \quad x \in \Re^n, \quad u \in \Re^m \\
&u_i = \begin{cases} u_i^+(x,t) & \sigma_i(x) > 0 \\ u_i^-(x,t) & \sigma_i(x) < 0 \quad i = 1,\dots,m \end{cases} \\
&x^T = \begin{bmatrix} x_1 & \dots & x_n \end{bmatrix} \\
&u^T = \begin{bmatrix} u_1 & \dots & u_m \end{bmatrix} \\
&\sigma^T = \begin{bmatrix} \sigma_1 & \dots & \sigma_m \end{bmatrix}
\end{aligned} \tag{13.1}$$

It is assumed that $f(x,t,u)$, u_i^+, u_i^- and σ_i are continuous functions of the system state. Motion of the system (13.1) may exhibit different behaviors depending on the changes in control input. It may stay on one of the discontinuity surfaces $\sigma_i(x) = 0$ or on their intersections. The motion will stay on surface $\sigma_i(x) = 0$, if the distance from surface and its rate of change are of the opposite signs [EM70]. The motion in discontinuity surfaces is so-called sliding mode motion. This motion is characterized by high frequency (theoretically infinite) switching of the control inputs. Additionally, due to changes in control input, function $f(x, t, u)$ on the different side of the discontinuity surface ($x^1 \neq x^2$) satisfy relation $f(x^1,t,u_i^+) \neq f(x^2,t,u_i^-)$ and consequently conditions for the uniqueness of solution of ordinary differential equations do not hold in this case. It has been shown that, in the cases in which the so-called regularization method [VI92] yields an unambiguous result, the equations of motion in discontinuity surface exist and can be derived from the equations of discontinuity surface and the system motion outside of the discontinuity surface. Regularization method consists in replacing ideal equations of motion (13.1) by a more accurate one $\dot{x} = f(x, t, \tilde{u})$, which takes into account nonidealities (like hysteresis, delay, dynamics nonidealities, etc.) in implementation of the control input $\tilde{u}$. Equations $\dot{x} = f(x, t, \tilde{u})$ have solution in conventional sense but motion is not anymore in manifold $\sigma_i(x) = 0$ but run in some Δ—vicinity (boundary layer) of that manifold. If thickness Δ of the boundary layer tends to zero, the motion in the boundary layer tends to the motion of system with ideal control. Equations of motion obtained as a result of such a limit process are regarded as ideal sliding mode. The procedure may lead to unique solution for the sliding mode equations as the function of the motion outside the manifold $\sigma_i(x) = 0$. For systems linear with respect to control, the regularization allows substantiation of so-called equivalent control method, which allows simple procedure of finding equations of motion in sliding mode.

A formal procedure is as follows: the projection of the system motion $\sigma(x, t) = 0$ in manifold $\sigma_i(x) = 0$ expressed as $\dot{\sigma}(x) = Gf(x, t, u_{eq}) = 0$ with $[\partial\sigma/\partial x] = G$ is solved for control $u = u_{eq}(G, f, x, t)$ and this solution is plug in equation $\dot{x} = f(x, t, u_{eq})$. Equations obtained as a result of such a procedure are accepted as sliding mode equations in manifold $\sigma_i(x) = 0$. By virtue of the equation $\dot{\sigma}(x) = Gf(x, t, u_{eq}) = 0$, it is obvious that motion in sliding mode will proceed along trajectories that lie in manifold $\sigma(x) = 0$, if initial conditions are $\sigma(x(0)) = 0$.

13.1.1 Equivalent Control and Equations of Motion

Further analysis will be restricted to the systems linearly dependent on control

$$\begin{aligned}
\dot{x} &= f(x) + B(x)u(x) + Dh(x,t) \quad x \in \Re^n, \quad u \in \Re^m, \quad h \in \Re^p \\
u_i &= \begin{cases} u_i^+(x,t) & \sigma_i(x) > 0 \\ u_i^-(x,t) & \sigma_i(x) < 0 \quad i = 1,\ldots,m \end{cases} \\
x^T &= \begin{bmatrix} x_1 & \ldots & x_n \end{bmatrix} \\
u^T &= \begin{bmatrix} u_1 & \ldots & u_m \end{bmatrix} \\
\sigma^T &= \begin{bmatrix} \sigma_1 & \ldots & \sigma_m \end{bmatrix} \\
h^T &= \begin{bmatrix} h_1 & \ldots & h_p \end{bmatrix}
\end{aligned} \tag{13.2}$$

In accordance with this equivalent control method, in (13.2) control should be replaced by the equivalent control, which is the solution to $\dot{\sigma}(x, t, u = u_{eq}) = 0$, with

$$\dot{\sigma}(x,t) = Gf + GBu + GDh, \quad G = \frac{\partial\sigma(x,t)}{\partial x} \in R^{m\times n} \tag{13.3}$$

The rows of matrix G are gradients of functions $\sigma_i(x)$. If $\det(G(x)B(x)) \neq 0$ for all x, the equivalent control for system (13.2) in manifold $\sigma(x) = 0$ can be from (13.3) calculated as

$$\dot{\sigma}(x) = 0 \quad \Rightarrow \quad u_{eq} = -(GB)^{-1} G(f + Dh) \tag{13.4}$$

The equivalent control (13.4) enforces the projection of the system velocity vector in the direction of gradients of surfaces defining sliding mode manifold to be zero. Substitution of equivalent control into (13.2) yields the equation

$$\begin{aligned}
\dot{x} &= f + Bu_{eq} + Dh \\
&= \left(I - B(GB)^{-1}G\right)\left(f + Dh\right)
\end{aligned} \tag{13.5}$$

with

$$\sigma(x) = 0 \tag{13.6}$$

Equation 13.5 along with equation of sliding mode manifold (13.6) describes the ideal sliding mode motion. Equation 13.6 allows to determine m components of the state vector x as a function of the rest of $n - m$ components. By inserting these m components into (13.5), the sliding mode equations can be obtained as $n - m$ dimensional system with trajectories in sliding mode manifold if initial conditions are $\sigma(x(0)) = 0$. Equation 13.5 depends on matrix G—thus the selection of the sliding mode manifold $\sigma(x) = 0$.

Projection matrix $P = (I - B(GB)^{-1}G)$ in Equation 13.5 satisfies $PB = (B - B(GB)^{-1}GB) = 0$. If vector $(f + Dh)$ can be partitioned as $f + Dh = B\lambda + \xi$, then Equation 13.5 can be rewritten in the following form

$$f + Dh = B\lambda + \xi \quad \Rightarrow \quad \dot{x} = (I - B(GB)^{-1}G)\xi \tag{13.7}$$

The sliding mode equations are invariant in components of vector $(f + Dh)$ that lie in the range space of system input matrix B—so-called matching conditions [DB69]. This result offers a way of assessing the invariance of the sliding mode motion to different elements of the system (13.2) dynamics (parameters, disturbances, etc.).

So far we have been discussing the equations of motion and their features under assumption that initial state lies in the sliding mode manifold. The sliding mode equations are found to be of the lower order than the order of the system and under certain conditions may be invariant to the certain components of the system dynamics. This shows potential benefits of establishing sliding mode in dynamic systems. In order to effectively use these potential in the control system design, the question of the sliding mode existence conditions should be clarified along the sliding mode design methods.

13.1.2 Existence and Stability of Sliding Modes

Existence of sliding mode has to do with stability of the motion in the sliding mode manifold. The stability of nonlinear dynamics (13.2) in sliding mode manifold can be analyzed in the Lyapunov stability framework [VI92,RD94,KD96] by finding a continuously differentiable positive definite function $V(\sigma, x)$ such that its time derivative along trajectories of system (13.2) is negative definite. Without loss of generality let Lyapunov function candidate be selected as

$$V(\sigma, x) = \frac{\sigma^T \sigma}{2} \tag{13.8}$$

In order to satisfy Lyapunov stability conditions, $\dot{V}(\sigma, x)$ must be negative definite on the trajectories of system (13.2). Let stability requirement, for appropriate function $\zeta(V, \sigma, x) > 0$, be expressed as

$$\dot{V}(\sigma, x) = \sigma^T \dot{\sigma} \le -\zeta(V, \sigma, x) \tag{13.9}$$

It is obvious that by selecting structure of $\zeta(V, \sigma, x)$, one can determine specific rate of change of $V(\sigma, x)$ and can thus define the transient from initial state to the sliding mode manifold. The equation (13.3) describing projection of the system motion in sliding mode manifold can be rearranged into

$$\dot{\sigma} = GB(u + (GB)^{-1}G(f + Dh)) = GB(u - u_{eq}) \tag{13.10}$$

By selecting control as in (13.11) with $M(x)$ being a scalar function

$$\begin{aligned} &u = u_{eq} - (GB)^{-1} M(x) sign(\sigma), \quad M(x) > 0 \\ &[sign(\sigma)]^T = [sign(\sigma_1) \quad \dots \quad sign(\sigma_m)] \end{aligned} \tag{13.11}$$

the derivative $\dot{V}(\sigma, x)$ can be expressed as

$$\dot{V}(\sigma,x) = -M(x)\sigma^T sign(\sigma) < 0 \tag{13.12}$$

With control (13.11), the stability conditions (13.9) are satisfied and consequently sliding mode exists in manifold $\sigma(x) = 0$. Control (13.11) is component-wise discontinuous and Lyapunov function reaches zero value in finite time. The selection of function $M(x)$ determines the rate of change of the Lyapunov function. It can be selected as linear or nonlinear function of state thus enforcing different transients in reaching the equilibrium.

13.2 Design

Decreasing dimensionality of the equations of motion in sliding mode permits to decouple the design problem into two independent steps involving systems of dimensionality lower than the original system:

a. In the first step, the desired dynamics is determined from (13.5) and (13.6) by selecting sliding mode manifold;
b. In the second step, control input should be selected to enforce sliding mode or stated differently to make projection of system motion in sliding mode manifold (13.3) stable.

Equations 13.5 and 13.6 describe $(n - m)$ order continuous dynamical system but selection of the sliding mode manifold providing desired behavior requires selection of matrix G. In order to make selection of the sliding mode manifold simpler, it is useful to transform system (13.2) with matched disturbance $Dh = B\lambda$ in the so-called regular form [LU81]

$$\begin{aligned} \dot{x}_1 &= f_1(x_1,x_2) & x_1 &\in \Re^{n-m} \\ \dot{x}_2 &= f_2(x_1,x_2) + B_2(x_1,x_2)(u+\lambda) & x_2 &\in \Re^{m} \end{aligned} \tag{13.13}$$

In writing (13.13), the matched disturbance and nonsingular matrix $B_2 \in R^{m\times m}$ are assumed.

The first equation is of $n - m$ order and it does not depend on control. In this equation, vector x_2 can be handled as virtual control. Assume x_2, satisfying some performance criterion, is given by

$$x_2 = -\sigma_0(x_1) \tag{13.14}$$

and the dynamics described by first equation in (13.13) may be rewritten in the following form:

$$\dot{x}_1 = f_1(x_1, -\sigma_0(x_1)) \tag{13.15}$$

The second equation in (13.13) describes mth order system with m-dimensional control vector and full rank gain matrix B_2. In the second step of design, discontinuous control u should be selected enforcing sliding mode in manifold

$$\sigma = x_2 + \sigma_0(x_1) = 0 \tag{13.16}$$

The projection of the system (13.13) motion in manifold (13.16) can be written as

$$\begin{aligned} \dot{\sigma} &= \dot{x}_2 + \dot{\sigma}_0(x_1) \\ &= f_2(x_1, x_2) + B_2(u+\lambda) + G_0 f_1(x_1, x_2) \\ G_0 &= \frac{\partial \sigma_0}{\partial x_1} \end{aligned} \tag{13.17}$$

By selecting Lyapunov function candidate as in (13.8), stability in manifold (13.16) will be enforced if control is selected as

$$\begin{aligned} u &= u_{eq} - B_2^{-1} M(x_1, x_2) sign(\sigma), \quad M(x_1, x_2) > 0 \\ u_{eq} &= -\lambda - B_2^{-1}(f_2 + G_0 f_1) \\ \left[sign(\sigma)\right]^T &= \left[sign(\sigma_1) \quad \ldots \quad sign(\sigma_m)\right] \end{aligned} \tag{13.18}$$

Problem with control (13.18) is the requirement to have information on equivalent control—thus full information on the system's structure and parameters. In most of the real control systems, some estimation of the equivalent control $\hat{u}_{eq}$ and matrix B_2 will be available. In this situation, the structure of the control can be selected in the form

$$u = \hat{u}_{eq} - M(x_1, x_2) sign(\sigma), \quad M(x_1, x_2) > 0 \tag{13.19}$$

In (13.19), the estimate of equivalent control is used instead of the exact value and discontinuous part does not depend on the unknown gain matrix. In this case, the stability of the sliding mode can be proved as follows. Let scalar function $L > 0$ be the component-vice upper bound of $B_2(\hat{u}_{eq} - u_{eq})$. Then the time derivative of the Lyapunov function (13.8) on the trajectories of system (13.13) with control (13.19) and with $M(x_1, x_2) \geq L$ is negative definite and consequently sliding mode will be enforced in manifold $\sigma = x_2 + \sigma_0(x_1) = 0$. Manifold will be reached in finite time and motion in sliding mode is described by (13.15)—thus satisfying the system specification of the dynamics of closed loop system.

13.2.1 Control in Linear System

The discontinuous control strategies become transparent when applied to linear systems. In this section, we will discuss the realization of enforcement of sliding mode in a linear system:

$$\begin{aligned} &\dot{x} = Ax + Bu + Dh \quad x \in \Re^n, \quad u \in \Re^m, \quad h \in \Re^p \\ &A \in \Re^{n\times n}, \quad B \in \Re^{n\times m}, \quad D \in \Re^{n\times p} \end{aligned} \tag{13.20}$$

Further we will assume that pair (A, B) is controllable, $rank(B) = m$, and disturbance satisfies matching conditions $Dh = B\lambda$. By reordering components of vector x, matrix B may be partitioned into two matrices $B_1 \in \Re^{(n-m)\times m}$ and $B_2 \in \Re^{m\times m}$ with $rank(B_2) = m$. As shown in [VI92,ED98], there exists nonsingular transformation $x_T = Tx$ such that

$$TB = \begin{bmatrix} B_1 \\ B_2 \end{bmatrix} = \begin{bmatrix} 0 \\ B_2 \end{bmatrix} \tag{13.21}$$

and, consequently transformed system can be written in the following form

$$\begin{aligned} \dot{x}_1 &= A_{11}x_1 + A_{12}x_2 & x_1 &\in \Re^{n-m} \\ \dot{x}_2 &= A_{21}x_1 + A_{22}x_2 + B_2(u + \lambda) & x_2 &\in \Re^m, \quad u \in \Re^m \end{aligned} \tag{13.22}$$

with $A_{ij}(i, j = 1, 2)$ being constant matrices and pair (A_{11}, A_{12}) being controllable.

Sliding mode design procedure allows handling x_2 as a virtual control input to the first equation in (13.22). By selecting $x_2 = -Gx_1$, first equation in (13.22) becomes $\dot{x}_1 = (A_{11} - A_{12}G)x_1$, and, due to the controllability of the pair (A_{11}, A_{12}), matrix G may be selected in such a way that $(n - m)$ eigenvalues are assigned desired values.

Control should be selected to enforce desired value of $x_2 = -Gx_1$ or to enforce sliding mode in manifold (13.23)

$$\sigma = x_2 + Gx_1 = 0 \tag{13.23}$$

Sliding mode may be enforced in manifold (13.23) by selecting control

$$\begin{aligned} u &= u_{eq} - \alpha B_2^{-1}\frac{\sigma}{\|\sigma\|}, \quad \alpha > 0, \|\sigma\| = \sqrt{(\sigma^T\sigma)} \\ u_{eq} &= -\lambda - B_2^{-1}\left[(A_{21} + GA_{11})x_1 + (A_{22} + GA_{12})x_2\right] \end{aligned} \tag{13.24}$$

The derivative of the Lyapunov function $V = \sigma^T\sigma/2$ is $\dot{V} = -\alpha B_2^{-1}\sigma^T\sigma/\|\sigma\| < 0$, thus guaranteeing the sliding mode. Control (13.24) has discontinuities in manifold (13.23) but not in each surface $\sigma_i = 0$, $(i = 1, \ldots, m)$.

13.2.2 Discrete-Time SMC

The direct application of discontinuous control in sampled systems will inevitably lead to chattering due to the finite sampling rate. The chattering appears due to the finite switching frequency limited by sampling rate and problem of realization of the continuity of the equivalent control. The salient feature of the sliding mode motion is that system trajectories reach selected manifold in finite time and remain in the manifold thereafter. With this understanding in discrete-time systems, the concept means that after reaching sliding mode manifold, the distance from it at each subsequent sampling interval should be zero. This requires to determine control input at sampling time kT such that $\sigma(kT + T) = 0$ with initial condition $\sigma(kT) = 0$.

The concept will be again explained on linear time invariant plant (13.20). By applying sample and hold process with sampling period T, and integrating solution over interval $t \in [kT, (k + 1)T]$ with $u(t) = u(kT)$ and disturbance constant during sampling interval, $h(t) = h(kT)$, $(k = 1, 2, \ldots)$, the discrete time model of plant (13.20) may be represented as

$$x_{k+1} = \bar{A}x_k + \bar{B}u_k + \bar{D}h_k \tag{13.25}$$

$$\bar{A} = e^{AT}, \quad \bar{B} = \int_0^T e^{A(T-\tau)}Bd\tau, \quad \bar{D} = \int_0^T e^{A(T-\tau)}Dd\tau$$

Let sliding mode manifold be defined as $\sigma_k = Gx_k$. The change of sliding mode manifold can be expressed as

$$\begin{aligned} \sigma_{k+1} &= Gx_{k+1} = G\bar{A}x_k + G\bar{B}u_k + G\bar{D}h_k \\ &= \sigma_k + G(\bar{A} - I)x_k + G\bar{D}h_k + G\bar{B}u_k \end{aligned} \tag{13.26}$$

If $\det(G\bar{B}) \neq 0$ from (13.26) control enforcing $\sigma_{k+1} = 0$, thus yielding motion in sliding mode manifold, is called equivalent control [DR89,VI93,KD96]

$$u_{keq} = -(G\bar{B})^{-1}\left(\sigma_k + G(\bar{A} - I)x_k + G\bar{D}h_k\right) \tag{13.27}$$

It is important to note that matching conditions now are defined in terms of matrices G, $\bar{B}$, $\bar{D}$ and have to be examined for discretized system. The required magnitude of control (13.27) may be large and limitation should be applied. If the control resources are defined by $\|u_{keq}\| < M$, then control can be implemented as in (13.28)

$$u_k = \begin{cases} u_{keq} & \text{if } \|u_{keq}\| \leq M \\ M\dfrac{u_{keq}}{\|u_{keq}\|} & \text{if } \|u_{keq}\| > M \end{cases} \tag{13.28}$$

It can be proven that such control will enforce sliding mode after finite number of sampling intervals. The most significant difference between continuous time and discrete time sliding modes is that motion in the sliding mode manifold may occur in discrete time systems with continuous (in the sense of discrete-time systems) control input.

13.2.3 Sliding Mode–Based Observers

In most of the control systems, design information on control plant is not complete and should be filled by estimating missing data from known structure of the system and the available measurements. The situation in control systems with sliding mode is not different. The realization of the control law requires at least information to evaluate distance from sliding mode manifold and in some realizations the equivalent control or at least some of its components. Estimation of the system state variables and estimation of the system disturbance are often part of the overall control system design. In general, observers are constructed based on available measurements and nominal—known—structure of the system. The idea of sliding mode application in observer design rests on designing control that forces some output of the nominal plant to track corresponding output of the real plant in sliding mode. Then equivalent control can be used to determine information reflected in the difference between nominal and real plant. The sliding mode–based state observers are well detailed in the available literature [VI92,ED98]. They are mostly based on the idea of Luenberger observers with control enforcing sliding mode thus achieving finite time convergence.

In order to illustrate some other ideas related to sliding mode–based observers let look at first-order system with scalar control input

$$\dot{x} = f(x,a) + u \tag{13.29}$$

Assume control input u and output x measured, $f(x, a)$ unknown continuous scalar function of state, time, and parameter a. Assume that nominal plant as integrator with input consisting of the known part of the system structure $f_o(\hat{x}, a_o)$, control input of actual plant, and the additional—observer control input u_o are to be determined:

$$\dot{\hat{x}} = f_o(\hat{x},a_o) + u + u_o \tag{13.30}$$

Let observer control input u_o be selected such that sliding mode is enforced in

$$\sigma = x - \hat{x} \tag{13.31}$$

From $\dot{\sigma} = \dot{x} - \dot{\hat{x}} = 0$, the equivalent control can be expressed as

$$\begin{aligned} u_{oeq} &= f(x,a) - f_o(\hat{x},a_o) \\ \hat{x} &= x \end{aligned} \tag{13.32}$$

In sliding mode for system (13.29), in manifold (13.31), the equivalent control is equal to the difference between real plant $f(x, a)$ and assumed plant $f_o(\hat{x}, a_o)$ structure. That would allows us to estimate the real plant unknown structure as

$$f(x,a) = f_o(\hat{x},a_o) + u_{oeq} \tag{13.33}$$

The estimation (13.33) is valid after sliding mode in manifold (13.31) is established. It is obvious that (13.33) allows estimation of the function $f(x, a)$ but not its components separately. In the case of MIMO systems, the procedure follows the same steps. The difference comes from dimensionality of the system and the unequal dimensions of the system and the dimension of the sliding mode manifold.

Let system (13.29) be linear with respect to unknown parameter a, thus may be expressed as $f(x, a) = f_x(x) + f_a(x)a$. Assume pair (x, u) is measured, structure of functions $f_x(x)$ and $f_a(x)$ known continuous functions. Let us construct a model as

$$\dot{\hat{x}} = f_x(\hat{x}) + f_a(\hat{x})u_o + u \tag{13.34}$$

System (13.34) is linear with respect to control u_o, thus sliding mode can be established in $\sigma = x - \hat{x} = 0$. In sliding mode for system (13.29), (13.34) leads to algebraic equation linear with respect to unknown parameter, and consequently for $x = \hat{x}$ unknown parameter can be determined:

$$\begin{gathered} \dot{x} - \dot{\hat{x}} = f_a(x)a - f_a(\hat{x})u_{oeq} = 0 \\ x = \hat{x} \quad \Rightarrow \quad f_a(x) = f_a(\hat{x}) \quad \Rightarrow \quad u_{oeq} = a \end{gathered} \tag{13.35}$$

Extension of this result to MIMO systems is possible under assumptions that the system dynamics can be linearly parametrized with some set of parameters. The selection of these parameters is neither simple nor unique. If number of such parameters is lower or equal to the order of system, then application is straight forward. In the case that number of parameters is higher than order of system, then additional constraint equations should be found. Interesting application of this idea for vision-based motion estimation is presented in [UN08].

Assume MIMO system (13.36), with measured x, u and unknown vector valued function $f(x,a)$, matched unknown and unmeasured disturbance $Dh(x,t) = B\lambda$:

$$\begin{gathered} \dot{x} = f(x,a) + Bu + Dh(x,t) \\ x \in \Re^n, \quad u \in \Re^m, \quad d \in \Re^p, \quad rank(B) = m, \quad m < n \end{gathered} \tag{13.36}$$

Let "nominal" plant of system (13.36) be given by (13.37)

$$\begin{gathered} \dot{\hat{x}} = f_o(\hat{x},a_o) + Bu + Bu_o \\ \hat{x} \in \Re^n, \quad u \in \Re^m, \quad u_o \in \Re^m, \quad rank(B) = m \end{gathered} \tag{13.37}$$

Let control u_o be selected to enforce sliding mode in manifold

$$\sigma = G(x - \hat{x}) = 0 \quad \sigma \in \Re^m, \quad rank(G) = m \tag{13.38}$$

Assuming det(GB) ≠ 0 equivalent control for system (13.36), (13.37) in manifold (13.38) determined as the solution of $\dot{\sigma}(G, x, \hat{x}) = 0$ is

$$\begin{aligned} u_{oeq} &= \lambda - (GB)^{-1}G(f(x,a) - f_o(\hat{x},a_o)) \\ G(x - \hat{x}) &= 0 \end{aligned} \tag{13.39}$$

Equation 13.39 allows developing different schemes in the sliding mode–based observers. For example, if the plant structure $f(x, a)$ is known then from (13.39) projection of the disturbance to the range space of input matrix can be fully estimated by structure (13.36) through (13.39). That offers a possibility to develop estimation of disturbance guarantying finite time convergence. By feeding disturbance estimation into the system input, one can obtain unknown input cancellation. For example, by partitioning system state as in (13.13), modeling block describing dynamics of x_2 by simple integrator $\dot{\hat{x}}_2 = B_2(u + u_o)$ and selecting u_o, which enforces sliding mode in manifold $\sigma = x_2 - \hat{x}_2$ equivalent control becomes $B_2 u_{oeq} = f_2(x_1, x_2) + B_2\lambda$. Feeding $-B_2 u_{oeq}$ to the input of second block in (13.13), the dynamics of overall system reduces to

$$\begin{aligned} \dot{x}_1 &= f_1(x_1, x_2) \\ \dot{x}_2 &= B_2 u \end{aligned} \tag{13.40}$$

Proposed structure effectively linearizes second block in system (13.13) by compensating matched disturbance and the part of the system dynamics. This result can be predicted by analyzing (13.40).

13.3 Some Applications of VSS

As shown sliding mode–based design can be applied to systems with and without discontinuities in control. That offers a wide scope for the application of sliding mode–based design to different dynamic systems. At one side are systems—like power electronics converters—in which switching is natural consequence of operational requirements and are not introduced due to the control system preferences. Application of sliding mode control in such systems is consequence of the system nature and at the same time illustrates very clearly the peculiarities of the sliding mode design. In this category, one can include power converters as electrical sources and electrical drives as example of the power conversion control.

On the other hand, complex mechanical systems like robots represent challenge in control system design due to their complexity and nonlinear nature. Despite the fact that in many such systems electrical machines are primary source of actuation and consequently power converters are used in their control structure, mechanical motion systems are mostly modeled with forces/torques as control input variables. In such systems, discontinuity of control may not be easy to implement or it may cause excitement of unmodeled dynamics or excessive wear of actuators. At the same time, robustness of the sliding mode with respect to matched disturbances and system parameters is very attractive feature. These systems are very good examples of application of sliding modes for systems with continuous inputs.

13.3.1 Control in Power Converters

The role of a power converter is to modulate electrical power flow between power sources. A converter should enable interaction of any input source to any output source of the power systems interconnected by converter (Figure 13.1). Disregarding wide variety of designs in most switching converters, control of power flow is accomplished by varying the length of time intervals for which one or more energy storage elements are connected to or disconnected from the energy sources. That allows analysis of converter as switching matrix enabling interconnections and that way creating systems with variable structure.

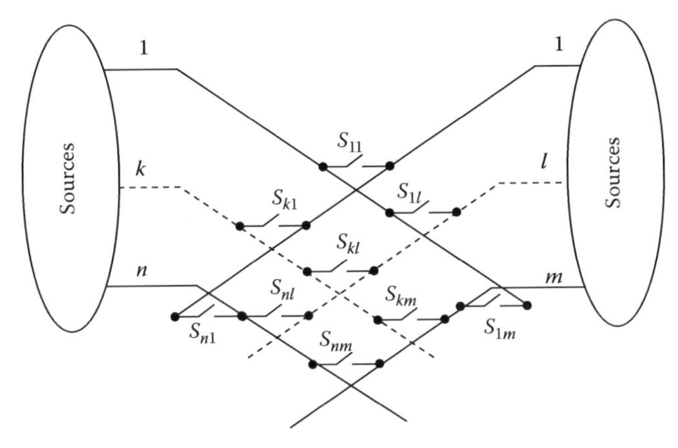

FIGURE 13.1 Converter as interconnecting element.

For purposes of mathematical modeling operation of a switch may be described by a two-valued parameter $u_{ij}(t)$, ($i = 1, \ldots, n$; $j = 1, \ldots, m$), having value 0 when the switch is open and value 1 when the switch is closed. The laws governing the operation of electrical circuits restrict operation of the switches in a sense that voltage source cannot be shortened and current source cannot be left out of the closed loop. These restrictions limit set of permissible combinations of the switches, states.

13.3.1.1 DC-to-DC Converters

In control of DC-to-DC converters, only "canonical structures"—buck, boost, and buck boost (Figure 13.2)—in continuous current mode will be analyzed. All three of these structures use inductance to transfer energy from the source to the sink. The capacitances are used as storage elements to smooth the output voltage, so basic structures are represented by two storage elements and switch realizing switching of the current path. The peculiarities of the structures are reflected on the mathematical

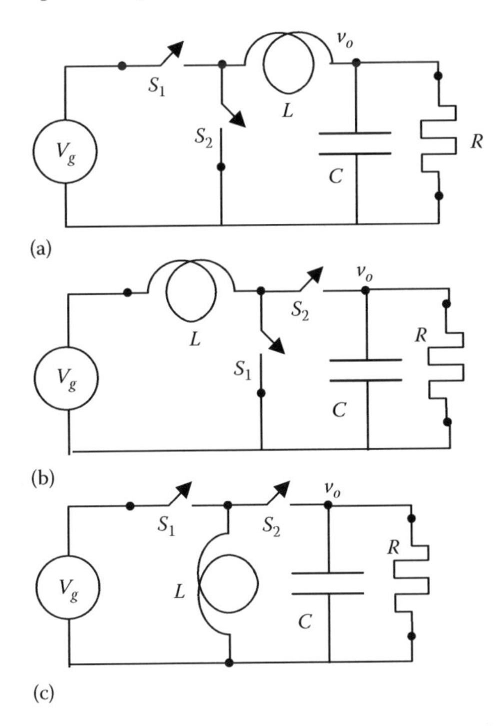

FIGURE 13.2 Topology of DC-to-DC converters: (a) buck, (b) boost, and (c) buck–boost.

description of the converters. All three structures (Figure 13.2) can be represented in regular form. Scalar parameter u, describing operation of the switch, is treated as the discontinuous control input and is defined as in (13.41).

$$u = \begin{cases} 1 & \text{if switch } S_1 = \text{ON} \quad \text{and} \quad \text{switch } S_2 = \text{OFF} \\ 0 & \text{if switch } S_1 = \text{OFF} \quad \text{and} \quad \text{switch } S_2 = \text{ON} \end{cases} \tag{13.41}$$

Operation of buck converter (Figure 13.2a) is described as in (13.42)

$$\begin{aligned} \frac{dv_o}{dt} &= \frac{i_L}{C} - \frac{v_o}{RC} & \frac{dv_o}{dt} &= f_v(i_L, v_o) \\ \frac{di_L}{dt} &= \frac{V_g}{L}u - \frac{v_o}{L} & \frac{di_L}{dt} &= f_i(v_o) + b(V_g)u \end{aligned} \tag{13.42}$$

For the boost converter (Figure 13.2b) both first and second block depend explicitly on control as shown in (13.43) with control parameter expressed as in (13.41)

$$\begin{aligned} \frac{dv_o}{dt} &= \frac{i_L}{C}(1-u) - \frac{v_o}{RC} & \frac{dv_o}{dt} &= f_v(v_o) + b_1(i_L)(1-u) \\ \frac{di_L}{dt} &= \frac{V_g}{L} - \frac{v_o}{L}(1-u) & \frac{di_L}{dt} &= f_i(V_g) + b(v_o)(1-u) \end{aligned} \tag{13.43}$$

The buck-boost converter (Figure 13.2c) are combination of the buck and boost structures and can be mathematically modeled as in (13.44) with control parameter expressed as in (13.41)

$$\begin{aligned} \frac{dv_o}{dt} &= \frac{i_L}{C}(1-u) - \frac{v_o}{RC} & \frac{dv_o}{dt} &= f_v(v_o) + b_1(i_L)(1-u) \\ \frac{di_L}{dt} &= \frac{V_g}{L}u - \frac{v_o}{L}(1-u) & \frac{di_L}{dt} &= f_i(V_g)u + b(v_o)(1-u) \end{aligned} \tag{13.44}$$

Despite simple structure with only two energy storing elements DC-to-DC converters represent complex dynamical systems due to the presence of switches and the variable structure. The peculiarities of the converters can be better understood from their dynamic structure. By describing each of the energy storing elements and inserting switch block the block diagram for each of the generic converters can be depicted as in Figure 13.3 (a) buck structure, (b) boost structure, and (c) buck–boost structures. The switching element in buck converters is placed in front of the overall dynamics and influences only one storage (inductance) second element is not directly influenced by control (as it can be see from the model (13.42)).

In boost structure, the switching element is placed in direct and feedback paths and acts on both storage elements—thus both blocks depend on the control input. The same is in buck–boost structures as clear combination of the buck and boost structures.

Dynamics of buck converters is written in regular form and design procedure may be done in two steps. First current should be selected from $\dot{v}_o = (i_L - v_o/R)/C$ to ensure specification in the voltage control. For example, by selecting $i_L = i_L^{ref} = v_o/R + C(\dot{v}_o^{ref} - k\Delta v_o)$ the error in voltage control loop $\Delta v_o = v_o^{ref} - v_o$ will be governed by $\Delta\dot{v}_o + k\Delta v_o = 0$, $k > 0$. In the second step, control u should be selected from $di_L/dt = (V_g u - v_o)/L$ and tracking requirements $\sigma = i_L^{ref} - i_L = 0$. If $|di_L^{ref}/dt + v_o| < |V_g|$ is satisfied and control is selected as $u = (sign(\sigma) - 1)/2$, the positive definite function $V = \sigma^2/2$ will have negative definite

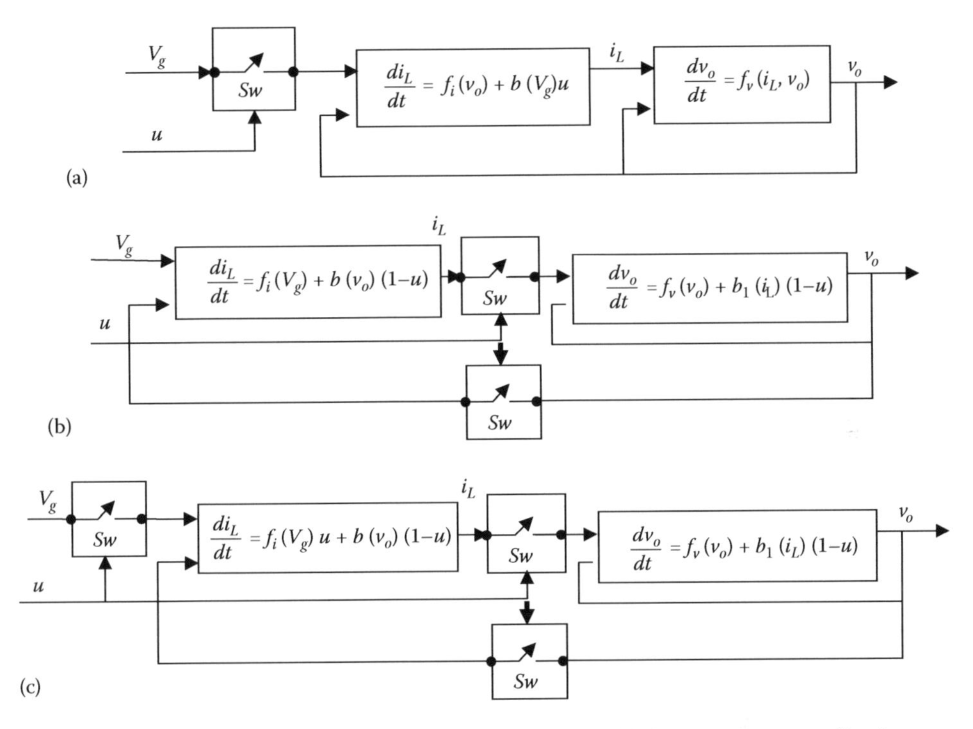

FIGURE 13.3 Dynamic structure of generic DC-to-DC converters: (a) dynamic structure of buck converters, (b) dynamic structure of boost converters, and (c) dynamic structure of buck–boost converters.

derivative and consequently sliding mode will be enforced in $\sigma = i_L^{ref} - i_L = 0$ thus transient in voltage loop will track dynamics dictated by $i_L = i_L^{ref}$. The operation of switching devices is directly defined by value of control and mapping (13.42).

Boost and buck–boost topologies are also written in block form and control input is entering both blocks thus motion separation as in buck converter is not so obvious. In real converters, the rate of change of the current is much faster than the rate of change of the output voltage. That allows separation of motion method to be applied for these converters and consequently splitting design into two steps. As shown in [BI83,VR85,VR86,MA04], the control system design procedure is slightly different than for the buck converters. For boost and buck boost converters in the first step, discontinuous control is selected to enforce sliding mode in the current loop $\sigma = i_L^{ref} - i_L$, assuming that current reference is continuous function of system state and converter parameters. In the second step, equivalent control u_{eq} derived in first step is plug in the equation describing output voltage and then current reference, as a control in obtained system, is determined to obtain desired transients in the voltage loop.

Let i_L^{ref} for a boost converter be continuous current reference to be tracked by the inductor current. The discontinuous control enforcing sliding mode in $\sigma = i_L^{ref} - i_L = 0$ can be derived from

$$\dot{\sigma} = \frac{di_L^{ref}}{dt} - \frac{di_L}{dt} = \frac{di_L^{ref}}{dt} - \frac{V_g - v_o}{L} - \frac{v_o}{L}u \tag{13.45}$$

If $|V_g| < |v_o|$ is satisfied and control is selected as $u = (sign(\sigma) - 1)/2$, the positive definite function $V = \sigma^2/2$ will have negative definite derivative and consequently sliding mode will be enforced in $\sigma = i_L^{ref} - i_L = 0$. The equivalent control obtained from (13.45) can be expressed as

$$\dot{\sigma} = 0 \;\Rightarrow\; 1 - u_{eq} = \frac{V_g}{v_o} - L\frac{1}{v_o}\frac{di_L^{ref}}{dt} \tag{13.46}$$

By inserting (13.46) into first equation in (13.43) and taking into account equality $i_L^{ref} = i_L$, one obtains

$$\frac{C}{2}\frac{dv_o^2}{dt} + \frac{v_o^2}{R} = V_g i_L^{ref} - \frac{L}{2}\frac{d\left(i_L^{ref}\right)^2}{dt} \tag{13.47}$$

Dynamics (13.47) describes the power in boost converters. Dynamics (13.47) is linear in v_o^2. That shows peculiarity in the dynamical behavior of the boost converters. If reference current is constant or slow, changing (13.47) can be simplified by discarding current derivative term. By doing that and selecting reference current as $i_L^{ref} = \left(v_d^{ref}\right)^2/(RV_g)$, the dynamics (13.47) will ensure stable transient to steady-state value $v_o = v_d^{ref}$.

For buck–boost converter, the same approach with sliding mode in $\sigma = i_L^{ref} - i_L = 0$ gives equivalent control $u_{eq} = L\left(di_L^{ref}/dt + v_o\right)/(V_g + v_o)$. By inserting this value of equivalent control in voltage dynamics yields

$$C\frac{dv_o}{dt} + \frac{v_o}{R} = \frac{V_g}{V_g + v_o}\left(i_L^{ref} - \frac{L}{2V_g}\frac{d\left(i_L^{ref}\right)^2}{dt}\right) \tag{13.48}$$

The dynamics (13.48) describes complex nonlinear behavior. Direct selection of reference current based on (13.48) is complicated but linearization around steady-state point (constant due to the fact that DC-to-DC converter is analyzed) may be easily applied. Based on linearization around $\left(V_o^{ref}, I_o^{ref}\right)$, reference current should be selected as $i_L^{ref} = V_o^{ref}\left(V_g + V_o^{ref}\right)/(RV_g)$.

The deliberate introduction of sliding mode in DC-to-DC converters control shows potential of such a design in the systems in which discontinuity of control is already present. The design follows well-established separation on subsystems of lower order. It has been shown how to deal with systems for which block control form does not lead to the separation on the blocks that do not depend on control and blocks that depend on control. The enforcement of sliding mode in current control loop allows calculation of equivalent control and effective derivation of the averaged—continuous—dynamics in the voltage loop after sliding mode is established in current control loop. That allows application of the methods of continuous system design in voltage control loop. In the following section, more complex, three-phase converters will be analyzed and control in sliding mode framework will be developed to deal with these systems.

13.3.1.2 Three-Phase Converters

The topology of switching converters that may have DC or three-phase AC sources on input and/or output side are shown in Figure 13.4a. Further we will limit analysis to voltage source converters and will discuss both inverters and rectifiers. In order to show some peculiarities in sliding mode, design buck inverter (Figure 13.4b) and boost rectifier (Figure 13.4c) will be analyzed in more detail.

The DC-to-AC (inverters) and AC-to-DC (rectifiers) three-phase converters have the same switching matrix. The state of a switching matrix in Figure 13.4a is described by the vector with elements being parameters u_{ij}, $i = 1, 2, \ldots, n$, $j = 1, 2, \ldots, m$, which define the ON–OFF state of switches $S = [u_{11} \; \ldots \; u_{ik} \; \ldots \; u_{nm}]$ [SI81,SOS92].

By selecting voltages and currents as state variables in the three-phase (a, b, c) frame of references, we are able to describe the three-phase converters dynamics. Assuming converters are supplied or are sources of three-phase balanced voltages or currents, zero-component is zero and can be neglected in the analysis. Simplification of the system description could be obtained by mapping corresponding equations into orthogonal stationary (α, β) frame of references. Further simplification may be obtained

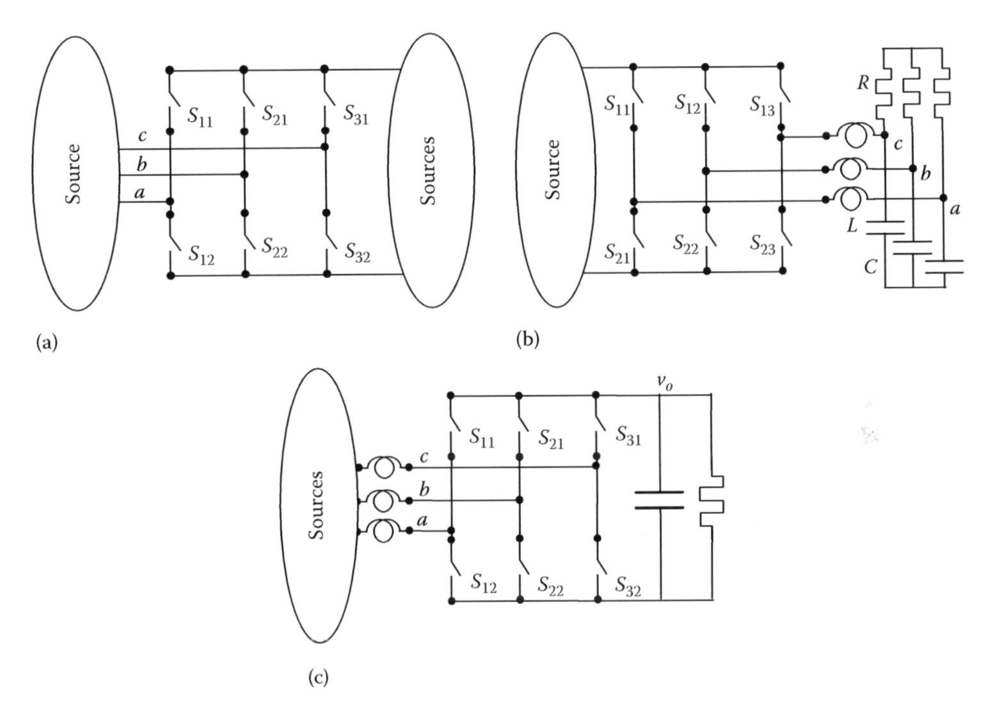

FIGURE 13.4 (a) Basic topology of three-phase converters, (b) buck inverter, (c) boost rectifier.

by mapping from (α, β) frame of references into rotating frame of references (*d*, *q*) with *d*-axis having θ_r position with respect to α-axis of stationary frame of references [SOS92]. Mapping between these frames of reference is defined by matrix $\mathbf{A}_{abc}^{\alpha\beta}$ for (*a*, *b*, *c*) to (α, β) and $\mathbf{A}_{\alpha\beta}^{dq}$ for (α, β) to (*d*, *q*). Matrix $\mathbf{F}(\theta_r) = \mathbf{A}_{\alpha\beta}^{dq}\mathbf{A}_{abc}^{\alpha\beta}$ describes transformation from three-phase (*a*, *b*, *c*) to synchronous orthogonal (*d*, *q*) frames of reference

$$
\begin{aligned}
&u_{dq} = F(\theta_r)\, u_{abc} = A_{\alpha\beta}^{dq} A_{abc}^{\alpha\beta} u_{abc} \\
&A_{\alpha\beta}^{dq} = \begin{bmatrix} \cos\theta_r & \sin\theta_r \\ -\sin\theta_r & \cos\theta_r \end{bmatrix}, \quad A_{abc}^{\alpha\beta} = \begin{bmatrix} 1 & \frac{-1}{2} & \frac{-1}{2} \\ 0 & \frac{\sqrt{3}}{2} & \frac{-\sqrt{3}}{2} \end{bmatrix} \\
&u_{abc}^T = \begin{bmatrix} u_a & u_b & u_c \end{bmatrix} \\
&u_{dq}^T = \begin{bmatrix} u_d & u_q \end{bmatrix} \\
&u_{\alpha\beta}^T = \begin{bmatrix} u_\alpha & u_\beta \end{bmatrix}
\end{aligned}
\tag{13.49}
$$

Notation used is as follows: $u^T = [u_d \quad u_q]$ is the control vector in (*d*, *q*); $u_{abc}^T = [u_a \quad u_b \quad u_c]$ is the control vector in (*a*, *b*, *c*) frame of references. Due to $rank(F(\theta_r)) = rank\left(A_{\alpha\beta}^{dq} A_{abc}^{\alpha\beta}\right) = 2$, mapping from (*a*, *b*, *c*) to either (α, β) or (*d*, *q*) frame of references is not unique. In the analysis, (*d*, *q*) frame of reference is selected to be synchronous with voltage vector on the three-phase side of a converter (input side for rectifiers and output side for inverters). The buck inverter and boost rectifier have eight permissible connections $S = \{S_1, S_2, S_3, S_4, S_5, S_6, S_7, S_8\}$ of the switches in the switching matrix. All eight permissible connections can generate six voltages different from zero and two zero values vectors as depicted in Figure 13.5.

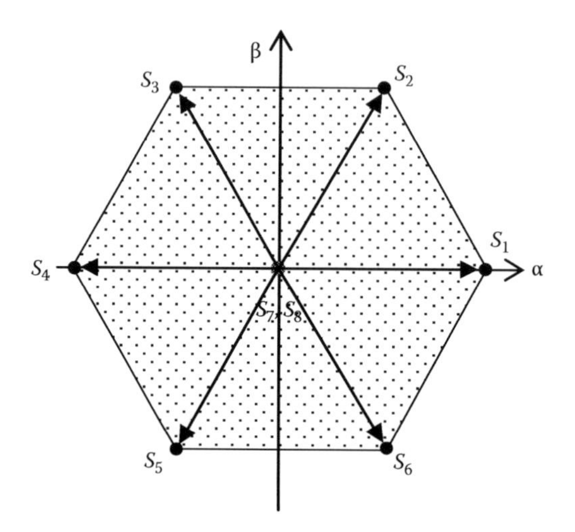

FIGURE 13.5 Control vectors for permissible states of the switching matrix.

$$S_1 = [1,0,0,0,1,1]$$

$$S_2 = [1,1,0,0,0,1]$$

$$S_3 = [0,1,0,1,0,1]$$

$$S_4 = [0,1,1,1,0,0]$$

$$S_5 = [0,0,1,1,1,0]$$

$$S_6 = [1,0,1,0,1,0]$$

$$S_7 = [1,1,1,0,0,0]$$

$$S_8 = [0,0,0,1,1,1]$$

13.3.1.2.1 *Dynamics of Three-Phase Converters*

Dynamics of three-phase buck inverter, in (d, q) frame of references with d-axis collinear with the reference output voltage vector, may be described as

$$\begin{bmatrix} \dfrac{dv_{od}}{dt} \\ \dfrac{dv_{oq}}{dt} \end{bmatrix} = \begin{bmatrix} -\dfrac{v_{od}}{RC} + \omega_r v_{oq} \\ -\omega_r v_{od} - \dfrac{v_{oq}}{RC} \end{bmatrix} + \frac{1}{C}\begin{bmatrix} 1 & 0 \\ 0 & 1 \end{bmatrix}\begin{bmatrix} i_{Ld} \\ i_{Lq} \end{bmatrix}$$
$$\frac{dv_o}{dt} = f_v(v_o, \omega_r) + \frac{1}{C} i_L \tag{13.50}$$

$$\begin{bmatrix} \dfrac{di_{Ld}}{dt} \\ \dfrac{di_{Lq}}{dt} \end{bmatrix} = \begin{bmatrix} -\dfrac{v_{od}}{L} + \omega_r i_{Lq} \\ -\dfrac{v_{oq}}{L} - \omega_r i_{Ld} \end{bmatrix} + \frac{V_g}{L}\begin{bmatrix} 1 & 0 \\ 0 & 1 \end{bmatrix}\begin{bmatrix} u_d \\ u_q \end{bmatrix}$$
$$\frac{di_L}{dt} = f_i(v_o, i_L, \omega_r) + \frac{V_g}{L} u_{dq} \tag{13.51}$$

where

$v_o^T = [v_{od} \quad v_{oq}]$ stands for the capacitance output voltage vector
$i_L^T = [i_{Ld} \quad i_{Lq}]$ stands for inductor current vector
V_g is amplitude of input voltage source
$\omega_r = \dot{\theta}_r$ stands for angular velocity of the frame of references
R, L, C stand for converter parameters

The structure is described in the regular form where current can be treated as virtual control in Equation 13.50.

Boost rectifier in (d, q) with d-axis collinear with the voltage supply vector V_g may be described as

$$\frac{dv_o}{dt} = -\frac{v_o}{RC} + \frac{i_{Ld}u_d + i_{Lq}u_q}{C} = -\frac{v_o}{RC} + \frac{1}{C} i_L^T u_{dq} \tag{13.52}$$

$$\begin{bmatrix} \dfrac{di_{Ld}}{dt} \\ \dfrac{di_{Lq}}{dt} \end{bmatrix} = \begin{bmatrix} \omega_r i_{Lq} + \dfrac{V_g}{L} \\ -\omega_r i_{Ld} \end{bmatrix} - \frac{v_o}{L} \begin{bmatrix} 1 & 0 \\ 0 & 1 \end{bmatrix} \begin{bmatrix} u_d \\ u_q \end{bmatrix} \tag{13.53}$$

$$\frac{di_L}{dt} = f_i(V_g, i_L, \omega_r) - \frac{v_o}{L} u_{dq}$$

Design may follow the same procedure as one shown for DC-to-DC converters. In the first step, sliding mode current control will be designed for the current control loop and in the second step voltage control loop will be designed. In real applications, design of control in (d, q) frame of references requires mapping of the control vector from (d, q) to (a, b, c) frame of references. Due to the fact that $rank(F(\theta_r)) = 2$, mapping from (d, q) to (a, b, c) frame of references does not have unique solution, so it has to be discussed separately. Application of sliding mode control to boost rectifier used as reactive power compensator can be found in [RS93].

13.3.1.2.2 Current Control in Three-Phase Converters

Current control is based on the sliding mode existence in the manifold $\sigma^T = [i^{ref}(t) - i] = 0$, where vector $\sigma^T = [\sigma_d \quad \sigma_q]$ with $\sigma_d = i_d^{ref}(t) - i_d$, $\sigma_q = i_q^{ref}(t) - i_q$. Components $i_d^{ref}(t), i_q^{ref}(t)$ of reference current vector are continuous functions to be determined later. From (13.51) and (13.53) follows that component-vise analysis may be applied and controls u_d, u_q may be determined from $\sigma_d\dot{\sigma}_d \langle 0$ and $\sigma_q\dot{\sigma}_q \langle 0$. By representing rate of change of the current errors as $d\sigma_j/dt = b_j\left(u_j - u_j^{eq}\right)$, $(j = d, q)$, $b_j \neq 0$, the sliding mode conditions will be satisfied if $sign\left(u_j - u_j^{eq}\right) = -sign(\sigma_j)$, $(j = d, q)$. Controls u_d, u_q should be mapped back to the (a, b, c) frame of references in order to decide the corresponding switches ON and OFF pattern. This mapping is not unique. In [SI81] solution, implicitly applied in most of the so-called space vector PWM algorithms, based on using transformation $u_{abc} = \left(A_{\alpha\beta}^{dq} A_{abc}^{\alpha\beta}\right)^T u_{dq}$ has been proposed. This solution has disadvantage that zero vector is not used and consequently switching is not optimized. Another problem of this algorithm is in the fact that transformation $u_{abc} = \left(A_{\alpha\beta}^{dq} A_{abc}^{\alpha\beta}\right)^T u_{dq}$ can be realized with the accuracy to the sign of the control. The amplitude of the control cannot be mapped correctly since number of available control vectors is limited as shown in Figure 13.5.

Another solution [SOS92,SOS93] can be found by expressing time derivative of $\sigma^T = [\sigma_d \quad \sigma_q]$ for systems (13.51) and (13.53) in the following form

$$\frac{d\sigma_{dq}}{dt} = B_{udq}\left[u_{dq}(S_i) - u_{dq}^{eq}\right], \quad i = (1, \ldots, 8) \tag{13.54}$$

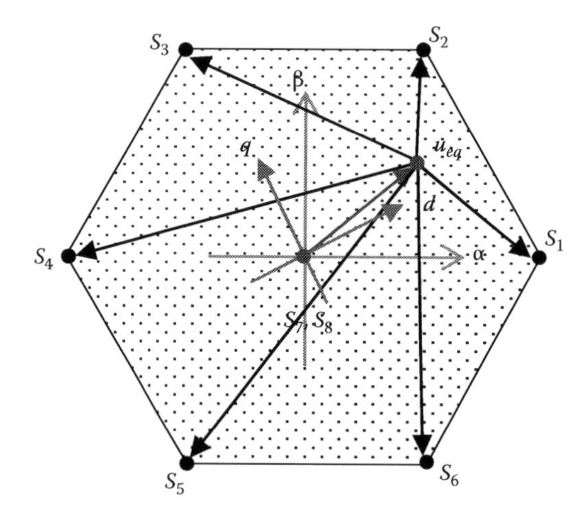

FIGURE 13.6 Effective rate of change of control errors for permissible control vectors.

Matrix B_{udq} for both converters under consideration is diagonal positive definite and equivalent control may be easily found from (13.54) and (13.51) or (13.53) depending on the converters to be analyzed. As shown in Figure 13.5, control vectors could take values from the discrete set $S = \{S_1, S_2, S_3, S_4, S_5, S_6, S_7, S_8\}$. For a particular combination of errors σ_d and σ_q all permissible vectors $S_i(i = 1,\ldots, l)$ that satisfy sliding mode existence conditions $\sigma_i\dot{\sigma}_i \langle 0(i = d, q)$ can be found from $sign\left(u_{dq}(S_i) - u_{dq}^{eq}\right) = -sign(\sigma_{dq})$. For some combinations of errors, there is more than one permissible vector (Figure 13.6). That leads to an ambiguous selection of the control and consequently existence of more than one solution for the selection of switching pattern. Ambiguity in selection of the control vector based on selected u_d and u_q allows a number of different PWM algorithms that satisfy sliding mode conditions.

In early works related to electrical machine control, the following solution was proposed: add an additional requirement $\vartheta(t) = 0$ to the control system specification, so that (13.54) is augmented to have the form

$$\begin{bmatrix} \dfrac{d\sigma_{dq}}{dt} \\ \dfrac{d\vartheta}{dt} \end{bmatrix} = \begin{bmatrix} \dfrac{di^{ref}}{dt} - f_{idq} \\ f_\vartheta \end{bmatrix} - \begin{bmatrix} B_{udq}F(\theta_r) \\ b_\vartheta^T \end{bmatrix} u_{abc}(S_i) \tag{13.55}$$

$$\frac{d\sigma_N}{dt} = f_N - B_N u_{abc}(S_i) \quad u_{abc}^T(S_i) = \begin{bmatrix} u_a & u_b & u_c \end{bmatrix} \tag{13.56}$$

vector b_ϑ should be selected so that $rank(B_N) = 3$. The simplest solution is for $\dot{\vartheta}(t) = u_a + u_b + u_c$ [SI81,RM86,VGS99], then matrix B_N will have full rank. To determine the switching pattern, the simplest way is to use the nonlinear transformation $\sigma_s = B_N^{-1}\sigma_N$. The sliding mode conditions are satisfied if the control is selected as

$$\begin{gathered} sign(u_j(S_i)) = -sign(\sigma_{sj}), \quad -1 \le u_{eqj} \le 1 \\ \left\| B_N^{-1} f_n \right\| < 1 \end{gathered} \tag{13.57}$$

This line of reasoning with some variations has been the most popular in designing the sliding mode–based switching pattern [RM86,VGS99]. Solution (13.57) enforces balanced condition for average values of phase controls $\int(u_a + u_b + u_c)\,d\tau = 0$. That generates unnecessary rise of switching frequency in the system real application. This algorithm can be easily extended by requiring the non-zero sum $\int(u_a + u_b + u_c)\,d\tau = z$ (actually standing for the average of the neutral point vector in star connection for there phase system). In this case, optimization of switching can be formulated as one of the requirement of the overall system design.

A new class of switching algorithms based on the simple requirement that control should be selected to give the minimum rate of change of control error. It may be designed, so the same current control error will be achieved with less switching effort [SOS92,SA02,PO07]. The algorithm can be formulated in the following form:

$$S_i = \left\{ \begin{array}{c} \min\left\|\mathbf{u}(S_i) - \mathbf{u}^{eq}(t)\right\| \\ \text{AND} \\ sign\left\{\left[u_d(S_i) - u_d^{eq}\right]\cdot\sigma_d(t)\right\} = -1 \\ \text{AND} \\ sign\left\{\left[u_q(S_i) - u_q^{eq}\right]\cdot\sigma_q(t)\right\} = -1 \end{array} \right\} \tag{13.58}$$

$$i = (1,2,\ldots,8)$$

This algorithm is universal in a sense that it works for all converters that have switching matrix as shown in Figure 13.4. Modification with introduction of the optimization of the total harmonic distortion (THD) had been successfully applied [CE97]. All of the above algorithms operate in so-called overmodulation as a normal operational mode without change of the algorithm.

In order to complete the design, the reference currents $i_d^{ref}(t), i_q^{ref}(t)$ shall be determined. In the sliding mode dynamics of current, control loop is reduced to $\sigma_d = i_d^{ref}(t) - i_d = 0$ and $\sigma_q = i_q^{ref}(t) - i_q = 0$. Buck inverter dynamical model (50), with components of the current as virtual controls, $dv_o/dt = f_v(v_o, \omega_r) + (1/C)i_L$ describes a two-dimensional system affine in control and design of the currents may follow the same procedure as for buck DC-to-DC converter.

For boost rectifier approach established for boost DC-to-DC converter can be applied. The equivalent control in $\sigma_d = i_d^{ref} - i_d = 0$ and $\sigma_q = i_q^{ref} - i_q = 0$ can be expressed as

$$\begin{aligned} u_d^{eq} &= \left(V_g + L\omega_r i_q - L\frac{di_d^{ref}}{dt}\right)\frac{1}{v_o} \\ u_q^{eq} &= -\left(L\omega_r i_d + L\frac{di_q^{ref}}{dt}\right)\frac{1}{v_o} \end{aligned} \tag{13.59}$$

By inserting (13.59) into (13.52), the dynamics of output voltage with sliding mode in current control loop becomes

$$\frac{C}{2}\frac{dv_o^2}{dt} + \frac{v_o^2}{R} = V_g i_d^{ref} - \frac{L}{2}\left(\frac{d\left(i_d^{ref}\right)^2}{dt} + \frac{d\left(i_q^{ref}\right)^2}{dt}\right) \tag{13.60}$$

Dynamics (13.60) is a first-order system in v_o^2—the same form as the dynamics of DC-to-DC converter. The q-component of the input current represents the reactive power flow in the system and in (13.60), it can be treated as a disturbance. Then i_d^{ref} can be selected the same way as in DC-to-DC boost converter.

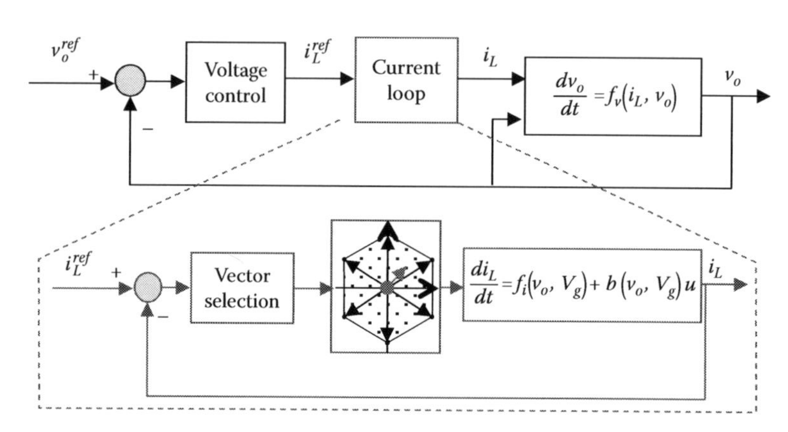

FIGURE 13.7 Structure of the sliding mode control for switching power converters.

The structure of the power converters control system can be presented as in Figure 13.7 where the current control loop operates in sliding mode with discontinuous control. The selected structure is only one of the several possible solutions.

13.3.2 SMC in Electrical Drives

Control of electrical drives requires usage of a switching matrix as interconnection between electrical machine and the power source. The topology of interconnection of the switching matrix and machines, for standard three-phase machines, is like three-phase buck inverters in which output filter is replaced by machine. Specifics of the sliding mode current controller design for three-phase electrical machines, in comparison with buck inverter, lies in the selection of the (d, q) frame of references. For three phase machines, d-axis should be selected collinear with vector of the flux (rotor or stator). With such selection sliding mode algorithms for current control developed for three-phase converters can be directly applied for electrical machines as well [SI81,VGS99].

Another field of application of sliding modes in electrical drives is estimation of states and parameters of electrical machines [VII93,SG95,SA02]. As an example, flux and velocity observer for Induction Machine (IM) will be shown.

13.3.2.1 Induction Machine Flux Observer

Description of the induction machine in the (α, β) frame of reference can be written as

$$\frac{d\psi_r}{dt} = -B_r\psi_r + \frac{R_r L_m}{L_r} i_s; \quad B_r = \begin{bmatrix} \frac{R_r}{L_r} & -\omega \\ \omega & \frac{R_r}{L_r} \end{bmatrix} \tag{13.61}$$

$$\frac{di_s}{dt} = \frac{1}{\sigma L_s}\left\{\frac{L_m}{L_r}\left(-B_r\psi_r + \frac{R_r L_m}{L_r} i_s\right) - R_s i_s + u_s\right\} \tag{13.62}$$

where

$\psi_r^T = [\psi_{r\alpha} \quad \psi_{r\beta}]$ stands for rotor flux vector
$i_s^T = [i_{s\alpha} \quad i_{s\beta}]$ stands vector of stator current
$u_s^T = [u_{s\alpha} \quad u_{s\beta}]$ stands for stator voltage
ω is velocity
R_s, R_r, L_m, L_r, L_s, σ stand for machine parameters

Formally, it is possible to design a stator current observer based on voltage and current measurements and with a rotor flux vector derivative as the control input:

$$\frac{d\hat{i}_s}{dt} = \frac{1}{\sigma L_s}\left(-\frac{L_m}{L_r}u_\psi - R_s\hat{i}_s + u_s\right) \tag{13.63}$$

This selection of the observer control input is different from the usually used current error feedback, application of sliding mode control based on current feedback, or selecting the unknown velocity as the control input [DI83,SA02]. The estimation error is determined as

$$\frac{d\varepsilon_i}{dt} = \frac{d(i_s - \hat{i})_s}{dt} = \frac{1}{\sigma L_s}\left(\frac{L_m}{L_r}\left(\left(-B_r\psi_r + \frac{R_r L_m}{L_r}i_s\right) + u_\psi\right) - R_s\varepsilon_i\right) \tag{13.64}$$

Selection of control $u_\psi^T = [u_{\psi\alpha} \quad u_{\psi\beta}]$ to enforce sliding mode in (13.64) is straight forward. After sliding mode is enforced from $d\boldsymbol{\varepsilon}_i/dt = 0$, $\boldsymbol{\varepsilon}_i = 0$ and under the assumption that the angular velocity is known, from (13.64) one can find

$$\hat{\psi}_r = B_r^{-1}\left(u_\psi^{eq} - \frac{R_r L_m}{L_r}i_s\right). \tag{13.65}$$

Extension of this idea to estimation of the flux and velocity is discussed in details in [SG95,ZU02].

13.3.3 Sliding Modes in Motion Control Systems

Control of multibody systems is generally analyzed in joint space (often called configuration space) and in task space (often called operational space). In this section, application of SMC will be shown for both joint space and task space.

13.3.3.1 Joint Space Control

Mathematical model of fully actuated mechanical system (number of actuators equal to number of the primary masses) in configuration space may be found in the following form

$$\begin{aligned} A(q)\ddot{q} + b(q,\dot{q}) + g(q) &= \tau \\ b(q,\dot{q}) + g(q) &= n(q,\dot{q}) \end{aligned} \tag{13.66}$$

where

$q \in \Re^n$ stands for vector of generalized positions
$\dot{q} \in \Re^n$ stands for vector of generalized velocities
$A(q) \in \Re^{n\times n}$ is the generalized positive definite inertia matrix with bounded parameters
$n(q,\dot{q}) \in \Re^{n\times 1}$ represents vector of coupling forces including gravity and friction
$\tau \in \Re^{n\times 1}$ is vector of generalized input forces

Vector $n(q,\dot{q}) \in \Re^{n\times 1}$ satisfies matching conditions. Model (13.66) can be rewritten in regular form

$$\begin{aligned} \dot{q} &= v \\ A(q)\dot{v} &= -b(q,\dot{q}) - g(q) + \tau \end{aligned} \tag{13.67}$$

In the first step, velocity vector v, taken as virtual control in (13.67), can be selected as $\Delta v = -G\Delta q$ with $\Delta v = v - v^{ref}$, $\Delta q = q - q^{ref}$ and q^{ref}, v^{ref} continuous differentiable functions. In the second step, control τ should be selected to enforce sliding mode in

$$\sigma = \Delta v + G\Delta q = 0 \tag{13.68}$$

The equivalent control for system (13.66) in manifold (13.68) takes the form

$$\begin{aligned} \dot{\sigma} &= A^{-1}(\tau - \tau_{eq}) \\ \tau_{eq} &= b(q, \dot{q}) + g(q) + A(\dot{v}^{ref} - G\Delta v) \end{aligned} \tag{13.69}$$

The simplest solution to enforce sliding mode is to select $\tau = -M_o\, sign(\sigma)$ with M_o large enough [YK78]. This solution will have large amplitude and in most cases will cause chattering. Another solution can be developed by merging disturbance observer method [OK96] and sliding mode technique. Since dimension of the vectors τ, b, g, $\lambda = A(\dot{v}^{ref} - G\Delta v)$ are the same then component-wise disturbance observers may be used. Then control enforcing sliding mode in manifold (13.68) may be written as

$$\begin{aligned} \tau &= \left(\hat{b}(q,\dot{q}) + \hat{g}(q) + \hat{A}(\dot{v}^{ref} - G\Delta v)\right) - \tau_0 sign(\sigma) \\ \tau_0 &> 0 \end{aligned} \tag{13.70}$$

In (13.70), $\tau_0 > 0$ is determined to compensate for the disturbance estimation error. Very similar result may be obtained for motion control in operation space.

Another way of looking at control (13.70) is to interpret the term $(\dot{v}^{ref} - G\Delta v) = \dot{v}^{des} = \ddot{q}^{des}$ as desired acceleration in the closed loop system (which is equal to the solution for acceleration of the $\dot{\sigma} = 0$ and in that sense stands for "equivalent acceleration"). Then (13.70) can be rewritten as

$$\begin{aligned} \tau &= \hat{A}\dot{v}^{des} + \left(\hat{b}(q,\dot{q}) + \hat{g}(q)\right) - \tau_0 sign(\sigma), \quad \tau_0 > 0 \\ \tau &= \hat{A}\ddot{q}^{des} + \left(\hat{b}(q,\dot{q}) + \hat{g}(q)\right) - \tau_0 sign(\sigma), \quad \tau_0 > 0 \end{aligned} \tag{13.71}$$

Physical interpretation of control is now apparent. It consists of the disturbance rejection term $(\hat{b}(q, \dot{q}) + \hat{g}(q))$, the weighted desired (equivalent) acceleration $A\dot{v}^{des} = A\ddot{q}^{des}$ and convergence term $\tau_0 sign(\sigma)$ enforcing finite-time convergence to manifold (13.68). In real systems, this term may be replaced by proportional term $\tau_0\sigma$, but then only exponential convergence to manifold (13.68) may be guaranteed and strictly speaking no sliding mode motion will be realized in the system and the motion will be in the vicinity of manifold (13.68). This situation is already encountered in the power converters and electrical drives control with sliding mode in inner current loop. In this case, sliding mode manifold (current error) is reached in finite time but since reference current is continuous the dynamics in outer loop (voltage or velocity or position) is at best exponentially stable. In that sense, if one assumes that in motion control systems the torque input is generated by electrical machines, then application of the (13.71) may cause problem and it may be that better way is to use just proportional term $\tau_0\sigma$ instead of discontinuous one $\tau_0 sign(\sigma)$ as a convergence term [SA08].

13.3.3.2 Operation Space Control

In the analysis of the motion system, a joint space description (13.66) is readily available. The dynamics of the mechanical system associated with performing a certain task defined by a minimal set of coordinated $x = x(q)$, $x \in \Re^{p\times 1}$, $p < n$ involves mapping of the configuration space dynamics (13.66) into new set of coordinates. In general, operation space coordinates may represent any set of coordinates defining kinematic mapping between joint space and the task space, but they are usually selected to represent a description of motion control task. For given task vector $x = x(q)$, $x \in \Re^{p\times 1}$, $p < n$, the following relationship is obtained $\dot{x} = (\partial x(q)/\partial q)\dot{q} = J(q)\dot{q}$, where $J(q) \in \Re^{p\times n}$ is task Jacobian matrix. Let internal configuration–posture consistent with the given task be defined by a minimal set of independent coordinates $x_P \in \Re^{(n-p)\times 1}$, with posture Jacobian $J_P = \left(\partial x_P/\partial q\right) \in \Re^{(n-p)\times n}$. Then posture velocity can be expressed as $\dot{x}_P = \Gamma\dot{q}$, with $\Gamma = (J_P\Gamma_1) \in \Re^{(n-p)\times n}$. $\Gamma_1 \in \Re^{n\times n}$ stands for the matrix to be selected in such a way that task and posture dynamics are decoupled. Fundamental relation linking forces $f \in \Re^{p\times 1}$ and $f_P \in \Re^{(n-p)\times 1}$ with joint spaces forces $\tau \in \Re^{n\times 1}$ is defined as $\tau = J^T f_x + \Gamma^T f_P$, [OK87].

The task–posture velocity vector can be written as

$$\dot{q}_{J\Gamma} = \begin{bmatrix} \dot{x} \\ \dot{x}_P \end{bmatrix} = \begin{bmatrix} J \\ \Gamma \end{bmatrix} \dot{q} = J_{JP}\dot{q} \tag{13.72}$$

The task and posture dynamics can be written in the following form:

$$\begin{bmatrix} \ddot{x} \\ \ddot{x}_P \end{bmatrix} = \begin{bmatrix} JA^{-1}J^T & JA^{-1}\Gamma^T \\ \Gamma A^{-1}J^T & \Gamma A^{-1}\Gamma^T \end{bmatrix} \begin{bmatrix} f_x \\ f_P \end{bmatrix} - \begin{bmatrix} JA^{-1}(b+g) \\ \Gamma A^{-1}(b+g) \end{bmatrix} + \begin{bmatrix} \dot{J}\dot{q} \\ \dot{\Gamma}\dot{q} \end{bmatrix} \tag{13.73}$$

If $\Gamma_1 \in \Re^{n\times n}$ is selected as task consistent null space projection matrix $\Gamma_1 = (1 - J^{\#} J)$ with $J^{\#} = A^{-1}J^T (JA^{-1}J^T)^{-1}$, then $JA^{-1}\Gamma_1^T J_P^T = 0$ holds. Then (13.73) describes two decoupled subsystems, which can be written as

$$\begin{aligned} \Lambda_x\ddot{x} &= f_x - \Lambda_x JA^{-1}(b+g) + \Lambda_x \dot{J}\dot{q} \\ \Lambda_\Gamma \ddot{x}_P &= f_P - \Lambda_\Gamma \Gamma A^{-1}(b+g) + \Lambda_\Gamma \dot{\Gamma}\dot{q} \\ \Lambda_x &= (JA^{-1}J^T)^{-1} \\ \Lambda_\Gamma &= (\Gamma A^{-1}\Gamma^T)^{-1} \end{aligned} \tag{13.74}$$

Design of control input for both task and internal configuration–posture is now straightforward having established consistent equations describing dynamical behavior of the system. The dynamical models (13.74) are in the same form as joint space dynamics (13.66). That opens a way of applying the same procedure in the sliding mode design. By selecting task space sliding mode manifold

$$\sigma_x = \Delta\dot{x} + G\Delta x = 0, \quad \Delta x = x - x^{ref} \tag{13.75}$$

task control enforcing sliding mode in manifold (13.75) can be expressed in the same form as (13.71) to obtain

$$\begin{aligned} f_x &= \Lambda_x \ddot{x}^{des} + \Lambda_x\left(JA^{-1}(\hat{b} + \hat{g}) - \dot{J}\dot{q}\right) - f_o sign(\sigma_x), \quad f_o > 0 \\ \ddot{x}^{des} &= \ddot{x}^{ref} - G\Delta\dot{x} \end{aligned} \tag{13.76}$$

Similarly by selecting sliding mode manifold

$$\sigma_P = \Delta\dot{x}_P + G\Delta x_P = 0, \quad \Delta x_P = x_P - x_P^{ref} \tag{13.77}$$

control enforcing desired internal configuration–posture can be expressed as

$$\begin{aligned} f_P &= \Lambda_\Gamma \ddot{x}_P^{des} + \Lambda_\Gamma\left(\Gamma A^{-1}(\hat{b} + \hat{g}) - \dot{\Gamma}\dot{q}\right) - f_{oP} sign(\sigma_P), \quad f_{oP} > 0 \\ \ddot{x}_P^{des} &= \ddot{x}_P^{ref} - G\Delta\dot{x}_P \end{aligned} \tag{13.78}$$

13.4 Conclusion

The aim of the chapter has been to outline the main mathematical methods and sliding mode control design techniques and to demonstrate its applicability in power electronics and motion control systems. Many interesting results from the wide range of results accumulated in the variable structure systems remain beyond the scope of this chapter.

References

[BI83] Bilalovic, F., Music, O., and Sabanovic, A., Buck converter regulator operating in sliding mode, *Proceeding of the Power Conversion Conference, PCI'83*, Orlando, FL, 1983, pp. 146–152.

[CE97] Chen, Y., Fujikawa, K., Kobayashi, H., Ohnishi, K., and Sabanovic, A., Direct instantaneous distortion minimization control for three phase converters, *Transactions of IEEE of Japan*, 117-D(7), July 1997 (in Japanese).

[DB69] Drazenovic, B., The invariance conditions in variable structure systems, *Automatica*, 5, 287–295, 1969.

[DI83] Izosimov, D., Multivariable nonlinear induction motor state identifier using sliding modes (in Russian), *Problems of Multivariable Systems Control*, Moscow, 1983.

[DR89] Drakunov, S. V. and Utkin, V. I., On discrete-time sliding modes, *Proceedings of the Nonlinear Control System Design Conference*, March, Capri, Italy, 1989, pp. 273–278.

[ED98] Edwards, C. and Spurgeon, S., *Sliding Mode Control: Theory and Applications* (Hardcover), 1st edn., CRC Press LLC, Boca Raton, FL, August 27, 1998.

[EM70] Emelyanov, S. V., *Theory of Variable Structure Systems*, Nauka, Moscow (in Russian), 1970.

[FL53] Fluge-Lotz, I., *Discontinuous Automatic Control*, Princeton University Press, New York, 1953.

[KD96] Young, K. D., Utkin, V. I., and Özgüner, Ü., A control engineer's guide to sliding mode control, *Proceedings of the 1996 IEEE International Workshop on Variable Structure Systems (VSS-96)*, Tokyo, Japan, December 5–6, 1996, pp. 1–14.

[LU66] Luenberger, D. G., Observers for multivariable systems, *IEEE Transactions*, AC-11, 190–197, 1966.

[LU81] Luk'yanov, A. G. and Utkin, V. I., Methods of reducing equations of dynamic systems to regular form, *Automatica Remote Control*, 42(4, Part 1), 5–14. 1981.

[MA04] Ahmed, M., Sliding mode control for switched power supplies, PhD thesis, University of Technology, Lappeenranta, Finland, December 2004.

[OK87] Khatib, O., A unified approach to motion and force control of robot manipulators: The operational space formulation, *International Journal of Robotics Research*, 3(1), 43–53, 1987.

[OK96] Ohnishi, K., Shibata, M., and Murakami, T., Motion control for advanced mechatronics, *IEEE/ASME Transactions on Mechatronics*, 1(1), 56–67, March 1996.

[PO07] Polic, A. and Jezernik, K., Event-driven current control structure for a three phase inverter, *International Review of Electrical Engineering (Testo stamp.)*, 2(1), 28–35, January–February 2007.

[RD94] De Carlo, R. and Drakunov, S., Sliding mode control design via Lyapunov approach, *Proceedings of the 33rd Conference on Decision and Control*, Lake Buena Vista, FL, December 1994.

[RM86] Mahadevan, R., Problems in analysis, control and design of switching inverters and rectifiers, PhD thesis, CALTECH, Pasadena, CA, 1986.

[RS93] Radulovic, Z., Masada, E., and Sabanovic, A., Sliding mode control strategy of the three phase reactive power compensator, *Conference Record of the Power Conversion Conference (PCC'93)*, Yokohama, April 19–21, 1993, pp. 581–586.

[SA02] Sabanovic, A., Sabanovic, N., and Jezernik, K., Sliding modes applications in power electronics and electrical drives, *Variable Structure Systems: Towards 21st Century, Lecture Notes in Control 274*, Eds. Yu, X. and Xu, J.-X., Springer-Verlag, Berlin, Germany, 2002.

[SA08] Sabanovic, A., Elitas, M., and Ohnishi, K., Sliding modes in constrained systems control, *IEEE Transactions on Industrial Electronics*, 55(9), 4055–4064, September 2008.

[SG95] Sahin, C., Sabanovic, A., and Gokasan, M., Robust position control based on chattering free sliding modes for induction motors, *Proceedings of the Industrial Electronics Conference (IECON 95)*, Orlando, FL, 1995, pp. 512–517.

[SI81] Šabanovic, A. and Izosimov, D. B., Application of sliding mode to induction motor control, *IEEE Transaction on Industrial Applications*, IA 17(1), 41–49,1981.

[SOS92] Šabanovic, N., Ohnishi, K., and Šabanovic, A., Sliding modes control of three phase switching converters, *Proceedings of the IECON'92 Conference*, San Diego, CA, 1992, pp. 319–325.

[SOS93] Sabanovic, A., Ohnishi, K., and Sabanovic, N., Control of PWM three phase converters: A sliding mode approach, *Conference Record of the Power Conversion Conference (PCC'93)*, Yokohama, April 1983, pp. 188–193.

[TS55] Tsypkin, Ya. Z., *Theory of Relay Control Systems*, Gostekhizdat, Moscow (in Russian), 1955.

[UN08] Unel, M., Sabanovic, A., Yilmaz, B., and Dogan, E., Visual motion and structure estimation using sliding mode observers, *International Journal of Systems Science*, 39(2), 149–161, February 2008.

[VGS99] Utkin, V., Guldner, J., and Shi, J., *Sliding Modes in Electromechanical Systems*, Taylor & Francis, Boca Raton, FL, 1999.

[VI92] Utkin, V. I., *Sliding Modes in Control and Optimization*, Springer-Verlag, Berlin, Germany, 1992.

[VI93] Utkin, V. I., Sliding mode control in discrete time and difference systems, *Variable Structure and Lyapunov Control*, Ed. Zinober A.S., Springer Verlag, London, U.K., 1993.

[VII93] Utkin, V. I., Sliding mode control design principles and applications to electric drives, *IEEE Transactions on Industrial Electronics*, 40(1), 421–434, 1993.

[VR85] Venkataramana, R., Sabanovic, A., and Cuk, S., Sliding mode control of DC-to-DC converters, *Proceedings of the IECON'85*, San Francisco, CA, 1985, pp. 267–273.

[VR86] Vankataramanan, R., Sliding mode control of switching converters, PhD thesis, CALTECH, Pasadena, CA, 1986.

[YK78] Young, K.-K. D., Controller design for a manipulator using theory of variable structure systems, *IEEE Transactions on Systems, Man and Cybernetics*, 8, 210–218, 1978.

[ZU02] Yan, Z. and Utkin, V., Sliding mode observers for electrical machines—An overview, *Proceedings of the IECON 2002*, Sevilla, Spain, 2002, Vol. 3, pp. 1842–1947.

14

Digital Control

Timothy N. Chang
New Jersey Institute of Technology

John Y. Hung
Auburn University

14.1 Introduction

Enabled by low-cost, high-speed signal acquisition and processing hardware, digital control has become an important tool for the analysis, design, and implementation of control systems [1,2,4]. A typical digital control system is shown in Figure 14.1 where the continuous-time signals are low-pass filtered by an anti-aliasing filter and sampled by an analog–digital converter (ADC) into discrete-time signals. The discrete-time signals are then processed by a microprocessor to generate the discrete-time control outputs. The digital–analog converter (DAC) in turn converts the discrete-time sequence into a continuous-time signal, generally by means of a zeroth-order hold circuit [2]. A smoothing filter is sometimes added to remove harmonics present in the passband. Other variations of this scheme are also possible [8] but the fundamental analytic and design methods presented here are generally applicable.

Shown in Figure 14.2 are the two paradigms for digital control design. On the left side is the discrete emulation approach where, based upon the continuous plant, a continuous-time controller is first designed and then discretized (digitized). However, a more direct approach is to first discretize the plant model and then design the corresponding digital control. This approach is shown on the right-hand side path in Figure 14.2. An advantage of this approach is that a number of additional tools, not available to continuous-time design, are available. For example, direct discrete-time design includes the ability to embed delay into the state-space model, achieve deadbeat behavior, linear phase [2], etc. Controllers directly designed from the discrete time usually can tolerate a lower sampling frequency as well. Finally, the emulation method [4] has been historically applied to obtain digital proportional–integral–derivative (PID) type of control. Details of this approach are given in Section 14.4.1 of this chapter.

In this chapter, conversion of a continuous-time model into discrete time is first discussed. Embedding delay into the discrete-time system is discussed in Section 14.3. In Section 14.4, the popular digital PID design is presented, followed by a discussion on "additional discrete-time controllers" that are unique to the direct discrete-time design path. A DC servo motor system serves as a running example in this chapter. Finally, readers interested in reviewing sampling and reconstruction are referred to [1,2] for details.

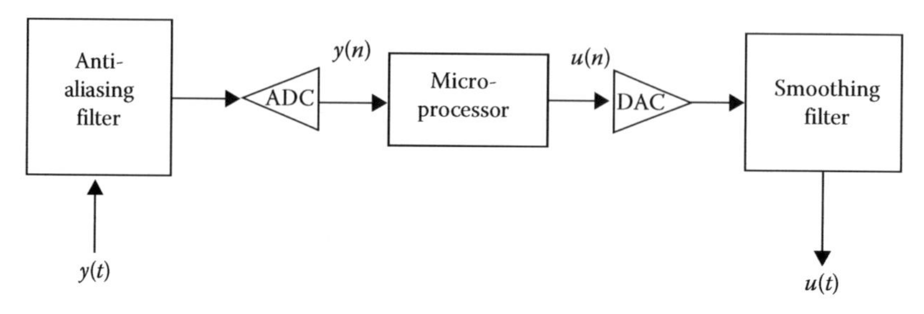

FIGURE 14.1 Typical digital control system.

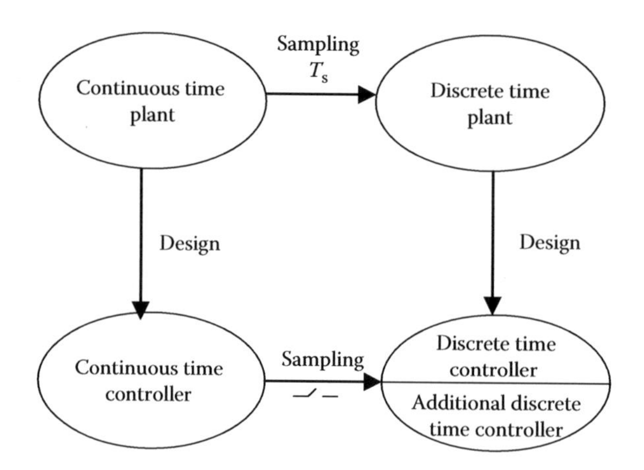

FIGURE 14.2 Digital control design paradigms.

14.2 Discretization of Continuous-Time Plant

Let the M-input, R-output, Nth order continuous-time plant be represented by the following linear, time-invariant state-space model:

$$\begin{aligned}\dot{\bar{x}}(t) &= \bar{A}\bar{x}(t) + \bar{B}\bar{u}(t)\\ \bar{y}(t) &= \bar{C}\bar{x}(t) + \bar{D}\bar{u}(t)\end{aligned} \tag{14.1}$$

where $\bar{x}(t)$, $\bar{u}(t)$, $\bar{y}(t)$ are the continuous-time state, input, and output vectors

$$\bar{x}(t) \triangleq \begin{pmatrix}\bar{x}_1(t)\\ \bar{x}_2(t)\\ \vdots\\ \bar{x}_N(t)\end{pmatrix}, \quad \bar{u}(t) \triangleq \begin{pmatrix}\bar{u}_1(t)\\ \bar{u}_2(t)\\ \vdots\\ \bar{u}_M(t)\end{pmatrix}, \quad \bar{y}(t) \triangleq \begin{pmatrix}\bar{y}_1(t)\\ \bar{y}_2(t)\\ \vdots\\ \bar{y}_R(t)\end{pmatrix} \tag{14.2}$$

(A line drawn over a variable denotes a continuous-time signal.). Therefore, $\bar{A}$, $\bar{B}$, $\bar{C}$, $\bar{D}$ are, respectively, N-by-N, N-by-M, R-by-N, and R-by-M matrices. Given a sampling period T_s and that the DAC is based on a zeroth-order hold device, the continuous-time plant (14.1) can be readily discretized so that the resultant discrete-time plant is given as

$$\begin{aligned}x(n+1) &= Ax(n) + Bu(n)\\ y(n) &= Cx(n) + Du(n)\end{aligned} \tag{14.3}$$

with

$$A \triangleq e^{\bar{A}T_s}, \quad B = \int_0^{T_s} e^{\bar{A}\tau}\bar{B}d\tau$$
$$C = \bar{C}, \quad D = \bar{D} \tag{14.4}$$

where it is understood that $x(n) \triangleq \bar{x}(nT_s)$, $n = 0, 1, 2, \ldots$, etc.

Dimensions of the matrices in discrete model are the same as for the continuous model. Calculation of A, B can be done in a number of ways. For low-order systems (typically $N \le 3$) or those with a sparse $\bar{A}$ matrix, the following inverse LaPlace transform formula can be handy:

$$e^{\bar{A}T_s} = L^{-1}\{(sI - \bar{A})^{-1}\}_{t=T_s} \tag{14.5}$$

On the other hand, for diagonalizable $\bar{A}$, $e^{\bar{A}T_s}$ can be readily computed as

$$e^{\bar{A}T_s} = V \begin{pmatrix} e^{\bar{\lambda}_1 T_s} & \cdots & 0 \\ \vdots & \ddots & \vdots \\ 0 & \cdots & e^{\bar{\lambda}_N T_s} \end{pmatrix} W \tag{14.6}$$

where

V, W are the right and left modal matrices

$\bar{\lambda}_1, \ldots, \bar{\lambda}_N$ are the continuous-time eigenvalues [7]

In general, the Cayley–Hamilton method can be applied to determine $e^{\bar{A}T_s}$

Example 14.1 Discretization of DC Servomotor

A DC motor [4] has the following simplified transfer function model:

$$T(s) = \frac{36}{s(s+3)} \tag{14.7}$$

It can be represented in state-space format as

$$\dot{\bar{x}}(t) = \begin{bmatrix} 0 & 1 \\ 0 & -3 \end{bmatrix} \bar{x}(t) + \begin{bmatrix} 0 \\ 36 \end{bmatrix} \bar{u}(t)$$
$$\bar{y}(t) = \begin{bmatrix} 1 & 0 \end{bmatrix} \bar{x}(t) \tag{14.8}$$

For a 100 Hz sampling rate, the discrete-time model is

$$x(n+1) = \begin{bmatrix} 1.0000 & 0.0099 \\ 0 & 0.9704 \end{bmatrix} x(n) + \begin{bmatrix} 0.0018 \\ 0.3547 \end{bmatrix} u(n)$$
$$y(n) = \begin{bmatrix} 1 & 0 \end{bmatrix} x(n) \tag{14.9}$$

It should be noted that a discrete-time model can be sensitive to numerical resolution, especially as the sampling frequency becomes large relative to the plant dynamics. Under those circumstances, coefficient rounding can result in significant errors. Modern computer software tools such as MATLAB®, which perform continuous-to-discrete conversions, are widely available. Some tools offer multiple approaches to discretization, each yielding different numerical characteristics. Alternatively, the discretization can also be carried out on the transfer function corresponding to (14.1):

$$\bar{T}(s) = \bar{D} + \bar{C}(sI - \bar{A})^{-1}\bar{B} = \begin{pmatrix} \bar{T}_{1,1}(s) & \cdots & \bar{T}_{1,M}(s) \\ \vdots & \ddots & \vdots \\ \bar{T}_{R,1}(s) & \cdots & \bar{T}_{R,M}(s) \end{pmatrix} \quad (14.10)$$

So that the corresponding discrete transfer function matrix is

$$T(z) = D + C(zI - A)^{-1}B = \begin{pmatrix} T_{1,1}(z) & \cdots & T_{1,M}(z) \\ \vdots & \ddots & \vdots \\ T_{R,1}(z) & \cdots & T_{R,M}(z) \end{pmatrix} \quad (14.11)$$

Elements of the discrete-time transfer function are given by

$$T_{i,j}(z) = Z\left\{\frac{\bar{T}_{i,j}(s)}{s}\right\}(1 - z^{-1}) \quad (14.12)$$

The expression $Z\{(\bar{T}_{i,j}(s))/s\}$ literally means (1) take the inverse LaPlace transform of $\bar{T}_{i,j}(s)/s$, (2) sample the corresponding impulse response at $t = nT_s$, and (3) take the Z transform. However, it is more customary to use table look up [2] or computer software.

When the continuous plant is preceded by a zeroth-order hold device (D/A converter), the resulting discrete model is also called the "pulse" transfer function, owing to the notion that the hold device produces rectangular pulses of duration T_s.

Example 14.2 Discretization of Transfer Function

Applying (14.12) on the DC motor transfer function (14.7), the discrete-time transfer function is

$$T(z) = \frac{4\left[\left(3T_s - 1 + e^{-3T_s}\right)z + \left(1 - e^{-3T_s} - 3T_s e^{-3T_s}\right)\right]}{(z-1)\left(z - e^{-3T_s}\right)} \quad (14.13)$$

For 100 Hz sampling frequency, $T_s = 0.01$ s,

$$T(z) = \frac{0.0018z + 0.0018}{(z-1)(z-0.9704)} \quad (14.14)$$

14.3 Discretization of Continuous-Time System with Delay

It is quite common that a delay τ_d is present in various design problems:

$$\begin{aligned} \dot{\bar{x}}(t) &= \bar{A}\bar{x}(t) + \bar{B}\bar{u}(t - \tau_d) \\ \bar{y}(t) &= \bar{C}\bar{x}(t) + \bar{D}\bar{u}(t - \tau_d) \end{aligned} \quad (14.15)$$

One approach to handling time delay in continuous-time design employs the Pade approximation, a transfer function approximation of the transcendental form $e^{-s\tau_d}$. A criticism of the Pade approximation is that a high-order transfer function may be required to reasonably model the time delay. In contrast, the design of controllers directly in the discrete time permits directly absorbing the delay in an equivalent discrete-time state-space model. Consider the delay $\tau_d = (d-1)T_s + \delta$ where $d \geq 1$ is an integer representing the integral delay factor and δ is the fractional delay, $\tau_d > \delta > 0$. The delay-embedded discrete-time model can be written as [1,2]:

$$\begin{aligned} \mathbb{Z}(n+1) &= \mathbb{A}\mathbb{Z}(n) + \mathbb{B}u(n) \\ y(n) &= \mathbb{C}\mathbb{Z}(n) + Du(n) \end{aligned} \tag{14.16}$$

where $\mathbb{Z}(n)$ is the augmented state vector. The model (14.16) takes on two forms, depending on the magnitude of the delay.

Case 1: $T_s > \tau_d > 0$, i.e., the delay is within one sampling period

$$\mathbb{Z}(n) \triangleq \begin{pmatrix} x(n) \\ u(n-1) \end{pmatrix}, \quad \mathbb{A} \triangleq \begin{pmatrix} A & B_1 \\ 0 & 0 \end{pmatrix}, \quad \mathbb{B} \triangleq \begin{pmatrix} B_0 \\ I_M \end{pmatrix}, \quad \mathbb{C} \triangleq (C \quad 0) \tag{14.17}$$

Case 2: $\tau_d > T_s$ so that $d > 1$

$$\mathbb{Z}(n) \triangleq \begin{pmatrix} x(n) \\ u(n-d+1) \\ \vdots \\ u(n-2) \\ u(n-1) \end{pmatrix}, \quad \mathbb{A} \triangleq \begin{pmatrix} A & B_1 & B_0 & 0 & \cdots & 0 \\ 0 & 0 & I_M & 0 & & 0 \\ \vdots & & & & & \vdots \\ 0 & 0 & 0 & 0 & \cdots & I_M \\ 0 & 0 & 0 & 0 & \cdots & 0 \end{pmatrix}, \quad \mathbb{B} \triangleq \begin{pmatrix} 0 \\ 0 \\ \vdots \\ 0 \\ I_M \end{pmatrix}, \quad \mathbb{C} \triangleq (C \quad 0 \quad 0 \quad \cdots \quad 0) \tag{14.18}$$

I_M represents the M by M identity matrix and

$$B_0 \triangleq \int_0^{T_s-\delta} e^{\bar{A}t}\bar{B}dt, \quad B_1 \triangleq e^{\bar{A}(T_s-\delta)} \int_0^{\delta} e^{\bar{A}t}\bar{B}dt \tag{14.19}$$

Example 14.3 DC Servomotor with Delay

The DC motor model (14.7) is augmented with a delay at the input:

$$T(s) = \frac{36}{s(s+3)} e^{-s\tau_d} \tag{14.20}$$

Consider the same sampling of 100 Hz, two cases are presented here.

Case 1: $T_s > \tau_d = 0.005$

The delay-embedded discrete-time model is given by (14.17), where

$$\mathbb{A} \triangleq \begin{pmatrix} 1 & 0.0099 & 0.0013 \\ 0 & 0.9704 & 0.1760 \\ 0 & 0 & 0 \end{pmatrix}, \quad \mathbb{B} \triangleq \begin{pmatrix} 0.0004 \\ 0.1787 \\ 1 \end{pmatrix}, \quad \mathbb{C} \triangleq (C \quad 0), \quad D = 0$$

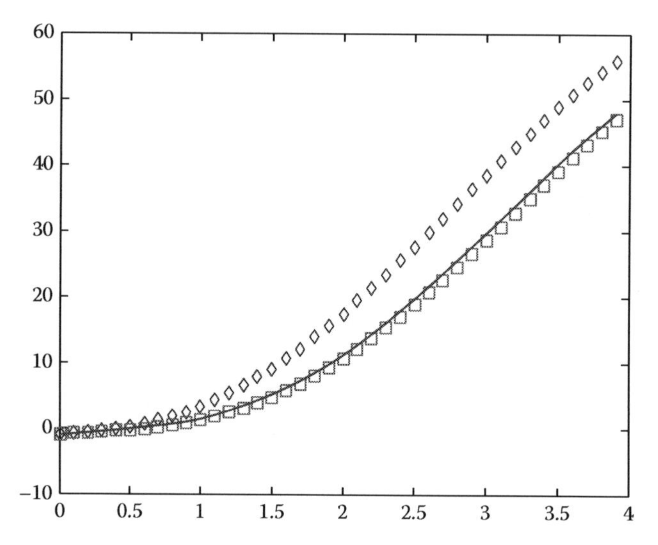

FIGURE 14.3 Response of DC motor (solid, continuous time; diamond, discrete time without embedding delay; square, discrete time with delay embedded).

Case 2: $\tau_d = 0.015 > T_s$ so that $d = 2$ and $\delta = 0.005$

Applying (14.18), the delay-embedded system is

$$\mathbb{A} \triangleq \begin{pmatrix} 1 & 0.0099 & 0.0013 & 0.0004 \\ 0 & 0.9704 & 0.1760 & 0.1787 \\ 0 & 0 & 0 & 1 \\ 0 & 0 & 0 & 0 \end{pmatrix}, \quad \mathbb{B} \triangleq \begin{pmatrix} 0 \\ 0 \\ 0 \\ 1 \end{pmatrix}, \quad \mathbb{C} \triangleq (C \quad 0 \quad 0), \quad D = 0$$

with $\mathbb{Z}(n) \triangleq \begin{pmatrix} x(n) \\ u(n-2) \\ u(n-1) \end{pmatrix}$

A simulation of the continuous-time plant with delay, discrete-time plant, and delay-embedded discrete-time plant is shown in Figure 14.3. The delayed-embedded discrete-time system correctly tracks the continuous-time response.

14.4 Digital Control Design

14.4.1 Digital PID Design

The PID control is a well-established technique in many industrial applications [3,5]. One continuous-time version is given as

$$\bar{u}(t) = K\left[\underset{P}{\bar{e}(t)} + \underset{I}{\frac{1}{T_I}\int_0^t \bar{e}(\tau)d\tau} + \underset{D}{T_D\frac{d\bar{e}(t)}{dt}}\right] \tag{14.21}$$

where

K is the controller gain
T_I is the integration time
T_D is the derivative time
$\bar{e}(t) = y^{ref}(t) - \bar{y}(t)$ is the process error
$y^{ref}(t)$ is the reference command

A common variation is that the derivative action is applied only on the process variable $y(t)$, so that changes in the reference command will not introduce excess control action—this is further explained below. Digital PID is obtained by discretizing (14.21) above:

a. Using rectangular rule to approximate integration and one-step finite difference for derivative

$$u(n) = K\left[e(n) + \frac{T_S}{T_I}\sum_{k=0}^{n-1} e(k) + \frac{T_D}{T_S}\left[e(n) - e(n-1)\right]\right] \tag{14.22}$$

Again, the notation here is that $e(n) = \bar{e}(nT_S)$, etc.

b. Using trapezoidal (bilinear or Tustin) rule instead of rectangular rule results in

$$u(n) = K\left[e(n) + \underbrace{\frac{T_S}{T_I}\left\{\sum_{k=1}^{n-1} e(k) + \frac{e(0)+e(n)}{2}\right\}}_{\text{better fit}} + \frac{T_D}{T_S}\{e(n) - e(n-1)\}\right] \tag{14.23}$$

Both Equations 14.22 and 14.23 are non-recursive and are referred to as "position algorithms." More commonly, the velocity algorithm (14.26) is used. It is derived by subtracting (14.25) from (14.24):

$$u(n) = K\left[e(n) + \frac{T_S}{T_I}\sum_{k=0}^{n-1} e(k) + \frac{T_D}{T_S}(e(n) - e(n-1))\right] \tag{14.24}$$

$$u(n-1) = K\left[e(n-1) + \frac{T_S}{T_I}\sum_{k=0}^{n-2} e(k) + \frac{T_D}{T_S}(e(n-1) - e(n-2))\right] \tag{14.25}$$

Resulting in the following recursive equation:

$$u(n) - u(n-1) = K\left[e(n) - e(n-1) + \frac{T_S}{T_I}e(n-1) + \frac{T_D}{T_S}\{e(n) - 2e(n-1) + e(n-2)\}\right] \tag{14.26}$$

or more conveniently,

$$u(n) = u(n-1) + q_0 e(n) + q_1 e(n-1) + q_2 e(n-2)$$

$$q_0 = K\left(1 + \frac{T_D}{T_S}\right)$$

$$q_1 = K\left[-1 + \frac{T_S}{T_I} - \frac{2T_D}{T_S}\right] \tag{14.27}$$

$$q_2 = K\left[\frac{T_D}{T_S}\right]$$

Velocity algorithms are simpler to implement. Similarly,

For trapezoidal rule integration, the parameters are

$$q_0 = K\left(1 + \frac{T_S}{2T_I} + \frac{T_D}{T_S}\right)$$

$$q_1 = K\left(-1 + \frac{T_S}{2T_I} - 2\frac{T_D}{T_S}\right) \tag{14.28}$$

$$q_2 = K\frac{T_D}{T_S}$$

Some implementation considerations:

a. Initialization for (14.26) or (14.27): Use position algorithm (14.22) to calculate $u(0)$:

$$u(0) = K\left[e(0) + \frac{T_D}{T_S}\left[e(0) - e(-1)\right]\right]$$

$$= q_0 e(0) - q_2 e(-1) \neq 0 \quad \text{in general}$$

where

$e(0) = y^{ref}(0) - y(0)$

$e(-1) = y^{ref}(-1) - y(-1)$

b. PID implementation: For the case that the reference command $y^{ref}(t)$ contains sharp jumps/steps, the set-point-on-I configuration shown in Figure 14.4 [1] is recommended, so that changes in reference y^{ref} will not induce a sharp PD reaction. This is achieved by replacing the error $e(n)$'s with the output $y(n)$'s in the P and D terms.

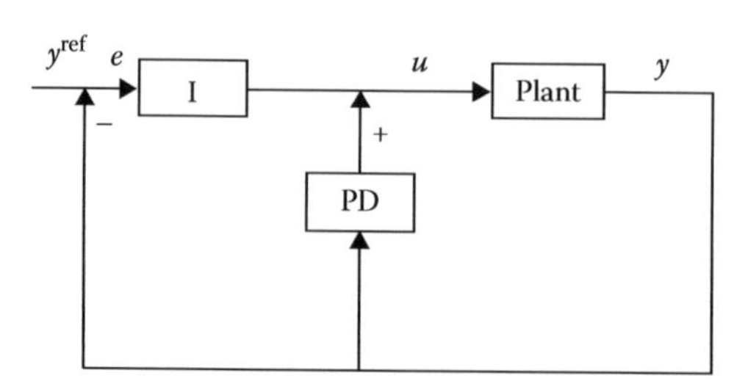

FIGURE 14.4 Set-point-on-I PID control.

c. Tuning Rules for PID controller: The Ziegler–Nichols tuning rules are applicable to digital PID tuning: the "transient response" method for well-damped systems and the "ultimate sensitivity" method for oscillatory systems. These methods are summarized as follows:

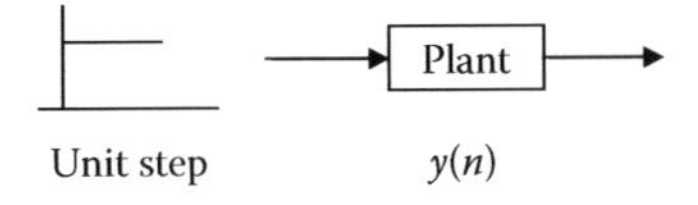

FIGURE 14.5 Set-up for the "transient response" method for tuning PID controller.

- Transient Response Method: A unit step is applied to the plant as shown in Figure 14.5 and the plant response is recorded.
  Upon measuring the steepest slope (L) and dead-time (L) from the unit step response shown in Figure 14.6, the estimated PID parameters are tabulated in Table 14.1.
- Ultimate Sensitivity Method: A proportional feedback is applied to the plant as shown in Figure 14.7.

Adjust the proportional control until a sustained oscillation appears at the output. Define critical gain as the value of the proportional gain at which the oscillation occurs and T_p as the period of oscillation as depicted in Figure 14.8. The corresponding PID parameters are listed in Table 14.2.

In both cases, it is necessary to fine tune K, T_I, and T_D. For systems with high complexity, the tuning procedures may fail. Furthermore, for most digital PID, the sampling rate should be 5X–20X of the unity gain cross over frequency of the open-loop plant (i.e., to minimize phase lag) [2].

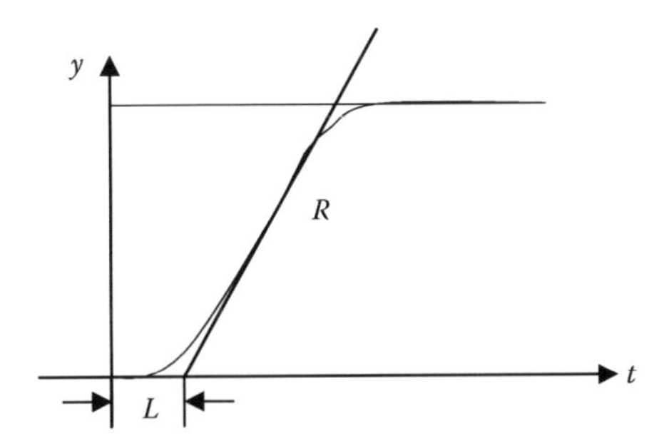

FIGURE 14.6 Measurement of steepest slope (R) and dead-time (L) for the "transient response" tuning method.

TABLE 14.1 Approximate PID Parameters Obtained from the Transient Response Method

Controller Type	K	T_I	T_D
P	$\frac{1}{RL}$		
PI	$\frac{0.9}{RL}$	3L	
PID	$\frac{1.2}{RL}$	2L	0.5L

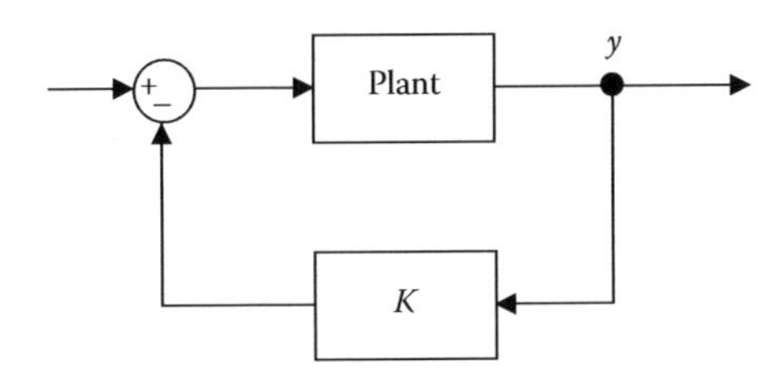

FIGURE 14.7 Set-up for the ultimate sensitivity method for tuning digital PID controller.

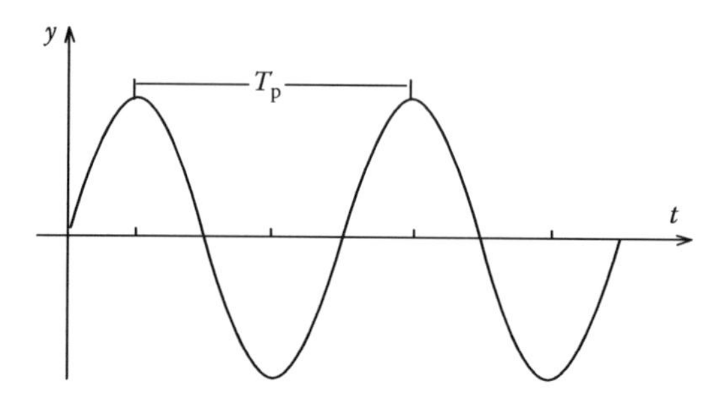

FIGURE 14.8 Measurement of period of oscillation for the "ultimate sensitivity" tuning method.

TABLE 14.2 Approximate PID Parameters Obtained from the Ultimate Sensitivity Method

Controller Type	K	T_I	T_D
P	$0.5K_C$		
PI	$0.45K_C$	$\frac{T_P}{1.2}$	
PID	$0.6K_C$	$\frac{T_P}{2}$	$\frac{T_P}{8}$

Ziegler–Nichols tuning is also employed in some modern autotuning controllers. One common method employs relay control to force the feedback system into an oscillation. The oscillation period is measured as described above. The ultimate gain can be computed from the amplitudes of the process variable and relay output.

14.4.2 Input Shaping Design

Input shaping is a feedforward technique to suppress command-induced vibrations [6]. The design is generally initiated in continuous time while leaving the implementation in discrete time due the requirement for relatively high amplitude and timing precisions. A brief overview of input shaping is presented [5,6] here based on a second-order system subjected to two-impulse excitation:

$$G(s) = \frac{\omega_n^2}{\left(s^2 + 2\varsigma\omega_n s + \omega_n^2\right)} \tag{14.29}$$

Let $\bar{y}_i(t)$ be the response to an impulse $A_i\delta(t - t_i)$, then the total response at a preselected settling time $t = t_N$, $t_N > t_1, t_2, \ldots, t_{N-1}$ can be written as

$$\bar{y}(t_N) = e^{-\zeta\omega_n t_N}\left\{\left[\sum_{i=1}^{m} B_i e^{\zeta\omega_n t_i}\cos(\omega_d t_i)\right]\sin(\omega_d t_N) - \left[\sum_{i=1}^{m} B_i e^{\zeta\omega_n t_i}\sin(\omega_d t_i)\right]\cos(\omega_d t_N)\right\} \tag{14.30}$$

where $B_i = (A_i\omega_n)/\sqrt{1-\varsigma^2}$ and $\omega_d = \omega_n\sqrt{1-\varsigma^2}$. By eliminating sin $(\omega_d t_N)$ and $\cos(\omega_d t_N)$ terms, the residual vibration V can therefore be expressed as

$$V(\omega_n,\varsigma,t_N) = |y(t_N)|$$

$$= e^{-\varsigma\omega_n t_N}\sqrt{\left[\sum_{i=1}^{m} B_i e^{\varsigma\omega_n t_i}\cos(\omega_d t_i)\right]^2 + \left[\sum_{i=1}^{m} B_i e^{\varsigma\omega_n t_i}\sin(\omega_d t_i)\right]^2} \quad (14.31)$$

In the case of $m = 2$, to zero out the residual vibration, the zero vibration (ZV) shaper is derived as [6]:

$$A_1 = \frac{1}{1+OS}, \quad A_2 = \frac{OS}{1+OS} \quad (14.32)$$

and

$$\Delta T = t_2 - t_1 = t_p \quad (14.33)$$

OS and t_p are the overshoot and peak time of the unit step response of second-order system (14.29), respectively,

$$OS = e^{\frac{-\pi\varsigma}{\sqrt{1-\varsigma^2}}}, \quad t_p = \frac{\pi}{\omega_n\sqrt{1-\varsigma^2}} \quad (14.34)$$

and the ZV shaper is given in (14.35).

$$u(t) = A_1\delta(t-t_1) + A_2\delta(t-t_2) \quad (14.35)$$

That is, the ZV-shaped input command is obtained by convolving the reference command with a sequence of two impulses. For step inputs, the result is a staircase command. The method is also known as Posicast control, which was originally studied in the 1960s [10].

Example 14.4 Input Shaping of DC Servomotor

For this example, consider closing a simple unity gain proportional loop on the DC Motor (14.7) as shown in Figure 14.9.

The closed-loop pole polynomial is $s^2 + 3s + 36$. It can be readily shown that

$$OS = e^{\frac{-\pi\varsigma}{\sqrt{1-\varsigma^2}}} = 0.4443, \quad t_p = \frac{\pi}{\omega_0\sqrt{1-\varsigma^2}} = 0.5408$$

Therefore, the shaper parameters are computed as shown in Table 14.3.

Unit step response of the DC Motor with unity gain feedback is shown in Figure 14.10. It is observed that the input shaper produces fast and nonoscillatory transients compared to the 44.43% overshoot in the regular step response. However, it should be noted that the ZV shaper is sensitive to parametric variations and other shaper designs should be considered for plants with higher degree of uncertainty [9]. For example,

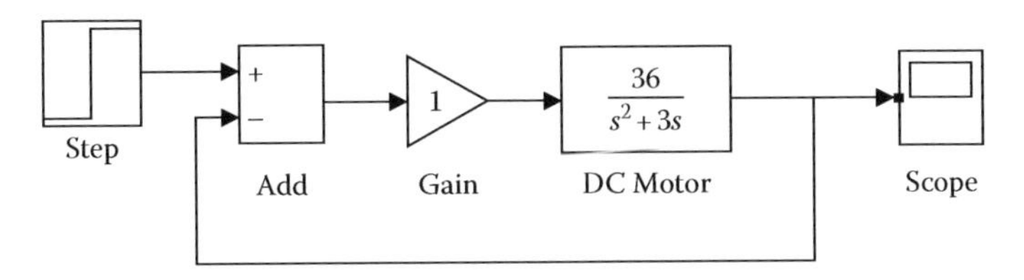

FIGURE 14.9 DC motor with simple proportional loop.

shaping within a feedback configuration (as opposed to feedforward shaping) has been proven in power electronic applications [10,11]. The implementation of the necessary delay functions can be accomplished with timers that are often integral subsystems of microcontrollers.

TABLE 14.3 ZV Shaper Parameters for Example 14.4

t_1	t_2	A_1	A_2
0	0.5408	0.6924	0.3076

14.4.3 State Feedback Design

State feedback control is among the most powerful control designs. It has an appealingly simple structure:

$$u(n) = -K_s x(n) \tag{14.36}$$

However, the plant (14.3) must be controllable and all of the state variables must be measured. The latter may not always be possible due to restriction on sensor deployment. In such case, an observer/state estimator may be used to estimate the state vector. The readers are referred to the corresponding chapters in this handbook for observer/state estimator designs. Consider the case of a single-input system so that $M = 1$ in (14.3), a number of pole placement algorithms are available to determine the feedback gain K_s [1,2,7]. The Bass–Gura method is considered as possessing the best numerical stability and is described below.

Let the open-loop characteristic polynomial of the discrete plant be given as

$$P_o = z^N + a_{N-1}z^{N-1} + \cdots + a_1 z + a_0 \tag{14.37}$$

Similarly, define the desired closed-loop characteristic polynomial as

$$P_c = z^N + \alpha_{N-1}z^{N-1} + \cdots + \alpha_1 z + \alpha_0 \tag{14.38}$$

Construct the Toeplitz matrix $\Upsilon = \begin{pmatrix} 1 & 0 & 0 & \cdots & 0 \\ a_{N-1} & 1 & 0 & & 0 \\ a_{N-2} & a_{N-1} & 1 & & 0 \\ \vdots & & & & \vdots \\ a_1 & a_2 & a_3 & \cdots & 1 \end{pmatrix}$

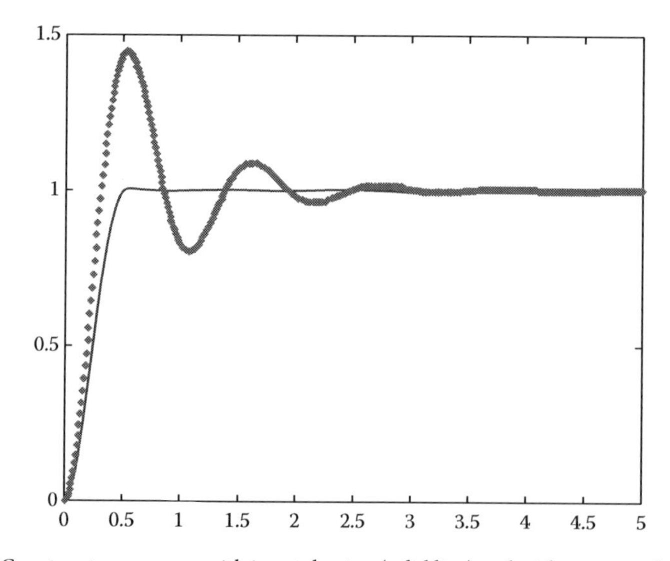

FIGURE 14.10 DC motor step response with input shaping (solid line) and without input shaping (dotted line).

then, the state feedback gain is given as

$$K_s = (\alpha_{N-1} - a_{N-1} \quad \alpha_{N-2} - a_{N-2} \quad \ldots \quad \alpha_0 - a_0)\Upsilon^{-1}M_c^{-1} \tag{14.39}$$

Here, M_c is the controllability matrix,

$$M_c = [B \quad AB \quad A^2B \quad \ldots \quad A^{N-1}B]$$

A unique property of the discrete-time domain is the so-called deadbeat behavior, that the system response can be driven to the set point levels in finite time. This is achieved by assigning all of the discrete-time poles to the center of the z-plane $z = 0$. The desired characteristic equation is given by $z^N = 0$. However, deadbeat control must be carried out with consideration of a reasonable speed of convergence. Overly aggressive designs may lead to saturation, inter-sample oscillation, and even instability [1]. The primary design parameter in deadbeat control is the sample period. The next example illustrates this effect.

Example 14.5 Deadbeat Control of DC Motor

Using the pole placement algorithm (14.35), the deadbeat feedback is given by

$$u(n) = -[281.9653 \quad 4.1391]x(n)$$

Response to an initial condition of $x(0) = \begin{bmatrix} 1 \\ 1 \end{bmatrix}$ is shown in Figure 14.11, where it is observed that the motor position converges to zero within two sampling periods (0.02 s). However, a closer examination of the velocity response in Figure 14.12 reveals that the velocity transient is unacceptably large.

Reducing the sampling rate to 10 Hz, the corresponding control is

$$u(n) = -\begin{bmatrix} 3.2152 & 0.3909 \end{bmatrix} x(n)$$

The responses are shown in Figures 14.13 and 14.14. Slowing down the system allows the energy to be dissipated in a more reasonable manner.

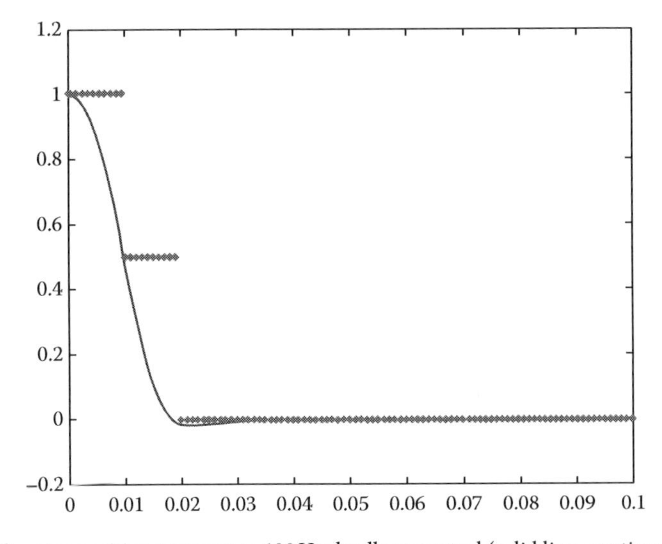

FIGURE 14.11 DC motor position response to 100 Hz deadbeat control (solid line, continuous time; dotted line, discrete time).

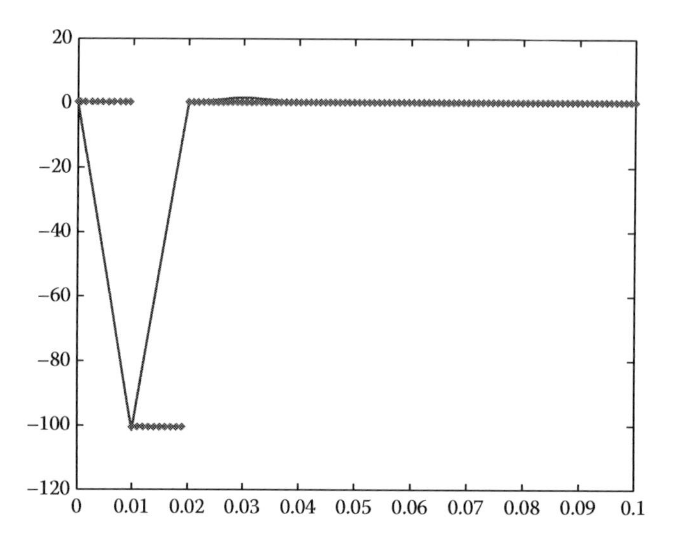

FIGURE 14.12 DC motor velocity response to 100 Hz deadbeat control (solid line, continuous time; dotted line, discrete time).

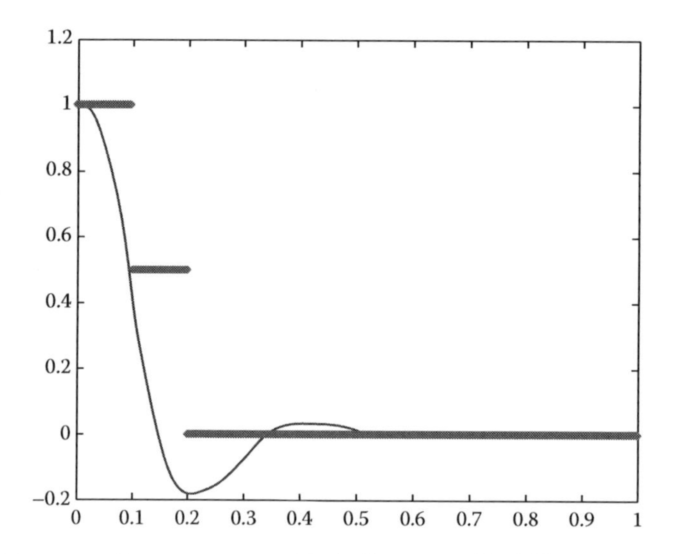

FIGURE 14.13 DC motor position response to 10 Hz deadbeat control (solid line, continuous time; heavy line, discrete time).

14.5 Multirate Controllers

Another approach to state feedback employs multirate sampling. Multirate sampling means that different sample periods are employed within the system. In general, three different sample periods could be employed in a digital controller: sampling of the output or process variable occurs at period T_y, the reference is sampled at period T_r, and the plant input zeroth-order hold is changed with period T_u. Different sampling periods may be justified due to hardware limitations of the plant or digital controller. More importantly, multirate sampling can achieve state feedback performance without the use of an observer. Excellent tracking performance and disturbance rejection can be achieved by multirate sampling control. The interested reader can explore some of the references [12–14].

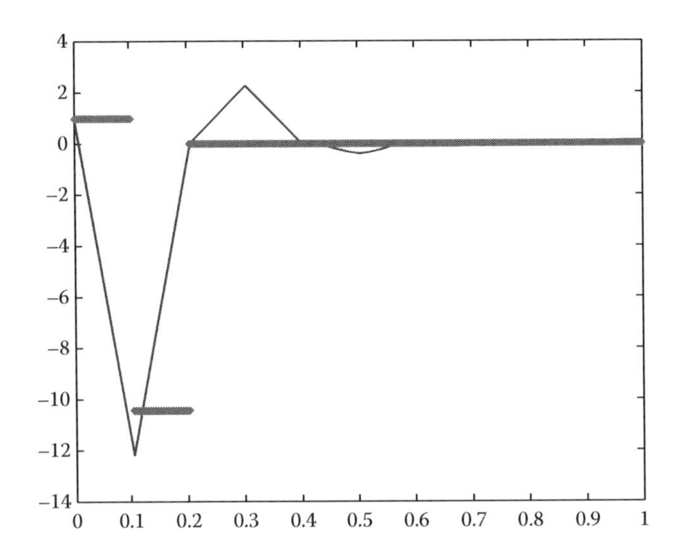

FIGURE 14.14 DC motor velocity response to 10 Hz deadbeat control (solid line, continuous time; dotted line, discrete time).

14.6 Conclusions

Digital control is considered in this chapter. Discretization of continuous-time process is first discussed. The process can either be a plant or a controller. Emulating continuous controllers in discrete time is very popular and requires minimal knowledge of discrete system theory, but a more flexible approach is to carry out direct control design in discrete time. The ability to embed delay, to implement control in great precision, and to carry out deadbeat response is unique to the direct approach and can be advantageous in many applications. Finally, the popular digital PID control is also presented in this chapter. Fundamentals of discrete control theory have been in existence for approximately 50 years, but advances such as multirate sampling continue to develop as electronic technologies evolve.

References

1. Astrom, K. and Wittenmark, B., *Computer Control Systems: Design and Theory*, Prentice-Hall, Englewood Cliffs, NJ, 1997.
2. Isermann, R., *Digital Control Systems*, Vols. 1 and 2, Springer-Verlag, Berlin, Germany, 1989.
3. Wittenmark, B., Nilsson, J., and Tormgren, M., Timing problems in real-time systems, *Proceedings of the American Control Conference*, Seattle, WA, June 1995, pp. 2000–2004.
4. Hung, J. and Trent, V., *Digital Control*, Chapter 37, IEEE Industrial Electronics Society handbook, 1997.
5. Chang, T.N., Servo control systems, *Encyclopedia of Life Support Systems*, United Nations Educational, Scientific, and Cultural, New York, 2004, Sec. 6.43.13.19.
6. Singer, N.C. and Seering, W.P. Preshaping command inputs to reduce system vibration, *Journal of Dynamic Systems, Measurement, and Control*, 112, 76–82, March 1990.
7. Kailath, T., *Linear Systems* (Prentice-Hall Information and System Science Series), Prentice-Hall, Englewood Cliffs, NJ, 1979.
8. Chang, T.N., Cheng, B., and Sriwilaijaroen, P., Motion control firmware for high speed robotic systems, *IEEE Transactions on Industrial Electronics*, 25(5), 1713–1722, 2006.
9. Chang, T.N., Hou, E., and Godbole, K., Optimal input shaper design for high speed robotic workcells, *Journal of Vibration and Control*, 9(12), 1359–1376, December 2003.

10. Hung, J.Y., Feedback control with Posicast, *IEEE Transactions on Industrial Electronics*, 50(1), 94–99, February 2003.
11. Feng, Q., Nelms, R.M., and Hung, J.Y., Posicast-based digital control of the buck converter, *IEEE Transactions on Industrial Electronics*, 53(3), 759–767, June 2006.
12. Hagiwara, T. and Araki, M., Design of a stable state feedback controller based on the multirate sampling of the plant output, *IEEE Transactions on Automatic Control*, 33(9), 812–819, September 1988.
13. Fujimoto, H., Hori, Y., and Kawamura, A., Perfect tracking control based on multirate feedforward control with generalized sampling periods, *IEEE Transaction on Industrial Electronics*, 48(3), 636–664, June 2001.
14. Fujimoto, H., Hori, Y., and Kondo, S., Perfect tracking control based on multirate feedforward control and applications to motion control and power electronics-a simple solution via transfer function approach, *Proceedings of the 2002 Power Conversion Conference*, Osaka, Japan, April 2–5, 2002, pp. 196–201.

15

Phase-Lock-Loop-Based Control

Guan-Chyun Hsieh
Chung Yuan Christian University

15.1 Introduction

An early description of the phase-locked loop (PLL) appeared in the papers by Appleton [1] in 1923 and de Bellescize [2] in 1932. The advent of the PLL has contributed to coherent communication systems without the Doppler shift effect. In the late 1970s, the theoretical description of the PLL was well established [3–5], but the PLL did not achieve widespread use until much later because of the difficulty in realization. With the rapid development of integrated circuits in the 1970s, the applications of the PLL came to be widely used in modern communication systems. Since then the PLL has made much progress and has turned its earlier professional use in high-precision apparatuses to its current use in consumers' electronic products. It has enabled modern electronic systems to improve performance and reliability, especially in common electronic appliances used daily. After 2000, owing to the fast development in IC design, PLLs have become the inevitably essential devices in many systems, such as microprocessors, digital signal processors, PIC controllers, mixed-mode integrated circuits and so on, because it can contribute the stable control function to increase system reliability in versatile applications. Presently, the PLL can be classified as the linear PLL (LPLL), the digital PLL (DPLL), the all-digital PLL (ADPLL), the software PLL (SPLL), etc., as shown in Figure 15.1 and Table 15.1 [6]. New topologies for increasing the performance of the PLL have been developed in succession, such as the enhanced PLL or the fast PLL, etc. [7,8,12]. Nowadays, PLLs are widely put in use in areas of communication, optics, control, power, and so on.

In the 1970s, researchers in the control field first turned their attention to the realization of the PLL for a synchronous motor [14]. Since then, the phase-locked servo system (PLS) has been rapidly developed for use in AC and DC motors' servomechanisms with the analog PLL ICs [14–17]. Over the past two decades, high-performance digital ICs and versatile microprocessors have resulted in a strong motivation for the implementation of PLLs in the digital domain, including communications, servomechanism, power electronics, power systems, and renewable energy systems. New types of controllers to increase the features of the PLL have been developed continuously so as to accomplish an easy-use and easy-control strategy for system applications.

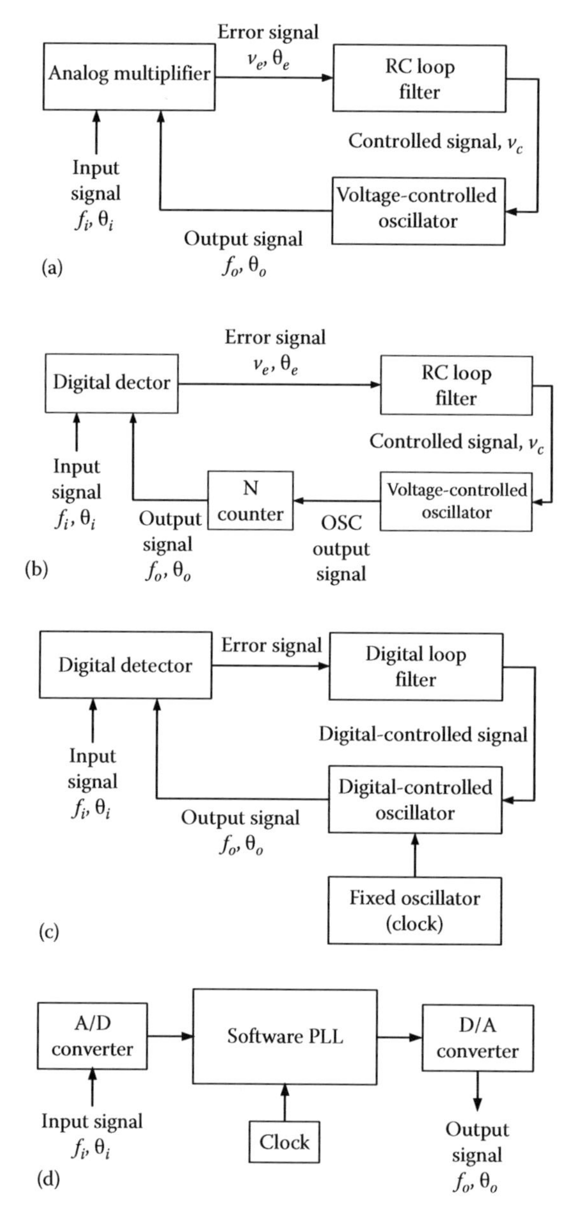

FIGURE 15.1 Configuration of PLLs: (a) linear PLL (LPLL), (b) digital PLL (DPLL), (c) all-digital PLL (ADPLL), and (d) software PLL (SPLL).

TABLE 15.1 Classification of PLLs

Type	Phase Detector	Loop Filter	Oscillator
Linear PLL (LPLL)	Analog multiplier	RC passive or active	Voltage-controlled
Digital PLL (DPLL)	Digital detector	RC passive or active	Voltage-controlled
All-digital PLL (ADPLL)	Digital detector	Digital filter	Digital-controlled
Software PLL (SPLL)	Software multiplier	Software filter	Software-controlled

15.2 Basic Concept of PLL

A phase-locked loop or phase lock loop (PLL) is a negative feedback control system that generates a signal that has a fixed relation to the phase of a reference signal. The PLL can cause a system to track with another one by keeping its output signal synchronizing with a reference input signal in both frequency and phase. More precisely, the PLL simply controls the phase of the output signal, locking the reference one by adaptively reducing their phase error to a minimum. The functional block diagram of a typical PLL is shown in Figure 15.2, which consists of a phase detector (PD), a loop filter (LF), and a voltage-controlled oscillator (VCO). Generally, a PLL initially compares the frequencies of two signals and produces an error signal that is proportional to the difference between the input frequencies. The error signal is then low-pass filtered through a RC or charge pump scheme to drive a VCO that creates an output frequency [9–11,13]. The output frequency is fed through a frequency divider back to the input of the system, producing a negative feedback loop. If the output frequency drifts, the error signal will increase, driving the frequency in the opposite direction so as to reduce the error. Thus, the output is locked to the frequency at the other input. This input is called the reference and is often derived from a crystal oscillator, which is very stable in frequency. We presume that x_i and x_o are, respectively, the input and the VCO signals, which can be expressed as [3–4]

$$x_i(t) = A\cos(\omega_i t + \theta_i) \tag{15.1}$$

$$x_o(t) = B\cos(\omega_o t + \varphi_o) \tag{15.2}$$

The angular frequency of the input signal is ω_i, and ω_o is the VCO central angular frequency. θ_i and φ_o are the phase constants. If the loop is initially unlocked and the PD has a sinusoidal characteristic, the significant output signal $v_e(t)$ at the output of PD is given by

$$v_e(t) = K_d\cos[(\omega_i - \omega_o)t + \theta_i - \varphi_o] \tag{15.3}$$

where K_d is the gain of the PD and the higher frequency item $\omega_i + \omega_o$ is negligible here due to the rejection of the LF. After a period of time sufficiently long for transient phenomena, it will be observed that the VCO output signal x_o has become synchronous with the input signal x_i. Signal x_o can then be expressed as

$$x_o(t) = B\sin(\omega_i t + \phi_o) \tag{15.4}$$

From (15.1) and (15.4), the quantity φ_o in (15.2) has become a linear function of time expressed as

$$\varphi_o = (\omega_i - \omega_o)t + \phi_o \tag{15.5}$$

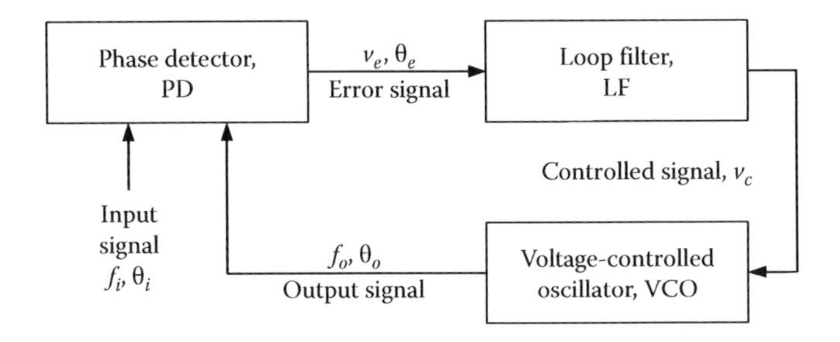

FIGURE 15.2 The functional block diagram of a typical PLL.

and the PD output signal $v_e(t)$ in (15.3) becomes a DC signal given by

$$v_e(t) = K_d \cos(\theta_i - \phi_o) \tag{15.6}$$

The LF is of the low-pass type, so that the controlled signal $v_c(t)$ is given by

$$v_c(t) = v_e(t) = K_d \cos(\theta_i - \phi_o) \tag{15.7}$$

The VCO is a frequency-modulated oscillator, whose instantaneous angular frequency ω_{inst} is a linear function of the controlled signal $v_c(t)$, around the central angular frequency ω_o, i.e.,

$$\begin{aligned}\omega_{inst} &= \frac{d}{dt}(\omega_o t + \varphi_o) \\ &= \omega_o + K_v v_c(t)\end{aligned} \tag{15.8}$$

and

$$\frac{d\varphi_o}{dt} = K_v v_c(t) \tag{15.9}$$

where K_v is the VCO sensitivity. From (15.5), (15.6), and (15.9), we have

$$\omega_i - \omega_o = K_d K_v \cos(\theta_i - \phi_o) \tag{15.10}$$

from which we have

$$\phi_o = \theta_i - \cos^{-1}\frac{\omega_i - \omega_o}{K_d K_v} \tag{15.11}$$

Substituting (15.11) into (15.6), we obtain

$$v_e = \frac{\omega_i - \omega_o}{K_v} \tag{15.12}$$

Equation 15.12 clearly shows that it is the DC signal $v_c = v_e$ that changes the VCO frequency from its central value ω_o to the input signal angular frequency ω_i, i.e.,

$$\omega_{inst} = \omega_o + K_v v_c = \omega_i \tag{15.13}$$

If the angular frequency difference $\omega_i - \omega_o$ is much lower than the product $K_d K_v$, Equation 15.11 becomes $\theta_i - \phi_o \approx \cos^{-1} 0 = \pi/2$. It indicates that the VCO signal is actually in phase quadrature with the input signal while the loop is exactly in lock. Strictly speaking, the phase quadrature actually corresponds to $\omega_i = \omega_o$. For this reason, we substitute the phase constant ϕ_o for the constant θ_o, so that $\theta_o = \phi_o - \pi/2$. Then

$$\begin{aligned}v_e &= K_d \cos(\theta_i - \phi_o) \\ &= K_d \sin(\theta_i - \theta_o)\end{aligned} \tag{15.14}$$

The difference $\theta_i - \theta_o$ is called the phase error between the two signals, which is null when the initial frequency offset is null. When the difference $\theta_i - \theta_d$ is sufficiently small, the following approximation is used, i.e.,

$$v_e \approx K_d(\theta_i - \theta_o) \tag{15.15}$$

Another interpretation for the signals in (15.1) and (15.2) can also be represented by

$$x_i = A\sin(\omega_i t + \theta_i) \tag{15.16}$$

$$x_o = B\cos(\omega_o t + \varphi_o) \tag{15.17}$$

The PD output can be represented by $v_e(t) = K_d \sin[(\omega_i - \omega_o)t + \theta_i - \varphi_o]$ when the loop is out of lock, and $v_e = K_d \sin(\theta_i - \theta_o)$ when it is working, with

$$\phi_o = \theta_i - \cos^{-1}\frac{\omega_i - \omega_o}{K_d K_v} \tag{15.18}$$

The product $K = K_d K_v$ is referred to as the loop gain. When the difference $|\omega_i - \omega_o|$ exceeds the loop gain K in a sinusoidal-characteristic PD, a proper θ_o for lock can no longer be found by means of (15.18). The synchronization no longer maintains and the loop falls out of lock.

15.2.1 Phase Detector

The PD in the PLL can be described by two categories, i.e., sinusoidal and square signal PDs. The sinusoidal PD inherently has a phase-detected interval $(-\pi/2, +\pi/2)$. It operates as a multiplier, which is a zero memory device. The square signal PDs are implemented by sequential logic circuits. Sequential PDs contain memory of past crossing events. They can generate PD characteristics that are difficult or impossible to obtain with multiplier circuits. Sequential PDs are usually built up from digital circuits and operate with binary, rectangular input waveforms. Accordingly, they are often called digital PDs. The characteristics of the square signal PDs are of the linear type over the phase-detected interval $(-\pi/2, +\pi/2)$ for a triangular PD, $(-\pi, +\pi)$ for a sawtooth PD, $(-\pi, +\pi)$ for a rectangular PD, and $(-2\pi, +2\pi)$ for a sequential phase/frequency detector (PFD). All aspects of the PD characteristics are depicted in Figure 15.3. In fact, the shape of the PD characteristic depends on the applied waveforms and not necessarily on the circuit. If the input signal is sinusoidal, the PD characteristic is also sinusoidal, with a maximum DC output equal to the peak signal amplitude. If rectangular waveforms are applied at both inputs of any multiplier or switch PD, the output characteristic becomes triangular. Remarkably, if waveforms are rectangular, digital circuits can be used instead of analog circuits. The digital-circuit equivalence of a multiplier PD is an exclusive-OR circuit. A sawtooth characteristic results from sampling a sawtooth input waveform, but that characteristic is more readily obtained from sequential PDs. A widely used sequential PD of greater complexity consists of four flip-flops plus additional logic and is available in several versions as single-chip ICs. It is called a phase/frequency detector (PFD) because it also provides an indication of frequency error when the loop is out of lock [4]. All curves of Figure 15.3, except the rectangular, are shown with the same slope at the phase error $\theta_e = \theta_i - \theta_o = 0$, which means that the different PDs all have the same factor K_d. An increased PD output capability provides a larger tracking range and a larger lock limit than those obtainable from a sinusoidal PD.

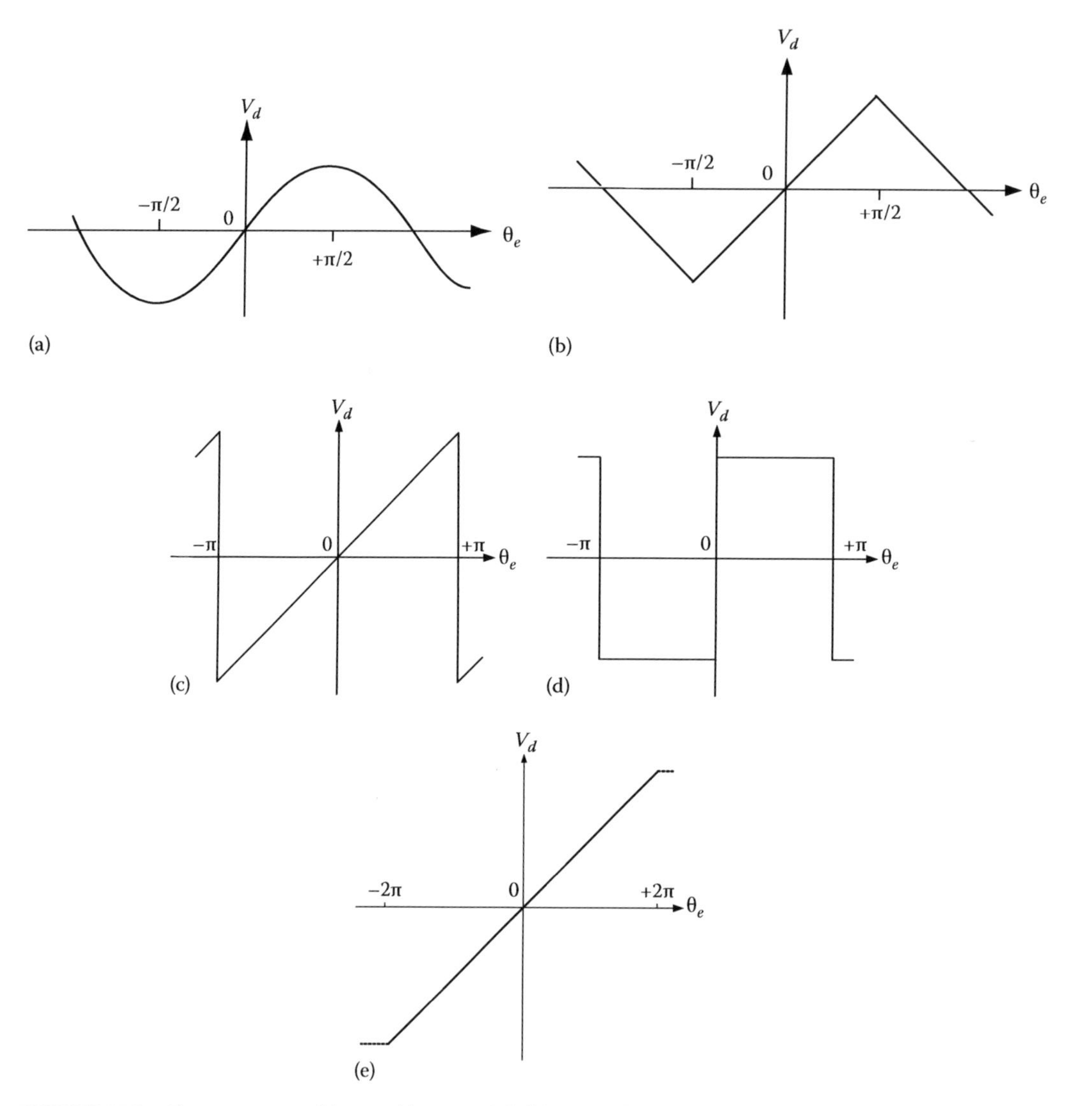

FIGURE 15.3 Characteristics of the PD: (a) sinusoidal, (b) triangular, (c) sawtooth, (d) rectangular, and (e) phase/frequency.

15.2.2 Voltage-Controlled Oscillator

The VCOs used in the PLL are basically not different from those employed for other applications, such as the modulation and the automatic frequency control. The main requirements for the VCO are as follows: (1) phase stability, (2) large frequency deviation, (3) high modulation sensitivity K_v, (4) linearity of frequency versus control voltage, and (5) capability for accepting wideband modulation. The phase stability is directly in opposition to all other four requirements. The four types of VCO commonly used are given in order of decreasing stability; i.e., crystal oscillators (VCXO), resonator oscillators, RC multivibrators, and YIG-tuned oscillators [3,4]. The phase stability can be enhanced by a number of factors: (1) high Q in the crystal and the circuit, (2) low noise in the amplifier portion, (3) temperature stability, and (4) mechanical stability. Remarkably, much of the phase jitter of an oscillator arises from noise in the associated amplifier. If a wider frequency range is required, an oscillator formed by inductor (L) and capacitor (C) must be used. In this application, the standard Hartley, Colpitts, and Clapp circuits make their appearance. The tuning is accomplished by means of a varactor. For microwave frequencies, YIG-tuned Gunn oscillators have become popular. Tuning of the YIG-tuned oscillator is accomplished by altering a magnetic field. In servomechanism, a servo motor replaces the VCO and can be controlled to maintain constant speed by detecting its phase with an encoder.

15.2.3 Loop Filter and Other Subsystems

The LF in the PLL is of a low-pass type. It is used to suppress the noise and high-frequency signal components from the phase error θ_e and provides a DC-controlled signal for the VCO. We assume that the loop is in lock, that the PD is linear, and the PD output voltage is proportional to the phase error; that is,

$$v_d \approx K_d(\theta_i - \theta_o) \tag{15.15a}$$

The phase error voltage v_e is filtered by the LF. The servo scheme of a typical PLL (or PLS) under the locking state is shown in Figure 15.4, in which *F*(*s*) is the loop transfer function. According to servo terminology, the type of loop is determined by the number of perfect integrators within the loop. Any PLL is at least, a type one loop because of the perfect integrator inherently in the VCO. If the LF contains one perfect integrator, then the loop is type two. A second-order PLL with a high-gain active filter can be approximately a type two loop, whereas a PLL with a passive filter is type one. The widely used passive and active filters for the PLL are shown in Figure 15.4a and b, respectively.

For the passive filter, the closed-loop transfer function *H*′(*s*) of the PLS (or PLL) is

$$H'(s) = \frac{K_o K_d(s\tau_2 + 1)/\tau_1}{s^2 + s(+1K_o + K_d\tau_2/\tau_1) + K_o K_d/\tau_1} \tag{15.19}$$

For the active filter, the closed-loop transfer function *H*″(*s*) is found to be

$$H''(s) = \frac{K_o K_d(s\tau_2 + 1)/\tau_1}{s^2 + s(K_o K_d/\tau_1) + K_o K_d/\tau_1} \tag{15.20}$$

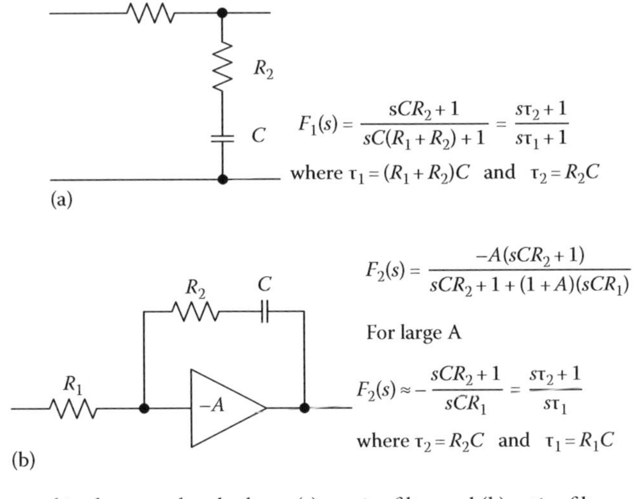

FIGURE 15.4 Filters used in the second-order loop: (a) passive filter and (b) active filter.

For convenience in description, (15.19) and (15.20) may be rewritten as

$$H'(s) = \frac{s\left(2\zeta\omega_n - \omega_n^2/K_oK_d\right) + \omega_n^2}{s^2 + 2\zeta\omega_n s + \zeta\omega_n^2} \tag{15.21}$$

and

$$H''(s) = \frac{2\zeta\omega_n s + \omega_n^2}{s^2 + 2\zeta\omega_n s + \omega_n^2} \tag{15.22}$$

in which ω_n is the natural frequency of the loop and ζ is the damping ratio. The relevant parameters for a passive filter are

$$\omega_n\left(\frac{K_vK_d}{\tau_1}\right)^{1/2}, \quad \zeta = \frac{1}{2}\left(\frac{K_vK_d}{\tau_1}\right)^{1/2}\left(\tau_2 + \frac{1}{K_vK_d}\right), \quad \tau_1 = (R_1 + R_2)C, \quad \text{and} \quad \tau_2 = R_2C;$$

and for an active filter are

$$\omega_n = \left(\frac{K_oK_d}{\tau_1}\right)^{1/2}, \quad \zeta = \frac{\tau_2}{2}\left(\frac{K_oK_d}{\tau_1}\right)^{1/2} = \frac{\tau_2\omega_n}{2}, \quad \tau_1 = R_1C, \quad \text{and} \quad \tau_2 = R_2C.$$

The two transfer functions are nearly the same if $1/K_vK_d \ll \tau_2$ in the passive filter. The open-loop transfer function of any PLL is given by

$$G(s) = \frac{K_oK_dF(s)}{s} \tag{15.23}$$

and the closed-loop transfer function can be given by

$$H(s) = \frac{G(s)}{1 + G(s)} \tag{15.24}$$

We define the DC gain of the loop as

$$K_D = K_vK_dF(0) \tag{15.25}$$

A large K_D is generally required for achieving a good performance of the loop [4]. We define a hold-in range of a loop as $\Delta\omega_H = K_D$. If the input frequency closes sufficiently to VCO frequency, a PLL locks up with just a phase transient; there is no cycle slipping prior to lock. The frequency range over which the loop acquires the phase to lock without slips is called the lock-in range of the PLL. In a first-order loop, the lock-in range is equal to the hold-in range; but for the second- or higher order loops, the lock-in range is still less than the hold-in range. Besides, there is a frequency interval, smaller than the hold-in interval and larger than the lock-in interval, over which the loop will acquire lock after slipping cycles for a while. This interval is called the pull-in range. To ensure stable tracking, it is common practice to build LFs with equal numbers of poles and zeros. At high frequencies, the loop is indistinguishable from a first-order loop with gain $K = K_o K_d F(\infty)$. As a fair approximation, we can say that the higher order loop has the same lock-in range as the equivalent-gain, first-order loop. The lock-in limit of a first-order loop is equal to the loop gain. We argue here, that a higher order loop has nearly the same lock limit. The lock-in range $\Delta\omega_L$ can be approximately estimated as

$$\Delta\omega_L \approx \pm K_d K_v F(\infty) \tag{15.26}$$

The acquisition of frequency in the PLL is more difficult, is slower, and requires more design attention than does the phase acquisition. Remarkably, the self-acquisition of the frequency is known as the frequency pull-in, or simply the pull-in; the self-acquisition of the phase as the phase lock-in, or lock-in.

15.3 Applications of PLL-Based Control

15.3.1 Phase-Locked Servo System

Integrated PLLs, developed since the 1970s, are capable of versatile functions, which are suitable for use in a variety of frequency-selective demodulations, for signal conditioning, or in frequency synthesis applications. The usual configurations of the PLL in communications are well developed and widely used for FM, AM, video, signal processing, commercial apparatus, control systems, telecommunication systems, etc. The use of the PLL, with the analog PLL IC—NE565, first developed by Signetics for synchronous and DC motors, began in the 1970s [14,15]. A PLS is certainly a frequency feedback control configuration that continually maintains the motor speed or the motor position by tracking the phase and the frequency of the incoming reference signal corresponding to the input command, such as speed or position. Basically, the PLS configuration is composed of a PD, an LF, and a servo motor, as shown in Figure 15.5, in which the servo motor operates like the VCO in the PLL as shown in Figure 15.1. From 1985 to 2009, a variety of PLL controllers with DSP, vector control, or intelligent strategies were developed in succession to increase the servo performance of a versatile servomechanism. They are, the voltage pump controller (VPC) and the adaptive digital pump controller (ADPC) for the DC motor speed servo control, the frequency-pumped controller (FPC), the two-phase-type detection, and the fuzzy control for the DC motor position servo control, the phase-controlled oscillator (PCO) for stepping up the motor speed servo control, the microcomputer-based variable slope pulse pump controller (VSPPC) for stepping up the motor position control, and the sensorless position control for the induction motor, etc., [14–34].

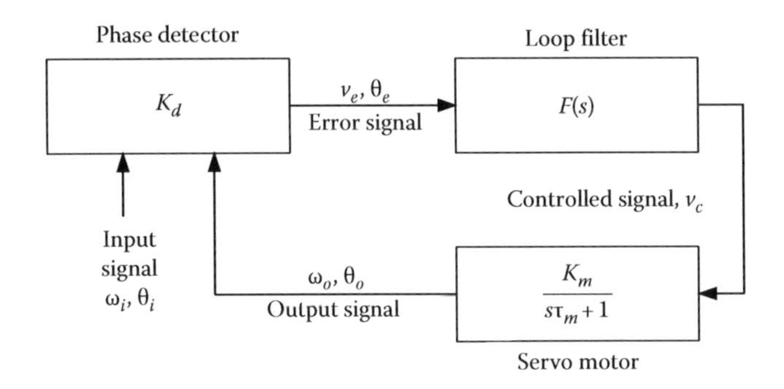

FIGURE 15.5 Typical phase-locked servo system (PLS).

15.3.2 Power Electronics

The high-power inverter, used in grid connections, with a power utility is being widely explored nowadays, in particular for renewable energy resources including fuel cell and solar photovoltaic cell systems. The high-precision control of a single- or a three-phase PWM inverter for constant voltage and constant frequency applications, such as uninterruptible power supply, is often controlled by employing a capacitor current feedback with a PLL compensator that minimizes the steady-state error of the output voltage. Furthermore, the power supply system with a synchronous rectifier frequently uses as DPLL technique to synchronize the secondary driver signal with the switching frequency so as to raise the power efficiency [35–55].

15.3.3 Active Power Filters

An approach using the PLL technique to realize the active power filter (APF) is widely applied to achieve high accuracy in tracking highly distorted current waveforms and to minimize the ripple in multiphase systems. The methodologies for the aforementioned achievement are realized with the aid of a current loop proportional–integral regulator, the hysteresis current-control method, a software implementation of a robust synchronizing circuit, a fully digital control algorithm with a field-programmable gate array (FPGA), and so on [56–61].

15.3.4 Power Quality

The phase angle of the utility voltage is a critical piece of information for the operation of most apparatuses, such as AC to DC converters, static VAR compensators, cycle-converters, active harmonic filters, and other energy storage systems coupled with electric utility. This information may be used to synchronize the turning on/off of power devices, calculate and control the flow of active/reactive power, or transform the feedback variables to a reference frame suitable for control purposes. The angle information is typically extracted using some form of a PLL. The quality of the lock directly affects the performance of the control loops in the aforementioned applications. Thus, an operation of a PLL system is widely adopted under distorted utility conditions. The PLL system is quite often recommended for the making or the tuning of the phase angle of the utility voltage, especially for an operation under common utility distortions, such as line notching, voltage unbalance/loss, and frequency variations [62–68].

15.3.5 Lighting

The electronic ballast, formed by a series-resonant topology, a class-D inversion, or a piezoelectric transformer with a current-equalization network, has been tried to be stabilized by a PLL, which also includes the detection of the phase signals of both the resonant-tank input voltage and the two leakage-inductor terminals in the transformer, and finally it is also used to track the resonant frequency. The primary objective is to increase ballast stabilization and efficiency, and to improve the temperature effect [69–72].

15.3.6 Battery Charger

Nowadays, battery chargers for versatile apparatuses have already come to play important roles in our daily lives. In the charger system, the more important charging processes include the bulk current charging process, the current charging process, and the float charging process. A lot of detections exist during the charging process, including charging current detection, voltage detection, the instant charging state, etc. The PLL applied in a charger is based on the charge trajectory, which is programmed to frequency to achieve the auto-tracking and the auto-locking of the charging status, so as to maintain high accuracy in the charging process [73–77].

15.4 Analog, Digital, and Hybrid PLLs

Due to the rapid progress in digital ICs in the 1970s, the integrated PLL was developed for filtering, frequency synthesis, motor-speed control, frequency modulation, demodulation, signal detection, and a variety of other applications. The PLL ICs have become one of the most important communication circuits. PLLs can be analog or digital, but most of them are composed of both analog and digital components. The first analog PLL ICs—NE565 and the CD4046 were developed by Signetics and RCA in the 1970s. They are integrated on a chip with a PD, an LF, and a VCO. The PD is a multiplier, whose phase-lock range is from $-\pi/2$ to $+\pi/2$. In order to extend the lock-in range of the PLL, the Motorola digital phase/frequency comparator (PFD) MC4044, capable of a phase-detection range from -2π to $+2\pi$ was proposed in 1972. A hybrid PLL realized by combining discrete analog and digital components was then achieved. Thereafter, versatile PLL components for raising the locking performance were developed in succession. Nowadays, it is a future-exploited trend to combine a PD, a prescaler, a programmable counter, and a VCO into a modular device to obtain a DPLL. Programmable interface controllers (PICs) are popular with developers and hobbyists alike due to their low cost and wide availability, a large user base, an extensive accessibility to application notes, the availability of low cost or free development tools, and the serial programming (and reprogramming with flash memory) capability. Easy programming PICs are easily available for PLL applications to help increase system functions.

References

General

1. E. V. Appleton, Automatic synchronization of triode oscillators, *Proc. Camb. Philos. Soc.*, 21(pt. III), 231, 1922–1923.
2. H. de Bellescize, La reception synchrone, *Onde Electron.*, 11, 230–240, June 1932.
3. A. Blanchard, *Phase-Locked Loops: Application to Coherent Receiver Design*, New York: John Wiley & Sons, Inc., 1976.
4. F. M. Garder, *Phaselock Techniques*, 2nd edn., New York: Wiley, 1979.
5. W. C. Lindsey and C. M. Chie, A survey of digital phase-locked loops, *Proc. IEEE*, 69(4), 410–431, April 1981.
6. B. Razavi (ed.), *Monolithic Phase-Locked Loops and Clock Recovery Circuits*, New York: IEEE Press, 1997.
7. Y. Fouzar, M. Sawan, and Y. Savaria, CMOS wide-swing differential VCO for fully integrated fast PLL, *Proc. IEEE Midwest Symp. Circuit Syst.*, 2, 948–950, August 2000.
8. M. Karimi-Ghartemani and M. R. Iravani, A new phase-locked loop (PLL) system, *Proc. IEEE MWSCAS 2007*, Montreal, Canada, August 2001, vol. 1, pp. 421–424.
9. K. H. Cheng, W. B. Yang, and C. M. Ying, A dual-slope phase frequency detector and charge pump architecture to achieve fast locking of phase-locked loop, *IEEE Trans. Circuits Syst. II: Analog Digit. Signal Process.*, 50(11), 892–896, November 2003.
10. P. K. Hanumolu, M. Brownlee, K. Mayaram, and Un-Ku Moon, Analysis of charge-pump phase-locked loops, *IEEE Trans. Circuits Syst. I: Regular Papers*, 51(9), 1665–1674, September 2004.
11. Z. Wang, An analysis of charge-pump phase-locked loops, *IEEE Trans. Circuits Syst. I: Regular Papers*, 52(10), 2128–2138, October 2005.
12. S. Gokcek, J. A. Jeltema, and C. Gokcek, Adaptively enhanced phase locked loops, *Proceedings of the IEEE Conference on Control Applications*, Toronto, Canada, August 2005, pp. 1140–1145.
13. C. Y. Yao and C. C. Yeh, An application of the second-order passive lead–lag loop filter for analog PLLs to the third-order charge-pump PLLs, *IEEE Trans. Ind. Electron.* 55(2), 972–974, February 2008.

Servo Control

14. G. T. Volpe, A phase-locked loop control system for a synchronous motor, *IEEE Trans. Automatic Contr.*, AC-15, 88–95, February 1970.
15. W. Moore, Phase-locked loops for motor-speed control, *IEEE Spectr.*, 10, 61–67, April 1973.
16. J. Tal, Speed control by phase locked servo system—New possibilities and limitations, *IEEE Trans. Ind. Electron. Contr. Instrum.*, IECI-24, 118–125, February 1977.
17. D. F. Geiger, *Phase Lock Loops for DC Motor Speed Control*, New York: Wiley, 1981.
18. N. Margaris and V. Petridis, Voltage pump phase-locked loops, *IEEE Trans. Ind. Electron.*, IE-32, 41–49, February 1985.
19. G. C. Hsieh, Y. P. Wu, C. H. Lee, and C. H. Liu, An adaptive digital pump controller for phase locked servo systems, *IEEE Trans. Ind. Electron.*, IE-34, 379–386, 1987.
20. G. C. Hsieh, A study on position servo control systems by frequency-locked technique, *IEEE Trans. Ind. Electron.*, IE-36, 365–373, 1989.
21. J. C. Li and G. C. Hsieh, A phase/frequency-locked controller for stepping servo systems, *IEEE Trans. Ind. Electron.*, 39(2), 379–386, April 1992.
22. M. F. Lai, G. C. Hsieh, and Y. P. Wu, Variable slope pulse pump controller for stepping position servo control using frequency-locked technique, *IEEE Trans. Ind. Electron.*, 42(3), 290–299, 1995.
23. C. A. Karybakas and T. L. Laopoulos, Analysis of unlocked and acquisition operation of a phase-locked speed control system, *IEEE Trans. Ind. Electron.*, 44(1), 138–140, February 1997.
24. J. Holtz, Sensorless position control of induction motors-an emerging technology, *IEEE Trans. Ind. Electron.*, 45(6), 840–851, December 1998.
25. T. Emura, L. Wang, M. Yamanaka, and H. Nakamura, A high-precision positioning servo controller based on phase/frequency detecting technique of two-phase-type PLL, *IEEE Trans. Ind. Electron.*, 47(6), 1298–1306, December 2000.
26. L. Ying and N. Ertugrul, A novel, robust DSP-based indirect rotor position estimation for permanent magnet AC motors without rotor saliency, *IEEE Trans. Power Electron.*, 18(2), 539–546, March 2003.
27. L. R. Chen, G. C. Hsieh, and H. M. Lee, A design of fast-locking without overshoot frequency-locking servo control system by fuzzy control technology, *J. Chin. Inst. Eng.*, 27(1), 147–151, January 2004.
28. H. Abu-Rub, J. Guzinski, Z. Krzeminski, and H. A. Toliyat, Advanced control of induction motor based on load angle estimation, *IEEE Trans. Ind. Electron.*, 51(1), 5–14, February 2004.
29. L. R. Chen, G. C. Hsieh, and H. M. Lee, A design for an adjustable fuzzy pulse pump controller in a frequency-locked servo system, *J. Chin. Inst. Eng.*, 28(3), 441–452, May 2005.
30. M. Comanescu and L. Xu, An improved flux observer based on PLL frequency estimator for sensorless vector control of induction motors, *IEEE Trans. Ind. Electron.*, 53(1), 50–56, February 2006.
31. J. X. Shen, and S. Iwasaki, Sensorless control of ultrahigh-speed PM brushless motor using PLL and third harmonic back EMF, *IEEE Trans. Ind. Electron.*, 53(2), 421–428, April 2006.
32. O. Wallmark and L. Harnefors, Sensorless control of salient PMSM drives in the transition region, *IEEE Trans. Ind. Electron.*, 53(4), 1179–1187, June 2006.
33. C. T. Pan and E. Fang, A phase-locked-loop-assisted internal model adjustable-speed controller for BLDC motors, *IEEE Trans. Ind. Electron.*, 55(9), 3415–3425, September 2008.
34. Y. W. Li, B. Wu, D. Xu, and N. R. Zargari, Space vector sequence investigation and synchronization methods for active front-end rectifiers in high-power current-source drives, *IEEE Trans. Ind. Electron.*, 55(3), 1022–1034, March 2008.

Power Electronics

35. S. Musumeci, A. Raciti, A. Testa, A. Galluzzo, and M. Melito, Switching-behavior improvement of insulated gate-controlled devices, *IEEE Trans. Power Electron.*, 12(4), 645–653, July 1997.
36. S. K. Chung, A phase tracking system for three phase utility interface inverters, *IEEE Trans. Power Electron.*, 15(3), 431–438, May 2000.

37. M. P. Chen, J. K. Chen, K. Murata, M. Nakahara, and K. Harada, Surge analysis of induction heating power supply with PLL, *IEEE Trans. Power Electron.*, 16(5), 702–709, September 2001.
38. S. K. Chung, H. B. Shin, and H. W. Lee, Precision control of single-phase PWM inverter using PLL compensation, *IEE Proc., Electric Power Appl.*, 152(2), 429–436, March 2005.
39. H. Miura, S. Arai, F. Sato, H. Matsuki, and T. Sato, A synchronous rectification using a digital PLL technique for contactless power supplies, *IEEE Trans. Magnetics*, 41(10), 3997–3999, October 2005.
40. J. W. Choi, Y. K. Kim, and H. G. Kim, Digital PLL control for single-phase photovoltaic system, *IEE Proc. Electric Power Appl.*, 153(1), 40–46, January 2006.
41. S. B. Yaakov and S. Lineykin, Maximum power tracking of piezoelectric transformer HV converters under load variations, *IEEE Trans. Power Electron.*, 21(1), 73–78, January 2006.
42. P. Rodriguez, J. Pou, J. Bergas, J. I. Candela, R. P. Burgos, and D. Boroyevich, Decoupled double synchronous reference frame PLL for power converters control, *IEEE Trans. Power Electron.*, 22(2), 584–592, March 2007.
43. J. T. Matysik, The current and voltage phase shift regulation in resonant converters with integration control, *IEEE Trans. Ind. Electron.*, 54(2), 1240–1242, April 2007.
44. M. Dai, M. N. Marwali, J. W. Jung, and A. Keyhani, Power flow control of a single distributed generation unit, *IEEE Trans. Power Electron.*, 23(1), 343–352, January 2008.
45. H. Karimi, A. Yazdani, and R. Iravani, Negative-sequence current injection for fast islanding detection of a distributed resource unit, *IEEE Trans. Power Electron.*, 23(1), 298–307, January 2008.
46. C. A. Busada, H. G. Chiacchiarini, and J. C. Balda, Synthesis of sinusoidal waveform references synchronized with periodic signals, *IEEE Trans. Power Electron.*, 23(2), 581–590, March 2008.
47. S. Shinnaka, A robust single-phase PLL system with stable and fast tracking, *IEEE Trans. Ind. Electron.*, 44(2), 624–633, March–April 2008.
48. M. A. Perez, J. R. Espinoza, L. A. Moran, M. A. Torres, and E. A. Araya, A robust phase-locked loop algorithm to synchronize static-power converters with polluted AC systems, *IEEE Trans. Ind. Electron.*, 55(5), 2185–2192, May 2008.
49. X. Peng, K. A. Corzine, and G. K. Venayagamoorthy, Multiple reference frame-based control of three-phase PWM boost rectifiers under unbalanced and distorted input conditions, *IEEE Trans. Power Electron.*, 23(4), 2006–2017, July 2008.
50. R. M. S. Filho, P. F. Seixas, P. C. Cortizo, L. A. B. Torres, and A. F. Souza, Comparison of three single-phase PLL algorithms for UPS applications, *IEEE Trans. Ind. Electron.*, 55(8), 2923–2932, August 2008.
51. G. Iwanski and W. Koczara, DFIG-based power generation system with UPS function for variable-speed applications, *IEEE Trans. Ind. Electron.*, 55(8), 3047–3054, August 2008.
52. H. Tao, J. L. Duarte, and M. A. M. Hendrix, Line-interactive UPS using a fuel cell as the primary source, *IEEE Trans. Ind. Electron.*, 55(8), 3012–3021, August 2008.
53. R. Teodorescu and F. Blaabjerg, Flexible control of small wind turbines with grid failure detection operating in stand-alone and grid-connected mode, *IEEE Trans. Power Electron.*, 19(5), 1323–1332, September 2004.
54. M. Wei and Z. Chen, A fast PLL method for power electronic systems connected to distorted grids, *Proc. IEEE IECON'07*, Taipei, Taiwan, November 2007, pp. 1702–1707.
55. J. Eloy-Garcia, S. Arnaltes, and J. L. Rodriguez-Amenedo, Direct power control of voltage source inverters with unbalanced grid voltages, *IET Power Electron.*, 1(3), 395–407, September 2008.

Active Power Filters

56. L. Malesani, P. Mattavelli, and P. Tomasin, High-performance hysteresis modulation technique for active filters, *IEEE Trans. Power Electron.*, 12(5), 876–884, September 1997.
57. H. Awad, J. Svensson, and M. J. Bollen, Tuning software phase-locked loop for series-connected converters, *IEEE Trans. Power Delivery*, 20(1), 300–308, January 2005.

58. M. Cichowlas, M. Malinowski, M. P. Kazmierkowski, D. L. Sobczuk, P. Rodriguez, and J. Pou, Active filtering function of three-phase PWM boost rectifier under different line voltage conditions, *IEEE Trans. Ind. Electron.*, 52(2), 410–419, April 2005.
59. L. G. B. Barbosa Rolim, D. R. Rodrigues da Costa Jr., and M. Aredes, Analysis and software implementation of a robust synchronizing PLL circuit based on the pq theory, *IEEE Trans. Ind. Electron.*, 53(6), 1919–1926, December 2006.
60. W. Stefanutti, P. Mattavelli, G. Spiazzi, and P. Tenti, Digital control of single-phase power factor preregulators based on current and voltage sensing at switch terminals, *IEEE Trans. Power Electron.*, 21(5), 1356–1363, September 2006.
61. Z. Shu, Y. Guo, and J. Lian, Steady-state and dynamic study of active power filter with efficient FPGA-based control algorithm, *IEEE Trans. Ind. Electron.*, 55(4), 1527–1536, April 2008.

Power Quality

62. V. Kaura and V. Blasko, Operation of a phase locked loop system under distorted utility conditions, *IEEE Trans. Ind. Appl.*, 33(1), 58–63, January–February 1997.
63. L. N. Arruda, S. M. Silva, and B. J. C. Filho, PLL structures for utility connected systems, *IEEE IAS*, 4(30 September 4), 2655–2660, October 2001.
64. M. Karimi-Ghartemani, H. Karimi, and M. R. Iravani, A magnitude/phase-locked loop system based on estimation of frequency and in-phase/quadrature-phase amplitudes, *IEEE Trans. Ind. Electron.*, 51(2), 511–517, April 2004.
65. B. Han and B. Bae, Novel phase-locked loop using adaptive linear combiner, *IEEE Trans. Power Deliv.*, 21(1), 513–514, January 2006.
66. A. Cataliotti, V. Cosentino, and S. Nuccio, A phase-locked loop for the synchronization of power quality instruments in the presence of stationary and transient disturbances, *IEEE Trans. Instrum. Meas.*, 56(6), 2232–2239, December 2007.
67. K. Young and R. A. Dougal, SRF-PLL with dynamic center frequency for improved phase detection, *IEEE International Conference on Clean Electrical Power*, Guilin, China, June 2009, Issue 9–11, pp. 212–216.
68. J. A. Suul, K. Ljøkelsøy, and T. Undeland, Design, tuning and testing of a flexible PLL for grid synchronization of three-phase power converters, *EPE 2009*, pp. 1–10, September 2009.

Lighting

69. C. H. Lin and J. Y. Chen, The tracking of the optimal operating frequency in a class E backlight inverter using the PLL technique, *IEICE Trans. Electron.*, E88-C(6), 1253–1262, June 2005.
70. R. L. Lin and C. H. Wen, PLL control scheme for the electronic ballast with a current-equalization network, *J. Display Technol.*, 2(2), 160–169, June 2006.
71. R. L. Lin and Y. T. Chen, Electronic ballast for fluorescent lamps with phase-locked loop control scheme, *IEEE Trans. Power Electron.*, 21(1), 254–262, January 2006.
72. C. H. Lin, Y. Lu, H. J. Chiu, and C. L. Ou, Eliminating the temperature effect of piezoelectric transformer in backlight electronic ballast by applying the digital phase-locked-loop technique, *IEEE Trans. Ind. Electron.*, 54(2), 1024–1031, April 2007.

Battery Charger

73. L. R. Chen, PLL-based battery charge circuit topology, *IEEE Trans. Ind. Electron.*, 51, 1244–1346, December 2004.
74. L. R. Chen, C. P. Chou, C. S. Liu, and B. G. Ju, A design of a digital frequency-locked battery charger for Li-ion batteries, *Proceedings of the IEEE International Symposium on Industrial Electronics (ISIE)*, Dubrovnik, Croatia, June 2005, pp. 1093–1098.

75. L. R. Chen, J. Y. Han, J. L. Jaw, C. P. Chou, and C. S. Liu, A resistance-compensated phase-locked battery charger, *Proceedings of the IEEE International Conference on Industrial Electronics and Applications (ICIEA)*, Singapore, May 2006.
76. L. R. Chen and C. S. Wang, Modeling, analyzing and designing of a phase-locked charger, *J. Chin. Inst. Eng.*, 30(6), 1037–1046, 2007.
77. L. R. Chen, J. J. Chen, N. Y. Chiu, and J. Y. Han, Current-pumped battery charger, *IEEE Trans. Ind. Electron.*, 55(6), 2482–2488, June 2008.

16

Optimal Control

Victor M. Becerra
University of Reading

16.1 Introduction

16.1.1 Optimal Control

Optimal control is the process of determining control and state trajectories for a dynamic system over a period of time to minimize a performance index. The performance index might include, for example, a measure of the control effort, a measure of the tracking error, a measure of energy consumption, a measure of the amount of time taken in achieving an objective, or any other quantity of importance to the designer of the control system.

16.1.2 Origins

Optimal control is closely related in its origins to the theory of *calculus of variations*. The theory of finding maxima or minima of functions can be traced to the isoperimetric problems formulated by Greek mathematicians such as Zenodorus (495–435 BC). In 1699, Johan Bernoulli (1667–1748) posed as a challenge to the scientific community the brachistochrone problem—the problem of finding a path of quickest descent between two points that are not on the same horizontal or vertical line. The problem was solved by Johan Bernoulli himself, his brother Jacob Bernoulli (1654–1705), Gottfried Leibniz (1646–1716), and (anonymously) Isaac Newton (1642–1727). Leonhard Euler (1707–1793) worked with Johan Bernoulli and made significant contributions, which influenced Ludovico Lagrange (1736–1813) who provided a way of solving such problems using the method of first variations, a set of necessary conditions for an extremum of a functional, which in turn led Euler to coin the term *calculus of variations*. Andrien

Legendre (1752–1833) and Carl Jacobi (1804–1851) developed sufficient conditions for the extremum of a functional. William Hamilton (1805–1865) did some remarkable work in mechanics, which, together with later work by Jacobi, resulted in the Hamilton–Jacobi equations, which had a strong influence in calculus of variations, optimal control, and dynamic programming. Karl Weierstrass (1815–1897) distinguished between strong and weak extrema, while Adolph Mayer (1839–1907) and Oskar Bolza (1857–1942) worked on generalizations of calculus of variations. Some important milestones in the development of optimal control in the twentieth century include the formulation dynamic programming by Richard Bellman (1920–1984) in the 1950s, the development of the minimum principle by Lev Pontryagin (1908–1988) and coworkers also in the 1950s, and the formulation of the linear quadratic regulator (LQR) and the Kalman filter by Rudolf Kalman (b. 1930) in the 1960s. Modern computational optimal control also has roots in nonlinear programming (parameter optimization with inequality and equality constraints), which was developed soon after the Second World War [BMC93]. See the review papers [SW97] and [Bry96] for further historical details.

16.1.3 Applications of Optimal Control

Optimal control and its ramifications (such as model predictive control) have found applications in many different fields, including aerospace, process control, robotics, bioengineering, economics, finance, and management science, and it continues to be an active research area within control theory. Before the arrival of the digital computer in the 1950s, only fairly simple optimal control problems could be solved. The arrival of the digital computer has enabled the application of optimal control theory and methods to many complex problems.

16.2 Formulation of Optimal Control Problems

There are various types of optimal control problems, depending on the performance index, the type of time domain (continuous, discrete), the presence of different types of constraints, and what variables are free to be chosen. The formulation of an optimal control problem requires the following:

- A mathematical model of the system to be controlled
- A specification of the performance index
- A specification of all boundary conditions on states and constraints to be satisfied by states and controls
- A statement of what variables are free

16.3 Continuous-Time Optimal Control Using the Variational Approach

16.3.1 Preliminaries

This section introduces some fundamental concepts in optimization theory and calculus of variations. Readers interested in further details may wish to consult [GF03,Lei81,Wan95].

Consider the following optimization problem:

$$\begin{aligned} &\text{minimize} \quad && f(\mathbf{x}) \\ &\text{subject to} \quad && x \in \Omega \end{aligned}$$

where

$f: \mathcal{R}^n \mapsto \mathcal{R}$ is a real-valued *objective function*

$\mathbf{x} = [x_1, x_2, \ldots, x_n]^T \in \mathcal{R}^n$ is called the *decision vector*

The set Ω, which is a subset of $\mathcal{R}^n$, is called the *admissible set*

Definition 16.1 (Local minimizer) *Suppose that* f: $\mathcal{R}^n \mapsto \mathcal{R}$ is a real-valued function defined on a set $\Omega \subset \mathcal{R}^n$. A point $\mathbf{x}^* \in \Omega$ is a local minimizer of f over Ω if there exists $\varepsilon > 0$ such that $f(\mathbf{x}) \geq f(\mathbf{x}^*)$ for all $\mathbf{x} \in \Omega$, $\mathbf{x} \neq \mathbf{x}^*$, and $\|\mathbf{x} - \mathbf{x}^*\| \leq \varepsilon$.

Consider now the case when there admissible set is defined by m equality constraints, such that $\Omega = \{\mathbf{x} \in \mathcal{R}^n: \mathbf{h}(\mathbf{x}) = 0\}$, where $\mathbf{h}$: $\mathcal{R}^n \mapsto \mathcal{R}^m$. Assume that $\mathbf{h}$ is continuously differentiable.

Definition 16.2 (Regular point) A point $\mathbf{x}^*$ satisfying the constraints $\mathbf{h}(\mathbf{x}^*) = 0$ is said to be a regular point of the constraints if rank$[\partial \mathbf{h}/\partial \mathbf{x}] = m$, that is, the Jacobian matrix of $\mathbf{h}$ has full rank.

The following well-known theorem establishes the first-order necessary condition for a local minimum (or maximum) of a continuously differentiable function $f(\mathbf{x})$ subject to equality constraints $\mathbf{h}(\mathbf{x}) = 0$. See [CZ96] for the proof.

Theorem 16.1 (Lagrange multiplier theorem) Let $\mathbf{x}^* \in \Omega$ be a local minimizer of f: $\mathcal{R}^n \mapsto \mathcal{R}$ subject to $\mathbf{h}(\mathbf{x}) = \mathbf{0}$, with $\mathbf{h}$: $\mathcal{R}^n \mapsto \mathcal{R}^m$, $m \leq n$. Assume that $\mathbf{x}$ is a regular point. Then there exists $\lambda^* \in \mathcal{R}^m$ such that

$$\nabla f(\mathbf{x}^*) + \left[\frac{\partial \mathbf{h}(\mathbf{x})}{\partial \mathbf{x}}\right]_{\mathbf{x}=\mathbf{x}^*}^T \lambda^* = \mathbf{0} \tag{16.1}$$

where

$\nabla f(\mathbf{x}^*) = [\partial f(\mathbf{x})/\partial \mathbf{x}]_{\mathbf{x}=\mathbf{x}^*}^T$ is the gradient vector of f

λ^* is the vector of Lagrange multipliers

Note that by augmenting the objective function $f(\mathbf{x})$ with the inner product of the vector of Lagrange multipliers λ and the constraint vector function $\mathbf{h}(\mathbf{x})$,

$$\bar{f}(\mathbf{x}, \lambda) = f(\mathbf{x}) + \lambda^T \mathbf{h}(\mathbf{x})$$

it is possible to express Equation 16.1 as follows:

$$\nabla \bar{f}(\mathbf{x}^*, \lambda^*) = \mathbf{0}$$

where $\nabla \bar{f}(\mathbf{x}^*, \lambda^*) = [\partial \bar{f}(\mathbf{x}, \lambda)/\partial \mathbf{x}]_{\mathbf{x}=\mathbf{x}^*, \lambda=\lambda^*}^T$.

Definition 16.3 (Functional) A functional J is a rule of correspondence that assigns a unique real number to each vector function $\mathbf{x}(t)$ in a certain class Ω^* defined over the real interval $t \in [t_0, t_f]$.

Definition 16.4 (Increment of a functional) The increment of a functional J is defined as follows:

$$\Delta J = J(\mathbf{x} + \delta \mathbf{x}) - J(\mathbf{x}) \tag{16.2}$$

where

$\delta \mathbf{x}$ is called the variation of $\mathbf{x}$

$\delta \mathbf{x}(t)$ is an incremental change in function $\mathbf{x}(t)$ when the variable t is held fixed

Definition 16.5 (Norm of a function) The norm of a function is a rule of correspondence that assigns to each function $\mathbf{x} \in \Omega^*$, defined for $t \in [t_0, t_f]$, a real number. The norm of $\mathbf{x}$, denoted as $\|\mathbf{x}\|$, satisfies the following properties:

- $\|\mathbf{x}\|>0$ if $x \neq 0$, $\|x\| = 0$ if and only if $\mathbf{x} = 0$ for all $t \in [t_0, t_f]$
- $\|\alpha\mathbf{x}\| = \|\alpha\| \, \|\mathbf{x}\|$ for all $\alpha \in \mathcal{R}$
- $\|\|\mathbf{x}^{(1)} + \mathbf{x}^{(2)}| \leq \|\|\mathbf{x}^{(1)}\| + \|\mathbf{x}^{(2)}\|$

Definition 16.6 The variation of a differentiable functional J is defined as follows:

$$\delta J = \lim_{\|\delta \mathbf{x}\| \mapsto 0} \Delta J = \lim_{\|\delta \mathbf{x}\| \mapsto 0} J(\mathbf{x} + \delta \mathbf{x}) - J(\mathbf{x}) \tag{16.3}$$

In general, the variation of a differentiable functional of the form

$$J = \int_{t_0}^{t_f} G(\mathbf{x}) \mathrm{d}t \tag{16.4}$$

can be calculated as follows:

$$\delta J = \int_{t_0}^{t_f} \left\{ \left[\frac{\partial G(\mathbf{x})}{\partial \mathbf{x}} \right] \delta \mathbf{x} \right\} \mathrm{d}t \tag{16.5}$$

Definition 16.7 (Local minimum of a functional) A functional J has a local minimum at $\mathbf{x}^*$ if for all functions $\mathbf{x}$ in the vicinity of $\mathbf{x}^*$ we have

$$\Delta J = J(\mathbf{x}) - J(\mathbf{x}^*) \geq 0 \tag{16.6}$$

The fundamental theorem of calculus of variations states that if a functional J has a local minimum or maximum at $\mathbf{x}^*$, then the variation of J must vanish on $\mathbf{x}^*$:

$$\delta J(\mathbf{x}^*, \delta \mathbf{x}) = 0 \tag{16.7}$$

Suppose now that we wish to find conditions for the minimum of a functional

$$J(\mathbf{x}) = \int_{t_0}^{t_f} W(\mathbf{x}, \dot{\mathbf{x}}, t) \mathrm{d}t$$

where $\mathcal{W}$: $\mathcal{R}^n \times \mathcal{R}^n \times \mathcal{R} \mapsto \mathcal{R}$, $\dot{\mathbf{x}} = \mathrm{d}\mathbf{x}/\mathrm{d}t$, subject to the constraints

$$\mathbf{g}(\mathbf{x}, \dot{\mathbf{x}}, t) = 0$$

where $\mathbf{g}$: $\mathcal{R}^n \times \mathcal{R}^n \times \mathcal{R} \mapsto \mathcal{R}^m$ is a vector function. The necessary conditions for a minimum of J are often found using Lagrange multipliers. Define an augmented functional

$$J_a(\mathbf{x}, \lambda) = J(\mathbf{x}) = \int_{t_0}^{t_f} \left\{ W(\mathbf{x}, \dot{\mathbf{x}}, t) + \lambda^T(t) \mathbf{g}(\mathbf{x}, \dot{\mathbf{x}}, t) \right\} \mathrm{d}t$$

where λ: $[t_0, t_f] \mapsto \mathcal{R}^m$ is a vector of time-dependent Lagrange multipliers. The necessary conditions for a minimum of $J_a(\mathbf{x}, \lambda)$ can be analyzed using the fundamental theorem of calculus of variations with $\mathbf{x}$ and λ as variables (see [Kir70]).

Note that if the constraints are satisfied, $\mathbf{g} = 0$, then $J(\mathbf{x}) = J_a(x, \lambda)$, so that the Lagrange multipliers can be chosen arbitrarily.

16.3.2 Case with Fixed Initial and Final Times and No Terminal or Path Constraints

If there are no path constraints on the states or the control variables, if the initial state conditions are fixed, and if the initial and final times are fixed, a fairly general continuous time optimal control problem can be defined as follows:

PROBLEM 16.1

Find the control vector trajectory $\mathbf{u}:[t_0,t_f]\subset\mathcal{R}\mapsto\mathcal{R}^{n_u}$ to minimize the performance index

$$J = \varphi(\mathbf{x}(t_f),t_f) + \int_{t_0}^{t_f} L(\mathbf{x}(t),\mathbf{u}(t),t)\,dt$$

subject to

$$\begin{aligned}\dot{\mathbf{x}}(t) &= \mathbf{f}(\mathbf{x}(t),\mathbf{u}(t),t)\\ \mathbf{x}(t_0) &= \mathbf{x}_0\end{aligned} \tag{16.8}$$

where $[t_0, t_f]$ is the time interval of interest, $\mathbf{x}:[t_0,t_f]\mapsto\mathcal{R}^{n_x}$ is the state vector, $\varphi:\mathcal{R}^{n_x}\times\mathcal{R}\mapsto\mathcal{R}$ is a terminal cost function, $L:\mathcal{R}^{n_x}\times\mathcal{R}^{n_u}\times\mathcal{R}\mapsto\mathcal{R}$ is an intermediate cost function, and $\mathbf{f}:\mathcal{R}^{n_x}\times\mathcal{R}^{n_u}\times\mathcal{R}\mapsto\mathcal{R}^{n_x}$ is a vector field. Note that Equation 16.8 represents the dynamics of the system and its initial state condition. Problem 16.1 as defined above is known as the Bolza problem. If $L(\mathbf{x}, \mathbf{u}, t) = 0$, then the problem is known as the Mayer problem; if $\varphi(\mathbf{x}(t_f), t_f) = 0$, it is known as the Lagrange problem. Note that the performance index $J = J(\mathbf{u})$ is a functional. In this section, calculus of variations is used to derive necessary optimality conditions for the minimization of $J(\mathbf{u})$.

Adjoin the constraints to the performance index with a time-varying Lagrange multiplier vector function $\lambda:[t_0,t_f]\mapsto\mathcal{R}^{n_x}$ (also known as the costate), to define an augmented performance index $\bar{J}$:

$$\bar{J} = \varphi(\mathbf{x}(t_f),t_f) + \int_{t_o}^{t_f}\left\{L(\mathbf{x},\mathbf{u},t) + \lambda^T(t)\left[\mathbf{f}(\mathbf{x},\mathbf{u},t) - \dot{\mathbf{x}}\right]\right\}dt$$

Define the Hamiltonian function H as follows:

$$H(\mathbf{x}(t),\mathbf{u}(t),\lambda(t),t) = L(\mathbf{x}(t),\mathbf{u}(t),t) + \lambda(t)^T\mathbf{f}(\mathbf{x}(t),\mathbf{u}(t),t)$$

such that $\bar{J}$ can be written as

$$\bar{J} = \varphi(\mathbf{x}(t_f), t_f) + \int_{t_0}^{t_f}\left\{H(\mathbf{x}(t),\mathbf{u}(t),\lambda(t),t) - \lambda^T(t)\dot{\mathbf{x}}\right\}dt$$

Assume that t_0 and t_f are fixed. Now consider an infinitesimal variation in $\mathbf{u}(t)$, which is denoted as $\delta\mathbf{u}(t)$. Such a variation will produce variations in the state history $\delta\mathbf{x}(t)$, and a variation in the performance index $\delta\bar{J}$, which can be expressed as follows:

$$\delta\bar{J} = \left[\left(\frac{\partial\varphi}{\partial\mathbf{x}} - \lambda^T\right)\delta\mathbf{x}\right]_{t=t_f} + \left[\lambda^T\delta\mathbf{x}\right]_{t=t_0} + \int_{t_0}^{t_f}\left\{\left(\frac{\partial H}{\partial\mathbf{x}} + \dot{\lambda}^T\right)\delta\mathbf{x} + \left(\frac{\partial H}{\partial\mathbf{u}}\right)\delta\mathbf{u}\right\}dt$$

Since the Lagrange multipliers are arbitrary, they can be selected to make the coefficients of $\delta\mathbf{x}(t)$ and $\delta\mathbf{x}(t_f)$ equal to zero, such that

$$\dot{\lambda}(t)^T = -\frac{\partial H}{\partial\mathbf{x}} \tag{16.9}$$

$$\lambda(t_f)^T = \left.\frac{\partial\varphi}{\partial\mathbf{x}}\right|_{t=t_f} \tag{16.10}$$

This choice of $\lambda(t)$ results in the following expression for $\bar{J}$, assuming that the initial state is fixed, so that $\delta\mathbf{x}(t_0) = 0$:

$$\delta\bar{J} = \int_{t_0}^{t_f}\left\{\left(\frac{\partial H}{\partial\mathbf{u}}\right)\delta\mathbf{u}\right\}dt$$

For a minimum, it is necessary that $\delta\bar{J} = 0$. This gives the stationarity condition

$$\frac{\partial H}{\partial\mathbf{u}} = 0 \tag{16.11}$$

Equations 16.8 through 16.11 are the first-order necessary conditions for a minimum of the performance index J. Equation 16.9 is known as the costate (or adjoint) equation. Equation 16.10 and the initial state condition represent the boundary (or transversality) conditions. These necessary optimality conditions, which define a two-point boundary-value problem, are very useful as they allow to find analytical solutions to special types of optimal control problems and to define numerical algorithms to search for solutions in general cases. Moreover, they are useful to check the extremality of solutions found by computational methods. Sufficient conditions for general nonlinear problems have also been established. Distinctions are made between sufficient conditions for weak local, strong local, and strong global minima. Sufficient conditions are useful to check if an extremal solution satisfying the necessary optimality conditions actually yields a minimum, and the type of minimum that is achieved. See [GF03,Lei81,Wan95] for further details.

The theory presented in this section does not deal with the existence of an optimal control that minimizes the performance index J. See the [Ces83] for theoretical issues on the existence of optimal controls. Moreover, a key point in the mathematical theory of optimal control is the existence of the Lagrange multiplier function $\lambda(t)$. See the book [Lue97] for details on this issue.

16.3.3 Case with Terminal Constraints

In case Problem 16.1 is also subject to a set of terminal constraints of the form

$$\psi(\mathbf{x}(t_f), t_f) = 0 \tag{16.12}$$

where $\psi : \mathcal{R}^{n_x} \times \mathcal{R} \mapsto \mathcal{R}^{n_\psi}$ is a vector function, it is possible to show by means of variational analysis [LS95] that the necessary conditions for a minimum of J are Equations 16.11, 16.9, 16.8, and the following terminal condition:

$$\left(\frac{\partial \varphi}{\partial x}^T + \frac{\partial \psi}{\partial x}^T \nu - \lambda\right)^T \Bigg|_{t_f} \delta \mathbf{x}(t_f) + \left(\frac{\partial \varphi}{\partial t} + \frac{\partial \psi}{\partial t}^T \nu + H\right)\Bigg|_{t_f} \delta t_f = 0 \quad (16.13)$$

where
$\nu \in \mathcal{R}^{n_\psi}$ is the Lagrange multiplier associated with the terminal constraint
δt_f is the variation of the final time
$\delta \mathbf{x}(t_f)$ is the variation of the final state

Note that if the final time is fixed, then $\delta t_f = 0$ and the second term vanishes. Also, if the terminal constraint is such that element j of $\mathbf{x}$ is fixed at the final time, then element j of $\delta \mathbf{x}(t_f)$ vanishes.

16.3.4 Example: Minimum Energy Point-to-Point Control of a Double Integrator

Consider the following optimal control problem

$$\min_{u(t)} J = \frac{1}{2}\int_0^1 u(t)^2 \, dt$$

subject to the differential constraints

$$\begin{aligned} \dot{x}_1(t) &= x_2(t) \\ \dot{x}_2(t) &= u(t) \end{aligned} \quad (16.14)$$

and the boundary conditions

$$\begin{aligned} x_1(0) &= 1, \quad x_2(0) = 1 \\ x_1(1) &= 0, \quad x_2(1) = 0 \end{aligned} \quad (16.15)$$

The first step to solve the problem is to define the Hamiltonian function, which is given by

$$H = \frac{1}{2}u^2 + \lambda_1 x_2 + \lambda_2 u$$

The stationarity condition (16.11) gives

$$u + \lambda_2 = 0 \quad \Rightarrow \quad u = -\lambda_2 \quad (16.16)$$

The costate Equation 16.9 gives

$$\begin{aligned} \dot{\lambda}_1 &= 0 \\ \dot{\lambda}_2 &= -\lambda_1 \end{aligned}$$

So that the costates can be expressed as follows:

$$\lambda_1(t) = a$$

$$\lambda_2(t) = -at + b$$

where a and b are integration constants to be found. Replacing the expression for λ_2 in (16.16), gives

$$u(t) = -at + b \tag{16.17}$$

The terminal constraint function in this problem is $\psi(\mathbf{x}(1)) = [x_1(1), x_2(1)]^T = [0, 0]^T$, noting that the final value of the state vector is fixed. It follows that $\delta\mathbf{x}(t_f) = 0$. Since the terminal time is fixed $t_f = 1$, then $\delta t_f = 0$, and the terminal condition (16.13) is satisfied. Replacing (16.17) in (16.14) and integrating on both sides gives

$$\begin{aligned} x_1(t) &= \frac{1}{6}at^3 - \frac{1}{2}bt^2 + ct + d \\ x_2(t) &= \frac{1}{2}at^2 - bt + c \end{aligned} \tag{16.18}$$

Evaluating (16.18) at the boundary points $t = 0$ and $t = 1$, and using the boundary conditions (16.15), results in the following values for the integration constants: $a = 18$, $b = 10$, $c = 1$, and $d = 1$, so that the optimal control is given by

$$u(t) = 18t - 10, \quad t \in [0,1]$$

16.3.5 Case with Input Constraints: Minimum Principle

Realistic optimal control problems often have inequality constraints associated with the input variables, so that the input variable $\mathbf{u}$ is restricted to be within an admissible compact region Ω, such that $\mathbf{u}(t) \in \Omega$.

It was shown by Pontryagin and coworkers [Pon87] that in this case, the necessary conditions (16.8), (16.9), and (16.10) still hold, but the stationarity condition (16.11) has to be replaced by

$$H(\mathbf{x}^*(t), \mathbf{u}^*(t), \lambda^*(t), t) \leq H(\mathbf{x}^*(t), \mathbf{u}(t), \lambda^*(t), t)$$

for all admissible $\mathbf{u}$, where * denotes optimal functions. This condition is known as Pontryagin's *minimum principle*. According to this principle, the Hamiltonian must be minimized over all admissible $\mathbf{u}$ for optimal state and costate functions. In their work, Pontryagin and coworkers also showed additional necessary conditions that are often useful when solving problems:

- If the final time t_f is fixed and the Hamiltonian function does not depend explicitly on time t, then the Hamiltonian must be constant when evaluated along the optimal trajectory:

$$H(x^*(t), \mathbf{u}^*(t), \lambda^*(t)) = \text{constant}, \quad t \in [t_0, t_f] \tag{16.19}$$

- If the final time t_f is free, and the Hamiltonian does not depend explicitly on time t, then the Hamiltonian must be zero when evaluated along the optimal trajectory:

$$H(\mathbf{x}^*(t), \mathbf{u}^*(t), \lambda^*(t)) = 0, \quad t \in [t_0, t_f] \tag{16.20}$$

16.3.6 Minimum Time Problems

A special class of optimal control problem involves finding the optimal input $\mathbf{u}(t)$ to satisfy a terminal constraint in minimum time. This kind of problem is defined as follows.

PROBLEM 16.2
Find t_f and $\mathbf{u}(t)$, $t \in [t_0, t_f]$, to minimize

$$J = \int_{t_0}^{t_f} 1\mathrm{d}t = t_f - t_0$$

subject to

$$\dot{\mathbf{x}}(t) = \mathbf{f}(\mathbf{x}(t), \mathbf{u}(t), t),$$

$$\mathbf{x}(t_0) = \mathbf{x}_o$$

$$\psi(\mathbf{x}(t_f), t_f) = \mathbf{0}$$

$$\mathbf{u}(t) \in \Omega$$

See [LS95] and [Nai03] for further details on minimum time problems.

16.3.7 Problems with Path- or Interior-Point Constraints

Sometimes it is necessary to restrict state and control trajectories such that a set of constraints is satisfied along the interval of interest $[t_0, t_f]$:

$$\mathbf{c}(\mathbf{x}(t), \mathbf{u}(t), t) \le \mathbf{0}$$

where $\mathbf{c} : \mathcal{R}^{n_x} \times \mathcal{R}^{n_u} \times [t_0, t_f] \mapsto \mathcal{R}^{n_c}$. Moreover, in some problems it may be required that the state satisfies equality constraints at some intermediate point in time t_1, $t_0 \le t_1 \le t_f$. These are known as interior-point constraints and can be expressed as follows:

$$\mathbf{q}(\mathbf{x}(t_1), t_1) = \mathbf{0}$$

where $\mathbf{q} : \mathcal{R}^{n_x} \times \mathcal{R} \mapsto \mathcal{R}^{n_q}$. See [BH75] for a detailed treatment of optimal control problems with path- and interior-point constraints.

16.3.8 Singular Arcs

In some optimal control problems, extremal arcs satisfying (16.11) occur where the matrix $\partial^2 H / \partial \mathbf{u}^2$ is singular. These are called singular arcs. Additional tests are required to verify if a singular arc is optimizing. A particular case of practical relevance occurs when the Hamiltonian function is linear in at least one of the control variables. In such cases, the control is not determined in terms of the state and costate by the stationarity condition (16.11). Instead, the control is determined by the condition that the time derivatives of $\partial H / \partial \mathbf{u}$ must be zero along the singular arc. In the case of a single control u, once the control is obtained by setting the time derivative of $\partial H / \partial u$ to zero, then additional necessary conditions, known as the generalized Legendre–Clebsch conditions, must be checked:

$$(-1)^k \frac{\partial}{\partial u} \left\{ \frac{\mathrm{d}^{(2k)}}{\mathrm{d}t^{(2k)}} \frac{\partial H}{\partial u} \right\} \ge 0, \quad k = 0, 1, 2, \ldots$$

The presence of singular arcs may cause difficulties to computational optimal control methods to find accurate solutions if the appropriate conditions are not enforced a priori [Bet01]. See [BH75] and [ST00] for further details on the handling of singular arcs.

16.3.9 The Linear Quadratic Regulator

A special case of Problem 16.1 that is of particular practical importance arises when the objective function is a quadratic function of **x** and **u**, and the dynamic equations are linear. The resulting feedback law in this case is known as the linear quadratic regualtor (LQR). The performance index is given by

$$J = \frac{1}{2}\mathbf{x}(t_f)^T \mathbf{S}_f \mathbf{x}(t_f) + \frac{1}{2}\int_{t_0}^{t_f} (\mathbf{x}(t)^T \mathbf{Q}\mathbf{x}(t) + \mathbf{u}(t)^T \mathbf{R}\mathbf{u}(t))\,dt \tag{16.21}$$

where $\mathbf{S}_f \in \mathcal{R}^{n_x \times n_x}$ and $Q \in \mathcal{R}^{n_x \times n_x}$ are positive semidefinite matrices, and $\mathbf{R} \in \mathcal{R}^{n_u \times n_u}$ is a positive definite matrix, while the system dynamics obey

$$\begin{aligned} \dot{\mathbf{x}}(t) &= \mathbf{A}\mathbf{x}(t) + \mathbf{B}\mathbf{u}(t) \\ \mathbf{x}(t_0) &= \mathbf{x}_0 \end{aligned} \tag{16.22}$$

where $\mathbf{A} \in \mathcal{R}^{n_x \times n_x}$ is the system matrix and $\mathbf{B} \in \mathcal{R}^{n_x \times n_u}$ is the input matrix.

In this case, using the optimality conditions (16.8), (16.9), (16.10), and (16.11), it is possible to find that the optimal control law can be expressed as a linear state feedback

$$\mathbf{u}(t) = -\mathbf{K}(t)\mathbf{x}(t) \tag{16.23}$$

where the state feedback gain is given by

$$\mathbf{K}(t) = \mathbf{R}^{-1}\mathbf{B}^T\mathbf{S}(t)$$

and $\mathbf{S}$: $[t_0, t_f] \mapsto \mathcal{R}^{n \times n}$ is the solution to the differential Ricatti equation

$$\begin{aligned} -\dot{\mathbf{S}} &= \mathbf{A}^T\mathbf{S} + \mathbf{S}\mathbf{A} - \mathbf{S}\mathbf{B}\mathbf{R}^{-1}\mathbf{B}^T\mathbf{S} + \mathbf{Q} \\ \mathbf{S}(t_f) &= \mathbf{S}_f \end{aligned}$$

In the particular case where $t_f \to \infty$, and provided the pair (**A**, **B**) is stabilizable,* the Ricatti differential equation converges to a limiting solution **S**, and it is possible to express the optimal control law as a state feedback as in (16.23) but with a constant gain matrix **K**. This is known in the literature as the *infinite horizon* or *steady-state* LQR solution. The steady-state gain **K** is given by

$$\mathbf{K} = \mathbf{R}^{-1}\mathbf{B}^T\mathbf{S}$$

where **S** is the positive definite solution to the algebraic Ricatti equation:

$$\mathbf{A}^T\mathbf{S} + \mathbf{S}\mathbf{A} - \mathbf{S}\mathbf{B}\mathbf{R}^{-1}\mathbf{B}^T\mathbf{S} + \mathbf{Q} = \mathbf{0} \tag{16.24}$$

* A linear dynamic system $\dot{\mathbf{x}} = \mathbf{A}\mathbf{x} + \mathbf{B}\mathbf{u}$ is stabilizable if the eigenvalues of the matrix $(\mathbf{A} - \mathbf{B}\mathbf{K})$ have (strict) negative real part for some matrix **K**.

Moreover, if the pair $(\mathbf{A}, \mathbf{C})$ is observable,* where $\mathbf{C}^T\mathbf{C} = \mathbf{Q}$, then the closed-loop system

$$\dot{\mathbf{x}}(t) = (\mathbf{A} - \mathbf{BK})\mathbf{x}(t)$$

is asymptotically stable. This is an important result, as the LQR provides a way of stabilizing any linear system that is stabilizable. In addition to its desirable closed-loop stability property, the steady-state LQR has certain guaranteed robustness properties that make it even more useful: an infinite gain margin and a phase margin of at least 60° [LS95].

It is worth pointing out that there are well-established methods and software for solving the algebraic Ricatti equation (16.24). This facilitates the design of LQRs. A useful extension of the LQR ideas involves modifying the performance index (16.21) to allow for a reference signal that the output of the system should track. Moreover, an extension of the LQR concept to systems with Gaussian additive noise, which is known as the linear quadratic Gaussian (LQG) controller, has been widely applied. The LQG controller involves coupling the LQR with the Kalman filter using the separation principle. See [LS95] for further details.

Example: Stabilization of a Rotary Inverted Pendulum

The rotary inverted pendulum is an underactuated mechanical system that is often used in control education (Figure 16.1). The servo motor drives an independent output gear whose angular position is measured by an optical encoder. The rotary pendulum arm is mounted to an output gear. At the end of the pendulum arm is a hinge instrumented with another encoder. The pendulum attaches to the hinge. The second encoder measures the angular position of the pendulum.

Notice that the pendulum in open loop has two equilibrium points. A stable equilibrium at the point where the rod is vertical and pointing down and an unstable equilibrium at the point where the rod is

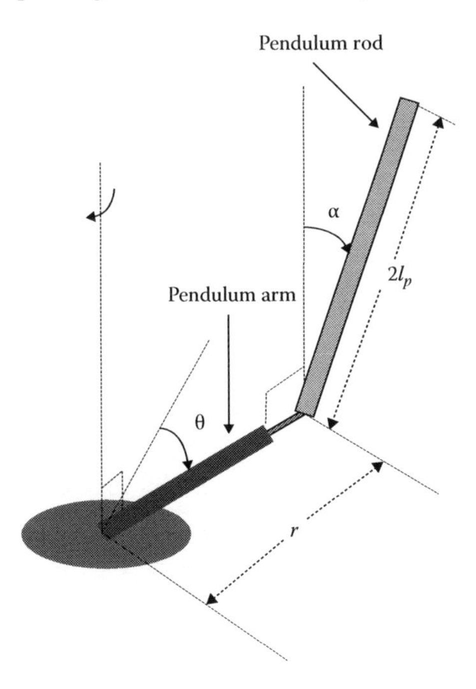

FIGURE 16.1 Illustration of the rotary inverted pendulum.

* A linear dynamical system $\dot{\mathbf{x}} = \mathbf{Ax} + \mathbf{Bu}$ with output observations $\mathbf{y} = \mathbf{Cx} + \mathbf{Du}$ is said to be observable if for any unknown initial state $x(0)$, there exists a finite t_1 such that knowledge of the input u and the output $\mathbf{y}$ over $[0, t_1]$ suffices to determine uniquely the initial state $\mathbf{x}(0)$.

vertical and pointing up. This example illustrates how a steady-state LQR controller can be designed to balance the pendulum around the unstable equilibrium point. Steady-state LQR controllers are often used in practice to stabilize the operation of systems around unstable equilibria.

A linearized dynamic model of a practical rotary inverted pendulum is given by

$$\begin{bmatrix} \dot{x}_1 \\ \dot{x}_2 \\ \dot{x}_3 \\ \dot{x}_4 \end{bmatrix} = \begin{bmatrix} 0 & 0 & 1 & 0 \\ 0 & 0 & 0 & 1 \\ 0 & -69.3642 & -29.0777 & 0 \\ 0 & 144.5070 & 36.6164 & 0 \end{bmatrix} \begin{bmatrix} x_1 \\ x_2 \\ x_3 \\ x_4 \end{bmatrix} + \begin{bmatrix} 0 \\ 0 \\ 54.1585 \\ -68.1996 \end{bmatrix} u$$

where

$x_1 = \dot{\theta}$ is the arm angle (rad)
$x_2 = \dot{\alpha}$ is the pendulum angle (rad)
$x_3 = \ddot{\theta}$ is the angular acceleration of the arm (rad/s^2)
$x_4 = \ddot{\alpha}$ is the angular acceleration of the pendulum (rad/s^2)
u is the voltage (V) applied to the servo motor

Note that the eigenvalues of the linearized model of the pendulum in open loop are

$$\text{eig}(\mathbf{A}) = \begin{bmatrix} 0 & -31.9716 & 8.8008 & -5.9069 \end{bmatrix}$$

so the local instability is apparent. To stabilize the pendulum around the upper equilibrium point ($\alpha = 0$), an LQR is designed using the following state and control weights, respectively:

$$Q = \begin{bmatrix} 5 & 0 & 0 & \\ 0 & 5 & 0 & 0 \\ 0 & 0 & 0.2 & 0 \\ 0 & 0 & 0 & 0.2 \end{bmatrix}, \quad R = 5$$

The Ricatti solution, which can be found easily using widely available software (including MATLAB®, Octave, and SciLab), results in the following state feedback gain:

$$K = \begin{bmatrix} -1 & -12.3967 & -1.4207 & -1.4682 \end{bmatrix}$$

Using this feedback gain **K**, the closed-loop eigenvalues are

$$\text{eig}(\mathbf{A} - \mathbf{BK}) = \begin{bmatrix} -36.5536 & -1.7692 & -7.8177 & -6.1229 \end{bmatrix}$$

which are all stable. The stabilizing controller can be implemented as follows:

$$u(t) = -\mathbf{Kx}(t) = \theta(t) + 12.3967\alpha(t) + 1.4207\dot{\theta}(t) + 1.4682\dot{\alpha}(t)$$

The implementation requires measurements (or good estimates) of the angular positions and velocities. Figure 16.2 shows an ideal simulation of the pendulum under closed-loop control, where the dynamics are simulated using the linearized dynamic model, and the initial state is given by $\mathbf{x}(0) = [0, 0.1, 0.0]^T$.

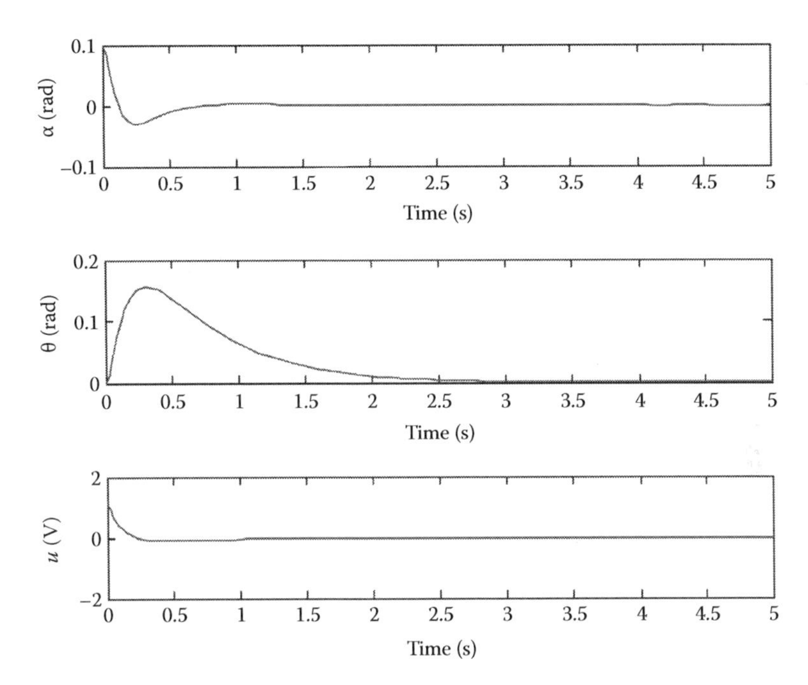

FIGURE 16.2 Simulation of the rotary inverted pendulum under LQR control.

16.4 Discrete-Time Optimal Control

Most of the problems defined above have discrete-time counterparts. These formulations are useful when the dynamics are discrete (for example, a multistage system) or when dealing with computer-controlled systems. In discrete-time, the dynamics can be expressed as a difference equation

$$\mathbf{x}(k+1)=\mathbf{f}(\mathbf{x}(k),\mathbf{u}(k),k) \quad \mathbf{x}(N_0)=\mathbf{x}_0$$

where

k is an integer index in the interval $[N_0, N_f - 1]$
$\mathbf{x}:[N_0,N_f]\mapsto R^{n_x}$ is the state vector
$\mathbf{u}:[N_0,N_f-1]\mapsto R^{n_u}$ is the control vector
$\mathbf{f}:R^{n_x}\times R^{n_u}\times[N_0,N_f-1]\mapsto R^{n_x}$ is a vector function

The objective is to find a control sequence $\mathbf{u}(k), k = N_0,\ldots, N_f - 1$, to minimize a performance index of the form

$$J=\varphi(\mathbf{x}(N_f),N_f)+\sum_{k=N_0}^{N_f-1}L(\mathbf{x}(k),\mathbf{u}(k),k)$$

where

$\varphi:R^{n_x}\times Z\mapsto \mathcal{R}$ is a terminal cost function
$L:\mathcal{R}^{n_x}\times\mathcal{R}^{n_u}\times[N_0,N_f-1]\mapsto\mathcal{R}$ is an intermediate cost function

See, for example, [LS95] or [Bry99] for further details on discrete-time optimal control.

16.5 Dynamic Programming

Dynamic programming is an alternative to the variational approach to optimal control. It was proposed by Bellman in the 1950s and is an extension of Hamilton–Jacobi theory. Bellman's *principle of optimality* can be stated as follows: "An optimal policy has the property that regardless of what the previous decisions have been, the remaining decisions must be optimal with regard to the state resulting from those previous decisions." This principle serves to limit the number of potentially optimal control strategies that must be investigated. It also shows that the optimal strategy must be determined by working backward from the final time.

Dynamic programming includes formulations for discrete-time systems as well as combinatorial systems, which are discrete systems with quantized states and controls. The original discrete dynamic programming method, however, suffers from the "curse of dimensionality," which causes the computations and memory requirements to grow dramatically with the problem size. This problem has been addressed in various ways, for example, see the work by Luus [Luu02]. The books [BH75,Kir70,LS95] and [Luu02] can be consulted for further details on dynamic programming.

16.6 Computational Optimal Control

The solutions to most practical optimal control problems cannot be found by analytical means. Over the years, many numerical procedures have been developed to solve general optimal control problems.

Indirect methods involve iterating on the necessary optimality conditions to seek their satisfaction. This usually involves attempting to solve nonlinear two-point boundary-value problems, through the forward integration of the plant equations and the backward integration of the costate equations. Examples of indirect methods include the gradient method and the multiple shooting method, both of which are described in detail in the book [Bry99].

With direct methods, optimal control problems are discretized and converted into nonlinear programming problems of the form:

PROBLEM 16.3

Find a decision vector $\mathbf{y} \in R^{n_y}$ to minimize

$$F(\mathbf{y})$$

subject to

$$\mathbf{G}_l \leq \mathbf{G}(\mathbf{y}) \leq \mathbf{G}_u$$

$$\mathbf{y}_l \leq \mathbf{y} \leq \mathbf{y}_u$$

where

$F : \mathcal{R}^{n_y} \mapsto \mathcal{R}$ is a twice continuously differentiable scalar function

$\mathbf{G} : \mathcal{R}^{n_y} \mapsto \mathcal{R}^{n_g}$ is a twice continuously differentiable vector function

Many direct methods involve the approximation of the control and/or states using basis functions, such as splines or Lagrange polynomials. Direct collocation methods involve the discretization of the differential equations using, for example, trapezoidal, Hermite–Simpson [Bet01], or pseudospectral approximations [EKR95], by defining a grid of N points covering the time interval $[t_0, t_f]$, $t_0 = t_1 < t_2 \cdots < t_N = t_f$. In this way, the differential equations become a finite set of equality constraints of the nonlinear programming problem. For example, in the case of trapezoidal collocation, the decision vector $\mathbf{y}$ contains the control and state variables at the grid points and possibly the initial and final times. The nonlinear optimization problems that arise from direct collocation methods may be very large, having possibly hundreds to tens of thousands of variables and constraints. It is, however, interesting that such large nonlinear programming problems are easier to solve than boundary-value problems. The reason for the relative ease of computation of direct methods (particularly direct collocation methods) is that the nonlinear programming problem is sparse, and well-known methods and software exist for its solution [Bet01,WB06]. Moreover, the region of convergence of indirect methods tends to be quite narrow, so that these methods require good initial guesses. Also, direct methods using nonlinear programming are known to deal more efficiently with problems involving path constraints. As a result, the range of problems that can be solved with direct methods is larger than the range of problems that can be solved via indirect methods. See [Bet01] for more details on computational optimal control using sparse nonlinear programming.

Some complex optimal control problems can be conveniently formulated as having multiple phases. Phases may be inherent to the problem (for example, a spacecraft drops a mass and enters a new phase). Phases may also be introduced by the analysist to allow for peculiarities in the solution of the problem such as discontinuities in the control variables or singular arcs. Additional constraints are added to the problem to define the linkages between interconnected phases. See [Bet01,RF04] for further details on multiphase optimal control problems.

16.7 Examples

16.7.1 Obstacle Avoidance Problem

Consider the following optimal control problem, which involves finding an optimal trajectory that minimizes energy expenditure for a particle to travel between two points on a plane, while avoiding two forbidden regions or obstacles [RE09]. Find $\theta(t) \in [0, t_f]$ to minimize the cost functional

$$J = \int_0^{t_f} \left[V^2 \sin^2(\theta(t)) + V^2 \cos^2(\theta(t)) \right] \mathrm{d}t \tag{16.25}$$

subject to the dynamic constraints

$$\begin{aligned} \dot{x} &= V\cos(\theta) \\ \dot{y} &= V\sin(\theta) \end{aligned} \tag{16.26}$$

The path constraints

$$\begin{aligned} (x(t)-0.4)^2 + (y(t)-0.5)^2 &\geq 0.1 \\ (x(t)-0.8)^2 + (y(t)-1.5)^2 &\geq 0.1 \end{aligned} \tag{16.27}$$

and the boundary conditions

$$\begin{aligned} x(0) &= 0 \\ y(0) &= 0 \\ x(t_f) &= 1.2 \\ y(t_f) &= 1.6 \end{aligned} \tag{16.28}$$

where $t_f = 1.0$ and $V = 2.138$.

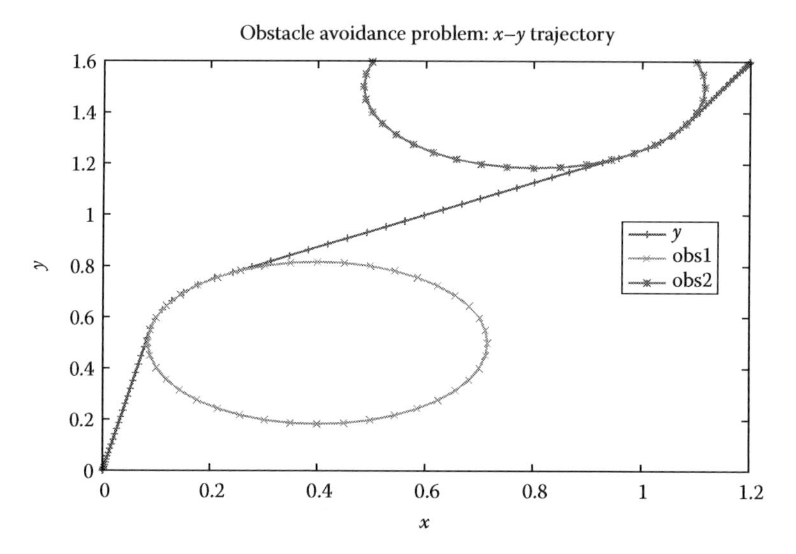

FIGURE 16.3 Optimal (x, y) trajectory for obstacle avoidance problem.

The problem was solved using PSOPT [Bec09], a computational optimal control solver that employs pseudospectral discretization methods, large-scale nonlinear programming, and automatic differentiation. The problem was solved initially by using 40 grid points, then the mesh was refined to 80 grid points, and an interpolation of the previous solution was employed as an initial guess for the new solution. Figure 16.3 shows the optimal (x, y) trajectory of the particle. Table 16.1 presents a summary of the solution to the problem. CPU time refers to a PC with an Intel Core 2 Quad Q6700 processor running at 2.66 GHz with 8 GB of RAM, under the Ubuntu Linux 9.04 operating system.

16.7.2 Missile Terminal Burn Maneuver

This example illustrates the computational design of a missile trajectory to strike a specified target from given initial conditions in minimum time [SZ09], so that the problem is to find t_f and $\mathbf{u}(t) = [\alpha(t), T(t)]^T$, $t \in [0, t_f]$, to minimize

$$J = t_f$$

TABLE 16.1 Solution Summary for the Obstacle Avoidance Problem

No. of grid points	{40, 80}
No. of NLP variables	{122, 242}
No. of NLP constraints	{165, 325}
No. of NLP iterations	{44, 22}
No. of objective evaluations	{295, 15}
No. of constraint evaluations	{294, 16}
No. of Jacobian evaluations	{46, 15}
Objective function	4.571044 s
CPU time	2.01 s

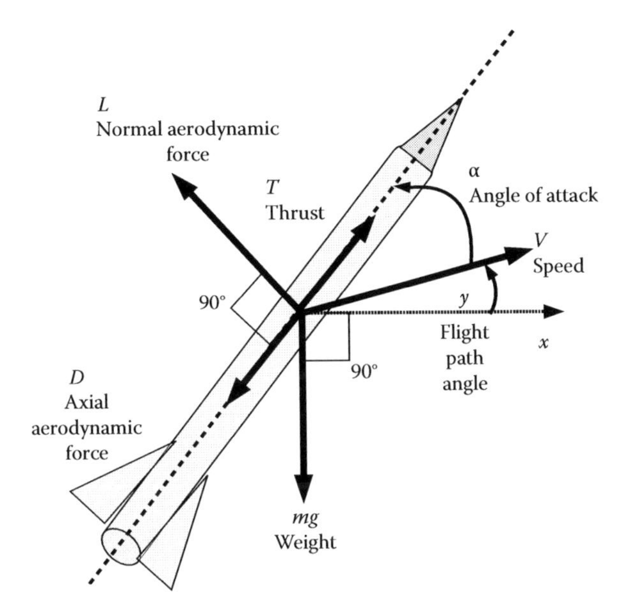

FIGURE 16.4 Illustration of the variables associated with the missile model.

Figure 16.4 shows the variables associated with the dynamic model of the missile employed in this example, where γ is the flight path angle, α is the angle of attack, V is the missile speed, x is the longitudinal position, h is the altitude, D is the axial aerodynamic force, L is the normal aerodynamic force, and T is the thrust.

The Equations of motion of the missile are given by

$$\dot{\gamma} = \frac{T-D}{mg}\sin\alpha + \frac{L}{mV}\cos\alpha - \frac{g\cos\gamma}{V}$$

$$\dot{V} = \frac{T-D}{m}\cos\alpha - \frac{L}{m}\sin\alpha - g\cos\gamma$$

$$\dot{x} = V\cos\gamma$$

$$\dot{h} = V\sin\gamma$$

where

$$D = \frac{1}{2}C_d\rho V^2 S_{ref}$$

$$C_d = A_1\alpha^2 + A_2\alpha + A_3$$

$$L = \frac{1}{2}C_l\rho V^2 S_{ref}$$

$$C_l = B_1\alpha + B_2$$

$$\rho = C_1h^2 + C_2h + C_3$$

TABLE 16.2 Parameter Values of the Missile Model

Parameter	Value	Units
M	1005	kg
G	9.81	m/s^2
S_{ref}	0.3376	M^2
A_1	−1.9431	
A_2	−0.1499	
A_3	0.2359	
B_1	21.9	
B_2	0	
C_1	3.312×10^{-9}	kg/m^5
C_2	-1.142×10^{-4}	kg/m^4
C_3	1.224	kg/m^3

where all the model parameters are given in Table 16.2. The initial conditions for the state variables are

$$\gamma(0) = 0$$

$$V(0) = 272 \text{ m/s}$$

$$x(0) = 0 \text{ m}$$

$$h(0) = 30 \text{ m}$$

The terminal conditions on the states are

$$\gamma(t_f) = -\frac{\pi}{2}$$

$$V(t_f) = 310 \text{ m/s}$$

$$x(t_f) = 10{,}000 \text{ m}$$

$$h(t_f) = 0 \text{ m}$$

The problem constraints are given by

$$200 \le V \le 310$$

$$1000 \le T \le 6000$$

$$-0.3 \le \alpha \le 0.3$$

$$-4 \le \frac{L}{mg} \le 4$$

$$h \ge 30 \quad (\text{for } x \le 7500 \text{ m})$$

$$h \ge 0 \quad (\text{for } x > 7500 \text{ m})$$

Note that the path constraints on the altitude are non-smooth. Given that non-smoothness causes problems with nonlinear programming, the constraints on the altitude were approximated by a single smooth constraint

$$\mathcal{H}_\varepsilon(x-7500))h(t)+[1-\mathcal{H}_\varepsilon(x-7500)][h(t)-30]\geq 0$$

where $\mathcal{H}_\varepsilon(z)$ is a smooth version of the Heaviside function, which is computed as follows:

$$\mathcal{H}_\varepsilon(z)=0.5\left(1+\tanh\left(\frac{z}{\varepsilon}\right)\right)$$

where $\varepsilon > 0$ is a small number.

The problem was solved using PSOPT [Bec09] with 40 grid points initially, then the mesh was refined to 80 grid points, and an interpolation of the previous solution was employed as an initial guess for the new solution. Figure 16.5 shows the missile altitude as a function of the longitudinal position. Figures 16.6 and 16.7 show, respectively, the missile speed and angle of attack as functions of time. Table 16.3 gives a summary of the solution.

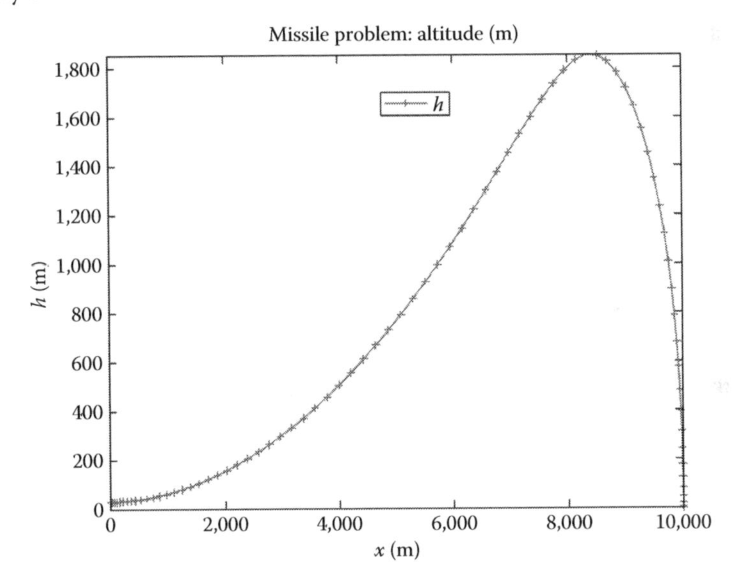

FIGURE 16.5 Missile altitude and a function of the longitudinal position.

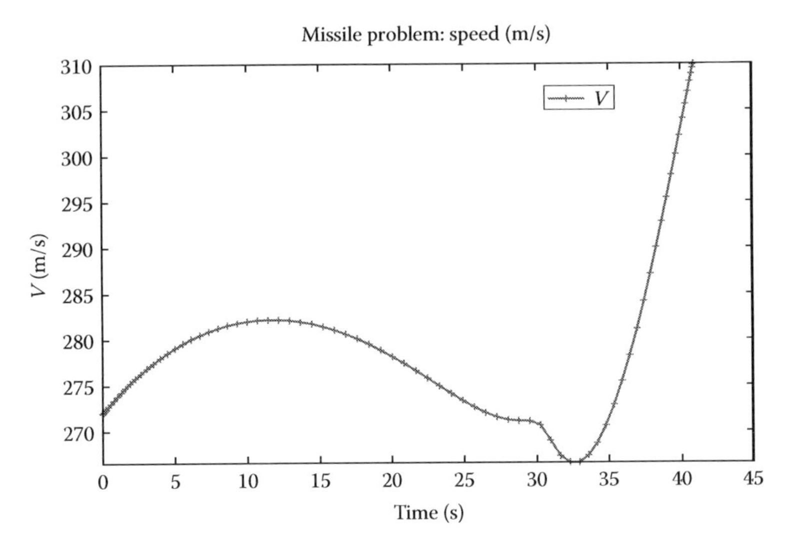

FIGURE 16.6 Missile speed as a function of time.

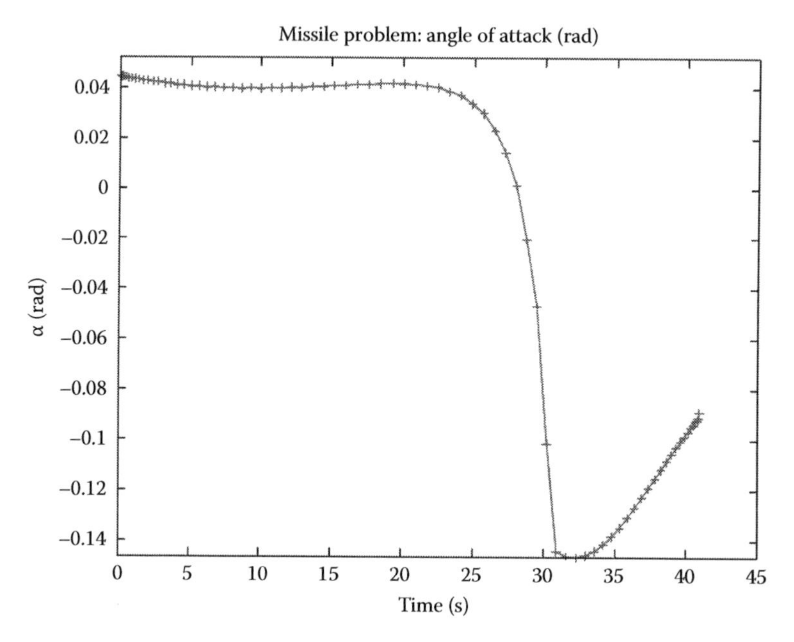

FIGURE 16.7 Missile angle of attack as a function of time.

TABLE 16.3 Solution Summary for the Missile Problem

No. of grid points	{40, 80}
No. of NLP variables	{242, 482}
No. of NLP constraints	{249, 489}
No. of NLP iterations	{22, 13}
No. of objective evaluations	{543, 22}
No. of constraint evaluations	{542, 23}
No. of Jacobian evaluations	{57, 22}
Objective function	40.9121 s
CPU time	6.83 s

References

[Bec09] V.M. Becerra. *PSOPT: Optimal Control Solver User Manual.* University of Reading, Reading, U.K., 2009.

[Bet01] J.T. Betts. *Practical Methods for Optimal Control Using Nonlinear Programming.* SIAM, Philadelphia, PA, 2001.

[BH75] A.E. Bryson and Y. Ho. *Applied Optimal Control.* Halsted Press, New York, 1975.

[BMC93] H.D. Sherali, M.S. Bazaraa, and C.M. Shetty, *Nonlinear Programming.* Wiley, New York, 1993.

[Bry96] A.E. Bryson. Optimal control—1950 to 1985. *IEEE Control Systems Magazine*, 16:23–36, 1996.

[Bry99] A.E. Bryson. *Dynamic Optimization.* Addison-Wesley, Menlo Park, CA, 1999.

[Ces83] L. Cesari. *Optimization-Theory and Applications: Problems with Ordinary Differential Equations.* Springer, Berlin, Germany, 1983.

[CZ96] E.K.P. Chong and S.H. Zak. *An Introduction to Optimization.* Wiley-Interscience, Hoboken, NJ, 1996.

[EKR95] G. Elnagar, M.A. Kazemi, and M. Razzaghi. The pseudospectral Legendre method for discretizing optimal control problems. *IEEE Transactions on Automatic Control*, 40:1793–1796, 1995.

[GF03] I.M. Gelfand and S.V. Fomin. *Calculus of Variations*. Dover Publications, New York, 2003.

[Kir70] D.E. Kirk. *Optimal Control Theory*. Prentice-Hall, Englewood Cliffs, NJ, 1970.

[Lei81] G. Leitmann. *The Calculus of Variations and Optimal Control*. Springer, New York, 1981.

[LS95] F.L. Lewis and V.L. Syrmos. *Optimal Control*. Wiley, New York, 1995.

[Lue97] D.G. Luenberger. *Optimization by Vector Space Methods*. Wiley, New York, 1997.

[Luu02] R. Luus. *Iterative Dynamic Programming*. Chapman & Hall/CRC, Boca Raton, FL, 2002.

[Nai03] D.S. Naidu. *Optimal Control Systems*. CRC Press, Boca Raton, FL, 2003.

[Pon87] L.S. Pontryagin. *The Mathematical Theory of Optimal Processes. Classics of Soviet Mathematics*. CRC Press, Boca Raton, FL, 1987.

[RE09] P.E. Rutquist and M.M. Edvall. *PROPT Matlab Optimal Control Software*. TOMLAB Optimization, Pullman, WA, 2009.

[RF04] I.M. Ross and F. Fahroo. Pseudospectral knotting methods for solving nonsmooth optimal control problems. *Journal of Guidance Control and Dynamics*, 27:397–405, 2004.

[ST00] S. Sethi and G.L. Thompson. *Optimal Control Theory: Applications to Management Science and Economics*. Kluwer, Dordrecht, the Netherlands, 2000.

[SW97] H.J. Sussmann and J.C. Willems. 300 years of optimal control: From the brachystochrone to the maximum principle. *IEEE Control Systems Magazine*, 17:32–44, 1997.

[SZ09] S. Subchan and R. Zbikowski. *Computational Optimal Control: Tools and Practice*. Wiley, New York, 2009.

[Wan95] F.Y.M. Wan. *Introduction to the Calculus of Variations and Its Applications*. Chapman & Hall, Boca Raton, FL, 1995.

[WB06] A. Wächter and L.T. Biegler. On the implementation of a primal-dual interior point filter line search algorithm for large-scale nonlinear programming. *Mathematical Programming*, 106:25–57, 2006.

17

Time-Delay Systems

Emilia Fridman
Tel Aviv University

17.1 Models with Time-Delay

Time-delay systems (TDS) are also called systems with aftereffect or dead-time, hereditary systems, equations with deviating argument, or differential-difference equations. They belong to the class of *functional differential equations* that are infinite-dimensional, as opposed to ordinary differential equations (ODEs). The simplest example of such a system is $\dot{x}(t) = -x(t-h)$, $x(t) \in R$, where $h > 0$ is the time-delay. Time-delay often appears in many control systems (such as aircraft, chemical, or process control systems, communication networks) either in the state, the control input, or the measurements. There can be transport, communication, or measurement delays. Actuators, sensors, and field networks that are involved in feedback loops usually introduce delays. Thus, delays are strongly involved in challenging areas of communication and information technologies: stability of networked control systems or high-speed communication networks [23]. We consider two examples of models with delay.

Model 1 [11]. Imagine a showering person wishing to achieve the desired value T_d of water temperature by rotating the mixer handle for cold and hot water. Let $T(t)$ denote the water temperature in the mixer output and let h be the constant time needed by the water to go from the mixer output to the tip of the person's head. Assume that the change of the temperature is proportional to the angle of the rotation of the handle, whereas the rate of rotation of the handle is proportional to $T(t) - T_d$. At time t the person feels the water temperature leaving the mixer at time $t - h$, which results in the following equation with constant delay h:

$$\dot{T}(t) = -k[T(t-h) - T_d], \quad k \in R. \tag{17.1}$$

Model 2. Consider a sampled-data control system

$$\dot{x}(t) = Ax(t) + BKx(t_k), \quad t \in [t_k, t_{k+1}), \quad k = 0, 1, 2, \ldots \tag{17.2}$$

where $x(t) \in R^n$, A, B, K are constant matrices and $\lim_{k\to\infty} t_k = \infty$. This system can be represented as a continuous system with time-varying delay $\tau(t) = t - t_k$ [16]:

$$\dot{x}(t) = Ax(t) + BKx(t - \tau(t)), \quad t \in [t_k, t_{k+1}), \tag{17.3}$$

where the delay is piecewise continuous with $\dot{\tau} = 1$ for $t \neq t_k$.

First equations with delay were studied by Bernoulli, Euler, and Concordet in the eighteenth century. Systematical study started in the 1940s with A. Myshkis and R. Bellman. Since 1960 there have appeared more than 50 monographs on the subject (see, e.g., [2,6,9,11,19] and the references therein).

In spite of their complexity, TDS often appear as *simple* infinite-dimensional models of more complicated partial differential equations (PDEs) [6]. Conversely, time-delay equations can be represented by a classical transport PDE. Thus, denoting in (17.1) $z(s, t) = T(t - hs)$, $s \in [0, 1]$, we arrive to the following boundary value problem for the transport equation:

$$\begin{aligned} &\frac{\partial}{\partial t} z(s,t) + h\frac{\partial}{\partial s} z(s,t) = 0, \quad z \in [0,1], \\ &\frac{\partial}{\partial t} z(0,t) = -k\left[z(1,t) - T_d\right]. \end{aligned} \tag{17.4}$$

17.2 Solution Concept and the Step Method

Consider the simple delay equation

$$\dot{x}(t) = -x(t-h), \quad t \geq 0. \tag{17.5}$$

In order to define its solution for $t \in [0, h]$, we have to define the right-hand side $x(t - h)$ for $t \in [0, h]$, which results in the initial value function

$$x(s) = \phi(s), \quad s \in [-h, 0], \tag{17.6}$$

instead of the initial value $x(0)$ for ODE with $h = 0$. In order to find a solution to this problem, we shall use the *step method* initiated by Bellman. First, we find a solution on $t \in [0, h]$ by solving

$$t \in [0,h], \quad \dot{x}(t) = -\phi(t-h), \quad x(0) = \phi(0).$$

Then we continue this procedure for $t \in [h, 2h]$, $t \in [2h, 3h]$, ... The resulting solutions for $h = 1$ and for the initial functions $\phi \equiv 1$ and $\phi \equiv 0.5t$ are given in Figure 17.1. We note that the step method can be applied for solving the initial value problem for general TDS with constant delay.

As it is seen from Figure 17.1, (17.5) has several solutions that achieve the same value $x(t^*)$ at some instants t^*. This is different from ODEs, e.g., from $\dot{x}(t) = -x(t)$, where through each $x(t^*)$ only one solution passes. Therefore, in TDS, a proper state is a *function*

$$x(t + \theta) = x_t(\theta), \quad \theta \in [-h, 0],$$

corresponding to the past time-interval $[t - h, t]$ (on Figure 17.1, there is only one solution passing through x_{t^*} for all $t^* \geq 0$). This is an infinite-dimensional system. The vector $x(t)$ is the *solution at time t*.

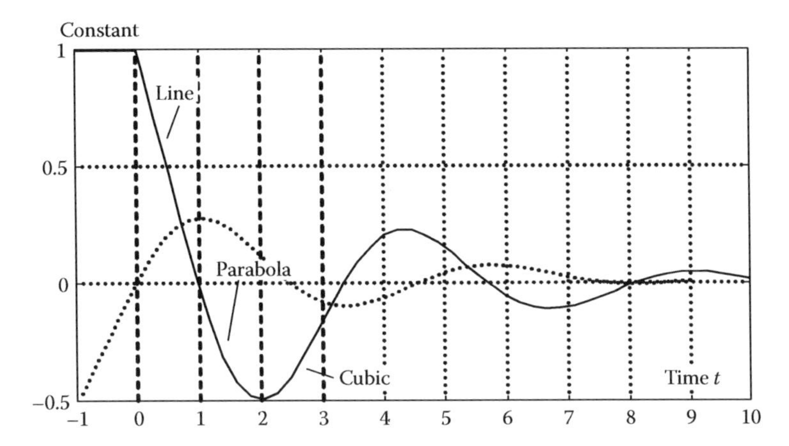

FIGURE 17.1 System (17.5) with $h = 1$, $\varphi \equiv 1$ (plain), or $0.5t$ (dotted).

17.3 Linear Time-Invariant Systems and Characteristic Equation

A linear time-invariant (LTI) system with K discrete delays and with a distributed delay has a form

$$\dot{x}(t) = \sum_{k=0}^{K} A_k x(t - h_k) + \int_{-h_d}^{0} A(\theta)x(t + \theta)d\theta, \tag{17.7}$$

where

$0 = h_0 < h_1 < \cdots < h_K$, $h_d \geq 0$, $x(t) \in R^n$
A_k are constant matrices
$A(\theta)$ is a continuous matrix function

The characteristic equation of this system is given by

$$det\left[\lambda I - \sum_{k=0}^{K} A_k e^{-\lambda h_k} - \int_{-h_d}^{0} A(s)e^{-\lambda s}ds\right] = 0. \tag{17.8}$$

Equation 17.8 is transcendental having infinite number of roots. This also reflects the *infinite-dimensional* nature of TDS. The LTI system has exponential solutions of the form $e^{\lambda t}c$, where λ is the characteristic root and $c \in R^n$ is an eigenvector of the matrix inside the *det* in (17.8). The latter can be verified by substitution of $e^{\lambda t}c$ into (17.7).

Location of the characteristic roots has a nice property (see Figure 17.2): there is a finite number of roots to the right of any vertical line.

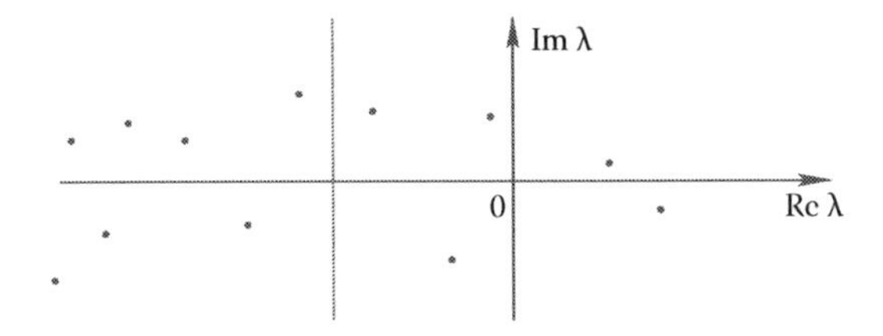

FIGURE 17.2 Location of the characteristic roots.

Delays are known to have complex effects on stability: they may be a source of instability. However, they may have also a stabilizing effect: a well-known example $\ddot{y}(t) + y(t) - y(t - h) = 0$ is unstable for $h = 0$ but is asymptotically stable (i.e., $y(t) \to 0$ for $t \to \infty$) for $h = 1$. The approximation $\dot{y}(t) \simeq [y(t) - y(t - h)]h^{-1}$ explains the damping effect. The stability and the performance of systems with delay are, therefore, of theoretical and practical importance.

Similar to linear ODEs, linear TDS (17.7) is asymptotically stable iff all the characteristic roots have negative real parts, i.e., iff $Re\lambda < -\alpha$ for some scalar $\alpha > 0$. Moreover, the stability of (17.7) is robust with respect to small delays: if (17.7) is asymptotically stable for $h_k = h_d = 0$, $k = 0,\ldots, K$, then it is asymptotically stable for all small enough values of delays.

For different root location tests see, e.g., [9,19].

Example 17.1

Consider a scalar TDS

$$\dot{x}(t) = -ax(t) - a_1x(t-h), \quad a + a_1 > 0. \tag{17.9}$$

The system without delay is asymptotically stable, and due to robustness it is stable for small enough h. For $a \geq |a_1|$ (17.9) is asymptotically stable for all delay, i.e., *delay-independently stable*. If $a_1 > |a|$, then the system is asymptotically stable for $h < h^*$ and becomes unstable for $h > h^*$, where

$$h^* = \frac{\arccos\left(-\frac{a}{a_1}\right)}{\sqrt{a_1^2 - a^2}}.$$

Thus, if $a = 0$, we have $h^* = \pi/2a_1$.

17.4 General TDS and the Direct Lyapunov Method

Consider the LTI system (17.7) and denote by $h = max\{h_K, h_d\}$. Let $C = C([-h, 0], R^n)$ be the space of continuous functions mapping the interval $[-h, 0]$ into R^n with the norm $\|x_t\|_C = max_{\theta\in[-h,0]} |x(t + \theta)|$. Then, (17.7) can be represented in the form $\dot{x}(t) = Lx_t$, where $L{:}C \to R^n$ is linear bounded functional defined by the right-hand side of (17.7). The scalar linear system $\dot{x}(t) = -x(t - \tau(t))$ with time-varying delay $\tau(t) \in [0, h]$ can be represented as $\dot{x}(t) = L(t)x_t$ with time-varying linear functional $L(t){:}C \to R^n$ defined by the right-hand side of the system.

The general form of a retarded functional differential equation is

$$\dot{x}(t) = f(t, x_t), \quad t \geq t_0, \tag{17.10}$$

where $x(t) \in R^n$ and f: $R \times C \to R^n$. Equation 17.10 indicates that the derivative of the state variable x at time t depends on t and on $x(\xi)$ for $t - h \leq \xi \leq t$. The initial condition φ must be prescribed as φ: $[-h, 0] \to R^n$ ($\varphi \in C$, or piecewise continuous)

$$x(t_0 + \theta) = x_{t_0}(\theta) = \phi(\theta), \quad \theta \in [-h, 0]. \tag{17.11}$$

The function x: $R \to R^n$ is a solution of (17.10) with the initial condition (17.11) if there exists a scalar $a > 0$ such that $x(t)$ is continuous on $[t_0 - h, t_0 + a)$, it is initialized by (17.11), and it satisfies (17.10) for $t \in (t_0, t_0 + a)$.

If f is continuous and is locally Lipschitz in its second argument (i.e., there exists $l = l(t_0, \phi) > 0$ such that $|f(t, \phi_1) - f(t, \phi_2)| < l\|\phi_1 - \phi_2\|_C$ for $\phi_1, \phi_2 \in C$ close to ϕ and for t close to t_0), then the solution is unique and it continuously depends on the initial data (t_0, ϕ). The existence, uniqueness, and continuous dependence of the solutions is similar to ODEs, except that the solution is considered in the *forward time direction*.

Consider (17.10), where f is continuous in both arguments and is locally Lipschitz continuous in the second argument. We assume that $f(t, 0) = 0$, which guarantees that (17.10) possesses a trivial solution $x(t) \equiv 0$.

Definition. The trivial solution of (17.10) is uniformly (in t_0) asymptotically stable if

(i) For any $\varepsilon > 0$ and any t_0, there exists $\delta(\varepsilon) > 0$ such that $\|x_{t_0}\|_C < \delta(\varepsilon)$ implies $|x(t)| < \varepsilon$ for $t \geq t_0$;
(ii) There exists a $\delta_a > 0$ such that for any $\eta > 0$ there exists a $T(\delta_a, \eta)$ such that $\|x_{t_0}\|_C < \delta_a$ implies $|x(t) < \eta$ for $t \geq t_0 + T(\eta)$ and $t_0 \in R$.

The trivial solution is *globally uniformly asymptotically stable* if in (ii) δ_a can be an arbitrary large, finite number. The system is uniformly asymptotically stable if its trivial solution is uniformly asymptotically stable.

As in systems without delay, an efficient method for stability analysis of TDS is the *direct Lyapunov method*. For TDS, there are two main direct Lyapunov methods for stability analysis: *Krasovskii* method of Lyapunov *functionals* [13] and *Razumikhin* method of Lyapunov *functions* [22]. The use of Lyapunov–Krasovskii functionals is a natural generalization of the direct Lyapunov method for ODEs, since a proper state for TDS is x_t. On the other hand, Lyapunov–Razumikhin functions $V(t, x(t))$ are much simpler to use.

We start with the Krasovskii method. If V: $R \times C \to R$ is a continuous functional, we define for x_t satisfying (17.10)

$$\dot{V}(t,x_t) = \limsup_{s \to 0^+} \frac{1}{s}\left[V(t+s, x_{t+s}) - V(t,x_t)\right]. \tag{17.12}$$

Lyapunov–Krasovskii Theorem. Suppose f: $R \times C \to R^n$ maps $R\times$ (bounded sets in C) into bounded sets of R^n and u, v, w: $R^+ \to R^+$ are continuous nondecreasing, positive for $s > 0$ functions and $u(0) = v(0) = 0$. The zero solution of $\dot{x}(t) = f(t, x_t)$ is uniformly asymptotically stable if there exists a continuous functional V: $R \times C \to R^+$, which is positive definite

$$u(|\phi(0)|) \leq V(t,\phi) \leq v(\|\phi\|_C), \tag{17.13}$$

such that its derivative along (17.10) is negative

$$\dot{V}(t,x_t) \leq -w(|x(t)|). \tag{17.14}$$

If, in addition, $\lim_{s\to\infty} u(s) = \infty$, then it is globally uniformly asymptotically stable.

Sometimes, the functionals depending on the state derivatives $V(t, x_t, \dot{x}_t)$ are useful [11]. Denote by W the Sobolev space of absolutely continuous functions ϕ: $[-h, 0] \to R$ with $\dot{\phi} \in L_2[-h, 0]$ (the space of square integrable functions) with the norm $\|\phi\|_W = |\phi(0)|^2 + \int_{-h}^{0} |\dot{\phi}(s)|^2\, ds$. Note that for (17.10), the stability results corresponding to continuous initial functions and to absolutely continuous initial functions are equivalent [2]. The above theorem is extended then to continuous functionals V: $R \times W \times L_2[-h, 0] \to R^+$, where inequalities (17.13) and (17.14) are modified as follows:

$$u(|\phi(0)|) \leq V(t,\phi,\psi) \leq v(\|\phi\|_W) \tag{17.15}$$

and

$$\dot{V}(t,x_t,\dot{x}_t) \le -w(|x(t)|). \tag{17.16}$$

Consider now a continuous *function* V: $R \times C \to R$ and define $\dot{V}(t,x(t))$ by the right-hand side of (17.12), where $x_{t+s} = x(t + s)$ and $x_t = x(t)$.

Lyapunov–Razumikhin Theorem. Suppose f: $R \times C \to R^n$ maps $R\times$ (bounded sets in C) into bounded sets of R^n and p, u, v, w: $R^+ \to R^+$ are continuous nondecreasing, positive for $s > 0$ functions and $p(s) > s$ for $s > 0$, $u(0) = v(0) = 0$. The zero solution of $\dot{x}(t) = f(t, x_t)$ is uniformly asymptotically stable if there exists a continuous function V: $R \times R^n \to R^+$, which is positive definite

$$u(|x|) \le V(t,x) \le v(|x|) \tag{17.17}$$

such that

$$\dot{V}(t,x(t)) \le -w(|x(t)|) \quad \text{if } V(t+\theta,x(t+\theta)) < p(V(t,x(t)), \quad \forall\theta \in [-h,0]. \tag{17.18}$$

If, in addition, $\lim_{s\to\infty} u(s) = \infty$, then it is globally uniformly asymptotically stable.

17.5 LMI Approach to the Stability of TDS

We consider a simple linear TDS

$$\dot{x}(t) = Ax(t) + A_1x(t-\tau(t)), \quad x(t) \in R^n \tag{17.19}$$

with time-varying bounded delay $\tau(t) \in [0, h]$.

For stability analysis of (17.19) with constant delay $\tau \equiv h$, the general form of Lyapunov–Krasovskii functional

$$V(t,x_t) = x(t)^T Px(t) + 2x^T(t)\int_{-h}^{0} Q(\xi)x(t+\xi)d\xi + \int_{-h}^{0}\int_{-h}^{0} x^T(t+s)R(s,\xi)x(t+\xi)dsd\xi \tag{17.20}$$

(that corresponds to necessary and sufficient conditions for stability) leads to a complicated system of PDEs with respect to P, Q, R. The choice of Lyapunov–Krasovskii functional is crucial for deriving stability criteria. Special forms of the functional lead to simpler *delay-independent* and *delay-dependent* conditions.

17.5.1 Delay-Independent Conditions

A simple Lyapunov–Krasovskii functional (initiated by N. N. Krasovskii) has the form

$$V(t,x_t) = x^T(t)Px(t) + \int_{t-\tau(t)}^{t} x^T(s)Qx(s)ds, \quad P > 0, \quad Q > 0, \tag{17.21}$$

where P and Q are $n \times n$ positive definite matrices. We further assume that the delay τ is a differentiable function with $\dot{\tau} \leq d < 1$ (this is the case of *slowly varying delays*). It is clear that V satisfies (17.13). Then, differentiating V along (17.19), we find

$$\dot{V}(t, x_t) = 2x^T(t)P\dot{x}(t) + x^T(t)Qx(t) - (1-\dot{\tau})x^T(t-\tau)Qx(t-\tau).$$

We further substitute for $\dot{x}(t)$ the right-hand side of (17.19) and arrive to

$$\dot{V}(t, x_t) \leq \left[x^T(t)x^T(t-\tau)\right] W \begin{bmatrix} x(t) \\ x(t-\tau) \end{bmatrix} < -\alpha |x(t)|^2,$$

for some $\alpha > 0$ if

$$W = \begin{bmatrix} A^TP + PA + Q & PA_1 \\ A_1^TP & -(1-d)Q \end{bmatrix} < 0. \tag{17.22}$$

LMI (17.22) does not depend on h and it is, therefore, delay-independent (but delay-derivative dependent). The feasibility of LMI (17.23) is sufficient condition for delay-independent uniform asymptotic stability for systems with slowly varying delays.

We will next derive stability conditions by using Razumikhin's method and the Lyapunov function $V(x(t)) = x^T(t)\, Px(t)$ with $P > 0$. Also here (17.17) holds. Consider the derivative of V along (17.19)

$$\dot{V}(x(t)) = 2x^T(t)P\left[Ax(t) + A_1x(t-\tau(t))\right].$$

Whenever Razumikhin's condition holds $V(x(t+\theta)) < pV(x(t))$ for some $p > 1$, we can conclude that, for any $q > 0$,

$$\begin{aligned} \dot{V}(x(t)) &\leq 2x^T(t)P\left[Ax(t) + A_1x(t-\tau(t))\right] + q\left[px^T(t)Px(t) - x(t-\tau(t))^T Px(t-\tau(t))\right] \\ &= \left[x^T(t)x^T(t-\tau(t))\right] W_R \begin{bmatrix} x(t) \\ x(t-\tau(t)) \end{bmatrix} < -\alpha |x(t)|^2, \end{aligned}$$

for some $\alpha > 0$ if

$$W_R = \begin{bmatrix} A^TP + PA + qpP & PA_1 \\ A_1^TP & -qP \end{bmatrix} < 0. \tag{17.23}$$

The latter matrix inequality does not depend on h. Therefore, the feasibility of (17.23) is sufficient for delay-independent uniform asymptotic stability for systems with *fast-varying* delays (without any constraints on the delay derivatives).

We note that the Krasovskii-based LMI (17.22) for small enough d is less restrictive than the Razumikhin-based condition (17.23): the feasibility of (17.23) implies the feasibility of (17.22) with the same P and with $Q = qpP$, where $d < 1 - 1/p$. However, till now only the Razumikhin method provides delay-independent conditions for systems with fast-varying delays.

Example 17.2

Consider [9]

$$\dot{x}(t)=\begin{bmatrix}-2 & 0\\ 0 & -0.9\end{bmatrix}x(t)+\beta\begin{bmatrix}-1 & 0\\ -1 & -1\end{bmatrix}x(t-\tau(t)),\quad \beta\geq 0. \tag{17.24}$$

By using LMI toolbox of MATLAB®, it was found that, for $\beta\leq 0.9$ and $d=0$, LMI (17.22) is feasible and, thus, (17.24) is uniformly asymptotically stable for constant delays. For $d>0$, LMI (17.22) is unfeasible and, thus, no conclusion can be made about the delay-independent stability of (17.24) via Krasovskii's approach. In this example, Razumikhin's approach leads to a less conservative result: matrix inequality (17.23) is feasible for $\beta\leq 0.9$, which guarantees the delay-independent stability of (17.24) for all time-varying delays. In this example (because of the triangular structure of system matrices), the stability analysis is reduced to analysis of two scalar systems: $\dot{x}_1(t)=-2x_1(t)-\beta x_1(t-\tau(t))$ and $\dot{x}_2(t)=-0.9x_2(t)-\beta x_2(t-\tau(t))$. As follows from Example 17.1, the second scalar equation with constant delay is not delay-independently stable for $\beta>0.9$. Thus, in Example 17.2 both approaches lead to the analytical results.

Feasibility of the delay-independent conditions (17.22) and (17.23) implies that A is Hurwitz. It means that delay-independent conditions cannot be applied for stabilization of unstable plants by the delayed feedback. Consider, e.g., stabilization of the scalar system $\dot{x}(t)=u(t-\tau(t))$ with the delayed control input via linear state feedback $u(t)=-x(t)$. The resulting closed-loop system $\dot{x}(t)=-x(t-\tau(t))$ is asymptotically stable for $\tau\equiv h<\pi/2$ and unstable for $h>\pi/2$ in the case of constant delay (see Example 17.1). In the case of fast-varying delay, it is uniformly asymptotically stable for $\tau(t)<1.5$ and unstable for $\tau(t)$ which may (for some values of time) become greater than 1.5 [11]. For such systems, *delay-dependent* (h-dependent) conditions are needed.

17.5.2 Delay-Dependent Conditions

The first delay-dependent (both, Krasovskii and Razumikhin-based) conditions were derived by using the relation $x(t-\tau(t))=x(t)-\int_{t-\tau(t)}^{t}\dot{x}(s)ds$ via different model transformations and by bounding the cross terms [3,12,14,20]. Most Lyapunov–Krasovskii functionals treated only the *slowly* varying delays, whereas the fast-varying delay was analyzed via Lyapunov–Razumikhin functions. For the first time, systems with *fast-varying delays* were analyzed by using Krasovskii method in [4] via the descriptor approach introduced in [3]. Most of the recent Krasovskii-based results do not use model transformations and cross terms bounding. They are based on the application of Jensen's inequality [9]

$$\int_{-h}^{0}\phi^T(s)R\phi(s)ds\geq\frac{1}{h}\int_{-h}^{0}\phi^T(s)dsR\int_{-h}^{0}\phi(s)ds,\quad \forall\phi\in L_2[-h,0],\quad R>0 \tag{17.25}$$

and of free weighting matrices technique [7]. Usually the Krasovskii method leads to less conservative delay-dependent results than the Razumikhin method.

The widely used by now Lyapunov–Krasovskii functional for delay-dependent stability is a state-derivative dependent one of the form [4,7]

$$V(t,x_t,\dot{x}_t)=x^T(t)Px(t)+\int_{t-h}^{t}x^T(s)Sx(s)ds+h\int_{-h}^{0}\int_{t+\theta}^{t}\dot{x}^T(s)R\dot{x}(s)dsd\theta$$

$$+\int_{t-\tau(t)}^{t}x^T(s)Qx(s)ds, \tag{17.26}$$

where $R \geq 0$, $S \geq 0$, $Q \geq 0$. This functional with $Q = 0$ leads to delay-dependent conditions for systems with fast-varying delays, whereas for $R = S = 0$, it leads to delay-independent conditions (for systems with slowly varying delays) and coincides with (17.21). Differentiating V, we find

$$\dot{V} \leq 2x^T(t)P\dot{x}(t) + h^2\dot{x}^T(t)R\dot{x}(t) - h\int_{t-h}^{t} \dot{x}^T(s)R\dot{x}(s)ds + x^T(t)[S+Q]x(t)$$
$$- x^T(t-h)Sx(t-h) - (1-\dot{\tau}(t))x^T(t-\tau(t))Qx(t-\tau(t)) \tag{17.27}$$

and employ the representation

$$-h\int_{t-h}^{t} \dot{x}^T(s)R\dot{x}(s)ds = -h\int_{t-h}^{t-\tau(t)} \dot{x}^T(s)R\dot{x}(s)ds - h\int_{t-\tau(t)}^{t} \dot{x}^T(s)R\dot{x}(s)ds. \tag{17.28}$$

Applying further Jensen's inequality (17.25)

$$\begin{aligned} \int_{t-\tau(t)}^{t} \dot{x}^T(s)R\dot{x}(s)ds &\geq \frac{1}{h}\int_{t-\tau(t)}^{t} \dot{x}^T(s)dsR\int_{t-\tau(t)}^{t} \dot{x}(s)ds, \\ \int_{t-h}^{t-\tau(t)} \dot{x}^T(s)R\dot{x}(s)ds &\geq \frac{1}{h}\int_{t-h}^{t-\tau(t)} \dot{x}^T(s)dsR\int_{t-h}^{t-\tau(t)} \dot{x}(s)ds, \end{aligned} \tag{17.29}$$

we obtain

$$\dot{V} \leq 2x^T(t)P\dot{x}(t) + h^2\dot{x}^T(t)R\dot{x}(t) - \left[x(t) - x(t-\tau(t))\right]^T R\left[x(t) - x(t-\tau(t))\right]$$
$$-\left[x(t-\tau(t)) - x(t-h)\right]^T R\left[x(t-\tau(t)) - x(t-h)\right] + x^T(t)[S+Q]x(t)$$
$$- x^T(t-h)Sx(t-h) - (1-d)x^T(t-\tau(t))Qx(t-\tau(t)). \tag{17.30}$$

We shall derive stability conditions in two forms. The first form is derived by substituting for $\dot{x}(t)$ in (17.30) the right-hand side of (17.19):

$$\begin{aligned} \dot{V} \leq\; & 2x^T(t)P\left[Ax(t) + A_1x(t-\tau(t))\right] \\ & + h^2\left[Ax(t) + A_1x(t-\tau(t))\right]^T R\left[Ax(t) + A_1x(t-\tau(t))\right] \\ & -\left[x(t) - x(t-\tau(t))\right]^T R\left[x(t) - x(t-\tau(t))\right] \\ & -\left[x(t-\tau(t)) - x(t-h)\right]^T R\left[x(t-\tau(t)) - x(t-h)\right] + x^T(t)[S+Q]x(t) \\ & - x^T(t-h)Sx(t-h) - (1-d)x^T(t-\tau(t))Qx(t-\tau(t)) \\ =\; & \eta^T\Phi\eta + h^2\left[Ax(t) + A_1x(t-\tau(t))\right]^T R\left[Ax(t) + A_1x(t-\tau(t))\right], \end{aligned} \tag{17.31}$$

where $\eta(t) = col\{x(t), x(t-h), x(t-\tau(t)),\}$ and where

$$\Phi = \begin{bmatrix} A^TP + PA + S + Q - R & 0 & PA_1 + R \\ * & -S - R & R \\ * & * & -(1-d)Q - 2R \end{bmatrix}. \tag{17.32}$$

Applying the Schur complements formula to the last term in (17.31), we find that (17.16) holds if the following LMI

$$\begin{bmatrix} A^TP + PA + S + Q - R & 0 & PA_1 + R & hA^TR \\ * & -S - R & R & 0 \\ * & * & -(1-d)Q - 2R & hA_1^TR \\ * & * & * & -R \end{bmatrix} < 0 \tag{17.33}$$

is feasible.

The second form is derived via the descriptor method, where the right-hand side of the expression

$$0 = 2[x^T(t)P_2^T + \dot{x}^T(t)P_3^T][Ax(t) + A_1x(t-\tau(t)) - \dot{x}(t)], \tag{17.34}$$

with some $P_2, P_3 \in R^{n\times n}$ is added to the right-hand side of (17.30). Setting $\eta_d(t) = col\{x(t), \dot{x}(t), x(t-h), x(t-\tau(t))\}$, we obtain that

$$\dot{V} \le \eta_d^T(t)\Phi_d\eta_d(t) \le 0 \tag{17.35}$$

is satisfied if the LMI

$$\Phi_d = \begin{bmatrix} \Phi_{11} & \Phi_{12} & 0 & P_2^TA_1 + R \\ * & \Phi_{22} & 0 & P_3^TA_1 \\ * & * & -(S+R) & R \\ * & * & * & -(1-d)Q - 2R \end{bmatrix} < 0 \tag{17.36}$$

holds, where

$$\begin{aligned} \Phi_{11} &= A^TP_2 + P_2^TA + S + Q - R, \\ \Phi_{12} &= P - P_2^T + A^TP_3, \\ \Phi_{22} &= -P_3 - P_3^T + h^2R. \end{aligned} \tag{17.37}$$

We note that the feasibility of the (delay-dependent) LMI (17.33) or (17.36) with $Q = 0$ guarantees uniform asymptotic stability of (17.19) for all fast-varying delays $\tau(t) \in [0, h]$. For $R \to 0, S \to 0$, the above LMIs guarantee the delay-independent stability. Thus, (17.33) coincides for $R \to 0, S \to 0$ with LMI (17.22).

Since LMIs (17.33), (17.36) are affine in the system matrices, they can be applied to the case where these matrices are uncertain. In this case, we denote $\Omega = [A\ A_1]$ and assume that $\Omega \in \mathcal{C}o\{\Omega_j, j = 1,\ldots,N\}$, namely,

$$\Omega = \sum_{j=1}^{N} f_j \Omega_j \quad \text{for some} \quad 0 \le f_j \le 1, \quad \sum_{j=1}^{N} f_j = 1, \tag{17.38}$$

where the N vertices of the polytope are described by $\Omega_j = \begin{bmatrix} A^{(j)} & A_1^{(j)} \end{bmatrix}$. In the case of time-varying uncertainty with $f_j = f_j(t)$, one has to solve the LMI (17.33) or (17.36) simultaneously for all the N vertices, applying the same decision variables (i.e., matrices P, P_2, P_3, S, Q, and R) for all vertices. In the case of time-invariant uncertainty, the conditions via the descriptor approach have advantages: one has to solve the descriptor-based LMI (17.36) simultaneously for all the N vertices, applying the same decision variables P_2. P_3 (since these matrices are multiplied by the system matrices) and different matrices $P^{(j)}$, $R^{(j)}$, $S^{(j)}$, and $Q^{(j)}$, $j = 1,\ldots,N$. By the first form of conditions (17.33), P and R should be common and only $S^{(j)}$ and $Q^{(j)}$ may be different in the vertices.

Comparing the above two forms of delay-dependent conditions, we see that the $1-st$ LMI (17.33) has less decision variables (which is an advantage). However, slack variables P_2 and P_3 of the descriptor method may lead to less conservative results in the analysis of uncertain systems and in design problems.

Example 17.3

Consider the simple scalar equation

$$\dot{x}(t) = -x(t - \tau(t)). \tag{17.39}$$

By verifying the feasibility of (17.33) and of (17.36), we arrive at the same results for uniform asymptotic stability of (17.39). Applying these conditions with $d = 0$, we find that the system is stable for (constant) $\tau \in [0, 1.41]$ (compared with the analytical result $\tau < 1.57$). Applying the above LMIs with $Q = 0$, we conclude that the system is stable for all fast-varying delays $\tau(t) \in [0, 1.22]$ (compared with the analytical result $\tau < 1.5$).

The above conditions can be easily extended to systems with multiple delays. We note that all delay-dependent conditions in terms of LMIs are *sufficient* only. Therefore, these conditions can be improved. Thus, the presented method conservatively applies Jensen's inequality (17.29), where in the right-hand side there could be written $1/\tau$ instead of its lower bound $1/h$. The improvements are usually achieved at the price of the LMI's complexity. Among the recent results, convex analysis of [21] has significantly improved the stability analysis of systems with time-varying delays (achieving in Example 17.3 the interval $\tau(t) \in [0, 1.33]$ instead of $\tau(t) \in [0, 1.22]$ for the stability of (17.39) with fast-varying delay). The presented above conditions guarantee the stability for small delay $\tau(t) \in [0, h]$. Recent results analyze also the stability for interval (or non-small) delay $\tau(t) \in [h_1, h_2]$ with $h_1 > 0$ (see, e.g., [7]).

A necessary condition for the application of simple Lyapunov–Krasovskii functionals is the asymptotic stability of (17.19) with $\tau = 0$. Consider, e.g., the following system with constant delay:

$$\dot{x}(t) = \begin{bmatrix} 0 & 1 \\ -2 & 0.1 \end{bmatrix} x(t) + \begin{bmatrix} 0 & 0 \\ 1 & 0 \end{bmatrix} x(t - h). \tag{17.40}$$

This system is unstable for $h = 0$ and is asymptotically stable for the constant delay $h \in (0.1002, 1.7178)$ [9]. For analysis of such systems, simple Lyapunov functionals are not suitable. One can use a discretized method of Gu [8], which is based on general Lyapunov–Krasovskii functional (17.20).

17.5.3 Exponential Bounds and L_2-Gain Analysis

System (17.19) is said to be *exponentially stable with a decay rate* $\delta > 0$ if there exists a constant $\beta \geq 1$ such that the following exponential estimate

$$|x(t)|^2 \leq \beta e^{-2\delta(t-t_0)} \|\phi\|_C^2 \quad \forall t \geq t_0, \tag{17.41}$$

holds for the solution of (17.19) initialized with $x_{t_0} = \phi \in C$. If LTI system with known matrices and constant delay is asymptotically stable, then it is also exponentially stable. The (uniform) asymptotic stability of the linear system with (time-varying) uncertain coefficients or delays does not imply exponential stability. In any case, Lyapunov–Krasovskii method gives sufficient conditions for finding the decay rate and the constant β in (17.41).

Consider, e.g., the following functional [17]

$$V(t, x_t) = x^T(t)Px(t) + \int_{t-\tau(t)}^{t} e^{2\delta(t-s)} x^T(s)Qx(s)ds, \quad P > 0, \quad Q > 0. \tag{17.42}$$

For $\delta = 0$, this functional coincides with (17.21) for delay-independent stability. If along (17.19)

$$\dot{V}(t, x_t) + 2\delta V(t, x_t) \leq 0, \tag{17.43}$$

then, by the comparison principle [10], it follows that

$$V(t, x_t) \leq V(t_0, \phi)e^{-2\delta(t-t_0)} \leq be^{-2\delta(t-t_0)} \|\phi\|_C^2 .$$

Differentiating V we find

$$\begin{aligned} \dot{V}(t, x_t) + 2\delta V(t, x_t) &\leq 2x^T(t)P\dot{x}(t) + 2\delta x^T(t)Px(t) + x^T(t)Qx(t) \\ &- (1-d)e^{-2\delta h} x^T(t-\tau(t))Qx(t-\tau(t)). \end{aligned} \tag{17.44}$$

Substituting further the right-hand side of (17.19) for $\dot{x}(t)$, we arrive to the following LMI, which guarantees (17.43):

$$\begin{bmatrix} A^T P + PA + 2\delta P + Q & PA_1 \\ A_1^T P & -(1-d)e^{-2\delta h} Q \end{bmatrix} < 0. \tag{17.45}$$

The latter LMI is delay independent when $\delta = 0$.

Modifying the delay-dependent Lyapunov–Krasovskii functional (17.26) by inserting $e^{2\delta(t-s)}$ into integral terms, one can arrive at delay-dependent-based conditions for (17.41) with $\phi \in W$ and with $\|\phi\|_W$ in the right-hand side instead of $\|\phi\|_C$.

Example 17.4

Consider [17]

$$\dot{x}(t) = \begin{bmatrix} -4 & 1 \\ 0 & -4 \end{bmatrix} x(t) + \begin{bmatrix} 0.1 & 0 \\ 4 & 0 \end{bmatrix} x(t-0.5). \tag{17.46}$$

For $d = 0$ and $\delta = 1.15$, it was found that LMI (17.45) is feasible and, thus, (17.46) is exponentially stable with the decay rate $\delta = 1.15$. It was found in [17] that the rightmost characteristic root corresponding to (17.46) is 1.15. The latter means that the derived LMI condition leads to the analytic result in this example.

Consider next a linear perturbed system

$$\dot{x}(t) = Ax(t) + A_1x(t - \tau(t)) + B_1w(t), \tag{17.47}$$

where $x(t) \in R^n$ is the state vector, $w(t) \in R^{n_w}$ is the disturbance with the controlled output

$$z(t) = C_0x(t) + C_1x(t - \tau(t)),\ z(t) \in R^{n_z}. \tag{17.48}$$

The matrices A, A_1, B_1, C_0, and C_1 are constant.

For a prechosen $\gamma > 0$, we introduce the following performance index:

$$J = \int_{t_0}^{\infty} \left[z^T(t)z(t) - \gamma^2 w^T(t)w(t) \right] dt. \tag{17.49}$$

We seek conditions that will lead to $J < 0$ for all $x(t)$ satisfying (17.47), (17.48) with zero initial condition and for all $0 \neq w(t) \in L_2[t_0, \infty)$. In this case the system (17.47), (17.48) has *L_2-gain less than* γ.

L_2-gain analysis for linear TDS can be performed by *using the Lyapunov–Krasovskii method* (Razumikhin's approach seems to be not applicable to this problem). We shall formulate sufficient conditions for L_2-gain analysis via the delay dependent Lyapunov functional (17.26). For a prescribed $\gamma > 0$, if along (17.47), (17.48)

$$\dot{V}(t, x_t, \dot{x}_t) + z^T(t)z(t) - \gamma^2 w^T(t)w(t) < 0, \quad \forall t \geq t_0, \quad \forall w \neq 0, \tag{17.50}$$

then (17.47), (17.48) has L_2-gain less than γ. Really, integration of (17.50) in t from t_0 to ∞ and taking into account that $V_{|t=t_0} = 0$ and $V_{|t=\infty} > 0$ leads to $J < 0$.

Similar to the stability analysis via the descriptor method, we add to $\dot{V}$ the right-hand side of the following expression:

$$0 = 2\left[x^T(t)P_2^T + \dot{x}^T(t)P_3^T \right]\left[Ax(t) + A_1x(t - \tau(t)) + B_1w(t) - \dot{x}(t) \right], \tag{17.51}$$

(with the additional term $B_1w(t)$ comparatively to (17.34)). Applying further the Schur complements formula to the term

$$z^T(t)z(t) = \left[C_0x(t) + C_1x(t - \tau(t)) \right]^T \left[C_0x(t) + C_1x(t - \tau(t)) \right],$$

we find that (17.50) holds if

$$\begin{bmatrix} & | & P_2^T B_1 & C_0^T \\ & | & P_3^T B_1 & 0 \\ \Phi_d & | & 0 & 0 \\ & | & 0 & C_1^T \\ - & - & - & - \\ * & | & -\gamma^2 I & 0 \\ * & | & * & -I \end{bmatrix} < 0, \tag{17.52}$$

where Φ_d is given by (17.36). We note that if LMI (17.52) is feasible, then $\Phi_d < 0$ and, thus, (17.47) with $w = 0$ is uniformly asymptotically stable.

17.6 Control Design for TDS

17.6.1 Predictor-Based Design

Consider the LTI system with the delayed input

$$\dot{x}(t) = Ax(t) + Bu(t-h), \tag{17.53}$$

where the constant $h > 0$ and the constant system matrices are supposed to be known. We seek a stabilizing state-feedback controller. Assume that there exists a gain K such that $A + BK$ is Hurwitz. Formally it means that $u(t) = Kx(t + h)$ stabilizes (17.53). Then, integrating (17.53) with the initial condition $x(t)$ we find

$$x(t+h) = e^{Ah}x(t) + \int_{t}^{t+h} e^{A(t+h-s)}Bu(s-h)ds. \tag{17.54}$$

Thus, the right-hand side of (17.54) predicts the value of $x(t + h)$. After the change of variable $\xi = s - t - h$ in the integral, we arrive to the following stabilizing controller:

$$u(t) = K\left[e^{Ah}x(t) + \int_{-h}^{0} e^{-A\xi}Bu(t+\xi)d\xi\right]. \tag{17.55}$$

This is the idea of the predictor-based controller introduced in [15], which reduces stabilization of TDS to that of its delay-free counterpart.

The predictor-based design has been extended to the linear quadratic regulator for LTI systems [11] and to other control problems with known constant or even known time-varying delay [25]. However, this approach faces difficulties in the case of uncertain systems and delays, as well as in the case of an additional state delay. For uncertain systems, the LMI approach leads to efficient design algorithms.

17.6.2 LMI-Based Design

Consider the system

$$\dot{x}(t) = Ax(t) + Bu(t-\tau(t)), \tag{17.56}$$

where

- $x(t) \in R^n$ is the state vector
- $u(t) \in R^{n_u}$ is the control input
- $\tau(t)$ is time-varying delay $\tau(t) \in [0, h]$

We seek a stabilizing state feedback $u(t) = Kx(t)$ that leads to a uniformly asymptotically stable closed-loop system (17.19), where $A_1 = BK$. Considering next the inequalities (17.33) or (17.36) for delay-dependent stability with $A_1 = BK$, we see that both of these inequalities are nonlinear: (17.33) because of the terms PBK, RBK and (17.36) because of $P_2^T BK$, $P_3^T BK$.

To find the unknown gain K, either an iterative procedure can be used (see, e.g., [18]) or some transformation of the nonlinear matrix inequality, which linearizes it. To linearize (17.36) denote $P_3 = \varepsilon P_2$, where ε is a scalar, $\bar{P}_2 = P_2^{-1}$, $\bar{P} = \bar{P}_2^T P\bar{P}$, $\bar{R} = \bar{P}_2^T R\bar{P}_2$, $\bar{S} = \bar{P}_2^T S\bar{P}_2$, $\bar{Q} = \bar{P}_2^T Q\bar{P}_2$. We multiply (17.36) by $diag\{\bar{P}_2^T, \bar{P}_2^T, \bar{P}_2^T, \bar{P}_2^T\}$ and its transpose, from the right and the left, respectively. Denoting $Y = K\bar{P}_2$, the following is obtained:

$$\begin{bmatrix} \bar{\Phi}_{11} & \bar{\Phi}_{12} & 0 & BY + \bar{R} \\ * & \bar{\Phi}_{22} & 0 & \varepsilon BY \\ * & * & -(\bar{S} + \bar{R}) & \bar{R} \\ * & * & * & -(1-d)\bar{Q} - 2\bar{R} \end{bmatrix} < 0 \tag{17.57}$$

where

$$\begin{aligned} \bar{\Phi}_{11} &= A\bar{P}_2 + \bar{P}_2^T A^T + \bar{S} + \bar{Q} - \bar{R}, \\ \bar{\Phi}_{12} &= \bar{P} - \bar{P}_2^T + \varepsilon \bar{P}_2 A^T, \\ \bar{\Phi}_{22} &= -\varepsilon \bar{P}_2 - \varepsilon \bar{P}_2^T + h^2 \bar{R}. \end{aligned} \tag{17.58}$$

Given ε, (17.57) is an LMI with respect to $\bar{P} > 0$, $\bar{R} \geq 0$, $\bar{Q} \geq 0$, $\bar{S} \geq 0$, and $\bar{P}_2$.

Various design problems for systems with state and input/output delays (including observer-based design) can be solved by using LMIs (see, e.g., [24]).

17.7 On Discrete-Time TDS

Consider a linear discrete-time system

$$x(k+1) = Ax(k) + A_1 x(k-h), \quad k = 0,1,2,\ldots, \quad h \in N, \quad x(k) \in R^n \tag{17.59}$$

with constant matrices and constant delay h. As in the continuous-time case, $x(k)$ can no longer be regarded as its state vector. It should be augmented by its finite history

$$x_{aug}(k) = col\{x(k-h), x(k-h+1), \ldots, x(k)\}, \tag{17.60}$$

resulting in the following augmented system without delay:

$$x_{aug}(k+1) = A_{aug} x_{aug}(k), \quad k = 0,1,2,\ldots, \quad x_{aug}(k) \in R^{(h+1)n}, \tag{17.61}$$

where

$$A_{aug} = \begin{bmatrix} 0 & I_n & \ldots & 0 \\ \ldots & \ldots & \ldots & \ldots \\ 0 & 0 & \ldots & I_n \\ A_1 & 0 & \ldots & A \end{bmatrix}.$$

Thus, the stability analysis of (17.59) can be reduced to the analysis of the higher-order system without delay (17.61). If we apply Lyapunov method to (17.59), then a necessary and sufficient condition for asymptotic stability of (17.59) is the existence of Lyapunov function $V(x_{aug}(k)) = x_{aug}^T(k) P_{aug} x_{aug}(k)$ with $P_{aug} > 0$ such that along (17.61) the following holds: $V_{aug}(x_{aug}(k+1)) - V_{aug}(x_{aug}(k)) < -\alpha \mid x_{aug}(k)|^2, \alpha > 0$.

Unlike the continuous-time case, the general Lyapunov function V_{aug} can be easily found by augmentation of the discrete-time LTI system. However, such augmentation suffers from the curse of dimensionality if h is high. Application of the augmentation becomes complicated if the delay is unknown or/and time varying. To derive simple sufficient stability conditions, discrete-time counterparts of simple Lyapunov–Razumikhin functions or Lyapunov–Krasovskii functionals can be applied [5]. Thus, for the linear system with time-varying delay

$$x(k+1) = Ax(k) + A_1x(k-\tau_k), \quad k = 0,1,2,\ldots, \quad \tau_k \in N, \quad 0 \le \tau_k \le h, \tag{17.62}$$

the discrete-time counterpart of (17.26) for delay-dependent analysis has the form

$$\begin{aligned} V(x(k)) &= x^T(k)Px(k) + h\sum_{m=-h}^{-1}\sum_{j=k+m}^{k-1} \bar{y}(j)^T R\bar{y}(j) + \sum_{j=k-h}^{k-1} x(j)^T Sx(j), \\ \bar{y}(k) &= x(k+1) - x(k), \quad P > 0, \quad R \ge 0, \quad S \ge 0. \end{aligned} \tag{17.63}$$

The technique for derivation of LMI conditions is then similar to the continuous-time case (see, e.g., [1]). We have

$$\begin{aligned} V(x(k+1)) - V(x(k)) &= 2x^T(k)P\bar{y}(k) + \bar{y}^T(k)P\bar{y}(k) + x^T(k)Sx(k) \\ &\quad - x^T(k-h)Sx(k-h) + h^2\bar{y}^T(k)R\bar{y}(k) - h\sum_{j=k-h}^{k-1} \bar{y}^T(j)R\bar{y}(j). \end{aligned} \tag{17.64}$$

We employ the representation

$$-h\sum_{j=k-h}^{k-1} \bar{y}^T(j)R\bar{y}(j) = -h\sum_{j=k-h}^{k-\tau_k-1} \bar{y}^T(j)R\bar{y}(j) - h\sum_{j=k-\tau_k}^{k-1} \bar{y}^T(j)R\bar{y}(j)$$

and apply further Jensen's (in fact, this is a particular case of Cauchy–Schwartz) inequality

$$\begin{aligned} h\sum_{j=k-h}^{k-\tau_k-1} \bar{y}^T(j)R\bar{y}(j) &\ge \left(\sum_{j=k-h}^{k-\tau_k-1} \bar{y}^T(j)\right) R \left(\sum_{j=k-h}^{k-\tau_k-1} \bar{y}(j)\right), \\ h\sum_{j=k-\tau_k}^{k-1} \bar{y}^T(j)R\bar{y}(j) &\ge \left(\sum_{j=k-\tau_k}^{k-1} \bar{y}^T(j)\right) R \left(\sum_{j=k-\tau_k}^{k-1} \bar{y}(j)\right). \end{aligned}$$

Then, by the descriptor method, where the right-hand side of the expression

$$0 = 2\left[x^T(k)P_2^T + \bar{y}^T(k)P_3^T\right]\left[(A-I)x(k) + A_1x(k-\tau_k) - \bar{y}(k)\right], \tag{17.65}$$

with some $P_2, P_3 \in R^{n\times n}$, is added into the right-hand side of (17.64), and setting $\eta_{dis}(k) = col\{x(k), \bar{y}(k), x(k-h), x(k-\tau_k)\}$, we obtain that

$$V(x(k+1)) - V(x(k)) \le \eta_{dis}^T(k)\Phi_{dis}\eta_{dis}(k) \le 0 \tag{17.66}$$

is satisfied, if the following LMI

$$\Phi_{dis} = \begin{bmatrix} \Phi_{11} & \Phi_{12} & 0 & P_2^T A_1 + R \\ * & \Phi_{22} & 0 & P_3^T A_1 \\ * & * & -(S+R) & R \\ * & * & * & -2R \end{bmatrix} < 0 \tag{17.67}$$

is feasible, where

$$\begin{aligned} \Phi_{11} &= (A-I)^T P_2 + P_2^T (A-I) + S - R, \\ \Phi_{12} &= P - P_2^T + (A-I)^T P_3, \\ \Phi_{22} &= -P_3 - P_3^T + P + h^2 R. \end{aligned} \tag{17.68}$$

Example 17.5

We consider

$$x(k+1) = \begin{bmatrix} 0.8 & 0 \\ 0 & 0.97 \end{bmatrix} x(k) + \begin{bmatrix} -0.1 & 0 \\ -0.1 & -0.1 \end{bmatrix} x(k-\tau_k). \tag{17.69}$$

For the constant $\tau_k \equiv h$, by using the augmentation, it is found that the system is asymptotically stable for all $h \leq 18$ (since the eigenvalues of the resulting A_{aug} are inside the unit circle). For time-varying τ_k, by verifying the feasibility of (17.67), we find that the system is asymptotically stable for all $0 \leq \tau_k \leq 12$.

References

1. W.-H. Chen, Z.-H. Guan, and X. Lu, Delay-dependent guaranteed cost control for uncertain discrete-time systems with delay, *IEE Proceedings Control Theory and Applications*, 150, 412–416, 2003.
2. L. El'sgol'ts and S. Norkin, *Introduction to the Theory and Applications of Differential Equations with Deviating Arguments*, Mathematics in Science and Engineering, Vol. 105, Academic Press, New York, 1973.
3. E. Fridman, New Lyapunov-Krasovskii functionals for stability of linear retarded and neutral type systems, *Systems & Control Letters*, 43, 309–319, 2001.
4. E. Fridman and U. Shaked, Delay dependent stability and H_∞ control: Constant and time-varying delays, *International Journal of Control*, 76(1), 48–60, 2003.
5. E. Fridman and U. Shaked, Stability and guaranteed cost control of uncertain discrete delay systems, *International Journal of Control*, 78(4), 235–246, 2005.
6. J.K. Hale and S.M.V. Lunel, *Introduction to Functional Differential Equations*, Springer-Verlag, New York, 1993.
7. Y. He, Q.G. Wang, C. Lin, and M. Wu, Delay-range-dependent stability for systems with time-varying delay, *Automatica*, 43, 371–376, 2007.
8. K. Gu, Discretized LMI set in the stability problem of linear time delay systems, *International Journal of Control*, 68, 923–934, 1997.

9. K. Gu, V. Kharitonov, and J. Chen, *Stability of Time-Delay Systems*, Birkhauser, Boston, MA, 2003.
10. H.K. Khalil, *Nonlinear Systems*, 3rd edn., Prentice Hall, Englewood Cliffs, NJ, 2002.
11. V. Kolmanovskii and A. Myshkis, *Applied Theory of Functional Differential Equations*, Kluwer, Dordrecht, the Netherlands, 1999.
12. V. Kolmanovskii and J.P. Richard, Stability of some linear systems with delays, *IEEE Transactions on Automatic Control*, 44, 984–989, 1999.
13. N. Krasovskii, *Stability of Motion*, Moscow, 1959 [Russian]. Translation, Stanford University Press, 1963.
14. X. Li and C. de Souza, Criteria for robust stability and stabilization of uncertain linear systems with state delay, *Automatica*, 33, 1657–1662, 1997.
15. A.Z. Manitius and A.W. Olbrot. Finite spectrum assignment problem for systems with delay. *IEEE Transactions on Automatic Control*, 24, 541–553, 1979.
16. Yu. Mikheev, V. Sobolev, and E. Fridman, Asymptotic analysis of digital control systems, *Automatic and Remote Control*, 49, 1175–1180, 1988.
17. S. Mondie and V. Kharitonov, Exponential estimates for retarded time-delay systems, *IEEE Transactions on Automatic Control*, 50(5), 268–273, 2005.
18. Y.S. Moon, P. Park, W.H. Kwon, and Y.S. Lee, Delay-dependent robust stabilization of uncertain state-delayed systems, *International Journal of Control*, 74, 1447–1455, 2001.
19. S.I. Niculescu, Delay effects on stability: A robust control approach, *Lecture Notes in Control and Information Sciences*, 269, Springer-Verlag, London, U.K., 2001.
20. P. Park, A delay-dependent stability criterion for systems with uncertain time-invariant delays, *IEEE Transactions on Automatic Control*, 44, 876–877, 1999.
21. P. Park and J.W. Ko, Stability and robust stability for systems with a time-varying delay, *Automatica*, 43, 1855–1858, 2007.
22. B. Razumikhin. On the stability of systems with a delay. *Prikladnaya Mathematika Mechanika*, 20, 500–512, 1956 [Russian].
23. J.P. Richard, Time-delay systems: an overview of some recent advances and open problems, *Automatica*, 39, 1667–1694, 2003.
24. V. Suplin, E. Fridman, and U. Shaked, Sampled-data H_∞ control and filtering: Nonuniform uncertain sampling, *Automatica*, 43, 1072–1083, 2007.
25. E. Witrant, D. Georges, C. Canudas-de-Wit, and M. Alamir, On the use of state predictors in networked control systems. In *Applications of Time-Delay Systems*, Eds. J. Chiasson and J.-J. Loiseau, pp. 315–327, Springer Verlag, Berlin, Germany, 2007.

18

AC Servo Systems

Yong Feng
RMIT University

Liuping Wang
RMIT University

Xinghuo Yu
RMIT University

18.1 Introduction

AC motors and AC servo systems have been playing more and more important roles in factory automation, household electrical appliances, computers, CNC machines, robots, high-speed aerospace drives, and high-technology tools used for outer space in the past decades because of their good performances, such as high power density, high efficiency, fast dynamics, and good compatibility [ZAATS06,JSHIS03]. AC motors include mainly permanent magnet synchronous motors (PMSMs) and induction motors (IMs). Both are multivariable, strong coupling, and uncertain nonlinear systems, in which uncertainties including internal parameters perturbation and external load disturbances exist. So far, many methods can improve the robustness and the dynamic performance of the system, such as the adaptive control, the fuzzy control, artificial neural networks, and the active disturbance rejection control [JL09,XVT03,KT07,RH98,SZD05]. Among them, the sliding-mode control method does not request a high accurate mathematical model and is strongly robust with respect to system parameter perturbation and external disturbances. Recently, due to the rapid improvement in power electronics and microprocessors, the performances of AC servo systems have reached and even exceeded that of traditional DC servo systems. This chapter will introduce the control of AC motors, including PMSMs and IMs.

18.2 Control System of PMSM

18.2.1 System Structure

A typical control system of PMSM is shown in Figure 18.1.

Generally, there are three controllers in the system: position controller, velocity controller, and current controller, which are used in the three closed loops, respectively. The task of the control system design for a PMSM is to design the three controllers: the position controller, the velocity controller, and the current controller, with a filter in the velocity loop. The design process of the three controllers is usually from the inner loop to the outer loop, that is, from the current controller to the position

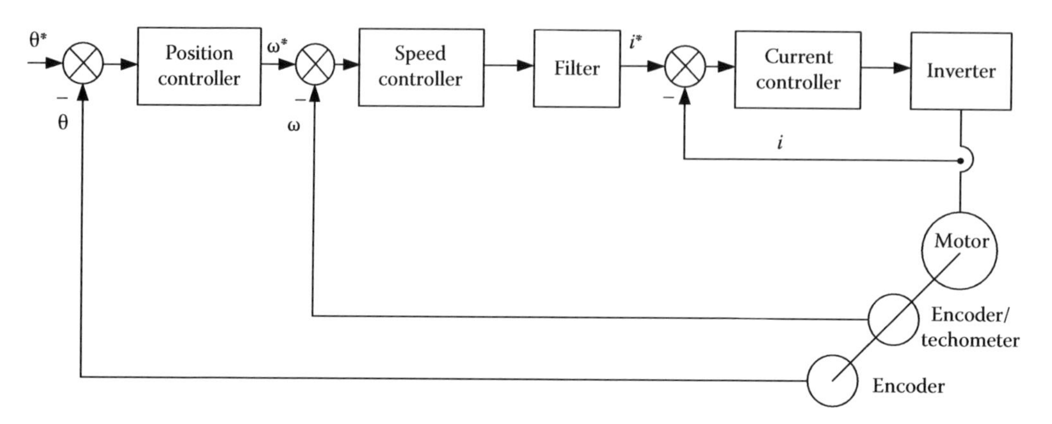

FIGURE 18.1 A typical control system of a PMSM.

controller. Compared to the velocity controller, the design tasks for the current controller and the position controller are relatively easy. The parameters of the current controller mainly depend on the system performance requirements and the physical parameters of the motor, such as the inductance and resistors, which can be readily measured. After completing the current controller and the velocity controller, the position controller can be designed in a relatively easy manner because the controlled plant has been well regulated to a second-order system. The design of the velocity controller is more challenging, mainly since mechanical resonances, backlash, and friction influence the performance and the stability of the velocity loop. The filter in the velocity loop is used to suppress mechanical resonance in the servo systems [ZF08].

18.2.2 Model of PSMS

Assume that the magnetic flux is not saturated, the magnetic field is sinusoidal, and the influence of the magnetic hysteresis is negligible. Taking the rotor coordinates (d-q axes) of the motor as the reference coordinates, a PMSM system can be described as follows [ED00,SZD05]:

$$\begin{cases} \dot{i}_d = -\dfrac{R_s}{L} i_d + p\omega i_q + \dfrac{u_d}{L} \\ \dot{i}_q = -p\omega i_d - \dfrac{R_s}{L} i_q - \dfrac{p\psi_f}{L}\omega + \dfrac{u_q}{L} \\ \dot{\omega} = \dfrac{1.5p\psi_f}{J} i_q - \dfrac{B}{J}\omega - \dfrac{T_L}{J} \\ \dot{\theta} = \omega \end{cases} \tag{18.1}$$

where

u_d and u_q are d-q axes stator voltages
i_d and i_q are d-q axes stator currents
L is the d-q axes inductance
R_s is the stator winding resistance
ψ_f is the flux linkage of the permanent magnets
p is the number of pole pairs
θ and ω are the angular position and the velocity of the motor, respectively

18.2.3 PID Control of PSMS

The vector-control method and PI controllers are widely used in the PMSM speed-control system. The structure of the PMSM vector-control system is shown in Figure 18.2.

According to the vector-control principle, three-phase currents of the motor stator are transformed into two rotor coordinates (d-q axes) of the motor. In order to realize the maximization of the driving torque, the phase angle between the stator flux and the rotor flux should be maintained at 90 electrical degrees, which can be realized by setting the d-axis current reference at zero. In practical applications, most PMSM systems utilize PID controllers till now.

In order to overcome the integral saturation of the PID controller in Figure 18.2, the anti-reset windup method can be used [S98]. The speed controller is shown in Figure 18.3 and expressed by the following equation:

$$i_q^* = k_p e_\omega + \left[e_\omega \frac{k_p}{\tau_i} - \frac{1}{k_c}\left(i_q^* - i_q^r\right) \right] \frac{1}{s} \tag{18.2}$$

where $k_{\omega m} > 0$ is the compensation coefficient.

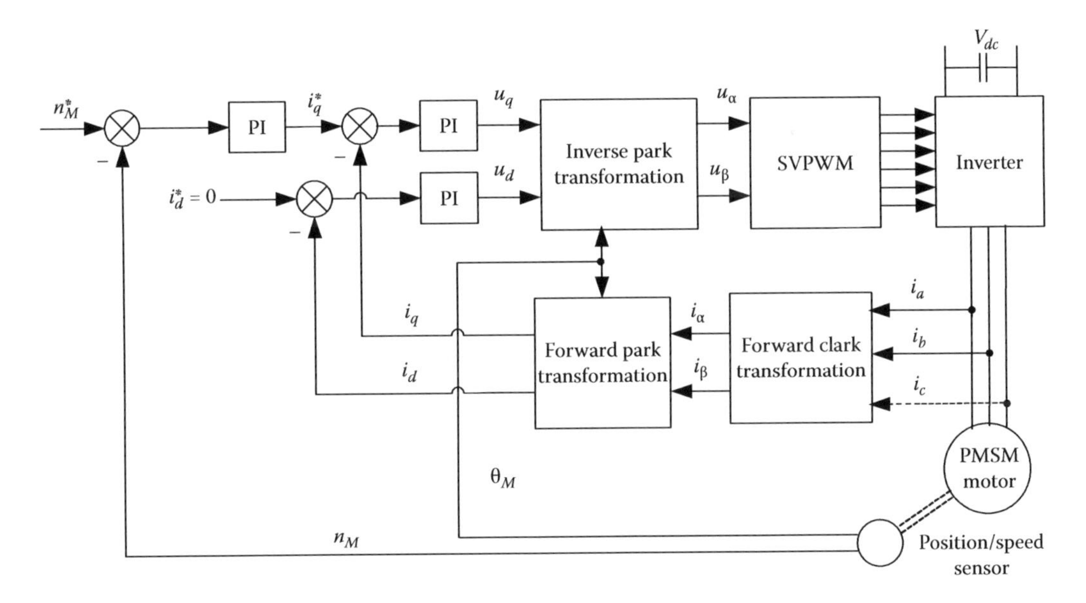

FIGURE 18.2 A PMSM vector control system based on PI controllers.

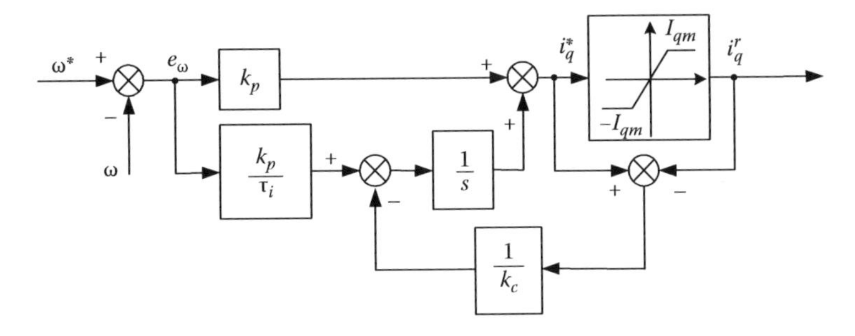

FIGURE 18.3 A PI controller using the anti-reset windup method.

18.2.4 High-Order TSM Control of PMSM

Besides PID control, many other control methods have been proposed for the control of the PMSM motor [BKY00,ES98]. Here, the high-order TSM control of the PMSM is introduced, which is shown in Figure 18.4.

Velocity controller design [ZFY08]

1. An NTSM manifold is chosen as follows [FYM02]:

$$l_{\omega} = e_{\omega} + \gamma_1 \dot{e}_{\omega}^{p_1/q_1} \tag{18.3}$$

   where
   $e_{\omega} = \omega^* - \omega$ is the motor speed error
   ω^* is the desired motor speed

2. The control strategies are chosen as follows:

$$i_q^* = i_{qeq} + i_{qn} \tag{18.4}$$

$$i_{qeq} = \frac{J}{1.5p\psi_f}\left(\dot{\omega}^* + \frac{B}{J}\omega\right) \tag{18.5}$$

$$i_{qn} = \frac{J}{1.5p\psi_f}\int_0^t \left[\frac{q_1}{\gamma_1 p_1}\dot{e}_{\omega}^{2-p_1/q_1} + (k_1 + \eta_{10})\operatorname{sgn}(l_{\omega}) + \eta_{11} l_{\omega}\right] d\tau \tag{18.6}$$

   where $k_1 > |\dot{T}_L/J|$, $\eta_{10} > 0$ and $\eta_{11} > 0$ are two design parameters. The error speed of the motor in Equation 18.6 will converge to zero in finite time.

In order to overcome the integral saturation in the control (18.6), the anti-reset windup method can be used. The speed controller is shown in Figure 18.5 and expressed by the following equation:

$$i_{qn} = \frac{J}{1.5p\psi_f}\int_0^t \left[\frac{q_1}{\gamma_1 p_1}\dot{e}_{\omega}^{2-p_1/q_1} + (k_1 + \eta_{10})\operatorname{sgn}(l_{\omega}) + \eta_{11} l_{\omega} - k_{\omega m}\left(i_q^* - i_q^r\right)\right] d\tau \tag{18.7}$$

where $k\omega_m > 0$ is the compensation coefficient.

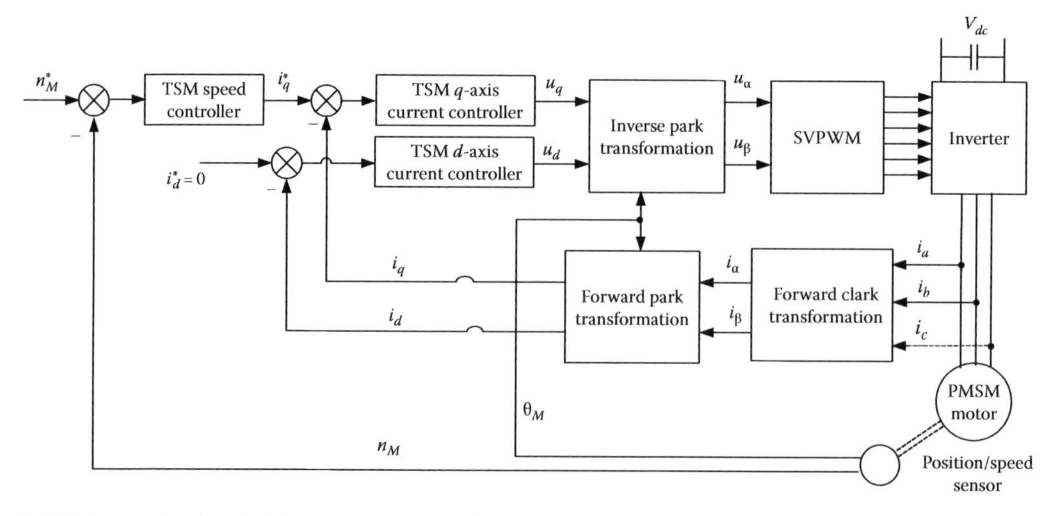

FIGURE 18.4 High-order TSM control system of a PMSM.

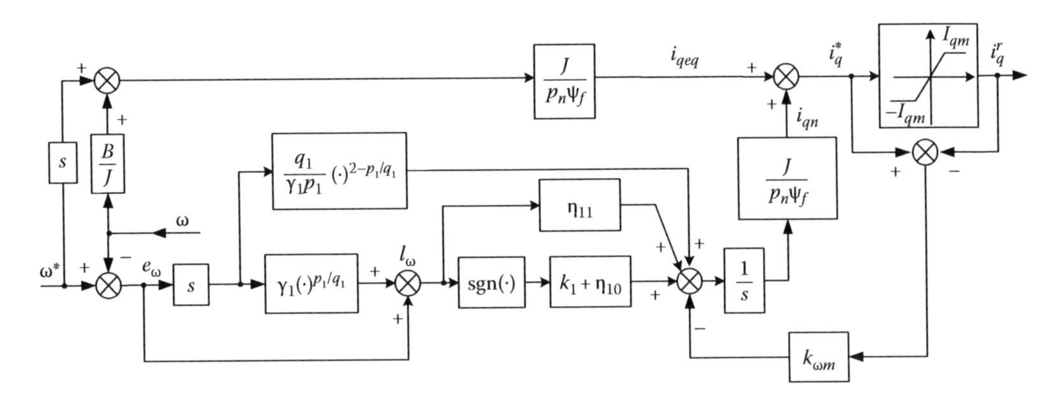

FIGURE 18.5 High-order TSM speed controller using the anti-reset windup method.

q-Axis current controller design [ZFY08]

1. An NTSM manifold is chosen as follows:

$$s_q = e_q + \gamma_2 \dot{e}_q^{p_2/q_2} \tag{18.8}$$

where
$\gamma_2 > 0$, p_2 and q_2 are odds, $1 < (p_2/q_2) < 2$
$e_q = i_q^* - i_q$ is the q-axis current error
i_q^* is the desired q-axis current

2. The control law is designed as follows:

$$u_q = u_{qeq} + u_{qn} \tag{18.9}$$

$$u_{qeq} = L\dot{i}_q^* + Lp\omega i_d + R_s i_q + p\psi_f \omega \tag{18.10}$$

$$u_{qn} = L\int_0^t \left[\frac{q_2}{\gamma_2 p_2} \dot{e}_q^{2-p_2/q_2} + k_{20}\,\mathrm{sgn}(s_q) + k_{21} s_q \right] \mathrm{d}\tau \tag{18.11}$$

where, $k_{20} > 0$, $k_{21} > 0$ are design parameters. The q-axis current controller can guarantee that the q-axis current converge to its desired reference in finite time.

d-Axis current controller design [ZFY08]

1. An NTSM manifold is chosen as follows:

$$s_d = e_d + \gamma_3 \dot{e}_d^{p_3/q_3} \tag{18.12}$$

where
$\gamma_3 > 0$, p_3 and q_3 are odds, $1 < p_3/q_3 < 2$
$e_d = i_d^* - i_d = -i_d$ is the current error
$i_d^* = 0$ is the desired current

2. The control law is designed as follows:

$$u_d = u_{deq} + u_{dn} \tag{18.13}$$

$$u_{deq} = -Lp\omega i_q + R_s i_d \tag{18.14}$$

$$u_{dn} = L\int_0^t \left[\frac{q_3}{\gamma_3 p_3} \dot{e}_d^{2-p_3/q_3} + k_3 \,\mathrm{sgn}(s_d) \right] \mathrm{d}\tau \tag{18.15}$$

where $k_3 > 0$ is the design parameter. The d-axis current controller can guarantee that the d-axis current converge to zero in finite time.

18.2.5 Simulations

The parameters of the PMSM are $P_N = 1.5\,\mathrm{kW}$, $n_N = 1000\,\mathrm{rpm}$, $I_N = 3.5$ A, $U_N = 380$ V, $p = 3$, $R_s = 2.875\ \Omega$, $L = 33\,\mathrm{mH}$, $J = 0.011\,\mathrm{kg\ m^2}$, $B = 0.002$ N ms, and $\psi_f = 0.8$ Wb. The q-axis current limit is $I_{qm} = 4$ A.

The parameters of the high-order TSM controller are designed as follows: $p_1 = 7, q_1 = 5, \gamma_1 = 0.002$, $k_1 = 910, \eta_{10} = 90, \eta_{11} = 5000, k_{\omega m} = 500; p_2 = 5, q_2 = 3, \gamma_2 = 0.01, k_{20} = 200, k_{21} = 0, \tau_0 = 0.001; p_3 = 5, q_3 = 3$, $\gamma_3 = 0.01$, and $k_3 = 0.1$. The simulation results of the two methods are shown in Figure 18.6.

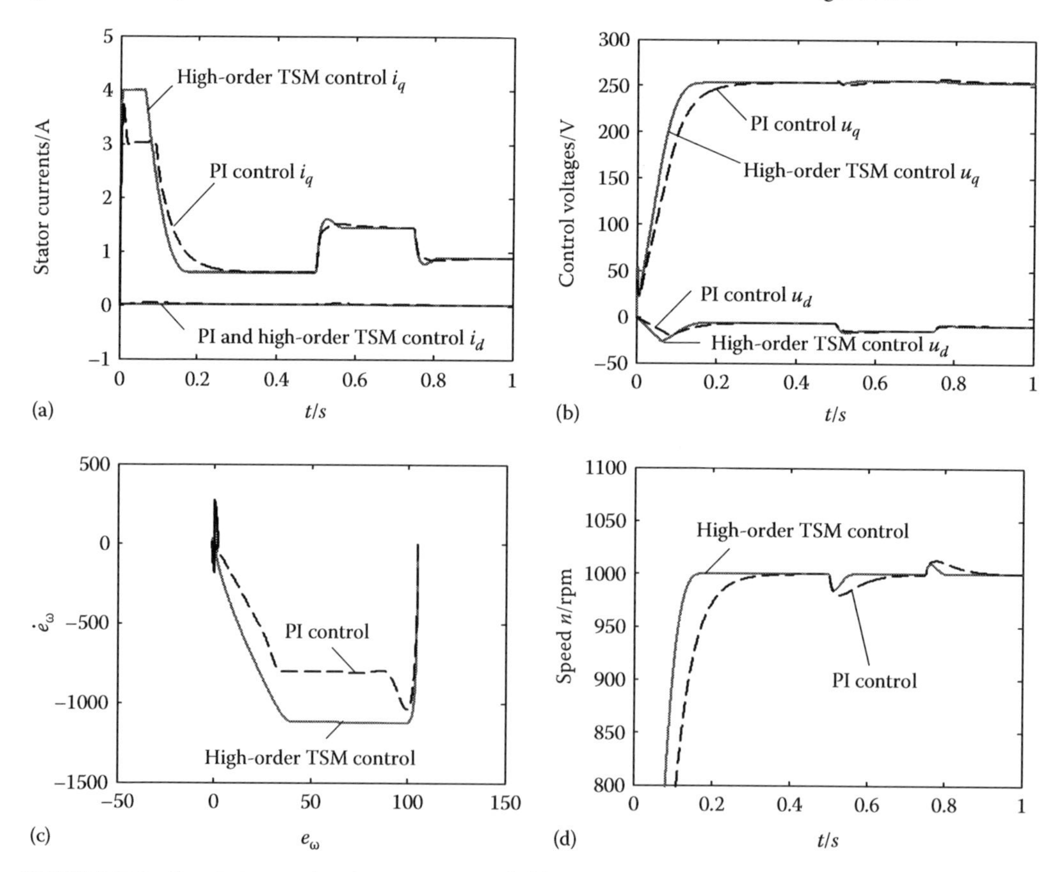

FIGURE 18.6 Simulation results when i_q is saturated: (a) stator currents, (b) control voltage, (c) phase diagram of e_ω and $\dot{e}_\omega$, and (d) motor speeds.

18.3 Observer for Rotor Position and Speed of PMSMs

In order to achieve high-performance in PMSMs, a vector-control strategy and a rotor position sensor are typically required. Usually, both the position and the speed of the motor are measured by mechanical sensors, such as encoders, resolvers, and so on. However, these high-precision sensors would increase the cost and size of the PMSM system, hence reducing the reliability and limiting the applications under certain conditions. The sensorless technique enables control of the PMSM without mechanical sensors, as well as estimation of the rotor position and the speed by the stator voltage and the current of the motor.

18.3.1 Conventional Sliding-Mode Observer

A typical structure of a sliding-mode-observer based PMSM vector-control system is shown in Figure 18.7 [UGS99,PK06,KKHK04]. In the system, a sliding-mode observer is utilized instead of the mechanical position sensor of the motor. The stator voltage and the current measurements of the motor are used to estimate the rotor position and the speed using the observer based on the PMSM model. Then, the estimated position and the speed can be used for the motor position or the speed control and the coordinate transformation between the stationary reference frame and the synchronous rotating frame. The position sensor in the motor system is replaced by a software algorithm, which can reduce the cost of hardware and improve the reliability of the motor system.

In the α-β stationary reference frame, the electrical dynamics of the PMSM can be described using the following equations [UGS99,LA06,BTZ03]:

$$\begin{cases} \dot{i}_\alpha = \dfrac{-R_s i_\alpha - e_\alpha + u_\alpha}{L} \\ \dot{i}_\beta = \dfrac{-R_s i_\beta - e_\beta + u_\beta}{L} \end{cases} \tag{18.16}$$

where

u_α and u_β are stator voltages
i_α and i_β are stator currents
L is the α-β axes inductance
R_s is the stator winding resistance
e_α and e_β are back-EMFs that satisfy the following equations,

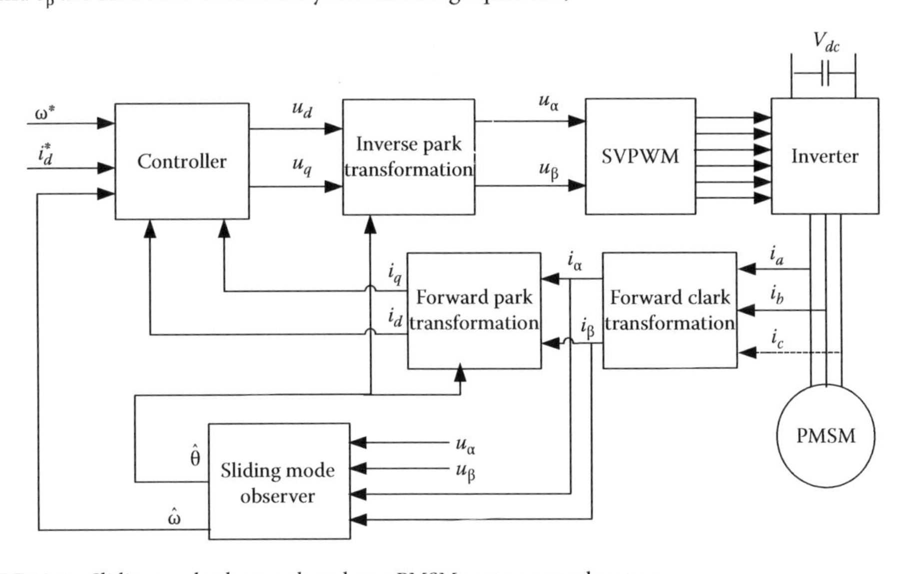

FIGURE 18.7 Sliding mode observer based on a PMSM vector-control system.

$$\begin{cases} e_\alpha = -\psi_f \omega \sin\theta \\ e_\beta = \psi_f \omega \cos\theta \end{cases} \tag{18.17}$$

where

ψ_f is the magnet flux linkage of the motor
ω is the rotor angular speed
θ is the rotor angular position

It can be seen from Equation 18.17 that the back-EMFs contain the information of both the rotor position and the speed of the motor. Therefore, the rotor position and the speed can be calculated from Equation 18.17, if the back-EMFs can be obtained accurately. From Equation 18.16, the conventional sliding-mode observer can be designed as follows [LE02]:

$$\begin{cases} \dot{\hat{i}}_\alpha = \dfrac{-\hat{R}_s \hat{i}_\alpha + u_\alpha + u_1}{\hat{L}} \\ \dot{\hat{i}}_\beta = \dfrac{-\hat{R}_s \hat{i}_\beta + u_\beta + u_2}{\hat{L}} \end{cases} \tag{18.18}$$

where

$\hat{i}_\alpha$ and $\hat{i}_\beta$ are two estimated stator current signals
$\hat{R}_s$ and $\hat{L}$ are the estimated values of R_s and L
u_1 and u_2 are two control inputs of the observer

Assume that the parameters satisfy $\hat{R}_s = R_s$ and $\hat{L} = L$. Then, the error equations of the stator current can be obtained by subtracting Equation 18.18 from Equation 18.16:

$$\begin{cases} \dot{\bar{i}}_\alpha = \dfrac{-\hat{R}_s \bar{i}_\alpha + e_\alpha + u_1}{\hat{L}} \\ \dot{\bar{i}}_\beta = \dfrac{-\hat{R}_s \bar{i}_\beta + e_\beta + u_2}{\hat{L}} \end{cases} \tag{18.19}$$

where $\bar{i}_\alpha = \hat{i}_\alpha - i_\alpha$, and $\bar{i}_\beta = \hat{i}_\beta - i_\beta$ are two stator current errors, respectively. The conventional sliding-mode observer for the rotor position and the speed can be shown in Figure 18.8.

18.3.2 Hybrid TSM Observer

Besides the conventional sliding-mode observer, several other observers for estimating the motor rotor position and speed have been proposed. Here, a hybrid TSM observer is introduced [FZYT09]. In this method, the chattering in the control signals can be eliminated using high order sliding mode technique [L03]. So, the low-pass filter is omitted, and the phase lag in the back-EMF signals can be avoided.

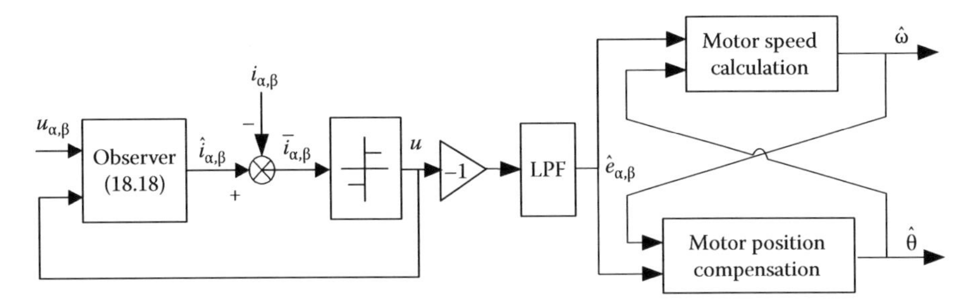

FIGURE 18.8 Conventional sliding mode observer for rotor position and speed.

The stator current error Equation 18.19 can be rewritten as follows:

$$\dot{\bar{\boldsymbol{i}}}_s = \frac{-\hat{R}_s \bar{\boldsymbol{i}}_s + \xi + \boldsymbol{u}}{\hat{L}} \tag{18.20}$$

where
$\bar{\boldsymbol{i}}_s = [\bar{i}_\alpha, \bar{i}_\beta]^T$ is the error vector of the stator current
$\xi = [e_\alpha, e_\beta]^T$ is the back EMF vector
$\boldsymbol{u} = [u_1, u_2]^T$ is the control of the observer

The observer controller can be designed as follows:

1. An NTSM manifold is chosen as follows:

$$\boldsymbol{s} = \bar{\boldsymbol{i}}_s + \boldsymbol{\gamma}\,\dot{\bar{\boldsymbol{i}}}_s^{\,p/q} \tag{18.21}$$

where
$\boldsymbol{s} = [s_\alpha, s_\beta]^T$
$\boldsymbol{\gamma} = \mathrm{diag}(\gamma_\alpha, \gamma_\beta)$, $\gamma_\alpha > 0$, $\gamma_\beta > 0$
p, q are all odd numbers, $1 < p/q < 2$
$\dot{\bar{\boldsymbol{i}}}_s^{\,p/q}$ is defined as

$$\dot{\bar{\boldsymbol{i}}}_s^{\,p/q} = \left[\dot{\bar{i}}_\alpha^{\,p/q}, \dot{\bar{i}}_\beta^{\,p/q}\right]^T \tag{18.22}$$

2. The control law is designed as

$$\boldsymbol{u} = \boldsymbol{u}_{eq} + \boldsymbol{u}_n \tag{18.23}$$

$$\boldsymbol{u}_{eq} = \hat{R}_s \bar{\boldsymbol{i}}_s \tag{18.24}$$

$$\boldsymbol{u}_n = -\int_0^t \left[\left(\frac{\hat{L}q}{p}\right)\boldsymbol{\gamma}^{-1}\dot{\bar{\boldsymbol{i}}}_s^{\,2-p/q} + (k' + \eta)\mathrm{sgn}(\boldsymbol{s}) + \mu \boldsymbol{s}\right] d\tau \tag{18.25}$$

where
$\mathrm{sgn}(\boldsymbol{s})$ is defined as $\mathrm{sgn}(\boldsymbol{s}) = [\mathrm{sgn}(s_\alpha), \mathrm{sgn}(s_\beta)]^T$
$k' > \|\dot{\xi}\|$, $\eta > 0$, $\mu > 0$ are the designed parameters

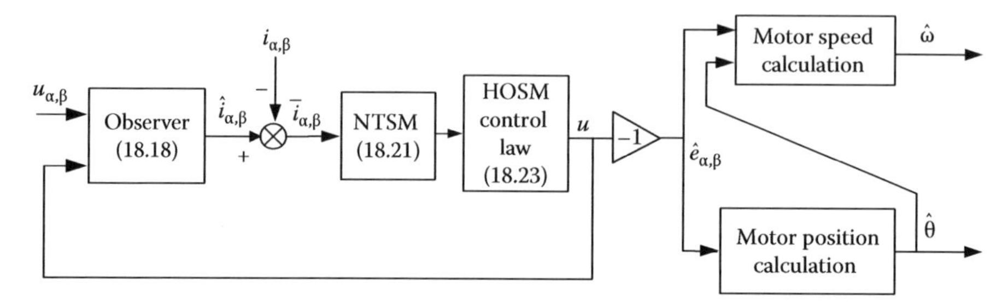

FIGURE 18.9 Hybrid TSM observer.

When the system (18.20) stays on the sliding mode, the dynamics of the system satisfy $\bar{\boldsymbol{i}}_s = \dot{\bar{\boldsymbol{i}}}_s = 0$. According to the equivalent control method [PK06], it can be obtained from Equation 18.20:

$$\hat{\xi} = [\hat{e}_\alpha, \hat{e}_\beta]^{\mathrm{T}} = -\boldsymbol{u} \tag{18.26}$$

The estimation of ω and θ can be obtained from Equation 18.17 as follows:

$$\hat{\theta} = \arctan\left(\frac{-\hat{e}_\alpha}{\hat{e}_\beta}\right) \tag{18.27}$$

$$\hat{\omega} = \frac{\sqrt{\hat{e}_\alpha^2 + \hat{e}_\beta^2}}{\psi_f}\operatorname{sgn}(\hat{e}_\beta \cos\hat{\theta} - \hat{e}_\alpha \sin\hat{\theta}) \tag{18.28}$$

The structure of the proposed observer can be described in Figure 18.9. Because of utilizing the second sliding-mode technique, it can be seen from Equations 18.23 through 18.25 that the control $\boldsymbol{u}$ is continuous and smooth, which can be used to estimate back-EMFs directly. The phase lag caused by the low-pass filter can be avoided effectively by omitting the low-pass filter in Figure 18.9.

18.4 Control of Induction Motors

An induction motor is a type of asynchronous AC motor that is designed to operate from a three-phase source of alternating voltage. The induction motor has many advantages, such as an intrinsically simple and a rugged structure, high reliability and low manufacturing cost, a wide speed range, high efficiencies, robustness, etc. Recently, due to the rapid improvements in power electronics and microprocessors, advanced control methods have made the induction motor for high-performance applications possible. More and more researchers have been attracted to induction motor control in the field of electric motor drives.

A lot of control methods for IMs have been proposed, among which the three important methods are the scalar control, the direct torque control (DTC), and the field-oriented control.

18.4.1 Scalar Control

The simplest scalar control of the induction motor is the constant Volts/Hertz (CVH) method, whose principle is to keep the stator flux as a constant via controlling both the amplitude and the frequency of the stator voltage, according to the following equation [T01]:

$$\Lambda_s \approx \frac{V_s}{\omega} = \frac{1}{2\pi}\frac{V_s}{f} \tag{18.29}$$

where

Λ_s denotes the stator flux
V_s is the stator voltage
f is the frequency of the stator voltage
ω is the angular velocity of the motor

The CVH method is generally called the variable-voltage variable-frequency (VVVF) control or simply the VF control. The main advantages of the VF control are that it is very simple, of low cost, and is suitable for applications without position-control requirements or the need for high accuracy of the speed control, such as in pumps, fans, and blowers. The main drawbacks are that field orientation is not used, the motor status is ignored, and the motor torque is not controlled.

The VF control can be used as an open-loop or a closed-loop control of IMs. An example of the VF-based closed-loop speed control of an induction motor is shown in Figure 18.10. The inverter in Figure 18.10 is a typical VVVF. There are a lot of VVVF products on the market. The control principle of IMs can be described using the following equation:

$$\omega_1 = \omega + \omega_s \tag{18.30}$$

where

ω is the actual velocity of the induction motor
ω_1 is the setpoint of the output frequency of the inverter
ω_s is the slip velocity

which can be expressed by

$$\omega_s = \omega^* - \omega \tag{18.31}$$

where ω^* is the desired motor velocity.

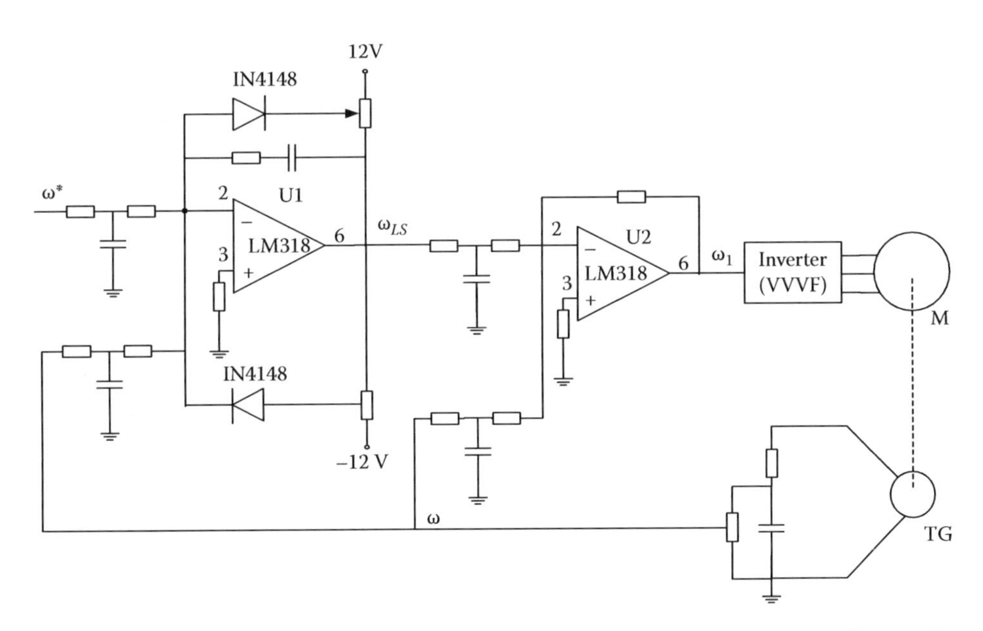

FIGURE 18.10 VF-based closed-loop speed control of an induction motor. (From Feng, Y. et al., Asynchronous motor closed-loop speed control using frequency converter (in Chinese), *Basic Automat.*, No. 2, 1994.)

The amplifier U1 in Figure 18.10 is used as a slip controller, which is a conventional PI controller and described by

$$\frac{-\omega_{sL}}{\omega^* - \omega} = 1 + \frac{1}{0.18s} \tag{18.32}$$

Another amplifier U2 in Figure 18.10 is used to perform addition, which can be described by

$$-\omega_1 = \omega_{sL} - \omega \tag{18.33}$$

18.4.2 Direct Torque Control

In recent years, the application fields of flux and torque decoupling control of induction machines have greatly increased in the areas of traction, paper and steel industry, and so on. The DTC is one of the actively researched control schemes that is based on the decoupled control of flux and torque. DTC provides a very quick and precise torque response without the complex field-orientation block and the inner-current regulation loop [HP92,T96,KK95].

The closed-loop direct torque speed-control system is shown in Figure 18.11. The torque and the stator flux calculator are used to estimate the torque and the stator flux from stator voltages and currents in the α, β coordinate. The stator flux in α, β coordinates can be expressed as follows [T01]:

$$\begin{cases} \psi_{s\alpha} = \int (u_{s\alpha} - R_s i_{s\alpha})\mathrm{d}t \\ \psi_{s\beta} = \int (u_{s\beta} - R_s i_{s\beta})\mathrm{d}t \end{cases} \tag{18.34}$$

Here, the amplitude and the phase angle are

$$\begin{cases} |\psi_s| = \sqrt{\psi_{s\alpha}^2 + \psi_{s\beta}^2} \\ \theta_s = \arctan\dfrac{|\psi_{s\beta}|}{|\psi_{s\alpha}|} \end{cases} \tag{18.35}$$

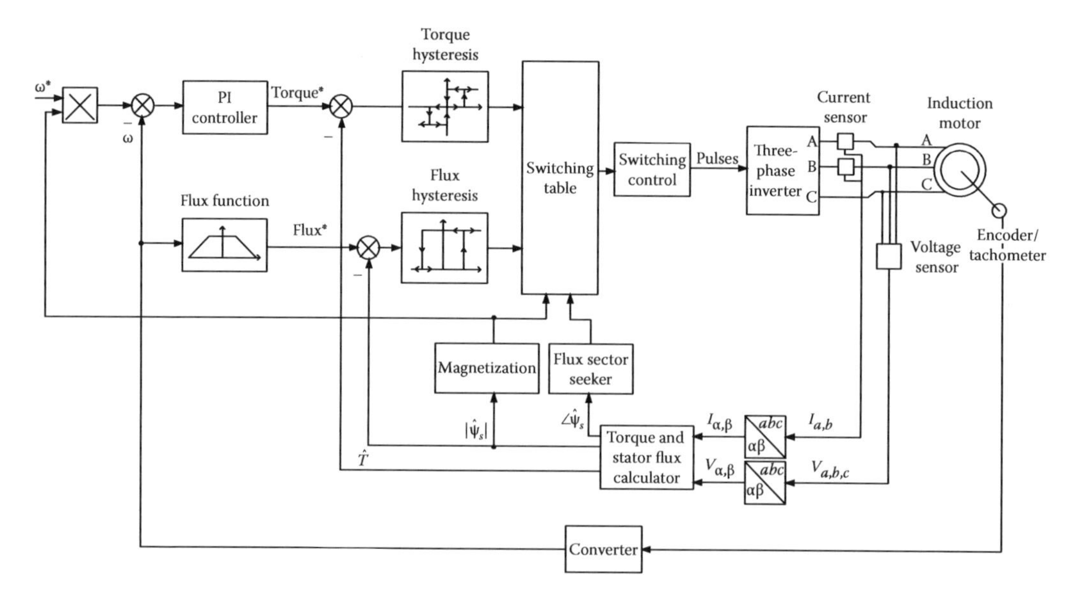

FIGURE 18.11 Closed-loop direct torque speed-control system.

where

R_s is the stator resistor

$u_{s\alpha}$, $u_{s\beta}$, and $i_{s\alpha}$, $i_{s\beta}$ are the stator voltages and the stator currents in the α,β coordinate

The motor torque can be calculated by using the following equation:

$$T_e = n_p L_m \left(i_{s\beta} i_{r\alpha} - i_{s\alpha} i_{r\beta} \right) \tag{18.36}$$

or

$$T_e = n_p \left(i_{s\beta} \psi_{s\alpha} - i_{s\alpha} \psi_{s\beta} \right) \tag{18.37}$$

where

n_p is the number of pole pairs

L_m is the magnetizing inductance

The magnetization in Figure 18.11 is used to create the stator flux. When the power is on, the switching table outputs v_1 and v_7 alternatively by a constant frequency, that is, v_1, v_7, v_1, v_7, …, till the stator flux reaches its desired value. Before the stator flux reaches its desired value, the magnetization blocks the motor speed reference. The block flux sector seeker is used to seek the sector that is the stator flux. The switching table utilizes eight possible stator voltage vectors to control the stator flux and the torque to follow the reference values within the hysteresis bands. The block switching control limits the switching frequency of six IGBTs not exceeding the maximum frequency.

18.4.3 Field-Oriented Control

In 1972 Blascke first developed the field-orientation control (FOC) method, which is a torque–flux decoupling technique applied to induction machine control [PTM99,T94,MPR02]. FOC is a well-established and widely used applied method, which can realize high-performance induction motor control. The control variables are i_{ds}, i_{qs}, the components of the stator currents, represented by a vector, in a rotating reference frame aligned to the rotor flux space vector (with a *d-q* coordinate system). Therefore, the information on the rotor flux is required.

FOC can be divided into two kinds of methods, direct field-oriented control (DFOC) and indirect field-oriented control (IFOC). In the former, the rotor flux is measured using sensors; in the latter, the rotor flux is estimated using algorithms.

A typical block diagram of the FOC-based induction motor control system is shown in Figure 18.12. For the FOC technique, the rotor speed is asymptotically decoupled from the rotor flux like the DC motor, and the speed depends linearly on the torque current [KK95,PW91]. However, the performance will be degraded due to motor parameter variations or unknown external disturbances [R96].

In Figure 18.12, the flux calculators are used to estimate the amplitude of the rotor flux for the flux closed-loop control and the position (angle) of the rotor flux for *DQ/dq* transformation. For the DFOC in Figure 18.12a, some sensors, such as the Hall-effect sensors are used to measure the air-gap flux in the *d*, *q* coordinate, λ_{dm}, λ_{qm}, which can be used to calculate the rotor flux.

For the IFOC in Figure 18.12b, the rotor flux can be calculated via the following steps:

Step 1: Stator flux vector [T01]

$$\psi_s = \int (u_s - R_s i_s) dt \tag{18.38}$$

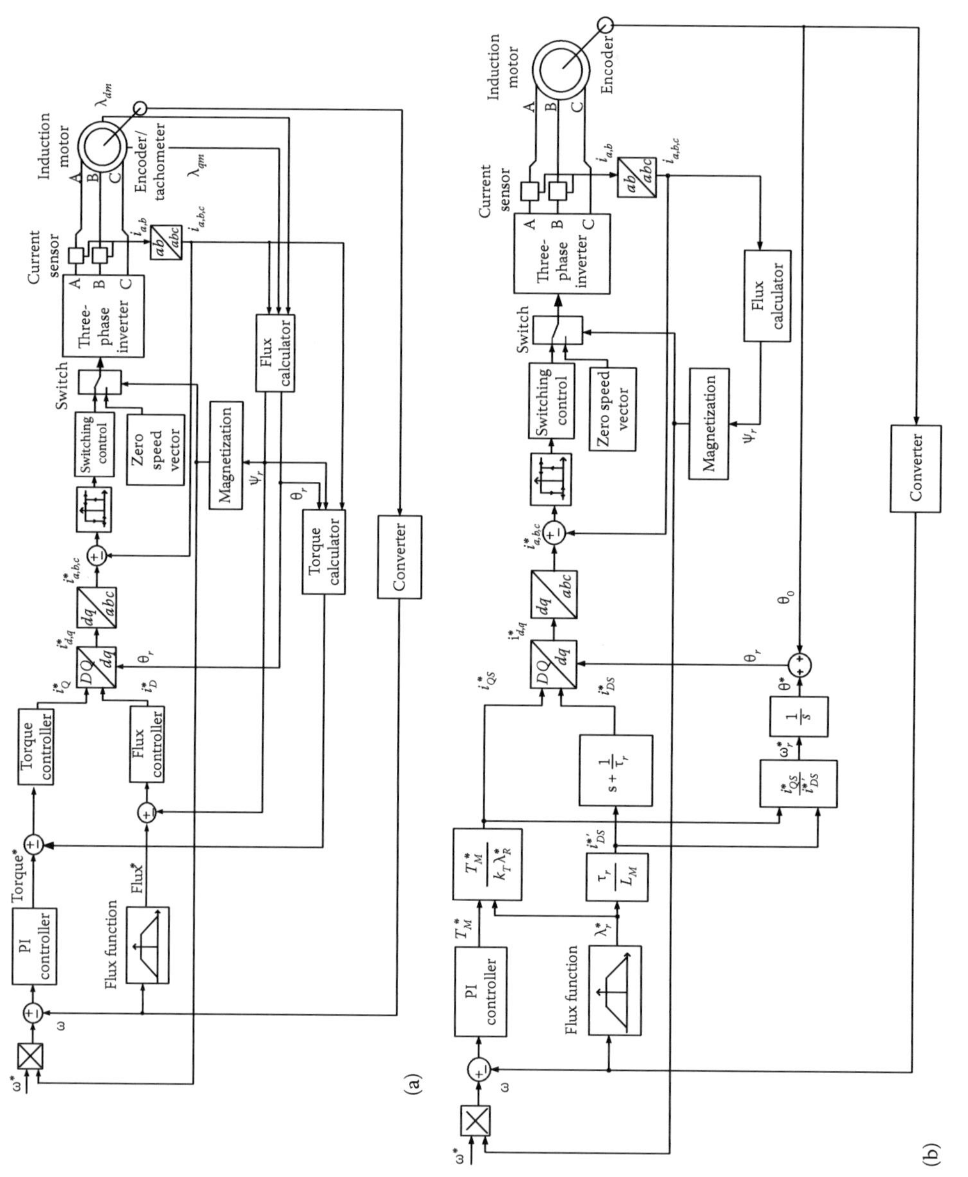

FIGURE 18.12 Closed-loop motor-speed control system with FOC: (a) DFOC and (b) IFOC.

Step 2: Air-gap flux vector

$$\psi_m = \psi_s - L_{1s} i_s \quad (18.39)$$

Step 3: Rotor flux vector

$$\psi_r = \frac{L_r}{L_m}\psi_m - L_{1s} i_s \quad (18.40)$$

where

R_s is the stator resistor
$\boldsymbol{u}_s$ and $\boldsymbol{i}_s$ are the stator voltage and the stator current, respectively
L_m is the magnetizing inductance
$L_r = L_{1r} + L_m$ is the sum of the rotor leakage inductance and the magnetizing inductance
L_{1s}, L_{1r} are the stator and the rotor leakage inductances, respectively

The main advantages of the FOC can be expressed as follows: good torque response, accurate speed control, full torque at zero speed, and performance approaching DC drive. The main drawbacks are feedback needed, costly, and modulator needed.

18.5 Conclusions

This chapter has described AC servo systems, mainly the control of PMSM (conventional PID control and high-order TSM control), an observer for rotor position and the speed of PMSMs (conventional sliding-mode observer and hybrid TSM observer), and the control of IMs (scalar control, DTC, and FOC). They are of substantial significance to both theoretical research and practical applications.

References

[BKY00] C. Baik, K.H. Kim, and M.J. Youn, Robust nonlinear speed control of PM synchronous motor using boundary layer integral sliding mode control technique, *IEEE Trans. Control Syst. Technol.*, 8(1), 47–54, 2000.

[BTZ03] S. Bolognani, L. Tubiana, and M. Zigliotto, Extended kalman filter tuning in sensorless PMSM drives, *IEEE Trans. Ind. Appl.*, 39(6), 1741–1747, 2003.

[ED00] G. Ellis and R.D. Lorenz, Resonant load control methods for industrial servo drives, in *Proceedings of the Industry Applications Conferences*, Rome, Italy, 2000, pp. 1438–1445.

[ES98] C. Edwards and S.K. Spurgeon, *Sliding Mode Control Theory and Applications*, Taylor & Francis Ltd., London, U.K., 1998.

[FWX94] Y. Feng, Y. Wang, and D. Xu, Asynchronous motor closed-loop speed control using frequency converter, (in Chinese), *Basic Automation*, no. 2, 1994.

[FYM02] Y. Feng, X.H. Yu, and Z.H. Man, Non-singular adaptive terminal sliding mode control of rigid manipulators, *Automatica*, 38(12), 2159–2167, 2002.

[FZYT09] Y. Feng, J. Zheng, X. Yu, and N.V. Truong, Hybrid terminal sliding mode observer design method for a permanent magnet synchronous motor control system, *IEEE Trans. Ind. Electron.*, 56(9), 3424–3431, 2009.

[HP92] T.G. Habetler and F. Profumo, Direct torque control of induction machines using space vector modulation, *IEEE Trans. Ind. Appl.*, 28, 1045–1052, 1992.

[JL09] H.Z. Jin and J.M. Lee, An RMRAC current regulator for permanent-magnet synchronous motor based on statistical model interpretation, *IEEE Trans. Ind. Electron.*, 56(1), 169–177, 2009.

[JSHIS03] J.H. Jang, S.K. Sul, J.I. Ha, K. Ide, and M. Sawamura, Sensorless drive of surface-mounted permanent-magnet motor by high-frequency signal injection based on magnetic saliency, *IEEE Trans. Ind. Appl.*, 39(4), 1031–1039, 2003.

[KK95] M.P. Kazmierkowski and A.B. Kasprowicz, Improved direct torque and flux vector control of PWM inverter-fed induction motor drives, *IEEE Trans. Ind. Electron.*, 42, 344–350, 1995.

[KKHK04] K.L. Kang, J.M. Kim, K.B. Hwang, and K.H. Kim, Sensorless control of PMSM in high speed range with iterative sliding mode observer, in *Proceedings of the IEEE APEC*, vol. 2, Anaheim, CA, 2004, pp. 1111–1116.

[KT07] Y.S. Kung and M.H. Tsai, FPGA-based speed control IC for PMSM drive with adaptive fuzzy control, *IEEE Trans. Power Electron.*, 22(6), 2476–2486, 2007.

[L03] A. Levant, Higher-order sliding modes, differentiation and output-feedback control, *Int. J. Control*, 76(9), 924–941, 2003.

[LA06] C. Lascu and G.D. Andreescu, Sliding-mode observer and improved integrator with DC-offset compensation for flux estimation in sensorless-controlled induction motors, *IEEE Trans. Ind. Electron.*, 53(3), 785–794, 2006.

[LE02] C.S. Li and M. Elbuluk, A robust sliding mode observer for permanent magnet synchronous motor drives, in *Proceedings of the IEEE IECON*, vol. 2, Sevilla, Spain, November 5–8, 2002, pp. 1014–1019.

[MPR02] H.A.F. Mohamed, H.W. Ping, and N.A. Rahim, Performance improvements to the vector control-based speed of induction motors using sliding mode control, *SICE 2002*, Osaka, Japan, August 5–7, 2002, pp. 258–263.

[PK06] K. Paponpen and M. Konghirun, An improved sliding mode observer for speed sensorless vector control drive of PMSM, in *Proceedings of the IEEE IPEMC*, vol. 2, Shanghai, China, August 14–16, 2006, pp. 1–5.

[PTM99] S. Peresada, A. Tonielli, and R. Morici, High-performance indirect field-oriented output-feedback control of induction motors, *Automatica*, 35, 1033–1047, 1999.

[PW91] M.-H. Park and C.-Y. Won, Time optimal control for induction motor servo system, *IEEE Trans. Power Electron.*, 6(3), 514–524, 1991.

[R96] X. Roboam, A special approach for the direct torque control of an induction motor, in 1st European Technical Scientific Report, Ansaldo-CRIS, Naples, Italy, February 1996, pp. 216–225.

[RH98] M.A. Rahman and M.A. Hoque, On-line adaptive artificial neural network based vector control of permanent magnet synchronous motors, *IEEE Trans. Energy Conversion*, 13(4), 311–318, 1998.

[S98] H.B. Shin, New antiwindup PI controller for variable-speed motor drives, *IEEE Trans. Ind. Electron.*, 45(3), 445–450, 1998.

[T01] M.A. Trzynadlowski, *Control of Induction Motors*, Academic Press, San Diego, CA, 2001.

[T94] A.M. Trzynadlowski, *The Field Orientation Principle in Control of Induction Motors*. Kluwer, Dordrecht, the Netherlands, 1994.

[T96] P. Tiitinen, The next motor control method—DTC direct torque control, in *Proceedings of the International Conference on Power Electronics, Drives and Energy System for Industrial Growth*, Delhi, India, 1996, pp. 37–43.

[UGS99] V. Utkin, J. Guldner, and J. Shi, *Sliding Mode Control in Electromechanical Systems*, 1st edn., Taylor & Francis, London, U.K., 1999.

[XVT03] Y. Xie, D.M. Vilathgamuwa, and K.J. Tseng, Observer-based robust adaptive control of PMSM with initial rotor position uncertainty, *IEEE Trans. Ind. App.*, 39(3), 645–656, 2003.

[ZAATS06] Y. Zhang, C.M. Akujuobi, W.H. Ali, C.L. Tolliver, and L.S. Shieh, Load disturbance resistance speed controller design for PMSM, *IEEE Trans. Ind. Electron.*, 53(4), 1198–1208, 2006.

[ZF08] J. Zheng and Y. Feng, High-order terminal sliding mode based mechanical resonance suppressing method in servo system, in *Proceedings of the Second International Symposium on Systems and Control in Aerospace and Astronautics*, Shenzhen, China, December 10–12, 2008, pp. 1–6.

[ZFY08] J. Zheng, Y. Feng, and X. Yu, Hybrid terminal sliding mode observer design method for permanent magnet synchronous motor control system, *IEEE 10th International Workshop on Variable Structure Systems, VSS'08*, Antalya, Turkey, June 8–10, 2008, pp. 106–111.

19

Predictive Repetitive Control with Constraints

Liuping Wang
RMIT University

Shan Chai
RMIT University

Eric Rogers
University of Southampton

19.1 Introduction

Control system applications in, for example, mechanical systems, manufacturing systems, and aerospace systems often require set-point following of a periodic trajectory. In this situation, the design of a control system that has the capability to produce zero steady-state error is paramount. It is well known from the internal model control principle that in order to follow a periodic reference signal with zero steady-state error, the generator for the reference must be included in the stable closed-loop control system [1]. Repetitive control systems follow a periodic reference signal and use the internal model control principle [2–5].

In the design of periodic repetitive control systems, the control signal is often generated by a controller that is explicitly described by a transfer function with appropriate coefficients [2,5]. If there are a number of frequencies contained in the exogenous signal, the repetitive control system will contain all periodic modes, and the number of modes is proportional to the period and inversely proportional to the sampling interval. As a result, if fast sampling is to be used, a very high-order control system may be obtained, which could lead to numerical sensitivity and other undesirable problems in a practical application.

Instead of retaining all the periodic modes in the periodic control system, an alternative is to embed fewer periodic modes at a given time, and when the frequency of the external signal changes, the coefficients of the controller change accordingly. This will effectively result in a lower order periodic control system through the use of switched linear controllers. In addition to knowing which periodic controller should be used, no bump occurring in the control signal, when switching from one controller to another, is paramount from implementation point of view.

This chapter develops a repetitive control algorithm using the framework of a discrete-time model predictive control. With a given reference signal, the frequency components of this signal are first analyzed and its reconstruction is performed using the frequency sampling filter model [6], from which the dominant frequencies are identified and error analysis is used to justify the selections. Second, the input disturbance model that contains all the dominant frequency components is formed and is embedded into an augmented state-space model. The augmented state-space model input signal is inversely filtered with the disturbance model, and consequently the current control signal is computed from past controls to ensure a bumpless transfer when the periodic reference signal changes. Third, with the augmented state-space model, the receding horizon control principle is used with an online optimization scheme [7] to generate the repetitive control law where any plant operational constraints are imposed.

This chapter is organized as follows. In Section 19.2, the frequency sampling filter model is used to construct the reference signal, which forms the basis for analysis of dominant frequency components and the quantification of errors when insignificant components are neglected. In Section 19.3, the disturbance model is formulated and embedded into a state-space model. In Section 19.4, the discrete-time model predictive control framework based on the Laguerre functions is used to develop the repetitive control law, with which constraints are imposed. Finally, in Section 19.5, simulation results from application to a gantry robot are given.

19.2 Frequency Decomposition of Reference Signal

We assume that the periodic reference signal of period T is uniformly sampled at a rate Δt, and within one period, the corresponding discrete sequence is denoted by $r(k)$, where $k = 0, 1, \ldots, M-1$. Here, $M = T/\Delta t$, and is assumed to be an odd integer for the reason explained below. From Fourier analysis [8], this discrete periodic signal can be uniquely represented by the inverse Fourier transform

$$r(k) = \frac{1}{M}\sum_{i=0}^{M-1}\left(e^{j(2\pi i/M)}\right)e^{j(2\pi ik/M)} \tag{19.1}$$

where $R(e^{j(2\pi i/M)})$ $(i = 0, 1, 2, \ldots, M-1)$ are the frequency components contained in the periodic signal. Note that the discrete frequencies are $0, 2\pi/M, 4\pi/M, \ldots, (M-1)2\pi/M$ for this periodic signal. For notational simplicity, we express the fundamental frequency as $\omega = 2\pi/M$.

The z-transform of the signal $r(k)$ is given by

$$R(z) = \sum_{k=0}^{M-1} r(k)z^{-k} \tag{19.2}$$

By substituting (19.1) into (19.2) and interchanging the summation, we obtain the z-transform representation for the periodic signal $r(k)$ as

$$R(z) = \sum_{l=-(M-1)/2}^{(M-1)/2} R(e^{jl\omega})H^{l}(z) \tag{19.3}$$

where $H^l(z)$ is termed the lth frequency sampling filter [6] and has the form

$$\begin{aligned} H^{l}(z) &= \frac{1}{M}\frac{1-z^{-M}}{1-e^{jl\omega}z^{-1}} \\ &= \frac{1}{M}\left(1+e^{jl\omega}z^{-1}+\cdots+e^{j(M-1)l\omega}z^{-(M-1)}\right) \end{aligned}$$

Here, we used the assumption that M is an odd number to include the zero frequency. The frequency sampling filters are bandlimited and centered at $l\omega$. For instance, at $z = e^{jl\omega}$, $H^l(z) = 1$. Equation 19.3 can also be written in terms of real and imaginary parts of the frequency component $R(e^{jl\omega})$ [9] as

$$R(z) = \frac{1}{M}\frac{1-z^{-M}}{1-z^{-1}}R(e^{j0}) + \sum_{l=1}^{(M-1)/2}\left[\mathrm{Re}(R(e^{jl\omega})F_R^l(z) + \mathrm{Im}(R(e^{jl\omega})F_I^l(z)\right] \tag{19.4}$$

where $F_R^l(z)$ and $F_I^l(z)$ are the lth second-order filters given by

$$F_R^l(z) = \frac{1}{M}\frac{2(1-\cos(l\omega)z^{-1})(1-z^{-M})}{1-2\cos(l\omega)z^{-1}+z^{-2}}$$

$$F_I^l(z) = \frac{1}{M}\frac{2\sin(l\omega)z^{-1}(1-z^{-M})}{1-2\cos(l\omega)z^{-1}+z^{-2}}$$

The following points are now relevant:

- The number of frequency components is M, which is determined by the period T and the sampling interval Δt. As Δt reduces, the number of frequencies increases. However, the number of dominant frequencies may not change with respect to a faster sampling rate, which is similar to the application of a frequency sampling filter in the modeling of dynamics.
- Given the z-transform representation of the reference signal, the time domain signal $r(k)$ can be constructed using the frequency components contained in the periodic signal with the input taken as a unit impulse function. As a direct result, the contribution of each frequency component can be considered in terms of the error arising when a specific frequency component is dropped. This approach is used in determining the dominant components of the periodic reference signal.
- Once the dominant frequency components are selected, they are embedded in the design of predictive repetitive control system.

19.3 Augmented Design Model

Suppose that the plant to be controlled is described by the state-space model

$$x_m(k+1) = A_m x_m(k) + B_m u(k) + \Omega_m \mu(k) \tag{19.5}$$

$$y(k) = C_m x_m(k) \tag{19.6}$$

where

$x_m(k)$ is the $n_1 \times 1$ state vector
$u(k)$ is the input signal
$y(k)$ the output signal
$\mu(k)$ represents the input disturbance

By the internal model principle, in order to follow a periodic reference signal with zero steady-state error, the generator for the reference must be included in the stable closed-loop control system [1]. From the identification of dominant frequency components in Section 19.2, we assume that the dominant frequencies are 0 and some $l\omega$ and that the input disturbance $\mu(k)$ is generated through the denominator of the disturbance model in the form

$$D(z) = \left(1 - z^{-1}\right)\Pi_l\left(1 - 2\cos(l\omega)z^{-1} + z^{-2}\right) \tag{19.7}$$

$$= 1 + d_1 z^{-1} + d_2 z^{-2} + d_3 z^{-3} + \cdots + d_\gamma z^{-\gamma} \tag{19.8}$$

In the time domain, the input disturbance $\mu(k)$ is described by the difference equation with the backward shift operator q^{-1}

$$D(q^{-1})\mu(k) = 0 \tag{19.9}$$

Define the following auxiliary variables using the disturbance model:

$$z(k) = D(q^{-1})x_m(k) \tag{19.10}$$

$$u_s(k) = D(q^{-1})u(k) \tag{19.11}$$

where these are the filtered state vector and control signals, respectively, in the inverse disturbance model.
Applying the operator $D(q^{-1})$ to both sides of the state equation in the system model (19.5) gives

$$D(q^{-1})x_m(k+1) = A_m D(q^{-1})x_m(k) + B_m D(q^{-1})u(k)$$

or

$$z(k+1) = A_m z(k) + B_m u_s(k) \tag{19.12}$$

where the relation $D(q^{-1})\mu(k) = 0$ has been used. Similarly, application of the operator $D(q^{-1})$ to both sides of the output equation in (19.6) gives

$$D(q^{-1})y(k+1) = C_m z(k+1) = C_m A_m z(k) + C_m B_m u_s(k) \tag{19.13}$$

or, on expanding the right-hand side,

$$\begin{aligned} y(k+1) = &-d_1 y(k) - d_2 y(k-1) - \cdots - d_{\gamma-1} y(k-\gamma+1) \\ &-d_\gamma y(k-\gamma) + C_m A_m z(k) + C_m B_m u_s(k) \end{aligned} \tag{19.14}$$

To obtain a state-space model that embeds the disturbance model, choose the new state vector as

$$x(k) = \left[z(k)^T \quad y(k) \quad y(k-1) \quad \cdots \quad y(k-\gamma)\right]^T$$

leading to the augmented state-space model

$$\begin{aligned} x(k+1) &= Ax(k) + Bu_s(k) \\ y(k) &= Cx(k) \end{aligned} \tag{19.15}$$

where

$$A=\begin{bmatrix} A_m & O & O & \cdots & O & O \\ C_mA_m & -d_1 & -d_2 & \cdots & -d_{\gamma-1} & -d_\gamma \\ O^T & 1 & 0 & \cdots & 0 & 0 \\ \cdots & \ddots & & & & \\ O^T & 0 & \cdots & 1 & 0 & 0 \\ O^T & 0 & \cdots & 0 & 1 & 0 \end{bmatrix}; \quad B=\begin{bmatrix} B_m \\ C_mB_m \\ 0 \\ \vdots \\ 0 \\ 0 \end{bmatrix}$$

$$C=\begin{bmatrix} O^T & 1 & 0 & \cdots & 0 & 0 \end{bmatrix}$$

and O denotes the $n_1 \times 1$ zero vector. The structure of the augmented model remains unchanged when the plant is multiple-input and multiple-output (or multivariable) except that the O vector is replaced by zero matrix with appropriate dimensions and the coefficients $-d_1, -d_2, \ldots$ become $-d_1I, -d_2I, \ldots$ matrices with appropriate dimensions, where I denotes the identity matrix with compatible dimensions.

19.4 Discrete-Time Predictive Repetitive Control

Having obtained the augmented state-space model, the next task is to optimize the filtered control signal $u_s(k)$ using the receding horizon control principle, where the auxiliary control signal $u_s(k)$ is modeled using a set of discrete-time Laguerre functions, where a summary of the main points is now given with complete details in Ref. [7].

Assuming that the current sampling instant is k_i, the control trajectory at the future time m is taken to be

$$u_s(k_i+m \mid k_i)=\sum_{i-1}^{N} l_i(m)c_i = L(m)^T\eta \tag{19.16}$$

where $\eta^T = \begin{bmatrix} c_1 & c_2 & \ldots & c_N \end{bmatrix}$ is the coefficient vector that contains the Laguerre coefficients and $L(\cdot)$ is the vector that contains the Laguerre functions $l_1(\cdot)$, $l_2(\cdot)$, …, $l_N(\cdot)$. Then the prediction of the future state variable, $x(k_i + m \mid k_i)$, becomes

$$x(k_i+m \mid k_i) = A^m x(k_i) + \sum_{i=0}^{m-1} A^{m-i-1}BL(i)^T\eta \tag{19.17}$$

and the objective for the predictive repetitive control is to find the Laguerre coefficient vector η such that the cost function

$$J=\sum_{m=1}^{N_p} x(k_i+m \mid k_i)^T Q x(k_i+m \mid k_i) + \eta^T R_L \eta \tag{19.18}$$

is minimized where $Q \geq 0$ and $R_L > 0$ are symmetric positive semi-definite and positive definite matrices, respectively, with identical elements. Substituting (19.17) into (19.18) gives the optimal solution as

$$\eta = -\left(\sum_{m=1}^{N_p} \phi(m)Q\phi(m)^T + R_L\right)^{-1}\left(\sum_{m=1}^{N_p} \phi(m)QA^m x(k_i)\right) \tag{19.19}$$

where

$$\phi^T(m) = \sum_{i=0}^{m-1} A^{m-i-1} BL(i)^T$$

Given the optimal parameter vector η, the filtered control signal using receding horizon control is realized as

$$u_s(k_i) = L(0)^T \eta \tag{19.20}$$

and the actual control signal to the plant is constructed using the relation

$$D(q^{-1})u(k) = u_s(k)$$

or, using the state-space realization,

$$U(k) = A_s U(k-1) + B_s u_s(k) = A_s U(k-1) + B_s L(0)^T \eta \tag{19.21}$$

where $U(k) = [u(k)\ u(k-1) \ldots u(k-\gamma)]^T$, and

$$A_s = \begin{bmatrix} -d_1 & -d_2 & \cdots & -d_{\gamma-1} & -d_\gamma \\ 1 & 0 & \cdots & 0 & 0 \\ \vdots & & & & \\ 0 & \cdots & 1 & 0 & 0 \\ 0 & \cdots & 0 & 1 & 0 \end{bmatrix}; \quad B_s = \begin{bmatrix} 1 \\ 0 \\ \vdots \\ 0 \\ 0 \\ 0 \end{bmatrix}$$

When the predictive repetitive controller is used for disturbance rejection, the control objective is to maintain the plant in steady-state operation with, in the steady state, zero-filtered state vector $x_s(k)$ and constant plant output. If the predictive repetitive controller is to be used to track a periodic input signal, the reference signal will enter the computation through the augmented output variables. Note that the state vector $x(k)$ contains the filtered state vector $x_s(k)$, and the output $y(k)$, $y(k-1)$, …, $y(k-\gamma)$, and hence at sampling instant k_i, the feedback errors are captured as

$$\begin{aligned} &[y(k_i) - r(k_i) y(k_i - 1) - r(k_i - 1) \ldots y(k_i - \gamma) - r(k_i - \gamma)]^T \\ &= [e(k_i) e(k_i - 1) \ldots e(k_i - \gamma)]^T \end{aligned}$$

These error signals then replace the original output elements in $x(k)$ to form state-feedback in the computation of the filtered control signal $u_s(k)$ using (19.19) and (19.20). The key strength of predictive repetitive control lies in its ability to systematically impose constraints on the plant input and output variables. The constrained control system performs the minimization of the objective function J (19.18) in real time subject to constraints. For instance, control amplitude constraints are translated into a set of linear inequalities as

$$U^{\min} \le A_s U(k-1) + B_s L(0)^T \eta \le U^{\max} \tag{19.22}$$

where $U^{\min}$ and $U^{\max}$ are of dimension γ and contain the required lower and upper limits of the control amplitude, respectively.

19.5 Application to a Gantry Robot Model

In this section, the design algorithm developed in this chapter is applied to the approximate model of one axis of a gantry robot undertaking a pick-and-place operation that mimics many industrial tasks where control is required.

19.5.1 Process Description

The gantry robot, shown in Figure 19.1, is a commercially available system found in a variety of industrial applications whose task is to place a sequence of objects onto a moving conveyor under synchronization. The sequence of operations is that the robot collects the object from a specified location, moves until it is synchronized (in terms of both position and speed) with the conveyor, places the object on the conveyor, and then returns to the same starting location to collect the next object and so on. This experimental facility has been extensively used in the benchmarking of repetitive and iterative learning control algorithms [10,11].

$$G(s) = \frac{(s+500.19)(s+4.90\times 10^5)(s+10.99\pm j169.93)(s+5.29\pm j106.86)}{s(s+69.74\pm j459.75)(s+10.69\pm j141.62)(s+12.00\pm j79.10)} \quad (19.23)$$

The gantry robot can be treated as three single-input single-output (SISO) systems (one for each axis) that can operate simultaneously to locate the end effector anywhere within a cuboid work envelope. The lowest axis, X, moves in the horizontal plane, parallel to the conveyor beneath. The Y-axis is mounted on the X-axis and moves in the horizontal plane, but perpendicular to the conveyor. The Z-axis is the shorter vertical axis mounted on the Y-axis. The X- and Y-axes consist of linear brushless dc motors, while the Z-axis is a linear ball-screw stage powered by a rotary brushless dc motor. All motors are energized by performance matched dc amplifiers. Axis position is measured by means of linear or rotary optical incremental encoders as appropriate.

To obtain a model for the plant on which to base controller design, each axis of the gantry was modeled independently by means of sinusoidal frequency response tests. From this data, it was possible to construct Bode plots for each axis and hence determine approximate transfer functions. These were then refined, by means of a least mean squares optimization technique, to minimize the difference between the frequency response of the real plant and that of the model. Here, we only consider the X-axis where a seven-order transfer function was used in design, and the Laplace transfer function is given by (19.23).

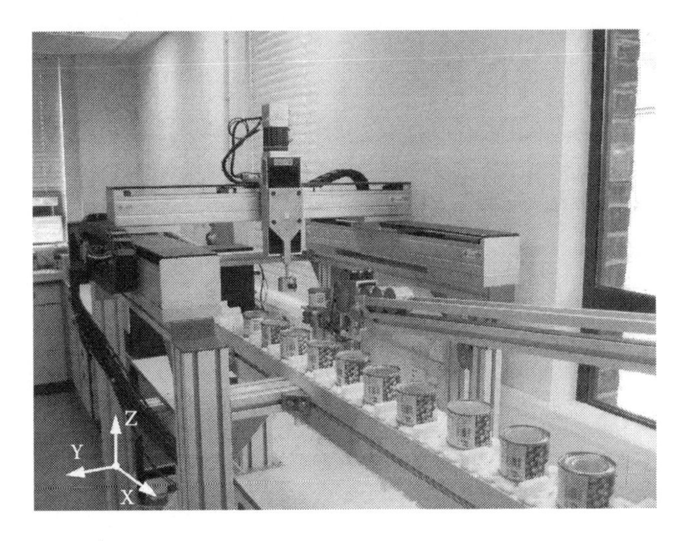

FIGURE 19.1 The gantry robot.

19.5.2 Frequency Decomposition of the Reference Signal

The reference signal with one period has 199 samples and is shown in Figure 19.2. With a sampling interval of 0.01 s, the period of this signal is $T = 2$ (s). Figure 19.3 shows the magnitude of the coefficients for all the frequency sampling filters, which clearly decay very fast as the frequency increases. Thus, it is reasonable to reconstruct the reference signal with just a few low-frequency components. Figure 19.2 compares the reconstructed signals using 0–1st, 0–3rd, and 0–6th frequency components, respectively. Table 19.1 shows the sum of squared errors between the actual reference signal and the reconstructed signal using a limited number of frequencies. It is obvious that the reference signal can be closely

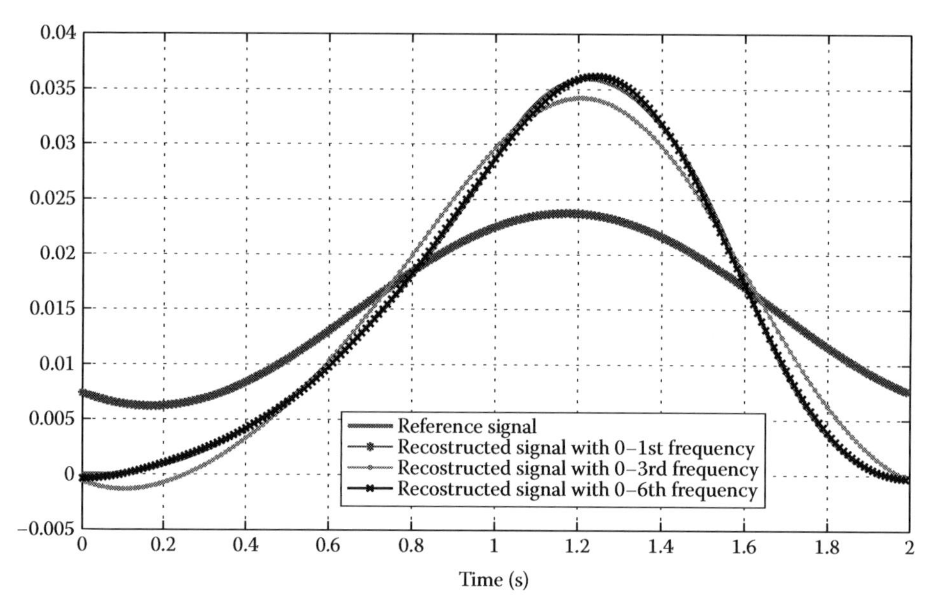

FIGURE 19.2 Reference signal and reconstructed signal with a limited number of frequencies.

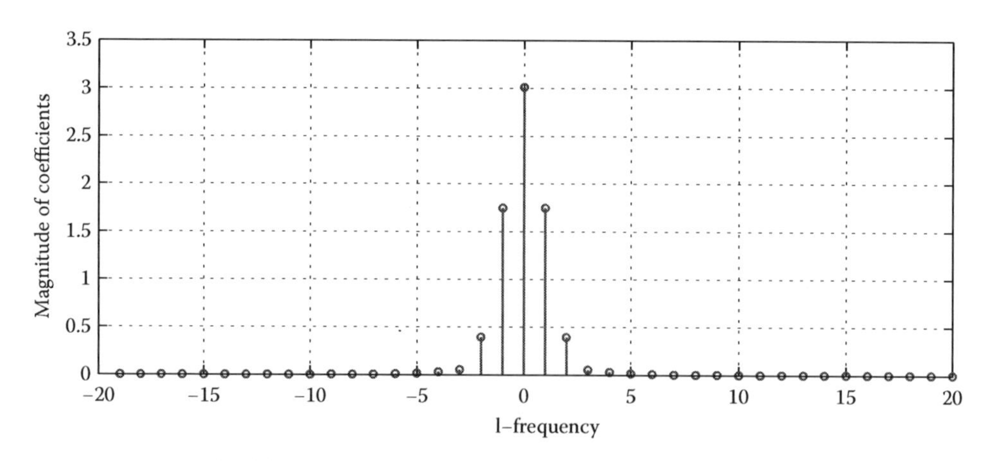

FIGURE 19.3 Magnitude of the coefficients in the frequency sampling filter.

TABLE 19.1 Sum of the Squared Errors of the Reconstructed Signal

Frequency	0	0–1	0–2	0–3	0–4	0–5	0–6
Error	0.0778	0.0325	0.0094	0.0017	0.0005	0.0001	0

approximated using 0–6th frequency components with only a small error. Thus, 0–6th frequency components are identified as the dominant frequencies for this example.

19.5.3 Closed-Loop Simulation Results

In this section, results from a simulation study on the predictive repetitive controller applied to the gantry robot model represented by the transfer function (19.23) are given. The denominator of the frequency sampling filter model with the 0–6th frequency sampling filter embedded has been used to design the controller. Figure 19.4a shows the perfect tracking performance of the design and the repetitive nature of input signal as shown in Figure 19.4b is as expected.

In practice, there are circumstances where the input of gantry robot could be saturated due to limitations of either its electrical or mechanical components. For example, if saturation occurs, the result could be a deterioration of the tracking performance of periodic reference signal when the input signal saturates. For example, given that $u^{max} = 0.6$ and $u^{min} = -0.8$, Figure 19.5a demonstrates that the system is unstable if the input signal is directly saturated as described by $u^{min} \leq u(k) \leq u^{max}$. Conversely, Figure 19.6a shows that proposed constraint algorithm, as given by (19.22), can maintain the tracking performance under the constraints by input signal. The mismatch in tracking happens when the input signal hits the limits and is saturated.

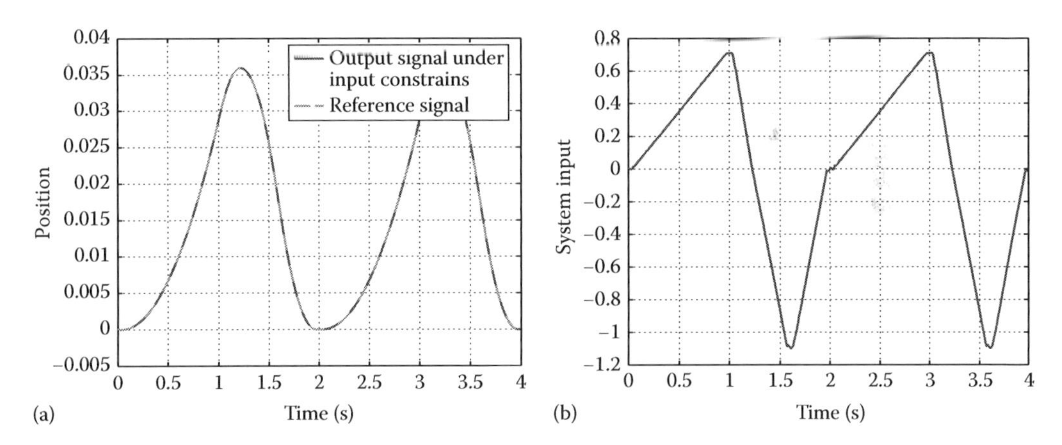

FIGURE 19.4 System simulation without input constraints: (a) system output and (b) input signal.

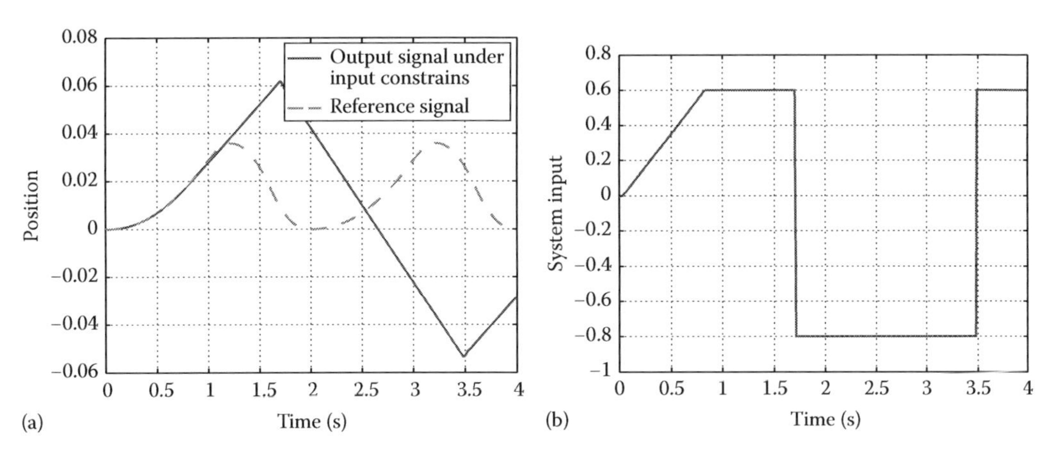

FIGURE 19.5 System simulation under input constraints: (a) system output under input constraints and (b) input signal under constraints.

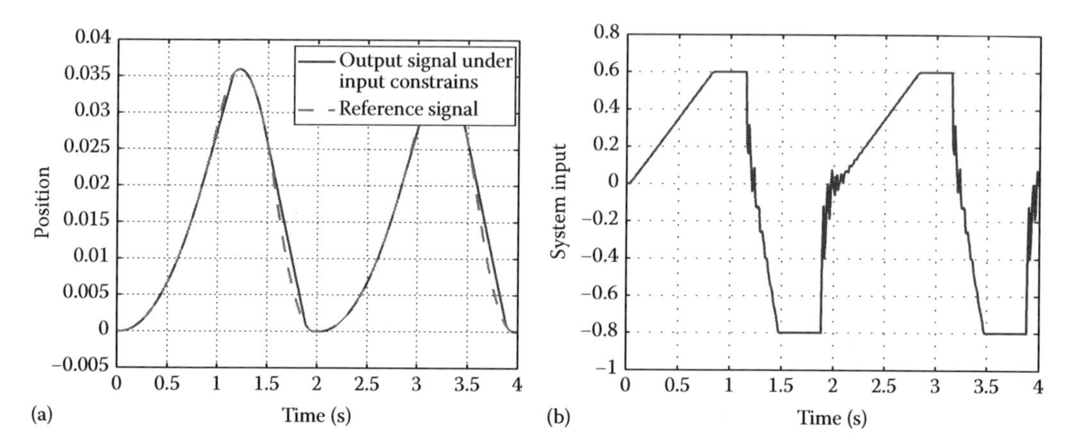

FIGURE 19.6 System simulation under input constraints with proposed method: (a) system output under input constraints and (b) input signal under constraints.

19.6 Conclusions

In this chapter, a predictive repetitive control algorithm using frequency decomposition has been developed. The augmented model is constructed by incorporating the dominant frequency components from the frequency sampling filter model. A model predictive control algorithm based on the receding horizon principle is employed to optimize the filtered input to the plant. The simulation results based on the gantry robot model show that the algorithm can achieve (1) perfect tracking of the periodic reference signal and (2) superior tracking performance when the input is subject to constraints.

References

1. B. A. Francis and W. M. Wonham, The internal model principle of control theory, *Automatica*, 5(12), 457–465, May 1976.
2. S. Hara, Y. Yamamoto, T. Omata, and M. Nakano, Repetitive control system: A new type servo system for periodic exogenous signals, *IEEE Transactions on Automatic Control*, 33(7), 659–668, July 1988.
3. T. Manayathara, T.-C. Tsao, and J. Bentsman, Rejection of unknown periodic load disturbances in continuous steel casting process using learning repetitive control approach, *IEEE Transactions on Control Systems Technology*, 4(3), 259–265, May 1996.
4. M. R. Bai and T. Y. Wu, Simulation of an internal model based active noise control system for suppressing periodic disturbances, *Journal of Vibration and Acoustics-Transaction of the ASME*, 120, 111–116, 1998.
5. D. H. Owens, L. M. Li, and S. P. Banks, Multi-periodic repetitive control system: A Lyapunov stability analysis for MIMO systems, *International Journal of Control*, 77, 504–515, 2004.
6. L. Wang and W. R. Cluett, *From Plant Data to Process Control: Ideas for Process Identification and PID Design*, 1st edn. Taylor & Francis, London, U.K., 2000.
7. L. Wang, *Model Predictive Control System Design and Implementation Using MATLAB*, 1st edn. Springer, London, U.K., 2009.
8. A. V. Oppenheim, R. W. Schafer, and J. R. Buck, *Discrete-Time Signal Processing*, 2nd edn. Prentice Hall, Englewood Cliffs, NJ, February 1999.
9. R. R. Bitmead and B. D. O. Anderson, Adaptive frequency sampling filters, in *19th IEEE Conference on Decision and Control including the Symposium on Adaptive Processes, 1980*, vol. 19, December 1980, pp. 939–944.

10. J. D. Ratcliffe, P. L. Lewin, E. Rogers, J. J. Hatonen, and D. H. Owens, Norm-optimal iterative learning control applied to gantry robots for automation applications, *IEEE Transactions on Robotics*, 22(6), 1303–1307, December 2006. [Online]. Available: http://eprints.ecs.soton.ac.uk/13227/
11. J. D. Ratcliffe, J. J. Hatonen, P. L. Lewin, E. Rogers, and D. H. Owens, Repetitive control of synchronized operations for process applications, *International Journal of Adaptive Control and Signal Processing*, 21(4), 300–325, May 2007. [Online]. Available: http://eprints.ecs.soton.ac.uk/13865/

20

Backstepping Control

Jing Zhou
International Research Institute of Stavanger

Changyun Wen
Nanyang Technological University

20.1 Introduction

In the beginning of 1990s, a new approach called "backstepping" was proposed for the design of adaptive controllers. The technique was comprehensively addressed by Krstic, Kanellakopoulos, and Kokotovic in [1]. Backstepping is a recursive Lyapunov-based scheme for the class of "strict-feedback" systems. In fact, when the controlled plant belongs to the class of systems transformable into the parametric-strict-feedback form, this approach guarantees global or regional regulation and tracking properties. An important advantage of the backstepping design method is that it provides a systematic procedure to design stabilizing controllers, following a step-by-step algorithm. With this method, the construction of feedback control laws and Lyapunov functions is systematic. Another advantage of backstepping is that it has the flexibility to avoid cancellations of useful nonlinearities and achieve stabilization and tracking. A number of results using this approach has been obtained [2–9].

This chapter reviews basic backstepping tools for state feedback control. The recursive backstepping design steps back toward the control input starting with a scalar equation and involves the systematic construction of feedback control laws and Lyapunov functions. To easily grasp the design and analysis procedures, we start with simple low-order systems. Then, the ideas are generalized to arbitrarily n-order systems. Application to the control of unstable oil wells is presented to illustrate the theory and the comparative simulation results illustrate the effectiveness of the proposed backstepping control algorithm.

20.2 Backstepping

The idea of backstepping is to design a controller recursively by considering some of the state variables as "virtual controls" and designing for them intermediate control laws. Backstepping achieves the goals of stabilization and tracking. The proof of these properties is a direct consequence of the recursive procedure because a Lyapunov function is constructed for the entire system including the parameter estimates. To give a clear idea of such development, we consider the following third-order strict-feedback system.

$$\dot{x}_1 = x_2 + x_1^2,$$
$$\dot{x}_2 = x_3 + x_2^2, \quad (20.1)$$
$$\dot{x}_3 = u,$$

where
x_1, x_2, and x_3 are system states
u is control input

The control objective is to design a state feedback control to asymptotically stabilize the origin.

Step 1. Start with the first equation of (20.1), we define $z_1 = x_1$ and derive the dynamics of the new coordinate

$$\dot{z}_1 = x_2 + x_1^2. \quad (20.2)$$

We view x_2 as a control variable and define a virtual control law for (20.2), say α_1, and let z_2 be an error variable representing the difference between the actual and virtual controls of (20.2), i.e.,

$$z_2 = x_2 - \alpha_1. \quad (20.3)$$

Thus, in terms of the new state variable, we can rewrite (20.2) as

$$\dot{z}_1 = \alpha_1 + x_1^2 + z_2. \quad (20.4)$$

In this step, our objective is to design a virtual control law α_1, which makes $z_1 \to 0$. Consider a control Lyapunov function

$$V_1 = \frac{1}{2} z_1^2. \quad (20.5)$$

The time derivative of which becomes

$$\dot{V}_1 = z_1(\alpha_1 + x_1^2) + z_1 z_2. \quad (20.6)$$

We can now select an appropriate virtual control α_1, which would make the first-order system stabilizable.

$$\alpha_1 = -c_1 z_1 - x_1^2, \quad (20.7)$$

$$\dot{\alpha}_1 = -(c_1 + 2x_1)(x_2 + x_1^2), \quad (20.8)$$

where c_1 is a positive constant. Then, the time derivative of V_1 becomes

$$\dot{V}_1 = -c_1 z_1^2 + z_1 z_2. \quad (20.9)$$

Clearly, if $z_2 = 0$, then $\dot{V}_1 = -c_1 z_1^2$ and z_1 are guaranteed to converge to zero asymptotically.

Step 2. We derive the error dynamics for $z_2 = x_2 - \alpha_1$.

$$\dot{z}_2 = \dot{x}_2 - \dot{\alpha}_1 = x_3 + x_2^2 + (c_1 + 2x_1)(x_2 + x_1^2), \tag{20.10}$$

in which x_3 is viewed as a virtual control input. Define a virtual control law a_2 and let z_3 be an error variable representing the difference between the actual and virtual controls

$$z_3 = x_3 - \alpha_2. \tag{20.11}$$

Then, (20.10) becomes

$$\dot{z}_2 = z_3 + \alpha_2 + x_2^2 + (c_1 + 2x_1)(x_2 + x_1^2). \tag{20.12}$$

The control objective is to make $z_2 \to 0$. Choose a control Lyapunov function

$$V_2 = V_1 + \frac{1}{2} z_2^2. \tag{20.13}$$

Taking time derivative gives

$$\begin{aligned} \dot{V}_2 &= -c_1 z_1 + z_1 z_2 + z_2 (z_3 + \alpha_2 + x_2^2 + (c_1 + 2x_1)(x_2 + x_1^2)) \\ &= -c_2 z_2^2 + z_2 (\alpha_2 + z_1 + x_2^2 + (c_1 + 2x_1)(x_2 + x_1^2)) + z_2 z_3. \end{aligned} \tag{20.14}$$

We can now select an appropriate virtual control α_2 to cancel some terms related to z_1, x_1, and x_2, while the term involving z_3 cannot be removed

$$\alpha_2 = -z_1 - c_2 z_2 - x_2^2 - (c_1 + 2x_1)(x_2 + x_1^2), \tag{20.15}$$

where c_2 is a positive constant. So the time derivative of V_2 becomes

$$\dot{V}_2 = -c_1 z_1^2 - c_2 z_2^2 + z_2 z_3 = -\sum_{i=1}^{2} c_i z_i^2 + z_2 z_3. \tag{20.16}$$

Clearly, if $z_3 = 0$, we have $\dot{V}_2 = -\sum_{i=1}^{2} c_i z_i^2$, and, thus, both z_1 and z_2 are guaranteed to converge to zero asymptotically.

Step 3. Proceeding to the last equation in (20.1), we derive the error dynamics for $z_3 = x_3 - a_2$.

$$\dot{z}_3 = u - \frac{\partial \alpha_2}{\partial x_1}(x_2 + x_1^2) - \frac{\partial \alpha_2}{\partial x_2}(x_3 + x_2^2). \tag{20.17}$$

In this equation, the actual control input u appears and is at our disposal. Our objective is to design the actual control input u such that z_1, z_2, z_3 converge to zero. Choose a Lyapunov function V_3, as

$$V_3 = V_2 + \frac{1}{2} z_3^2. \tag{20.18}$$

Its time derivative is given by

$$\dot{V}_3 = -\sum_{i=1}^{2} c_i z_i^2 + z_3\left(u + z_2 - \frac{\partial \alpha_2}{\partial x_1}(x_2 + x_1^2) - \frac{\partial \alpha_2}{\partial x_2}(x_3 + x_2^2)\right). \tag{20.19}$$

We are finally in the position to design control u by making $\dot{V}_3 \le 0$ as follows:

$$u = -z_2 - c_3 z_3 + \frac{\partial \alpha_2}{\partial x_1}(x_2 + x_1^2) + \frac{\partial \alpha_2}{\partial x_2}(x_3 + x_2^2), \tag{20.20}$$

where c_3 is a positive constant. Then, the derivative of Lyapunov function of V_3 is

$$\dot{V}_3 = -\sum_{i=1}^{3} c_i z_i^2. \tag{20.21}$$

Then, the Lasalle theorem in [10] guarantees the global uniform boundedness of z_1, z_2, and z_3. It follows that $z_1, z_2, z_3 \to 0$ as $t \to \infty$. Since $x_1 = z_1$, x_1 is also bounded and $\lim_{t\to\infty} x_1 = 0$. The boundedness of x_2 follows from boundedness of α_1 in (20.7) and the fact that $x_2 = z_2 + \alpha_1$. Similarly, the boundedness of x_3 then follows from boundedness of α_2 in (20.15) and the fact that $x_3 = z_3 + \alpha_2$. Combining this with (20.20), we conclude that the control $u(t)$ is also bounded.

With the above example, the idea of backstepping has been illustrated. In the following, we will consider parametric uncertainties and achieve both boundedness of the closed-loop states and asymptotic tracking.

20.3 Adaptive Backstepping

In this section, we will consider unknown parameters which appear linearly in system equations. An adaptive controller is designed by combining a parameter estimator, which provides estimates of unknown parameters, with a control law. The parameters of the controller are adjusted during the operation of the plant. In the presence of such parametric uncertainties, the adaptive controller is able to ensure the boundedness of the closed-loop states and asymptotic tracking.

To illustrate the idea of adaptive backstepping, let us first consider the following second-order system:

$$\begin{aligned} \dot{x}_1 &= x_2 + \phi_1^{\mathrm{T}}(x_1)\theta, \\ \dot{x}_2 &= u + \phi_2^{\mathrm{T}}(x_1, x_2)\theta, \end{aligned} \tag{20.22}$$

where

$\theta \in R^r$ is an unknown vector constant

$\phi_1 \in R^r$ and $\phi_2 \in R^r$ are known nonlinear functions

Our problem is to globally stabilize the system and also to achieve the asymptotic tracking of x_r by x_1. The reference signal x_r and its first second-order derivative are piecewise continuous and bounded.

The design procedure is elaborated in the following. Introduce the change of coordinates

$$z_1 = x_1 - x_r, \tag{20.23}$$

$$z_2 = x_2 - \alpha_1 - \dot{x}_r, \tag{20.24}$$

where α_1 is called virtual control and will be determined in later discussion.

Step 1. We start with the first equation of (20.22) by considering x_2 as control variable. The derivative of tracking error z_1 is given as

$$\dot{z}_1 = z_2 + \alpha_1 + \phi_1^T\theta. \tag{20.25}$$

Since θ is unknown, this task is fulfilled with an adaptive controller consisting of a control law and an update law to obtain an estimate of θ. Design the first stabilizing function α_1 and parameter updating law $\dot{\hat{\theta}}$ as

$$\alpha_1 = -c_1 z_1 - \phi_1^T\hat{\theta}_1, \tag{20.26}$$

$$\dot{\hat{\theta}} = \Gamma\phi_1 z_1, \tag{20.27}$$

where
- $\hat{\theta}_1$ is an estimate of θ
- c_1 is a positive constant
- Γ is a positive definite matrix

Our task in this step is to stabilize (20.25) with respect to the Lyapunov function

$$V_1 = \frac{1}{2}z_1^2 + \frac{1}{2}\tilde{\theta}_1^T\Gamma^{-1}\tilde{\theta}_1, \tag{20.28}$$

where $\tilde{\theta}_1 = \theta - \hat{\theta}_1$. Then the derivative of V_1 is given by

$$\begin{aligned}\dot{V}_1 &= z_1\dot{z}_1 - \tilde{\theta}_1^T\Gamma^{-1}\dot{\hat{\theta}}_1 \\ &= -c_1 z_1^2 + z_1 z_2 - \tilde{\theta}_1^T\Gamma^{-1}\left(\dot{\hat{\theta}}_1 - \Gamma\phi_1 z_1\right) \\ &= -c_1 z_1^2 + z_1 z_2.\end{aligned} \tag{20.29}$$

Step 2. The derivative of z_2 with (20.26) and (20.27) is now expressed as

$$\begin{aligned}\dot{z}_2 &= \dot{x}_2 - \dot{\alpha}_1 - \ddot{x}_r \\ &= u - \frac{\partial\alpha_1}{\partial x_1}x_2 + \left(\phi_2 - \frac{\partial\alpha_1}{\partial x_1}\phi_1\right)^T\theta - \frac{\partial\alpha_1}{\partial\hat{\theta}_1}\Gamma\phi_1 z_1 - \frac{\partial\alpha_1}{\partial x_r}\dot{x}_r - \ddot{x}_r.\end{aligned} \tag{20.30}$$

In this equation, the actual control input u appears and is at our disposal. At this point, we need to select a control Lyapunov function and design u to make its derivative nonpositive. We use a Lyapunov function

$$V_2 = V_1 + \frac{1}{2}z_2^2, \tag{20.31}$$

whose derivative is

$$\dot{V}_2 = -c_1 z_1^2 + z_2\left[u + z_1 - \frac{\partial\alpha_1}{\partial x_1}x_2 + \left(\phi_2 - \frac{\partial\alpha_1}{\partial x_1}\phi_1\right)^T\theta - \frac{\partial\alpha_1}{\partial\hat{\theta}_1}\Gamma\phi_1 z_1 - \frac{\partial\alpha_1}{\partial x_r}\dot{x}_r - \ddot{x}_r\right]. \tag{20.32}$$

The control u is able to cancel the rest six terms in (20.32) to ensure $\dot{V}_2 \le 0$. To deal with the term containing the unknown parameter θ, we try to use the estimate $\hat{\theta}_1$ designed in the first step

$$u = -z_1 - c_2 z_2 + \frac{\partial \alpha_1}{\partial x_1} x_2 - \left(\phi_2 - \frac{\partial \alpha_1}{\partial x_1} \phi_1 \right)^{\mathrm{T}} \hat{\theta}_1 + \frac{\partial \alpha_1}{\partial \hat{\theta}_1} \Gamma \phi_1 z_1 + \frac{\partial \alpha_1}{\partial x_{\mathrm{r}}} \dot{x}_{\mathrm{r}} + \ddot{x}_{\mathrm{r}}, \tag{20.33}$$

where c_2 is a positive constant. The resulting derivatives of V_2 is given as

$$\dot{V}_2 = -c_1 z_1^2 - c_2 z_2^2 + \left(\phi_2 - \frac{\partial \alpha_1}{\partial x_1} \phi_1 \right)^{\mathrm{T}} (\theta - \hat{\theta}_1). \tag{20.34}$$

It can be observed that term $\left(\phi_2 - (\partial \alpha_1 / \partial x_1) \phi_1\right)^{\mathrm{T}} (\theta - \hat{\theta}_1)$ cannot be canceled. To eliminate this term, we need to treat θ in the equation (20.30) as a new parameter vector and assign to it a new estimate $\hat{\theta}_2$ by selecting u as

$$u = -z_1 - c_2 z_2 + \frac{\partial \alpha_1}{\partial x_1} x_2 - \left(\phi_2 - \frac{\partial \alpha_1}{\partial x_1} \phi_1 \right)^{\mathrm{T}} \hat{\theta}_2 + \frac{\partial \alpha_1}{\partial \hat{\theta}_1} \Gamma \phi_1 z_1 + \frac{\partial \alpha_1}{\partial x_{\mathrm{r}}} \dot{x}_{\mathrm{r}} + \ddot{x}_{\mathrm{r}}. \tag{20.35}$$

With this choice, (20.30) becomes

$$\dot{z}_2 = -z_1 - c_2 z_2 + \left(\phi_2 - \frac{\partial \alpha_1}{\partial x_1} \phi_1 \right)^{\mathrm{T}} (\theta - \hat{\theta}_2). \tag{20.36}$$

Our task in this step is to stabilize the (z_1, z_2) system. The presence of the new parameter estimate $\hat{\theta}_2$ suggests the following form of the Lyapunov function:

$$V_2 = V_1 + \frac{1}{2} z_2^2 + \frac{1}{2} \tilde{\theta}_2^{\mathrm{T}} \Gamma^{-1} \tilde{\theta}_2, \tag{20.37}$$

where $\tilde{\theta}_2 = \theta - \hat{\theta}_2$. Then the derivative of Lapunov function of V_2 is

$$\begin{aligned} \dot{V}_2 &= \dot{V}_1 + z_2 \dot{z}_2 - \tilde{\theta}_2^{\mathrm{T}} \Gamma^{-1} \dot{\hat{\theta}}_2 \\ &= -c_1 z_1^2 + z_2 \left(-c_2 z_2 + \left(\phi_2 - \tilde{\theta}_2^{\mathrm{T}} \frac{\partial \alpha_1}{\partial x_1} \phi_1 \right) \right) - \tilde{\theta}_2^{\mathrm{T}} \Gamma^{-1} \dot{\hat{\theta}}_2 \\ &= -c_1 z_1^2 - c_2 z_2^2 - \tilde{\theta}_2^{\mathrm{T}} \Gamma^{-1} \left(\dot{\hat{\theta}}_2 - \Gamma \left(\phi_2 - \frac{\partial \alpha_1}{\partial x_1} \phi_1 \right) z_2 \right). \end{aligned} \tag{20.38}$$

We choose the update law

$$\dot{\hat{\theta}}_2 = \Gamma \left(\phi_2 - \frac{\partial \alpha_1}{\partial x_1} \phi_1 \right) z_2. \tag{20.39}$$

Then, the derivative of V_2 gives

$$\dot{V}_2 = -c_1 z_1^2 - c_2 z_2^2. \tag{20.40}$$

By using the Lasalle's theorem, this Lyapunov function (20.40) guarantees the global uniform boundedness of $z_1, z_2, \hat{\theta}_1, \hat{\theta}_2$, and z_1, $z_2 \to 0$ as $t \to \infty$. It follows that asymptotic tracking is achieved, such that $\lim_{t\to\infty}(x_1 - x_r) = 0$. Since z_1 and x_r are bounded, x_1 is also bounded from $x_1 = z_1 = x_r$. The boundedness of x_2 follows from boundedness of $\dot{x}_r$ and a_1 in (20.26) and the fact that $x_2 = z_2 + \alpha_1 + \dot{x}_r$. Combining this with (20.35), we conclude that the control $u(t)$ is also bounded.

In conclusion, the above adaptive backstepping employs the overparametrization estimation, i.e., two estimates for the same parameter vector θ in this case. This means that the dynamic order of the controller is not of minimal order. In the next section, a new backstepping design is presented to avoid such a case from happening, which employs the minimal number of parameter estimates.

20.4 Adaptive Backstepping with Tuning Functions

To give a clear idea of tuning function design, we consider the same system in (20.22) with the same control objective, namely globally stabilization and also asymptotic tracking of x_r by x_1, and use the same change of coordinates

$$z_1 = x_1 - x_r, \tag{20.41}$$

$$z_2 = x_2 - \alpha_1 - \dot{x}_r, \tag{20.42}$$

where a_2 is virtual control. The design procedure is elaborated as follows.

Step 1. We start with the first equation of (20.22) by considering x_2 as control variable. The derivative of tracking error z_1 is given as

$$\dot{z}_1 = z_2 + \alpha_1 + \phi_1^T\theta. \tag{20.43}$$

Our task in this step is to stabilize (20.43). Choose the control Lyapunov function

$$V_1 = \frac{1}{2}z_1^2 + \frac{1}{2}\tilde{\theta}^T\Gamma^{-1}\tilde{\theta}, \tag{20.44}$$

where Γ is a positive definite matrix, $\tilde{\theta} = \theta - \hat{\theta}$. Then, the derivative of V_1 is

$$\dot{V}_1 = z_1\left(z_2 + \alpha_1 + \phi_1^T\hat{\theta}\right) - \tilde{\theta}^T\left(\Gamma^{-1}\dot{\hat{\theta}} - \phi_1 z_1\right). \tag{20.45}$$

We may eliminate $\tilde{\theta}$ by choosing $\dot{\hat{\theta}} = \Gamma\phi_1 z_1$. If x_2 is the actual control and let $z_2 = 0$, we choose a_1 to make $\dot{V}_1 \leq 0$.

$$\alpha_1 = -c_1 z_1 - \phi_1^T\hat{\theta}, \tag{20.46}$$

where

c_1 is a positive constant
$\hat{\theta}$ is an estimate of θ

To overcome the overparametrization problem caused by the appearance of θ, as shown in the previous subsection, we define a function τ_1, named tuning function, as follows:

$$\tau_1 = \phi_1 z_1. \tag{20.47}$$

The resulting derivative of V_1 is

$$\dot{V}_1 = -c_1 z_1^2 + z_1 z_2 - \tilde{\theta}^{\mathrm{T}}\left(\Gamma^{-1}\dot{\hat{\theta}} - \tau_1\right). \tag{20.48}$$

Step 2. We derive the second tracking error for z_2

$$\dot{z}_2 = u - \frac{\partial \alpha_1}{\partial x_1} x_2 + \left(\phi_2 - \frac{\partial \alpha_1}{\partial x_1}\phi_1\right)^{\mathrm{T}} \theta - \frac{\partial \alpha_1}{\partial \hat{\theta}}\dot{\hat{\theta}} - \frac{\partial \alpha_1}{\partial x_r}\dot{x}_r - \ddot{x}_r. \tag{20.49}$$

In this equation, the actual control input u appears and is at our disposal. The control Lyapunov function is selected as

$$V_2 = V_1 + \frac{1}{2} z_2^2 = \frac{1}{2} z_1^2 + \frac{1}{2} z_2^2 + \frac{1}{2}\tilde{\theta}^{\mathrm{T}}\Gamma^{-1}\tilde{\theta}. \tag{20.50}$$

Our task is to make $\dot{V}_2 \leq 0$.

$$\begin{aligned} \dot{V}_2 = {} & -c_1 z_1^2 + z_2\left(u + z_1 - \frac{\partial \alpha_1}{\partial x_1} x_2 + \hat{\theta}^{\mathrm{T}}\left(\phi_2 - \frac{\partial \alpha_1}{\partial x_1}\phi_1\right) - \frac{\partial \alpha_1}{\partial \hat{\theta}}\dot{\hat{\theta}} - \frac{\partial \alpha_1}{\partial x_r}\dot{x}_r - \ddot{x}_r\right) \\ & + \tilde{\theta}^{\mathrm{T}}\left(\tau_1 + \left(\phi_2 - \frac{\partial \alpha_1}{\partial x_1}\phi_1\right) z_2 - \Gamma^{-1}\dot{\hat{\theta}}\right). \end{aligned} \tag{20.51}$$

Finally, we can eliminate the $\tilde{\theta}$ term from (20.51) by designing the update law as

$$\dot{\hat{\theta}} = \Gamma \tau_2, \tag{20.52}$$

where τ_2 is called the second tuning function and is selected as

$$\tau_2 = \tau_1 + \left(\phi_2 - \frac{\partial \alpha_1}{\partial x_1}\phi_1\right) z_2. \tag{20.53}$$

Then,

$$\dot{V}_2 = -c_1 z_1^2 + z_2\left(u + z_1 - \frac{\partial \alpha_1}{\partial x_1} x_2 + \hat{\theta}^{\mathrm{T}}\left(\phi_2 - \frac{\partial \alpha_1}{\partial x_1}\phi_1\right) - \frac{\partial \alpha_1}{\partial \hat{\theta}}\Gamma\tau_2 - \frac{\partial \alpha_1}{\partial x_r}\dot{x}_r - \ddot{x}_r\right). \tag{20.54}$$

To stabilize the system (20.49), the actual control input is selected to remove the residual term and make $\dot{V}_2 \leq 0$

$$u = -z_1 - c_2 z_2 + \frac{\partial \alpha_1}{\partial x_1} x_2 - \hat{\theta}^{\mathrm{T}}\left(\phi_2 - \frac{\partial \alpha_1}{\partial x_1}\phi_1\right) + \frac{\partial \alpha_1}{\partial \hat{\theta}}\Gamma\tau_2 + \frac{\partial \alpha_1}{\partial x_r}\dot{x}_r + \ddot{x}_r, \tag{20.55}$$

where c_2 is a positive constant. The resulting derivative of V_2 is

$$\dot{V}_2 = -c_1 z_1^2 - c_2 z_2^2. \tag{20.56}$$

This Lyapunov function provides the proof of uniform stability and the proof of asymptotic tracking $x_1(t) - x_r(t) \to 0$.

The controller designed in this section also achieves the goals of stabilization and tracking. By using tuning functions, only one update law is used to estimate unknown parameter θ. This avoids the overparametrization problem and reduces the dynamic order of the controller to its minimum.

20.5 State Feedback Control

The controller designed in this section achieves the goals of stabilization and tracking. The proof of these properties is a direct consequence of the recursive procedure because a Lyapunov function is constructed for the entire system including the parameter estimates. The overparametrization problem is overcomed by using tuning functions. The number of parameter estimates are equal to the number of unknown parameters.

The adaptive backstepping design with tuning functions is now generalized to a class of nonlinear system as in the following parametric strict-feedback form

$$\begin{aligned}
\dot{x}_1 &= x_2 + \phi_1^{\mathrm{T}}(x_1)\theta + \psi_1(x_1) \\
\dot{x}_2 &= x_3 + \phi_2^{\mathrm{T}}(x_1, x_2)\theta + \psi_2(x_1, x_2) \\
&\vdots \\
\dot{x}_{n-1} &= x_n + \phi_{n-1}^{\mathrm{T}}(x_1, \ldots, x_{n-1})\theta + \psi_n(x_1, \ldots, x_{n-1}) \\
\dot{x}_n &= bu + \phi_n^{\mathrm{T}}(x)\theta + \psi_n(x),
\end{aligned} \tag{20.57}$$

where $x = [x_1, \ldots, x_n]^{\mathrm{T}} \in R^n$, the vector $\theta \in R^r$ is constant and unknown, $\phi_i \in R^r$, $\psi_1 \in R$, $i = 1, \ldots, n$ are known nonlinear functions, and the high-frequency gain b is an unknown constant. The control objective is to force the output x_1 to asymptotically track the reference signal x_r with the following assumptions.

Assumption 20.1 The parameters θ and b are unknown and the sign of b is known.

Assumption 20.2 The reference signal x_r and its n-order derivatives are piecewise continuous and bounded.

For system (20.57), the number of design steps required is equal to n. At each step, an error variable z_i, a stabilizing function a_i, and a tuning function τ_i are generated. Finally, the control u and parameter estimate $\hat{\theta}$ are developed. The scheme is now concisely summarized in Table 20.1, where c_i, $i = 1, \ldots, n$ and γ are positive constants, Γ is a positive definite matrix, $\hat{\theta}$, $\hat{p}$ are estimates of θ, $1/b$.

TABLE 20.1 Adaptive Backstepping Control

Change of Coordinates:
$z_1 = x_1 - x_r$ (20.58)
$z_i = x_i - \alpha_{i-1} - x_r^{(i-1)}, \quad i = 2,3,\ldots,n$ (20.59)
Adaptive Control Laws:
$u = \hat{p}(\alpha_n + x_r^{(n)})$ (20.60)
$\alpha_1 = -c_1 z_1 - \phi_1^T\hat{\theta} - \psi_1$ (20.61)
$\alpha_2 = -z_1 - c_2 z_2 - \psi_2 + \frac{\partial\alpha_1}{\partial x_1}(x_2 + \psi_1) - \hat{\theta}^T\left(\phi_2 - \frac{\partial\alpha_1}{\partial x_1}\phi_1\right) + \frac{\partial\alpha_1}{\partial\hat{\theta}}\Gamma\tau_2 + \frac{\partial\alpha_1}{\partial x_r}\dot{x}_r$ (20.62)
$\alpha_i = -c_i z_i - z_{i-1} - \psi_i + \sum_{j=1}^{i-1}\frac{\partial\alpha_{i-1}}{\partial x_j}(x_{j+1} + \psi_j) - \hat{\theta}^T\left(\phi_i - \sum_{j=1}^{i-1}\frac{\partial\alpha_{i-1}}{\partial x_j}\phi_j\right)$
$+\frac{\partial\alpha_{i-1}}{\partial\hat{\theta}}\Gamma\tau_i + \left(\sum_{j=2}^{i-1} z_j\frac{\partial\alpha_{j-1}}{\partial\hat{\theta}}\right)\Gamma\left(\phi_i - \sum_{j=1}^{i-1}\frac{\partial\alpha_{i-1}}{\partial x_j}\phi_j\right) + \sum_{j=2}^{i-1}\frac{\partial\alpha_{i-1}}{\partial x_r^{(j-1)}}x_r^{(j)}$ (20.63)
Tuning Functions:
$\tau_1 = \phi_1 z_1$ (20.64)
$\tau_2 = \tau_1 + \left(\phi_2 - \frac{\partial\alpha_1}{\partial x_1}\phi_1\right)z_2$ (20.65)
$\tau_i = \tau_{i-1} + \left(\phi_i - \sum_{j=1}^{i-1}\frac{\partial\alpha_{i-1}}{\partial x_j}\phi_j\right)z_i$ (20.66)
Parameter Update Laws:
$\dot{\hat{\theta}} = \Gamma\tau_n$ (20.67)
$\dot{\hat{p}} = -\gamma\,\text{sign}(b)\,(\alpha_n + x_r^{(n)})z_n$ (20.68)

We choose the Lyapunov function

$$V_n = \sum_{i=1}^{n}\frac{1}{2}z_i^2 + \frac{1}{2}\tilde{\theta}^T\Gamma^{-1}\tilde{\theta} + \frac{|b|}{2\gamma}\tilde{p}^2. \tag{20.69}$$

Then, its derivative is given by

$$\dot{V}_n = -\sum_{i=1}^{n}c_i z_i^2 + \left(\tilde{\theta}^T + \sum_{j=2}^{n}z_j\frac{\partial\alpha_{j-1}}{\partial\hat{\theta}}\Gamma\right)\left(\tau_n - \Gamma^{-1}\dot{\hat{\theta}}\right) - \frac{|b|}{\gamma}\tilde{p}\left(\dot{\hat{p}} + \gamma\text{sign}(b)\left(\alpha_n + x_r^{(n)}\right)z_n\right)$$

$$= -\sum_{i=1}^{n}c_i z_i^2 \le 0. \tag{20.70}$$

From the Lasalle's theorem, this Lyapunov function provides the proof of uniform stability, such that $z_1, z_2, \ldots, z_n$, $\hat{\theta}$, $\hat{p}$ are bounded and z_i 0, $i = 1,\ldots,n$. This further implies that $\lim_{t\to\infty}(x_1 - x_r) = 0$.

Since $x_1 = z_1 + x_r$, x_1 is also bounded from the boundedness of z_1 and x_r. The boundedness of x_2 follows from boundedness of $\dot{x}_r$ and α_1 in (20.61) and the fact that $x_2 = z_2 + \alpha_1 + \dot{x}_r$. Similarly, the boundedness of $x_i (i = 3, \ldots, n)$ can be ensured from the boundedness of $x_r^{(i-1)}$ and α_i in (20.63) and the fact that $x_i = z_i + \alpha_{i-1} + x_r^{(i-1)}$. Combining this with (20.60), we conclude that the control $u(t)$ is also bounded. Since V_n is nonincreasing, we have

$$\| z_1 \|_2^2 = \int_0^\infty | z_1(\tau) |^2 \, d\tau \le \frac{1}{c_1} (V_n(0) - V_n(\infty)) \le \frac{1}{c_1} V_n(0). \tag{20.71}$$

Therefore, boundedness of all signals and asymptotic tracking are ensured as formally stated in the following theorem.

Theorem 20.1 Consider the closed-loop adaptive system (20.57) under Assumptions 20.1 and 20.2, the adaptive controller (20.60), virtual control laws (20.61), (20.62), and (20.63), and parameter updating laws (20.67) and (20.68), the following statements hold:

- The resulting closed-loop system is globally stable.
- The asymptotic tracking is achieved, i.e., $\lim_{t \to \infty} [x_1(t) - x_r(t)] = 0$.
- The transient displacement tracking error performance is given by

$$\| x_1(t) - x_r(t) \|_2 \le \frac{1}{\sqrt{c_1}} \sqrt{V_n(0)}. \tag{20.72}$$

Remark 20.1 The transient tracking error performance for $\|x(t) - x_r(t)\|_2$ depends on the initial states $x_i(0)$, the initial estimated errors $\tilde{\theta}(0)$, $\tilde{p}(0)$, and the explicit design parameters. It is an explicit function of design parameters, and thus computable. We can decrease the effects of the initial error estimates on the transient performance by increasing the adaptation gains $c_1\Gamma$ and γ.

20.6 Backstepping Control of Unstable Oil Wells

This section illustrates the potential of backstepping control applied for stabilization of unstable flow in oil wells. A simple empirical model is developed that describes the qualitative behavior of the downhole pressure during severe riser slugging. Two nonlinear controllers are designed by integrator backstepping approach, and stabilization for open-loop unstable pressure set points is demonstrated. If the parameters are unknown, an adaptive controller is designed by adaptive backstepping approach. The proposed backstepping controller is shown in simulations to perform better than proportional-integral (PI) and proportional-derivative (PD) controllers for low pressure set points. Operation at a low pressure set point is desirable since it corresponds to a high production flow rate.

20.6.1 Unstable Multiphase Flow

Multiphase-flow instabilities present in all phases of the lifetime of a field; however, the likelihood for multiphase-flow instabilities increase when entering oil production. The stabilization is related to the purpose of attenuating an oscillation phenomenon, called severe slugging, which exists in pipelines carrying multiphase flow. In oil production, unstable multiphase flow from wells or severe slugging is an increasing problem. In particular, unstable flow causes reduced production and oil recovery as the well must be choked down for the downstream processing equipment on the platforms to be able to handle the resulting variations in liquid and gas flow rates. The diagram of riser slug rig is shown in Figure 20.1.

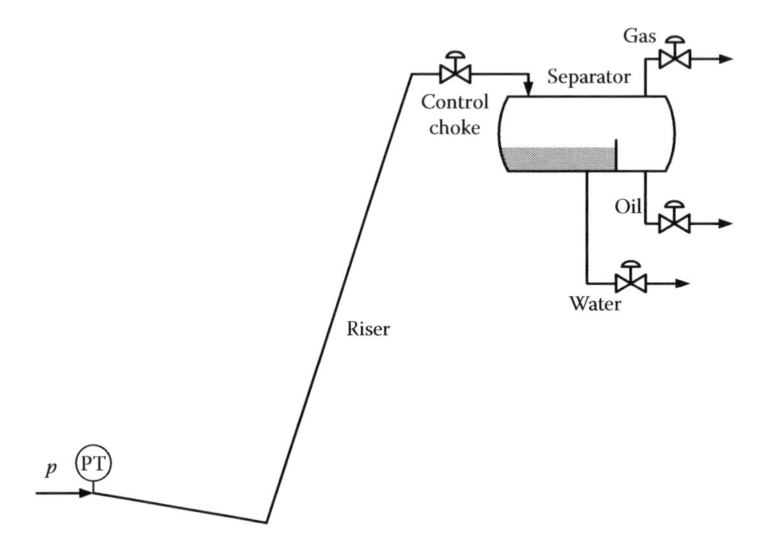

FIGURE 20.1 Schematic diagram of riser slug rig.

Active control of the production choke at the well head can be used to stabilize or reduce these instabilities. Conventionally, this is done by applying PI control to a measured downhole pressure to stabilize this at a specified set point, thus, stabilizing the flow. However, PI control is not robust and requires frequent retuning, or it does not achieve proper stabilization at all. Consequently, improved methods for stabilization of slugging wells have significant potential for increased production and recovery. Research on handling severe slugging in unstable wells has received much attention in the literature and in the industry, such as [11–22]. There is significant potential to improve performance of the control methods.

20.6.2 Dynamic Model

In order to understand the underlaying instabilities and to predict the controllability of slugging, a simple model is needed that captures the fundamental dynamics of the system, which can be used to develop a model-based stabilizing control law to counteract the destabilizing mechanisms of slugging. The schematic of the severe slugging cyclic behavior is shown in Figure 20.2, which includes four phases: slug formation phase, slug production phase, blow out phase, and liquid fallback phase. The oscillating

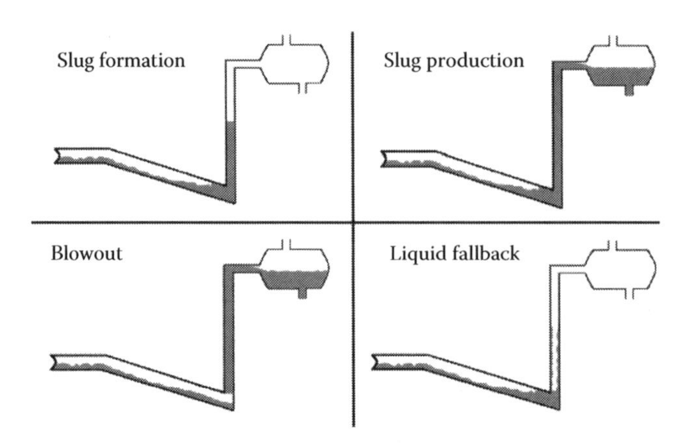

FIGURE 20.2 Schematics of the severe slug cycle in flowline riser systems.

behavior of the downhole pressure of a slugging well can be characterized as a stable limit cycle, which exhibits qualitatively the same behavior as the slightly modified van der Pol equation

$$\dot{p} = w, \tag{20.73}$$

$$\dot{w} = a_1(\beta - p) + a_2\left(\zeta - w^2\right)w, \tag{20.74}$$

$$\dot{q} = -\frac{1}{\tau}q + \frac{1}{\tau}\delta, \tag{20.75}$$

where the states p and w are the downhole pressure in the riser and its time derivative, respectively, q is related to the effect of the differential pressure over the production choke, δ represents the control input and is a strictly decreasing function of the production choke opening. The active control of the production choke at the well head is used to stabilize or reduce these instabilities. The coefficients in (20.73) through (20.75) can be explained as follows:

- a_1: frequency or stiffness of the system
- τ: a time lag due to the transportation delay
- β: steady-state pressure. Assuming constant reservoir influx such that β can be given in

$$\beta(q) = b_0 + b_1 q, \tag{20.76}$$

  where b_0 and b_1 are positive constants.
- a_2, ζ: local "degree of the stability/instability" and amplitude of the oscillation. Assuming constant flow rates of liquid and gas from the reservoir, ζ can be given as

$$\zeta(q) = c_0 - c_1 q, \tag{20.77}$$

  where c_0/c_1 denotes the bifurcation point and c_0, c_1 are positive constants.

20.6.2.1 Simplified Model of Riser Slugging

Based on (20.76) and (20.77), the system dynamics (20.73) through (20.75) can be assembled into

$$\dot{p} = w \tag{20.78}$$

$$\dot{w} = -a_1 p + h(w) + g(w)q + a_1 b_0 \tag{20.79}$$

$$\dot{q} = -\frac{1}{\tau}q + \frac{1}{\tau}\delta, \tag{20.80}$$

where the functions h and g are defined as

$$h(w) = a_2 c_0 w - a_2 w^3 = h_0 w - h_1 w^3 \tag{20.81}$$

$$g(w) = a_1 b_1 - a_2 c_1 w = g_0 - g_1 w. \tag{20.82}$$

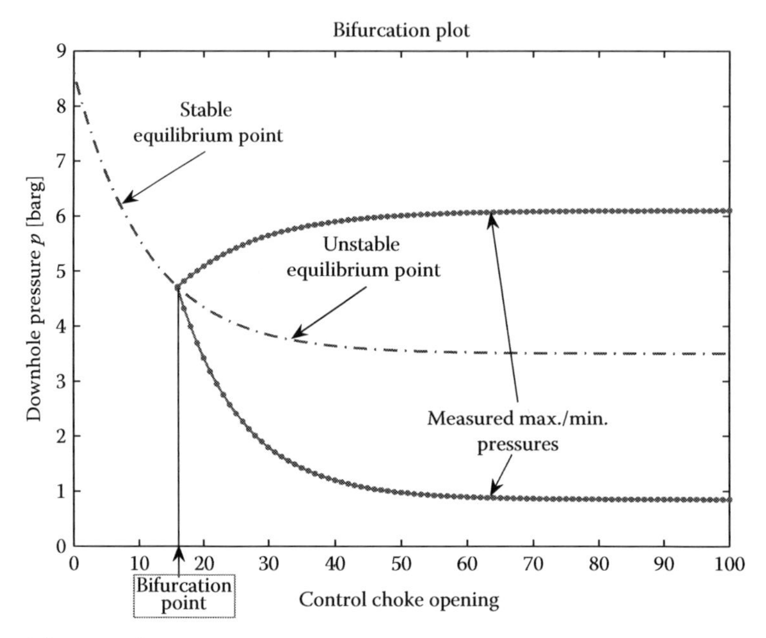

FIGURE 20.3 Bifurcation plot.

The positive constants a_i, b_i, and c_i are empirical parameters that are adjusted to produce the right behavior of the downhole pressure p. The system (20.78) through (20.80) can capture the qualitative properties in the downhole pressure during riser slugging.

- Decreasing control gain: A characteristic property of riser slugging is that the static gain decreases with choke opening.
- Bifurcation: The model exhibits the characteristic bifurcation that occurs at a certain choke opening c_0/c_1, i.e., the steady-state response of the downhole pressure exhibits changes from a stable point when choke opening is smaller than c_0/c_1 to a stable limit cycle when choke opening is larger than c_0/c_1 (see Figure 20.3).
- Time lag: The transportation delay between a change in choke opening to the resulting change in downhole pressure p is modeled by simple first-order lag.

Our objective is to design a control law for $\delta(t)$, which stabilizes downhole pressure $p(t)$ at the desired set point $p_{\text{ref}} > b_0$.

20.6.3 Backstepping Control Design

In this section, we assume all the parameters in dynamic model (20.78) through (20.80) are known, and we design stabilizing controllers using integrator backstepping.

20.6.3.1 Control Scheme I

We iteratively look for a change of coordinates in the form

$$z_1 = p - p_{\text{ref}}, \tag{20.83}$$

$$z_2 = w - \alpha_w, \tag{20.84}$$

$$z_3 = q - \alpha_q, \tag{20.85}$$

where functions α_w and α_q are virtual controls to be determined

$$\alpha_w = -C_1 z_1, \tag{20.86}$$

$$\alpha_q = \frac{1}{g(w)}(-C_2 z_2 - z_1 + a_1(z_1 + p_{\text{ref}} - b_0) - h(w) - a_1 b_0 + \dot{\alpha}_w), \tag{20.87}$$

$$\delta = -\tau C_3 z_3 - \tau g(w) z_2 + \alpha_q + \tau \dot{\alpha}_q, \tag{20.88}$$

where C_1, C_2, C_3 are positive constants. The time derivative of the control Lyapunov function (CLF) $U_3 = \sum_{i=1}^{3} \frac{1}{2} z_i^2$ becomes

$$\dot{U}_3 \leq -C_1 z_1^2 - C_2 z_2^2 - C_3 z_3^2, \tag{20.89}$$

The LaSalle–Yoshizawa theorem guarantees that the equilibrium $(z_1, z_2, z_3) = 0$ is globally exponentially stable, and in particular $p(t)$ is regulated to the set point p_{ref}. Since U_3 is nonincreasing, we have

$$\| z_1 \|_2^2 = \int_0^\infty | z_1(\tau) |^2 \, d\tau \leq \frac{1}{C_1}(U_3(0) - U_3(\infty)) \leq \frac{1}{C_1} U_3(0). \tag{20.90}$$

Lemma 20.1 Consider the riser slugging system (20.78) through (20.80). With the application of the controller (20.88) and virtual control laws (20.86) and (20.87), the following statements hold:

- The resulting closed-loop system is globally stable.
- The asymptotic tracking is achieved, i.e., $\lim_{t \to \infty}[p(t) - p_{\text{ref}}] = 0$.
- The transient displacement tracking error performance is given by

$$\| p(t) - p_{\text{ref}} \|_2 \leq \frac{1}{\sqrt{C_1}} \sqrt{U_3(0)}. \tag{20.91}$$

Remark 20.2 We refer to this choice of α_q as an exact canceling design because we simply cancel existing dynamics and replace it with some desirable linear feedback terms: $-C_1 z_1$ and $-C_2 z_2$. Note that this design is not necessarily the best choice of control law because stabilizing nonlinearities may be canceled, potentially wasting control effort, losing robustness to modeling errors, and making the control law overly complicated. It is desirable to obtain a simpler control law by avoiding cancellation of useful nonlinearities.

20.6.3.2 Control Scheme II

By inspection of the second step of backstepping in the previous section, we recognize that the terms $-h_1 w^3$ and $-g_1 wq$ are expected to be stabilizing, since physically $q > 0$. Hence, canceling these terms is not necessary at this point in the design. We select

$$\alpha_w = 0 \tag{20.92}$$

$$\alpha_q = -\frac{C_2 + h_0}{g_0} z_2 + \frac{a_1}{g_0}(p_{\text{ref}} - b_0), \tag{20.93}$$

$$\delta = -\tau C_3 z_3 - \tau g_0 z_2 + q + \tau \dot{\alpha}_q. \tag{20.94}$$

Consider now the CLF $U_3 = \sum_{i=1}^{3} \frac{1}{2} z_i$ the time derivative is

$$\dot{U}_3 = -(C_2 + g_1 q) z_2^2 - h_1 z_2^4 - C_3 z_3^2 \leq -C_2 z_2^2 - C_3 z_3^2. \tag{20.95}$$

The stabilizing terms $-h_1 z_2^4$ and $-g_1 q z_2^2$ increase negativity of $\dot{U}_3$. LaSalle's invariance principle now implies that the origin is asymptotically stable. The following result formalizes this.

Lemma 20.2 Consider the riser slugging system (20.78) through (20.80). With the application of the controller (20.94) and the virtual control laws (20.92) and (20.93), the resulting closed-loop system is stable and the asymptotic tracking is achieved, such as $\lim_{t \to \infty} [p(t) - p_{ref}] = 0$.

20.6.3.3 Adaptive Backstepping Control

Consider the parameters a_1, b_0, h_0, h_1, g_0, g_1 are unknown constants. Define

$$\theta_1 = [a_1, a_1 b_0, h_0]^{\mathrm{T}}, \tag{20.96}$$

$$\theta_2 = [a_1, a_1 b_0, h_0, h_1, g_0, g_1]^{\mathrm{T}}, \tag{20.97}$$

$$\vartheta = \frac{1}{g_0}, \tag{20.98}$$

and the regressor errors

$$\phi_1 = [-p, 1, w]^{\mathrm{T}}, \quad \phi_2 = [-p, 1, w, -w^3, q, -wq]^{\mathrm{T}}. \tag{20.99}$$

Thus, we use the change of coordinates in (20.83) through (20.85) and adaptive backstepping control technique. The final control δ and virtual controls α_w and α_q determined as

$$\alpha_w = 0, \tag{20.100}$$

$$\alpha_q = \hat{\vartheta}\left(-C_2 z_2 - \hat{\theta}_1^{\mathrm{T}} \phi_1 - z_1\right), \tag{20.101}$$

$$\delta = \tau\left(-C_3 z_3 + \frac{1}{\tau} q + \frac{\partial \alpha_q}{\partial p} w + \frac{\partial \alpha_q}{\partial \hat{\theta}_1} \dot{\hat{\theta}}_1 + \frac{\partial \alpha_q}{\partial \hat{\vartheta}} \dot{\hat{\vartheta}} + \frac{\partial \alpha_q}{\partial w} \hat{\theta}_2^{\mathrm{T}} \phi_2 - \hat{\theta}_2^{\mathrm{T}} e_5 z_2\right), \tag{20.102}$$

and the parameter updating laws are given

$$\dot{\hat{\theta}}_1 = \Gamma_1 \phi_1 z_2, \tag{20.103}$$

$$\dot{\hat{\theta}}_2 = -\Gamma_2 z_3 \left(\frac{\partial \alpha_q}{\partial w} \phi_2 - e_5 z_2\right), \tag{20.104}$$

$$\dot{\hat{\vartheta}} = -\gamma z_2 \left(-C_2 z_2 - \hat{\theta}_1^{\mathrm{T}} \phi_1 - z_1\right), \tag{20.105}$$

where C_2, C_3, γ are positive constants and Γ_1, Γ_2 are positive definite matrix, and $e_5 = [0,0,0,0,1,0]^T$. The system (20.78) through (20.80) can be rewritten as

$$\dot{p} = w \tag{20.106}$$

$$\dot{w} = \theta_1^{\mathrm{T}}\phi_1 - h_1 w^3 + g_0 q - g_1 w q \tag{20.107}$$

$$\dot{q} = -\frac{1}{\tau}q + \frac{1}{\tau}\delta, \tag{20.108}$$

Note that $q > 0$. Consider the CLF $U_3 = \sum_{i=1}^{3}\frac{1}{2}z_i^2 + \sum_{i=1}^{3}\frac{1}{2}\tilde{\theta}_i^{\mathrm{T}}\Gamma_i^{-1}\tilde{\theta}_i + \frac{g_0}{2\gamma}\tilde{\vartheta}^2$. The time derivative of U_3 follows (20.103) through (20.105)

$$\begin{aligned}\dot{U}_3 = & -C_2 z_2^2 - C_3 z_3^2 - \tilde{\theta}_1^{\mathrm{T}}\Gamma_1^{-1}\left(\dot{\hat{\theta}}_1 - \Gamma_1\phi_1 z_2\right) - \frac{g_0}{\gamma}\tilde{\vartheta}\left(\dot{\hat{\vartheta}} + \gamma\left(-C_2 z_2 - \hat{\theta}_1^{\mathrm{T}}\phi_1 - z_1\right)\right) \\ & -\tilde{\theta}_2^{\mathrm{T}}\Gamma_2^{-1}\left(\dot{\hat{\theta}}_2 + \Gamma_2 z_3 \frac{\partial \alpha_q}{\partial w}\phi_2 - \Gamma_2 z_3 e_5 z_2\right) - h_1 w^4 - g_1 w^2 q \\ & \leq -C_2 z_2^2 - C_3 z_3^2,\end{aligned} \tag{20.109}$$

which is negative semidefinite and proves that the system is stable.

Lemma 20.3 With the application of the adaptive control law (20.102), virtual control laws (20.100) through (20.101), and the parameter update laws (20.103) through (20.105), the resulting closed-loop system is stable and the asymptotic tracking is achieved as $\lim_{t\to\infty}[p(t) - p_{\mathrm{ref}}] = 0$.

20.6.4 Simulation Results

In this section, the proposed backstepping controllers are tested for stabilization of unstable flow in oil well, compared with traditional PI and PD controllers. For simulation studies, the following values are selected as "true" parameters for the system: $h_0 = 1$, $h_1 = 50$, $g_0 = 0.125$, $g_1 = 5$, $a_1 = 0.025$, $b_0 = 3.5$, and $\tau = 0.1$. The choke function is $\delta(t) = e^{-10u(t)}$, where u is the choke opening.

20.6.4.1 Backstepping Controller

We test our proposed backstepping controller (20.94) for stabilization of unstable flow in oil well. The design objective is to stabilize p at the desired set point $p_{\mathrm{ref}} = 3.51$. With the proposed backstepping control Scheme II, we take the following set of design parameters: $C_2 = 0.2$ and $C_3 = 5$. The initials are set as $p(0) = 3.51$, $w(0) = q(0) = 0$ and the choke opening $u(0) = 0.10$, respectively. Figure 20.4 illustrates the backstepping controller applied for stabilization in the unstable region at reference pressure $p_{\mathrm{ref}} = 3.51$.

20.6.4.2 PI Control

The conventional way to stabilize riser slugging is by applying a simple control law u_{PI}

$$u_{\mathrm{PI}} = u_I - K_p\left(p - p_{\mathrm{ref}}\right) \tag{20.110}$$

$$\dot{u}_I = -\frac{K_i}{T_i}\left(p - p_{\mathrm{ref}}\right), \tag{20.111}$$

where K_p, K_i, T_i are positive constants. Figure 20.5 illustrates PI controller applied for stabilization in the region at reference pressure $p_{\mathrm{ref}} = 4.498$. Figure 20.6 shows that the system loses stability at the pressure

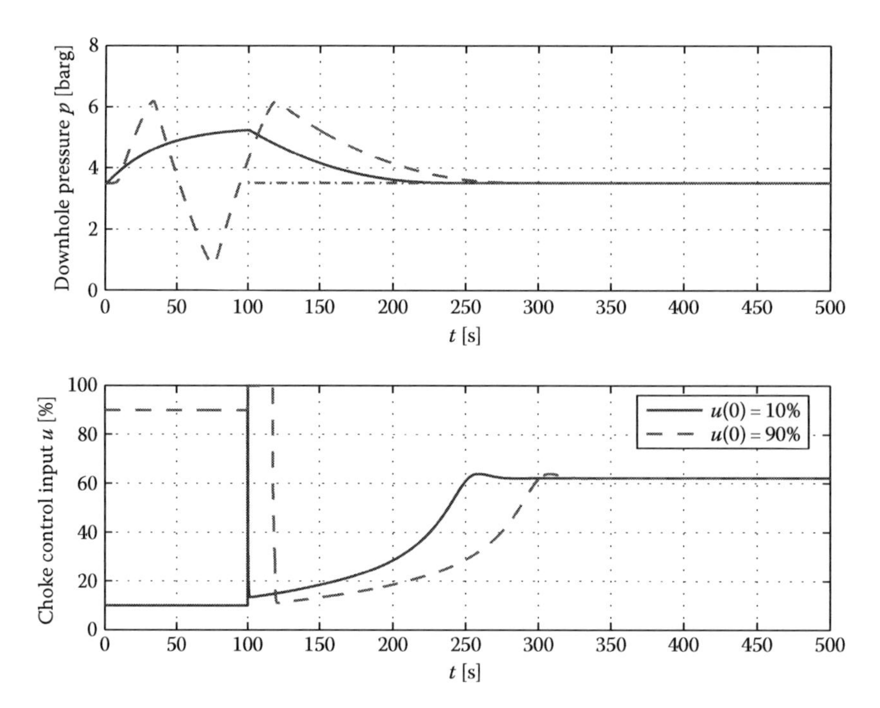

FIGURE 20.4 Simulations of stabilization at $p_{ref} = 3.51$ using backstepping control Scheme II.

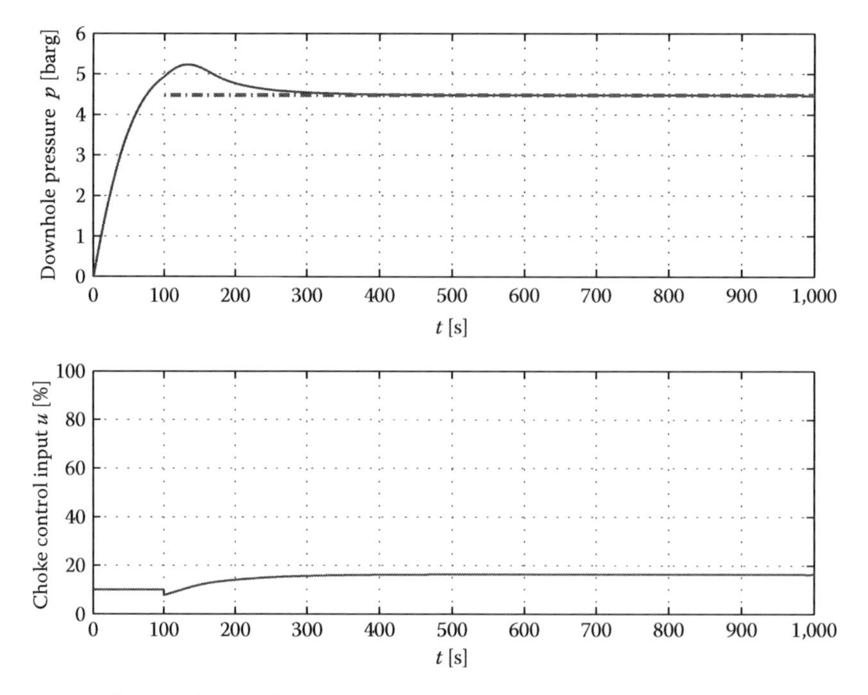

FIGURE 20.5 Simulations of PI stabilization at $p_{ref} = 4.498$ using PI controller.

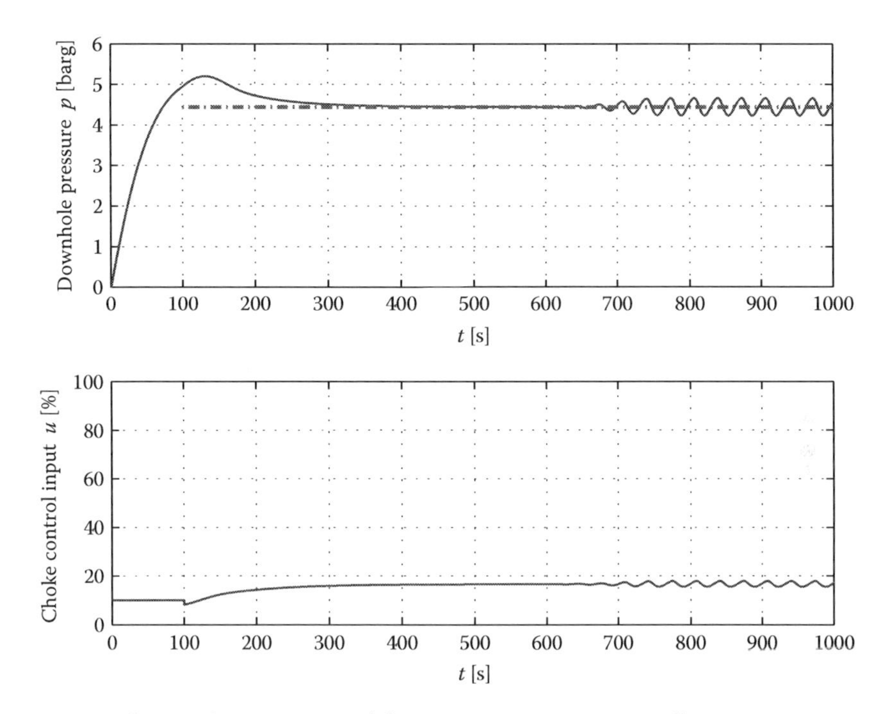

FIGURE 20.6 Simulations of an attempt to stabilize at $p_{ref} = 4.45$ using PI controller.

$p_{ref} = 4.45$, which is higher than the required pressure reference $b_0 = 3.5$. The design parameters are chosen as $K_p = 0.1$, $K_i = 0.1$, and $T_i = 25$.

20.6.4.3 PD Control

Another way to stabilize riser slugging is by applying a simple control law u_{PD}

$$u_{PD} = -K_p\left(p - p_{ref}\right) + u_D, \tag{20.112}$$

$$u_D = -K_d \frac{d(p - p_{ref})}{dt} = -K_d w, \tag{20.113}$$

where K_p, K_d are positive constants. Figure 20.7 illustrates PD controller applied for stabilization at reference pressure $P_{ref} = 4.6$. The design parameters are chosen as $K_p = 2$ and $K_d = 2$, which satisfy the stability conditions. Figure 20.8 shows that the system loses stability at the pressure $P_{ref} = 3.51$. The design parameters are chosen as $K_p = 0.02$ and $K_d = -1$, which satisfy the stability conditions. When the pressure is small, feasible K_p and K_d, according to the Hurwitz criterium, give an aggressive actuation that the choke saturates repeatedly and stabilization is not achieved.

20.6.4.4 Adaptive Control

If some parameters are unknown, the adaptive backstepping control (20.100) through (20.102) with parameter updating laws (20.103) through (20.105) is applied to stabilize the pressure p at the desired set point $p_{ref} = 3.8$. The design parameters are chosen as $C_2 = 0.2$, $C_3 = 5$, and $\Gamma_1 = diag\{0.01, 0.5, 10\}$, $\Gamma_2 = diag\{0.01, 0.5, 10, 0.01, 0.01, 0.01\}$, $\gamma = 0.1$. The initials of states are set as $p(0) = w(0) = q(0) = 0$, and the initials of estimates $\hat{a}_1(0) = 0.8a_1$, $\hat{b}_1(0) = 0.5b_1$, $\hat{h}_0(0) = 1.2h_0$, $\hat{h}_1(0) = 0.5h_1$, $\hat{g}_0(0) = 1.2g_0$, $\hat{g}_1(0) = 1.2g_1$, respectively. Figure 20.9 illustrates the adaptive backstepping controller applied for stabilization at $p_{ref} = 3.8$.

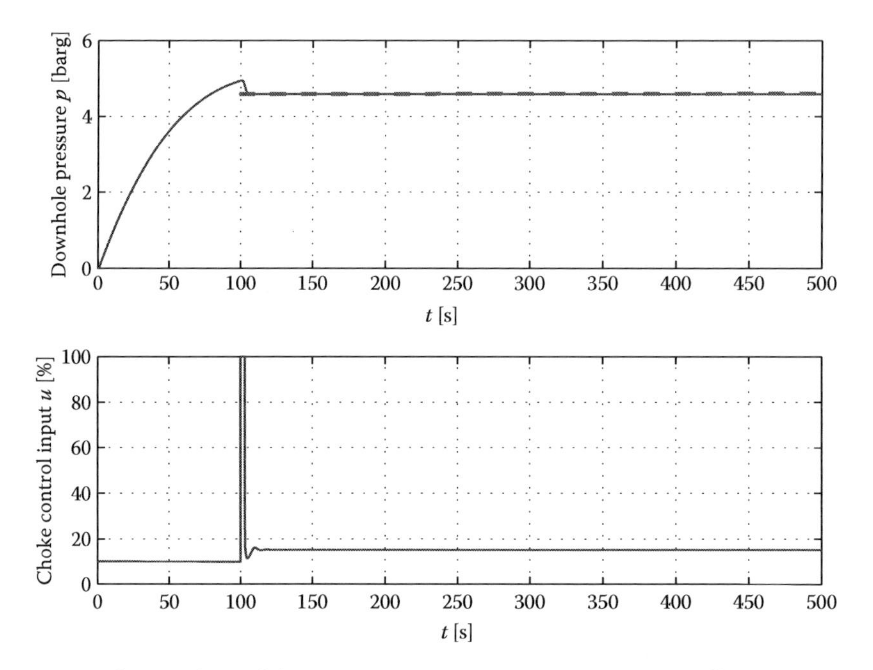

FIGURE 20.7 Simulations of PD stabilization at a pressure $p_{ref} = 4.60$ using PD controller.

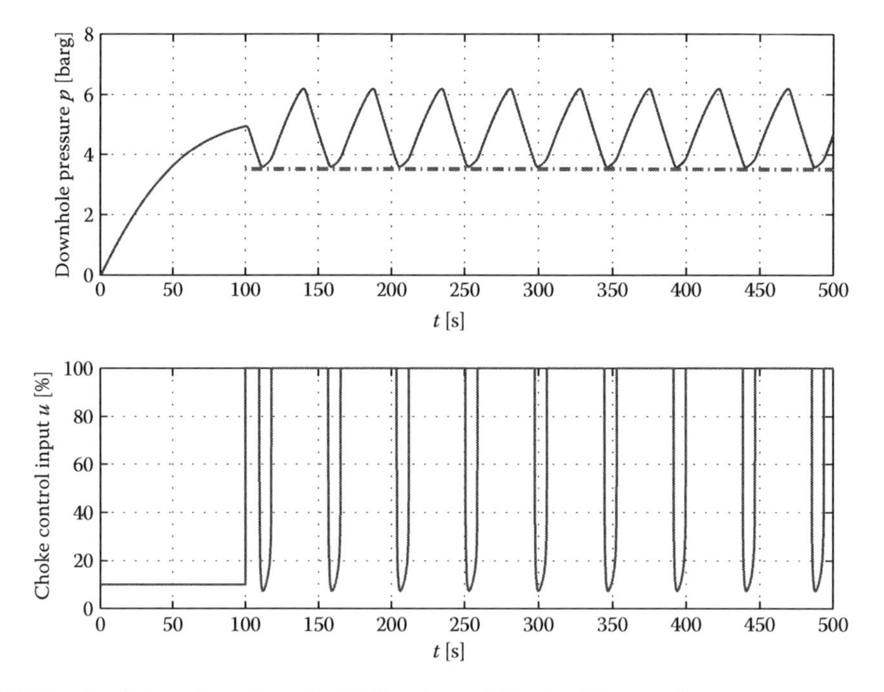

FIGURE 20.8 Simulations of an attempt to stabilize at $p_{ref} = 3.51$ using PD controller.

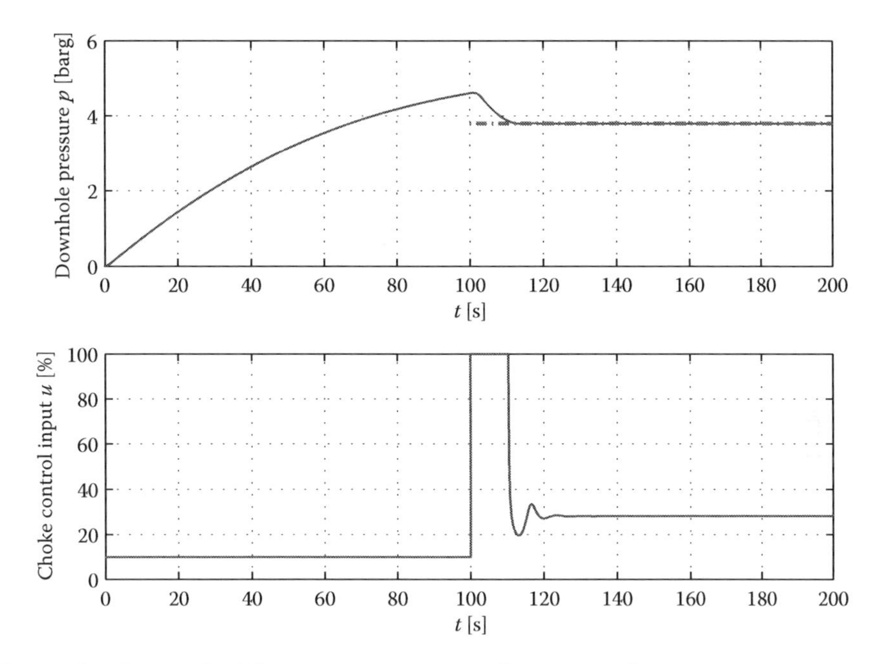

FIGURE 20.9 Simulations of stabilization at $p_{ref} = 3.8$ using adaptive control.

References

1. M. Krstic, I. Kanellakopoulos, and P. V. Kokotovic. *Nonlinear and Adaptive Control Design*. Wiley, New York, 1995.
2. R. Lozano and B. Brogliato. Adaptive control of a simple nonlinear system without a priori information on the plant parameters. *IEEE Transactions on Automatic Control*, 37(1):30–37, 1992.
3. C. Wen. Decentralized adaptive regulation. *IEEE Transactions on Automatic Control*, 39:2163–2166, 1994.
4. C. Wen, Y. Zhang, and Y. C. Soh. Robustness of an adaptive backstepping controller without modification. *Systems & Control Letters*, 36:87–100, 1999.
5. C. Wen and J. Zhou. Decentralized adaptive stabilization in the presence of unknown backlash-like hysteresis. *Automatica*, 43:426–440, 2007.
6. Y. Zhang, C. Wen, and Y. C. Soh. Adaptive backstepping control design for systems with unknown high-frequency gain. *IEEE Transactions on Automatic Control*, 45:2350–2354, 2000.
7. J. Zhou, C. Wen, and Y. Zhang. Adaptive backstepping control of a class of uncertain nonlinear systems with unknown backlash-like hysteresis. *IEEE Transactions on Automatic Control*, 49:1751–1757, 2004.
8. J. Zhou, C. Wen, and Y. Zhang. Adaptive output control of a class of time-varying uncertain nonlinear systems. *Journal of Nonlinear Dynamics and System Theory*, 5:285–298, 2005.
9. J. Zhou, C. J. Zhang, and C. Wen. Robust adaptive output control of uncertain nonlinear plants with unknown backlash nonlinearity. *IEEE Transactions on Automatic Control*, 52:503–509, 2007.
10. H. K. Khalil. *Nonlinear Systems*, 3rd edn. Prentice-Hall, Englewood Cliffs, NJ, 2002.
11. M. Dalsmo, E. Halvorsen, and O. Slupphaug. Active feedback control of unstable wells at the brage field. *SPE Annual Technical Conference and Exhibition*, SPE 77650, San Antonio, TX, September 29–October 2, 2002.
12. T. Drengstig and S. Magndal. Slugcontrol of production pipeline. In *Proceedings of SIMS2001*, Porsgrunn, Norway, 2002, pp. 361–366.

13. J.-M. Godhavn, M. P. Fard, and P. H. Fucks. New slug control strategies, tuning rules and experimental results. *Journal of Process Control*, 15:454–463, 2005.
14. V. Henriot, A. Courbot, E. Heintzé, and L. Moyeux. Simulation of process to control severe slugging: Application to dunbar pipeline. *SPE Annual Technical Conference and Exhibition*, SPE 56461, Houston, TX, 1999.
15. G. Kaasa, V. Alstad, J. Zhou, and O. M. Aamo. Attenuation of slugging in unstable oil wells by nonlinear control. In *17th International Federation of Automatic Control World Congress*, Korea, 2008, pp. 6251–6256.
16. G.-O. Kaasa, V. Alstad, J. Zhou, and O. M. Aamo. Nonlinear control for a riser-slugging system in unstable wells. *Journal of Modeling, Identification and Control*, 28:69–79, 2007.
17. J. P. Kinvig and P. Molyneux. Slugging control. US patent US6716268, 2001.
18. P. Molyneux, A. Tait, and J. Kinvig. Characterisation and active control of slugging in a vertical riser. In *Proceedings from BHR Group Conference: Multiphase Technology*, Banff, Canada, 2000.
19. P. F. Pickering, G. F. Hewitt, M. J. Watson, and C. P. Hale. The prediction of flows in production risers—Truth & myth? In *IIR Conference*, Atlanta, GA, 2001.
20. H. B. Siahaan, O. M. Aamo, and B. A. Foss. Suppressing riser-based slugging in multiphase flow by state feedback. In *Proceedings of the 44th IEEE Conference on Decision and Control, and the European Control Conference*, Seville, Spain, 2005.
21. E. Storkaas. Stabilizing control and controllability: Control solutions to avoid slug flow in pipeline. PhD thesis, Norwegian University of Science and Technology, Trondheim, Norway, 2005.
22. J. Zhou, G. Kaasa, and O. M. Aamo. Nonlinear adaptive observer control for a riser slugging system in unstable wells. In *Proceedings of American Control Conference*, Seattle, WA, 2008, pp. 2951–2956.

21

Sensors

Tiantian Xie
Auburn University

Bogdan M. Wilamowski
Auburn University

21.1 Introduction

A sensor is a device that provides an electrical output responding to a stimulation of a nonelectrical signal. It converts a physical parameter into a signal suitable for processing. Sensors are widely used in various places such as automobiles, airplanes, radios, and countless other applications. In this chapter, various types of sensors for measuring different physical parameters are described.

21.2 Distance and Displacement Sensors

In a control system, displacement or distance is an important parameter to be measured. Depending on the distance range, various types of sensors are used.

21.2.1 Sensors for Large Distances

The most popular sensors for large distance measurement are optical sensors [ABLMR01]. An optical sensor usually requires at least three essential components: a light source, a photodetector, and light guidance devices, which may include lenses, mirrors, optical fiber, etc. In a basic optical position sensor, light is guided toward a target by focusing lenses and it is diverted back to detectors by the reflector.

The time cost for the light traveling through the target and back to the receiver is proportional to the distance. These sensors are used for sensing distances from hundreds of meters to thousands of meters.

Laser triangulation sensors are specially designed optical sensors for more precision position measurements over short and long ranges. As shown in Figure 21.1, the position of the target is determined by measuring reflection from the target surface. A laser diode projects a spot of light to the target, and its reflection is focused via an optical lens on a light-sensitive array. If the target changes its position, the position of the reflected spot of light on the detector changes as well. The most critical element in this arrangement is the receiver/detector. It is commonly a charge-coupled device (CCD). The CCD is registering bright on pixel array and then reading it by transferring information across array. The intensity distribution of the imaged spot is viewed, and then image processing is incorporated for the linear triangulation measurement.

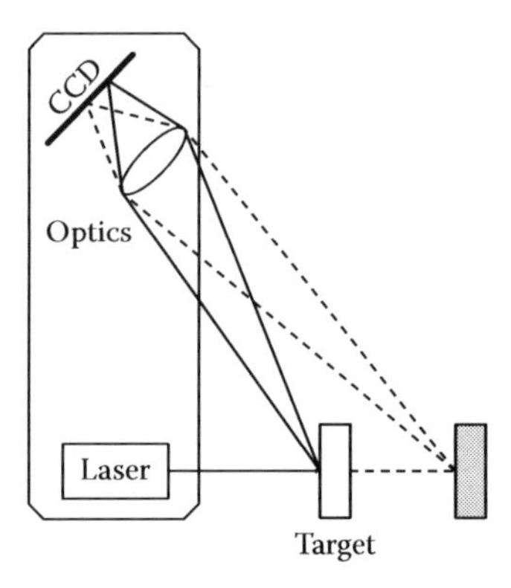

FIGURE 21.1 Laser distance sensor using triangulation principle.

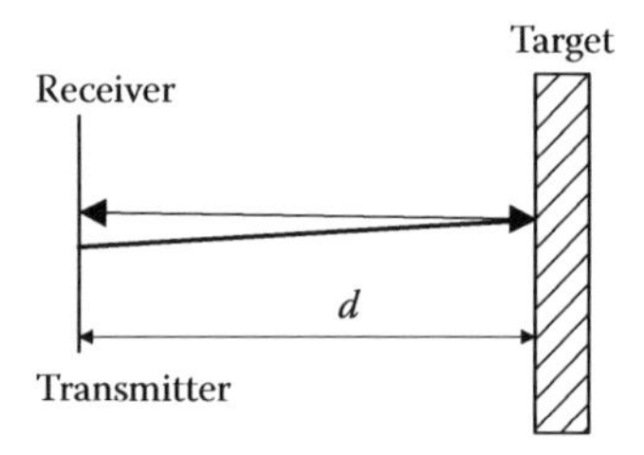

FIGURE 21.2 Schematic structure of ultrasonic distance sensor.

21.2.2 Sensors for Medium Distances

Ultrasonic distance sensors measure the traveling of ultrasonic pulse. The schematic structure of an ultrasonic distance sensor is shown in Figure 21.2. When the waves are incident on the target, part of their energy is reflected in a diffuse manner. A distance d to the target can be calculated though the speed v of the ultrasonic waves in the media:

$$d = \frac{vt}{2} \quad (21.1)$$

where t is the time cost for the ultrasonic waves traveling from the source to the target and back to the receiver. To generate the ultrasonic, piezoelectric materials are used. Ultrasonic sensors are used for distance ranging from several centimeters up to a few meters.

Another displacement sensor is linear variable differential transformer (LVDT). The transformer comprises three coils: a primary center coil and two outer secondary coils. The transfer of current between the primary and the secondary coils of the transducer depends on the position of a magnetic core shown in Figure 21.3. At the center of the position measurement stroke, the two secondary voltages of the displacement transducer are equal; since they are connected oppositely, the output from the sensor is zero. As the core moves away from the center, the result is an increase position sensor in one of the secondary coils and a decrease in the other, which results in an output from the measurement sensor. The LVDT is used to measure displacement ranging from fractions of a millimeter to several centimeters. They can be manufactured to meet stringent accuracy and resolution requirements, for example, accuracy better than 0.2%.

21.2.3 Sensors for Small Distances

Capacitive position sensors are based on the principle that the capacitance between two plates is proportional to the area A and inversely proportional to the distance d between them:

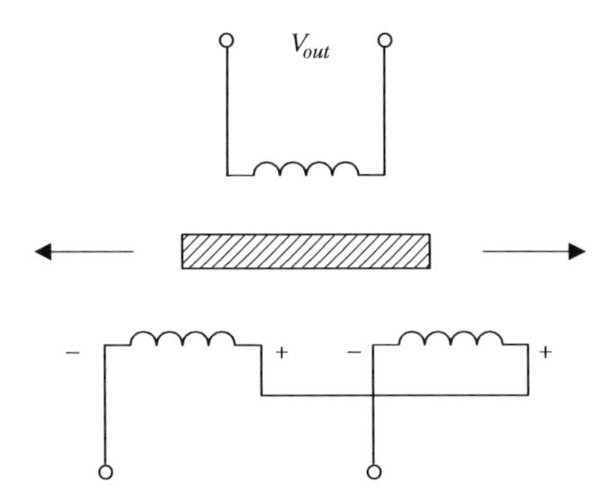

FIGURE 21.3 Structure of LVDT variation.

$$C = \frac{A\varepsilon_0\varepsilon_r}{d} \tag{21.2}$$

where

ε_r is the electric permittivity of the material
ε_0 is the electric permittivity of the vacuum

In capacitive position sensors, one plate of a capacitor is the base and fixed, while the other plate is movable [B97]. The capacitance position sensor can operate over a range of a few millimeters, but it is sensitive to temperature and humidity, especially if the dielectric is air.

There are two fundamental types of capacitive position sensors: spacing and area variations as shown in Figure 21.4.

In the first type of sensor (Figure 21.4a), distance between two plates are being measured. The capacitance is inversely proportional to the distance between plates. This relationship is nonlinear. However, if the circuit in Figure 21.5 is used for measurement, the nonlinear relationship can be eliminated. In this circuit, the output voltage is calculated by

$$V_{out} = -\frac{1/sC_2}{1/sC_1}V_{in} = -\frac{C_1}{C_2}V_{in} = -C_1V_{in}\frac{d}{A\varepsilon_r\varepsilon_0} \tag{21.3}$$

The nonlinear dependence on d (see Equation 21.2) can be transferred to linear sensor using this method.

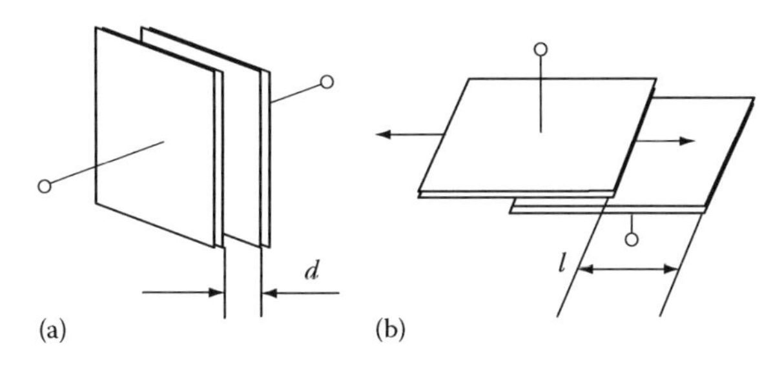

FIGURE 21.4 Two types of capacitive position sensors: (a) spacing variation and (b) area.

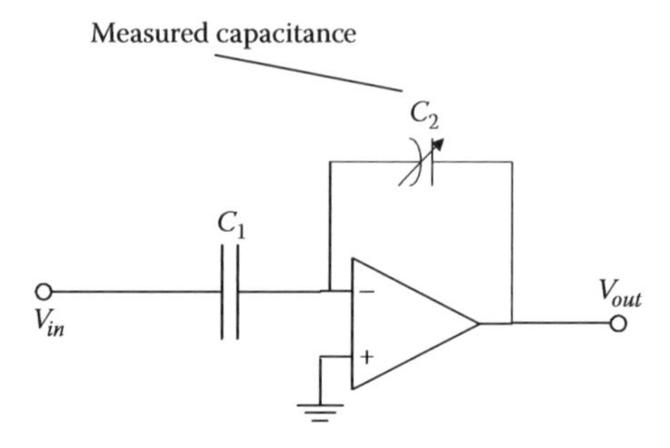

FIGURE 21.5 Circuit of linear spacing variation capacitive distance sensor.

For larger distances, the area variation (Figure 21.4b) is preferred. As the plates slide transversely, the capacitance of the area variation changes as a linear function of sliding length.

For small and medium distances, the grating displacement sensor can be used as shown in Figure 21.6. Grating capacitive displacement sensor, which is based on the principle resembling that of caliper, is used for precision displacement measurement [PST96]. The sensor is fabricated with two overlapping gratings with slightly different pitch, which serve as a capacitive modulator. One grating is fixed while the other is movable. When the opaque sector of the moving grating is maximum aligned with the transmitting sector, the overlapping area is the least (Figure 21.6a). The overlapping area is periodic variable as shown in Figure 21.7, which induces the capacitance modulated. The displacement of the moving grating can be obtained by calculating the number of hanging periods

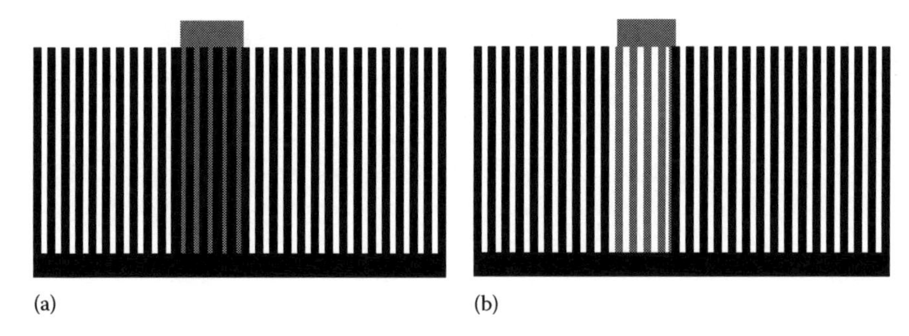

FIGURE 21.6 Grating type displacement sensor: (a) minimum capacitance area and (b) maximum capacitance area.

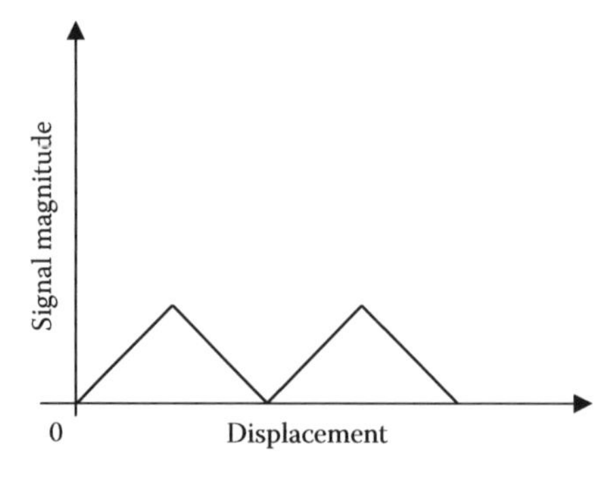

FIGURE 21.7 Transfer function of grating type displacement sensor.

of the capacitance. For high sensitivity, the grating pitch can be made very small, so that very short movements of the grating will result in a large output signal.

21.3 Pressure Sensors

Pressure sensors are the devices used to measure pressure. They can be used to measure the flow of liquid, the weight or force exerted by one object on another, atmospheric pressure, and anything else involving force [N97]. Many modern pressure sensors are much more sensitive than scales and give an accurate output that can be measured electronically.

21.3.1 Piezoresistive Pressure Sensors

Piezoresistivity is a material property where bulk resistivity is influenced by mechanical stress applied to material. The common piezoresistors can be silicon, poly silicon, silicon dioxide, zinc oxide, etc.

A piezoresistive pressure sensor consists of a thin silicon membrane supported by a thick silicon rim as shown in Figure 21.8a and b. The diaphragm acts as a mechanical stress amplifier. When a pressure difference is applied across the device, the thin diaphragm will bend downward or upward, indicating traction or compression on the piezoresistors. The resistance change caused by this stress can be measured by Wheatstone bridge (Figure 21.8c).

Figure 21.8c shows the four piezoresistors connected in the Wheatstone bridge configuration. By placing one piezoresistor parallel to two edges of the diaphragm and the other perpendicular to the other two edges, the resistance change of the two piezoresistors will always be opposite. When the diaphragm is bent downward, causing the tensile stress on the diaphragm surface at the edges, the parallel resistors are under lateral stress and show a decrease in resistance while the perpendicular ones are under longitudinal stress and show an increase in resistance. The applied stress is proportional to the change of resistance, which is precisely measured.

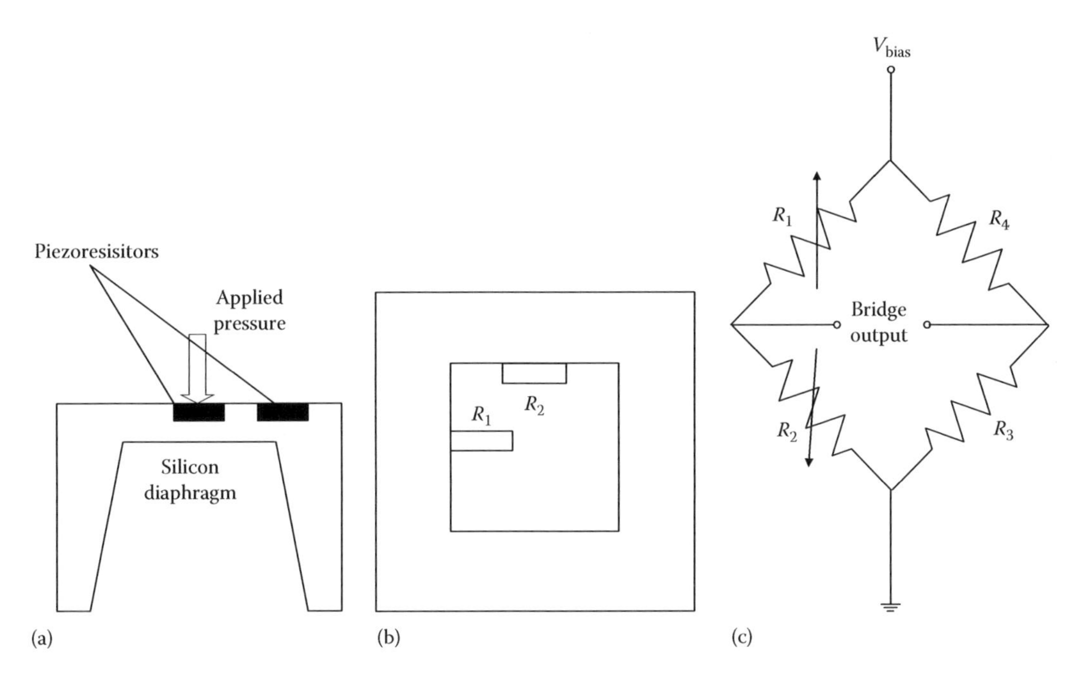

FIGURE 21.8 Piezoresistive pressure sensors: (a) cross section, (b) top view, and (c) Wheatstone bridge.

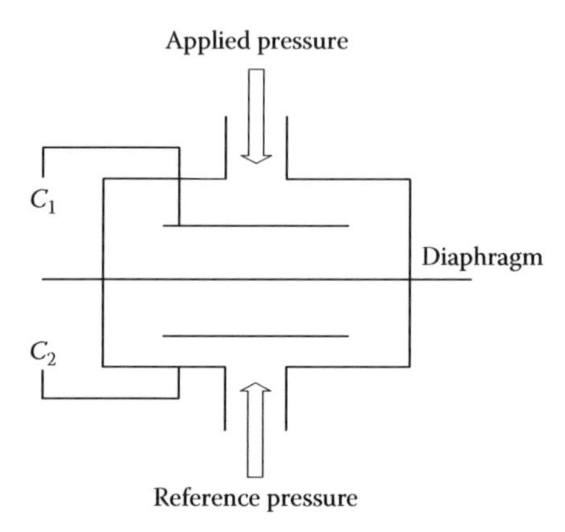

FIGURE 21.9 Schematic of a capacitive pressure sensor.

21.3.2 Capacitive Pressure Sensors

The capacitive pressure sensors use capacitance sensitive components to transform the pressures to electrical signals. The structure of a capacitive pressure sensor is shown in Figure 21.9. A diaphragm is suspended between two parallel metallic plates so as to form two capacitances C_1 and C_2. The capacitances of C_1 and C_2 will be changed if the diaphragm is deflected due to a pressure difference between its two sides. This type of sensor is mostly used to measure small changes of a fairly low static pressure. The capacitive pressure sensor is more sensitive and less temperature dependent than piezoresistive sensor.

21.3.3 Piezoelectric Pressure Sensors

A piezoelectric pressure sensor is a device that uses the piezoelectric effect to measure force by converting it to an electrical signal [SC00]. The scheme of piezoelectric pressure sensor is shown in Figure 21.10. The pressure is applied on the piezoelectric material and makes it deformed, so as to generate charge. The circuit part in Figure 21.10 converts the input current into voltage. Therefore, the pressure can be calculated by measuring output voltage. The electrical signal generated by the piezoelectric material decays rapidly after the application of force. This makes these devices unsuitable for the detection of static force. Depending on the application requirements, dynamic force can be measured as compression, tensile, or torque force.

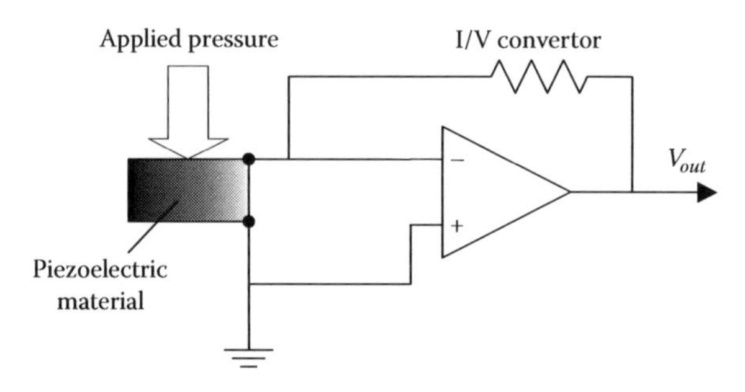

FIGURE 21.10 Scheme of piezoelectric pressure sensor.

21.4 Accelerometer

An accelerometer is an electromechanical device that will measure acceleration forces. These forces may be static or dynamic. The accelerometer has been used in engineering, building and structural monitoring, medical application, etc. An accelerometer requires a component whose movement lags behind that of the accelerometer's housing, which is coupled to the object under study. This component is usually called inertial mass. No matter how the sensors are designed or what is the conversion technique, an ultimate goal of the measurement is the detection of the mass displacement with respect to the housing. Hence, any suitable displacement sensor capable of measuring microscopic movements under strong vibrations or linear acceleration can be used as an accelerometer.

21.4.1 Piezoelectric Accelerometer

Piezoelectric accelerometer utilizes the piezoelectric effect of materials to measure dynamic changes in mechanical variables. The structure of piezoelectric accelerometer is shown in Figure 21.11. When a physical force is exerted on the accelerometer, the seismic mass loads the piezoelectric element according to Newton's second law of motion ($F = ma$). The force exerted on the piezoelectric material can be observed corresponding to the change of the voltage generated by the piezoelectric material. Therefore, the acceleration can be obtained. Single crystal, like quartz, and ceramic piezoelectric materials, such as barium titanate and lead-zirconate-lead-titanate, can be used for the purposes of accelerometer. These sensors are suitable from frequency as low as 2 Hz and up to about 5 kHz. They possess high linearity and a wide operating temperature range.

21.4.2 Piezoresistive Accelerometer

The structure of piezoresistive accelerometer is the same as piezoelectric accelerometer. The piezoelectric material is substituted by a Wheatstone bridge of resistors incorporating one or more legs that change value when strained. The piezoresistive material's resistance value decreases when it is subjected to a compressive force and increases when a tensile force is applied. A seismic mass is fabricated at the center of the sensor die. This mass behaves like a pendulum, responding to acceleration and causing deflection of the diaphragm.

21.4.3 Capacitance Accelerometer

Capacitance accelerometers are usually designed as parallel-plate air-gap capacitors in which motion is perpendicular to the plates. The sensor of a typical capacitance accelerometer is constructed of three silicon elements bonded together to form a hermetically sealed assembly shown in Figure 21.12a. Two of

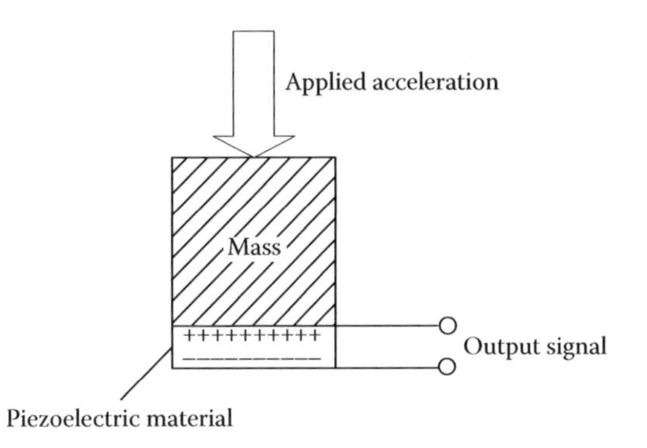

FIGURE 21.11 Structure of piezoelectric accelerometer.

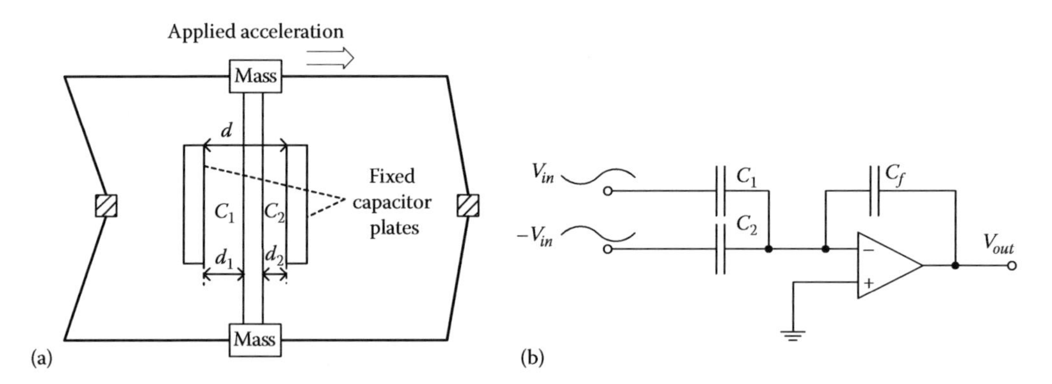

FIGURE 21.12 Capacitance accelerometer: (a) structure and (b) circuit.

the elements are the electrodes of an air dielectric, parallel-plate capacitor. The middle element is a rigid central mass. Changes in capacitance due to acceleration are sensed by a pair of current detectors that convert the changes into voltage output.

Using the circuit in Figure 21.12b, the relationship between output voltage and capacitance change can be described as

$$V_{out} = \frac{C_2 - C_1}{C_f} V_{in} \tag{21.4}$$

By combining Equations 21.2 and 21.4,

$$V_{out} = \frac{A\varepsilon_0\varepsilon_r}{C_f} V_{in}\left(\frac{1}{d_2} - \frac{1}{d_1}\right) \tag{21.5}$$

With the Newton's second law, it can be obtained that

$$\frac{d_1 - d_2}{2} = \frac{at^2}{2} \tag{21.6}$$

where

t is the time cost for changing

a is the accelerator

Since

$$d_1 + d_2 = d \tag{21.7}$$

where d is the distance between the two fixed capacitor plates.

By combining Equations 21.6 and 21.7, the accelerator can be calculated as

$$a = \frac{\sqrt{4 + \left(V_{out}/V_{in}\right)^2 \left(C_f^2 d^2 / A^2 \varepsilon_r^2 \varepsilon_0^2\right)} - 2}{(V_{out}/V_{in})(C_f/A\varepsilon_r\varepsilon_0)t^2} \tag{21.8}$$

21.5 Temperature Sensors

The temperature of electronic systems needs to be measured at regular intervals. The temperature testing must be interpreted and processed properly, so that actions can be taken to counteract the unwanted temperature change. There are an abundance of applications where temperature must be monitored and controlled. Many devices have been developed to match the widely varying technical and economic requirements, including thermistor, thermocouple, and integrated circuit (IC) temperature sensors.

21.5.1 Thermistor

Thermistors are special solid temperature sensors that behave like temperature-sensitive electrical resistors that are fabricated in forms of droplets, bars, cylinders, rectangular flakes, and thick films. There are basically two types: negative temperature coefficient (NTC) thermistors, used mostly in temperature sensing; and positive temperature coefficient (PTC) thermistors, used mostly in electrical current control.

The resistances of the NTC thermistors decrease with the increase of temperature. They are composed of metal oxides. The most commonly used oxides are those of manganese, nickel, cobalt, iron, copper, and titanium. The relationship between the resistance and temperature is highly nonlinear. The most popular equation is the exponential form

$$R_t = R_{t0} \exp\left(\beta\left(\frac{1}{T} - \frac{1}{T_0}\right)\right) \tag{21.9}$$

where

T_0 is the calibrating temperature
R_{t0} is the resistance at calibrating temperature
β is a material's characteristic temperature

The NTC α can be found by

$$\alpha = \frac{1}{R}\frac{\partial R}{\partial T} = -\frac{\beta}{T^2} \tag{21.10}$$

One may notice that NTC thermistors depend on both β and T. An NTC thermistor is much more sensitive at lower temperatures and its sensitivity drops fast with a temperature increase. In the reality, β is not constant and depends on temperature. The sensitivity α varies over the temperature range from 2% to 8%/°C, which implies that this is a very sensitive device.

PTC thermistors are generally made by introducing small quantities of semiconducting material into a polycrystalline ceramic, usually barium titanate or solid solutions of barium and strontium titanate. These ceramic PTC materials in a certain temperature range are characterized by very large temperature dependence. Above the Curie temperature of a composite material, the ferroelectric properties change rapidly, resulting in a rise in resistance. The coefficient changes vary significantly with temperature and may be as large as 2/°C.

Thermistors have high sensitivity and accuracy, and fast response. However, they have limited temperature range and a nonlinear resistance–temperature relationship. Errors may also be generated from self-excitation currents being dissipated by the thermistors.

21.5.2 Thermocouple

Thermocouples operate under the principle of Seebeck. Thermocouples for practical measurement of temperature are junctions of specific alloys that have a predictable and repeatable relationship

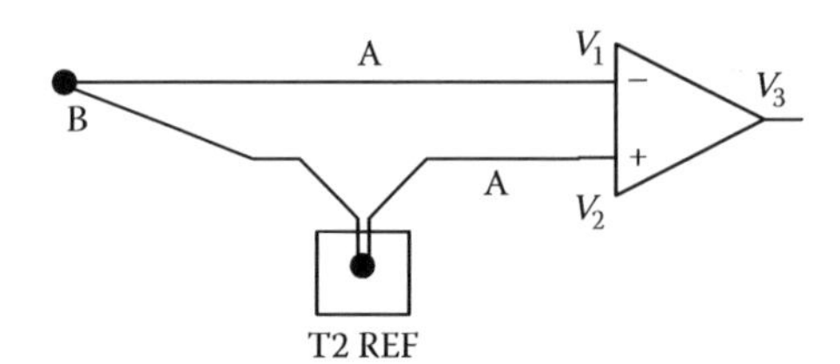

FIGURE 21.13 The structure of thermocouple.

between temperature and voltage. The typical structure of thermocouple is shown in Figure 21.13. The Seebeck voltage developed by the two junctions is $V_1 - V_2$. Absolute temperature measurements can be made once one of the junctions is held at a known temperature, or an electronic reference junction is used. The three most common thermocouple materials for moderate temperatures are Iron-Constantan (Type J), Copper-Constantan (Type T), and Chromel-Alumel (Type K).

Thermocouple can be made in very tough designs; they are very simple in operation and measure temperature at a point. Over different types, they cover from −250°C to +2500°C.

21.5.3 IC Temperature Sensors

IC temperature sensors employ the principle that a bipolar junction transistor (BJT)'s base-emitter voltage to collector current varies with temperature:

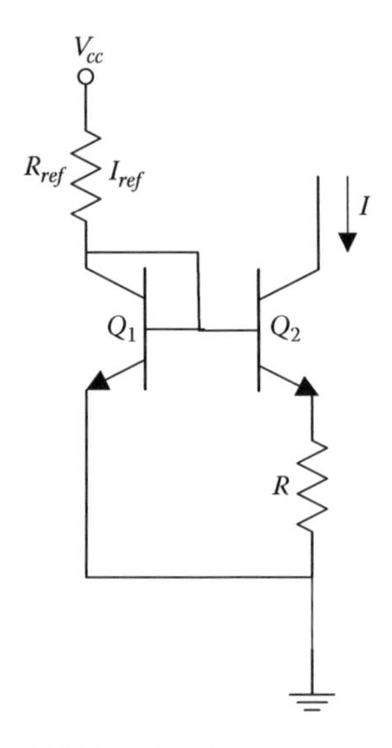

FIGURE 21.14 Circuit of silicon temperature sensor.

$$V_{BE} = \frac{kT}{q} \ln\left(\frac{I_{CE}}{I_S}\right) \tag{21.11}$$

Figure 21.14 shows an example of a circuit utilizing this principle. If currents in a fixed ratio or equal currents flow through one transistor and a set of N identical paralleled transistors, the current change can be calculated by

$$I = \frac{\Delta V_{BE}}{R} = \frac{(kT/q)\ln\left(I_{ref}/I_S\right) - (kT/q)\ln\left(I/I_S\right)}{R} = \frac{kT}{qR} \ln\left(\frac{I_{ref}}{I}\right) \tag{21.12}$$

One may notice that the current change I is proportional to absolute temperature.

The fundamental design of IC temperature sensor results from the diode and silicon temperature sensor [FL02]. The use of IC temperature sensors is limited to applications where the temperature is within a −55°C to 150°C range. The measurement range of IC temperature sensors is small compared to that of thermocouples, but they have several advantages: they are small, accurate, and inexpensive and are easy to interface with other devices such as amplifiers, regulators, DSPs, and microcontrollers.

21.6 Radiation Sensors

Radiation sensors convert incident radiant signals into standard electrical output signals. Radiation, which can be classified by their frequency from low to high, is microwave, infrared, visible, and ultraviolet radiation, respectively. The common source of radiation can be thermal radiation, which includes the solar, earth, atmosphere, human body, incandescent, gas-discharge light source, solid-state laser, and semiconductor light source. There are mainly two fundamental detectors for radiation detecting: bolometer and photon detector.

21.6.1 Bolometer

A bolometer is a device that heats up by absorbing radiation. The change in temperature is sensed in some way, resulting in the signal. It can be used to measure the energy of incident electromagnetic radiation. There are two fundamental components in the bolometer: a sensitive thermometer and high cross-section absorber. Both thermometer and absorber are connected by a weak thermal link to a heat sink. The incoming energy is converted to heat in the absorber. Then, the temperature increase decays as power in absorber flows out to the heat sink. The temperature increase is proportional to the incoming energy.

There are several types of bolometer due to different sensitive thermometer, such as bulk resistor, diode, and pyroelectric material.

The bolometer uses high thermo-resistance material such as a suitable metal, VO_x, amorphous silicon, and semiconductors. In order to determine the amount of absorbed radiation, the resistor has to be biased by a pulse with its duration limited to tens of microseconds. Otherwise, the bolometer would be damaged by excessive Joule heat dissipated within the bolometer membrane.

A semiconductor diode made of single-crystal silicon has been applied to replace resistive temperature sensor due to its excellent stability and low noise. Diodes as temperature sensors can be operated in various modes, such as constant current or constant voltage either forward biased or reverse biased. The major problem is thermal isolation of a silicon island containing the diode from the substrate required for the bolometer operation. However, the problem could be solved by electrochemical etch stop technique or utilizing silicon-on insulator wafers.

Pyroelectric bolometer is based on a pyroelectric crystal covered by absorbing layer (silver or silver blackened with carbon). Once the radiation is absorbed by pyroelectric bolometer and its temperature rises, the polarization of dipolar domains inside the crystal changes to more chaotic, which results in an electric current flowing across the crystal. The current can be amplified by current-to-voltage converter based on operational amplifier. The amplifier should be placed as close as possible to the detector to reduce the noise. Pyroelectric sensors have a flat response over wide spectral range.

21.6.2 Photon Detectors

In photon detectors, light energy quantum produces free electrons and causes the change of electric signal. The photon must have sufficient energy to exceed some thresholds. The wavelength must be shorter than the cutoff wavelength. Photodiode, photoresistor, and photomultiplier are three typical photon detectors that utilize photoelectric effect.

21.6.3 Photoresistor

Photoresistor is a photoconductive device. It requires a power source as it does not generate resistance because the photoconductive effect is manifested in change in the material's resistance. Figure 21.15 shows a schematic diagram of a photoresistor. In darkness, the resistance of the material is high. Hence, applied voltage V results in small dark current due to temperature effect. When light is incident on the surface, current flows.

Consider an optical beam of power P and frequency ν, which is incident on a photoconductive detector. Taking the probability for excitation of a carrier by an incident photon, the quantum efficiency η, the carrier generation rate is

$$G = \frac{P\eta}{h\nu} \tag{21.13}$$

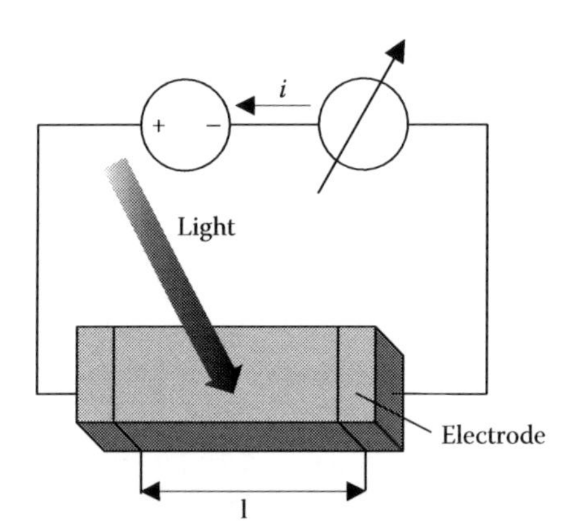

FIGURE 21.15 Schematic diagram of a photoresistor.

If the carriers last on the average seconds τ_0 before recombining, the average number of carriers N_c is found by equating the generation rate to the recombination rate:

$$N_c = G\tau_0 = \frac{P\eta\tau_0}{hv} \tag{21.14}$$

To a current in the external circuit,

$$i_e = \frac{e\bar{v}}{l} \tag{21.15}$$

where l is the length of semiconductor.

The total current

$$\bar{i} = N_c i_e = \frac{P\eta\tau_0 e\bar{v}}{hvl} = \frac{e\eta}{hv}\left(\frac{\tau_0}{\tau_l}\right)P \tag{21.16}$$

where $\tau_l = l/\bar{v}$ is the drift time for a carrier across the length d.

The resistor is obtained by

$$R = \frac{V}{i} = \frac{Vhv\tau_l}{e\eta\tau_0 P}$$

where V is the bias voltage. This equation describes the response of a photoconductive detector to a constancy optical flux. It can be seen that

$$R \infty \frac{1}{P} \tag{21.17}$$

It can be shown that for large sensitivity, the carrier life time τ_0 should be larger. Photoresistor has nonlinear responsibility and cannot be used to detect short pulse. In addition, it is very sensitive to temperature variation.

Depending on their spectral responsively function, photoresistors are divided into photoconductive detectors for the visible wavelength range (such as cadmium sulfide or CdS photoconductive detectors), the near-infrared wavelength range (such as lead sulfide or PbS photoconductive detectors), and the infrared wavelength range (such as silicon doped with arsenide or Si:As photoconductive detectors and the mercury–cadmium–telluride or HgCdTe photoconductive detector).

21.6.4 Photodiode

Photodiode is a type of photodetector device converting light into either current or voltage depending on the mode of operation. When a photon of sufficient energy strikes the diode, it excites an electron, thereby creating a mobile electron and a positively charged electron hole. If the absorption occurs in the junction's depletion region, or one diffusion length away from it, these carriers are swept from the junction by the built-in field of the depletion region. Thus, holes move toward the anode, and electrons toward the cathode, and a photocurrent is produced.

If a p-n junction is forward biased and exposed to light of proper frequency, the current increase will be very small with respect to a dark current. If the junction is reversely biased, the current will increase quite noticeably. Because the increase of the reverse bias can widen the depletion layer, decrease the capacitance, and increase the drift velocity, the efficiency of the direct conversion of optical power into electric power becomes quite low. The p-n photodiode possess 1–3 μm wide depletion layer and is usable for visible light with Si and near-infrared light with Ge.

There are two general operating modes for a photodiode: the photoconductive and the photovoltaic.

For photovoltaic mode, no bias voltage is applied. The result is that there is no dark current, so there is only thermal noise present. This allows much better sensitivities at low light levels; however, the speed response is worse due to an increase in internal capacitance. For photoconductive operating mode, a reverse bias voltage is applied to the photodiode. The result is a wider depletion region, lower junction capacitance, lower series resistance, shorter rise time, and linear response in photocurrent over a wider range of light intensities. However, as the reverse bias is increased, shot noise increases as well due to increase in a dark current.

In the sensor technologies, an additional high-resistivity intrinsic layer is present between p and n types of the material, which is called PIN photodiode. PIN has wider depletion region due to very lightly doped region between a p-type semiconductor and n-type semiconductor. The wide intrinsic region makes the PIN diode largely improve the response time and thus enhance sensitivity for longer wavelengths.

The photodiode directly converts photons into charge carriers. The phototransistors can do the same and provide additional current again, resulting in a much higher sensitivity but slow response. The collector-base junction is a reverse biased diode that functions as described above. If the transistor is connected into a circuit containing a battery, a photoinduced current flows through the loop, which includes the based-emitter region. This current is amplified by the transistor as in a conventional transistor, resulting in a significant increase in the collector current.

21.6.5 Photomultiplier

Photomultiplier tube is a typical photoemissive detector whose structure is shown in Figure 21.16. Photomultipliers are constructed from a glass envelope with a high vacuum inside, which houses a photocathode, several dynodes, and an anode. Incident photons strike the photocathode material with electrons being produced as a consequence of the photoelectric effect. The electron multiplier consists of a number of dynodes. Each dynode is held at a more positive voltage than the previous one. The electrons leave the photocathode, having enough incoming energy. As the electrons move toward the first dynode, they are accelerated by the electric field and arrive with much greater energy. Upon striking the first dynode, more low-energy electrons are emitted, and these electrons in turn are accelerated toward the second dynode. The number of the electrons is multiplied. When the electrons reach the anode, the accumulation of charge results in a sharp current pulse. This current pulse indicates the arrival of a photon at the photocathode.

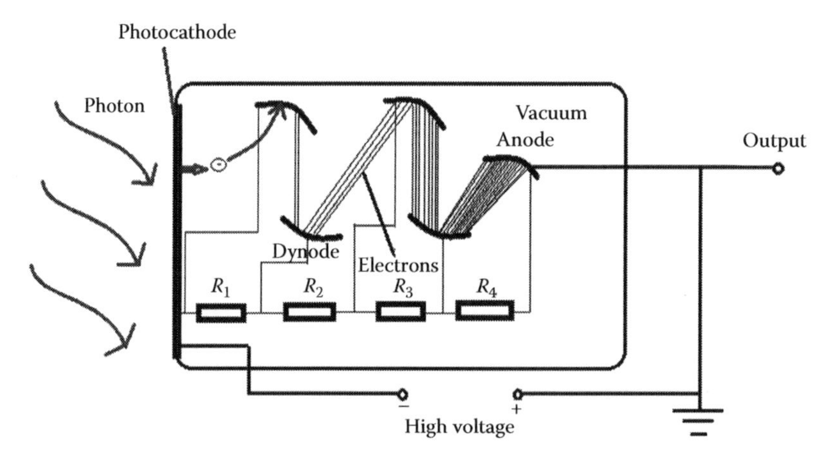

FIGURE 21.16 Structure of photomultiplier tube.

21.7 Magnetic Field Sensors

Magnetic field sensors are used for detecting magnetic field or other physical parameter via magnetic field. Based on their principle, there are four fundamental types of magnetic field sensors: induction coil sensor, Hall effect sensor, fluxgate sensor, and magnetoresistive sensor.

21.7.1 Induction Coil Sensors

The induction coil sensor is one of the oldest and well-known magnetic sensors. Its transfer function results from the fundamental Faraday's law of induction:

$$V = -n\frac{d\Phi}{dt} = -nA\frac{dB}{dt} = -\mu_0 nA\frac{dH}{dt} \tag{21.18}$$

where

Φ is the magnetic flux passing through a coil with an area A

n is a number of turns

The output signal V of a coil sensor depends on the rate of change of flux density dB/dt.

In order to improve the sensitivity, the coil should have large number of turns and large active area. The coil sensors are widely used in detecting displacement, magnetic field, etc.

21.7.2 Hall Effect Sensors

The Hall element is another basic magnetic field sensor. When a current-carrying conductor is placed into a magnetic field, a voltage will be generated perpendicular to both the current and the field. This voltage is the Hall voltage and is given by

$$V_{out} = \frac{IB}{qnd} \tag{21.19}$$

where

d is the thickness of the Hall plate

n is the carrier density

q is the charge of the electron

From the above equation, in order to increase the sensitivity of the Hall device, small thickness d and low carrier concentration n are needed.

This principle is known as the Hall effect, which is shown in Figure 21.17. If the current changes direction or the magnetic field changes direction, the polarity of the Hall voltage flips. The Hall effect sensor can be used to measure magnitude and direction of a field. The Hall voltage is a low-level signal on the order of 30 mV/T. Hall effect can also be used for measuring angular displacement, power, etc.

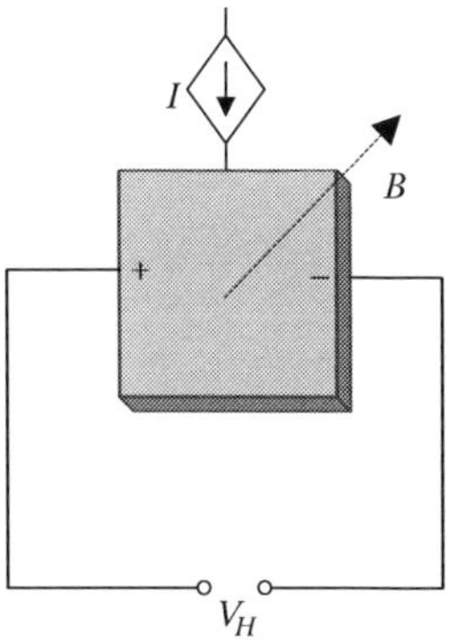

FIGURE 21.17 Principle of Hall effect.

21.7.3 Fluxgate Sensors

Fluxgate sensor is a device measuring the magnitude and direction of the DC or low-frequency AC magnetic field in the range 10^{-10}–10^{-4} T with 10 pT resolution [R92].

Fluxgate sensor exploits the hysteresis of soft magnetic materials. The basic ring-core fluxgate sensor configuration is shown in Figure 21.18. The core is made of soft magnetic material. Two coils are wrapped around these cores: the drive coil and pick-up coil. The drive coil is thought to be two separate half cores. As the current flows through the drive coil, one half of the core will generate a field in the same direction as external magnetic field and the other will generate a field in the opposite direction. When the external measured magnetic field is present, the hysteresis loop of half of the core is distorted. For some critical value of magnetic field, half of the core in which the excitation and measured fields have the same direction is saturated. At this moment, the magnetic resistance of the circuit rapidly increases and the effective permeability of the other half core decreases. As a result, the other half core comes out in saturation later. There is a net change in flux in the pick-up coil. According to Faraday's law, this net change in flux induces a voltage in the same direction as external magnetic field. Consequently, there are two spikes in voltage for each transition in the drive and the induced voltage is at twice the drive frequency.

21.7.4 Magnetoresistive Sensors

The magnetoresistive effect is the change of the resistivity of a material due to a magnetic field. There are two basic principles of magnetoresistive effect [M97]. The first which is similar to Hall elements is

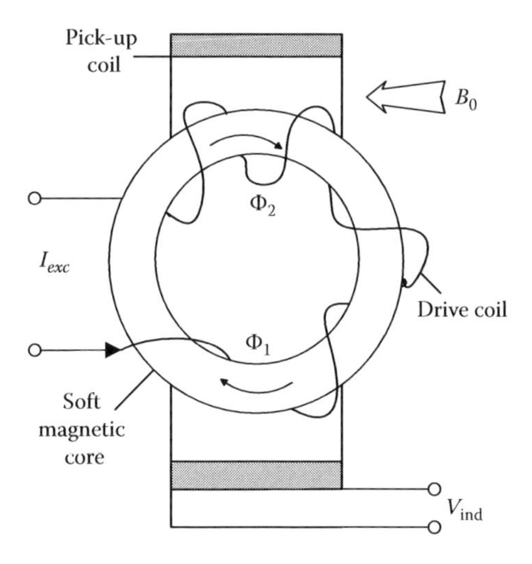

FIGURE 21.18 Ring core fluxgate sensor.

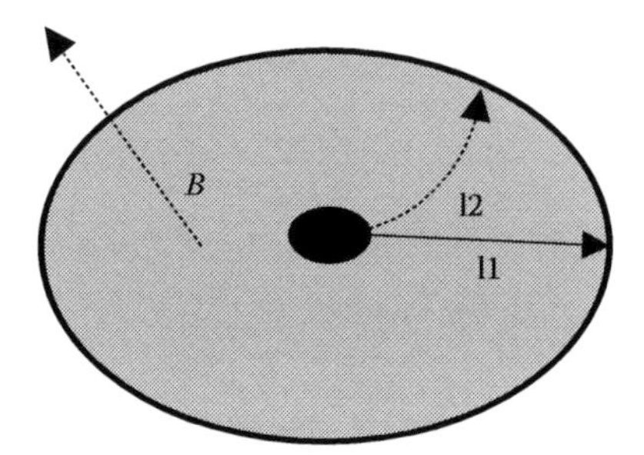

FIGURE 21.19 Scheme of a Corbino ring.

known as the geometrical magnetoresistive effect. The magnetic field forces electrons to take a longer path and thus the resistance of the material increases. A relationship between magnetic field and current is established. The resistance of the device becomes a measure of the field. The relation between field and current is proportional to B^2 for most configurations. It is dependent on carrier mobility in the material used. The Corbino ring shown in Figure 21.19 is a particularly useful configuration for magnetoresistor. Without the magnetic field, the radial current flows in the conducting annulus due to the battery connected between the conductivity rims. When a magnetic field along the axis is turned on, the Lorentz force drives a circular component of current and the resistance between the inner and outer rims goes up. Geometrical magnetoresistor is used in a manner similar to Hall elements or magnetoresistive read heads where Hall elements cannot be used. It is much more sensitive than Hall elements.

The second principle is based on those materials with highly anisotropic properties, such as ferromagnetics, whose resistances are changed in the presence of a magnetic field when a current flows through them. Unlike geometrical magnetoresistor, the effect is due to change of their magnetization direction due to application of the field, thus called anisotropic magnetoresistor. A magnetoresistive material is exposed to the magnetic field to be sensed. A current passes through the magnetoresistive material at the same time. Magnetic field is applied perpendicular to the current. The sample has an internal magnetization vector parallel to the flow of current. When the magnetic field is applied, the internal magnetization changes direction by an angle α. Magnetoresistor offers several advantages as compared with other magnetic sensors. First, their mathematical model is a zero-order system. This differs from inductive sensors in which response depends on the time derivative of magnetic flux density. When compared with Hall effect sensors, which also have a first-order model, magnetoresistor shows increased sensitivity, temperature range, and frequency passband. Magnetoresistor can be used for position sensor, pressure sensor, accelerometer, etc.

21.8 Sensorless Control System

Instead of measuring a parameter that is difficult to measure, other parameters are being measured and the value of this differs to measuring parameters that are evaluated by countless physical effects. The sensorless control system estimates the parameters without using a detector. For example, inductive motor utilizes the sensorless control system to estimate torque, speed, and flux without sensor.

Permanent-magnet synchronous (PMS) motor with high-energy permanent magnet materials provides fast dynamics, efficient operation, and very good compatibility. The electric windings of PMS motors are spaced on the armature at regular angles. When excited with current, they each produce magnetic fluxes that add vectorially in space to produce the stator flux vector. By controlling the proportions of currents in the windings, flux magnitude and orientation are determined.

There are two competing control strategies for PMS motors: vector control (VC) and direct torque control (DTC) technology. Sensorless VC is a variable frequency drive (VFD) control strategy that improves motor performance by regulating the VFD output based on a mathematical determination of motor characteristics and operating conditions. Operating conditions are estimated from measurements of electrical parameters. Both types of VC provide improved performance over the basic control strategy called "V/Hz control." With the simplest type of V/Hz, control, the drive simply acts as a power supply that provides an adjustable output frequency with output voltage proportional to frequency. To provide the maximum possible motor torque with the minimum possible current, the motor voltage needs to be "fine tuned" to suit the exact motor characteristics and to compensate for changes in those characteristics due to load changes and motor temperature changes.

For DTC, the required measurements for this control technique are only the input currents. The input of the motor controller is the reference speed, which is directly applied by the pedal of the vehicle. The DTC technique is based on the direct stator flux and torque control. The input voltage (v_s) and current (i_s) of the motor on the stationary reference frame can be expressed as

$$v_s = v_{s\alpha} + jv_{s\beta}$$

$$i_s = i_{s\alpha} + ji_{s\beta}$$

The actual stator flux can be estimated from the equivalent circuit of the motor as follows:

$$\psi_s = \int (v_s - R_s i_s)dt + \psi_{s0}$$

$$\phi_s = \sqrt{\psi_{s\alpha}^2 + \psi_{s\beta}^2}$$

where
- ψ_s is flux vector
- ψ_{s0} is the initial flux vector
- ϕ_s is the flux vector rms value
- R_s is the stator resistance

The electromagnetic torque of the motor is

$$T_e = \phi_{s\alpha} i_{s\beta} - \phi_{s\beta} i_{s\alpha}$$

The control command for the system is speed. The flux reference can be calculated based on the speed. Below the rated speed, rated flux is used as a reference (constant torque region). Above the rated speed, a relatively large rated flux may imply the need to exceed the supply voltage limits to maintain speed. Therefore, a flux-weakening method generates the flux reference for higher-than-rated speed. The reference flux is selected, proportional to the inverse of the reference speed. The reference torque can be calculated using the difference between reference speed and instantaneous speed.

These inductive motors can be widely used in industries such as oil well diagnosis [B00]. The information from the terminal characteristics of the induction motor could be combined with the fault identification of electrical motors by neural networks. With this approach, the quality of the oil well could be monitored continuously and proper adjustments could be made.

References

[ABLMR01] M. C. Amann, T. Bosch, M. Lescure, R. Myllylä, M. Rioux, Laser ranging: A critical review of usual techniques for distance measurement. *Optical Engineering*, 40, 10–19, 2001.

[B00] B. M. Wilamowski, O. Kaynak, Oil well diagnosis by sensing terminal characteristics of the induction motor, *IEEE Transactions on Industrial Electronics*, 47, 1100–1107, 2000.

[B97] L. K. Baxter, *Capacitive Sensors: Design and Applications*. Institute of Electrical & Electronics Engineering, New York, pp. 61–80, 1997.

[FL02] I. M. Filanovsky, Su Tam Lim, Temperature sensor application of diode-connected MOS transistors. *IEEE International Symposium on Circuits and Systems*, 2, 149–152, 2002.

[M97] D. J. Mapps, Magnetoresistive Sensors. *Sensors and Actuators A*, 59, 9–19, 1997.

[N97] N. T. Nguyen, Micromachined flow sensors—A review. *Flow Measurement and Instrumentation*, 8, 7–16, 1997.

[PST96] O. Parriaux, V. A. Sychugov, A. T. V. Tishchenko, Coupling gratings as waveguide functional elements. *Pure and Applied Optics*, 5, 453–469, 1996.

[R92] P. Ripka, Review of fluxgate sensors. *Sensors and Actuators A*, 33, 129–141, 1992.

[SC00] J. Sirohi, I. Chopra, Fundamental understanding of piezoelectric strain sensors. *Journal of Intelligent Material Systems and Structures*, 11, 246–257, 2000.

22

Soft Computing Methodologies in Sliding Mode Control

Xinghuo Yu
RMIT University

Okyay Kaynak
Bogazici University

22.1 Introduction

Sliding mode control (SMC) is a special class of the variable structure systems (VSSs) [34], which has been studied extensively for over 50 years and widely used in practical applications due to its simplicity and robustness against parameter variations and disturbances [35,46]. The essence of SMC is that in a vicinity of a prescribed switching manifold, the velocity vector of the controlled state trajectories always points toward the switching manifold. Such motion is induced by imposing disruptive (discontinuous) control actions, commonly in the form of switching control strategies. An ideal sliding mode exists only when the system state satisfies the dynamic equation that governs the sliding mode for all time. This requires an infinite switching in general to ensure the sliding motion. For details about the fundamentals of SMC, readers are referred to Chapter 13 [27] of this handbook.

Despite the sustained active research on SMC over the last 50 years, the key technical problems such as chattering, the removal of the effects of unmodeled dynamics, disturbances and uncertainties, adaptive learning, and improvement of robustness remain to be the key research challenges that have attracted continuing attention. Various approaches have been developed to address these problems, though there has not been a perfect solution. The integration of other research methodologies has been considered, and many new technologies have been introduced, such as those that will be discussed in this chapter. One key objective of the recent SMC research is to make it more intelligent. Naturally, this leads to the introduction of intelligent agents into SMC paradigms.

Soft computing ([SC], as opposed to hard computing), coined by Professor Lotif Zadeh as one of the key future intelligent systems technologies, has been researched and applied in solving various practical problems extensively. SC methodologies include neural networks (NNs), fuzzy logic (FL), and probabilistic reasoning (PR), which are paradigms for mimicking human intelligence and smart optimization mechanisms observed in the nature [1]. The diffusion of SC methodologies into SMC architectures has received numerous successes and has become a major research topic in SMC theory and applications.

This chapter introduces the basic SC methodologies used in SMC. Section 22.2 outlines the key technical issues associated with it. Section 22.3 briefs the basics of SC techniques that are relevant to SMC. Section 22.4 presents key developments of SC methodologies in SMC. Section 22.5 concludes the chapter.

22.2 Key Technical Issues in SMC

22.2.1 Fundamentals of SMC and Its Design Methods

In recent years, the majority of research in SMC has been done with regards to multi-input and multi-output (MIMO) systems; therefore, we will use MIMO system framework as a platform for discussing the integration of SC methodologies in SMC. Readers can also refer to the comprehensive tutorial [6] on the fundamentals of MIMO SMC systems.

Commonly, the MIMO SMC systems considered are of the form

$$\dot{x} = f(x,t) + B(x,t)u + \xi(x,t) \tag{22.1}$$

where

$x \in R^n$ is the system state vector

$u \in R^m$

$\xi(x, t) \in R^n$ represents all the factors that affect the performance of the control system, such as disturbances and uncertainties in the parameters of the system

If $\xi \in$ range B, then there exists a control u_t, such as $\xi = Bu_\xi$. It is well known that when this condition (the matching condition) is satisfied, the celebrated invariance property of SMC stands [7].

As detailed in [27], the design procedure of SMC includes two major steps encompassing two main phases of SMC:

1. Reaching phase, where the system state is driven from any initial state to reach the switching manifolds (the anticipated sliding modes) in finite time.
2. Sliding mode phase, where the system is induced into the sliding motion on the switching manifolds, that is, the switching manifolds become an attractor.

These two phases correspond to the following two main design steps:

1. Switching manifold selection: A set of switching manifolds are selected with prescribed desirable dynamical characteristics. Common candidates are linear hyperplanes.
2. Discontinuous control design: A discontinuous control strategy is formed to ensure the finite time reachability of the switching manifolds. The controller may be either local or global, depending upon specific control requirements.

In the context of the system (1), the switching manifolds can be denoted as $s(x) = 0 \in R^m$, where $s = (s_1, \ldots, s_m)^T$ is a m-dimensional vector underpinned by the desired dynamical properties.

The SMC $u \in R^m$ is characterized by the control structure defined by

$$u_i = \begin{cases} u_i^+(x) & \text{for } s_i(x) > 0 \\ u_i^-(x) & \text{for } s_i(x) < 0 \end{cases} \tag{22.2}$$

where $i = 1, \ldots, m$.

It is well known [27] that when the sliding mode occurs, a so-called equivalent control (u_{eq}) is induced which drives the system dynamics $\dot{s} = 0$ and can be written as (assuming $(\partial s/\partial x)B(x,t)$ is nonsingular):

$$u_{eq} = -\left(\frac{\partial s}{\partial x}B(x,t)\right)^{-1}\left(\frac{\partial s}{\partial x}\left(f(x,t)+\xi(x,t)\right)\right) \tag{22.3}$$

And in the sliding mode, the system dynamics is immune to the matched uncertainties and external disturbances.

There are several SMC controller types seen in the literature; the choice of the type of controller to be used is dependent upon the specific problems to be dealt with. However, central to the SMC design is the use of the Lyapunov stability theory, in which the Lyapunov function of the form

$$V = \frac{1}{2}s^T s \tag{22.4}$$

is commonly taken. The control design task is then to find a suitable discontinuous control such that $\dot{V} < 0$ in the neighborhood of the equilibrium $s = 0$.

If one would like to embed learning and adaptation in SMC to enhance its performance, one common alternative Lyapunov function may be constructed as

$$V = \frac{1}{2}s^T Qs + \frac{1}{2}z^T Rz \tag{22.5}$$

where

$z = 0$ represents a desirable outcome

R and Q are symmetric nonnegative definite matrices of appropriate dimensions

For example, if one wants to enable system parameter learning while controlling, say for a set of system parameters $p \in R^l$ to converge to its true values p^*, one may set $z = p - p^*$. Therefore, $\dot{V} < 0$ in the neighborhood of $V = 0$ would lead to the convergence of the system states during which certain parameters would be learned simultaneously. However, it should be pointed out that the set of parameters are not necessarily convergent toward their true values, that is, z may converge to some nonzero constants.

Typical SMC strategies for MIMO systems include

1. Equivalent control-based SMC: This is a control of the type

$$u = u_{eq} + u_s \tag{22.6}$$

where u_s is a switching control component that may have two types of switching

- $u_{s_i} = -\alpha_i(x)\mathrm{sgn}(s_i)$ for $\alpha i(x) > 0$ or
- $u_{s_i} = -\beta\frac{s_i}{||s_i||}, \quad \beta > 0$

and there are other variants.

2. Bang-bang type SMC: This is a control of direct switching type

$$u_{s_i} = -M\,\mathrm{sgn}(s_i)$$

where $M > 0$ should be large enough to suppress all bounded uncertainties and unstructured systems dynamics. Design of such control relies on the Fillipov method, by which, a sufficient large local attraction region is created to suck in all system trajectories.

3. Enforce

$$\dot{s} = -R\text{sgn}s - K\sigma(s) \tag{22.7}$$

to realize finite time reachability [18], where $R > 0$ and $K > 0$ are diagonal matrices and $s_i\sigma(s_i) < 0$.

4. Embedding adaptive estimation and learning of parameters or uncertainties in (1), (2), and (3), which largely follows the methodologies of adaptive control, for example, [16] and iterative learning control mechanism, for example, [38].

22.2.2 Key Technical Issues in SMC and Applications

There are several key issues that are commonly seen as obstacles affecting the widespread applications of SMC, which have prompted the extensive use of SC methodologies in SMC systems in recent years. Some of these are listed in the following.

22.2.2.1 Chattering

As has been mentioned above, SMC is a special class of VSSs, in which the sliding modes are induced by disruptive control forces. As demonstrated in the previous sections, despite the advantages of simplicity and robustness, SMC generally suffers from the well-known problem, namely *chattering*, which is a motion oscillating around the predefined switching manifold(s). Two causes are commonly conceived [42].

- The presence of parasitic dynamics in series with the control systems causes a small amplitude, high-frequency oscillation. These parasitic dynamics represent the fast actuator and sensor dynamics that are normally neglected during the control design.
- The switching nonidealities can cause high-frequency oscillations. These may include small time delays due to sampling (e.g., zero-order-hold [ZOH]), and/or execution time required for calculation of control, and more recently transmission delays in networked control systems.

Various methods have been proposed to "soften" the chattering, for example,

- The boundary layer control in which the sign function is replaced by $\text{sgn}_b(s_i) = \text{sgn}s_i$, if $|s_i| > o_i$, $o_i > 0$, and $\text{sgn}_b(s_i) = \rho(s_i)$ if $|s_i| \leq o_i$, where o_i is small and $\rho(s_i)$ is a function of s_i, which could be a constant or a function of states as well [29].
- The continuous approximation method in which the sign function is replaced by a continuous approximation $\text{sgn}_a(s_i) = (s_i/(|s_i| + \varepsilon))$, where ε is a small positive number [3]. This, in fact, gives rise to a high gain control when the states are in the close neighborhood of the switching manifold.

22.2.2.2 Matched and Unmatched Uncertainties

It is known that the celebrated invariance property of SMC lies in its insensitivity to matched uncertainties when in the sliding mode [7]. However, if the matching condition is not satisfied, the sliding mode (motion) is dependent on the uncertainties ξ, which is not desirable because the condition for the well-known robustness of SMC does not hold. Research efforts in this aspect have been focussed on restricting the influence of the uncertainties ξ within a desired bound.

22.2.2.3 Unmodeled Dynamics

Mathematically, it is impossible to model a practical system perfectly. There always exists some unmodeled dynamics. The situation may get worsened if the unmodeled dynamics contains high-frequency oscillatory dynamics that may be excited by the high-frequency control switching of SMC. There are several aspects of dealing with this unmodeled dynamics.

- The unmodeled dynamics can refer to the presence of parasitic dynamics in series with the control system as shown above.
- Due to the complexity of practical systems, the unmodeled dynamics that may have different dimensions and characteristics cannot be modeled at all. In this case, the unmodeled dynamics can be commonly lumped with uncertainties and a sufficiently large switching control has to be imposed. Certainly, the more the knowledge of the uncertainties is, the less crude will the SMC force be.

More broadly, in the complex systems environment, the object to be controlled is difficult to model without a significant increase in the dimensions of the problem space. Often, it is more feasible to study the component subsystems linked through aggregation. The challenge is how to design an effective SMC without knowing the dynamics of the entire system (apart from the aggregative links and local models). This research topic has been very popular in recent years in the intelligent control community using, for example, NNs, FL, and evolutionary computation (EC).

22.3 Key SC Methodologies

In this section, we will give a brief introduction to the key SC techniques that are commonly used in dynamical systems and control. Components of SC methodologies are NNs, FL, and PR that subsume belief networks, EC, chaos theory, and parts of learning theory [1].

22.3.1 Neural Networks

An NN is a computational structure that is inspired by observed processes in natural networks of biological neurons in the brain [11]. It consists of a mass of neurons, each of which is a simple computational unit, yet the intersections between the neurons emulate the enormous learning capability of the brain. This learning is done by adjusting the so-called weights that represent the interconnection strength of neurons, according to certain learning algorithms. In general, the learning can be supervised or unsupervised. In a supervised learning algorithm, learning is guided by specifying, for each training input pattern, the class to which the pattern is supposed to belong. The adjustments of weights are done to minimize the difference between the desired and actual outputs incrementally. A common structure is the feedforward neural networks (FNNs) [11] where the information flow in the network is directional.

Supervised learning in FNNs appears to be the most popular research area in dynamical systems and control applications. Generally, it can be formulated as follows. Given the inputs, weights, desired outputs, and actual outputs of an NN as $x \in R^n$, $w \in R^l$, $y_d \in R^m$, and $y \in R^m$, respectively, the target, for learning purposes, is to minimize the performance index, which can be generically written as

$$J = \frac{1}{\tau} \int_{t-\tau}^{t} \frac{1}{2} \| y(\theta) - y_d(\theta) \|^2 \tag{22.8}$$

where

τ is the length of time window

$\|\cdot\|$ represents the Euclidean norm

Most supervised learning (backpropagation, BP) algorithms for NNs can be considered as finding zeros of $\partial J / \partial w$, which correspond to their local as well as global minima. The search performance of this class of learning algorithms somehow relies on initial weights and, frequently, it traps into local minima. As shown in [43], all FNN learning can be generally regarded as finding minimum for the Lyapunov function

$$V = J + \frac{1}{2}\mu\left\|\frac{\partial J}{\partial w}\right\|^2 \tag{22.9}$$

where the parameter μ determines the relative importance of each term in the function.

One particular FNN that offers a much faster way of learning is the so-called radial basis function-based neural network (RBFNN) [28]. In this setting, the network structure is much simpler, that is, $y_i(t) = \sum_{j=1}^{m} w_j^i \phi_j(x - \mu_j, \theta_j)$, where $\phi(x - \mu_j, \theta_j)$ is a Gaussian function. The learning of the weights becomes much simpler as well, using the conventional FNN learning paradigm, although extensive computation is still required and efficient algorithms can be derived [36].

Another class of NNs is the so-called recurrent NNs (RNNs) that has a feedback mechanism [11], similar to dynamic feedback systems. This class of NNs has also been popular in the area of dynamical systems.

22.3.2 Fuzzy Systems

A fuzzy system is any system whose variables (or, at least, some of them) range over states that are fuzzy sets based on FL, which is a form of multi-valued logic derived from fuzzy set theory to deal with reasoning, that is, vague rather than precise. In FL the degree of truth of a statement can range between 0 and 1.

For each variable, the fuzzy sets are defined on some relevant universe of discourses, which are often an interval of real numbers. In this special but important case, the fuzzy sets are fuzzy numbers, and the associated variables are linguistic variables. Representing states of variables by fuzzy sets is a way of quantifying the variables. Fuzzy systems provide an alternative representation framework to express problems that cannot easily be described using deterministic and probabilistic mathematical models.

In the domain of control systems, the output of a fuzzy system can be computed by a mechanism of Takagi–Sugeno–Kang (TSK) type IF-THEN rules [31,32], which can be described as

$$\begin{aligned} &R^i: \text{IF} \quad z_1 \text{ is } F_1^i \quad \text{AND} \quad \ldots z_n \text{ is } F_n^i \\ &\text{THEN} \quad y^i = c^i \end{aligned} \tag{22.10}$$

for $i = 1,2,\ldots,m$, where R^i represents the ith fuzzy inference rule, m is the number of inference rules, $F_i (i = 1,\ldots, n)$ are the fuzzy sets, $z(t)$ is the system state, u is the system input, and y_i is the system output that could be a function c^i. Denote $w^i(z(t))$ as the normalized fuzzy membership function of the inferred fuzzy set F^i where $F^i = \bigcap_{j=1}^{n} F_j^i$; the output of the fuzzy system is formulated as $y = \left(\sum_{i=1}^{m} w^i y^i\right) \Big/ \left(\sum_{i=1}^{m} w^i\right)$. Another well-known fuzzy control method is the Mamdani method [19] that makes use of much intuitive nature of human expert control. However, the TSK method is more suitable for model-based fuzzy control systems.

In the domain of control systems, dynamical systems are commonly dealt with. Hence, in this case, the output function $y^i = c^i$ can be replaced with a dynamical equation, which can be viewed as a local dynamics.

22.3.3 Evolutionary Computation

EC techniques have been a most widely used technology for optimization in SMC. The EC paradigm attempts to mimic the evolution processes observed in nature and utilize them for solving a wide range of optimization problems. EC technologies include genetic algorithms (GAs), genetic programming, evolutionary algorithms, and strategies. In general, EC performs directed random searches using mutation and/or crossover operations through evolving populations of solutions with the aim of finding the best solutions (the fittest survives). The criterion that is expressed in terms of an objective function is usually referred to as a fitness function [10].

22.3.4 Integration of SC Methodologies

The aforementioned SC paradigms offer different advantages. The integration of these paradigms would give rise to powerful tools for solving difficult practical problems. It should be noted that NNs and EC are about a process that enables learning and optimization while the FL systems are a representation tool. Emerging technologies such as swarm optimization, chaos theory, and complex network theory also provide alternative tools for optimization and for explaining emerging behaviors that are predicted to be widely used in the future.

22.4 Sliding Mode Control with SC Methodologies

There have been extensive research done in using SC methodologies for SMC systems. We will use the same categories as in Section 22.3 to summarize the developments. The integration of SMC and SC has two dimensions. One is the application of SC methodologies in SMC to make it "smarter" and the other using SMC to enhance SC capabilities. This chapter will be confined within the scope of the former, that is, the application of SC methodologies in SMC.

22.4.1 SMC with NNs

As has been outlined before, there are three essential interconnected aspects that affect usability and applications of SMC, namely, chattering, matched and unmatched uncertainties, and unmodeled dynamics. NNs offer a "model-free" mechanism to learn from examples the underlying dynamics. Conventional NNs learn by employing BP-type of learning as stipulated in [11,43]. This learning process is usually very time consuming.

Unlike conventional learning of constant parameters or functions, learning speed is very important in dynamical environments. Reducing learning errors rather than obtaining exact parameter values is of paramount importance. In control theory, the application of NNs is done in the context of dynamical learning focusing on achieving stability and fast convergence, rather than reducing learning errors. The convergence of learning parameters to their true values becomes secondary, though, under certain conditions, for example, in adaptive learning and control, when the persistence excitation condition is satisfied, such convergence is guaranteed [29].

One of the early attempts in using NNs in SMC [20] aimed to estimate the equivalent control. An online NN estimator is constructed and the NN is trained using the alternative function $V = 0.5(\dot{s} + ds)^2$ instead of (22.4). The same approach is also contemplated in [14]. In another work [8], two NNs are used to approximate the equivalent control and the correction control, respectively. NNs in control systems adopt the same topological structures but the learning is done through real-time estimation.

In terms of learning mechanisms, while the conventional BP-based NN learning paradigm is usable, there is another trend of using the control design techniques to derive the learning algorithms with convergence to the true parameter values being a secondary requirement. Instead of using the conventional NN performance index (22.9) for learning, a performance index of adaptive control type (22.5) is used, where z is defined as functions that enable the convergence of NN weights to deliver the convergence $V \to 0$.

In the literature, there are several research works using similar ideas. For example, in [2], a dynamical NN approach is used to estimate the unmodeled dynamics to reduce the requirement of large switching control magnitude using the Lyapunov function (22.5), the stability as well as modeling of the uncertainties are guaranteed for an electronic throttle. In [40], the switching manifolds are used as performance criteria and adaptive algorithms are derived for learning and control. In [21], a NN estimator is used to estimate the modeling errors to compensate the SMC to reduce the tracking errors in discrete nonlinear systems. In [24], a NN is used to estimate the lumped unmatched uncertainties with time delayed states. Adaptive algorithms are derived for estimating parameters and guarantee stability. It is indicated that the normal requirement of bounded uncertainties is not required in the control scheme. This approach is proved to be effective in reducing chattering and practically implemented on a see-saw system.

22.4.2 SMC with FL

The strength of fuzzy systems based on FL has been taken advantage of in SMC design due to its simple representation, underpinned by the heuristic nature of human reasoning. The rationale central to the use of fuzzy systems with SMC is similar to those for NNs, that is, to alleviate the problems associated with chattering, matched and unmatched uncertainties, and unmodeled dynamics. The key advantages include introducing heuristics underpinned by human expert experience to reduce (or soften) chattering, to model the unmodeled uncertainties based on partial knowledge of experts through their years of experience, and to control complex systems that are modeled through aggregated linear models via fuzzy logic. This section will outline major developments in this respect.

22.4.2.1 Dealing with Chattering

One of the early works in this respect uses FL in a low pass filter to "smooth" SMC signals, which would reduce chattering [13]. This is done by forming a specifically constructed fuzzy rule table.

Another approach to dealing with chattering is to fuzzify SMC. Fuzzifying the sliding-mode concept and the reaching phase to the sliding manifold involves replacing the otherwise well-defined mathematical derivations with heuristic rules underpinned by FL [25,37].

One of the earliest works in which the fuzzy concepts are introduced is [22]. In this paper, instead of using the well-defined SMC functions, the control action is derived from an FL-based control table, compounded by fuzzy membership functions. In such a way, the landing on the switching manifold and overshooting it is softened, and this eases the chattering. In [26], the softening FL idea is introduced in tuning the switching magnitude according to the distance from the state to the switching manifold.

The main idea behind the use of SMC with FL is to use human control intuition and logic to enhance SMC performance. For example, one simple and direct way to design a fuzzified SMC is by using the actual position to the switching manifold and the velocity toward it as two inputs and to derive rules as to what control action is required for the state to reach the switching manifold. Consider the case when the following standardized fuzzy values are used, that is, NB = "negative big," NS = "negative small," ZO = "zero," PB = "positive big," and PS = "positive small". If the state of the trajectory to the switching manifold is NB and the velocity toward the switching manifold is NS, an intuitive control action should be PB since only a drastic action can reverse the speedy trend of the trajectory leaving the switching manifold. However, if the state of the trajectory to the switching manifold is PB and the velocity toward the switching manifold is NB, meaning that the system dynamics has a useful tendency toward the switching manifold, one may choose NS or NB as the control output to further accelerate the convergence. There are many choices, depending upon what control performance is required. Certainly, the control choices are not unique. The key to fuzzified SMC lies in the selection of membership functions associated with the fuzzy values. The above example serves as a typical case for fuzzified SMC and similar ideas have been used in a number of papers. For example, in a recent work [39], a robust fuzzy sliding-mode controller is proposed for active suspension of a nonlinear half-car model in which the ratio of the derivative of the error to the error is used as the input to the controller.

22.4.2.2 Dealing with Uncertainties

Uncertainties are another cause of chattering. Efforts to reduce chattering are directed to ensure that information about the uncertainties and the unmodeled dynamics be obtained as much as possible in order to reduce the requirements of high enough (crude) switching gains to suppress them and to secure the existence of sliding modes and hence the robustness. The estimation of uncertainties using fuzzy systems has been quite a popular approach. In this section, we also discuss uncertainties associated with unmodeled dynamics, that is, the uncertainties also include lumped unmodeled dynamics, which may refer to the presence of parasitic dynamics in series of the control systems that causes small amplitude high-frequency oscillations. These parasitic dynamics are normally neglected during the control design. It may also refer to other residual dynamics that cannot be modeled. However, the latter is difficult to handle due to the uncontrollable nature of the unmodeled dynamics. An example of using FL in SMC includes a fuzzy system architecture [4] to adaptively model the nonlinearities and the uncertainties. In [41], schemes are proposed to estimate the unknown functions in order to reduce the magnitudes of the switching controls.

22.4.2.3 Dealing with Complex Systems

More broadly, in the complex systems environments, the object to be controlled is difficult to model without a significant increase in the dimension of the problem space. The feasible solutions in such cases are often to study the component subsystems, networked through a topological structure. The challenge is how to design an effective SMC without knowing the dynamics of the entire system (apart from the topological links). This direction of research has become very popular in recent years in the intelligent control community using, for example, NNs, FL, and EC technologies.

For complex dynamical systems, it is possible to linearize them around some given operating points such that the well-developed linear control theory can be applied in the local region. Such a treatment is quite common in practice. However, there may exist a number of operating points in complex nonlinear systems that should be considered during control. To aggregate the locally linearized models into a global model to represent the complex system is not an easy task. One effective approach is the FL approach, by which the set of linearized mathematical models is aggregated into a global model that is equivalent to the complex system. Various fuzzy models and their controls have been discussed by, for example, Sugeno and Kang [30], Takagi and Sugeno [32], and Feng et al. [9]. Such a formulation can be easily accommodated by replacing $y^i = c^i$ by a dynamic linear system [9]

$$\dot{x}(t) = A_i x(t) + B_i u(t), \quad y_i(t) = D_i x(t) + E_i$$

where
the matrices A_i, B_i, D_i are of appropriate dimensions
E_i is a constant vector

Using the standard fuzzy inference approach, the global fuzzy state space model can be obtained as

$$\dot{x}(t) = Ax(t) + Bu(t), \quad y = Dx(t) + E \tag{22.11}$$

where

$$A = \sum_{i=1}^{m} \mu_i A_i$$

$$B = \sum_{i=1}^{m} \mu_i B_i$$

$$D = \sum_{i=1}^{m} \mu_i D_i$$

$$E = \sum_{i=1}^{m} \mu_i E_i$$

While the global fuzzy model can approximate the complex system, the aggregated SMC cannot guarantee the global stability. This is because the globally asymptotical stability is not guaranteed in the overlapping regions of fuzzy sets where several subsystems are activated at the same time, unless it satisfies a sufficient condition [44]. In [33], the dynamics of (22.7) is used to enforce the reachability of the sliding modes for MIMO nonlinear systems represented in a similar form as shown in above. A fuzzy adaptive control scheme is then developed to deal with the model following a control problem. In another work [12], robust adaptive fuzzy SMC is proposed for nonlinear systems approximated by the Takagi–Sugeno fuzzy model.

22.4.2.4 SMC with EC

The optimization of the control parameters and the model parameters are vitally important in reducing chattering and improving robustness. In this respect, various optimization techniques can be used [11], such as the gradient-based search, Levenberg–Marquart algorithm, etc.; however, a significant amount of information about the tendency of search parameters toward optima (i.e., the derivative information) is required. The complexity of the problem and the sheer number of parameters make the application of the conventional search methods rather hard. The key process of using EC is to treat the set of parameters (often a large number) as the attributes of an individual in a population and generate enough number of individuals randomly to ensure a rich diversity of "genes" in the population, through bio-inspired operations such as mutation and/or crossover.

The advantages mentioned above have motivated various researchers to use EC as an alternative to find "optimal" solutions for SMC. For example, Lin et al. [17] use a GA to reach an online estimation of the magnitude of switching control to reduce chattering. In [23], a GA is used to optimize the matrix selection to obtain better membership functions for smoother SMC.

22.4.3 SMC with Integrated NN, FL, and EC

It is recognized that NNs and EC are processes that enable learning and optimization while the fuzzy systems are a representation tool [15]. For complex systems that are difficult to model and represent in an analytic form, it is certainly advantageous to use fuzzy systems as a paradigm to represent the complex systems as an aggregated set of simpler models, just like the role local linearization plays in nonlinear function approximation by piecing together locally linearized models. NNs on the other hand can be used as a learning mechanism to learn the dynamics while the EC approaches can be used to optimize the representation and learning. Such ideas have been used quite extensively in dynamic systems and control areas, for example, in [5], two fuzzy NNs are used to learn the control as well as to identify the uncertainties to eliminate chattering in distributed control systems. The conventional BP-based learning mechanism is employed for learning.

22.5 Conclusions

In this chapter, the key developments of SC methodologies in SMC have been discussed. For a comprehensive coverage of literature on this topic, readers are referred to the survey paper [45].

References

1. The Berkeley Institute in Soft Computing. http://www-bisc.cs.berkeley.edu.
2. M. Baric, I. Petrovic, and N. Peric. Neural network-based sliding mode control of electronic throttle. *Engineering Applications of Artificial Intelligence*, 18:951–961, 2005.
3. J. A. Burton and A. S. I. Zinober. Continuous approximation of variable structure control. *International Journal of Systems Science*, 17(6):875–885, 1986.

4. C. C. Chen and W. L. Chen. Robust adaptive sliding mode control using fuzzy modeling for an inverted-pendulum system. *IEEE Transactions on Industrial Electronics*, 45:297–306, 1998.
5. F. Da. Decentralized sliding mode adaptive controller design based on fuzzy neural networks for interconnected uncertain nonlinear systems. *IEEE Transactions on Neural Networks*, 11(6):1471–1480, 2000.
6. R. A. DeCarlo, S. H. Zak, and G. P. Matthews. Variable structure control of nonlinear multivariable systems: A tutorial. *Proceedings of the IEEE*, 76:212–232, 1988.
7. B. Drazenovic. The invariance conditions in variable structure systems. *Automatica*, 5:287–295, 1969.
8. M. Ertugrul and O. Kaynak. Neuro sliding mode control of robotic manipulators. *Mechatronics*, 10:239–263, 2000.
9. G. Feng, S. G. Cao, N. W. Rees, and C. K. Chak. Design of fuzzy control systems based on state feedback. *Journal of Intelligent and Fuzzy Systems*, 3:203–304, 1995.
10. D. Fogel. *Evolutionary Computation—Toward a New Philosophy of Machine Intelligence*. IEEE Press, New York, 1995.
11. S. Haykin. *Neural Networks—A Comprehensive Foundation*. Prentice-Hall, Englewood Cliffs, NJ, 1999.
12. C.-L. Hwang. A novel takagi-sugeno based robust adaptive fuzzy sliding mode controller. *IEEE Transactions on Fuzzy Systems*, 12(5):676–687, 2004.
13. Y. R. Hwang and M. Tomizuka. Fuzzy smoothing algorithms for variable structure systems. *IEEE Transactions on Fuzzy Systems*, 2:277–284, 1994.
14. K. Jezernik, M. Rodic, M. Saaric, and B. Curk. Neural network sliding mode robot control. *Robotica*, 15:23–30, 1997.
15. O. Kaynak, K. Erbatur, and M. Ertugrul. The fusion of computationally intelligent methodologies and sliding-mode control: A survey. *IEEE Transactions on Industrial Electronics*, 48:4–17, 2001.
16. T. P. Leung, Q.-J. Zhou, and C. Y. Su. An adaptive variable structure model following control design for robot manipulators. *IEEE Transactions on Automatic Control*, 36(3):347–353, 1991.
17. F.-J. Lin, W.-D. Chou, and P.-K. Huang. Adaptive sliding-mode controller based on real-time genetic algorithm for induction motor servo drive. *IEEE Proceedings—Electrical Power Applications*, 150(2):1–13, 2003.
18. X. Y. Lu and S. Spurgeon. Robust sliding mode control of uncertain nonlinear systems. *Systems and Control Letters*, 32:75–90, 1997.
19. E. H. Mamdani and S. Assilian. An experiment in linguistic synthesis with a fuzzy logic controller. *International Journal of Man-Machine Studies*, 7(1):1–13, 1975.
20. H. Morioka, K. Wada, A. Sabanovic, and K. Jezernik. Neural network based chattering free sliding mode control. In *Proceedings of the 34th SICE Annual Conference*, pp. 1303–1208, 1995.
21. D. Munoz and D. Sbarbaro. An adaptive sliding mode controller for discrete nonlinear systems. *IEEE Transactions on Industrial Electronics*, 43(3):574–581, 2000.
22. M. De Neyer and R. Gorez. Use of fuzzy concepts in adaptive sliding mode control. In *Proceedings of IEEE Conference on Systems, Man and Cybernetics*, San Antonio, TX, 1994.
23. K. C. Ng, Y. Li, D. J. Murray-Smith, and K. C. Sharman. Genetic algorithms applied to fuzzy sliding mode controller design. In *Proceedings of the First International Conference on Genetic Algorithms in Engineering Systems: Innovations and Applications*, pp. 220–225, 1995.
24. Y. Niu, J. Lam, X. Wang, and D. W. C. Ho. Sliding mode control for nonlinear state-delayed systems using neural network approximation. *IEE Proceedings: Control Theory Applications*, 150(3):233–239, 2003.
25. R. Palm. Sliding mode fuzzy control. In *Proceedings of 1992 IEEE Conference on Fuzzy Systems*, pp. 519–526, San Diego, CA, 1992.
26. R. Palm. Robust control by sliding mode. *Automatica*, 30(9):1429–1437, 1994.
27. A. Sabanovic and N. Sabanovic-Behlilovic. Variable structure control techniques. *Control and Mechatronics*. Taylor & Francis, Boca Raton, FL, 2010.

28. K. P. Seng, Z. Man, and H. Wu. Lyapunov theory based radial basis function networks for adaptive filtering. *IEEE Transactions on Neural Networks*, 49(8):1215–1220, 2002.
29. J.-J. Slotine and W. Li. *Applied Nonlinear Control*. Prentice Hall, Englewood. Cliffs, NJ, 1991.
30. M. Sugeno and G. T. Kang. Fuzzy modeling and control of multilayer incinerator. *Fuzzy Sets and Systems*, 18:329–346, 1986.
31. M. Sugeno and G. T. Kang. Fuzzy identification of fuzzy model. *Fuzzy Sets and Systems*, 28:15–33, 1988.
32. T. Takagi and M. Sugeno. Fuzzy identification of systems and its application to modeling and control. *IEEE Transactions on Systems, Man, and Cybernetics*, 15:110–32, 1985.
33. S. Tong and H. Li. Fuzzy adaptive sliding mode control of mimo nonlinear systems. *IEEE Transactions on Fuzzy Systems*, 11(3):354–360, 2003.
34. V. I. Utkin. Variable structure systems with sliding modes. *IEEE Transactions on Automatic Control*, 22:212–222, 1977.
35. V. I. Utkin. *Sliding Modes in Control and Optimization. Communications and Control Engineering Series*. Springer Verlag, Berlin, Germany, 1992.
36. B. M. Wilamowski, N. J. Cotton, O. Kaynak, and G. Dundar. Computing gradient vector and jacobian matrix in arbitrarily connected neural networks. *IEEE Transactions on Industrial Electronics*, 55(10):3784–3790, 2008.
37. J. C. Wu and T. S. Ham. A sliding mode approach to fuzzy control design. *IEEE Transactions on Control Systems Technology*, 4:141–151, 1996.
38. J.-X. Xu and Y. Tan. *Linear and Nonlinear Iterative Learning Control*. Lecture Notes in Control and Information Science. Springer, Berlin, Germany, 2003.
39. N. Yagiz, Y. Hacioglu, and Y. Taskin. Fuzzy sliding-mode control of active suspensions. *IEEE Transactions on Industrial Electronics*, 55(10), 2008.
40. Y. Yildiz, A. Sabanovic, and K. S. Abidi. Sliding mode neuro controller for uncertain systems. *IEEE Transactions on Industrial Electronics*, 54(3):1676–1685, 2007.
41. B. Yoo and W. Ham. Adaptive sliding mode control of nonlinear systems. *IEEE Transactions on Fuzzy Systems*, 6:315–321, 1998.
42. K. D. Young, V. I. Utkin, and U. Ozguner. A control engineer's guide to sliding mode control. *IEEE Transactions on Control Systems Technology*, 7:328–342, 1999.
43. X. Yu, O. Efe, and O. Kaynak. A general backpropagation algorithm for feedforward neural networks learning. *IEEE Transactions on Neural Networks*, 13:251–259, 2002.
44. X. Yu, Z. Man, and B. Wu. Design of fuzzy sliding mode control systems. *Fuzzy Sets and Systems*, 95(3):295–306, 1998.
45. X. Yu and O. Kaynak. Sliding mode control with soft computing: A survey. *IEEE Transactions on Industrial Electronics*, 56:3275–3285, 2009.
46. X. Yu and J.-X. Xu. *Variable Structure Systems: Towards the 21st Century*, volume 274 of *Lecture Notes in Control and Information Sciences*. Springer Verlag, Berlin, Germany, 2002.

III

Estimation, Observation, and Identification

23

Adaptive Estimation

Seta Bogosyan
University of Alaska

Metin Gokasan
Istanbul Technical University

Fuat Gurleyen
Istanbul Technical University

23.1 Introduction

The process of estimation involves the mathematical determination of unavailable or hard to access system states and uncertain parameters with the use of measurable states and a system model that may be available in analytic form or may be derived recursively as part of the estimation process. The diverse applications of estimation theory range from the design of state observers and/or parameter identifiers for control systems, as well as sensorless control systems. Most control systems use output feedback, while some also require partial- or full-state feedback systems. In both cases, it is obvious that the control performance very much depends on the proper use of sensor information. For state feedback, theoretically, the number of states and sensors has to match, which calls for the acquisition and processing of a large number of sensors and related data. However, realistic considerations, such as cost, space, noise, system robustness, as well as availability are major reasons leading to the replacement of sensors wherever possible. Estimator (or observer) schemes serve as mathematical substitutes to sensors and are the most widely used solutions against these limitations.

The use of estimators is not limited to control systems, but is extended to system identification and modeling with diverse applications in monitoring and fault detection. Estimators are also extended for prediction and decision support systems in a multitude of areas, where estimation schemes may be needed to calculate a single state, fuse data from multiple sensors, or to predict a certain future state of the system or environment.

In this chapter, we will first group estimators in two main categories based on their estimation domain, that is, parameter, state, or both parameter and state. While all the methods to be discussed below can be applied for both linear and nonlinear systems (if necessary, after linearizing the given model), some estimator schemes are specifically designed for deterministic and some for stochastic systems:

The first category, mostly referred as "**adaptive parameter estimators (APE) or parameter identifiers**," involves the estimation of unknown parameters of a system model based on the principles of optimization or stability theory. Methods based on optimization theory require a system model given as an input–output transfer function, or in phase-canonical state-space representation and use the system output and/or measurement noise, and/or input signals to solve for parameter values that will minimize a given performance index. Least-squares estimation (LSE) and recursive LSE (RLSE) schemes are the commonly known estimation techniques in this category. There are also APE schemes developed based on the principles of stability. One such estimator is the model reference adaptive scheme (MRAS) in which the parameter vector is updated while minimizing the error between the actual system output and the output of a desired system based on the Lyapunov's stability theorem. While linear system models must be developed for these two schemes, adaptive estimation may also be performed for nonlinear systems provided the system satisfies a certain parameter-regressor vector form to be discussed further on in this chapter. Once again, the parameter update law is obtained based on Lyapunov's theory and estimates the system parameter vector, θ, online, using the error between the output signal, $y(t)$, and control reference signal, $r(t)$, as input for the parameter update law.

APE schemes to be discussed in this chapter can be applied for both constant and varying system parameters and used in combination with model-based adaptive control schemes that require updates from system states and parameters, that is, self-tuning regulators (STR). LSE and Lyapunov-based estimation schemes can also be used in combination with model-based nonlinear control schemes, such as PD+, feedback linearization, and inverse dynamic control approaches. An important condition for accurate estimation with methods in this category is *persistency of excitation*, which is the requirement that either the system or the input signal be rich in terms of frequency content to avoid parameter drift and assure convergence in steady state.

- Estimation schemes in the second category are developed based on linear, nonlinear, or recursive system models, and can perform the estimation of states (i.e., Kalman filter, KF), or simultaneous estimation of both system states and unknown system parameters (i.e., stochastic extended Kalman filter, EKF; deterministic extended Luenberger observer, ELO). Although estimation schemes in this category are referred both as "estimators" or "observers" in the literature, the term "observer" is used for estimators performing state estimation only. In this chapter, we will comply with this common approach.

The estimated states and parameters derived from observers or estimators can further be used to update a control law, or a system model online or offline. Hence, estimators and observers in this latter category may be used to predict the future values of a system state or output with the use of the estimated model and measured signals, that is, LSE applications in sensor networks or KF applications in sensor fusion.

An essential requirement for observer and estimator schemes is the fast convergence of the estimated states to their actual values. Normally, an open-loop observer structure warranting the determination of the system output, y, based on system input, u, may appear to be good enough. Such an open-loop structure may be sufficient to calculate the state values based on u when the system at hand is linear and time invariant (LTI) and has a well-defined model, with known system parameters and initial conditions. However, when the system's initial conditions are not known, the estimated states will take a longer time to converge to their actual values and will not even converge if the model and system parameters do not match. Fast convergence of the estimated states in this case can only be ensured with the use of closed-loop observers/estimators, in which the output estimation error (deviation between measured and calculated output) is also taken into consideration to update the observer control input. Such observer configurations are more widely used than their open-loop counterparts and are known as closed-loop estimators or observers.

An important consideration in the design of observers is the unknown or varying system parameters in the observer model, which will naturally affect the estimation performance negatively. To remedy this problem, these unknown/variable parameters could be estimated and used to update the observer model online in the same process or in a separate estimation process running in parallel. Similarly, model-based control laws may also require accurate knowledge of system parameters and disturbances (i.e., load observers). While such estimators can easily be developed for systems with analytical models, in the more likely event that system parameters and properties are time varying, or where it is impossible to perform separate identification experiments for the system to establish an observer model, a recursive identification approach can be conducted in an online manner to infer the properties of the system. Consequently, schemes for online determination of suitable regulators/predictors/filters (mostly associated with the first category) can inherently be considered *adaptive* and hence, could be called *adaptive estimators*. In this chapter, estimators performing parameter estimation or identification only, or simultaneous state and parameter estimation performing parameter estimation will be referred to as *adaptive observers* and the term "observer" will be used only for state estimators. It is also worth mentioning that there are numerous examples in the literature that refer to all online estimators (and observers) as adaptive [B07], or use the terms *adaptive estimators* or *adaptive observers* interchangeably.

To summarize, observers and estimators can be regarded as "model-based, measurement-based, closed-loop, information reconstructors," as they rely on a model using available measurements for online adaptation based on available measurements. Usually, the model is given as a state-space representation, in continuous-time or discrete-time, deterministic or stochastic, finite-dimensional or infinite-dimensional, smooth or "with singularities" [B07].

This chapter aims to address the most widely used linear, nonlinear, deterministic, and stochastic type estimation and observer schemes with corresponding simulation and experimental examples wherever possible.

23.2 Adaptive Parameter Estimation Schemes

Adaptive parameter estimation (APE) schemes, also called parameter identifiers, are online estimation schemes that guarantee the convergence of the estimated parameters to their unknown actual values. The design of APEs includes the selection of the plant input so that the signal vector $\boldsymbol{W}(t)$ (also called the regressor vector), which contains input and output measurements, is persistently exciting (PE). The PE property of $\boldsymbol{W}(t)$ is necessary for the convergence of the estimated parameters to the unknown parameter values. The PE property is

$$k_1\mathbf{I} \geq \int_t^{t+T} \boldsymbol{W}(\tau)\boldsymbol{W}^T(\tau)d\tau \geq k_2\mathbf{I}, \quad \text{where } k_1 \text{ and } k_2 > 0$$

A lack of this property may bring about a parameter drift in steady state, which is one of the most important problems encountered with APEs in practice.

In this chapter, most commonly used APEs will be briefly discussed. While most schemes can be applied for both linear and nonlinear systems after linearization or application of a proper parameterization approach, in this chapter, methods based on LSE and MRAS will be discussed in relation with linear system models, while the Lyapunov-based parameter estimation scheme will be addressed within the context of nonlinear system applications to comply with the common application trends in the literature.

23.2.1 Least-Squares Estimation

The least-squares method, created by Gauss, is probably the oldest and the most used estimation method. The method serves the purpose of finding the parameters of a mathematical model using the measured inputs and outputs of the system, hence, calls for the mathematical model to be given in certain forms. The implementation can be done either offline or online depending on the application or system; for example, with a system with slow-varying system parameters, it makes sense to find the parameters offline, and inversely, online, when dealing with a system that has fast-varying parameters in an operation range. The approach is based on finding a parameter set that yields the minimum sum for the square of errors found as the difference between the measured output and the output calculated based on the system model with the calculated parameters. This error minimization approach is known as the LSE, which can be applied to a very wide range of systems with discrete or continuous-time models of stochastic or deterministic nature, and even nonlinear systems. However, in most applications, the mathematical model is in linear form, either as an input–output transfer function or as a state-space representation. This chapter will discuss LSE methods for both continuous-time and discrete-time dynamical system [IF06,IS96,AW95].

23.2.1.1 Least-Squares Estimation Algorithms for Discrete-Time Systems

The discrete-time model of a system can be given in two forms:

1. As a transfer function:

$$y(z)=\left[\frac{B(z^{-1})}{A(z^{-1})}\right]u(z) \tag{23.1}$$

   where
   z^{-1} is the backward shift operator
   u is the control input
   y is the output [AW95]

   The parameters to be calculated are given in

$$A(z^{-1})=1+a_1z^{-1}+\cdots+a_{n_a}z^{-n_a},\ B(z^{-1})=b_0+B(z^{-1})=b_0+b_1z^{-1}+\cdots+b_{n_b}z^{-n_b}$$

2. As a difference equation, when the system output, $y(t)$, is given in terms of output values at previous instances, $(t-n_a)$ and input values at the current instant (t) as well as previous instances, $(t-n_b)$, as given in (23.2). Since the discrete model in (23.1) can also be represented as (23.2), our derivations will take that model as basis:

$$y(t)=a_1y(t-1)+\cdots+a_{n_a}y(t-n_a)+b_0u(t)+b_1u(t-1)+\cdots+b_{n_b}u(t-n_b) \tag{23.2}$$

   Here, a_i $(i=1,\ldots,n_a)$ and b_j $(j=0,\ldots,n_b)$ with a_i ve b_j being the unknown parameters.

   Next, these parameters are combined as a vector, $\boldsymbol{P}=\left[a_1\ldots a_{n_a}b_0\ldots b_{n_b}\right]^T$ while measured inputs and outputs, $u(t-j)$ and $y(t-i)$ are combined in $\boldsymbol{W}^T(t)=[-y(t-1)\ldots-y(t-n_a)\ u(t)\ u(t-1)\ldots u(t-n_b)]$. As a result, the model can now be given in the following parameter linearizable form:

$$y(t)=\boldsymbol{W}^T\boldsymbol{P} \tag{23.3}$$

   Here, $\boldsymbol{P}=[P_1\quad P_2\quad\ldots\quad P_n]^T$ is known as the parameter vector with dimension $n=n_a+n_b$ and $\boldsymbol{W}^T(i)=[w_1(i)\quad w_2(i)\quad\ldots\quad w_n(i)]$ is known as the regression vector.

This leads to the calculation of the output, $y(i)$, for each instant i based on the following model:

$$y(i)=\boldsymbol{W}^T(i)\boldsymbol{P},\quad i=1,2,\ldots,t,\ldots \tag{23.4}$$

The deviation between the estimated output, $\hat{y}(i)$ calculated at an instant, i, and its measured value, $y(i)$, gives the error

$$e(i) = y(i) - \hat{y}(i) = y(i) - \mathbf{W}(i)^T \mathbf{P}(i) \tag{23.5}$$

The calculation of the unknown parameter vector $\boldsymbol{P}$ can be achieved by solving the least-squares cost function below, in terms of the parameter vector, $\boldsymbol{P}$, or similarly, by finding P that will minimize the cost function, $J(\boldsymbol{P}, t)$ [AW95]. The cost function can be expressed as below:

$$J(\mathbf{P},t) = \frac{1}{2}\sum_{i=1}^{t}\left(y(i) - \mathbf{W}(i)^T \mathbf{P}(i)\right)^2 \tag{23.6}$$

$$J(\mathbf{P},t) = \frac{1}{2}\sum_{i=1}^{t} e(i)^2 \tag{23.7}$$

$$J(\mathbf{P},t) = \frac{1}{2}\left\{e(1)^2 + \cdots + e(t)^2\right\} = J\left[\mathbf{P}(1)\right] + \cdots + J\left[\mathbf{P}(t)\right] \tag{23.8}$$

(23.8) can be further organized by making the following definitions:

$$\mathbf{Y}(t) = \begin{bmatrix} y(1) & y(2) & \cdots & y(t) \end{bmatrix}^T, \quad \mathbf{E}(t) = \begin{bmatrix} e(1) & e(2) & \cdots & e(t) \end{bmatrix}^T, \quad \mathbf{P} = \begin{bmatrix} P_1 & P_2 & \cdots & P_n \end{bmatrix}^T$$

$$\mathbf{E} = \mathbf{Y} - \hat{\mathbf{Y}} = \mathbf{Y} - \mathbf{W}\mathbf{P}, \quad \mathbf{W}(t) = \begin{bmatrix} w_1(1)w_2(1)\ldots w_n(1) \\ \vdots \\ w_1(t)w_2(t)\ldots w_n(t) \end{bmatrix} \tag{23.9}$$

$$J(\mathbf{P},t) = \frac{1}{2}\mathbf{E}^T\mathbf{E} = \frac{1}{2}\mathbf{P}^T(\mathbf{W}^T(t)\mathbf{W}(t))\mathbf{P} - \mathbf{P}^T\mathbf{W}^T(t)\mathbf{Y} + \frac{1}{2}\mathbf{Y}^T\mathbf{Y} \tag{23.10}$$

We will now discuss some commonly used methods for calculating the unknown parameter vector that will minimize the above cost function. These methods can be listed as the gradient method, recursive least squares, and recursive least squares with forgetting factor. We will start our discussion with the gradient methods.

23.2.1.1.1 Gradient Methods

Standard gradient methods, such as steepest descent, Newton's method, etc., are commonly used methods for the minimization of J. As can be seen from (23.8), the minimization of the cost function, $J(\boldsymbol{P}, t)$, can be also achieved by minimizing each error-square term. The cost function in (23.8) is rewritten below by replacing $\boldsymbol{P}$ with $\hat{\boldsymbol{P}}$, as the deviation at each i is actually calculated in terms of the estimated parameter vector as

$$J\left(\hat{\mathbf{P}}(t)\right) = \frac{1}{2}e^2(t) = \frac{1}{2}\left(y(t) - \mathbf{W}^T(t)\hat{\mathbf{P}}(t)\right)^2 \tag{23.11}$$

The cost function in (23.11) should decrease in the direction of its negative gradient as expressed with the relationship below:

$$\Delta\hat{\boldsymbol{P}}(t) = \hat{\boldsymbol{P}}(t) - \hat{\boldsymbol{P}}(t-1) = -\boldsymbol{\Gamma}(t)\frac{\partial J(\hat{\boldsymbol{P}}(t))}{\partial\hat{\boldsymbol{P}}(t)} \tag{23.12}$$

where $\boldsymbol{\Gamma}(t) = \boldsymbol{\Gamma}^T(t) > 0$ is a symmetric positive definite matrix. This is also equivalent to

$$\frac{\partial J(\hat{\boldsymbol{P}}(t))}{\partial\hat{\boldsymbol{P}}(t)} = \frac{\partial e(t)}{\partial\hat{\boldsymbol{P}}(t)}e(t) = -\boldsymbol{W}(t)e(t) \tag{23.13}$$

Hence, by substituting (23.13) in Equation 23.12, a gradient descent-like iteration is obtained, which is also known as the MIT rule:

$$\hat{\boldsymbol{P}}(t) = \hat{\boldsymbol{P}}(t-1) - \boldsymbol{\Gamma}(t)\frac{\partial e(t)}{\partial\hat{\boldsymbol{P}}(t)}e(t) = \hat{\boldsymbol{P}}(t-1) + \boldsymbol{\Gamma}(t)\boldsymbol{W}(t)e(t) \tag{23.14}$$

With the assumption that the parameter changes are slower in comparison to those of the other variables in the system, the derivative $(\partial e(t)/\partial\hat{\boldsymbol{P}}(t))$ can be evaluated under the assumption that $\boldsymbol{P}$ is constant.

There are many alternatives to the cost function given by Equation 23.10. If it is chosen to be

$$J(\hat{\boldsymbol{P}}(t)) = |e(t)|, \tag{23.15}$$

the gradient method gives

$$\hat{\boldsymbol{P}}(t) = \hat{\boldsymbol{P}}(t-1) - \boldsymbol{\Gamma}(t)\frac{\partial e(t)}{\partial\hat{\boldsymbol{P}}(t)}\operatorname{sgn}[e(t)] = \hat{\boldsymbol{P}}(t-1) + \boldsymbol{\Gamma}(t)\boldsymbol{W}(t)\operatorname{sgn}[e(t)] \tag{23.16}$$

As can be seen in the above equations, the parameter convergence rate can be adjusted by the proper selection of $\boldsymbol{\Gamma}$ based on trial and error; however, a recursive optimization approach can also be taken to this aim, which will be discussed next.

23.2.1.1.2 Recursive Least-Squares Estimation

In adaptive processes, the observations are obtained sequentially in real time. Accordingly, computations of the least-squares estimate should be carried out in such a way that the results obtained at time $t - 1$ can be used to get the estimates at time t (Figure 23.1).

To get the least-squares estimate recursively, we can directly apply iterative methods. The cost function, $J(\boldsymbol{P}, t)$, given in (23.10) will be minimal, if the Euclidean norm of the residual vector, $\boldsymbol{E} = \boldsymbol{Y} - \boldsymbol{WP}$, is minimal for all parameters, $\boldsymbol{P}$, which make the gradient vector equal to be zero;

$$\nabla J(\boldsymbol{P}, t) = \frac{1}{2}\frac{\partial J(\boldsymbol{P}, t)}{\partial\boldsymbol{P}} = \left(\frac{\partial\boldsymbol{E}}{\partial\boldsymbol{P}}\right)^T\boldsymbol{E} = -\boldsymbol{W}^T(t)(\boldsymbol{Y} - \boldsymbol{WP}) = 0 \tag{23.17}$$

such that a vector $\boldsymbol{P}$ is a least-squares solution of $\mathbf{W}\boldsymbol{P} = \boldsymbol{Y}$ if and only if $\boldsymbol{P}$ is a solution of $\mathbf{W}^T(t)\mathbf{W}(t)\boldsymbol{P} = \mathbf{W}^T(t)\boldsymbol{Y}$. We may assume that $\mathbf{W}^T(t)\mathbf{W}(t)$ is positive definite for $t \geq n$. In that case, the least-squares estimate is unique and can be formulated as

$$\hat{\boldsymbol{P}}(t) = \left(\mathbf{W}^T(t)\mathbf{W}(t)\right)^{-1}\mathbf{W}^T(t)\boldsymbol{Y}(t) = \mathbf{S}(t)\mathbf{W}^T(t)\boldsymbol{Y}(t) \tag{23.18}$$

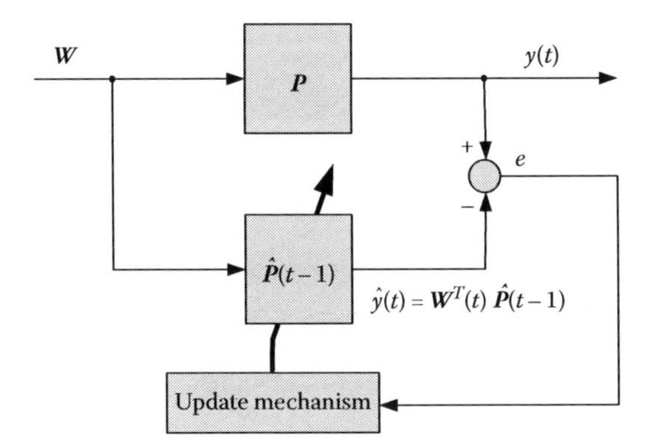

FIGURE 23.1 Schematic representation of the RLSE.

where

$$\mathbf{S}(t) = \left(\mathbf{W}^T(t)\mathbf{W}(t)\right)^{-1} = \left(\sum_{i=1}^{t} \boldsymbol{W}(i)\boldsymbol{W}^T(i)\right)^{-1}, \quad t \ge n \tag{23.19}$$

In order to calculate $\hat{\boldsymbol{P}}(t)$ iteratively using $\hat{\boldsymbol{P}}(t-1)$, t must be placed by $t-1$ in (23.18):

$$\hat{\boldsymbol{P}}(t-1) = \mathbf{S}(t-1)\mathbf{W}^T(t-1)\boldsymbol{Y}(t-1) \tag{23.20}$$

Using the definitions below,

$$\mathbf{W}^T(t)\mathbf{W}(t) = [\mathbf{W}^T(t-1)\ \boldsymbol{W}(t)]\begin{bmatrix}\mathbf{W}(t-1)\\ \boldsymbol{W}^T(t)\end{bmatrix} = \mathbf{W}^T(t-1)\mathbf{W}(t-1) + \boldsymbol{W}(t)\boldsymbol{W}^T(t) \tag{23.21}$$

$$\mathbf{W}^T(t)\boldsymbol{Y}(t) = [\mathbf{W}^T(t-1)\ \boldsymbol{W}(t)]\begin{bmatrix}\boldsymbol{Y}(t-1)\\ y(t)\end{bmatrix} = \mathbf{W}^T(t-1)\boldsymbol{Y}(t-1) + \boldsymbol{W}(t)y(t) \tag{23.22}$$

Hence, (23.19) can be written using (23.21) as

$$\mathbf{S}^{-1}(t) = \boldsymbol{W}^T(t)\boldsymbol{W}(t) = \mathbf{S}^{-1}(t-1) + \boldsymbol{W}(t)\boldsymbol{W}^T(t) \tag{23.23}$$

To obtain $\mathbf{S}(t)$ from (23.23), the Matrix Inversion Lemma [AW95] is used, which gives

$$\mathbf{S}(t) = \left(\mathbf{S}^{-1}(t-1) + \boldsymbol{W}(t)\boldsymbol{W}^T(t)\right)^{-1} = \mathbf{S}(t-1) - \mathbf{S}(t-1)\boldsymbol{W}(t)\left(1 + \boldsymbol{W}^T(t)\mathbf{S}(t-1)\boldsymbol{W}(t)\right)^{-1}\boldsymbol{W}^T(t)\mathbf{S}(t-1) \tag{23.24}$$

Finally, the recursion for $\mathbf{S}(t)$ will be combined with the recursion for $\boldsymbol{W}^T(t)\boldsymbol{Y}(t)$ to give a direct recursion for $\hat{\boldsymbol{P}}(t)$ from $\hat{\boldsymbol{P}}(t-1)$. The most common way is to define the residual error variable $e(t)$ as defined in Equation 23.5 by

$$e(t) = y(t) - \hat{y}(t) = y(t) - \boldsymbol{W}^T(t)\hat{\boldsymbol{P}}(t-1) \tag{23.25}$$

and substitute for $y(t)$ in Equation 23.21. This gives

$$\boldsymbol{W}^T(t)\boldsymbol{Y}(t) = \boldsymbol{W}^T(t-1)\boldsymbol{Y}(t-1) + \boldsymbol{W}(t)\boldsymbol{W}^T(t)\hat{\boldsymbol{P}}(t-1) + \boldsymbol{W}(t)e(t) \tag{23.26}$$

Substituting for $\boldsymbol{W}^T(t-1)\boldsymbol{Y}(t-1)$, $\boldsymbol{W}^T(t)\boldsymbol{Y}(t)$ using Equations 23.18, 23.22, and 23.24 gives

$$\hat{\boldsymbol{P}}(t) = \hat{\boldsymbol{P}}(t-1) + \mathbf{S}(t)\boldsymbol{W}(t)e(t) \tag{23.27}$$

As an alternative procedure to update the parameter $\hat{\boldsymbol{P}}(t)$ in (23.27), the definition can also be taken as $\boldsymbol{L}(t) = \mathbf{S}(t)\boldsymbol{W}(t)$ in Equation 23.27 as a column vector of adjustment gains. Substituting the vector, $\boldsymbol{L}(t)$ into (23.27) we obtain

$$\hat{\boldsymbol{P}}(t) = \hat{\boldsymbol{P}}(t-1) + \boldsymbol{L}(t)e(t) \tag{23.28}$$

By combining Equation 23.22 and the expression for $\boldsymbol{P}(t)$ in (23.24), we can obtain the following equation for $\boldsymbol{L}(t)$:

$$\boldsymbol{L}(t) = \frac{\mathbf{S}(t-1)\boldsymbol{W}(t)}{1 + \boldsymbol{W}^T(t)\mathbf{S}(t-1)\boldsymbol{W}(t)} \tag{23.29}$$

Consequently, $\mathbf{S}(t)$ can be expressed in the following recursive form:

$$\mathbf{S}(t) = \mathbf{S}(t-1) - \boldsymbol{L}(t)(\mathbf{S}(t-1)\boldsymbol{W}(t))^T \tag{23.30}$$

Below is a flowchart for the procedure of calculating $\hat{\boldsymbol{P}}(t)$ recursively (Figure 23.2).

23.2.1.1.3 Recursive Least-Squares Estimation with Exponential Forgetting Factor [AW95]

Forgetting factor λ is a number between 0 and 1, which is used to progressively reduce the emphasis placed on the past values of error, e. The use of this factor becomes necessary especially when high amounts of measurement and/or input data are involved, that is, with online applications of RLSE.

$$J(\boldsymbol{P},t) = \frac{1}{2}\sum_{i=1}^{t} \lambda^{t-i}\left(y(i) - \boldsymbol{W}^T\boldsymbol{P}\right)^2 \tag{23.31}$$

Below we present the recursive estimation algorithm with the inclusion of the forgetting factor. While the parameter update laws in (23.27) and (23.28) remain the same, the $\mathbf{S}$ and $\boldsymbol{L}$ expressions reflect the inclusion of the forgetting factor:

$$\mathbf{S}(t) = \lambda^{-1}\mathbf{S}(t-1)\left[\mathbf{I}_n - \frac{\boldsymbol{W}(t)\boldsymbol{W}^T(t)\mathbf{S}(t-1)}{\lambda + \boldsymbol{W}^T(t)\mathbf{S}(t-1)\boldsymbol{W}(t)}\right] \tag{23.32}$$

Thus, the expression for $\mathbf{S}(t)$ and $\boldsymbol{L}(t)$ are given as follows.

$$\boldsymbol{L}(t) = \frac{\mathbf{S}(t-1)\boldsymbol{W}(t)}{\left(\lambda + \boldsymbol{W}^T(t)\mathbf{S}(t-1)\boldsymbol{W}(t)\right)}; \quad \hat{\boldsymbol{P}}(t) = \hat{\boldsymbol{P}}(t-1) + \boldsymbol{L}(t)e(t) \tag{23.33}$$

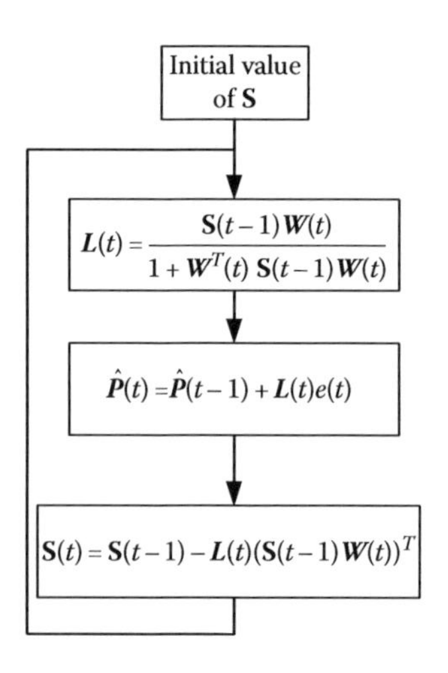

FIGURE 23.2 Flowchart for calculating $\hat{\boldsymbol{P}}(t)$ recursively with RLSE.

$$\mathbf{S}(t) = \frac{\left(\mathbf{I}_n - \boldsymbol{L}(t)\boldsymbol{W}^T(t)\right)\mathbf{S}(t-1)}{\lambda}$$

Below is a flowchart for the procedure of calculating $\hat{\boldsymbol{P}}(t)$ recursively (Figure 23.3).

23.2.1.2 Extended Least-Squares Estimation Method

In this section, we will also discuss the LSE approach when the output signals of the system in consideration are corrupted by noise. The approach is based on the system model given below

$$A(q)y(t) = B(q)u(t) + C(q)\varepsilon(t) \tag{23.34}$$

where $A(q)$, $B(q)$, and $C(q)$ are polynomials in forward shift operator q, and $\varepsilon(t)$ is white noise. Let us consider that the degrees of the $\boldsymbol{A}$, $\boldsymbol{B}$, and $\mathbf{C}$ polynomials are confined to be n = degree($\boldsymbol{A}$) = degree($\mathbf{C}$) = degree($\boldsymbol{B}$) + 1.

$$A(q) = q^n + a_1 q^{n-1} + \cdots + a_n, \quad B(q) = b_1 q^{n-1} + b_2 q^{n-2} + \cdots + b_n, \quad C(q) = q^n + c_1 q^{n-1} + c_2 q^{n-2} + \cdots + c_n$$

Due to the infeasibility of measuring past random noise values, we can make the assumption that when $[\hat{y}(t)\hat{y}(t-1)\ldots\hat{y}(t-n)]$ calculated by the LSE algorithm converges to the actual $[y(t)y(t-1)\ldots y(t-n)]$ values, the modeling error, $[e(t)e(t-1)\ldots e(t-n)]$, will also converge to the random noise, $[\varepsilon(t)\varepsilon(t-1)\ldots\varepsilon(t-n)]$. This assumption leads to the extended regressor vector given below, augmented by the unknown noise coefficients. The resulting recursive extended least-squares algorithm can be given as follows:

$$\hat{\boldsymbol{P}}_e^T = [\hat{a}_1 \quad \ldots \quad \hat{a}_n \quad \hat{b}_1 \quad \ldots \quad \hat{b}_n \quad \hat{c}_1 \quad \ldots \quad \hat{c}_n\,] \tag{23.35}$$

$$\mathbf{W}_e^T(t-1) = \left[-y(t-1)\cdots - y(t-n)\; u(t-1)\ldots u(t-n)\varepsilon(t-1)\ldots\varepsilon(t-n)\right] \tag{23.36}$$

$\varepsilon(t)$ can be written in the following form:

$$e(t) = y(t) - \mathbf{W}_e^T(t-1)\hat{\boldsymbol{P}}_e^T(t-1) \tag{23.37}$$

The variables $\varepsilon(t)$ are approximated by the modeling errors $e(t)$. The model can then be approximated by

$$y(t) = \mathbf{W}_e^T(t-1)\hat{\boldsymbol{P}}_e + \varepsilon(t) \tag{23.38}$$

The resulting approach is called the extended least squares (ELS). The equations for updating the estimates are given by

$$\hat{\boldsymbol{P}}_e(t) = \hat{\boldsymbol{P}}_e(t-1) + \mathbf{S}_e(t)\mathbf{W}_e(t-1)e(t) \tag{23.39}$$

$\mathbf{S}(t)$ can also be given in the following form as in RLSE as

$$\mathbf{S}_e(t) = \mathbf{S}_e(t-1)\left[\mathbf{I}_n - \frac{\mathbf{W}_e(t)\mathbf{W}_e^T(t)\mathbf{S}_e(t-1)}{1 + \mathbf{W}_e^T(t)\mathbf{S}_e(t-1)\mathbf{W}_e(t)}\right] \tag{23.40}$$

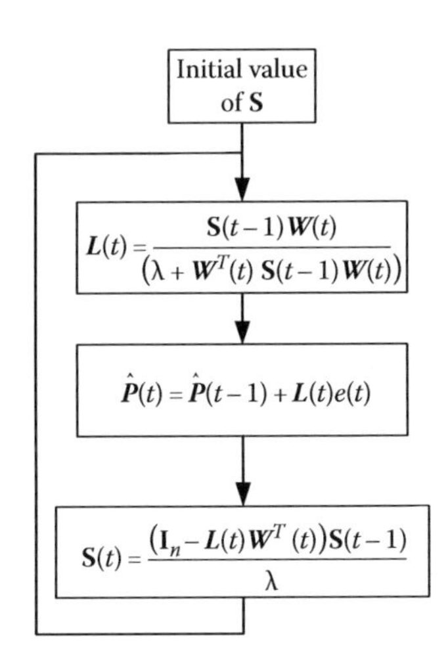

FIGURE 23.3 Flowchart for calculating $\hat{\boldsymbol{P}}(t)$ recursively for RLSE with forgetting factor.

$$L_e(t) = \frac{S_e(t-1)W_e(t)}{\left(1 + W_e^T(t)S_e(t-1)W_e(t)\right)}$$

$$\hat{P}_e(t) = \hat{P}_e(t-1) + L_e(t)e(t)$$

$$S_e(t) = S_e(t-1) - L_e(t)W_e(t)S_e(t-1)$$

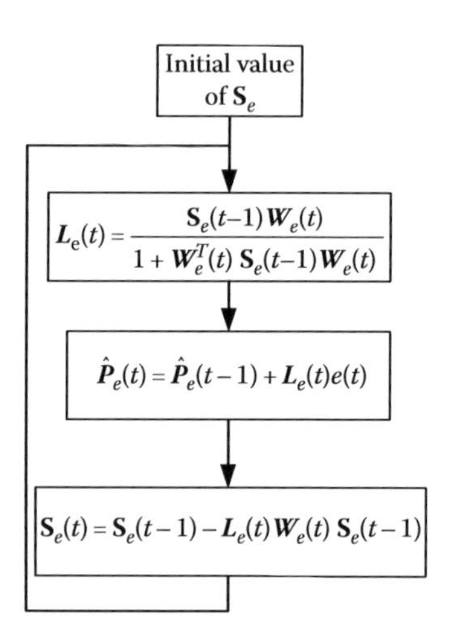

FIGURE 23.4 Flowchart for calculating $\hat{P}(t)$ recursively for ELSE.

The ELSE algorithm also uses the flowchart given for RLSE in Figure 23.4

23.2.1.3 Adaptive Parameter Estimation for Continuous-Time Systems

Single-input, single-output (SISO) linear, time invariant systems can be modeled with an input–output transfer function as below, even when all the system states may not be available:

$$y = \frac{N_m(s)}{D_n(s)}u \tag{23.41}$$

$N_m(s)$ and $D_n(s)$ have no common factors and can be given as below:

$$N_m(s) = b_m s^m + b_{m-1}s^{m-1} + \cdots + b_1 s + b_0 \tag{23.42}$$

$$D_n(s) = s^n + a_{n-1}s^{n-1} + \cdots + a_1 s + a_0 \tag{23.43}$$

$N_m(s)$ is of degree, m, while $D_n(s)$ is of degree, n, with $m \le n - 1$. The unknown parameters of $N_m(s)$ and $D_n(s)$, constituting the parameter vector, P, of dimension $(n + m + 1)$ can be written as

$$P = \begin{bmatrix} b_m & b_{m-1} & \ldots & b_1 & b_0 & a_{n-1} & a_{n-2} & \ldots & a_1 & a_0 \end{bmatrix} \tag{23.44}$$

We can start with (23.41) to obtain the continuous-time parametric model as an nth-order differential equation given by

$$y^{(n)} + a_{n-1}y^{(n-1)} + \cdots + a_0 y = b_m u^{(m)} + b_{m-1}u^{(m-1)} + \cdots + b_0 u \tag{23.45}$$

We can lump all the parameters in (23.45) in the parameter vector P and all the corresponding input–output signal derivatives in the signal vector as below:

$$X^T = \begin{bmatrix} u^{(m)} \; u^{(m-1)} \cdots u - y^{(n-1)} - y^{(n-2)} \cdots - y \end{bmatrix} = \begin{bmatrix} \alpha_m^T(p)u - \alpha_{n-1}^T(p)y \end{bmatrix} \tag{23.46}$$

Here, $\alpha_i^T(p) = \begin{bmatrix} p^i \; p^{i-1} \ldots 1 \end{bmatrix}$ and $p = d/dt$. The definitions in (23.44) and (23.46) can further be used in the representation of $y^{(n)}$ in the following compact form:

$$y^{(n)} = P^T X \tag{23.47}$$

It should be noted that the calculation of the parameter vector in (23.47) requires the m and nth-order derivatives of u and y, respectively. However, problems related to differentiation may cause limitations in practical applications. To address this limitation, low-pass filters (LPF) may be designed with orders that are the same or higher than that of the differential equation describing the system. To this aim, the input–output equation in (23.41) can be rewritten with the inclusion of $\Lambda(s)$ with degree n, in the following way:

$$y = \frac{N_m(s)}{D_n(s)} \frac{\Lambda(s)}{\Lambda(s)} u \tag{23.48}$$

Here, $\dfrac{1}{\Lambda(p)}$ is the nth-order low-pass filter, with $\Lambda(p) = p^n + \boldsymbol{\lambda}^T \boldsymbol{\alpha}_{n-1}(p)$, $\boldsymbol{\lambda}^T = [\lambda_{n-1}\ \lambda_{n-2} \ldots \lambda_0]$.

In order to avoid differentiation, both sides of (23.47) are divided by $\Lambda(p)$ to yield

$$\frac{1}{\Lambda(p)} y^{(n)} = \frac{p^n}{\Lambda(p)} y \quad \text{and} \quad \frac{1}{\Lambda(p)} \boldsymbol{X}^T(t) = \left[\frac{\boldsymbol{\alpha}_m^T(p)}{\Lambda(p)} u(t) \quad - \frac{\boldsymbol{\alpha}_{n-1}^T(p)}{\Lambda(p)} y(t) \right]$$

Following the use of this stable filter and z and W definitions below, the plant in (23.41) will take on the parameterization model below to develop adaptive laws for estimating, P.

$$z(t) = W^T P \tag{23.49}$$

where

$$W^T P = \frac{1}{\Lambda(p)} \boldsymbol{X}^T(t) = \left[\frac{\boldsymbol{\alpha}_m^T(p)}{\Lambda(p)} u(t) - \frac{\boldsymbol{\alpha}_{n-1}^T(p)}{\Lambda(p)} y(t) \right]$$

$$z(t) = \left[\frac{p^n}{\Lambda(p)} \right] y(t) = \left[\frac{\Lambda(p) - \boldsymbol{\lambda}^T \boldsymbol{\alpha}_{n-1}(p)}{\Lambda(p)} \right] y(t) = y(t) + \boldsymbol{\lambda}^T W_2 \tag{23.50}$$

Below we will discuss the LSE approach developed for the estimation of parameters for a system model given with a continuous input–output transfer function. To develop the LSE for continuous-time observation, the unknown parameters are assumed to appear in a linear form, such as in the linear parametric model given below:

$$z(t) = \boldsymbol{P}^T \boldsymbol{W}(t), \quad t \in \left[0, \infty\right) \tag{23.51}$$

The estimate $\hat{z}$ of z in Equation 23.51, $\hat{z} = \hat{\boldsymbol{P}}^T \boldsymbol{W}$ results in

$$e = z - \hat{z} = z - \hat{\boldsymbol{P}}^T \boldsymbol{W} \tag{23.52}$$

In the next section, we will briefly discuss the gradient method as a commonly applied approach to perform the least-square minimization of (23.52). Other methods, such as RLSE and RLSE with forgetting factor are also widely applied and can be found in [IF06,IS96,AW95] in further detail.

23.2.1.3.1 Gradient Methods for Minimization of J in Continuous-Time Systems

We start with the instantaneous quadratic cost function in the form

$$J(\hat{\boldsymbol{P}}) = \frac{e^2}{2} = \frac{(z - \hat{\boldsymbol{P}}^T \boldsymbol{W})^2}{2} \tag{23.53}$$

which will be minimized with respect to $\hat{\boldsymbol{P}}$.

$$\dot{\hat{P}} = -\Gamma \nabla J(\hat{P}) \tag{23.54}$$

where $\Gamma = \Gamma^T > 0$ is a scaling matrix, also known as adaptive gain. From (23.53), we have

$$\nabla J(\hat{P}) = -(z - \hat{P}^T W)W = -eW \tag{23.55}$$

Hence, the adaptive law for calculating $P(t)$ can be obtained as

$$\dot{\hat{P}} = \Gamma e W \tag{23.56}$$

23.2.1.4 APE Schemes for System Models in State-Space Form

APE methods discussed so far are based on relationships between input–output transfer functions and noise signals given as transfer function models. We will now discuss APE methods developed for system models given in state-space form. Now, consider an LTI plant represented by the state equation for these systems as

$$\begin{aligned} \dot{x} &= \mathbf{A}\dot{x} + \mathbf{B}u; \\ x(0) &= x_0; \\ y &= \mathbf{C}x \end{aligned} \tag{23.57}$$

where $x \in R^n$, $u \in R^m$, $y \in R^l$ and $n \geq m, l$.

The plant input is piecewise continuous and is a uniformly bounded function of time. The plant parameters $\mathbf{A} \in R^{n\times n}$, $\mathbf{B} \in R^{n\times m}$, and $\mathbf{C} \in R^{l\times n}$ are unknown constant parameter matrices. We assume that the plant (23.57) is completely controllable and observable. Once again, the objective of this section is to design online parameter identifiers (or parameter estimation schemes) that guarantee convergence of the estimated plant parameters to their actual values. The plant parameters to be estimated are the constant matrices **A, B, C** in (23.57). The parameter identifier can be designed using Gradient and Lyapunov methods.

23.3 Model Reference Adaptive Schemes

An LTI system with unknown plant parameters can be modeled by the LTI differential equation in (23.57). **A** and **B** are constant matrices with unknown components, while the state vector $x \in R^n$ may be fully or partially available for measurements. We also assume that **A** is stable, and u is bounded.

With these assumptions and also using the previously described parameter estimators, we will now discuss parameter identification techniques based on MRAS (with full- or partial-state measurements) to estimate **A** and **B** from the measurements of x and u.

23.3.1 MRAS with Full-State Measurements

MRAS involves the calculation of unknown system parameters, first with the selection of a plant model, the parameters of which are updated based on output measurements in a way that the deviation between the actual output and the model output is minimized. Similar to all other identification methods, MRAS also requires the control input to be PE. Now, let us assume that the actual plant and the estimated plant have the following models, respectively:

$$\dot{x} = \mathbf{A}x + \mathbf{B}u, \tag{23.58}$$

$$\dot{\hat{x}} = \hat{\mathbf{A}}\hat{x} + \hat{\mathbf{B}}u \tag{23.59}$$

We will use the models in (23.58) and (23.59) to discuss the Lyapunov method in order to derive the parameter update rules for $\hat{\mathbf{A}}$, $\hat{\mathbf{B}}$.

The Lyapunov theory–based identification algorithm [IF06,IS96,NA05] requires the derivation of the state error dynamics equation as

$$\dot{e}(t) = \dot{\hat{x}} - \dot{x} = \mathbf{A}e(t) + \tilde{\mathbf{A}}(t)\hat{x}(t) + \tilde{\mathbf{B}}(t)u(t) \tag{23.60}$$

where the parameter error matrices are defined by

$$\tilde{\mathbf{A}}(t) = \hat{\mathbf{A}}(t) - \mathbf{A} \quad \text{and} \quad \tilde{\mathbf{B}}(t) = \hat{\mathbf{B}}(t) - \mathbf{B} \tag{23.61}$$

Next, a Lyapunov function candidate is proposed to guarantee the asymptotic stability of the error dynamics in (23.60).

$$V(e,\tilde{\mathbf{A}},\tilde{\mathbf{B}}) = \frac{1}{2}\left[e^T\mathbf{T}e + tr\left(\tilde{\mathbf{A}}^T\mathbf{\Gamma}_1^{-1}\tilde{\mathbf{A}}\right) + tr\left(\tilde{\mathbf{B}}^T\mathbf{\Gamma}_2^{-1}\tilde{\mathbf{B}}\right)\right] \tag{23.62}$$

and its derivative

$$\dot{V}(e,\tilde{\mathbf{A}},\tilde{\mathbf{B}}) = e^T\mathbf{T}\dot{e} + tr\left\{\tilde{\mathbf{A}}^T\mathbf{\Gamma}_1^{-1}\dot{\tilde{\mathbf{A}}}\right\} + tr\left\{\tilde{\mathbf{B}}^T\mathbf{\Gamma}_2^{-1}\dot{\tilde{\mathbf{B}}}\right\} = -\frac{1}{2}e^T\mathbf{Q}e \le 0 \tag{23.63}$$

where $\mathbf{\Gamma}_1 = \mathbf{\Gamma}_1^T > 0$, $\mathbf{\Gamma}_2 = \mathbf{\Gamma}_2^T > 0$ are positive definite gain matrices. After substituting $\dot{e}(t)$ from (23.60) into (23.63), to guarantee negative definiteness of (23.63), $\mathbf{T}$, the symmetric positive definite matrix of order, n, must satisfy the following matrix equation:

$$\mathbf{A}^T\mathbf{T} + \mathbf{T}\mathbf{A} = -\mathbf{Q}, \quad \mathbf{Q} = \mathbf{Q}^T > 0 \tag{23.64}$$

while the rest of the terms in (23.63), which include the parameter errors, should equal zero. This condition yields the parameter adaptive laws

$$\dot{\tilde{\mathbf{A}}}(t) = \dot{\hat{\mathbf{A}}} = -\mathbf{\Gamma}_1\mathbf{T}e(t)x^T(t), \quad \dot{\tilde{\mathbf{B}}} = \dot{\hat{\mathbf{B}}} = -\mathbf{\Gamma}_2\mathbf{T}e(t)u^T(t) \tag{23.65}$$

When the plant is stable and u is bounded and sufficiently rich, then $\hat{x}(t)$, $\hat{\mathbf{A}}(t)$, $\hat{\mathbf{B}}(t)$ are bounded and $e(t)$ and the elements of $\dot{\hat{\mathbf{A}}}(t), \dot{\hat{\mathbf{B}}}(t)$ asymptotically converge to zero as $t \to \infty$. Therefore, the Lyapunov function based identification laws (23.65) guarantee convergence of the parameter estimates to their true values with uniform asymptotic stability, if the input $u(t)$ is PE.

23.3.2 Generalized Error Models in Estimation with Full-State Measurements [NA05]

This approach involves unifying the $\mathbf{A}$, $\mathbf{B}$ parameter matrices of system in (23.58) in one matrix $\mathbf{P}$, while x and u in (23.58) are unified in one regressor vector W.

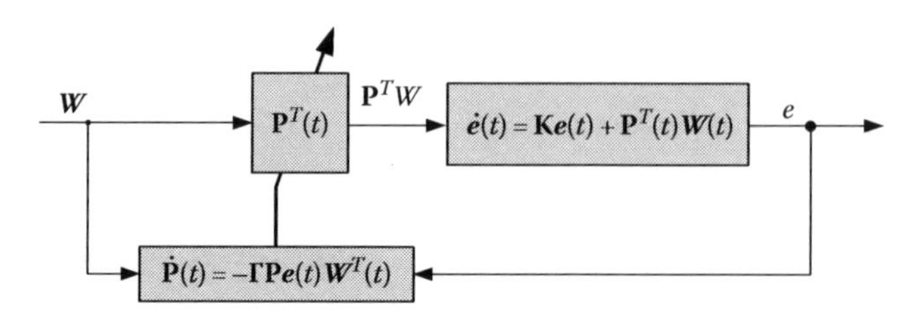

FIGURE 23.5 Generalized error models in estimation and control with full-state measurements.

$$\dot{x} = \mathbf{P}^T \boldsymbol{W}, \quad \dot{\hat{x}} = \hat{\mathbf{P}}^T \hat{\boldsymbol{W}}$$

where $(\mathbf{P}^T(t) = [\mathbf{A}(t)\ \mathbf{B}(t)])$ and $\boldsymbol{W}(t)$ $(\boldsymbol{W}^T(t) = [\boldsymbol{x}^T(t)\ \boldsymbol{u}^T(t)])$ is an s-vector.

Hence, the resulting dynamics of the error between the adjustable model and plant states can be given as below:

$$\dot{e}(t) = \mathbf{K}e(t) + \tilde{\mathbf{P}}^T(t)\boldsymbol{W}(t) \tag{23.66}$$

where $\mathbf{K}(\mathbf{K} = \mathbf{A})\tilde{\mathbf{P}}^T(t) = [\tilde{\mathbf{A}}(t)\ \tilde{\mathbf{B}}(t)]$ is a stable matrix of order n;

$\tilde{\mathbf{P}}(t)$ is unknown but $\dot{\tilde{\mathbf{P}}}(t)$ can be adjusted in such a way that $\lim_{t\to\infty} \boldsymbol{e}(t) = 0$ and $\lim_{t\to\infty} \tilde{\mathbf{P}}(t) = 0$.

This adaptive law can be found via the Lyapunov function $V(\boldsymbol{e}, \tilde{\mathbf{P}}) = (1/2)[\boldsymbol{e}^T\mathbf{T}\boldsymbol{e} + tr(\tilde{\mathbf{P}}(t)\Gamma^{-1}\tilde{\mathbf{P}}^T(t))]$, as a result of which

$$\dot{\tilde{\mathbf{P}}}^T(t) = -\Gamma\mathbf{T}e(t)v^T(t) \tag{23.67}$$

where $\Gamma = \Gamma^T > 0; \quad \mathbf{K}^T\mathbf{T} + \mathbf{T}\mathbf{K} = -\mathbf{Q}, \quad \mathbf{Q} = \mathbf{Q}^T > 0$.

Figure 23.5 illustrates the MRAS identification scheme with full-state measurement.

23.3.3 Generalized Error Models in Estimation with Partial-State Measurements [NA05]

Figure 23.6 illustrates the MRAS scheme with partial-state measurement, which arises when only some of the outputs of the plant are accessible. In such case, the error between the plant and model outputs can be described by

$$\dot{e}(t) = \mathbf{K}e(t) + \mathbf{D}\tilde{\mathbf{P}}^T(t)\boldsymbol{v}(t), \quad \boldsymbol{\eta}(t) = \mathbf{H}e(t) \tag{23.68}$$

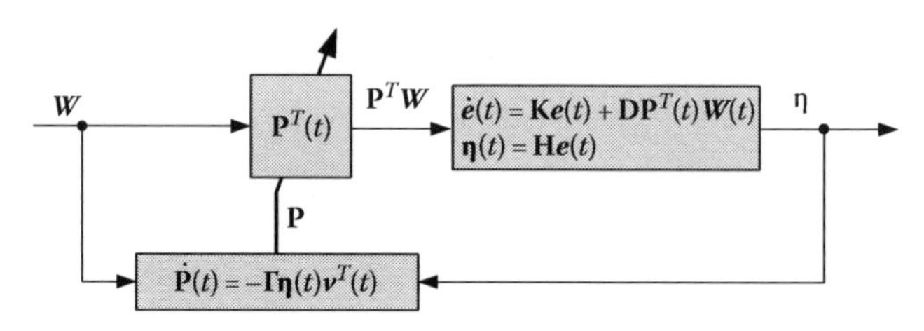

FIGURE 23.6 Generalized error models in estimation and control with partial-state measurements.

In (23.68), it is assumed that the output vector belongs to R^m and that the transfer matrix $\mathbf{H}(s\mathbf{I} - \mathbf{K})^{-1}\mathbf{D}$ is a strictly positive real such that $\mathbf{K}^T\mathbf{T} + \mathbf{TK} = -\mathbf{Q}$, $\mathbf{Q} = \mathbf{Q}^T > 0$, $\mathbf{T} = \mathbf{T}^T > 0$ and $\mathbf{H} = \mathbf{D}^T\mathbf{T}$. The adaptive law for updating the $m \times s$ matrix $\tilde{\mathbf{P}}^T(t)$ in this case is found to be

$$\dot{\tilde{\mathbf{P}}}^T(t) = -\mathbf{\Gamma}\mathbf{D}^T\mathbf{T}e(t)\boldsymbol{v}^T(t) = -\mathbf{\Gamma}\boldsymbol{\eta}(t)\boldsymbol{v}^T(t) \tag{23.69}$$

As in the previous case, $V(\boldsymbol{e}, \tilde{\mathbf{P}}) = (1/2)[\boldsymbol{e}(t)^T\mathbf{T}\boldsymbol{e}(t) + tr(\tilde{\mathbf{P}}(t)\mathbf{\Gamma}^{-1}\tilde{\mathbf{P}}^T(t))]$ is a Lyapunov function, hence, $\boldsymbol{e}(t)$ and $\tilde{\mathbf{P}}(t)$ tend to be zero, if the input $\boldsymbol{u}(t)$ and hence, $\boldsymbol{v}(t)$ is sufficiently rich.

For a single output system (23.69) specializes to $\dot{\boldsymbol{e}}(t) = \mathbf{K}\boldsymbol{e}(t) + \boldsymbol{d}\tilde{\mathbf{P}}^T(t)\boldsymbol{v}(t)$; $e_1(t) = \boldsymbol{h}e(t)$, where $\boldsymbol{h}$ is as row, $\dot{\tilde{\boldsymbol{P}}}(t) = -e_1(t)\boldsymbol{v}(t)$. If the transfer function $\boldsymbol{h}(s\mathbf{I} - \mathbf{K})^{-1}\boldsymbol{d}$ is a strictly positive real and the input $\underline{u}(t)$ is sufficiently rich, then $\boldsymbol{e}(t)$ and $\tilde{\boldsymbol{P}}(t)$ tend to zero.

23.4 Lyapunov-Based Adaptive Parameter Estimation for Nonlinear Systems

Besides its use in systems with linear models, Lyapunov-based APE proves to be a very effective approach for adaptive linearization in nonlinear systems. A common use of the approach is in combination with the feedback linearization technique, which aims to transform a nonlinear system into a (fully or partially) linear system, thereby allowing the use of linear techniques to complete the control design. The approach has been used to solve a number of practical nonlinear control problems involved with important classes of nonlinear systems, such as input–output or, input-state linearizable, or minimum-phase systems. However, feedback linearization does not guarantee robustness in the face of parameter uncertainty or disturbances, and requires (preferably) online parameter and/or load updates for accurate tracking of reference inputs. A brief discussion of the Lyapunov-based adaptive linearization scheme will be provided below.

A large class of nonlinear control systems can be made to have linear input–output behavior through a choice of nonlinear state feedback control laws. Consider the SISO system given in control-affine form as

$$\dot{\boldsymbol{x}}(t) = \boldsymbol{f}(\boldsymbol{x}) + \boldsymbol{g}(\boldsymbol{x})u \tag{23.70}$$

$$y(t) = h(\boldsymbol{x})$$

with $\boldsymbol{x} \in \Re^n$ and smooth $\boldsymbol{f}, \boldsymbol{g}, h$. Differentiating y with respect to time until the control input, u, appears in the equation, one obtains

$$y^{(r)} = L_f^r h + L_g L_f^{r-1} h u \tag{23.71}$$

Here, $L_f h$, $L_g h$ stand for the Lie derivatives of h w.r.t. $\boldsymbol{f}, \boldsymbol{g}$, respectively. Equation 23.83 thus satisfies the so-called matching condition defined in Taylor et al. [TKM89], which, with the condition $L_g L_f^{r-1} h(\boldsymbol{x}) \neq 0 \ \forall \boldsymbol{x} \in \Re^n$ yields the control law below:

$$u = \frac{1}{L_g L_f^{r-1} h}\left(-L_f^r h + \mathrm{v}\right) \tag{23.72}$$

Feedback linearization is based on the idea of using change of coordinates and feedback control to cancel all or most of the nonlinear terms so that the system behaves as a linear or partially linear system. Exact linearization in practical implementations poses some limitations, due to the fact that any uncertainty in the knowledge of the nonlinear functions f and g, will prevent the cancellation from being exact resulting in an input–output equation, which is not linear. The use of parameter adaptive control to be discussed in this section can help achieving exact cancellation, at least asymptotically.

Input–output feedback linearizable systems often have the structure

$$\dot{\boldsymbol{x}} = \begin{bmatrix} \dot{x}_1 \\ \dot{x}_2 \\ \vdots \\ \dot{x}_n \end{bmatrix} = \begin{bmatrix} x_2 \\ x_3 \\ \vdots \\ f_n(x_1, x_2, \ldots, x_{n-1}) \end{bmatrix} + \begin{bmatrix} 0 \\ 0 \\ \vdots \\ g_n(x_1, x_2, \ldots, x_{n-1}) \end{bmatrix} u = \boldsymbol{f}(\boldsymbol{x}) + \boldsymbol{g}(\boldsymbol{x})u \tag{23.73}$$

$$y = x_1 = h(\boldsymbol{x})$$

$$\dot{x}_n = f_n(\boldsymbol{x}) + g_n(\boldsymbol{x})\, u = \boldsymbol{W}_f^T(\boldsymbol{x})\boldsymbol{P}_f + \boldsymbol{W}_g^T(\boldsymbol{x})\boldsymbol{P}_g u \tag{23.74}$$

A system in the form above can be linearized by the proper application of a feedback control law, with the assumption of accurate knowledge of parameters in the control law. There are two existing approaches in the design of the control law.

23.4.1 Standard Parameterization Approach

The control input assumes the following form in [SI89]:

$$u = \frac{1}{\boldsymbol{W}_g^T(\boldsymbol{x})\boldsymbol{P}_g}\left(-\boldsymbol{W}_f^T(\boldsymbol{x})\boldsymbol{P}_f + \mathrm{v}\right) \tag{23.75}$$

where v is the tracking controller designed for the desired system dynamics. This controller will determine the behavior of the linear system arising after the proper cancellation of nonlinearities.

The performance of the linearizing controller in (23.75) depends on how well the parameter vector, $\boldsymbol{P}$ is known, which may naturally be variable or uncertain, hence calling for online parameter estimation schemes. Whatever the source of these parameters, it would make better sense to present (23.75) with the use of estimated parameter values, $\bar{\boldsymbol{P}}_f$, $\bar{\boldsymbol{P}}_g$ as below:

$$u = \frac{1}{\boldsymbol{W}_g^T(\boldsymbol{x})\bar{\boldsymbol{P}}_g}\left(-\boldsymbol{W}_f^T(\boldsymbol{x})\bar{\boldsymbol{P}}_f + \mathrm{v}\right) \tag{23.76}$$

With the substitution of (23.76) into (23.74)

$$\dot{x}_n = \mathrm{v} - \boldsymbol{W}_f^T\tilde{\boldsymbol{P}}_f - u\boldsymbol{W}_g^T\tilde{\boldsymbol{P}}_g \tag{23.77}$$

Thus, it is clear that if the parameters converge to their true values, the system becomes asymptotically linearized. Now, v must be chosen for tracking purposes and the update law for parameter estimation must be determined.

For simplicity, consider a second-order system such as

$$\dot{\boldsymbol{x}} = \begin{bmatrix} \dot{x}_1 \\ \dot{x}_2 \end{bmatrix} = \begin{bmatrix} x_2 \\ f_2(x_1, x_2) \end{bmatrix} + \begin{bmatrix} 0 \\ g_2(x_1, x_2) \end{bmatrix} u$$

$$y = x_1 = h(x_1, x_2)$$

Considering this second-order system, ν can be chosen as a simple proportional-derivative (PD) controller,

$$\nu = \dot{x}_2^{ref} + K_D\left(x_2^{ref} - x_2\right) + K_P\left(x_1^{ref} - x_1\right) = \dot{x}_2^{ref} + K_D\dot{e} + K_P e \quad (23.78)$$

Usually, the convergence rate σ of the asymptotic behavior is determined to provide critical damping. Thus, with

$$K_P = \sigma^2; \quad K_D = 2\sigma \quad (23.79)$$

the error equation will be obtained as

$$\ddot{e} + 2\sigma\dot{e} + \sigma^2 e - \boldsymbol{W}_g^T\tilde{\boldsymbol{P}}_g - i_q\boldsymbol{W}_f^T\tilde{\boldsymbol{P}}_f = 0 \quad (23.80)$$

Now, for the determination of the parameter update law a Lyapunov function is chosen as below:

$$V = \left(\dot{e} + \sigma e\right)^2 + \sigma^2 e^2 + \tilde{\boldsymbol{P}}_f^T\Gamma_1^{-1}\tilde{\boldsymbol{P}}_f + \tilde{\boldsymbol{P}}_g^T\Gamma_2^{-1}\tilde{\boldsymbol{P}}_g \quad (23.81)$$

which gives the $\dot{V}$ as

$$\dot{V} = -\sigma\left[\left(\dot{e} + \sigma e\right)^2 + \sigma^2 e^2\right] - \sigma\dot{e}^2 + 2\left[\dot{\tilde{\boldsymbol{P}}}_f^T\Gamma_1^{-1} + \boldsymbol{W}_f^T\left(\dot{e} + \sigma e\right)i_q\right]\tilde{\boldsymbol{P}}_f + 2\left[\dot{\tilde{\boldsymbol{P}}}_g^T\Gamma_2^{-1} + \boldsymbol{W}_g^T\left(\dot{e} + \sigma e\right)\right]\tilde{\boldsymbol{P}}_g \quad (23.82)$$

with Γ_1, Γ_2 chosen as positive definite adaptation gain matrices. Clearly, for the satisfaction of Lyapunov stability criteria, the parameter update laws must satisfy the equations

$$\begin{aligned} \dot{\bar{\boldsymbol{P}}}_f &= \dot{\tilde{\boldsymbol{P}}}_f = -\Gamma_1\left(\dot{e} + \sigma e\right)\boldsymbol{W}_f u \\ \dot{\bar{\boldsymbol{P}}}_g &= \dot{\tilde{\boldsymbol{P}}}_g = -\Gamma_2\left(\dot{e} + \sigma e\right)\boldsymbol{W}_g \end{aligned} \quad (23.83)$$

23.4.2 New Parameterization Approach

The main drawback of the parameterization approach in (23.75) and (23.76) is that changes in $\boldsymbol{P}_g$ may affect the numerator terms negatively, hence deteriorating the convergence of the algorithm, especially when $\boldsymbol{P}_g$ values become very small. To remedy this condition, another parameterization approach has been proposed [CP93,BG95], in which the parameters multiplying the control input are separated as those that are constant or known and those that are uncertain as below:

$$\boldsymbol{W}_g^T(x)\bar{\boldsymbol{P}}_g u = \bar{\boldsymbol{P}}_{gn}u + \boldsymbol{W}_g^T(x)\bar{\boldsymbol{P}}_g u = -\boldsymbol{W}_f^T(x)\bar{\boldsymbol{P}}_f + \nu$$

The resulting control structure is as below:

$$u = \frac{1}{\bar{\boldsymbol{P}}_{gn}}\left(-\boldsymbol{W}_g^T(x)\bar{\boldsymbol{P}}_g u - \boldsymbol{W}_f^T(x)\bar{\boldsymbol{P}}_f + \nu\right) \quad (23.84)$$

$\bar{P}_f$, $\bar{P}_g$ being parameter estimates; P_f, P_g are the actual parameters and $\tilde{P}_f$, $\tilde{P}_g$ represent the parameter errors. Now, by equating

$$\tilde{P}_f = \bar{P}_f - P_f; \quad \tilde{P}_g = \bar{P}_g - P_g \tag{23.85}$$

$$\dot{x}_n = \nu - W_g^T \tilde{P}_g - W_f^T \tilde{P}_f u \tag{23.86}$$

is obtained, where

$$W^T = \begin{bmatrix} W_f^T & W_g^T \end{bmatrix} \quad P = \begin{bmatrix} \tilde{P}_f^T & \tilde{P}_g^T \end{bmatrix} \tag{23.87}$$

Thus, it is clear that if the parameters converge to their true values, the system becomes asymptotically linearized. Now, ν must be chosen for tracking purposes and the update law for parameter estimation must be determined.

The new parameterization approach is performed by rewriting the equation of motion (23.77) in the following form:

$$\dot{x}_n = P_{gn} u + W^T P \tag{23.88}$$

As can be seen from the above equation, the torque terms are no longer classified according to multiplication with the control variable, u. However, the control variable, u, is now derived in a different form, as can also be observed from (23.88):

$$u = \frac{1}{P_{gn}} \left(\nu - W^T \bar{P} \right) \tag{23.89}$$

P_{gn}, can be replaced by the constant input parameter if it is known for the system, or it can be given some constant value.

Example 23.1: Experimental Implementation of Parameter Identification for Torque Ripple Minimization in Direct Drive Systems [CP93]

Below is a description of the experimental implementation of the Lyapunov-based parameter estimation for the estimation of load and friction uncertainties as well as the cancellation of undesirable nonlinear torque-ripple terms in permanent magnet synchronous motors (PMSM) [BG95] in a 1-degree-of-freedom (DOF) robotic arm configuration under gravity effects. PMSMs find wide range of applications in direct-drive robotics (DDR) applications, which require that load torque variations and torque pulsations (i.e., torque ripple and cogging torque) be eliminated mechanically, or via control. This example considers the latter approach. The PMSM in this example is driving a single-link arm with variable load. The motor has 12 poles on its permanent magnet rotor and 39 teeth on its stator surface. With those assumptions, the mathematical model of the system can be described by the following equations:

$$\frac{d\theta}{dt} = \omega \tag{23.90}$$

$$\frac{d\omega}{dt} = \underbrace{K_{c39}\cos 39\theta + K_{s39}\sin 39\theta - T_L \sin\theta + \xi}_{f(x)} + \underbrace{\left(K_t + K_{c6}\cos 6\theta + K_{s6}\sin 6\theta\right)}_{g(x)}\underbrace{i_q}_{u} \tag{23.91}$$

where

K_t; K_{c6}; K_{s6} are the torque constant, pulsation torque constants of 6th components (the biggest pulsation torque components), respectively, and all torque constants divided by total of inertia ($1/A \cdot s^2$)

K_{c39}; K_{s39} are the cogging torque constants of 39th components (the biggest cogging torque components), respectively, and all torque constants divided by total of inertia ($1/s^2$)

T_L is the load torque/total inertia due to the gravity effect ($1/s^2$)

ξ is the residues of the pulsation torques and cogging torques ($1/s^2$), this effect is small when compared with the other components

p is the number of pole pairs and $p = 6$

i_q is the torque current (A)

Implementation of the standard parameterization scheme

The torque pulsations in the torque function can be eliminated by an appropriately chosen current input with current harmonics to cancel the effects due to geometric harmonics, but it would require accurate knowledge of the parameters (coefficients of the load and torque pulsations in (23.91). To solve this problem, we will exploit the flexibility offered by the two parameter estimation schemes described above and perform the adaptive online identification of the unknown load and torque-ripple coefficients. Both methods are implemented for the direct-drive system in consideration with the use of the DS1104 motion controller.

Both parameterization approaches require the determination of ν for the desired tracking dynamics as well as the parameter update law as given in (23.78) and (23.83), respectively. First, for both schemes, ν is selected as below:

$$\nu = \dot{\omega}_d + K_D(\omega_d - \omega) + K_P(\theta_d - \theta) \tag{23.92}$$

where θ_d and ω_d are desired angular position and velocity, respectively. To further the implementation of the two schemes on the given direct-drive system, the following substitutions are made in (23.78) and (23.83):

$$x_1 = \theta, \quad x_2 = \omega, \quad u = i_q.$$

Clearly, with these substitutions, the parameter update laws will assume

$$\dot{\tilde{P}}_f = \dot{\hat{P}}_f = -\Gamma_1(\dot{e} + \sigma e)Wi_q \tag{23.93}$$

$$\dot{\tilde{P}}_g = \dot{\hat{P}}_g = -\Gamma_2(\dot{e} + \sigma e)W \tag{23.94}$$

where Γ_1, Γ_2 are arbitrary positive definite adaptation gain matrices,

$$\tilde{\boldsymbol{P}}^T = \left[\tilde{K}_t\, \tilde{K}_{c6}\, \tilde{K}_{s6}\, \tilde{K}_{c39}\, \tilde{K}_{s39}\, \tilde{T}_L\right]: \text{ parameter vector to be estimated} \tag{23.95}$$

$$\boldsymbol{W}^T = \left[1 \quad \cos 6\theta \quad \sin 6\theta \quad \cos 39\theta \quad \sin 39\theta \quad -\sin\theta\right]: \text{ regressor vector} \tag{23.96}$$

With the parameters updated according to (23.93) and (23.94), the control variable i_q will be obtained in the following form for the standard parameterization scheme:

$$i_q = \frac{\ddot{\theta}_d + K_P e + K_D \dot{e} + \bar{T}_L \sin\theta - \bar{K}_{c39}\cos 39\theta - \bar{K}_{s39}\sin 39\theta}{\bar{K}_t + \bar{K}_{c6}\cos 6\theta + \bar{K}_{s6}\sin 6\theta} \tag{23.97}$$

Inspection of (23.97) reveals the drawback of the standard adaptive linearization as applied to the direct-drive system. Due to the variables in the denominator of i_q, the boundedness of $\ddot{e}$ and consequently, the convergence of $\dot{e}$ and e cannot actually be guaranteed unless special measures are taken on the variation of those terms in the denominator. Parameter convergence is also not ensured due to this reason, which results in the lack of persistency of excitation. A theoretical coverage of this issue can be found in [CP93].

Implementation of the new parameterization scheme

To improve the convergence performance, we will now implement the new parameterization approach by rewriting the equation of motion (23.88) and (23.89) in the following form:

$$\dot{\omega} = K_t i_q + \boldsymbol{W}^T \boldsymbol{P} \tag{23.98}$$

As can be seen from the above equation, the torque terms are no longer grouped according to multiplication with the control variable, i_q. The parameter error and regressor vector have the same form as in (23.95) and (23.96). However, the control variable, i_q, is now derived in a different form, as can also be observed from (23.97), with only constant K_{tn} in the denominator. The derived control variable, i_q, is

$$i_q = \frac{1}{K_{tn}}\left(\nu - \boldsymbol{W}^T \bar{\boldsymbol{P}}\right) \tag{23.99}$$

K_{tn} can be replaced by the nominal torque constant if it is known for the system or it can be given some constant value, since K_t adaptation is also included in the control.

This new parameterization approach also allows the choice of a simpler Lyapunov structure, such as

$$V = \left(\dot{e} + \sigma e\right)^2 + \sigma^2 e^2 + \tilde{\boldsymbol{P}}^T \Gamma^{-1} \tilde{\boldsymbol{P}} \tag{23.100}$$

where $\tilde{\boldsymbol{P}} = \bar{\boldsymbol{P}} - \boldsymbol{P}$ is the parameter error vector with its update law obtained with the same procedure as before;

$$\dot{\boldsymbol{P}} = \dot{\bar{\boldsymbol{P}}} = -\left(\dot{e} + \sigma e\right)\Gamma \boldsymbol{W} \tag{23.101}$$

The only difference is that the update law has the same form for all parameters unlike the standard approach.

The new parameterization, which basically eliminates all the problems induced by the division of variables in i_q, also establishes guaranteed stability, robustness, and parameter convergence [CB93]. The improved performance through its application to the direct-drive system can be noted in the experimental results.

The experimental results for both parameterization approaches are given below. Parameter estimation is performed to determine the unknown coefficients of the torque pulsations as well as the unknown load mass, which is varied from zero to $M = 150\,\text{g}$. Comparative results are provided for both schemes. Figure 23.7a through d present the performance for a constant speed reference trajectory. Finally, for a better demonstration of the performance of the two schemes, torque-ripple spectra are also provided for both schemes, under both no load and load. Figure 23.8a and b depict the torque spectrum for no load and for an $M = 150\,\text{g}$, respectively, while Figure 23.8c and d present the torque-ripple spectrum for standard and new parameterization schemes, respectively.

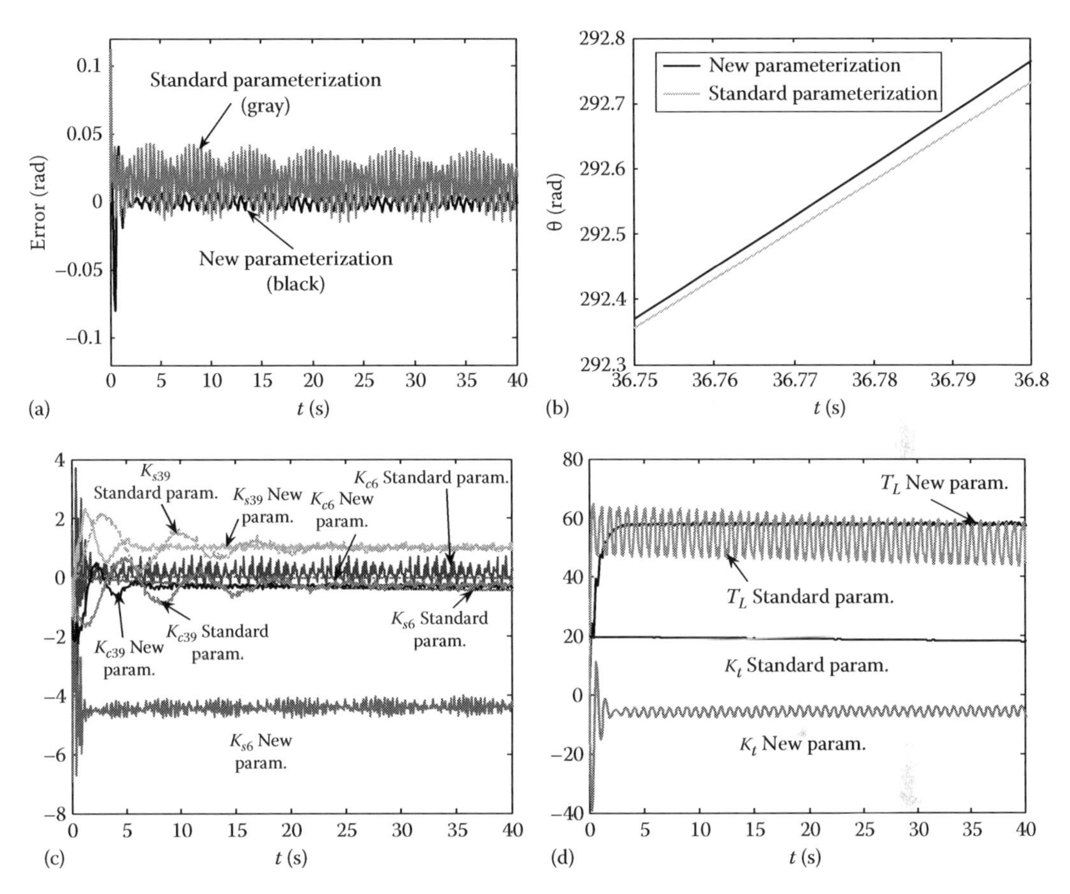

FIGURE 23.7 Comparative performance of standard and new parameterization scheme for constant speed ($\omega_d = 2$): (a) variation of error; (b) variation of θ (angular position); (c) variation of K_{c6}, K_{s6}, K_{c39}, K_{s39}; and (d) Variation of T_L and K_t.

A careful inspection of the figures demonstrates that, while parameters and load variations in the numerator of the control law are estimated with high performance for both schemes, that is, load parameters in the denominator drift or do not converge at all. Hence, the representation of all the variable parameters in the numerator achieved by the new parameterization approach results in an overall improved performance in terms of accuracy and robustness.

23.5 Adaptive Observers

The main drawback of the schemes described in the previous sections is that they depend on the accessibility of all state variables or the availability of a strictly positive real transfer function model with partial-state measurements of the plant to be identified. However, in most practical situations, it should be expected that only some of the state variables of the plant can be measured. Hence, it is necessary to construct a scheme that estimates both the parameters of the plant (23.58), that is, **A**, **B**, **C**, as well as the state vector $\boldsymbol{x}$ using only **input–output** measurements. In this chapter, we refer to such schemes as adaptive observer or adaptive estimator. A good starting point for choosing the structure of adaptive observer is the state observer, known as the Luenberger observer, used to estimate only the state vector $\boldsymbol{x}$ of the plant (23.58) when the parameters **A**, **B**, **C** are known.

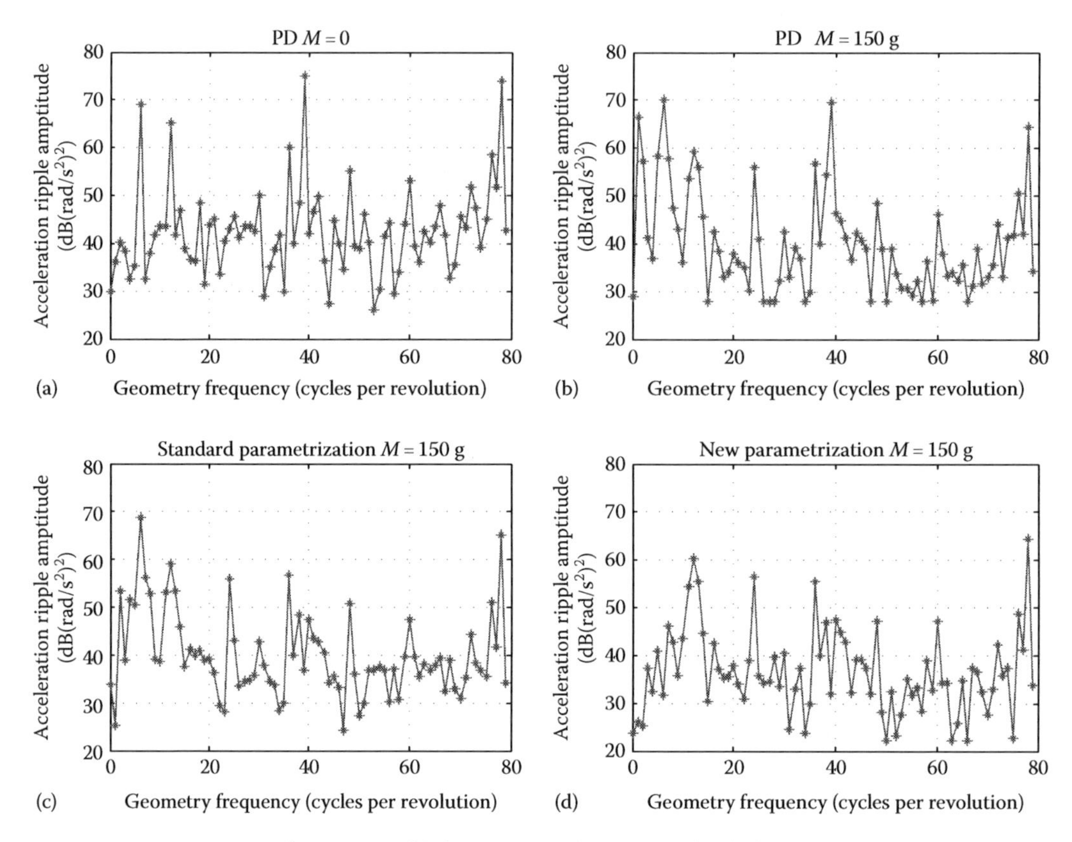

FIGURE 23.8 Torque ripple spectrum (a) for $M = 0$ with PD control, (b) for $M = 150$ g with PD control, (c) with adaptive linearization using standard parameterization, and (d) with adaptive linearization using new parameterization.

23.5.1 Luenberger Observer

When the initial condition $\boldsymbol{x}_0$ is unknown and $\mathbf{A}$ is not a stable matrix, or $\mathbf{A}$ is stable but state observation error $\boldsymbol{e} = \hat{\boldsymbol{x}} - \boldsymbol{x}$ is required to converge to zero faster than the rate with which $\|e^{\mathbf{A}t}\|$ goes to zero. The following observer, known as the Luenberger full-state observer is used:

$$\dot{\hat{\boldsymbol{x}}} = \mathbf{A}\hat{\boldsymbol{x}} + \mathbf{B}\boldsymbol{u} + \boldsymbol{K}(y - \hat{y}), \quad \hat{\boldsymbol{x}}(0) = \hat{\boldsymbol{x}}_0 \tag{23.102}$$

$$\hat{y} = \boldsymbol{C}\hat{\boldsymbol{x}}$$

where $\boldsymbol{K}$ is a column matrix to be chosen by the designer. The Luenberger observer is seen to have a feedback term that depends on the output observation error $\tilde{y} = \hat{y} - y$. The state observation error $\boldsymbol{e} = \hat{\boldsymbol{x}} - \boldsymbol{x}$ satisfies

$$\dot{\boldsymbol{e}} = (\mathbf{A} - \boldsymbol{KC})\boldsymbol{e}, \quad \boldsymbol{e}_0 = \hat{\boldsymbol{x}}_0 - \boldsymbol{x}_0 \tag{23.103}$$

Because $(\boldsymbol{C},\mathbf{A})$ is an observable pair and $(\mathbf{A}^T, \boldsymbol{C}^T)$ is a controllable pair, we can choose $\boldsymbol{K}$ so that the characteristic polynomial $|s\mathbf{I} - \mathbf{A} + \boldsymbol{KC}| = |s\mathbf{I} - \mathbf{A}^T + \boldsymbol{C}^T \boldsymbol{K}^T|$ is a stable polynomial with desired eigenvalues of $\mathbf{A} - \boldsymbol{KC}$, so the rate of convergence of the observation error $\boldsymbol{e}(t)$ to zero can arbitrarily be chosen by designing $\boldsymbol{K}$ appropriately.

23.5.2 Adaptive Estimation in Deterministic Nonlinear Systems Using the Extended Luenberger Observer

As previously discussed, the Luenberger observer has a linear structure and is used to estimate the system states with the use of input–output signals and under the assumption of known system parameters. However, it is a well-known fact that generally, system parameters are not known or may vary during system operation. To address such problems in addition to the commonly encountered nonlinearities in practical systems, the ELO has been developed as an extension of Luenberger observers. ELOs can perform the simultaneous estimation of states and unknown parameters, using deterministic system models based on the assumption of slow parameter variations. The ELO is designed for nonlinear systems in the following form:

$$\dot{\boldsymbol{x}}(t) = \boldsymbol{f}(\boldsymbol{x}(t), \boldsymbol{P}(t)) + \boldsymbol{B}u(t) \tag{23.104}$$

$$y(t) = \boldsymbol{C}\boldsymbol{x}(t)$$

$\boldsymbol{P}(t)$: variable parameter vector.

To perform the estimation of slowly varying system parameters, an "extended" system model is obtained as below:

$$\underbrace{\begin{bmatrix} \dot{\boldsymbol{x}}(t) \\ \dot{\boldsymbol{P}}(t) \end{bmatrix}}_{\dot{\boldsymbol{x}}_e} = \underbrace{\begin{bmatrix} \boldsymbol{f}(\boldsymbol{x}(t), \boldsymbol{P}(t)) \\ \boldsymbol{0} \end{bmatrix}}_{\boldsymbol{F}(\boldsymbol{x}_e(t))} + \underbrace{\begin{bmatrix} \boldsymbol{B} \\ \boldsymbol{0} \end{bmatrix}}_{\boldsymbol{B}_e} u(t) \tag{23.105}$$

$$y(t) = \underbrace{\begin{bmatrix} \boldsymbol{C} & \boldsymbol{0} \end{bmatrix}}_{\boldsymbol{C}_e} \underbrace{\begin{bmatrix} \boldsymbol{x}(t) \\ \boldsymbol{P}(t) \end{bmatrix}}_{\boldsymbol{x}_e} \tag{23.106}$$

Next, the nonlinear system in (23.104) is linearized around the estimated state vector, $\hat{x}_e(t)$ as

$$\dot{\boldsymbol{x}}_e(t) = \boldsymbol{F}(\hat{\boldsymbol{x}}_e(t)) + \left.\frac{\partial \boldsymbol{F}(\boldsymbol{x}_e)}{\partial \boldsymbol{x}_e}\right|_{\hat{\boldsymbol{x}}_e} (\boldsymbol{x}_e - \hat{\boldsymbol{x}}_e) + \boldsymbol{B}_e u(t) \tag{23.107}$$

to yield the equations below for the ELO scheme, similar to that of LO. It should be noted that in the ELO scheme, the gain matrix, $\boldsymbol{L}(t)$ will vary as the estimated system parameters are updated.

$$\dot{\hat{\boldsymbol{x}}}_e(t) = \boldsymbol{F}(\hat{\boldsymbol{x}}_e(t)) + \boldsymbol{B}_e u(t) + \boldsymbol{L}(t)(y(t) - \hat{y}(t))$$

$$\hat{y}(t) = \boldsymbol{C}_e \hat{\boldsymbol{x}}_e(t) \tag{23.108}$$

$$\dot{\hat{\boldsymbol{x}}}_e(t) = \boldsymbol{F}(\hat{\boldsymbol{x}}_e(t)) + \boldsymbol{B}_e u(t) + \boldsymbol{L}(t)(y(t) - \boldsymbol{C}_e \hat{\boldsymbol{x}}_e(t)) \tag{23.109}$$

Using the relationships

$$\mathbf{A}_e(t) = \left.\frac{\partial \boldsymbol{F}(\boldsymbol{x}_e(t))}{\partial \boldsymbol{x}_e(t)}\right|_{\boldsymbol{x}_e = \hat{\boldsymbol{x}}_e} \tag{23.110}$$

$$\tilde{\boldsymbol{x}}_e(t) = \boldsymbol{x}_e(t) - \hat{\boldsymbol{x}}_e(t) \tag{23.111}$$

the state error vector can be represented as below, with the design of an appropriate observer gain matrix, $\boldsymbol{L}(t)$ to yield the desired estimation dynamics:

$$\dot{\tilde{\boldsymbol{x}}}_e(t) = \left[\mathbf{A}_e(t) - \boldsymbol{L}(t)\boldsymbol{C}_e\right]\tilde{\boldsymbol{x}}_e(t) \tag{23.112}$$

The gain matrix, $\boldsymbol{L}(t)$ is calculated at every time step, t, by matching the actual characteristic polynomial with a characteristic polynomial representing the desired observer dynamics, with $s^n + \gamma_1 s^{n-1} + \cdots + \gamma_n = 0$ (γ_i: constant) as can be seen below:

$$\left|\mathbf{sI} - \mathbf{A}_e(\boldsymbol{P}) + \boldsymbol{L}(\boldsymbol{P})\boldsymbol{C}_e\right| = s^n + \alpha_1(\boldsymbol{L}(\boldsymbol{P}))s^{n-1} + \cdots + \alpha_n(\boldsymbol{L}(\boldsymbol{P})) = 0 \tag{23.113}$$

Now, we consider the problem where both the state $\boldsymbol{x}$ and parameters $\mathbf{A}$, $\mathbf{B}$, $\boldsymbol{C}$ are to be estimated online simultaneously using an adaptive observer.

23.5.3 Adaptive Estimation in Stochastic Nonlinear Systems Using the Extended Kalman Filter

The extended Kalman filter (EKF) scheme is another well-established adaptive estimation scheme for the simultaneous estimation of states and unknown parameters in nonlinear systems with stochastic models. Similar to the approach taken in the previous section, we will first discuss the basics of the Kalman filter (KF), which is the optimal "nonadaptive" state estimation scheme, upon which the "non-optimal" EKF scheme is built.

23.5.3.1 Kalman Filter

The KF is a well-known recursive algorithm that takes the stochastic linear state-space model of the system into account, together with measured outputs to achieve the optimal estimation of states [BGH01] in multi-input, multi-output systems. The optimization is performed based on the minimization of the mean of the squared error. The filter is capable of providing estimations of past, present, and future states even when the model of the system is not accurately known. The design of the KF is based on the following stochastic linear difference model of the system for which state estimation will be performed:

$$\boldsymbol{x}(k) = \mathbf{A}\boldsymbol{x}(k-1) + \mathbf{B}\boldsymbol{u}(k-1) + \boldsymbol{\zeta}(k) \tag{23.114}$$

with a measurement signal that can be given as

$$z(k) = \mathbf{H}\boldsymbol{x}(k) + \boldsymbol{\varepsilon}(k) \tag{23.115}$$

The random variables in (23.114) and (23.115) represent the process and measurement noise, respectively, w and v are considered to be white noise and with normal probability distributions $p(\boldsymbol{w}) \sim N(0, \mathbf{Q})$; $p(\boldsymbol{v}) \sim N(0, \mathbf{R})$. $\mathbf{Q}$ and $\mathbf{R}$ are the *process noise covariance* and *measurement noise covariance* matrices, respectively. The KF algorithm involves two main steps: *prediction step*, in which predictions of the state and error covariance estimates, namely the a priori estimates are calculated for the next time step, and the *correction step*, in which an improved a posteriori estimate is obtained by incorporating a new measurement into the a priori estimate. The *prediction–correction* algorithm can thus be summarized with the following equations:

Prediction

$$\begin{aligned} \hat{\boldsymbol{x}}(k\,|\,k-1) &= \mathbf{A}\hat{\boldsymbol{x}}(k-1) + \mathbf{B}\boldsymbol{u}(k-1) \\ \mathbf{U}(k\,|\,k-1) &= \mathbf{A}\mathbf{U}(k-1\,|\,k-1)\mathbf{A}^T + \mathbf{Q} \end{aligned} \tag{23.116}$$

Correction

$$\begin{aligned} \mathbf{K}(k) &= \mathbf{U}(k\,|\,k-1)\mathbf{H}^T(\mathbf{H}\mathbf{U}(k\,|\,k-1)\mathbf{H}^T + \mathbf{R})^{-1} \\ \hat{\boldsymbol{x}}(k\,|\,k) &= \hat{\boldsymbol{x}}(k\,|\,k-1) + \mathbf{K}(k)\,(z(k) - \mathbf{H}\hat{\boldsymbol{x}}(k\,|\,k-1)) \\ \mathbf{U}(k\,|\,k) &= (\boldsymbol{I} - \boldsymbol{K}(k)\mathbf{H})\mathbf{U}(k-1\,|\,k-1) \end{aligned} \tag{23.117}$$

23.5.3.2 Extended Kalman Filter

The EKF is an extension of KF scheme designed specifically for nonlinear systems. With this method, it is possible to make the online estimation of states while simultaneously performing identification of parameters in a relatively short time interval [SST93,BGH01,BBG02], also taking system/process errors and measurement noises directly into account. The method ensures a high convergence rate, while the system noise taken into account in the design process inherently fulfills the persistency of excitation requirement for accurate estimation. The recursive nature of the algorithm is yet another attraction of the algorithm as it provides accurate estimation based on data from previous time steps, which also makes the algorithm suitable for practical implementation. These are the main reasons why EKF has found wide application in the industrial control systems, in spite of its computational complexity, which has now ceased to be a burden, thanks to the developments in high-performance processor technology.

Similar to the design of the ELO, the design of the EKF scheme is also based on the extended model of the given nonlinear system as in (23.108), which is first discritized in the following form:

$$\boldsymbol{x}_e(k) = \boldsymbol{x}_e(k-1) + Tf_e(\boldsymbol{x}_e(k-1), \boldsymbol{u}(k-1)) = \boldsymbol{F}(\boldsymbol{x}_e(k-1), \boldsymbol{u}(k-1)) \tag{23.118}$$

Next, the quasi linearization of (23.121) is performed around the estimated $\hat{\boldsymbol{x}}_e(k-1)$ and $\boldsymbol{u}(k-1)$ of the previous step, yielding

$$\boldsymbol{x}_e(k) = \boldsymbol{F}[\hat{\boldsymbol{x}}_e(k-1), \hat{\boldsymbol{u}}(k-1)] + \mathbf{F}_x[\boldsymbol{x}_e(k-1) - \hat{\boldsymbol{x}}_e(k-1)] + \mathbf{F}_u[\boldsymbol{u}(k-1) - \hat{\boldsymbol{u}}(k-1)] + \boldsymbol{\zeta}(k-1) \tag{23.119}$$

Here,

$$\mathbf{F}_x = \left[\frac{\partial \boldsymbol{F}}{\partial \boldsymbol{x}_e}\right]_{\hat{\boldsymbol{x}}_e(k-1), \hat{\boldsymbol{u}}(k-1)}, \quad \mathbf{F}_u = \left[\frac{\partial \boldsymbol{F}}{\partial u}\right]_{\hat{\boldsymbol{x}}_e(k-1), \hat{\boldsymbol{u}}(k-1)} \tag{23.120}$$

$\boldsymbol{w}$: Gaussian system noise with zero mathematical expectation and variance, $\mathbf{Q}$

The system output, y in (23.108) is replaced with the measured output, z, which includes the measured signal as well as the measurement noise.

$$z(k) = \mathbf{H}\boldsymbol{x}_e(k) + \boldsymbol{\varepsilon}(k) \tag{23.121}$$

Here,

$\boldsymbol{H}$ is the measurement matrix

ε is the Gaussian measurement noise with zero mathematical expectation and variance, $\mathbf{R}$

The estimation of $\boldsymbol{x}_e(k)$ is performed based on the Bayes approach, using the fact that a posteriori probability density, $p[\boldsymbol{x}_e(k)/\boldsymbol{z}^k]$, contains all the statistical information on the value of $\hat{\boldsymbol{x}}_e(k)$ and can be expressed using the Bayes formula as

$$p\left[\frac{\boldsymbol{x}_e(k)}{\boldsymbol{z}^k}\right]=p\left[\frac{\boldsymbol{x}_e(k)}{\boldsymbol{z}^{k-1},z(k)}\right]=\frac{p\left[\boldsymbol{x}_e(k)/\boldsymbol{z}^{k-1}\right]p\left[z(\boldsymbol{k})/\boldsymbol{x}_e(k),\boldsymbol{z}^{k-1}\right]}{p\left[z(k)/\boldsymbol{z}^{k-1}\right]} \tag{23.122}$$

Here, $\boldsymbol{z}^k = \{z(1), z(2),\ldots, z(k)\}$, $\boldsymbol{z}^{k-1} = \{z(1), z(2),\ldots, z(k-1)\}$

The accuracy of the value will be determined by the covariance matrix of this distribution. Now, the parameter identification algorithm can be given as

$$\hat{\boldsymbol{x}}_e(k)=\boldsymbol{F}[\hat{\boldsymbol{x}}_e(k-1),\hat{\boldsymbol{u}}(k-1)]+\mathbf{U}(k)\boldsymbol{H}^T\mathbf{R}^{-1}\{z(k)-\boldsymbol{HF}[\hat{\boldsymbol{x}}_e(k-1),\hat{\boldsymbol{u}}(k-1)]\} \tag{23.123}$$

$$\mathbf{U}(k)=\mathbf{N}(k)-\mathbf{N}(k)\boldsymbol{H}^T\left[R+\boldsymbol{H}\mathbf{N}(k)\boldsymbol{H}^T\right]^{-1}\boldsymbol{H}\mathbf{N}(k) \tag{23.124}$$

$$\mathbf{N}(k)=\mathbf{F}_x\mathbf{U}(k-1)\mathbf{F}_x^T+\mathbf{F}_u\mathbf{D}_u(k-1)\mathbf{F}_u^T+\mathbf{Q} \tag{23.125}$$

Here,

- $\mathbf{N}$ is the covariance matrix of extrapolation error
- $\mathbf{U}$ is the covariance matrix of value error
- $\mathbf{D}_u$ is the variance of the input signal

The EKF given by (23.123) through (23.125) provides the optimal value of the vector x_e based on the optimization criteria and under measurement noise and external noise effecting the states.

Two practical examples will be provided below to further clarify the implementation of the EKF scheme: The first example involves the online estimation of the variable load for the high-performance control of a nonlinear single-link arm, while the second example is for the estimation of velocity, flux, load, and rotor resistance for the control of squirrel-cage induction motors (IMs) without the use of mechanical sensors, commonly known as sensorless control of IMs. The control and estimation algorithms for both examples are performed on the high-performance motion control board, DS1104.

Example 23.2: Parameter Identification for the Control of a Single Link Arm [BGH01]

The example below is a practical implementation of the EKF scheme for the high-performance control of a 1 DOF arm. A PD+ controller is designed for the control of the arm and compensation of the variable load; hence, an EKF scheme will be developed to accurately estimate the load. The estimator will be specifically designed to calculate the total inertia of the system in the calculation of the load, while also estimating the angular position, θ, and velocity, ω, at the same time. PMSM-driven single-link arm in Figure 23.9 is modeled as follows:

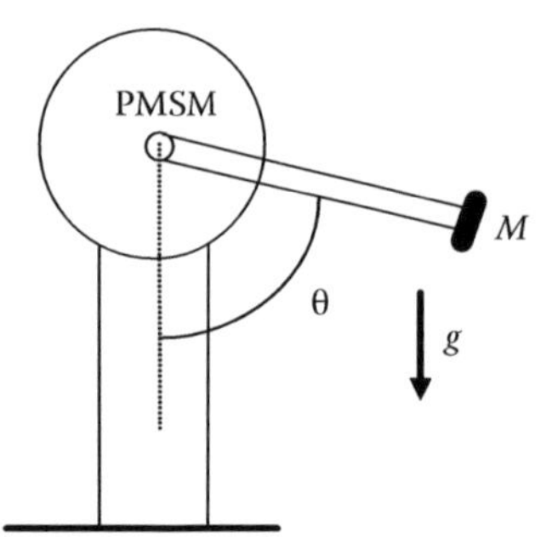

FIGURE 23.9 Single link arm.

$$\dot{\theta}=\omega,$$

$$\dot{\omega}=-\frac{B}{J}\omega+\frac{k_t}{J}i_q-\frac{Mgl}{J}\sin\theta, \tag{23.126}$$

where

- θ is the angular position of the arm (rad)
- ω is the angular velocity of the arm (rad/s)
- J is the total mass of inertia reduced to motor shaft (kg/m^2)
- B is the coefficient of viscous friction (Nm/rad/s)
- k_t is the torque constant (Nm/A)
- M is the mass of the payload (kg)
- g is the acceleration of gravity (m/s^2)
- l is the length of arm (m)
- i_q is the or u control input (A)

$$J = J_m + Ml^2$$

J_m is the moment of inertia of the motor

As discussed in the previous section, the EKF scheme given by (23.125) will provide the near-optimal value of the vector x under measurement and system noise. The estimation algorithm will be used to estimate $\hat{\theta}(k)$, $\hat{\omega}(k)$, and $\hat{J}(k)$ using actual measurements of $\theta(k)$. The estimated total inertia value, $\hat{J}(k)$, will then be used to calculate the unknown load mass using the relationship

$$\hat{M}(k) = \frac{\hat{J}(k) - J_m}{l^2} \tag{23.127}$$

The PD+ controller is designed as below, aiming for the cancellation of the variable load term while providing the desired tracking dynamics:

$$i_q = K_p e + K_d \dot{e} + \frac{\hat{M} gl}{k_t} \sin\theta \tag{23.128}$$

Here,

- $e = \theta^{ref} - \theta$
- K_p is the proportional gain
- K_d is the derivative gain
- θ^{ref} is the reference angular position value

The system parameters are given in Table 23.1.

The experimental results obtained with the EKF-based PD+ controller for different load masses of $M = 0.6$ and 0.2 kg load for are presented in the Figure 23.10, which demonstrates the actual θ values, while Figure 23.11 depicts the tracking error, $\theta_{ref} - \hat{\theta}$, between the reference and angular position. Figure 23.11 also provides zoomed-in sections of the output in the transient and steady state for both load values under a step-type reference trajectory. A careful inspection of the diagrams demonstrates steady

TABLE 23.1 System Parameters

$J = 0.0268$ kg m^2	$k_t = 0.2$ Nm/A	$\theta^{ref} = \pi/2$
$J_{min} = 0.0042$ kg m^2	$B = 0.0007$ Nm s	$l = 0.14$ m
$K_p = 20$	$K_d/T = 5000$	$D_\sigma = 2 \times 10^{-5}$
$T = 0.0008$ s	$M = 0.2$ and 0.6 kg	$D_{iq} = 4.88 \times 10^{-4}$

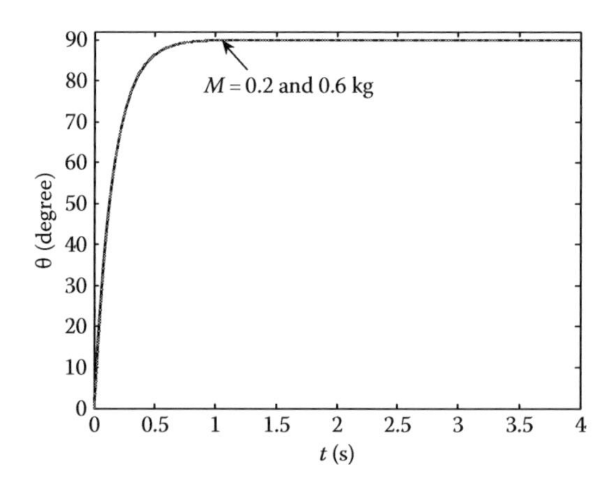

FIGURE 23.10 Angular position, θ, with EKF and PD+ for $M = 0.6 - 0.2\,\text{kg}$.

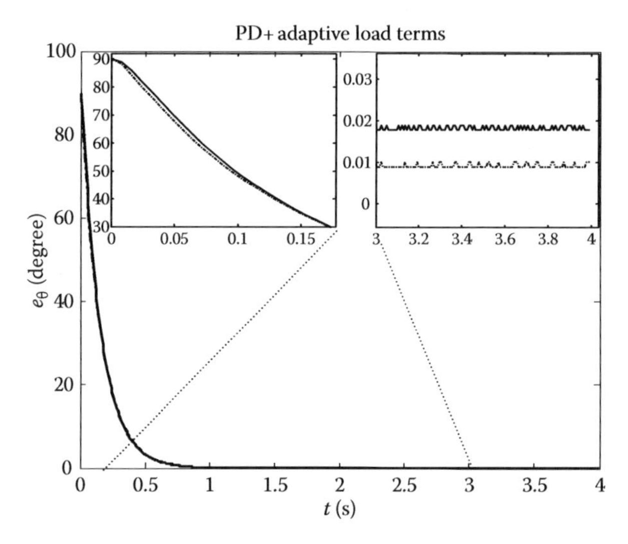

FIGURE 23.11 Error between reference and estimated angular positions $\theta^{ref} - \theta$ with EKF and PD+ for $M = 0.6$ and $0.2\,\text{kg}$.

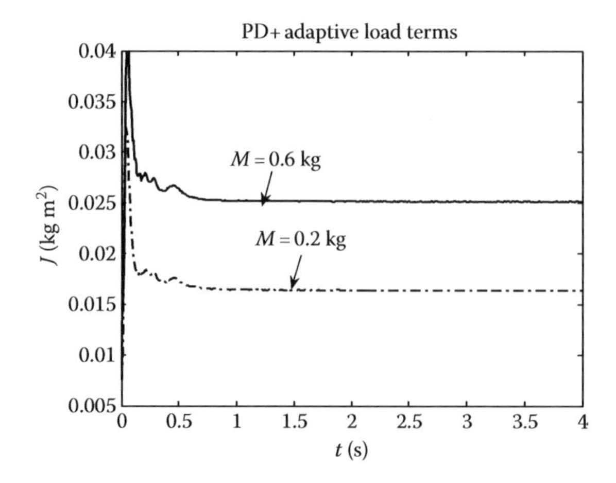

FIGURE 23.12 Estimated total inertia with EKF and PD+ for $M = 0.6$ and $0.2\,\text{kg}$.

state errors of 0.01, 0.02 degrees in the position control of the single-link arm with unknown load variations of mass, $M = 0.2$ and $0.6\,\text{kg}$, respectively. For both cases, the initial value of the inertia is $\hat{J}(k) = J_m$. The results indicate accuracy and robustness both in the transient and steady state, in spite of unknown load variations in Figure 23.12.

Example 23.3: EKF Based Estimation for the Sensorless Control in Induction Motors [BBG07_2,BBG07_3]

Below is another practical implementation of the EKF scheme, namely for the mechanical sensorless estimation of velocity in squirrel-cage IMs. IMs serve as benchmark problems for many sophisticated control and estimation schemes, due to the nonlinear and time-varying nature of the system. IM parameters vary significantly with operating conditions. Besides the load torque that can vary from no load to full load, stator and rotor resistances change with temperature and frequency, while inductances tend to saturate at high current levels. Sensorless control in electromechanical systems is a measure taken for increased reliability, elimination of vibration effects etc., in rugged environments. However, with the elimination of mechanical sensors, the effects of parameter and model uncertainties become even more relevant in IMs, calling for sophisticated methods for the estimation of flux and velocity.

Sensorless control schemes developed for IMs require the estimation of rotor flux components, $\psi_{r\alpha}$, $\psi_{r\beta}$, angular velocity, ω_m, and stator current components $i_{s\alpha}$ and $i_{s\beta}$, which are also measured as output. However, the accurate estimation of these states is very much dependent on how well the system parameters are known, particularly the rotor resistance, R_r. The detailed matrix representation of the two IM models can be given as below:

$$\frac{d}{dt}\underbrace{\begin{bmatrix} i_{s\alpha} \\ i_{s\beta} \\ \psi_{r\alpha} \\ \psi_{r\beta} \\ \omega_m \\ T_L \\ R_r \end{bmatrix}}_{\dot{\mathbf{x}}_e} = \underbrace{\begin{bmatrix} -\left(\frac{R_s}{L_\sigma}+\frac{L_m^2 R_r}{L_\sigma L_r^2}\right) & 0 & \frac{L_m R_r}{L_\sigma L_r^2} & \frac{L_m}{L_\sigma L_r}p\omega_m & 0 & 0 & 0 \\ 0 & -\left(\frac{R_s}{L_\sigma}+\frac{L_m^2 R_r}{L_\sigma L_r^2}\right) & -\frac{L_m}{L_\sigma L_r}p\omega_m & \frac{L_m R_r}{L_\sigma L_r^2} & 0 & 0 & 0 \\ \frac{R_r}{L_r}L_m & 0 & -\frac{R_r}{L_r} & -p\omega_m & 0 & 0 & 0 \\ 0 & \frac{R_r}{L_r}L_m & p\omega_m & -\frac{R_r}{L_r} & 0 & 0 & 0 \\ -\frac{3pL_m}{2JL_r}\psi_{r\beta} & \frac{3pL_m}{2JL_r}\psi_{r\alpha} & 0 & 0 & -\frac{B}{J} & 0 & 0 \\ 0 & 0 & 0 & 0 & 0 & 0 & 0 \\ 0 & 0 & 0 & 0 & 0 & 0 & 0 \end{bmatrix}}_{\mathbf{A}_e} \underbrace{\begin{bmatrix} i_{s\alpha} \\ i_{s\beta} \\ \psi_{r\alpha} \\ \psi_{r\beta} \\ \omega_m \\ T_L \\ R_r \end{bmatrix}}_{\mathbf{x}_e}$$

$$+\underbrace{\begin{bmatrix} \frac{1}{L_\sigma} & 0 & 0 & 0 & 0 & 0 & 0 \\ 0 & \frac{1}{L_\sigma} & 0 & 0 & 0 & 0 & 0 \\ 0 & 0 & 0 & 0 & 0 & 0 & 0 \\ 0 & 0 & 0 & 0 & 0 & 0 & 0 \\ 0 & 0 & 0 & 0 & 0 & 0 & 0 \\ 0 & 0 & 0 & 0 & 0 & 0 & 0 \\ 0 & 0 & 0 & 0 & 0 & 0 & 0 \end{bmatrix}}_{\mathbf{B}_e} \underbrace{\begin{bmatrix} V_{s\alpha} \\ V_{s\beta} \\ 0 \\ 0 \\ 0 \\ 0 \\ 0 \end{bmatrix}}_{\mathbf{u}_e} + \underbrace{\begin{bmatrix} 0 \\ 0 \\ 0 \\ 0 \\ -\frac{1}{J} \\ 0 \\ 0 \end{bmatrix}}_{\mathbf{d}} T_L + \zeta(t) \tag{23.129}$$

$$\mathbf{z}(t) = \mathbf{H}\mathbf{x}_e(t) + \mathbf{v}(t) \text{(measurement equation)} \ \mathbf{z} = \underbrace{\begin{bmatrix} 1 & 0 & 0 & 0 & 0 & 0 & 0 \\ 0 & 1 & 0 & 0 & 0 & 0 & 0 \end{bmatrix}}_{\mathbf{H}} \begin{bmatrix} i_{s\alpha} \\ i_{s\beta} \\ \psi_{r\alpha} \\ \psi_{r\beta} \\ \omega_m \\ t_L \\ R_r \end{bmatrix} + \boldsymbol{\varepsilon}(t) \quad (23.130)$$

where

p is the number of pole pairs
$L_\sigma = \sigma L_s$ is the stator transient inductance
$\sigma = 1 - L_m^2/(L_s L_r)$ is the leakage or coupling factor
L_s, R_s is the stator inductance and resistance, respectively
L_r, R_r is the rotor inductance and resistance, referred to the stator side, respectively
$v_{s\alpha}$ $v_{s\beta}$ is the stator stationary axis components of stator voltages
$i_{s\alpha}$ $i_{s\beta}$ is the stator stationary axis components of stator currents
$\psi_{r\alpha}$ $\psi_{r\beta}$ is the stator stationary axis components of rotor flux
J is the total inertia of the IM and load
ω_m is the angular velocity
$i_{s\alpha}$ and $i_{s\beta}$ are the measured variables

The load torque and rotor resistances are assumed to have a slow variation with time and, therefore, are taken into consideration as constant parameters.

23.6 Experimental Results

The experimental setup used includes the induction motor under consideration is 3 phase, 8 pole, 3 HP/2.238 kW, with its specification details given in Table 23.2. The EKF algorithm and all analog signals are developed and processed on a Power PC–based DS1104 Controller Board. A torque transducer rated at 50 [Nm] and an encoder with 3600 [counts/rev], are also used for the verification of the load torque and velocity estimation and hence, for the performance evaluation of the EKF. The phase voltages and currents are measured with high band voltage and current sensors from LEM Inc.

The variable load is generated through a DC machine operating in generator mode coupled to the IM. An array resistor connected to the armature terminals of the dc machine is used to vary the load torque applied to the IM based on the relationship as $T_L = k_t^2\omega/R$, where k_t is the torque constant of the dc machine; ω is angular velocity. The parameters for the IM and DC generator used in experiments are listed in Tables 23.2 and 23.3.

TABLE 23.2 Rated Values and Parameters of the Induction Motor Used in the Experiments

$R_s = 0.6619\ \Omega$	$R_r = 0.7322\ \Omega$	$L_s = 0.0375$ H	$L_r = 0.0376$ H	$L_m = 0.0334$ H	$V = 230$ V	$I = 12$ A
$P = 2.238$ kW	$f = 60$ Hz	$n_m = 850$ rpm	$T_L = 25.1$ Nm	$J = 0.2595$ kg m^2		$p = 4$

TABLE 23.3 Rated Values and Parameters of the DC Machine Used in the Experiments

P(kW)	V(V)	I(A)	N_m (rpm)	t_L (Nm)
3	125	24	1150	24.91

The initial values of the P and Q in the EKF algorithms are found by trial and error to achieve a rapid initial convergence as well as the desired transient and steady-state performance for the estimated states and parameters. The R matrix, on the other hand, is determined taking into account the measurement errors of the current and voltage sensors and the quantization errors of the analog-to-digital converters (ADCs).

As can be seen from the experimental results in Figures 23.13 through 23.19, EKF estimation has several advantages, like fast convergence, simultaneous estimation of states and parameters, and accurate estimation performance due to the PE property inherently satisfied by the system noise. However, this accuracy performance is clearly compromised when a high number of parameters/states are to be estimated with a single EKF algorithm, with a limited number of output measurements. To address this limitation, modified implementations of the EKF also exist in the literature, that is, the Braided EKF approach for demanding estimation applications with a high number of unknown parameters and states [BBG07_1,BBG08].

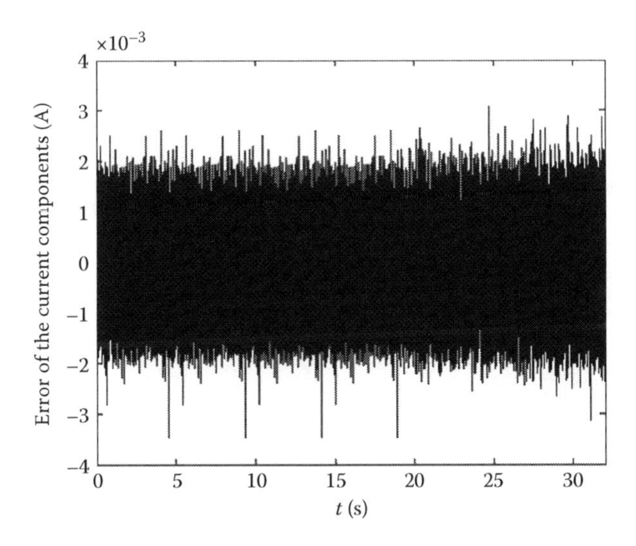

FIGURE 23.13 Variations of the error of the current components.

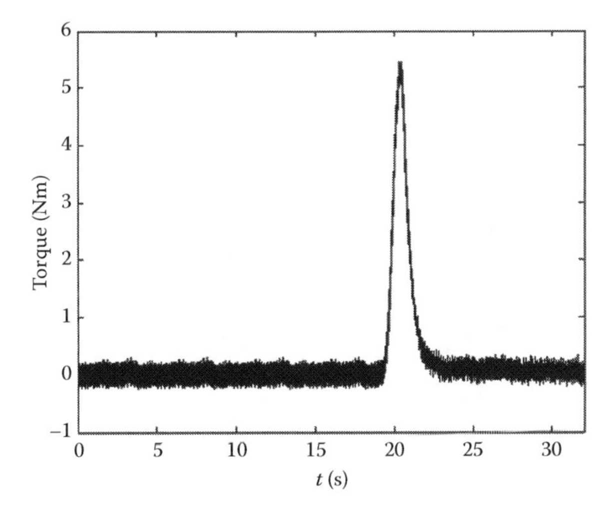

FIGURE 23.14 Variation of the estimated motor torque.

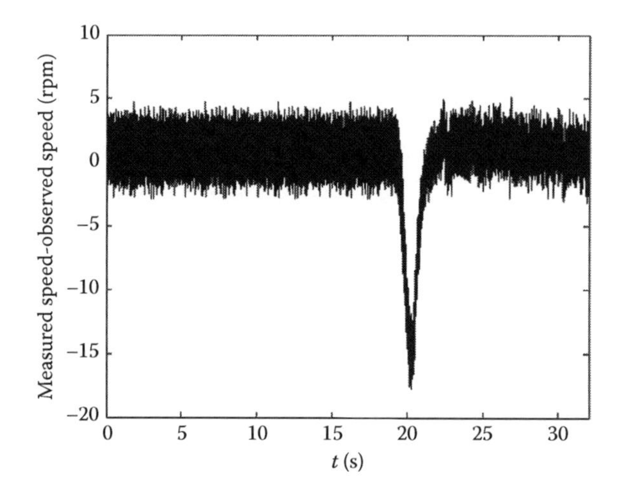

FIGURE 23.15 Error variation between measured and observed speeds.

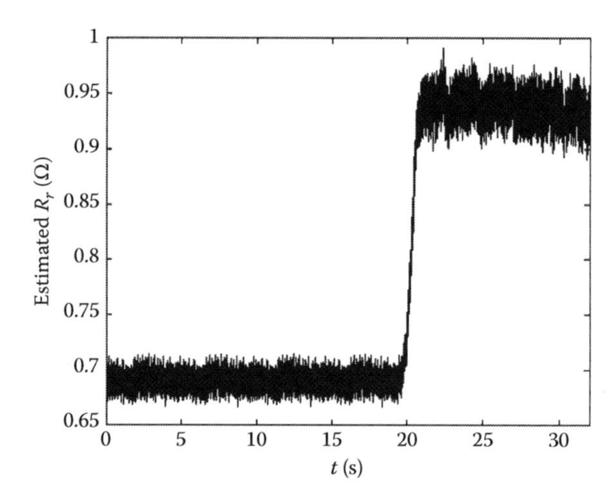

FIGURE 23.16 Estimated rotor resistance variation.

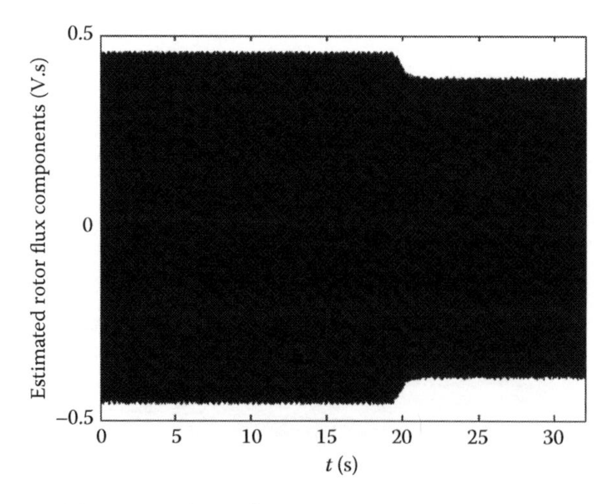

FIGURE 23.17 Variations of the estimated rotor flux components.

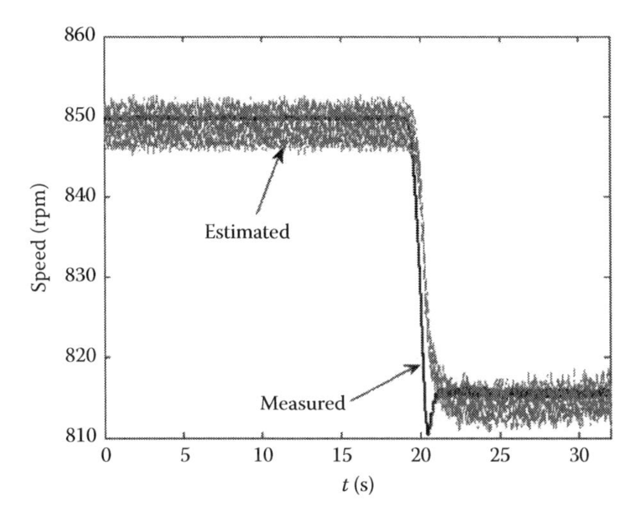

FIGURE 23.18 Estimated and measured motor speed.

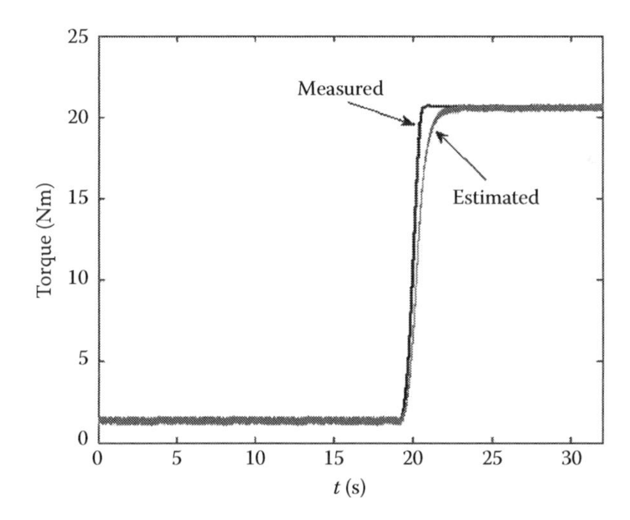

FIGURE 23.19 Estimated and measured torque variations.

References

[AW95] Aström K.J. and Wittenmark B., *Adaptive Control*, 2nd edn., Addison-Wesley Publishing Company, Reading, MA, 1995.

[B07] Besancon G. (Ed.), *Nonlinear Observers and Applications*, Springer, Berlin, Germany, 2007.

[BBG02] Barut M., Bogosyan O.S., and Gokasan M., EKF based estimation for direct vector control of induction motors, *Proceedings of the IEEE-IECON'02 Annual Meeting*, vol. 2, pp. 1710–1715, Sevilla, Spain, 2002.

[BBG07_1] Bogosyan S., Barut M., and Gokasan M., Braided extended Kalman filters for sensorless estimation in induction motors at high-low/zero speed, *IET Proc. Control Theory Appl.*, 1(4), 987–998, July 2007.

[BBG07_2] Barut M., Bogosyan S., and Gokasan M., Switching EKF technique for rotor and stator resistance estimation in speed sensorless control of IMs, *Energy Conversion Manage.*, 48(12), 3120–3134, December 2007.

[BBG07_3] Barut M., Bogosyan S., and Gokasan M., Speed-sensorless estimation for induction motors using extended Kalman filters, *IEEE Trans. Ind. Electron.*, 54(1), 272–280, February 2007.

[BBG08] Barut M., Bogosyan S., and Gokasan M., Experimental evaluation of braided EKF for sensorless control of induction motors, *IEEE Trans. Ind. Electron.*, 55(2), 620–632, February 2008.

[BG95] Bogosyan O.S. and Gokasan M., Adaptive torque ripple minimization of PMSM for direct drive applications, *IEEE IAS Annual Meeting, IAS 95*, pp. 231–237, Orlando, FL, 8–10 October 1995.

[BGH01] Bogosyan O.S., Gokasan M., and Hajiyev C., An application of Ekf for the position control of a single link arm, *Proceedings of the IEEE-IECON'01 Annual Meeting*, vol. 1, pp. 564–569, Denver, CO, 2001.

[CP93] Chen D. and Paden B., Adaptive linearization of hybrid step motors: Stability analysis, *IEEE Trans. Autom. Control*, 38(60), 874–887, 1993.

[IF06] Ioannou P.A. and Fidan B., *Adaptive Control Tutorial*, Siam, Philadelphia, PA, 2006.

[IS96] Ioannou P.A. and Sun J., *Robust Adaptive Control*, Prentice Hall PTR, Upper Saddle River, NJ, 1996.

[NA05] Narendra K.S. and Annaswamy A.M., *Stable Adaptive Systems*, Dover Books On Engineering, Mineola, NY, 2005.

[SI89] Sastry S. and Isidori A., Adaptive control of linearizable systems, *IEEE Trans. Autom. Control*, 34(11), 1123–1131, 1989.

[SST93] Salvatore L., Stasi S., and Tarchioni L., A new EKF-based algorithm for flux estimation in induction machines, *IEEE Trans. Ind. Electron.*, 40(5), 496–504, October 1993.

[TKM89] Taylor D.G., Kokotovic P.V., Marino, R., and Kanellakopoulos, I., Adaptive regulation of nonlinear systems with unmodeled dynamics, *IEEE Trans. Autom. Control*, 34(4), 405–412, 1989.

24

Observers in Dynamic Engineering Systems

Christopher Edwards
University of Leicester

Chee Pin Tan
Monash University

24.1 Introduction

The modeling of systems in so-called state-space form has had a significant impact on a wide number of scientific areas—but perhaps most especially in physics and engineering [23]. In this paradigm, the system is modeled as an interconnected set of *first-order* differential equations, specifically

$$\dot{x}(t) = f(t, x) \tag{24.1}$$

The quantity x is typically a vector and the function $f(\cdot)$ an appropriate mapping between vector spaces. The vector x is termed the *state* of the system and (at least for the form of Equation 24.1) contains all the necessary information to uniquely determine the future evolution of the system. In many engineering systems, certain aspects of the system have specific interpretations, such as tangible measurements generated from sensors termed *outputs* (e.g., angular velocity measured by a gyro), and certain quantities that can be manipulated and result in changes to outputs, termed *inputs* (e.g., the rudder on an aircraft or ship). A common special subclass of state-space systems are ones that possess the property of superposition of solutions and are termed *linear systems*. These can be modeled by the following pair of equations:

$$\dot{x}(t) = Ax(t) + Bu(t) \tag{24.2}$$

$$y(t) = Cx(t) \tag{24.3}$$

where

$x \in \mathbb{R}^n$ are the states
$y \in \mathbb{R}^p$ are the outputs
$u \in \mathbb{R}^m$ are the inputs

The state x may or may not have physical meaning. It is also an "internal" variable to the system, in the sense that the system in (24.2) and (24.3) can be represented in input–output form by taking Laplace transforms of (24.2) and (24.3) to obtain the expression

$$Y(s) = C(sI - A)^{-1}BU(s) \tag{24.4}$$

where $Y(s)$ and $U(s)$ are the Laplace transforms of the signals $y(t)$ and $u(t)$. It can be seen that the state x plays no role in (24.4) and has been effectively "eliminated."

However, information about the internal states is often useful and sometimes essential for solving a range of problems, such as stabilizing a system using state-feedback control (which will be demonstrated later in Section 24.9.1) or monitoring the health of the system.

Consider for example the double integrator

$$\ddot{\theta}(t) = u(t) \tag{24.5}$$

where

θ represents the position of a particle
u is the input in the form of an applied force

If θ is a measured output, then this input–output system can be written in state-space form, as in (24.2) and (24.3), if

$$A = \begin{bmatrix} 0 & 1 \\ 0 & 0 \end{bmatrix} \quad B = \begin{bmatrix} 0 \\ 1 \end{bmatrix} \quad C = \begin{bmatrix} 1 & 0 \end{bmatrix} \tag{24.6}$$

In this realization $x = \text{col}(\theta, \dot{\theta})$ and so the state contains information about the velocity of the particle, even though only position is measured. It is easy to verify that when A, B, and C are given in (24.6), the input–output form from (24.4) becomes

$$\Theta(s) = \frac{1}{s^2}U(s)$$

In many engineering systems, direct measurement of all the states is not always possible. First, sometimes it is not feasible to measure certain quantities as they are associated with very harsh environmental conditions or there are physical constraints that make it impossible for them to be measured. Second, sensors are costly and could potentially be the most expensive components in the system. Third, as alluded to earlier, sometimes the state has no physical meaning, and is purely an artifact of the modeling process.

The general solution to (24.2) is well known to be

$$x(t) = e^{At}x(0) + \int_0^t e^{A(t-s)}Bu(s)\,ds \tag{24.7}$$

where $x(0)$ represents the state at time $t = 0$. Consequently, if the state of system (24.2) and (24.3) is known at any point in time and the subsequent input signal $u(t)$ is known, the state can be evaluated from (24.7) for all subsequent time. The difficulty of predicting how $x(t)$ evolves when $u(t)$ is known is that if the state is unmeasurable and has no physical meaning, then $x(0)$ is unknown. However, it should be noted that in (24.7) the term $x(0)$ appears only once and is pre-multiplied by the matrix exponential term e^{At}. If the system in (24.2) is stable, i.e., the eigenvalues of A all have negative real parts, then $e^{At} \to 0$ as $t \to \infty$, and therefore, the influence of $x(0)$ "disappears" as time progresses. Consequently, the state provided by (24.7) asymptotically converges to the required value. This can be interpreted as creating a copy of the original system (24.2) and (24.3) and solving the system

$$\dot{z}(t) = Az(t) + Bu(t) \tag{24.8}$$

driven by the same inputs $u(t)$. Solving (24.8) yields $z(t)$ as a copy of the state $x(t)$ (at least asymptotically). However, such a scheme will *not* work if the system matrix A is unstable, and furthermore may also be impractical if the decay $e^{At} \to 0$ is "slow."

A *state observer*, as originally proposed in [20], provides an attractive solution to circumvent the above mentioned problems. It is essentially a replica of the system to be "observed," and processes the system inputs and outputs (which are measurable and available) to provide a real-time estimate of the unmeasurable internal state. An observer can be software based, and hence easily implemented on a computer—which is a relatively low-cost solution.

This chapter discusses the basic principles and theory of observers and associated design methods.

24.2 Linear Observers

Consider initially a linear dynamical engineering system that can be modeled as in (24.2) and (24.3), where $x \in \mathbb{R}^n$ are the states, $y \in \mathbb{R}^p$ are the outputs, and $u \in \mathbb{R}^m$ are the inputs. Assume that *only the inputs and outputs are measurable.*

An *observer* [20] for system (24.2) and (24.3) is

$$\dot{\hat{x}}(t) = A\hat{x}(t) + Bu(t) - G(C\hat{x}(t) - y(t)) \tag{24.9}$$

where $\hat{x} \in \mathbb{R}^n$ is the estimate of the state x and the gain $G \in \mathbb{R}^{n \times p}$ is a design matrix to be selected. Note the first two terms on the right-hand side constitute a model of the plant. The third term involving G is a corrective one; it feeds-back the difference between the "output" of the observer $C\hat{x}$ and the measurement y. Notice that the system in (24.9) requires only knowledge of y and u (which are assumed to be available). To solve (24.9), an initial condition $\hat{x}(0)$ must be provided. It will now be argued that the solution of (24.9), denoted by $\hat{x}(t)$, has the property that $\hat{x}(t) \to x(t)$, whatever initial value of $\hat{x}(0)$ is chosen, provided "an appropriate" choice of G can be found.

Define the state estimation error $\tilde{x} = \hat{x} - x$, then, from (24.2), (24.3), and (24.9) the following expression, which governs the error state, can be obtained:

$$\dot{\tilde{x}}(t) = (A - GC)\tilde{x}(t) \tag{24.10}$$

Notice that (24.10) shows a new unforced state-space system where the "state" is $\tilde{x}$. If the gain matrix G is chosen such that the eigenvalues of $(A - GC)$ have negative real parts, then the solution to (24.10) is

$$\tilde{x}(t) = e^{(A-GC)t}\tilde{x}(0) \tag{24.11}$$

and $\tilde{x}(t) \to 0$ as $t \to \infty$. Consequently, $\hat{x}(t) \to x(t)$ as $t \to \infty$. Notice that although $\hat{x}(t)$ is known for all time, the error $\tilde{x}(t)$ is unknown (because it depends on $x(t)$), and furthermore $\tilde{x}(0)$ is unknown. Nevertheless, because $\tilde{x}(t)$ satisfies (24.11), then it is guaranteed to vanish over time.

Example 24.1

Consider the double integrator system in (24.6). Suppose $G = \text{col}(g_1, g_2)$ where g_1, g_2 are scalars. Then,

$$A - GC = \begin{bmatrix} -g_1 & 1 \\ -g_2 & 0 \end{bmatrix}$$

The characteristic equation of this matrix is

$$\det(\lambda I - (A - GC)) = 0 \quad \Rightarrow \quad \lambda^2 + g_1\lambda + g_2 = 0 \qquad (24.12)$$

It is therefore clear that the scalars g_1 and g_2 can always be chosen to ensure that $(A - GC)$ has stable eigenvalues. If the eigenvalues are specified to be $\{-1, -2\}$, then the appropriate choice of gains is $g_1 = 3$, $g_2 = 2$.

Assume that the system position θ has an initial condition of −0.2 and is at rest ($\dot{\theta}(0) = 0$). Suppose the observer has zero initial conditions, in the following simulation, a sine-wave input is applied to the system.

Figures 24.1 and 24.2 show θ and $\dot{\theta}$ (dashed lines) together with their estimates (solid lines) obtained from the observer. It is clear that despite the initial mismatch (due to the difference in initial conditions), the observer estimates converge toward the actual state. It is important to note that the design and implementation of the observer is performed ***without any prior knowledge of the mismatch in initial conditions***. Furthermore, the observer in (24.9) provides asymptotic estimates of the states $(\theta, \dot{\theta})$ from the measurement of only θ.

A block diagram illustration of the observer, based on (24.9), is shown in Figure 24.3. The observer is within the rectangular box. Note that the concept of the observer is to use the measurable signals (u and y), and information about the model (A, B, C), to estimate the unmeasurable signal x.

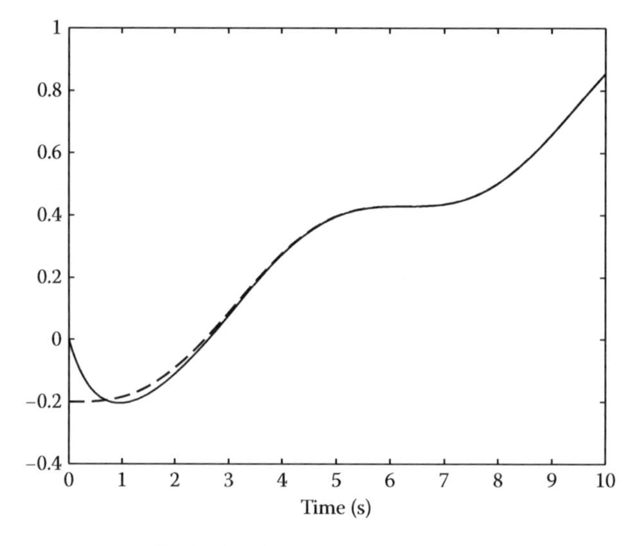

FIGURE 24.1 The system position θ (dashed) and its estimate (solid).

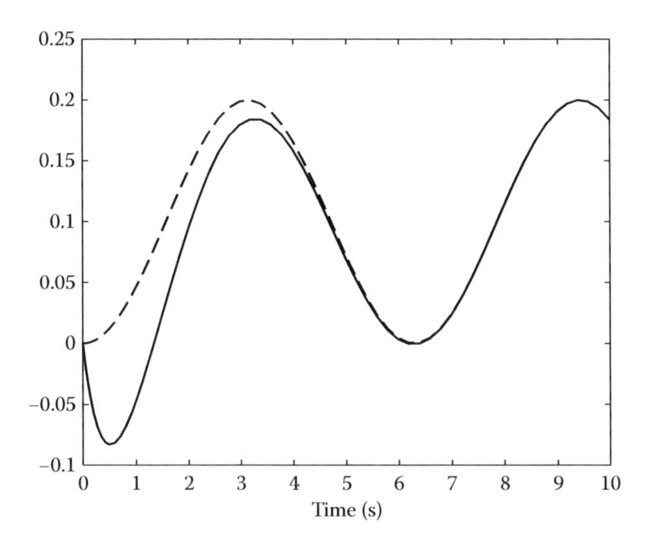

FIGURE 24.2 The system velocity $\dot{\theta}$ (dashed) and its estimate (solid).

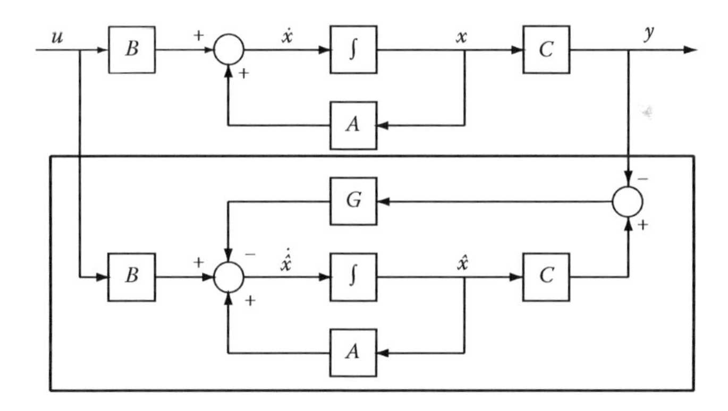

FIGURE 24.3 The block diagram representation of the linear observer (24.9).

Also notice that the observer structure in (24.9) has a filter interpretation since it is easy to see that (24.9) can be equivalently written as

$$\dot{\hat{x}}(t) = (A - GC)\hat{x}(t) + Bu(t) + Gy(t) \tag{24.13}$$

If $\hat{x}(t)$ is regarded as the output of the filter, then

$$\hat{X}(s) = (sI - (A - GC))^{-1}BU(s) + (sI - (A - GC))^{-1}GY(s)$$

and the state estimate $\hat{x}$ is a filtered combination of the signals $y(t)$ and $u(t)$.

24.2.1 Observability

The gain matrix G represents design freedom in (24.9). An obvious question is that under what conditions on the matrix pair (A, C), which are part of the model of the dynamic system, can a matrix G be found such that *all* eigenvalues of $(A - GC)$ can be placed in arbitrarily desired locations. In fact a purely algebraic condition can be established namely: the eigenvalues of $(A - GC)$ can be "placed" arbitrarily by choice of G if and only if the *observability* matrix

$$O := \begin{bmatrix} C \\ CA \\ CA^2 \\ \vdots \\ CA^{n-1} \end{bmatrix}$$

is full column rank (i.e., rank n).

Calculating the observability matrix for the system in (24.6) yields

$$O = \begin{bmatrix} C \\ CA \end{bmatrix} = \begin{bmatrix} 1 & 0 \\ 0 & 1 \end{bmatrix}$$

which is clearly rank 2 (which confirms the earlier explicit calculations).

Now consider the system with state-space matrices

$$A = \begin{bmatrix} 0 & 1 \\ 0 & 0 \end{bmatrix} \quad B = \begin{bmatrix} 0 \\ 1 \end{bmatrix} \quad C = \begin{bmatrix} 0 & 1 \end{bmatrix} \tag{24.14}$$

which represents the double integrator system with velocity rather than position measured. It can be shown that the observability matrix

$$O = \begin{bmatrix} 0 & 1 \\ 0 & 0 \end{bmatrix}$$

which clearly is rank deficient, and so the eigenvalues of $(A - GC)$ cannot be placed arbitrarily. To confirm this, once again suppose $G = \text{col}(g_1, g_2)$, then by repeating the earlier calculations, the eigenvalues of $(A - GC)$ are now given by the roots of the equation $\lambda(\lambda + g_2) = 0$. Clearly, one of the eigenvalues will always be at $\lambda = 0$. Consequently, the eigenvalues of $(A - GC)$ cannot be assigned arbitrarily.

24.3 Reduced-Order Observers

The observer described in Section 24.2 is known as a *full-order observer* as it seeks to estimate *all* states. However, this is not always necessary as the outputs, which can be thought of as the measurable components of the state, do not need to be estimated. An observer that estimates fewer than n states is known as a *reduced-order observer*. By a change of coordinates (which will be discussed in the sequel), the system (24.2) and (24.3) can be partitioned as follows:

$$\begin{bmatrix} \dot{x}_1(t) \\ \dot{y}(t) \end{bmatrix} = \begin{bmatrix} A_{11} & A_{12} \\ A_{21} & A_{22} \end{bmatrix} \begin{bmatrix} x_1(t) \\ y(t) \end{bmatrix} + \begin{bmatrix} B_1 \\ B_2 \end{bmatrix} u(t) \tag{24.15}$$

$$y(t) = \begin{bmatrix} 0 & I_p \end{bmatrix} \begin{bmatrix} x_1(t) \\ y(t) \end{bmatrix} \tag{24.16}$$

where $x_1 \in \mathbb{R}^{n-p}$ is the unmeasurable state to be estimated. Notice that in this form, the last p states are the outputs, giving the matrix C the special structure in (24.16). The change of coordinates $x \mapsto T_c x$ to achieve the form in (24.15) and (24.16) can easily be computed as

$$T_c = \begin{bmatrix} \bar{C} \\ C \end{bmatrix}$$

where $\bar{C} \in \mathbb{R}^{(n-p)\times n}$ is any matrix that makes T_c full rank.

Define a new variable $x_r := x_1 + Ly$ where L is a design matrix chosen such that $(A_{11} + LA_{21})$ has stable eigenvalues. Differentiating x_r and substituting for $\dot{x}_1$, $\dot{y}$ from (24.15) results in

$$\dot{x}_r(t) = (A_{11} + LA_{21})x_r(t) + (A_{12} + LA_{22} - (A_{11} + LA_{21})L)y(t) + (B_1 + LB_2)u(t) \tag{24.17}$$

A suitable reduced-order observer for (24.17) is

$$\dot{\hat{x}}_r(t) = (A_{11} + LA_{21})\hat{x}_r(t) + (A_{12} + LA_{22} - (A_{11} + LA_{21})L)y(t) + (B_1 + LB_2)u(t) \tag{24.18}$$

Define $\tilde{x}_r := \hat{x}_r - x_r$; then subtracting (24.17) from (24.18) results in

$$\dot{\tilde{x}}_r(t) = (A_{11} + LA_{21})\tilde{x}_r(t) \tag{24.19}$$

Since L has been chosen such that $(A_{11} + LA_{21})$ has stable eigenvalues, then as argued earlier, $\tilde{x}_r(t) \rightarrow 0$ as $t \rightarrow \infty$ and an asymptotic estimate of x_1 (written as $\hat{x}_1$) can be obtained from

$$\hat{x}_1 := \hat{x}_r - Ly \tag{24.20}$$

The reduced-order observer relies upon the calculation of a matrix L such that $(A_{11} + LA_{21})$ has stable eigenvalues. It can be shown that the fictitious pair (A_{11}, A_{21}) is observable if and only if (A, C) is observable (see the observability rank test in Lemma 5.6 of [9]).

Example 24.2

Consider the observable double integrator system (24.5) whose state-space representation is given in (24.6). Note that (24.6) is not in the form of (24.15) and (24.16) as the matrix C is not in the prescribed form. However, this can be overcome by interchanging the state variables so that $x := \mathrm{col}(\dot{\theta}, \theta)$, which results in

$$A = \begin{bmatrix} 0 & 0 \\ 1 & 0 \end{bmatrix}, \quad B = \begin{bmatrix} 1 \\ 0 \end{bmatrix}, \quad C = \begin{bmatrix} 0 & 1 \end{bmatrix} \tag{24.21}$$

Comparing the matrices above with the structure of the partitioned form in (24.15) and (24.16), it is clear that $A_{11} = 0, A_{21} = 1$, and (A_{11}, A_{21}) is observable. Choosing $L = -2$ makes $(A_{11} + LA_{21}) = -2$. The reduced-order observer is then given by

$$\dot{\hat{x}}_r(t) = -2\hat{x}_r(t) - 4y(t) + u(t) \tag{24.22}$$

and from (24.20) the estimate of $\dot{\theta}$ is provided by $\hat{x}_r + 2y$. The same initial conditions, as those used in Section 24.2, have been considered here and the following results have been obtained, as shown in Figure 24.4.

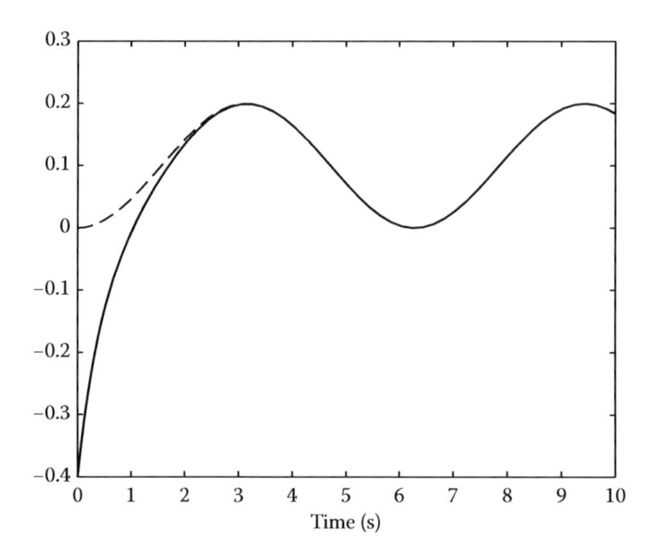

FIGURE 24.4 The system velocity $\dot{\theta}$ (dashed) and its estimate (solid).

24.4 Noise and Design Tradeoffs

Now consider the linear state-space system representation from (24.2) to (24.3), but suppose the measurements are corrupted. Specifically suppose

$$\dot{x}(t) = Ax(t) + Bu(t) \tag{24.23}$$

$$y(t) = Cx(t) + v \tag{24.24}$$

where $v \in \mathbb{R}^p$ is an unknown signal which additively corrupts the measurement y. Now employing the same observer configuration, as used in (24.9), and denoting $\tilde{x} = \hat{x} - x$, yields

$$\dot{\tilde{x}}(t) = (A - GC)\tilde{x}(t) + Gv \tag{24.25}$$

Clearly, unless $v \equiv 0$, asymptotic stability in (24.25) is lost, and $\tilde{x} \nrightarrow 0$. This exposes design issues associated with the choice of the gain matrix G. At least intuitively, a "large" G is likely to force the eigenvalues of $(A - GC)$ further into the left half of the complex plane. [Or put another way, in order to specify eigenvalues for $(A - GC)$ which have large and negative real parts, large values of G may be necessary.] Large and negative eigenvalues for $(A - GC)$ are advantageous from the point of view of ensuring a fast decay associated with $\tilde{x}(t)$ since

$$\| \tilde{x}(t) \| \le M \| \tilde{x}(0) \| e^{\alpha t}$$

where M is a scalar and $\alpha = \max_{i=1...N}\{Re(\lambda_i)\} < 0$ where the $\{\lambda_1, \ldots, \lambda_n\}$ are the eigenvalues of $(A - GC)$. However, in the presence of corrupted measurements (e.g., by noise), the gain term G in (24.25) directly amplifies the effects of v on the error system, and therefore big is no longer necessarily best. A more subtle problem formulation is therefore required.

24.5 Kalman–Bucy Filtering

Consider a linear system in the presence of internal plant noise and measurement noise:

$$\dot{x}(t) = Ax(t) + Bu(t) + Dw \quad (24.26)$$

$$y(t) = Cx(t) + v \quad (24.27)$$

The error system associated with the observer in (24.9) is

$$\dot{\tilde{x}}(t) = (A - GC)\tilde{x}(t) + Dw + Gv \quad (24.28)$$

If the noise vectors v and w are assumed to be white with zero mean and covariance matrices V and W, respectively, then a design problem can be posed as follows: synthesize the gain G in the observer, such that the standard deviation of the state estimation error $\tilde{x}$ is minimized. This is known as a Kalman–Bucy filtering problem [14].

The optimal gain $\hat{G}(t)$, which is time varying, is given by

$$\hat{G}(t) = P(t)C^T V^{-1} \quad (24.29)$$

where the time-varying symmetric positive definite matrix $P(t)$ is obtained by solving the matrix differential equation

$$\dot{P}(t) = AP(t) + P(t)A^T + DWD^T - P(t)C^T V^{-1} CP(t)$$

The initial condition $P(0)$ is the covariance matrix associated with the initial plant condition $x(0)$.

The steady-state version of (24.29) is

$$\hat{G} = PC^T V^{-1} \quad (24.30)$$

where the symmetric positive definite P solves the Riccati equation

$$AP + PA^T + DWD^T - PC^T V^{-1} CP = 0 \quad (24.31)$$

As an example consider the double integrator problem with

$$A = \begin{bmatrix} 0 & 1 \\ 0 & 0 \end{bmatrix} \quad B = \begin{bmatrix} 0 \\ 1 \end{bmatrix} \quad D = \begin{bmatrix} 0 \\ 1 \end{bmatrix} \quad C = \begin{bmatrix} 1 & 0 \end{bmatrix}$$

and $W = V = 0.01$. It can be shown that the steady-state solution to (24.31) is

$$P = \frac{1}{100}\begin{bmatrix} \sqrt{2} & 1 \\ 1 & \sqrt{2} \end{bmatrix} \Rightarrow G = \begin{bmatrix} \sqrt{2} \\ 1 \end{bmatrix}$$

In the following simulations $x(0) = \text{col}(1, 0)$ and $\hat{x}(0) = 0$. The state estimation error is shown in Figure 24.5. Figure 24.6 shows the measured output $y + v$ and the output of the observer $\hat{y}$. From a filtering perspective, $\hat{y}$ is an estimate of the true output $y = Cx$.

The filtering property with respect to the noise is clearly visible in Figure 24.6.

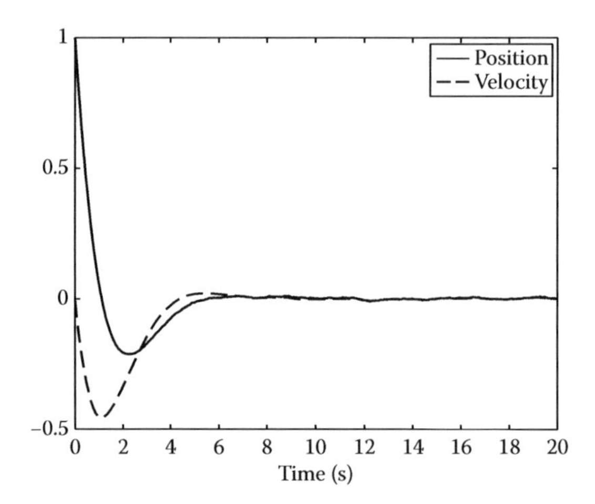

FIGURE 24.5 State estimation error.

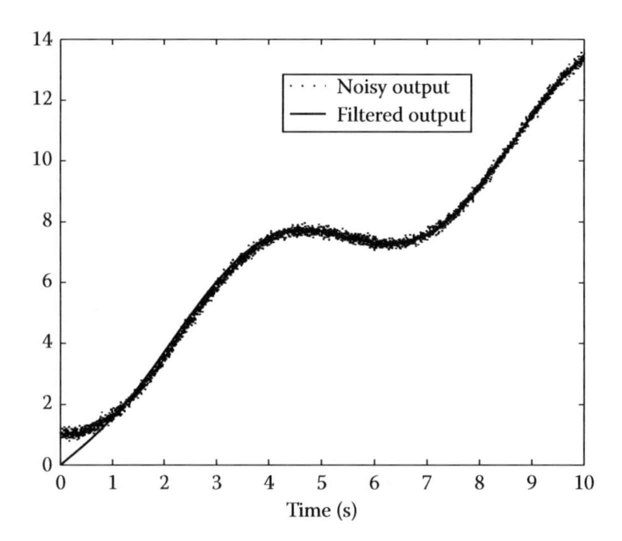

FIGURE 24.6 Measured output and the (filtered) output from the Kalman filter.

24.6 Nonlinear Observers: Thau Observer

Now consider the more general nonlinear state-space system represented in (24.1), that is,

$$\dot{x}(t) = f(x) \tag{24.32}$$

where the quantity x is typically a vector, and the function $f(\cdot)$ an appropriate (non) mapping between vector spaces. Suppose a linear combination of the states is measured so that as in the linear case

$$y(t) = Cx(t)$$

Further, suppose that the function on the right-hand side of (24.32) is decomposed as

$$f(x) = Ax + D\phi(x) \tag{24.33}$$

where

Ax represents the linear component of the function $f(\cdot)$

$\phi(\cdot)$ typically represents the higher order terms

Consider as an observer for (24.32) the nonlinear system

$$\dot{\hat{x}}(t) = f(\hat{x}) + G(y - C\hat{x}) \tag{24.34}$$

If $\tilde{x} = \hat{x} - x$, it follows from (24.32) through (24.34) that

$$\begin{aligned}\dot{\tilde{x}}(t) &= f(\hat{x}) - f(x) + G(y(t) - C\hat{x}(t)) \\ &= (A - GC)\tilde{x}(t) + D(\phi(\hat{x}) - \phi(x))\end{aligned} \tag{24.35}$$

Assume that the function $\phi(\cdot)$ in (24.33) is Lipschitz continuous [26], then

$$\|\phi(\hat{x}) - \phi(x)\| < l_\phi \|\hat{x} - x\|$$

for some positive scalar l_ϕ. Further suppose G is chosen so that $(A - GC)$ has stable eigenvalues and that P is a symmetric positive definite matrix that solves the matrix inequality

$$P(A - GC) + (A - GC)^T P + PDD^T P + l_\phi^2 I_n < 0 \tag{24.36}$$

where I_n is the identity matrix of the same order as A. Then, it can be proved* that $\tilde{x}(t) \to 0$ as $t \to \infty$ and the state estimate in (24.34) tracks the state in (24.32).

Example 24.3

Consider the pendulum-like equation

$$\ddot{\theta} = -\sin\theta \tag{24.37}$$

This can be written in the form of (24.32) if $x = \text{col}(\theta, \dot{\theta})$. Furthermore, the resulting function $f(\cdot)$ can be factorized in the form of (24.33) if

$$A = \begin{bmatrix} 0 & 1 \\ 0 & 0 \end{bmatrix} \quad D = \begin{bmatrix} 0 \\ 1 \end{bmatrix} \tag{24.38}$$

and $\phi(\cdot) = -\sin\theta$. Now also suppose that the output distribution matrix† is

$$C = \begin{bmatrix} 1 & 1 \end{bmatrix} \tag{24.39}$$

---

* Using Lyapunov theory, $V(t) = \tilde{x}^T P \tilde{x}$ can be used to show that $\tilde{x}(t) \to 0$ as $t \to \infty$.

† Here the output is chosen as the sum of position and velocity. This is done for pedagogical reasons. If $y = \theta$, the nonlinear term in (24.40) can be compensated for because $\sin(\theta)$ is known (since θ is measured). Alternatively, choosing $y = \dot{\theta}$ results in an unobservable system from a linear perspective as seen earlier, hence the choice of y in (24.39).

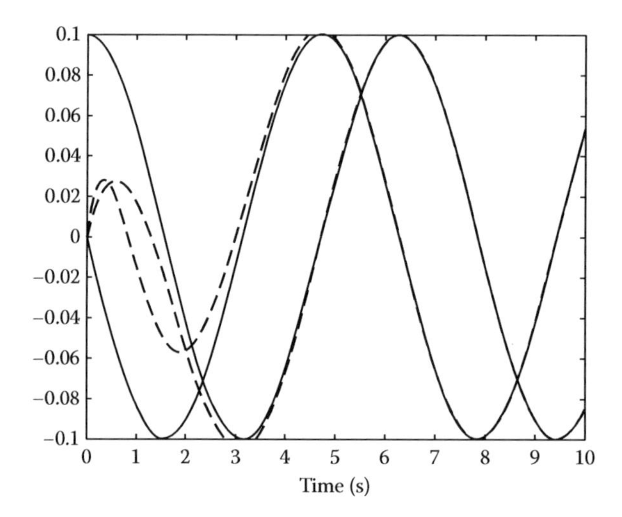

FIGURE 24.7 States of the plant (solid) and the observer (dashed).

From the mean value theorem [26], for any θ_1 and θ_2

$$\sin\theta_1 - \sin\theta_2 = m(\theta_1 - \theta_2) \tag{24.40}$$

where $m = \cos\bar{\theta}$ for some value of $\bar{\theta}$ belonging to the interval formed by θ_1 and θ_2. It follows from (24.40) that

$$|\sin\theta_1 - \sin\theta_2| \le |\theta_1 - \theta_2|$$

since $|m| = |\cos(\bar{\theta})| \le 1$ and consequently $l_\phi = 1$ is an appropriate choice for the Lipschitz gain.

It can be shown for $G = \text{col}(1, 2)$ if

$$P = \begin{bmatrix} 3.2435 & 0.3739 \\ 0.3739 & 0.5920 \end{bmatrix}$$

then (24.36) is satisfied for $l_\phi = 1$.

In the following simulations, the initial state of the system is $x(0) = \text{col}(0.1, 0)$, while the state of the observer $\hat{x}(0) = \text{col}(0, 0)$. As Figure 24.7 shows, the states $\hat{x}(t)$ track $x(t)$ asymptotically.

24.7 High-Gain Observers

Consider the nonlinear system in (24.37) written in state-space form as

$$\dot{x}(t) = Ax(t) + B\sin x_1(t) \tag{24.41}$$

where only the first component of x, written as x_1, is measured. It follows the output $y(t) = Cx(t)$, where A, B, and C are given in (24.6).

Consider the *linear observer* for the nonlinear system in (24.41)

$$\dot{z}(t) = Az(t) + G(y(t) - Cz(t)) \tag{24.42}$$

where

$$G := \begin{bmatrix} \alpha_1/\varepsilon \\ \alpha_2/\varepsilon^2 \end{bmatrix} \tag{24.43}$$

and the positive scalars α_1 and α_2 are such that

$$s^2 + \alpha_1 s + \alpha_2 = 0 \tag{24.44}$$

is a Hurwitz polynomial. The quantity ε in (24.43) is a small positive scalar to be determined. Define two new variables as $\eta_1 = (x_1 - z_1)/\varepsilon$ and $\eta_2 = x_2 - z_2$. Then, if $\eta := \text{col}(\eta_1, \eta_2)$, it is easy to show that

$$\varepsilon\dot{\eta}(t) = A_0\eta(t) + \varepsilon B \sin x_1(t) \tag{24.45}$$

where

$$A_0 := \begin{bmatrix} -\alpha_1 & 1 \\ -\alpha_2 & 0 \end{bmatrix}$$

The matrix A_0 is Hurwitz by construction since its characteristic equation is given by (24.44). As $\varepsilon \to 0$, the effect of the nonlinear term in (24.45) vanishes and therefore $\eta(t) \to 0$ as $t \to \infty$ and the state estimates $z(t)$ asymptotically track $x(t)$. Of course as $\varepsilon \to 0$, then the entries in the gain matrix G from (24.43) become large, hence the term high-gain observer. The scalar ε must be chosen as a tradeoff between performance (ε small) and the size of the entries in the gain G in (24.43).

In the following design $\alpha_1 = 2$ and $\alpha_2 = 1$, giving a characteristic equation with poles at $\{-1, -1\}$. Three different values of ε are compared: $\varepsilon = 0.25$, $\varepsilon = 0.1$, and $\varepsilon = 0.05$, respectively (Figures 24.8 through 24.10). As ε becomes smaller, as predicted by the theory, the quality of the state reconstruction improves. Another feature of this class of observer manifests itself in Figures 24.8 through 24.10, namely the so-called "peaking" phenomenon. It can be seen that in the initial stages of the estimation, there is a transient peak in the estimation error which, disappears quickly, but which temporarily results in poor estimation.

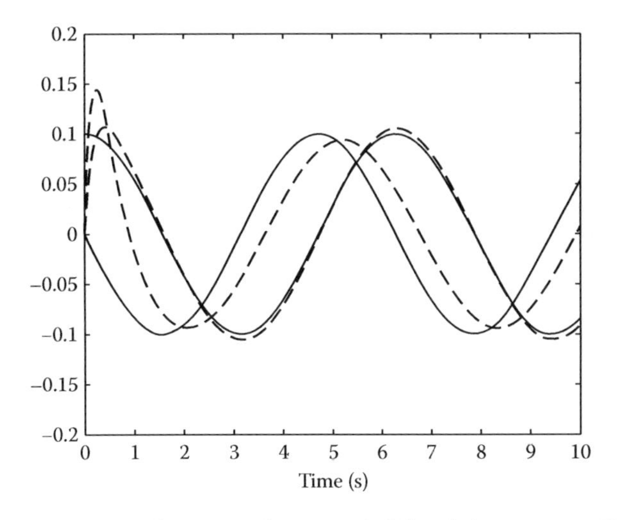

FIGURE 24.8 High-gain observer with $\varepsilon = 0.25$: plant states (solid) and observer states (dashed).

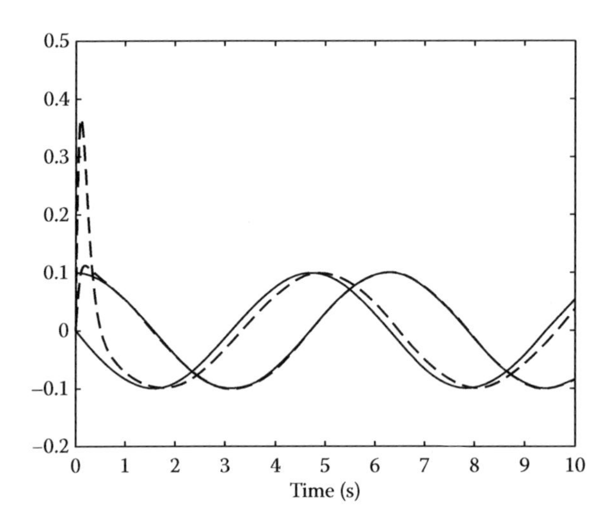

FIGURE 24.9 High-gain observer with $\varepsilon = 0.1$: plant states (solid) and observer states (dashed).

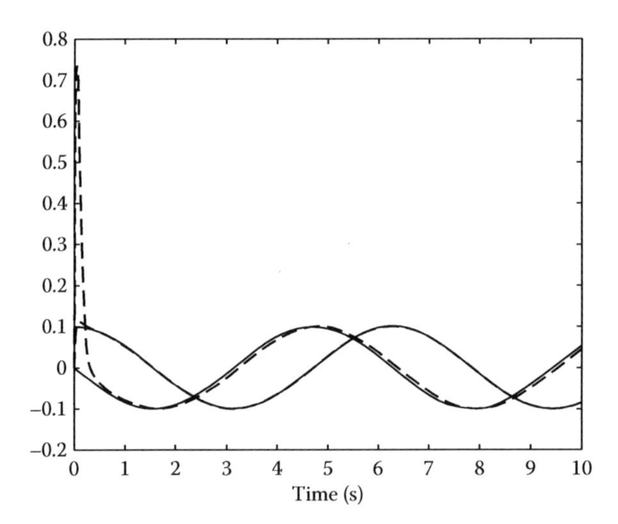

FIGURE 24.10 High-gain observer with $\varepsilon = 0.05$: plant states (solid) and observer states (dashed).

24.8 Sliding-Mode Observers

In the previous section, the nonlinear system was decomposed as

$$\dot{x}(t) = Ax + D\phi(x) \tag{24.46}$$

where $\phi(\cdot)$ is a nonlinear function, *but known*. This nonlinear term was incorporated as part of the observer in (24.34). However, if ϕ is imprecisely known, this approach may not be successful.

Consider instead the observer

$$\dot{\hat{x}} = A\hat{x} + G_n \nu \tag{24.47}$$

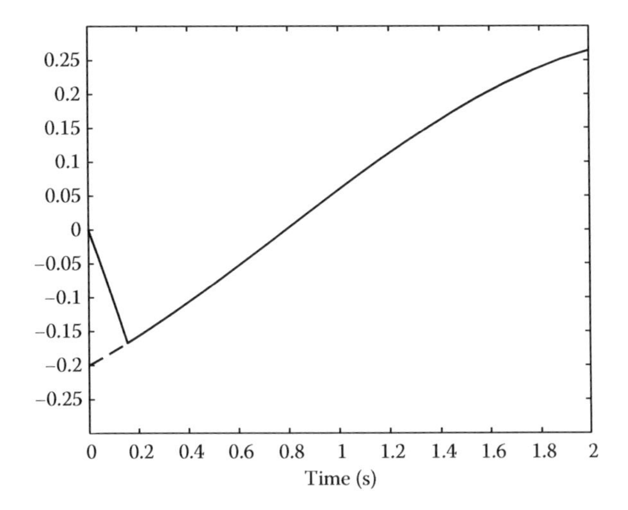

FIGURE 24.11 The system output y (dashed) and its estimate (solid).

where G_n is a gain to be chosen and

$$v = -\text{sign}(\tilde{y}) \tag{24.48}$$

where $\tilde{y} = C\hat{x} - y$. Note that v represents a *discontinuous injection* term.

Suppose for the system given in Section 24.6, the gain $G_n = \text{col}(0, 1)$ is chosen. The following simulation uses initial conditions $\theta(0) = -0.2$, $\dot{\theta}(0) = 0$ in the plant, and assumes zero initial conditions for the observer.

Figure 24.11 shows the system output (dashed) and its estimate (solid). It can be seen that at approximately $t = 0.15$ s, the output estimate converges to the output in *finite time* (compared to the linear observer scenario where only asymptotic convergence takes place). For all subsequent time, a so-called *sliding motion* takes place on the surface $\{\tilde{x}: C\tilde{x} = 0\}$, where $\tilde{x} = \hat{x} - x$. This phenomenon of finite time convergence is due to the nonlinear injection term $G_n\text{sign}(\tilde{y})$ (for a detailed mathematical proof of how the so-called sliding motion is achieved; see [30] for example). Figure 24.12 shows the states (dashed) and their respective estimates. At the instant when the sliding motion occurs ($t = 0.15$ s), there is an abrupt change in the error system trajectory, and the dynamics of $\tilde{y}$ are eliminated and $\tilde{y} \equiv 0$ for all subsequent time. Suppose $\tilde{x} = \text{col}(\tilde{x}_1, \tilde{x}_2)$, then from the structure of A and D it follows

$$\dot{\tilde{x}}_1 = \tilde{x}_2 \tag{24.49}$$

As shown in Figure 24.11, in finite time $\tilde{y} \equiv 0$ and therefore

$$0 = \tilde{y} = \tilde{x}_1 + \tilde{x}_2 \;\Rightarrow\; \tilde{x}_2 = -\tilde{x}_1$$

Substituting for $\tilde{x}_2$ in (24.49) yields

$$\dot{\tilde{x}}_1 = -\tilde{x}_1 \tag{24.50}$$

and hence the reduced-order motion is governed by a first-order system. From Equation 24.50, $\tilde{x}_1(t) \to 0$ as $t \to \infty$. The subsequent motion is known as the *reduced-order motion*.

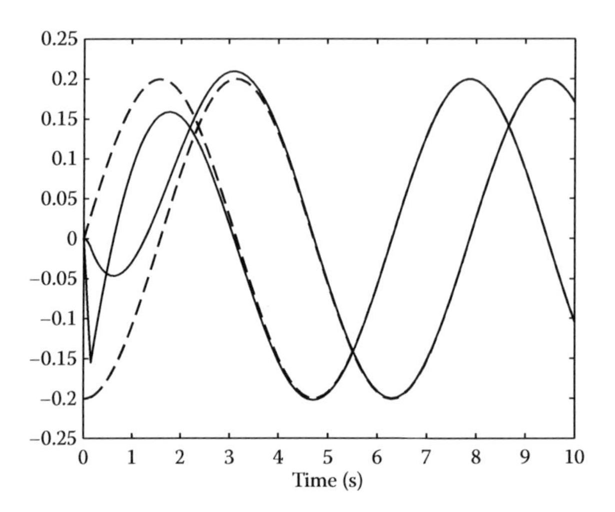

FIGURE 24.12 The states of the system θ, $\dot{\theta}$ (dashed) and their estimates (solid).

The discontinuous term ν can be replaced by the sigmoidal expression

$$\nu_\delta := -\frac{\tilde{y}}{|\tilde{y}|+\delta} \tag{24.51}$$

where δ is a small positive constant that governs the degree of accuracy to which an ideal sliding motion is obtained [30]. Notice that this asymptotic recovery of the states has been achieved without precise knowledge about the term $\phi(\cdot)$. In fact, the only knowledge which is required here is that $|\phi| < 1$.

Furthermore, it can be seen from Figure 24.13 that ν starts to track the disturbance $\phi(x) = \sin\theta$ after the sliding motion takes place. It can be seen that (1) in finite time, output estimation error convergence

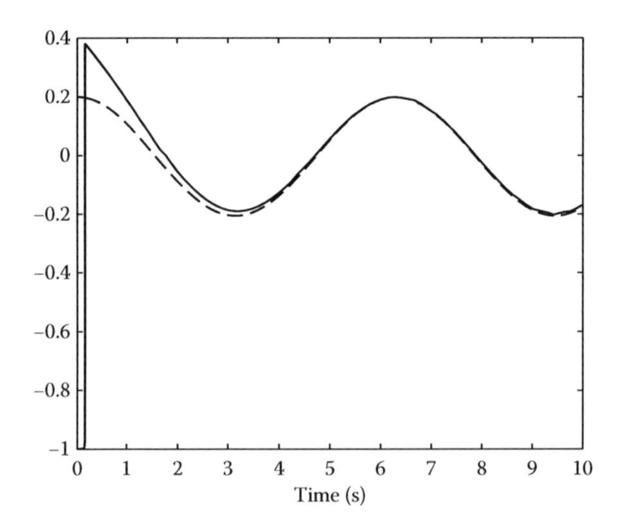

FIGURE 24.13 The "disturbance" $\sin\theta$ (dashed) and the equivalent output error injection ν (solid).

takes place; (2) the sliding motion (and hence the state estimates) are independent of the disturbance; and (3) information about the nonlinearity/disturbance (in this case sin θ) can be obtained from the nonlinear injection term v.

24.9 Applications

24.9.1 State-Feedback Control

A state-feedback controller for (24.2) arises when the control input is defined as $u = -Kx$, where $K \in \mathbb{R}^{m\times n}$. From (24.2), the resulting closed loop system is

$$\dot{x}(t) = (A - BK)x(t)$$

and K must be chosen such that the matrix $(A - BK)$ has stable eigenvalues. However, this situation is usually not realistic as it requires the state x to be measurable. This problem can be circumvented by implementing an observer and defining the control input as $u = -K\hat{x}$, where $\hat{x}$ is the estimate of the state from the observer. Using the definition of u and the definition of $\tilde{x}$, the augmented state-space system for the *observer-controller* closed loop system can be written as

$$\begin{bmatrix} \dot{x}(t) \\ \dot{\tilde{x}}(t) \end{bmatrix} = \underbrace{\begin{bmatrix} A - BK & -BK \\ 0 & A - GC \end{bmatrix}}_{A_{aug}} \begin{bmatrix} x(t) \\ \tilde{x}(t) \end{bmatrix} \tag{24.52}$$

Notice that A_{aug} has an upper "triangular" structure and therefore the eigenvalues of A_{aug} are the eigenvalues of $(A - BK)$ together with those of $(A - GC)$. This shows that the controller and observer can both be designed independently, and as long as both $(A - BK)$ and $(A - GC)$ are stable, the overall observer-controller structure will also be stable.

Example 24.4

It is well known that the double integrator in (24.5) cannot be stabilized by static output feedback $u = -ky$. Instead, state feedback will be employed based on state estimates from an observer. From (24.6), choosing $u = -K\hat{x}$, where $K = [8\ 4]$ results in $\lambda(A - BK) = -2 \pm j2$. The observer in Section 24.2 is used here to estimate the states for use in the control law. (Again initially there is a plant/observer state mismatch.)

From Figures 24.14 and 24.15, it can be seen that the observer tracks the states and that the states have been successfully regulated to zero using the state estimates combined with the state-feedback controller.

Note that the overall control law can be written as

$$\dot{\hat{x}}(t) = (A - GC - BK)\hat{x}(t) + Gy(t) \tag{24.53}$$

$$u(t) = -K\hat{x}(t) \tag{24.54}$$

and so the controller can be thought of as the compensator

$$\mathcal{K}(s) = K(sI - (A - GC - BK))^{-1}G$$

in negative feedback connection.

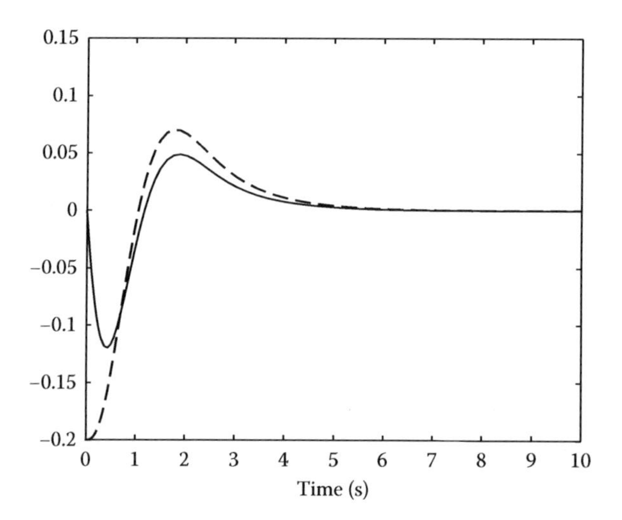

FIGURE 24.14 The system output y (dashed line) and its estimate (solid line).

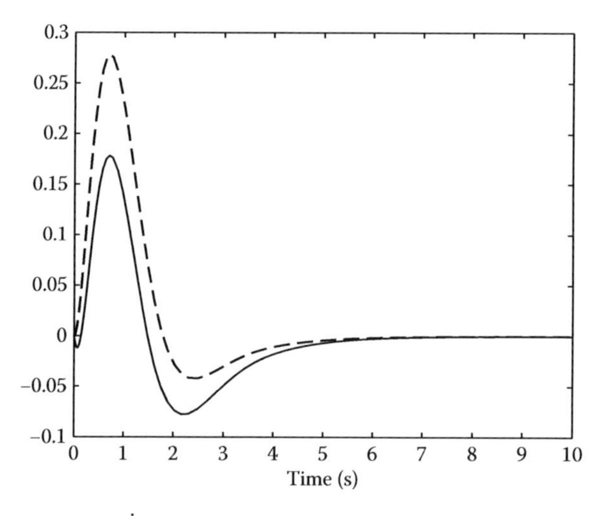

FIGURE 24.15 The system velocity $\dot{\theta}$ (dashed) and its estimate (solid).

In this example, it can be shown that

$$K(s) = \frac{3s + 14}{s^2 + 7s + 22}$$

Now suppose the reduced-order observer is used together with the same value of K. Since the estimate of $\dot{y}$ is given by $\hat{x}_r + 2y$, it can be shown that

$$u(t) = -4\hat{x}(t) - 16y(t) \tag{24.55}$$

where

$$\dot{\hat{x}}_r(t) = -6\hat{x}_r(t) - 20y(t) \tag{24.56}$$

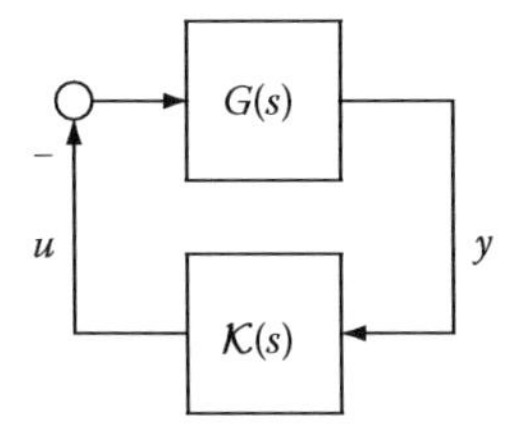

or equivalently, in transfer function form, in the block diagram above

$$\hat{\mathcal{K}}(s) = 16\frac{(s+1)}{(s+6)}$$

Notice that this is a phase advance compensator structure.

Figures 24.16 and 24.17 show the performance of the system and the reduced-order observer. Once again, it can be seen that the observer successfully tracks the unmeasurable velocity, and as a result, the states are regulated as desired.

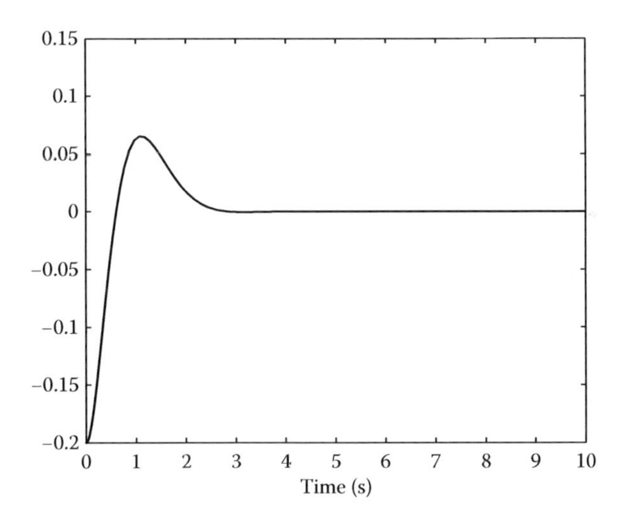

FIGURE 24.16 The system output y.

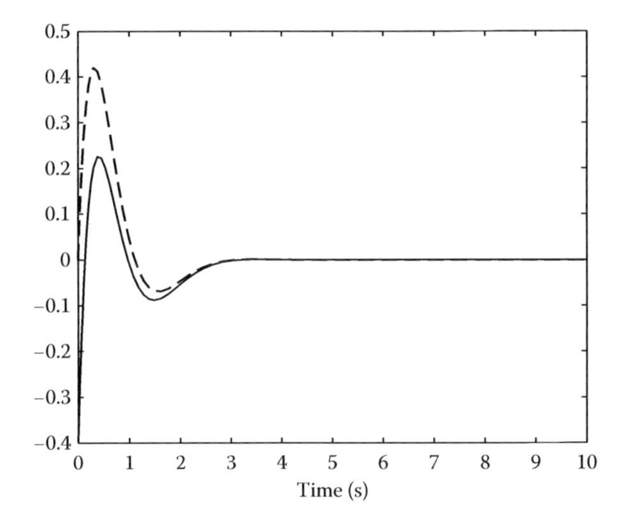

FIGURE 24.17 The system velocity $\dot{\theta}$ (dashed) and its estimate (solid).

24.9.2 Fault Detection and Isolation

Another useful application of observers is in the field of fault detection and isolation [4]. A fault is defined as a malfunction in the system of interest or a situation in which an abnormality occurs. A system with faults can be modeled in the following way:

$$\dot{x}(t) = Ax(t) + Bu(t) + Mf(t) \tag{24.57}$$

where

M is the "fault distribution matrix" which is constant

f is the vector of faults which is unknown

When $f(t) = 0$, the system is said to be fault free and (24.57) is identical to (24.2). When $f(t) \neq 0$, the effect of the fault propagates through the system. By implementing the observer (24.9), the following error dynamical system is obtained

$$\dot{\tilde{x}}(t) = (A - GC)\tilde{x}(t) - Mf(t) \tag{24.58}$$

Note that the signal $\tilde{y} = C\hat{x} - y = C\tilde{x}$ is a "measurable" output of the error system (24.58). In the presence of a persistent fault, $\tilde{x}$ does not decay to zero and $\tilde{y} \neq 0$. The signal $\tilde{y}$ is taken to be a *residual* signal, which can be used as an alarm to *detect* the presence of a fault when $\tilde{y} \neq 0$ and to certify that the system is healthy when $\tilde{y} = 0$. By designing the gain G, additional information about the fault can be obtained, such as identifying the location or *isolating* the fault.

Example 24.5

Consider the double integrator in (24.5), and suppose that the actuation device is fault prone. This situation can be modeled as

$$\ddot{\theta}(t) = u(t) + f(t) \tag{24.59}$$

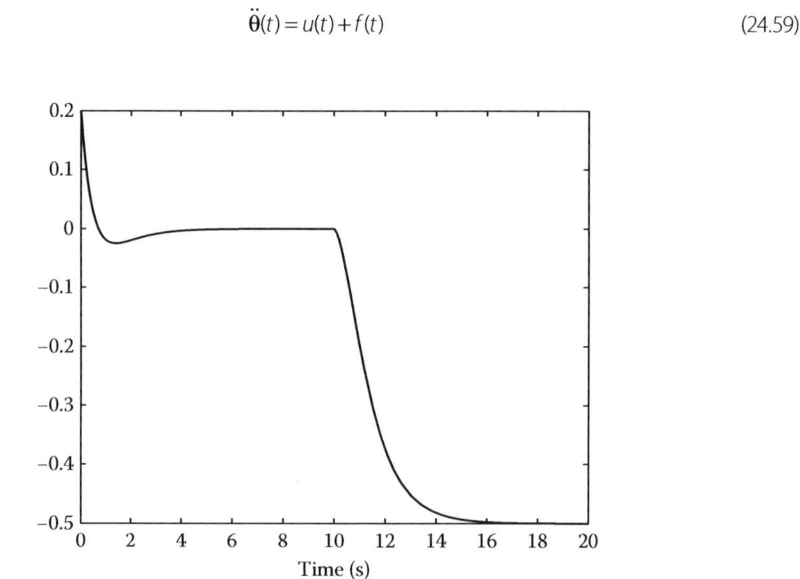

FIGURE 24.18 The output estimation error $\tilde{y}$ when a fault occurs at $t = 10$ s.

Let the fault be a unit step at time $t = 10$ s. Using the same observer and initial conditions, as described in Section 24.2, the results shown in Figure 24.18 are obtained. The signal $\tilde{y}$ has a nonzero initial condition due to the mismatch in initial conditions. This decays exponentially to zero as expected and in the period of time between 5 and 10 s, $\tilde{y} = 0$ at least visually (indicating fault-free performance). When the fault occurs at $t = 10$ s, $\tilde{y}$ immediately reacts and becomes nonzero, indicating the presence of the fault.

24.10 Conclusions

This chapter described a specific class of dynamical systems—so-called observers—used primarily to estimate the unmeasurable state vector in a state-space system based only on information about the inputs and outputs. The focus has been largely on linear systems, although some straightforward extensions to nonlinear systems have been described. These ideas, since their inception in the 1960s, have found a wide range of applications: as part of state-feedback control schemes, filtering of noisy data, as a basis for fault detection schemes for dynamical systems, as part of disturbance estimation schemes for improved control, and as a technology for cryptography. Although the linear case is well studied and understood, generic solutions for nonlinear systems is still an open problem and an area of active research.

24.11 Further Reading

The observers described in the chapter have been largely based on linear models. In engineering systems, other unmeasurable and unmodeled factors impact on the plant dynamics, and if unaccounted for, will corrupt the state estimation. This leads to a so-called "unknown input" observer problem, whereby the observer must robustly estimate the states and be insensitive to these disturbances. Based on the unknown input disturbance distribution matrix, researchers have exploited various observer structures and design methods in order to generate a robust state estimate; the most widely discussed approaches are so-called *unknown input observers* [5,7,15], adaptive techniques [18,21], and sliding-mode techniques [8,31]. Crucial to the success in obtaining a state estimate that is robust against the unknown inputs, is the *modeling* of the unknown inputs. A detailed discussion on this topic can be found in [24].

In particular these robust observer designs have been exploited in the area of *fault detection and isolation* [4,12,16]. In this paradigm, the unmeasurable faults can be modeled as an unknown input. Specific conditions on the system, thought of as a transfer function matrix between the unknown input signals and the measurements, have been established to identify when it is possible to create observers whose state estimation error dynamics are *independent* of the unknown inputs. Typically, these amount to relative degree one and minimum phase conditions [7].

The ability of the equivalent nonlinear injection signals in sliding-mode observers to track unknown inputs, as discussed in Section 24.8, has been exploited to achieve fault reconstruction systems [10,11,27,28]. Recently, the "relative degree one" assumption has been obviated in the work in [22] by using cascaded observers. So-called higher order sliding-mode observers have also been proposed to circumvent the relative degree condition [6,13]. High-gain observers expanding on the discussions in Section 24.7 can also be employed in systems with relative degree greater than one [2,3,17].

The nonlinear observer described in Section 24.6 is usually known as a Thau observer [25,29]. There is extensive subsequent work which has built on these ideas—especially in terms of how to design the gain to maximize the magnitude of the Lipschitz gain that can be tolerated.

The observers in this chapter have focused on linear systems (or at best quasi-linear systems). The work of Isidori and coworkers to convert a general class of nonlinear systems to linear ones, via nonlinear coordinate transformation, has also been employed to design observers for a reasonably general class of nonlinear systems [1,19].

References

1. M. Arcak and P. Kokotovic. Nonlinear observers: A circle criterion design and robustness analysis. *Automatica*, 37:1923–1930, 2001.
2. A.N. Atassi and H.K. Khalil. A separation principle for the stabilization of a class of nonlinear systems. *IEEE Trans. Automatic Control*, 44:1672–1687, 1999.
3. A.N. Atassi and H.K. Khalil. Separation results for the stabilization of nonlinear systems using different high-gain observer designs. *Syst. Control Lett.*, 39:183–191, 2000.
4. J. Chen and R.J. Patton. *Robust Model-Based Fault Diagnosis for Dynamic Systems*. Kluwer Academic Publishers, Norwell, MA, 1999.
5. J. Chen, R.J. Patton, and H.Z. Zhang. Design of unknown input observers and robust fault detection filters. *Int. J. Control*, 63:85–105, 1996.
6. W. Chen and M. Saif. Actuator fault diagnosis for uncertain linear systems using a high-order sliding-mode robust differentiator (HOSMRD). *Int. J. Robust Nonlinear Control*, 18:413–426, 2008.
7. M. Darouach. On the novel approach to the design of unknown input observers. *IEEE Trans. Automatic Control*, 39:698–699, 1994.
8. C. Edwards and S.K. Spurgeon. On the development of discontinuous observers. *Int. J. Control*, 59:1211–1229, 1994.
9. C. Edwards and S.K. Spurgeon. *Sliding Mode Control: Theory and Applications*. Taylor & Francis, London, U.K., 1998.
10. C. Edwards and S.K. Spurgeon. A sliding mode observer based FDI scheme for the ship benchmark. *Eur. J. Control*, 6:341–356, 2000.
11. C. Edwards, S.K. Spurgeon, and R.J. Patton. Sliding mode observers for fault detection and isolation. *Automatica*, 36:541–553, 2000.
12. P.M. Frank. Analytical and qualitative model-based fault diagnosis—A survey and some new results. *Eur. J. Control*, 2:6–28, 1996.
13. L. Fridman, J. Davila, and A. Levant. High-order sliding-mode observation of linear systems with unknown inputs. *Proceedings of the IFAC World Congress*, Seoul, Korea, pp. 4779–4790, 2008.
14. A. Golovan and A. Matasov. Kalman-Bucy filter in the guaranteed estimation problem. *IEEE Trans. Automatic Control*, 39:1282–1286, 1994.
15. Y. Guan and M. Saif. A novel approach to the design of unknown input observers. *IEEE Trans. Automatic Control*, 36:632–635, 1991.
16. R. Isermann. *Fault-Diagnosis Systems: An Introduction from Fault Detection to Fault Tolerance*. Springer-Verlag, Berlin, Germany, 2005.
17. H. Khalil and A. Saberi. Adaptive stabilization of a class of nonlinear systems using high-gain feedback. *IEEE Trans. Automatic Control*, 32:1031–1035, 1987.
18. G. Kreisselmeier. Adaptive observers with exponential rate of convergence. *IEEE Trans. Automatic Control*, 22:2–8, 1977.
19. A.J. Krener and A. Isidori. Linearization by output injection and nonlinear observers. *Syst. Control Lett.*, 3:47–52, 1983.
20. D.G. Luenberger. An introduction to observers. *IEEE Trans. Automatic Control*, 16:596–602, 1971.
21. R. Marino and P. Tomei. Adaptive observers with arbitrary exponential rate of convergence for nonlinear systems. *IEEE Trans. Automatic Control*, 40:1300–1304, 1995.
22. K.Y. Ng, C.P. Tan, C. Edwards, and Y.C. Kuang. New results in robust actuator fault reconstruction in linear uncertain systems. *Int. J. Robust Nonlinear Control*, 17:1294–1319, 2007.
23. K. Ogata. *Modern Control Engineering*, 4th edn. Pearson, New York, 2002.
24. R.J. Patton and J. Chen. Optimal unknown input distribution matrix selection in robust fault diagnosis. *Automatica*, 29:837–841, 1993.
25. R. Rajamani. Observers for lipschitz nonlinear systems. *IEEE Trans. Automatic Control*, 43:397–401, 1998.

26. M. Spivak. *Calculus*. Cambridge University Press, Cambridge, U.K., 2006.
27. C.P. Tan and C. Edwards. Sliding mode observers for detection and reconstruction of sensor faults. *Automatica*, 38:1815–1821, 2002.
28. C.P. Tan and C. Edwards. Sliding mode observers for robust detection and reconstruction of actuator and sensor faults. *Int. J. Robust Nonlinear Control*, 13:443–463, 2003.
29. F.E. Thau. Observing the state of nonlinear dynamic systems. *Int. J. Control*, 3:471–479, 1973.
30. V.I. Utkin. *Sliding Modes in Control Optimization*. Springer-Verlag, Berlin, Germany, 1992.
31. B.L. Walcott and S.H. Zak. State observation of nonlinear uncertain dynamical systems. *IEEE Trans. Automatic Control*, 32:166–170, 1987.

25

Disturbance Observation–Cancellation Technique

Kouhei Ohnishi
Keio University

25.1 Why Estimate Disturbance?

One simple and effective robust control technique is disturbance observation-cancellation. It is necessary to have modes of disturbance in the controller for proper regulation. The internal model principle assures only the steady-state convergence of error. It is effective to run parallel with feed-forward compensation for faster response. Feed-forward compensation needs future disturbance signal beforehand. However, from a practical control viewpoint, the signal of only one or two steps in future are sufficient. The low-order disturbance observer estimates an equivalent disturbance (or a modified disturbance) only several steps ahead. The estimated disturbance added to input cancels out the disturbance. This function is the same as feed-forward compensation and improves the transient performance to the disturbance, as well as, the steady-state operation of the plant. Since the equivalent disturbance includes parameters variation, the entire controller is expected to be robust against not only the disturbance but also the parameters variation. As a result, the controlled plant seems as if it had nominal parameters and no disturbances. That is why it is worthwhile to estimate the equivalent disturbance. Also, the controller is simple and applicable to practical controllers. Let us consider the system as simply as possible.

25.2 Plant and Disturbance

The first goal is to estimate the additive disturbance in a linear system. For this purpose, we assume that the linear system has a single-input and single-output (SISO). Without losing generality, such a linear system is represented in the companion forms in Equation 25.1. Here, d is an additive disturbance as a scalar function in the companion form

$$\begin{aligned}\dot{\mathbf{x}} &= \mathbf{A}\mathbf{x} + \mathbf{b}u + \mathbf{e}d \\ y &= \mathbf{c}\mathbf{x}\end{aligned} \tag{25.1}$$

Here, **x** is a state vector of plant in the form

$$\mathbf{x} = \begin{bmatrix} x_1 \\ x_2 \\ \vdots \\ x_n \end{bmatrix}$$

and y is an output and u is an input of the system, respectively.

A is a system matrix in the form

$$\mathbf{A} = \begin{bmatrix} 0 & 1 & 0 & \cdots & 0 \\ 0 & 0 & 1 & \cdots & 0 \\ \vdots & \vdots & \vdots & \ddots & \vdots \\ 0 & 0 & 0 & \cdots & 1 \\ -a_1 & -a_2 & -a_3 & \cdots & -a_n \end{bmatrix}$$

b is a distribution vector in the form

$$\mathbf{b} = \begin{bmatrix} 0 \\ 0 \\ \vdots \\ 0 \\ K \end{bmatrix}$$

e is a distribution vector of disturbance in the form

$$\mathbf{e} = \begin{bmatrix} 0 \\ 0 \\ \vdots \\ 0 \\ 1 \end{bmatrix}$$

c is an observation vector in the form

$$\mathbf{c} = \begin{bmatrix} c_1 & c_2 & \cdots & c_m & 0 & \cdots & 0 \end{bmatrix}$$

If d is zero, then the system is controllable and observable and is represented in the following transfer function:

$$\frac{Y(s)}{U(s)} = K \frac{c_m s^m + c_{m-1} s^{m-1} + \cdots + c_2 s + c_1}{s^n + a_n s^{n-1} + a_{n-1} s^{n-2} + \cdots + a_2 s + a_1} \tag{25.2}$$

Generally the system dynamics matrix **A** and the distribution vector **b** include the variable parameters in their elements. They are the sum of the nominal values and their variations, respectively.

$$\begin{aligned}\mathbf{A} &= \mathbf{A}_0 + \Delta\mathbf{A}\\ \mathbf{b} &= \mathbf{b}_0 + \Delta\mathbf{b}\end{aligned} \tag{25.3}$$

Here, $\mathbf{A}_0$ is a nominal system matrix in the form

$$\mathbf{A}_0 = \begin{bmatrix} 0 & 1 & 0 & \cdots & 0 \\ 0 & 0 & 1 & \cdots & 0 \\ \vdots & \vdots & \vdots & \ddots & \vdots \\ 0 & 0 & 0 & \cdots & 1 \\ -a_{01} & -a_{02} & -a_{03} & \cdots & -a_{0n} \end{bmatrix}$$

$\Delta\mathbf{A}$ is a variation of **A** in the form

$$\Delta\mathbf{A} = \begin{bmatrix} 0 & 0 & 0 & \cdots & 0 \\ 0 & 0 & 0 & \cdots & 0 \\ \vdots & \vdots & \vdots & \ddots & \vdots \\ 0 & 0 & 0 & \cdots & 0 \\ -\Delta a_{01} & -\Delta a_{02} & -\Delta a_{03} & \cdots & -\Delta a_{0n} \end{bmatrix}$$

$\mathbf{b}_0$ is a nominal distribution vector in the form

$$\mathbf{b}_0 = \begin{bmatrix} 0 \\ 0 \\ \vdots \\ 0 \\ K_0 \end{bmatrix}$$

$\Delta\mathbf{b}$ is a variation of **b** in the form

$$\Delta\mathbf{b} = \begin{bmatrix} 0 \\ 0 \\ \vdots \\ 0 \\ \Delta K \end{bmatrix}$$

It is noted that the variation of dynamic matrix **A** is equal to the variations of the coefficients in the lowest column of **A**. Also, the variation of **b** is substantially equal to the variation of forward gain K. The system equation is transformed into Equation 25.4.

$$\begin{aligned}\dot{\mathbf{x}} &= (\mathbf{A}_0 + \Delta\mathbf{A})\mathbf{x} + (\mathbf{b}_0 + \Delta\mathbf{b})u + \mathbf{e}d\\ &= \mathbf{A}_0\mathbf{x} + \mathbf{b}_0 u + (\Delta\mathbf{A}\mathbf{x} + \Delta\mathbf{b}u + \mathbf{e}d)\end{aligned} \tag{25.4}$$

The third term is the sum of the disturbance and the parameter variation effect. It is possible to define a scalar function as an equivalent disturbance

$$\tilde{d} = d + \mathbf{e}^t(\Delta\mathbf{A}\mathbf{x} + \Delta\mathbf{b}u) \tag{25.5}$$

$\tilde{d}$, termed equivalent disturbance, includes not only the unknown disturbance but also the unknown parameter variations. By substituting Equation 25.5 into Equation 25.1, Equation 25.6 holds:

$$\begin{aligned}\dot{\mathbf{x}} &= \mathbf{A}_0\mathbf{x} + \mathbf{b}_0 u + \mathbf{e}\tilde{d}\\ y &= \mathbf{c}\mathbf{x}\end{aligned} \tag{25.6}$$

Since $\tilde{d}$ is a function of time, it is expanded into power series of time. If $\tilde{d}$ is slower compared to system dynamics, it is approximated in the form

$$\frac{\mathrm{d}^{(p)}\tilde{d}}{\mathrm{d}t^p} = 0 \tag{25.7}$$

Equation 25.7 is easily combined to Equation 25.6. The results are

$$\begin{aligned}\dot{\tilde{\mathbf{x}}} &= \tilde{\mathbf{A}}_0\tilde{\mathbf{x}} + \tilde{\mathbf{b}}_0 u\\ y &= \tilde{\mathbf{c}}_0\tilde{\mathbf{x}}\end{aligned} \tag{25.8}$$

Here, $\tilde{\mathbf{x}}$ is an augmented state vector in the form

$$\tilde{\mathbf{x}} = \begin{bmatrix} x_1 \\ x_2 \\ \vdots \\ x_u \\ \tilde{d} \\ \dot{\tilde{d}} \\ \ddot{\tilde{d}} \\ \vdots \\ \tilde{d}^{(p-1)} \end{bmatrix}$$

$\tilde{\mathbf{A}}_0$ is an augmented system matrix in the form

$$\tilde{\mathbf{A}}_0 = \overbrace{\begin{bmatrix} 0 & 1 & 0 & \cdots & 0 & 0 & 0 & \cdots & 0 \\ 0 & 0 & 1 & \cdots & 0 & 0 & 0 & \vdots & 0 \\ \vdots & \vdots & \vdots & \ddots & \vdots & \vdots & 0 & \vdots & 0 \\ 0 & 0 & 0 & \cdots & 1 & 0 & 0 & \vdots & 0 \\ -a_{01} & -a_{02} & -a_{03} & \cdots & -a_{0n} & 1 & 0 & \vdots & 0 \\ 0 & 0 & 0 & \cdots & 0 & 0 & 0 & \vdots & 0 \\ \vdots & \vdots & \vdots & \vdots & \vdots & \vdots & \vdots & \vdots & \vdots \\ 0 & 0 & 0 & \cdots & 0 & 0 & 0 & \cdots & 0 \end{bmatrix}}^{n+p}$$

$\tilde{\mathbf{b}}_0$ is an augmented distribution vector in the form

$$\tilde{\mathbf{b}}_0 = \left.\begin{bmatrix} 0 \\ 0 \\ \vdots \\ 0 \\ K_0 \\ 0 \\ \vdots \\ 0 \end{bmatrix}\right\} n+p$$

$\tilde{\mathbf{c}}_0$ is an augmented observation vector in the form

$$\tilde{\mathbf{c}}_0 = \overbrace{\begin{bmatrix} c_1 & c_2 & \cdots & c_m & 0 & \cdots & 0 \end{bmatrix}}^{n+p}$$

In Equation 25.8, an equivalent disturbance is treated as if it were a state variable. This is the key point in the design process. Equation 25.8 is the same as Equation 25.1. However, Equation 25.8 does not seem to have any disturbance or any parameter variations. The difference is the size of dimension. Clearly, the controllability is lost; fortunately, however, the observability is preserved. It is possible to construct an observer that estimates an equivalent disturbance $\tilde{d}$. Such an observer is called a disturbance observer. Once the equivalent disturbance is estimated or identified, it is possible to synthesize an input u to include a signal to cancel the equivalent disturbance. This is the principle of the disturbance observation-cancellation technique. Sometimes this technique is called "zeroing" or "cancellation." The details of the above explanation are developed one by one in the following.

25.3 Higher-Order Disturbance Approximation

To regard disturbance as a state variable, the disturbance should have certain dynamics. From an analysis in the previous chapter, an equivalent disturbance is a function of time.

Since we do not know $\tilde{d}$ a priori, we should estimate it as closely as possible. From the point of digital control, we need $\tilde{d}$ only a few steps ahead of every control sampling time. This means that we will apply an approximation by a low-order polynomial of time to $\tilde{d}$ in every sampling time. For example, if $\tilde{d}$ is approximated by piecewise rectangular lines, the derivatives are zero almost everywhere. This is the case of $p = 1$ in Equation 25.7. Similarly, if the function is approximated by piecewise straight lines, the second derivatives are zero. By increasing the fitness of the function by a $(p - 1)$-order polynomial, we will get Equation 25.7, which means the equivalent disturbance of $(p - 1)$ steps ahead is estimated. From a practical standpoint, p less than 3 gives a good enough approximation. In the case of Equation 25.7, the augmented states and assorted matrix are corresponding to Equation 25.8.

25.4 Disturbance Observation

Since the linear system with additive disturbance is represented in the form of Equation 25.8, it is possible to construct the (reduced-order) observer that estimates $\tilde{d}$ whose pth derivative is zero. Using Equation 25.8, various observers can be constructed. A reduced-order observer is designed by Gopinath's method whose order is $n + p - m$. A full-order observer whose order is $n + p$ is also applicable. There are several design procedures for the design of the above observer. Most of them, including Gopinath's method, are found in other sections of the book or in the references. An example applied to motion system is shown later. Please note that the dynamics of any observer are specified arbitrarily; good results are obtained by careful thought of pole allocation of the observer.

25.5 Disturbance Cancellation

Once an equivalent disturbance is estimated, the input will be designed to have two parts.

$$u = u^{ref} + u^{dis} \tag{25.9}$$

The first term is a driving input for a nominal plant that has only nominal parameters without disturbance. The second term is a compensation input to regard the original plant as a nominal plant without disturbance. The second term is synthesized, so that the equivalent disturbance is canceled by the feedback of the estimated equivalent disturbance $\tilde{d}$. Clearly, u^{dis} is determined by the following equation.

$$\begin{aligned} u^{dis} &= -(\mathbf{b}_0^t \mathbf{b}_0)^{-1} \mathbf{b}_0^t \mathbf{e} \hat{\tilde{d}} \\ &= -\frac{1}{K_0} \hat{\tilde{d}} \end{aligned} \tag{25.10}$$

Equation 25.10 cancels out the real unknown equivalent disturbance. Since $\hat{\tilde{d}}$ is estimated with lag elements inside, the difference between the real value and the estimated value of the equivalent disturbance will converge to *zero* in steady state. The convergence velocity depends on the identification process, that is, the poles of the disturbance observer. Equation 25.10 is the direct result of this section.

25.6 Examples of Application

Disturbance observation and cancellation techniques were realized originally in motion control systems. The mechanical system driven by the dc motor has the following dynamic equation.

$$J = \frac{d\omega}{dt} = K_t I_a - T_l \tag{25.11}$$

Here
- J is the inertia about motor shaft
- K_t is the torque constant
- T_l is the load torque

Suppose only the position of the motor shaft is detected by the rotary encoder, then the output is written in the form

$$y = \theta = \int \omega dt \tag{25.12}$$

The companion form with the equivalent disturbance, shown in Equation 25.13, has the following elements:

$$\bar{d} = -\frac{T_l}{J} + \left(\frac{K_t}{J} - \frac{K_{t0}}{J_0} \right) I_a \tag{25.13}$$

$$\mathbf{A} = \begin{bmatrix} 0 & 1 \\ 0 & 0 \end{bmatrix}$$

$$\mathbf{b} = \begin{bmatrix} 0 \\ \dfrac{K_t}{J} \end{bmatrix}$$

$$\mathbf{c} = \begin{bmatrix} 1 & 0 \end{bmatrix}$$

Suppose the disturbance is sufficiently slow in one sampling time, it is possible to assume the following equation according to the previous consideration.

$$\frac{\mathrm{d}\bar{d}}{\mathrm{d}t} = 0 \tag{25.14}$$

The augmented equation is

$$\frac{\mathrm{d}}{\mathrm{d}t}\begin{bmatrix} \theta \\ \omega \\ \bar{d} \end{bmatrix} = \begin{bmatrix} 0 & 1 & 0 \\ 0 & 0 & 1 \\ 0 & 0 & 0 \end{bmatrix}\begin{bmatrix} \theta \\ \omega \\ \bar{d} \end{bmatrix} + \begin{bmatrix} 0 \\ \dfrac{K_{t0}}{J_0} \\ 0 \end{bmatrix} I_a \tag{25.15}$$

A disturbance observer that estimates $\tilde{d}$ is derived by Gopinath's algorithm as follows:

$$\hat{\tilde{d}} = k_l\theta + z_l$$

here z_l satisfies

$$\frac{\mathrm{d}}{\mathrm{d}t}\begin{bmatrix} z_1 \\ z_2 \end{bmatrix} = \begin{bmatrix} 0 & -k_1 \\ 1 & -k_2 \end{bmatrix}\begin{bmatrix} z_1 \\ z_2 \end{bmatrix} + \begin{bmatrix} -k_1k_2\theta \\ (k_1 - k_2^2)\theta + \dfrac{K_{t0}}{J_0} I_a \end{bmatrix} \tag{25.16}$$

Here, the two poles of the observer α and β that are arbitrarily allocated in the complex plane satisfy the equation

$$\begin{aligned} \alpha + \beta &= -k_2 \\ \alpha\beta &= k_1 \end{aligned} \tag{25.17}$$

Modifying Equation 25.16, we get

$$\begin{aligned} \hat{\tilde{d}} &= \frac{k_1}{s^2 + k_2 s + k_1}\left(s^2\theta - \frac{K_{t0}}{J_0} I_a\right) \\ &= \frac{k_1}{s^2 + k_2 s + k_1}\tilde{d} \end{aligned} \tag{25.18}$$

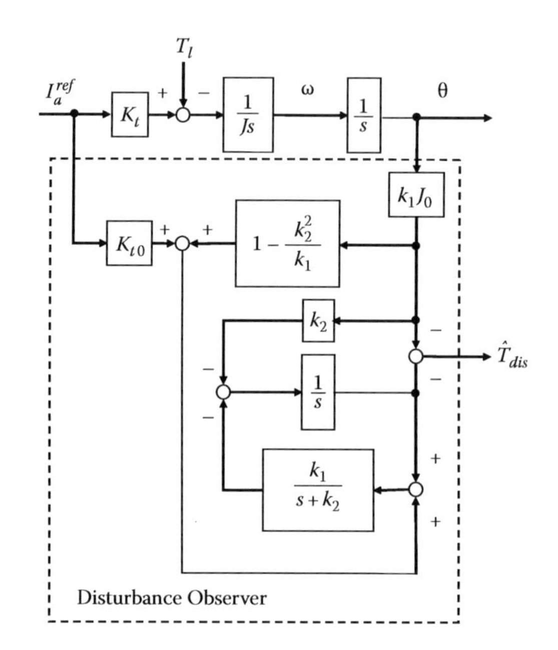

FIGURE 25.1 A disturbance observer realization in motion system.

$\hat{\tilde{d}}$ is estimated through a second-order lag system whose poles are α and β. The physically extended load disturbance T_{dis} is defined as

$$T_{dis} = -T_l - \Delta J \frac{d\omega}{dt} + \Delta K_t I_a \tag{25.19}$$

After some manipulations, we get

$$T_{dis} = J_0 \tilde{d} \tag{25.20}$$

Equation 25.19 means that the equivalent disturbance substantially has the terms of the following three terms, respectively:

- Mechanical load ($-T_l$)
- Varied self-inertia torque ($-\Delta J\, d\omega/dt$)
- Torque pulsation generated by the motor ($\Delta K_t I_a$)

Instead of $\tilde{d}$, it is possible to use T_{dis} in the motion system. One example of the disturbance cancellation technique realized in the motion system is shown in Figure 25.1. Here, the tachometer is used instead of the encoder. The direct solution of disturbance cancellation in the case of using a rotary encoder is left to be solved by readers. By comparing Figure 25.1 with the direct solution, the readers will appreciate the physical meaning of the disturbance cancellation in the motion system.

25.7 Conclusions

The disturbance observation and cancellation technique is now widely used in industrial controllers including position servo controller, video head controller, UPS, industrial robots, pneumatic servo controller, automobile active controller, and so on. The main reason is that higher performance is achieved by a simple and inexpensive controller. For example, the inverse dynamics in the controller of a robotic manipulator is saved by using Figure 25.1 at every joint. It is possible to extend this technique to other industrial plants. In most cases, the performance of robustness and linearization are expectedly superior to other controllers; however, it is necessary to use a fast CPU as a controller to realize a short sampling time.

Bibliography

1. K. Ohnishi et al. 1982. Torque control of DC motor observer, *Proceedings of the IEEJ Technical Meeting on Rotating Machinery*, Vol. RM-82–83, Tokyo, Japan.
2. K. Ohnishi et al. 1987. New development of servo technology in mechatronics, *Transactions on IEEJ*, 107-D(1), 83–86.
3. K. Ohnishi, N. Matsui, and Y. Hori. 1994. Estimation, identification, and sensorless control in motion control system, *Proceedings of the IEEE*, 82(8), 1253–1265.
4. K. Ohnishi et al. 1996. Motion control for advanced mechatronics, *IEEE/ASME Transactions on Mechatronics*, 1-1, 56–57.
5. M. Mizuochi, T. Tsuji, and K. Ohnishi. 2007. Multirate sampling method for acceleration control system, *IEEE Transactions on Industrial Electronics*, 54(3), 1462–1471.
6. H. Kobayashi, S. Katsura, and K. Ohnishi. 2007. An analysis of parameter variations of disturbance observer for motion control, *IEEE Transactions on Industrial Electronics*, 54(6), 3413–3421.

26

Ultrasonic Sensors

Lindsay Kleeman
Monash University

26.1 Overview

Ultrasonic sensors work with acoustic frequencies above 20 kHz, which is at the upper limit of audible human hearing. Ultrasonic sensing in robotics is popular due to the ability to directly achieve range sensing cheaply, simply, unobtrusively, and with low power consumption. Ultrasonic sensing is sometimes called SONAR derived from SOund NAvigation and Ranging. The basic principle of ranging is the measurement of the time-of-flight, *tof*, between transmission and reception of ultrasonic acoustic emissions. From the speed of sound, c, and *tof*, the range is $c\ tof/2$. The speed of sound varies with atmospheric conditions of temperature, humidity, and pressure as discussed below. The time-of-flight (TOF) can be measured from transmitting a pulse and processing the echo or using a continuous swept frequency transmission and examining the frequency of the echo. The former pulse echo technique is more common due to its processing and hardware simplicity. The frequency, f, of transmission can range from 25 to 500 kHz with corresponding wavelengths of ($\lambda = c/f$) approximately 14 mm down to 1 mm, respectively. Finer discrimination of targets is obtained using smaller wavelengths with the disadvantage of reduced maximum range due to the greater absorption losses during propagation in air as described below. Range measurement in the 25–60 kHz band is limited to around 10 m due geometric beam spreading and absorption losses in air.

There are several properties of ultrasonic sensing that need to be well understood for effective and accurate use of the sensors. Due to the long acoustic wavelength in comparison to imperfections in smooth surfaces, reflectors often behave in a specular acoustic manner analogous to an optical mirror where the angle of incidence and reflectance are equal. When the receiver and transmitter are in the same position, the sonar echo from a smooth wall will be reflected from a position on the wall that results in a direction of propagation perpendicular to the wall. When a sonar sensor is scanned across a wall, echoes will be obtained from this same perpendicular point with reducing amplitude as the sensor points away from this perpendicular point. The range reported by the sonar sensor pointing at a particular angle can be due to a reflection from the perpendicular point that differs in angle by half the beamwidth of the sensor. Thus, range readings obtained can be misleading due to angle errors up to the half beamwidth to the pointing direction.

The situation can be improved by using two receivers and sophisticated processing of the echo signals. This allows the bearing and ranges to specular targets to be estimated with accuracies typically better than 0.2 and 0.1 mm as described below. Using two transmitters and two receivers allows targets to be localized and classified into primitives such as planes, corners, and edges. Three-dimensional targets can be similarly classified. The use of phased arrays of transducers introduces the possibility of focused beams, improved signal-to-noise ratio, and greater discrimination of targets at similar range and bearing. Transducers can also be organized into a circular ring to allow complete coverage around the sensor rather than being restricted to the beamwidth of one transducer.

This chapter starts with the underlying physical properties of ultrasound speed, attenuation in air, and scattering from targets. Properties of the transducers are then discussed. Estimating range, bearing, and target type are then examined. Finally, arrays and rings are briefly discussed. References to more detailed treatment of these topics are provided.

26.2 Speed of Sound

The speed of sound varies significantly with atmospheric conditions, such as temperature, humidity, pressure, and altitude above sea level. The following formulae are derived from [1,2]. The speed of sound in dry air at sea-level air density and one atmosphere pressure is given by

$$c_T = 20.05\sqrt{T_C + 273.16}\ \text{ms}^{-1} \tag{26.1}$$

where T_C is the temperature in degrees Celsius. Equation 26.1 is accurate to 1% for most conditions. A more accurate estimate, c_H, can be given if the relative humidity of air, h_r, in percent is known:

$$c_H = c_T + h_r\left\{1.0059\times10^{-3} + 1.7776\times10^{-7}(T_C + 17.78)^3\right\}\text{ms}^{-1} \tag{26.2}$$

Equation 26.2 is accurate to around 0.1% for temperatures of −30°C to 43°C and most atmospheric pressure conditions at sea level. When atmospheric pressure p_s is known, Equation 26.3 is more accurate:

$$c_P = 20.05\sqrt{\frac{T_C + 273.16}{1 - 3.79\times10^{-3}(h_r p_{sat}/p_s)}}\ \text{ms}^{-1} \tag{26.3}$$

where the saturation pressure of air, p_{sat}, is a function of temperature defined by Equation 26.8 below.

26.3 Attenuation of Ultrasound due to Propagation

Due to geometric spreading of the wavefront from a point source, the power of an ultrasonic pulse reduces with the square of the distance it travels in a lossless open medium. This translates to a *linear* reduction of pressure with distance, and hence voltage on a receiver, since power is proportional to the square of pressure or voltage. This geometric spreading applies in the far field (i.e., beyond $a^2/4\lambda$, where a is the diameter of the transduce and λ is the wavelength), which is beyond 20 mm for the Polaroid 7000 electrostatic transducer.

In reality, air absorbs energy from a propagating wave in the form of heat. Losses are affected by many factors, including air viscosity, heat conduction, molecular vibration modes, and composition of air in terms of nitrogen, oxygen, carbon dioxide, and water vapor. The attenuation, α, is a function primarily of frequency, temperature, and relative humidity and is expressed in nepers per meter. Thus, the pressure wavefront after propagating a distance d (in meters) is further multiplied by $e^{-\alpha d}$ in addition to geometric spreading. The attenuation is most accurately determined by calibration. There are, however, empirical formulae available from the American National Standards [2], as given below in (26.4) through (26.8), which are accurate to 10% for temperatures 0°C–40°C, 10%–100% relative humidity, pressure less than 2 atmospheres, and frequency-to-pressure ratio 40–10^6 Hz atm^{-1}. The empirical formulae predict the attenuation coefficient, α (in nepers m^{-1}), in air for signal frequency f (in Hz), atmospheric pressure p_s (in Pa), temperature T (in K), and relative humidity h_r (in %) is

$$\alpha = f^2 \left[\begin{array}{l} 1.84\times10^{-11}\left(\dfrac{p_s}{p_{s0}}\right)^{-1}\left(\dfrac{T}{T_0}\right)^{1/2} \\ +\left(\dfrac{T}{T_0}\right)^{-5/2}\left\{ \begin{array}{l} 1.278\times10^{-2}\dfrac{\left[\exp(-2239.1/T)\right]}{\left(f_{r,O}+(f^2/f_{r,O})\right)} \\ +1.068\times10^{-1}\dfrac{\left[\exp(-3352/T)\right]}{\left(f_{r,N}+(f^2/f_{r,N})\right)} \end{array} \right\} \end{array} \right], \quad \text{in Nepers m}^{-1} \tag{26.4}$$

where the following are defined:

$$f_{r,O} = \left(\frac{p_s}{p_{s0}}\right)\left\{24 + 4.41\times10^4 h \times \left[\frac{(0.05+h)}{(0.391+h)}\right]\right\}, \quad \text{in Hz} \tag{26.5}$$

$$f_{r,N} = \left(\frac{p_s}{p_{s0}}\right)\left(\frac{T}{T_0}\right)^{-1/2}\left(9 + 350h\exp\left\{-6.142\left[\left(\frac{T}{T_0}\right)^{-1/3} - 1\right]\right\}\right), \quad \text{in Hz} \tag{26.6}$$

$$h = h_r \frac{(p_{sat}/p_{s0})}{(p_s/p_{s0})}, \quad \text{in \%} \tag{26.7}$$

$$\begin{aligned} \log_{10}\left(\frac{p_{sat}}{p_{s0}}\right) &= 10.795861\left[1-\left(\frac{T_{01}}{T}\right)\right] - 5.02808\log_{10}\left(\frac{T}{T_{01}}\right) \\ &\quad + 1.50474\times10^{-4}\times\left\{1-10^{-8.29692[(T/T_{01})-1]}\right\} \\ &\quad + 0.42873\times10^{-3}\times\left\{-1+10^{4.76955[1-(T_{01}/T)]}\right\} \\ &\quad - 2.2195983 \end{aligned} \tag{26.8}$$

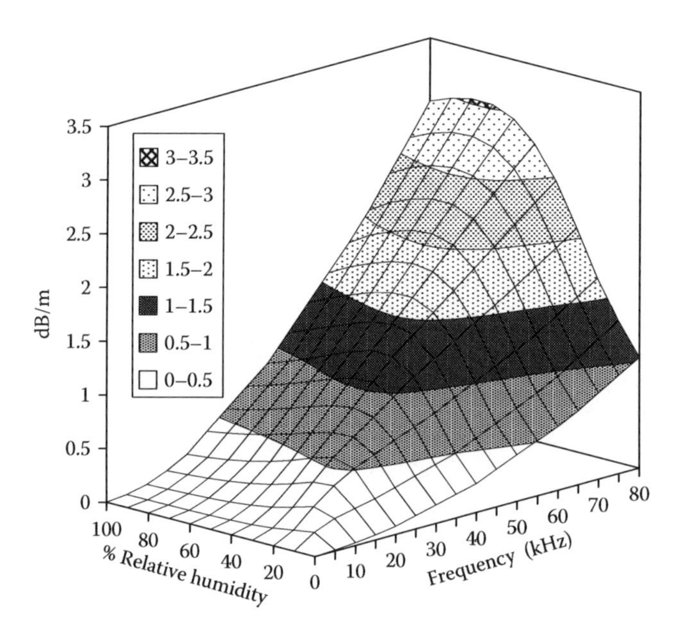

FIGURE 26.1 Absorption loss (dB m^{-1}) as a function of frequency and humidity in still air at 20°C. (Data derived from Weast, R.C. and Astle, M.J., *CRC Handbook of Chemistry and Physics*, 59th edn., CRC Press, Boca Raton, FL, 1978.)

where

p_{s0} is the reference atmospheric pressure (101.325 kPa)
T_0 is the reference ambient atmospheric temperature of 293.15 K (20°C)
T_{01} is the triple-point isotherm temperature with the exact value of 273.16 K

Figure 26.1 shows the dependency of absorption loss of air at 20°C on relative humidity and frequency [3]. Note the steeply increasing losses as frequency increases and also the peak losses occurring at intermediate relative humidities.

26.4 Target Scattering and Reflection

Acoustic wave propagation is disturbed by changes in the acoustic impedance of the medium, which is defined as the product of medium density and speed of sound. Discontinuities of the acoustic impedance can occur when the lower impedance of air meets solid objects with higher impedance and scattering of the acoustic wave results. When scattered waves make their way back to a receiver, an echo is registered. There are two basic types of scattering: *reflection* and *diffraction*. Reflection occurs from smooth surfaces with the angle of incidence to the normal of the surface equaling the angle of reflection. *Smooth* is defined by the size of rough features of the surface being much smaller than the acoustic wavelength. Reflection from smooth surfaces is often called *specularity*. Diffraction occurs due to discontinuities of the surface, such as edges (i.e., where a smooth surface ends or changes direction). Reflection from smooth planes and corners, and diffraction from edges will be considered here. More complex target profiles may be analyzed using approximations developed by Freedman [4].

26.4.1 Reflections from a Plane

Smooth planes are considered here; rough planes have been investigated in [5]. The plane is assumed to be free of losses due to reflection. Solid materials such as glass, timber, perspex, and plaster are practically lossless reflectors. Softer materials such as cardboard, cloth, felt, and foam can absorb ultrasonic energy.

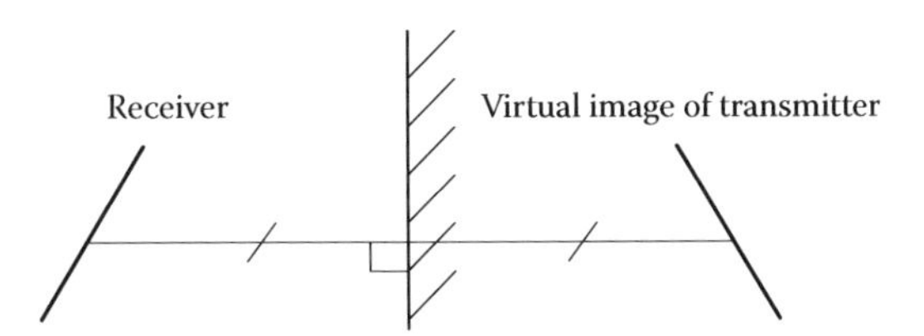

FIGURE 26.2 Virtual image formed by a plane.

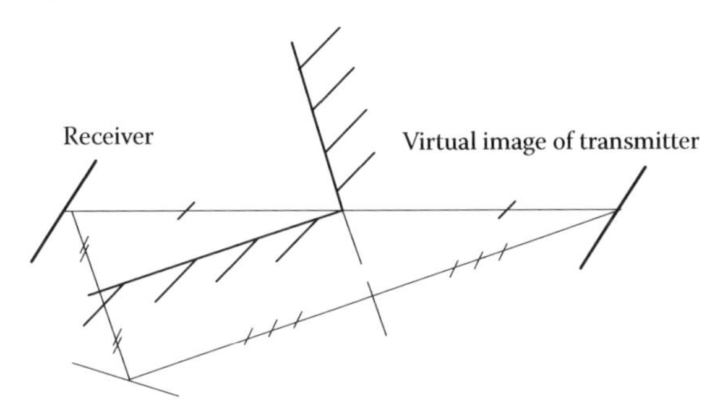

FIGURE 26.3 Virtual image formed by a corner.

In a lossless reflector, the plane reflector can be replaced by a virtual image of the transmitter or receiver. Shown in Figure 26.2 is a transducer that acts as both transmitter and receiver. The transmitter is considered separately as a virtual image reflected in the plane as with a mirror in optics. The effect of the plane reflector is then identical to moving the transmitter to the position of the virtual transmitter in Figure 26.2, fixing the original position of the receiver and removing the plane in between. Equally valid is to move the receiver to the virtual image position and to keep the transmitter in the original position.

26.4.2 Reflections from a Concave Right-Angled Corner

The corner is made up of two planes at right angles to each other. The virtual image of the transmitter in Figure 26.3 can be constructed by creating a virtual image through one plane and then taking the virtual image of that image through the other plane. This amounts to reversing the image with respect to a plane reflection by reflecting through the line of intersection of the two perpendicular planes. The image reversal property can be exploited to allow an ultrasonic system to discriminate corners from planes as described below.

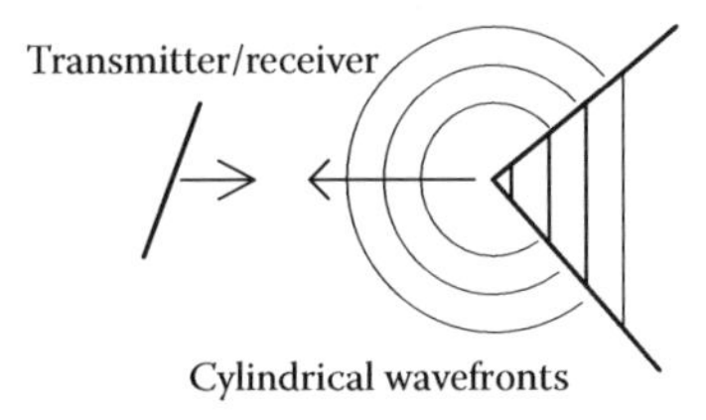

FIGURE 26.4 Scattering from an edge.

26.4.3 Diffraction from Edges

The edge as shown in Figure 26.4 scatters the acoustic wave in the form of cylindrical wavefronts with center line along the edge. This results in more rapid power spreading than that of a plane at the same range. This can be modeled as the amplitude reducing by $1/\sqrt{\text{range}}$ times that of the plane reflector. Moreover, an approaching wave front will see a smaller area when encountering an edge versus a plane. This further reduces the reflected amplitude of an edge in comparison to the plane.

Reverse edges, as shown in Figure 26.5, produce a sign reversed echo over the front on edge in Figure 26.4. This is due to the transition from higher acoustic impedance to lower. Moreover, Kuc and Viard [6] report an approximately half amplitude echo from the reverse edge compared to the front on edge.

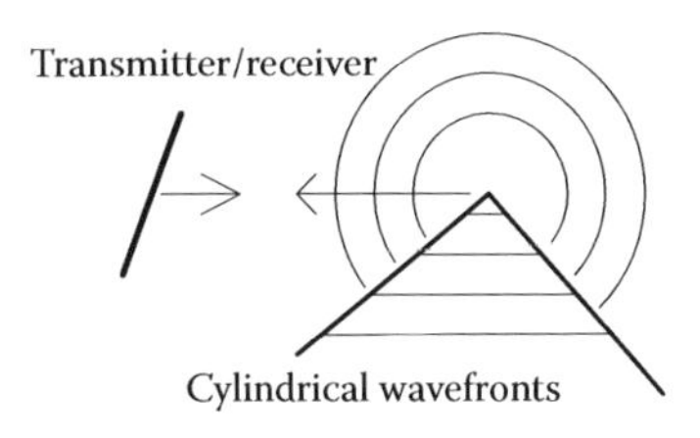

FIGURE 26.5 Scattering from a reversed edge.

26.5 Beamwidth—Round Piston Model

Many ultrasonic transducers are round and can be modeled by a vibrating circular disk in an infinite baffle. The acoustic pressure, p, in the far field can then be derived [7] in terms of the Bessel function J_1 as a function of the angle to the transducer normal, θ, wavelength, λ, and the radius of the disk, a:

$$p(\theta) = \frac{2J_1\left(ka\sin(\theta)\right)}{ka\sin(\theta)} \quad \text{where } k = \frac{2\pi}{\lambda} \tag{26.9}$$

The radiation pattern of the Polaroid instrument grade transducer has side lobes at 20° and 40° and nulls at 15° and 30°. Note that the physical principle of reciprocity between transmitter and receiver implies that the same beam pattern applies to each. The half-width beam angle, θ_0, is the angle to the first off-axis zero in the Bessel function and is given by

$$\theta_0 = \sin^{-1}\left(0.61\frac{\lambda}{a}\right) \tag{26.10}$$

This equation shows the relationship between transducer radius in wavelengths and the effective beamwidth. The larger the number of wavelengths across the transducer, the narrower the beamwidth is.

Note that the radiation pattern in Equations 26.9 and 26.10 above correspond to exciting the transducer with a *continuous* sine wave at 50 kHz. In practice, *short* pulses can be transmitted and thus contain a wider spectrum, which blurs the radiation pattern. In the extreme case where the pulse is as narrow as practical, the side lobes and nulls disappear altogether. This is illustrated in Figure 26.6, which shows the pulse, obtained by driving the transmitter with a 10 μs 300–0–300 V pulse, and the *corresponding* combined transmitter–receiver radiation patterns for a Polaroid 7000 electrostatic transducer. A Gaussian approximation can be employed as an effective approximation for the beam characteristics [6]. The beamwidth of a transducer is defined by the angle difference for −3 dB attenuation relative to that at zero angle. Values other than −3 dB are also used in the literature. Targets outside the beamwidth may still return a weak echo, which may or may not be discernible above the background acoustic and amplifier noise.

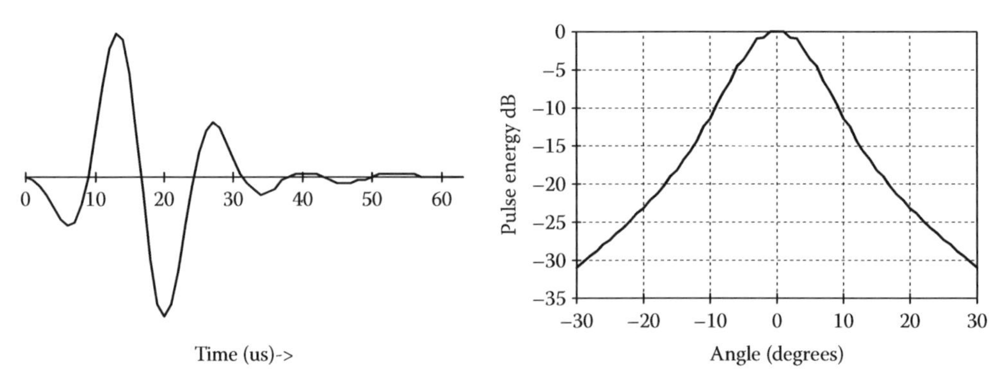

FIGURE 26.6 Received pulse shape from a plane reflector at 1 m range using Polaroid 7000 transducer and corresponding beam pattern for combined transmitter and receiver.

26.6 Transducers

Two types of transducers are commonly employed: electrostatic and the piezoelectric. An example of an electrostatic transducer is the Polaroid transducer constructed from a gold-coated plastic foil membrane stretched across a round grooved aluminum backplate. The membrane and the backplate form a varying capacitor. As a transmitter, the transducer membrane is vibrated by applying 0–300 V pulses across this capacitor. The charge induced by the 300 V on the capacitor causes an electrostatic attraction force between the membrane and the backplate. The same transducer also acts as a receiver when a 150 V DC bias voltage is applied to store opposing charges on the membrane and backplate. Incoming sound vibrates the membrane, changing the effective separation between the plates of the capacitor and hence the capacitance. The charge on the plates of the varying capacitance induce a voltage related to the incoming acoustic wave. The grooves on the backplate enhance the sensitivity of the transducer. Mounted on the front of the transducer is a protective grill—the author's experience suggests removing this, should the environmental conditions allow, in order to reduce losses and reverberation between the grill and the membrane. The −3 dB beamwidth of the transducer is of the order of 8°. The frequency bandwidth of a transmitter combined with a receiver is of the order of 20 kHz for −3 dB, centered at 50 kHz. The sensitivities of the transmitter and receiver allow range measurement upto approximately 10 m for plane reflectors.

Piezoelectric transducers can also act as both transmitters and receivers, although, often, manufacturers optimize the performance by offering separate devices. A piezoelectric resonant crystal mechanically vibrates when a voltage is applied across the crystal and in reverse generates a voltage when mechanically vibrated. A conical concave horn is mounted on the crystal to acoustically match the crystal acoustic impedance to that of air. An example is the Murata MA40A5R/S receiver and sender transducers, which operate at 40 kHz. This device has a diameter of 16 mm and a 60° beam angle. The attenuation at 60° off-axis for pulse echo is 20 dB down on the on-axis attenuation for a transmitter and receiver pointing in the same direction. The effective bandwidth of transmitter and receiver is only a few kilohertz due to the resonant nature of the crystals. This limits the envelope rise time of pulses to around 300 μs. An advantage is the ability to drive piezoelectric devices with low voltages, for example, by connecting each terminal to complementary CMOS logic outputs. Piezoelectric transducers are also available at resonant frequencies from 25 to 300 kHz.

In terms of performance, the electrostatic transducers offer the better sensitivity and bandwidth. Piezoelectric transducers, on the otherhand, are simpler to drive due to the lower voltages necessary.

26.7 Polaroid Ranging Module

The Polaroid ranging module (PRM) was originally developed as a range finder for auto-focus cameras. It has found extensive application in robotics due to its simplicity, low cost, and availability [8].

The 6500 series sonar ranging module consists of a single Polaroid electrostatic transducer that acts as transmitter and receiver, driving electronics, receiver time-gain compensated amplifier and filter, threshold circuitry, and echo output. A pulse train of 16 pulses at 49.4 kHz is transmitted and the echo is amplified with a gain that is stepped up with time. The signal is bandpass filtered and a decaying integrator applied before a threshold—so that short noise spikes do not trigger the threshold. The echo output is asserted once the threshold is triggered and the range to the first obstacle can be calculated from the elapsed time and the speed of sound.

There are some important limitations: (1) The beamwidth of the transducer must be taken into account when estimating the bearing of the obstacle. (2) Only the nearest returns are usually logged, thus masking of further obstacles occurs even though echoes are returned (it is, however, possible to log multiple echoes if they are spaced far enough apart and the receiver is re-enabled after each echo). (3) Weak echoes can produce errors in range due to the delay in charging the integration circuit.

(4) The time-gain compensation can only be approximate due to the variability of absorption of ultrasound with temperature and humidity and also variability of scattering efficiency of different target types. (5) Multipath echoes can cause obstacles to appear in completely the wrong direction [9]. Multiple (typically 16–24) PRMs are often employed in a ring around a robot and simultaneously fired in noninterfering clusters. More sophisticated sonar ring structures are discussed below.

26.8 Estimating the Echo Arrival Time

Various techniques to estimate the arrival time of an echo in an ultrasonic sensor are considered in this section in order of processing complexity.

26.8.1 Thresholding

A TOF is estimated by recording the time at which an echo exceeds a threshold. Consequently, low-amplitude echoes can give delayed estimates, due to the rise time of the signal envelope. Attempts to eliminate this amplitude dependency include the use of time-gain compensation as in the PRM. Since edges and planes, for example, return different echo amplitudes, time-gain compensation may not always work correctly [10]. Instead, computer sampling of the echo envelope can then be used to apply a threshold proportional to the echo amplitude. Variability may still occur due to the pulse shape dependence on bearing angle. Since the arrival time is essentially based on one sample, noise on that sample can unduly corrupt the arrival time. Other estimation techniques based on multiple samples improve the noise performance and are discussed next.

26.8.2 Curve Fitting to the Envelope

Discrete-time samples of the echo envelope can be extracted either using (a) Nyquist rate sampling and then processing or (b) analogue full-wave rectification, low-pass filtering, and then lower-rate digital sampling. The rising edge of the envelope is then used to fit curves to, for example, (a) an increasing parabola, (b) a polynomial multiplied by a decaying exponential, or (c) maximum slope straight line of a constant number of consecutive samples. The rising edge is used since it offers the maximum slope and is least likely to be corrupted by overlapping pulses.

26.8.3 Matched Filtering

A matched filter is obtained by examining the cross correlation of the potentially noisy echo with a predicted pulse shape. The TOF corresponds to the time-shifted position of the predicted pulse that gives a maximum in the cross correlation. Matched filtering can be applied if the sample rate of the echo exceeds twice the bandwidth of the echo signal. Theory of RADAR [11] shows that the Maximum Likelihood Estimate of the arrival time of the echo corrupted by additive white Gaussian noise is the matched filter. This means that it is usually the "best" estimator in practice. The matched filter needs a prediction or model of the echo pulse shape that can depend on the bearing angle to the target, dispersion in air of ultrasound, scattering properties of targets, and transmitter and receiver characteristics. Linear models exist that accurately predict pulse shape, and matched filtering has been implemented successfully [12].

26.9 Estimating the Bearing to Targets

The range of ultrasonic reflectors can be determined by examining the echo waveform from one receiver. However, accurate *bearing* estimation requires two or more receivers. Figure 26.7 shows two receivers observing the range to P, which may be an edge reflector or a virtual image of a transmitter in a plane or corner. The bearing to P is given by

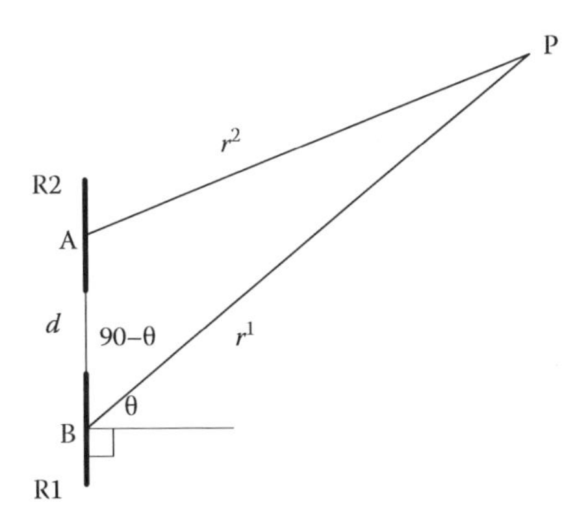

FIGURE 26.7 Two receivers used to estimate bearing to P.

$$\theta = \sin^{-1}\left(\frac{d^2 + r_1^2 - r_2^2}{2dr_1}\right) \tag{26.11}$$

where

d is the receiver separation

r_1 and r_2 are the ranges to P from each receiver

In a cluttered environment with many closely spaced echoes, it may be difficult to determine corresponding pairs of arrival times on each receiver. This correspondence problem, as it is known, is minimized by employing closely spaced receivers, provided a sufficiently accurate arrival time estimator (e.g., matched filter) has been employed to prevent bearing inaccuracy.

26.10 Specular Target Classification

It is possible to classify targets into planes, concave right angle corners, and edges using at least two transmitters and two receivers. Such an arrangement is shown in Figure 26.8, where T represents a transmitter and R a receiver.

The algorithm for classification is based on the virtual image of a plane being reversed to that of a corner. An edge produces fixed bearing independent of the transmitter position. As illustrated in Figure 26.9, classification can be performed based on the difference, β, in bearing angles to the reflector from the two

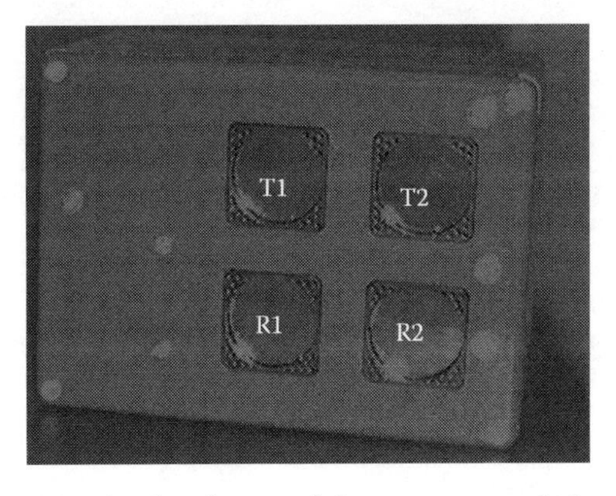

FIGURE 26.8 Sensor arrangement for identification of planes, corners, and edges—Note the two transmitters T1 and T2 above the two receivers R1 and R2.

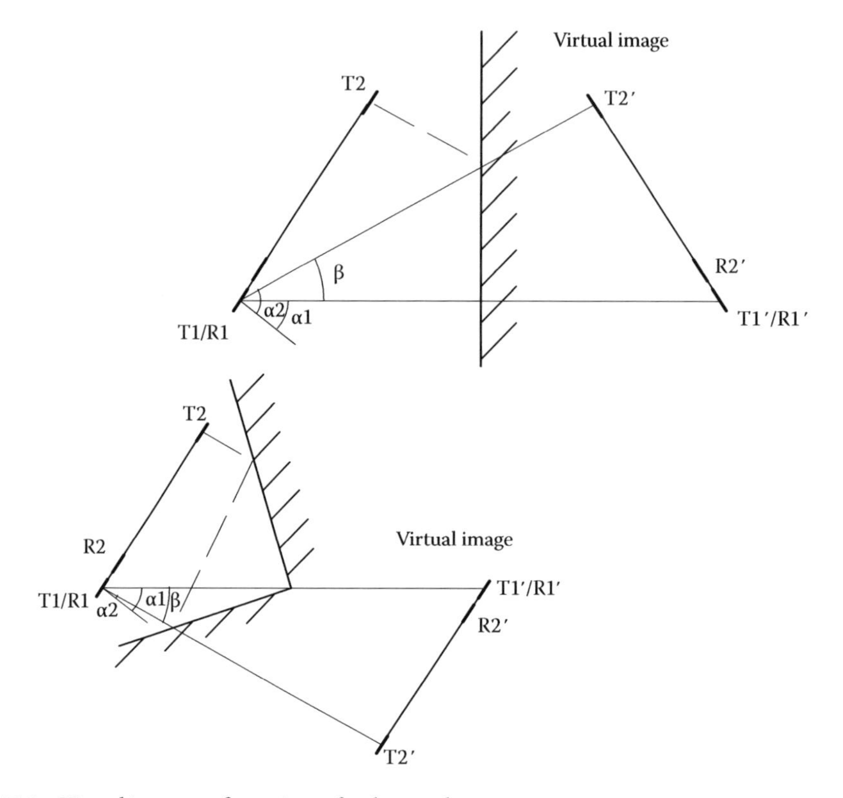

FIGURE 26.9 Virtual image configurations of a plane and a corner.

different transmitters fired in succession. The value of β is positive for a plane, negative for a corner, and zero for an edge. The implementations of this approach [12,13] can successfully classify reflectors up to ranges of 8 m to accuracies of 0.2 mm range and 0.2° bearing. Three-dimensional targets can be classified similarly [14].

26.11 Ultrasonic Beam-Forming Arrays

The sensor array shown in Figure 26.8 uses the full beam pattern of both the transmitter and receivers to collect echoes. Using beamforming techniques, it is possible to restrict the effective beamwidth of either the transmitter or the receiver, so that targets at the same range but with different bearings can be discriminated. Targets falling within the beamwidth and at the same range cannot be discriminated with the sensor of Figure 26.8 since pulse overlap occurs. By using phase delays on an array of transmitters or phase delays and summation on an array of receivers as in Figure 26.10, the beam can be narrowed or even focused to a point. The delay and summing can be performed electronically or with software processing of digitized echo waveforms. The advantage of phasing the transmitters is that more energy can be focused into a narrow beam, thus increasing the range of the sensor. However, covering a wide bearing range requires many measurements, slowing the sensor. The advantage of receiver arrays is that one measurement can cover a large angle and processing can then be done on the same data for different bearing angles. For narrow band systems, such as is the case when piezoelectric transducers are employed, the separation of the elements of the array must be less than half a wavelength to avoid ambiguity in bearing estimation of receivers or grating lobes in transmitters. Wide-band systems can exploit envelope shape to overcome this limitation. An example of an ultrasonic array implementation can be found in [14]. One practical difficulty is that the sensitivity of the transducers need to be matched either by a calibration procedure or precise manufacturing techniques.

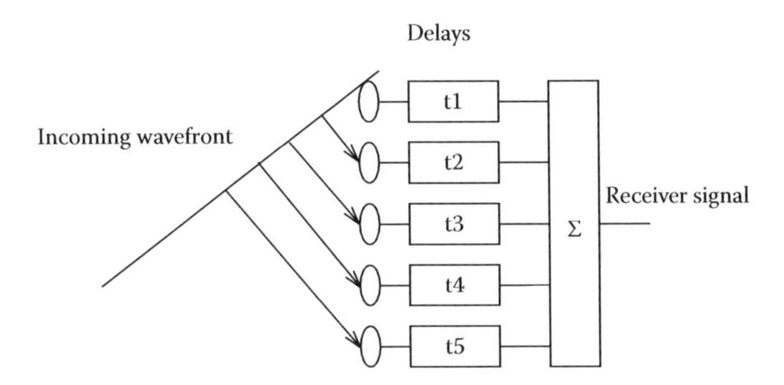

FIGURE 26.10 Phased receiving array.

26.12 Advanced Sonar Sensing

The term "advanced" sonar sensing in this section refers to sensing that provides range, bearing, target classification, and interference rejection in the one sensor. This section describes how these features can be achieved and the minimum sensor requirements. Range estimation can be achieved with a single transducer acting as a transmitter and a single receiver. Estimating the bearing angle in two dimensions of an echo can be achieved by two approaches.

The first approach uses just one receiver and estimates the angle based on the amplitude of the echo [15]. The echo amplitude at the receiver is a symmetrical function of the angle to the normal direction of the transducer, and consequently, only one side of the transducer beamwidth can be employed to unambiguously measure the angle. By modeling the amplitude response of a circular transducer as a function of angle and range, an estimate can be achieved of the angle of reception from high-reflectance targets. This approach is only practical for ranges less than 2 m because of the variation of absorption losses in air due to variations in humidity and temperature. Errors as low as 1°C have been reported with this approach [15]. When the reflectance is low, it is difficult to differentiate the reduction in echo amplitude due to reflectance from angle reduction, unless an angular scan of the environment is taken and the amplitude model fitted to the scan [16].

The second approach for bearing estimation uses two receivers and accurate TOF estimation on each receiver as described above. From the difference in arrival times of an echo, the angle of arrival can be accurately estimated. The accuracy is limited by the air conditions, such as turbulence and temperature gradients, and typical errors of 0.1° standard deviation have been reported in office-type environments [12,13]. The estimation of accurate arrival time is achieved by sampling the analogue echo signal at 1 MHz with a 12-bit analogue-to-digital converter (ADC) and applying matched filtering digital signal processing. Matched filtering consists of calculating the correlation of the echo with an echo template for different positions of the template relative to the echo. The position of maximum correlation corresponds to the arrival time. The resolution of the arrival time can be improved over the signal sample time by fitting a parabola to the correlation as a function of time shift and using the peak of the parabola as the arrival time [12]. The template echo shape can be derived from a calibration procedure performed a priori that collects an echo from a plane reflector perpendicular to the transducer normal. This zero degree echo shape is distorted to account for different angles of arrival by double convolution with an elliptical impulse response with width proportional to the angle from normal [12]. A set of templates for different angles are generated and used in the matched filter of receiver data. The template shape also varies with range due to the dispersion of sound in air and this can be modeled to generate different template sets for each meter of range. When many echoes are received on the two receivers, the correspondence problem must be solved in order to extract bearing measurements. The correspondence problem is how to associate an echo on one receiver to an echo on the other receiver. The problem can be alleviated by positioning the receivers as close together as the transducer dimensions allow, so that echoes on different receivers will be as close as possible in time, thus reducing the number of associations possible. Echo amplitude, template match angle, and other features can be used as attributes to assist in solving the correspondence problem. Nevertheless, certain reflector configurations can be constructed to produce ambiguous correspondences, but in practice, these occurrences are rarely found. A pragmatic approach is to flag any ambiguous correspondence and allow the resolution of the correspondence at a higher level in the data interpretation or even ignore these readings altogether.

Target classification with advanced sonar can be achieved with the following target types: plane, concave right angle corner, and point feature (often called an edge). A plane is analogous to a flat mirror in that a virtual sonar image is generated. Similarly with a corner, two plane reflections occur and a reversed sonar image is formed. With a point feature, the echo appears to come from one position independent of the transmitter position and can correspond to high-curvature objects such as table legs, or correspond to small concave corner reflectors where the dimensions of the corner are smaller than the transducer diameter. In order to classify targets into plane, corner, and point types, two transmitter positions and two receivers are required [12]. The classification process is analogous to observing one's image in a flat mirror, a corner mirror, and a shiny table leg. The image is reversed between the flat and corner mirrors, and the image is compressed into a small width in the table leg. In terms of sonar classification, one transmitter is fired and the angle of the echo is noted. Then, the other transmitter is fired and the difference in angle is calculated. This angle difference can be used to determine the target classification. Large-scale simultaneous mapping and localization with a mobile robot has been achieved using this classifying sonar system [17,18].

Interference rejection allows multiple advanced sonar systems to operate in the same acoustic environment. The transmitter source of an echo needs to be identifiable. A double-pulse coding approach is adopted in [13] whereby two pulses are transmitted separated in time by typically 150–500 μs. Should two sonar systems interfere with each other, by examining the echo TOF separations, the other sonar's transmissions can be rejected if they operate with different double-pulse separations. One obvious problem is that echoes may overlap in time causing the arrival time estimation to fail. The echo energy has been concentrated into approximately 80 μs or five or six cycles of 60 kHz to minimize this problem.

Classification and interference rejection have been combined into one sensing cycle [13], whereby two transmitters fire as the first and second pulses, respectively, of the double-pulse code. In practice, the receiver arrival times can be associated with a transmitter by examining the relative timing. Moreover, double-pulse spacing outside the expected range can be deemed to be interference and rejected. This approach has been implemented on a digital signal processor (DSP) and runs in close to real time at 25 Hz repetition rate for ranges out to 5 m [13].

26.13 Sonar Rings

26.13.1 Simple Ranging Module Rings

Since a single sonar sensor only detects objects lying within its beamwidth, a popular way to scan the environment is to use a ring of sonar sensors [19]. The most common configuration is the Denning ring that contains 24 sonars equally spaced around the robot periphery. This 15° spacing allows some overlap in the sonar beams so at least one of the sonars will detect a strong reflecting object. The sonar transducers in the ring are typically fired sequentially. Using a 50 ms probing pulse period, to reduce false readings, a complete environmental scan is accomplished every 1.2 s. This sample time is adequate for a translate-and-stop operation in research settings but may be too slow for a continually moving robot. A robot moving at 1 ms^{-1} may not detect an object with sufficient time to prevent a collision. Some researchers propose simultaneously employing sonar sensors on opposite ends of the ring to speed up measurement times, while others also reduce the probing pulse period and attempt to identify artifacts.

26.13.2 Advanced Rings

Yata et al. [20] have produced a design of a 32 cm diameter sonar ring composed of 30 transmitters interleaved with 30 receivers. Murata piezoelectric MA40S4R wide-angle transducers give rise to overlapping reception of echoes in neighboring receivers when all transmitters are fired simultaneously. An axial symmetrical exponential horn structure is used to vertically narrow the beam shape of the transmitters to avoid reflections from the floor. Received signals are compared with a decaying threshold to produce a 1 bit digitized sampled signal without rectification. Bearing is estimated from the leading edge of echoes and an error standard deviation of 0.4° is reported to a range of 1.5 m.

26.13.3 DSP Sonar Ring

A sonar ring has been developed [21,22] that is organized into 24 pairs of 7000 series Polaroid transducers. Each pair consists of a transceiver (i.e., can transmit and receive on the same transducer) and a receiver. The system includes six DSP processor boards to perform signal processing and another DSP board to coordinate these processing boards. The DSP sonar ring is shown in Figure 26.11. Each pair of transceiver and receiver can derive range and accurate bearing information using template matching as described above. A 12-bit ADC samples each receiver with a sample rate of 250 kHz. Each DSP board processes eight receiver channels. All 24 transceivers are fired simultaneously to sense the surrounding environment 11 times a second, to 6 m range. The system has experimentally validated range and bearing errors for plane targets of 0.6 mm and 0.2°. Interference between neighboring pairs is rejected by using two different transmitted pulse shapes in an interleaved fashion around the perimeter of the ring. One pulse shape has two cycles of 65 kHz excitation and the other has three cycles. The DSP sonar ring allows for rapid and accurate wall following, map building, and obstacle avoidance due to the high repetition and accurate range and bearing sensing. The beamwidth of the transducer pairs allows full 360° coverage with respect to smooth specular targets to a range of 3 m.

26.13.4 Single Transmitter Ring Horn

Multiple transmitter sonar rings are composed of many ultrasonic receivers arranged in a ring and have traditionally used many transmitters as described in the previous sections. The transmitters have been fired either serially or together and the receivers measure TOF to reflectors in the environment.

FIGURE 26.11 DSP sonar ring mounted on a mobile robot.

FIGURE 26.12 A two-dimensional omnidirectional transmitter with a conical parabolic reflector suitable for a sonar ring.

Serial transmission reduces the measurement rate around the entire ring, whilst parallel transmission can incur interference or cross talk between transmitters. An alternative approach [23] is to employ a single high-powered transmitter, for example, a Murata (ESTD01 Super Tweeter) and a parabolic conical reflector [24] to overcome these issues. The reflector is designed to narrow the vertical beamwidth of the transmitter and to provide good dispersion of the sound around the ring. The transmitter then behaves like a point source, insonifying all directions equally in two dimensions. A prototype of such a system is shown in Figure 26.12.

26.14 Conclusion

This chapter has discussed the important properties of ultrasound and its interaction with objects in the environment that cause an echo to be generated. The design of ultrasonic sensors has been discussed in terms of transducers, their configuration, signal processing, and feature extraction. Sonar sensing has the capability to not only find the range to objects but also determine their bearing angle and perform classification into various target types such as a plane, corner, edge, or curved surface.

References

1. H H Poole, *Fundamentals of Robotics Engineering*, 1989, New York: Van Nostrand.
2. J E Piercy et al., American National Standard: Method for calculation of the absorption of sound by the atmosphere, ANSI SI-26-1978, 1978, New York: American National Standards, Acoustical Society of America.
3. R C Weast and M J Astle, *CRC Handbook of Chemistry and Physics*, 59th edn., 1978, Boca Raton, FL: CRC Press.
4. A Freedman, 1962. A mechanism of acoustic echo formation, *Acustica*, 12: 10–21.
5. O Bozma and R Kuc. Characterizing the environment using echo energy, duration, and range: the ENDURA method, in *IEEE/RSJ International Conference on Intelligent Robots and Systems*, Raleigh, NC, 1992, pp. 813–820.
6. R Kuc and V B Viard, 1991. A physically based navigation strategy for sonar-guided vehicles, *International Journal of Robotics Research*, 10(2): 75–85.
7. P M Morse and K U Ingard, *Theoretical Acoustics*, 1968, New York: McGraw-Hill.
8. P J McKerrow, *Introduction to Robotics*, 1991, Reading, MA: Addison Wesley.
9. J J Leonard and H F Durrant-Whyte, *Directed Sonar Sensing for Mobile Robot Navigation*, 1992, Boston, MA: Kluwer Academic Publishers.
10. R Kuc, 1990. A spatial sampling criterion for sonar obstacle detection, *IEEE Transactions on Pattern Analysis and Machine Intelligence*, 12(7): 686–690.

11. P M Woodward, 1964. *Probability and Information Theory with Applications to Radar*, 2nd edn., Oxford: Pergamon Press.
12. L Kleeman and R Kuc, August 1995. Mobile robot sonar for target localization and classification, *International Journal of Robotics Research*, 14(4), 295–318.
13. L Kleeman, On-the-fly classifying sonar with accurate range and bearing estimation, *IEEE/RSJ International Conference on Intelligent Robots and Systems*, Lausanne, Switzerland, 2002, pp. 178–183.
14. M L Hong and L Kleeman, A low sample rate ultrasonic sensor system for the differentiation and location of common three-dimensional room features, MECSE-94-2, 1994, Monash University, Clayton, Victoria, Australia.
15. P F Webb, An ultrasonics based system for the extraction of range and bearing data for multiple targets, PhD thesis, 1994, The University of Nottingham, England, U.K.
16. T Yata, L Kleeman, and S Yuta, November 1998. Fast bearing measurement with a single ultrasonic transducer, *International Journal of Robotics Research*, 17(11), 1202–1213.
17. L Kleeman, Scanned monocular sonar and the doorway problem, *IEEE/RSJ International Conference on Intelligent Robots and Systems*, Osaka, November 1996, pp. 96–103.
18. K S Chong and L Kleeman, January 1999. Feature-based mapping in real, large scale environments using an ultrasonic array, *International Journal Robotics Research*, 18(1), 3–19.
19. K S Chong and L Kleeman, January 1999. Mobile robot map building for an advanced sonar array and accurate odometry, *International Journal Robotics Research*, 18(1), 20–36.
20. S A Walter. The sonar ring: Obstacle detection for a mobile robot. In *Proceedings IEEE International Conference on Robotics Automation*, Raleigh, NC, March 31–April 3, 1987, pp. 1574–1578.
21. T Yata, A Ohya, and S Yuta. A fast and accurate sonar-ring sensor for a mobile robot. In *Proceedings of the IEEE International Conference on Robotics and Automation*, pp. 630–636, 1999.
22. S Fazli and L Kleeman, May 2007. Simultaneous landmark classification, localisation and map building for an advanced sonar ring, *Robotica*, 25(3), 283–296.
23. S Fazli and L Kleeman, July 2006. Sensor design and signal processing for an advanced sonar ring, *Robotica*, 24(4), 433–446.
24. L Kleeman and A Ohya, The design of a transmitter with a parabolic conical reflector for a sonar ring, *Australasian Conference on Robotics and Automation*, Auckland, New Zealand, December 2006.

27

Robust Exact Observation and Identification via High-Order Sliding Modes

Leonid Fridman
National Autonomus University of Mexico

Arie Levant
Tel Aviv University

Jorge Angel Davila Montoya
National Polytechnic Institute

27.1 Introduction

Consider the dynamic system

$$\dot{x} = f(x), \quad x \in \mathbb{R}^n. \tag{27.1}$$

Let a constraint be given by the equation $\sigma(x) = 0$, $\sigma: \mathbb{R}^n \to \mathbb{R}$, with $\sigma(x)$ being a smooth constraining function. It is also supposed that total time derivatives of σ along the trajectories exist and are single-valued functions of x, which is not trivial for discontinuous dynamic systems. In other words, this means that discontinuities do not appear in the first $r - 1$ total time derivatives of the constraining function σ. Then, the rth-order sliding set we called an integral manifold of the dynamic system (27.1) in the Filippov sense [5] is defined as the intersection of the constraints

$$\sigma = \dot{\sigma} = \cdots = \sigma^{(r-1)} = 0. \tag{27.2}$$

Definition 27.1 Let the rth-order sliding set (27.2) be nonempty and assume that it is a locally integral set of (27.1) in the Filippov sense (i.e., it consists of Filippov trajectories of the discontinuous dynamical system (27.1)). The equations are understood in the Filippov sense [5] in order to provide for the possibility of using discontinuous signals in observers. Note that Filippov solutions coincide with the usual solutions when the right-hand sides are continuous. Then the corresponding motion of (27.1) satisfying (27.2) is called an rth-order sliding mode with respect to the constraining function $\sigma(x)$ if the shift operator for solutions (27.1) from this set is non-invertible.

One of the main results of the high order sliding mode (HOSM) theory is that a number of predefined standard r-sliding controllers have been developed, defined for each given relative degree r, which solve the problem of keeping $\sigma = 0$ in finite time [12–14]. Actually, such controllers only require the knowledge of the system relative degree r. The produced control is a discontinuous function of σ and of its real-time-calculated successive derivatives $\dot{\sigma},\ldots,\sigma^{(r-1)}$. In the absence of measurement noises such controllers provide for the accuracy

$$\sigma = O(\tau^r) \tag{27.3}$$

with the sampling interval τ [13]. In practice, the values of the derivatives are to be independently measured or produced by some real-time differentiator (see below).

Suppose that the control system is affine in control, and both the system and the output σ are smooth in time and state. Recall that r is the relative degree [9] if $\frac{\partial}{\partial u}\dot{\sigma} = 0,\ldots,\frac{\partial}{\partial u}\sigma^{(r-1)} = 0, \frac{\partial}{\partial u}\sigma^{(r)} \neq 0$. The following is a list of r-sliding controllers for $r = 1, 2, 3, 4$ (quasi-continuous controllers, [14]).

1. $u = -\alpha \operatorname{sign} \sigma$
2. $u = -\alpha\left(\dot{\sigma} + |\sigma|^{1/2} \operatorname{sign} \sigma\right)/\left(|\dot{\sigma}| + |\sigma|^{1/2}\right)$
3. $-\alpha\left[\ddot{\sigma} + 2\left(|\dot{\sigma}| + |\sigma|^{2/3}\right)^{-1/2}\left(\dot{\sigma} + |\sigma|^{2/3} \operatorname{sign}\sigma\right)\right]/\left[|\ddot{\sigma}| + 2\left(|\dot{\sigma}| + |\sigma|^{2/3}\right)^{1/2}\right]$
4. $\varphi = \dddot{\sigma} + 3\left[\ddot{\sigma} + \left(|\dot{\sigma}| + 0.5|\sigma|^{3/4}\right)^{-1/3}\left(\dot{\sigma} + 0.5|\sigma|^{3/4} \operatorname{sign}\sigma\right)\right]\left[|\ddot{\sigma}| + \left(|\dot{\sigma}| + 0.5|\sigma|^{3/4}\right)^{2/3}\right]^{1/2}$

   $N = |\dddot{\sigma}| + 3\left[|\ddot{\sigma}| + \left(|\dot{\sigma}| + 0.5|\sigma|^{3/4}\right)^{2/3}\right]^{1/2}, u = -\alpha\varphi/N$

The absolute value $|\alpha|$ has the sense of the control magnitude, and it has to be taken sufficiently large in order to provide for $\sigma = 0$ in finite time. The larger $|\sigma|$, the larger the region of the control effectiveness. The sign of α and of $\frac{\partial}{\partial u}\sigma^{(r)}$ should coincide. Note that the produced control is a continuous function everywhere except the r-sliding set $\sigma = \dot{\sigma} = \cdots = \sigma^{(r-1)} = 0$. Another important result of the HOSM theory is the construction of arbitrary-order robust exact differentiators with finite-time convergence [11,12]. The above asymptotics (27.3) is preserved, when such robust exact differentiator of the order $r - 1$ is applied as a standard part of the output-feedback r-sliding controller. The accuracy $|\sigma^{(i)}| = O(\varepsilon^{(r-i)/r})$ is assured if the output σ is available with an error magnitude ε [13].

27.1.1 Arbitrary-Order Robust Exact Differentiators

27.1.1.1 Recursive Form of the Differentiator

Let the input signal $f(t)$ be a function defined on $[0, \infty)$ and consisting of a bounded Lebesgue-measurable noise with unknown features, and of an unknown base signal $f_0(t)$, whose kth derivative has a known Lipschitz constant $L > 0$. The problem is to find real-time robust estimations of $\dot{f}_0(t), \ddot{f}_0(t),\ldots, f_0^{(k)}(t)$ being exact in the absence of measurement noises. This is solved by the following differentiator [12]

$$\begin{aligned}
&\dot{z}_0 = v_0, v_0 = -\lambda_k L^{1/(k+1)} |z_0 - f(t)|^{k/(k+1)} \operatorname{sign}(z_0 - f(t)) + z_1,\\
&\dot{z}_1 = v_1, v_1 = -\lambda_{k-1} L^{1/k} |z_1 - v_0|^{(k-1)/k} \operatorname{sign}(z_1 - v_0) + z_2,\\
&\vdots\\
&\dot{z}_{k-1} = v_{k-1}, v_{k-1} = -\lambda_1 L^{1/2} |z_{k-1} - v_{k-2}|^{1/2} \operatorname{sign}(z_{k-1} - v_{k-2}) + z_k,\\
&\dot{z}_k = -\lambda_0 L \operatorname{sign}(z_k - v_{k-1}).
\end{aligned} \tag{27.4}$$

Provided the parameters $\lambda_1, \lambda_2, \ldots, \lambda_k > 0$ are properly chosen, after a finite-time transient the equalities $z_i = f_0^{(i)}(t), i = 0,1,\ldots,k$, hold in the absence of input noises. Note that these equalities hold in two sliding modes. The convergence takes place for any initial conditions, and by enlarging L, it can be made arbitrarily fast. Such differentiator can be used in any feedback, trivially providing for the separation principle [1,13].

Note that the differentiator has a recursive structure. Once the parameters $\lambda_1, \lambda_2, \ldots, \lambda_{k-1}$ are properly chosen for the $(k-1)$th-order differentiator with the Lipschitz constant L, only one parameter k needs to be tuned for the kth-order differentiator with the same Lipschitz constant. The parameter k just has to be taken sufficiently large. Any $\lambda_0 > 1$ can be used to start this process.

Thus, an infinite positive sequence $\{\lambda_n\}$ can be chosen so that, for each natural k, parameters $\lambda_1, \lambda_2, \ldots, \lambda_k$ provide for the finite-time convergence of the kth-order differentiator (27.4). A possible choice for the differentiator parameters with $k \le 5$ is $\lambda_0 = 1.1$, $\lambda_1 = 1.5$, $\lambda_2 = 2$, $\lambda_3 = 3$, $\lambda_4 = 5$, $\lambda_5 = 8$, [13].

The asymptotic accuracy of the differentiator is provided by its homogeneity features [13]. Let the measurement noise be any Lebesgue-measurable function with a magnitude not exceeding ε. Then the accuracy

$$|z_i(t) - f_0^{(i)}(t)| = O(\varepsilon^{(k+1-i)/(k+1)}) \tag{27.5}$$

is obtained. That accuracy is shown to be the best one possible [10,11]. In the absence of noises, with discrete time measurements and the sampling interval τ, the accuracy

$$|z_i(t) - f_0^{(i)}(t)| = O(\tau^{k+1-i}) \tag{27.6}$$

is obtained.

27.1.1.2 Non-Recursive Form of the Differentiator

Substituting v_0 from the first equation in (27.4) into the second one, then substituting v_1 from the second equation in (27.4) into the third one, etc., the following non-recursive form of the differentiator is obtained:

$$\begin{aligned}
\dot{z}_0 &= -\tilde{\lambda}_k L^{1/(k+1)} |z_0 - f(t)|^{k/(k+1)} \operatorname{sign}(z_0 - f(t)) + z_1, \\
\dot{z}_1 &= -\tilde{\lambda}_{k-1} L^{2/(k+1)} |z_0 - f(t)|^{(k-1)/(k+1)} \operatorname{sign}(z_0 - f(t)) + z_2, \\
&\vdots \\
\dot{z}_{k-1} &= -\tilde{\lambda}_1 L^{k/(k+1)} |z_0 - f(t)|^{1/2} \operatorname{sign}(z_0 - f(t)) + z_k, \\
\dot{z}_k &= -\tilde{\lambda}_0 L \operatorname{sign}(z_0 - f(t)).
\end{aligned} \tag{27.7}$$

It is easy to see that $\tilde{\lambda}_k = \lambda_k, \tilde{\lambda}_i = \lambda_i \tilde{\lambda}_i^{i/(i+1)}, i = k-1, k-2, \ldots, 0$. In particular, $k=1$ yields $\tilde{\lambda}_0 = 1.1$, $\tilde{\lambda}_1 = 1.5$; $k = 2$ yields $\tilde{\lambda}_0 = 1.1, \tilde{\lambda}_1 = 2.12, \tilde{\lambda}_2 = 2$; $k = 3$ yields $\tilde{\lambda}_0 = 1.1, \tilde{\lambda}_1 = 3.06, \tilde{\lambda}_2 = 4.16, \tilde{\lambda}_3 = 3$; $k = 4$ yields $\tilde{\lambda}_0 = 1.1, \tilde{\lambda}_1 = 4.57, \tilde{\lambda}_2 = 9.3, \tilde{\lambda}_3 = 10.03, \tilde{\lambda}_4 = 5$; $k = 5$ yields $\tilde{\lambda}_0 = 1.1, \tilde{\lambda}_1 = 6.93, \tilde{\lambda}_2 = 21.4, \tilde{\lambda}_3 = 34.9$, $\tilde{\lambda}_4 = 26.4, \tilde{\lambda}_5 = 8$. Note that these parameter values can be rounded to two meaningful digits without any loss of convergence.

27.1.1.3 Implementation Issues

Let the input signal $f_0(t)$ be corrupted by a noise with a maximal magnitude 10^{-6}, and $|f_0^{(6)}| \leq 1$. Suppose that we need to estimate five derivatives of $f_0(t)$. The first possibility is to successively use the first-order differentiator. Then we need to ensure that all derivatives of order higher than one be bounded by known constants. Suppose that all of them are bounded by one in their absolute value. Then the error of the first derivative estimation will be at least $10^{-6/2} = 10^{-3}$, and the error of the second derivative will be of order $10^{-3/2}$. Continuing this reasoning, an error on the fifth derivative estimation of order $10^{-3/16} = 0.65$ is obtained, which is rather bad. Contrary to this, using the fifth-order differentiator, obtain an error of order $10^{-6/6} = 0.1$, which is still significant but cannot be improved [10,11]. Note that in this case, no conditions are needed for the lower derivatives of $f_0(t)$. Thus one needs a special differentiator for each differentiator order.

Differentiator (27.7) needs to be realized using computer chips. Since the dynamic system is discontinuous, standard integration methods like the popular Runge–Kutta methods are ineffective, and will not allow for asymptotic accuracy (27.5), (27.6). Indeed, they are based on the local Taylor expansion of the right-hand side, which does not exist in our case.

The only appropriate way to integrate system (27.7) is to use the simple Euler method. Recall that provided the dynamic system be of the form

$$\dot{z} = \zeta(t,z), \quad z \in \mathbb{R}^{k+1}, \tag{27.8}$$

and that the time integration step be $h > 0$, then an Euler integration step of (27.8) is

$$\hat{z}(t+h) = \hat{z}(t) + h\zeta(t,\hat{z}(t)), \tag{27.9}$$

where $\hat{z}(t)$ is the Euler approximation of a solution of (27.8).

During each integration step (27.9), the sampled value of $f(t)$ should obviously be constant. Therefore, in most cases it is reasonable to require $h \leq \tau$, where τ is the sampling interval of $f(t)$. Often $h = \tau$ is used. That is the case, for example, in image-processing applications, when inevitably $h = \tau = 1$. With large L and $h \ll \tau$ differentiator convergence is very fast, and the sampled input $f(t)$ actually turns out to be piece-wise constant. As a result, $z_1,\ldots, z_k$ converge to zero on each sampling interval, producing significant chattering of the differentiator outputs. In that case, $h = \tau$ is a good choice.

It is convenient to directly implement the recursive form (27.4) of the differentiator. A program code can be written in that case, which is valid for any k. Also in that case the Euler scheme is to be used. In order to avoid integration mistakes, one has to first calculate vector $v(t)$ as a function of the current differentiator state $\hat{z}(t)$ and the current last sampled value of f. Then the calculated right-hand sides of (27.4) are used for the integration step of the form (27.9).

While the differentiator converges with any initial value of $\hat{z}$, it is reasonable to choose the initial value of $\hat{z}_0$ equal to the initial sampled value of the input f. The initial values of the other coordinates of $\hat{z}$ can be taken as zero or, alternatively, can be numerically evaluated as the corresponding time derivatives. In the latter case, the differentiator transient can be completely avoided with small sampling noises.

Due to the natural sensitivity of the high-order differentiation to noises, the number of meaningful digits turns to be a restriction for computer implementation. In particular, one cannot observe the accuracy (27.6) of the fifth derivative if the internal computer accuracy is determined by the popular long double precision.

27.1.1.4 Examples

27.1.1.4.1 Signal Processing: Simulation of Real-Time Differentiation

The following is the fifth-order differentiator in the recursive form (27.4):

$$\dot{z}_0 = v_0, \quad v_0 = -8L^{1/6}|z_0 - f(t)|^{5/6}\,\mathrm{sign}(z_0 - f(t)) + z_1,$$

$$\dot{z}_1 = v_1, \quad v_1 = -5L^{1/5}|z_1 - v_0|^{4/5}\,\mathrm{sign}(z_1 - v_0) + z_2,$$

$$\dot{z}_2 = v_2, \quad v_2 = -3L^{1/4}|z_2 - v_1|^{3/4}\,\mathrm{sign}(z_2 - v_1) + z_3,$$

$$\dot{z}_3 = v_3, \quad v_3 = -2L^{1/3}|z_3 - v_2|^{2/3}\,\mathrm{sign}\,(z_3 - v_2) + z_4,$$

$$\dot{z}_4 = v_4, \quad v_4 = -1.5L^{1/2}|z_4 - v_3|^{1/2}\,\mathrm{sign}(z_4 - v_3) + z_5,$$

$$\dot{z}_5 = -1.1L\mathrm{sign}\,(z_5 - v_4); \quad f^{(6)} \le L.$$

Note that it recursively contains differentiators of lower orders. For example, the last two equations describe a first-order differentiator with the input v_3. It is applied with $L = 1$ for differentiation of the function

$$f(t) = \sin 0.5t + \cos 0.5t, \quad |f^{(6)}| \le L = 1.$$

The initial values of the differentiator variables are taken to be zero in order to demonstrate the transient process. Choosing the initial value of z_0 equal to the initial sampled value of $f(t)$ would significantly shorten the transient. Convergence of the differentiator is demonstrated in Figure 27.1. Note that the fifth derivative estimation z_5 is not exact due to software restrictions (the number of digits of long double-precision variables). Higher order differentiation requires special software development. With $h = \tau = 5 \times 10^{-4}$ the accuracies $|z_0 - f_0| \le 1.46 \times 10^{-13}$, $|z_1 - \dot{f}_0(t)| \le 7.16 \times 10^{-10}$, $|z_2 - \ddot{f}_0(t)| \le 9.86 \times 10^{-7}$, $|z_3 - \dddot{f}_0(t)| \le 3.76 \times 10^{-4}$, $|z_4 - f_0^{(4)}| \le 0.0306$, $|z_5 - f_0^{(5)}| \le 0.449$ were obtained.

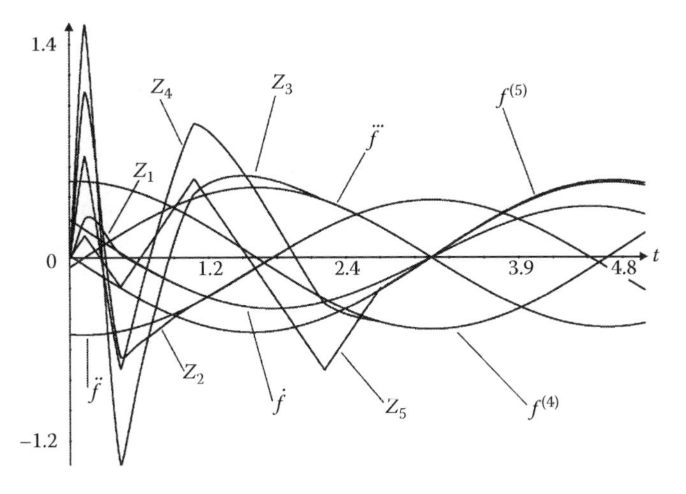

FIGURE 27.1 Fifth-order differentiation.

27.1.1.4.2 Output-Feedback Control

Consider a simple kinematic model of car control

$$\dot{x} = v\cos\varphi, \quad \dot{y} = v\sin\varphi, \quad \dot{\varphi} = \frac{v}{l}\tan\theta, \quad \dot{\theta} = u,$$

where

x and y are Cartesian coordinates of the rear-axle middle point
φ is the orientation angle
v is the longitudinal velocity
l is the length between the two axes
θ is the steering angle (i.e., the real input) (Figure 27.2)

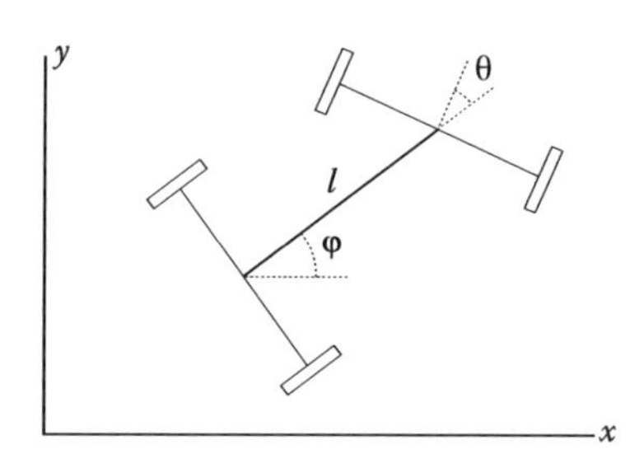

FIGURE 27.2 Kinematic car model.

The task is to steer the car from a given initial position to the trajectory $y = g(x)$, where $g(x)$ and y are assumed to be available in real time.

Define $\sigma = y - g(x)$. Let v = const = 10 m/s, l = 5 m, $x = y = \varphi = \theta = 0$ at $t = 0$, $g(x) = 10\sin(0.05x) + 5$. The relative degree of the system is 3 and the quasi-continuous 3-sliding controller (Section 27.1) solves the problem. The control magnitude $\alpha = 1$ is found by simulation and the parameter $L = 400$ of the second-order differentiator is chosen deliberately large in order to provide for a better performance in the presence of measurement errors ($L = 25$ is also sufficient but is much worse with sampling noises). The control was applied only from $t = 1$, in order to provide some time for the differentiator convergence.

The resulting output-feedback controller is

$$u = -\frac{\left[z_2 + 2\left(|z_1| + |z_0|^{2/3}\right)^{-1/2}\left(z_1 + |z_0|^{2/3}\operatorname{sign} z_0\right)\right]}{\left[|z_2| + 2\left(|z_1| + |z_0|^{2/3}\right)^{1/2}\right]},$$

$$\dot{z}_0 = v_0, \quad v_0 = -14.7361|z_0 - \sigma|2/3\operatorname{sign}(z_0 - \sigma) + z_1,$$

$$\dot{z}_1 = v_1, \quad v_1 = -30|z_1 - v_0|^{1/2}\operatorname{sign}(z_1 - v_0) + z_2,$$

$$\dot{z}_2 = -440\operatorname{sign}(z_2 - v_1).$$

Integration was carried out according to Euler's method, the sampling step being equal to the integration step $\tau = 10^{-4}$. In the absence of noises, the 3-sliding tracking accuracies $|\sigma| \leq 5.4 \times 10^{-7}$, $|\dot{\sigma}| \leq 2.4 \times 10^{-4}, |\ddot{\sigma}| \leq 0.042$ were obtained. After changing to $\tau = 10^{-5}$, the accuracies $|\sigma| \leq 5.6 \times 10^{-10}$, $|\dot{\sigma}| \leq 1.4 \times 10^{-5}, |\ddot{\sigma}| \leq 0.0042$ were in turn obtained, which mainly corresponds to the asymptotics (27.3). The car trajectory, 3-sliding tracking errors, steering angle θ, and its derivative u are shown in Figure 27.3a through d, respectively [13]. It is seen from Figure 27.3c that control u remains continuous until the very entrance into the 3-sliding mode. The steering angle θ remains rather smooth and is quite feasible.

In the presence of output noise with magnitude 0.01 m, the tracking accuracies $|\sigma| \leq 0.02$, $|\dot{\sigma}| \leq 0.14, |\ddot{\sigma}| \leq 1.3$ were obtained. With a measurement noise of magnitude $0.1m$ the accuracies changed to $|\sigma| \leq 0.20, |\dot{\sigma}| \leq 0.62, |\ddot{\sigma}| \leq 2.8$, which corresponds to the theoretical asymptotic accuracy $|\sigma^{(i)}| = O(\varepsilon^{(r-i)/r})$, $r = 3$. The performance of the controller with the measurement error magnitude 0.1 m is shown in Figure 27.4. It is seen from Figure 27.4c that control u is a continuous function of t. The steering-angle vibrations have a magnitude of about 7° and frequency 1, which is also quite feasible. The performance does not change significantly when the noise frequency varies in the range 100–100,000.

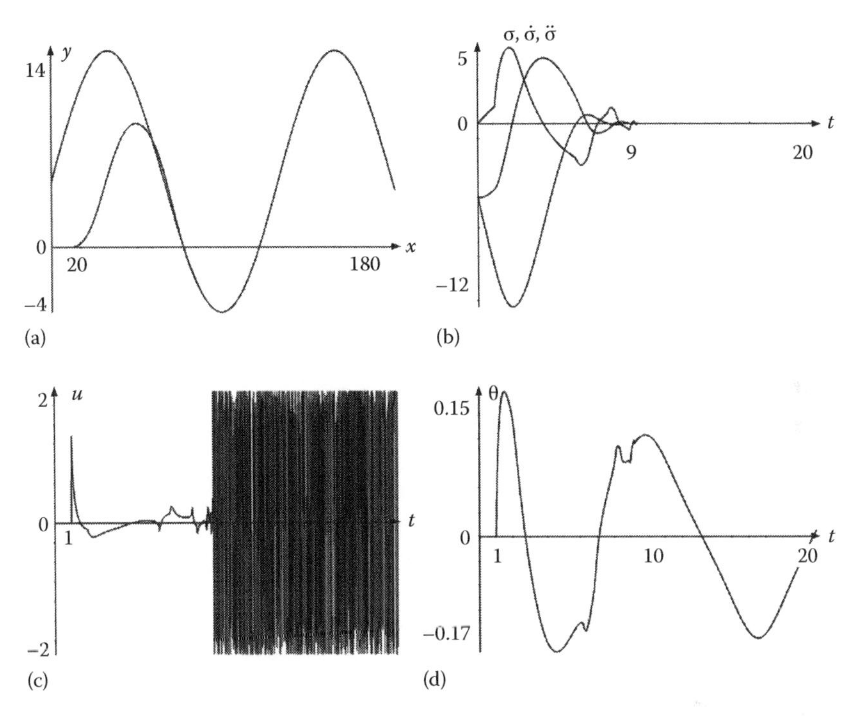

FIGURE 27.3 Noise magnitude $\varepsilon = 0$. (a) Car trajectory, (b) 3-sliding deviations, (c) steering-angle derivative, and (d) steering angle.

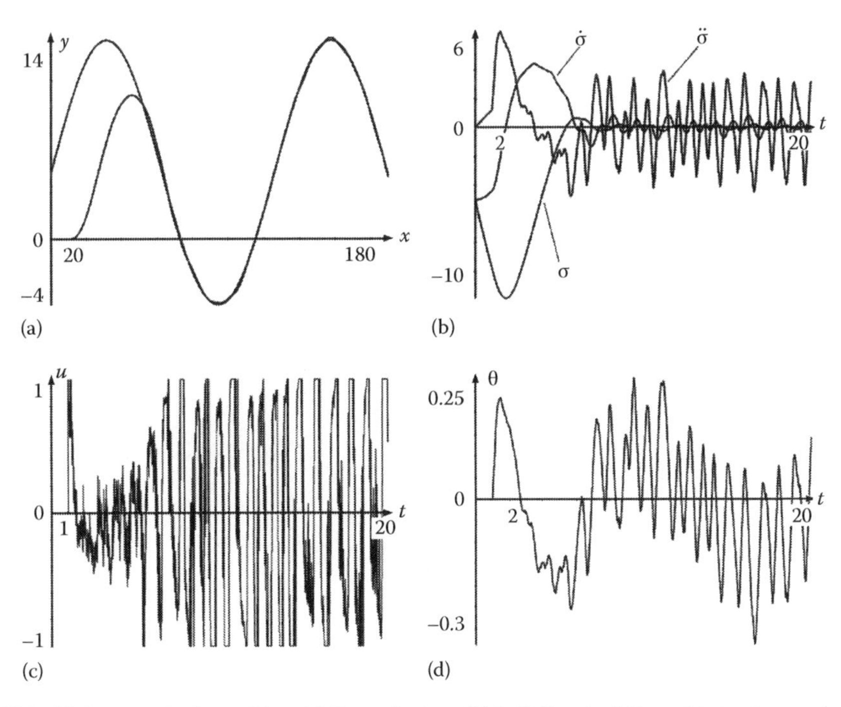

FIGURE 27.4 Noise magnitude $\varepsilon = 0.1$ m. (a) Car trajectory, (b) 3-sliding deviations, (c) steering-angle derivative, and (d) steering angle.

27.2 High-Order Sliding-Modes State Observation

27.2.1 Single Output–Single Unknown Input Case

Consider an LTI system

$$\dot{x} = Ax + Bu + D\zeta(t), \quad D \neq 0,$$
$$y = Cx, \tag{27.10}$$

where

$x \in \mathbb{R}^n$, $y \in \mathbb{R}$ are the system state and the output
$\zeta \in \mathbb{R}$ is the unknown input (disturbance)
$u \in \mathbb{R}$ is the known control and the known matrices A, B, C, D have suitable dimensions

It is assumed also that all considered inputs allow the existence and extension of solutions to the whole semi-axis $t \geq 0$.

The task is to build an observer providing for asymptotic (preferably finite-time convergent and exact) estimation of the states and the unknown input. Obviously, it can be assumed without loss of generality that the known input u is equal to zero (i.e., $u(t) = 0$).

The presence of unknown inputs makes introducing the concept of strong observability, which is a generalization of the classical observability concept, necessary.

System (27.10) is called (strongly) observable if for any initial state $x(0)$ and $\zeta(t) \equiv 0$ (any input $\zeta(t)$), $y(t) \equiv 0$ with $\forall t \geq 0$ implies that also $x \equiv 0$ (see [8]).

The relation of the relative degree with the state reconstruction problem in the presence of unknown inputs was studied in [6]. Here, the fundamental relationship is presented.

System (27.10) is strongly observable if and only if the output of system (27.10) has relative degree n with respect to the unknown input $\zeta(t)$.

27.2.1.1 State Reconstruction

Let us make a couple of assumptions about system (27.10).

System (27.10) has relative degree n with respect to the unknown input $\zeta(t)$ (this assumption ensures that the system be strongly observable).

The unknown input $\zeta(t)$ is a bounded Lebesgue-measurable function, $|\zeta(t)| \leq \zeta^+$.

Under last assumptions, the observer is built in the form

$$\dot{z} = Az + Bu + L(y - Cz), \tag{27.11}$$

$$\hat{x} = z + Kv, \tag{27.12}$$

$$\dot{v} = W(y - Cz, v), \tag{27.13}$$

where z, $\hat{x} \in \mathbb{R}^n$, $\hat{x}$ is the estimation of x, and the column matrix $L = [l_1, l_2, \ldots, ln]^T \in \mathbb{R}^n$ is a correction factor chosen such that the eigenvalues of the matrix $A - LC$ have negative real parts. Such L exists due to the fact that system is strongly observable. The selection of matrix K, vector v, and the nonlinear discontinuous function W will be presented in this section.

The proposed observer is actually composed of two parts. Equation 27.11 is a traditional Luenberger observer providing for the boundedness of the difference $z - x$ in the presence of the unknown bounded input ζ. System (27.13) is based on high-order sliding modes and ensures finite-time convergence of the resulting estimation error to zero.

Note that in the simplest case, when $n = 1$, the only observable coordinate coincides with the measured output and, therefore, only the input estimation problem makes sense, requiring additional assumptions. The latter problem is considered in the next section. Thus assume that $n > 1$.

Since the pair (C, A) is observable, arbitrary stable values are assigned to the eigenvalues of matrix $(A - LC)$, choosing an appropriate column gain matrix L [3]. Obviously, the pair $(C, A - LC)$ is also observable. Its observability matrix

$$\tilde{P} = \begin{bmatrix} C \\ C(A-LC) \\ C(A-LC)^2 \\ \vdots \\ C(A-LC)^{n-1} \end{bmatrix} \tag{27.14}$$

is not singular. Set the gain matrix $K = \tilde{P}^{-1}$ and assign

$$\hat{x} = z + \tilde{P}^{-1}v. \tag{27.15}$$

The nonlinear part of observer (27.13) is chosen using the high-order sliding mode differentiator described before and given by the following equation:

$$\begin{aligned}
\dot{v}_1 &= w_1 = -\alpha_n M^{1/n} |v_1 - y + Cz|^{(n-1)/n} \operatorname{sign}(v_1 - y + Cz) + v_2, \\
\dot{v}_2 &= w_2 = -\alpha_{n-1} M^{1/(n-1)} |v_2 - w_1|^{(n-2)/(n-1)} \operatorname{sign}(v_2 - w_1) + v_3, \\
&\vdots \\
\dot{v}_{n-1} &= w_{n-1} = -\alpha_2 M^{1/2} |v_{n-1} - w_{n-2}|^{1/2} \operatorname{sign}(v_{n-1} - w_{n-2}) + v_n, \\
\dot{v}_n &= -\alpha_1 M \operatorname{sign}(v_n - w_{n-1}),
\end{aligned} \tag{27.16}$$

where v_i, z_i, and w_i are the components of the vectors v, $z \in \mathbb{R}^n$, and $w \in \mathbb{R}^{n-1}$, respectively. Parameter M is chosen sufficiently large, in particular $M > |d|\ \zeta^+$, where $d = CA^{n-1}D$. The constants α_i are chosen recursively sufficiently large as in [12].

Recall that $x_O = Px$ is the canonical observation coordinates vector, and $e_O = P(\hat{x} - x)$ is the canonical observation error.

Summary: If system (27.10) is strongly observable and the unknown input is bounded then, with properly chosen α_j and M sufficiently large, the state x of the system is estimated exactly in finite time by the observer (27.11), (27.14), and (27.15).

Noise effects: Assume once again that the system is strongly observable and that the unknown input is bounded. Let the output be measured with a noise which is a Lebesgue-measurable function of time with a maximal magnitude ε. Then with properly chosen α_j and M sufficiently large, the system state x is estimated in finite time by the observer (27.11), (27.14), and (27.15). With sufficiently small ε the observation errors $e_{Oi} = \hat{x}_{Oi} - x_{Oi} = CA^{i-1}(\hat{x}_i - x_i)$ are of the order $\varepsilon^{(n-i+1)/n}$, i.e., the observation errors satisfy the inequalities $|e_{Oi}| \le \gamma_i \varepsilon^{(n-i+1)/n}$ for some constants $\gamma_i > 0$ depending only on the observer, system parameters, and the input upper bound. The accuracy of the order of $\varepsilon^{1/n}$ is obtained in non-canonical coordinates due to the mix of coordinates.

Let now the output γ be sampled at discrete times with the constant time step τ, $t \in [t_j, t_{j+1})$, $t_{j+1} - t_j = \tau$. Substituting $\bar{e}_1(t_j) = y(t_j) - Cz(t_j)$ for the term $\bar{e}_1 = y - Cz$ in (27.11), (27.13), a discrete-sampling observer is obtained. Note that the simultaneous sampling of the components of $\bar{e}_1$ is important.

Discretization effects: Let the system be strongly observable and the unknown input be bounded. Also, let the observer designed with a proper selection of parameters and the output be measured without measurement errors at discrete sampling times with a sufficiently small sampling interval τ. Then, after a finite-time transient, the canonical observation errors $e_{Oi} = CA^{i-1}(x_i - x_i)$ are of the order τ^{n-i+1}.

27.2.1.2 Identification of the Unknown Input

For the reconstruction of the unknown input $\zeta(t)$, an additional smoothness assumption is required. Let $\zeta(t)$ be a bounded function with successive derivatives up to order k bounded by the same constant ζ_1^+. The kth derivative is a Lipschitzian function with a Lipshitz constant not exceeding ζ_1^+. Thus, $\zeta^{(k+1)}(t)$ exists almost everywhere and it is a bounded Lebesgue-measurable function, $|\zeta^{(k+1)}(t)| \leq \zeta_1^+$.

If the unknown input is bounded and has a kth Lipschitz derivative, the observer with an extended version of the sliding mode term can be applied to reconstruct the state and the unknown input. Let now $v \in \mathbb{R}^{n+k}$ satisfy the nonlinear differential equation (27.13) of the form

$$\begin{aligned}
\dot{v}_1 &= w_1 = -\alpha_{n+k+1}M^{1/(n+k+1)}\,|v_1 - y(t) + Cz|^{(n+k)/(n+k+1)}\,\text{sign}(v_1 - y(t) + Cz) + v_2,\\
\dot{v}_2 &= w_2 = -\alpha_{n+k}M^{1/(n+k)}\,|v_2 - w_1|^{(n+k-1)/(n+k)}\,\text{sign}(v_2 - w_1) + v_3,\\
&\vdots\\
\dot{v}_n &= w_n = -\alpha_{k+2}M^{1/(k+2)}\,|v_n - w_{n-1}|^{(k+1)/(k+2)}\,\text{sign}(v_n - w_{n-1}) + v_{n+1},\\
&\vdots\\
\dot{v}_{n+k} &= w_{n+k} = -\alpha_2 M^{1/2}\,|v_{n+k-1} - w_{n+k-2}|^{1/2}\,\text{sign}(v_{n+k-1} - w_{n+k-2}) + v_{n+k},\\
\dot{v}_{n+k+1} &= -\alpha_1 M\,\text{sign}(v_{n+k+1} - w_{n+k}),
\end{aligned} \tag{27.17}$$

where M is a sufficiently large constant. As previously, (27.17) has a recursive form, and the parameters α_i are chosen in the same way [12]. In particular, one of the possible choices is $\alpha_1 = 1.1$, $\alpha_2 = 1.5$, $\alpha_3 = 2$, $\alpha_4 = 3$, $\alpha_5 = 5$, $\alpha_6 = 8$, which is sufficient for $n + k \leq 5$. In any computer realization one has to calculate the internal auxiliary variables w_j, $j = 1, \ldots, n + k$, using only the simultaneously sampled current values of y, z_1, and v_j.

The equality $\bar{e} = \omega$ is established in finite time, where ω is the truncated vector

$$\omega = (v_1, \ldots, v_n)^T. \tag{27.18}$$

Thus, the corresponding observer equation

$$\hat{x} = z + \tilde{P}^{-1}\omega \tag{27.19}$$

is obtained instead of (27.12) and $\tilde{P}$ is defined by (27.14). The estimation of the input ζ is defined as

$$\begin{aligned}
\hat{\zeta} &= \frac{1}{d}\left(v_{n+1} - (a_1 v_1 + a_2 v_2 + \cdots + a_n v_n)\right),\\
s^n - a_n s^{n-1} &- \cdots - a_1 = (-1)^n \det(A - LC - sI)
\end{aligned} \tag{27.20}$$

where the second line defines the characteristic polynomial of the matrix $A - LC$.

Summary: If the system is strongly observable and the unknown input is bounded and smooth (in the terms explained above), then with properly chosen α_j and M sufficiently large, the state x_O of the system is estimated exactly in finite time by the observer (27.11) and (27.17) through (27.20).

Noise effect: Let the output be measured with a noise that is a Lebesgue-measurable function of time with a maximal magnitude ε. With any sufficiently small ε the canonical observation errors e_{Oi} obtained are of order $\varepsilon^{(n-i+k+2)/(n+k+1)}$. The estimation error of the input ζ is of order $\varepsilon^{k/(n+k+1)}$.

27.2.2 Multiple Outputs–Multiple Unknown Inputs Case

Consider the system with m outputs and the same number of unknown inputs

$$\dot{x} = Ax + Bu + D\zeta, \quad y = Cx, \tag{27.21}$$

where $x \in \mathbb{R}^n$, $u \in \mathbb{R}^q$, ζ, $y \in \mathbb{R}^m$, the matrices are of the suitable dimensions. The observation matrix for system (27.21) takes on the form

$$P = \begin{bmatrix} c_1 \\ c_1A \\ \vdots \\ c_1A^{n-1} \\ \vdots \\ c_m \\ c_mA \\ \vdots \\ c_mA^{n-1} \end{bmatrix}$$

where c_i, $i = 1, \ldots, m$ are the rows of matrix C. Recall that the vector output $y = Cx$ is said to have vector relative degree $(r_1, \ldots, r_m)$ with respect to the input ζ, if

$$c_iA^sD_j = 0 \quad i,j = 1,2,\ldots,m, \quad s = 0,1,\ldots,r_i - 2 \tag{27.22}$$

and

$$\det Q \neq 0, \quad Q = \begin{bmatrix} c_1A^{r_1-1}D_1 & \ldots & c_1A^{r_1-1}D_m \\ & \ldots & \\ c_mA^{r_m-1}D_1 & \ldots & c_mA^{r_m-1}D_m \end{bmatrix}. \tag{27.23}$$

An important property of (27.21), when the output y has vector relative degree $(r_1, \ldots, r_m)$ with respect to the unknown input ζ, is that the vectors $c_1,\ldots,c_1A^{r_1-1},\ldots,c_m,\ldots,c_mA^{r_m-1}$ are linearly independent.

Recall that $s_0 \in \mathbb{C}$ is called an invariant zero of the triplet $\{A, D, C\}$ if rank $R(s_0) < n + \text{rank}(D)$, where R is the Rosenbrock matrix of system (27.21)

$$R = \begin{bmatrix} sI - A & -D \\ C & 0 \end{bmatrix}. \tag{27.24}$$

The system (27.21) is strongly observable if the triple $\{A, C, D\}$ has no invariant zeros [8].

The complete relative degree $r = r_1 + \cdots + r_m$ with respect to the unknown input ζ does not exceed the dimension of the system n.

To design the observer, assume the unknown input (perturbation) $\zeta_i(t)$ is a bounded function, $|\zeta_i(t)| \le \zeta_i^+$, with bounded $(r_M - r_i + k)$ successive derivatives, the last one being Lipschitzian. That means that its derivative $\zeta_i^{(r_M - r_i + k+1)}(t)$ exists almost everywhere and it is a bounded Lebesgue-measurable function, $|\zeta_i^{(r_M - r_i + k+1)}(t)| \le \zeta_{1i}^+$. Other derivatives are also supposed bounded by the same constant.

Choose a matrix L such that $A - LC$ be a Hurwitz matrix. As previously, the linear Luenberger part of the observer takes on the form

$$\dot{z} = Az + Bu + L(y - Cz). \tag{27.25}$$

The corresponding error system is

$$\dot{e} = (A - LC)e + D\zeta,$$

where $e = x - z$. Then in some new coordinates $\begin{bmatrix} e_C^T & e_N^T \end{bmatrix}^T = [T_C^T T_N^T]^T e = Te$. This system takes on the standard form

$$\begin{aligned} \dot{e}_C &= A_C e_C + A_{CN} e_N + D_C \zeta \\ \dot{e}_N &= A_{NC} e_C + A_N e_N \end{aligned}, \quad y = C_C e_C,$$

where the canonical observation errors $e_{Ci} = (e_{Ci1}^T, \ldots, e_{Cir_i}^T)^T \in \mathbb{R}^{r_i}$ are calculated as $e_{Cij} = c_i A^{j-1}(x - z)$, and $e_N \in \mathbb{R}^{n-r}$. The corresponding matrices have the form

$$\begin{bmatrix} A_C & A_{CN} \\ A_{NC} & A_N \end{bmatrix} = T(A - LC)T^{-1}, \tag{27.26}$$

$$A_C = \begin{bmatrix} A_{11} & \cdots & A_{1,m} \\ & \ddots & \\ A_{m,1} & \cdots & A_{mm} \end{bmatrix}, \quad A_{CN} = \begin{bmatrix} A_{C1} \\ \vdots \\ A_{Cm} \end{bmatrix} \quad D_C = \begin{bmatrix} D_{11} & \ldots & D_{m1} \\ & \ddots & \\ D_{1m} & \ldots & D_{mm} \end{bmatrix}, \tag{27.27}$$

$$A_{ii} = \begin{bmatrix} 0 & 1 & \cdots & 0 \\ \vdots & \vdots & & \vdots \\ 0 & 0 & \cdots & 1 \\ a_{ii,1} & a_{ii,2} & \cdots & a_{ii,r_i} \end{bmatrix}, \tag{27.28}$$

$$A_{i,j} = \begin{bmatrix} 0 & 0 & \cdots & 0 \\ \vdots & & \vdots & \vdots \\ 0 & 0 & \cdots & 0 \\ a_{ij,1} & a_{ij,2} & \cdots & a_{ij,r_j} \end{bmatrix}, \quad i \ne j, \quad A_{C,j} = \begin{bmatrix} 0 & 0 & \cdots & 0 \\ \vdots & & \vdots & \vdots \\ 0 & 0 & \cdots & 0 \\ a_{Cj,1} & a_{Cj,2} & \cdots & a_{Cj,n-r} \end{bmatrix}, \tag{27.29}$$

$$y_j = C_{Cj} e_{Cj}, \quad C_{Ci} = \begin{bmatrix} 1 & 0 & \cdots & 0 \end{bmatrix}, \quad D_{ij} = \begin{bmatrix} 0 & \cdots & 0 & d_{ij} \end{bmatrix}^T. \tag{27.30}$$

Here $d_{ij} = c_i A^{r_i - 1} D_j$ and other matrices do not have any specific form. Matrix A_N is Hurwitz. The nonlinear observer part takes on the form

$$\dot{v}_i = W_i(v_i, y_i(t) - C_i z), \quad v = (v_1, \ldots, v_m)^T \in \mathbb{R}^{m(r_M + k)}. \tag{27.31}$$

The auxiliary variable v_i is a solution of the discontinuous vector differential equation

$$\begin{aligned}
\dot{v}_{i,1} = w_{i,1} &= -\alpha_{r_M+k+1} M_i^{1/(r_M+k+1)} \left| v_{i,1} - y_i(t) + C_i z \right|^{(r_M+k)/(r_M+k+1)} \\
&\quad \times \operatorname{sign}(v_{i,1} - y_i(t) + z_{i,1}) + v_{i,2}, \\
\dot{v}_{i,2} = w_{i,2} &= -\alpha_{r_M+k} M_i^{1/(r_M+k)} \left| v_{i,2} - w_{i,1} \right|^{(r_M+k-1)/(r_M+k)} \operatorname{sign}(v_{i,2} - w_{i,1}) + v_{i,3}, \\
&\ \ \vdots \\
\dot{v}_{i,r_M+k} = w_{i,r_M+k-1} &= -\alpha_2 M_i^{1/2} \left| v_{i,r_M+k} - w_{i,r_M+k-1} \right|^{1/2} \\
&\quad \times \operatorname{sign}(v_{i,r_M+k-1} - w_{i,r_M+k-2}) + v_{i,r_M+k}, \\
\dot{v}_{i,r_M+k+1} &= -\alpha_1 M_i \operatorname{sign}(v_{i,r_M+k} - w_{i,r_M+k}),
\end{aligned} \tag{27.32}$$

where M_i are sufficiently large constants, and the parameters α_i are chosen in the same way as in [12].

Denote

$$\begin{aligned}
&\omega_{1i} = (v_{i,1}, \ldots, v_{i,r_i})^T, \quad \omega_1 = (\omega_{11}^T, \ldots, \omega_{1m}^T)^T \in \mathbb{R}^r, \quad \omega_2 \in \mathbb{R}^{n-r}, \\
&\bar{v} = (v_{1,r_1+1}, \ldots, v_{m,r_m+1})^T.
\end{aligned} \tag{27.33}$$

then the system for the observation of e_N takes on the form

$$\dot{\omega}_2 = A_{CN}\omega_1 + A_N\omega_2. \tag{27.34}$$

The coordinates are estimated as

$$\hat{x} = z + T^{-1} \begin{bmatrix} \omega_1 \\ \omega_2 \end{bmatrix}. \tag{27.35}$$

The unknown inputs ζ_i, $i = 1, 2, \ldots, m$, are estimated by $\hat{\zeta}_i$,

$$\hat{\zeta} = D_C^{-1} \left(\bar{v} - \begin{bmatrix} \bar{A}_C & \bar{A}_N \end{bmatrix} \begin{bmatrix} \omega_1 \\ \omega_2 \end{bmatrix} \right) \tag{27.36}$$

where the matrices $\bar{A}_C$, $\bar{A}_N$ are composed of the bottom rows of $A_{i,j}$ and $A_{C,j}$ composing the block matrices A_C and A_{CN}, respectively:

$$\begin{aligned}
\bar{A}_C &= \begin{bmatrix} a_{11} & \cdots & a_{1m} \\ & \ddots & \\ a_{m1} & \cdots & a_{mm} \end{bmatrix}, \quad a_{ij} = (a_{ij,1}, \ldots, a_{ij,r_i})^T, \\
\bar{A}_N &= \begin{bmatrix} a_{C1,1} & \cdots & a_{C1,n-r} \\ & \ddots & \\ a_{C,m,1} & \cdots & a_{C,m,n-r} \end{bmatrix}.
\end{aligned} \tag{27.37}$$

Recall once more that $x_O = P_n x$ is the vector of the canonical observation coordinates (not all of them are independent). Vector $e_O = P_n(\hat{x} - x)$ is naturally called the canonical observation error. Its linearly independent part naturally contains the coordinates $e_{Oij} = C_i A^{j-1}(\hat{x} - x), j = 1, \ldots, r_i$.

Summary: With a properly chosen parameter observer (27.25) through (27.37) provides, after finite-time transient, for the accuracies

$$|e_{Oij}| \sim \varepsilon^{(r_M+k+2-j)/(r_M+k+1)}, \quad j=1,2,\ldots,r_i, \quad \|\hat{x}_N - x_N\| \sim \varepsilon^{2/(r_M+k+1)}$$

with ε being the magnitude of the measurement noise, and for the accuracies

$$|e_{Oij}| \sim \tau^{(r_M+k+2-j)}, \quad j=1,2,\ldots,r_i, \quad \|\hat{x}_N - x_N\| \sim \tau^{k+2}$$

with the sampling interval τ in the absence of noises. All other coordinates are estimated asymptotically with accuracies $\varepsilon^{(k+2)/(r_M+k+1)}$ and τ^{k+2}, respectively. The inputs ζ_i are asymptotically estimated with accuracies of order $\varepsilon^{(r_M+k+1-r_i)/(r_M+k+1)}$ and $\tau^{(r_M+k+1-r_i)}$, respectively. Finite-time convergence of the coordinates e_C is achieved and asymptotically exact convergence to zero of the coordinates e_N and the unknown-input estimations are obtained with continuous measurements and $\varepsilon = 0$.

27.2.3 Example

Consider the system

$$\begin{aligned} \dot{x} &= Ax + D\zeta(t) + Bu, \\ y &= Cx \end{aligned} \tag{27.38}$$

with matrices

$$A = \begin{bmatrix} 0 & 1 & 0 & 0 \\ 0 & 0 & 1 & 0 \\ 0 & 0 & 0 & 1 \\ 6 & 5 & -5 & -5 \end{bmatrix},$$

$$B = D = \begin{bmatrix} 0 & 0 & 0 & 1 \end{bmatrix}^T, \quad C_O = \begin{bmatrix} 1 & 0 & 0 & 0 \end{bmatrix}$$

and initial conditions $x_O(0) = \begin{bmatrix} 1 & 0 & 1 & 1 \end{bmatrix}$. The system has eigenvalues −3, −2, −1, 1. The system is of relative degree $r = 4$ with respect to the unknown input and, in consequence, is strongly observable. Note that A is not stable. The "unknown" input

$$\zeta = \cos 0.5t + 0.5\sin t + 0.5$$

is taken, being obviously a bounded smooth function with bounded derivatives. For the sake of simplicity, it is assumed that $u = 0$.

27.2.3.1 State Estimation

In this section, only the state observation problem will be considered. The correction factor $L=\begin{bmatrix}5 & 5 & 5 & 5\end{bmatrix}^T$ provides for the eigenvalues –1, –2, –3, –4 of the matrix $A - LC$. The gain matrix K in (27.12) is chosen as

$$K = \begin{bmatrix} C \\ C(A-LC) \\ C(A-LC)^2 \\ C(A-LC)^3 \end{bmatrix}^{-1} = \begin{bmatrix} 1 & 0 & 0 & 0 \\ 5 & 1 & 0 & 0 \\ 5 & 5 & 1 & 0 \\ 5 & 5 & 5 & 1 \end{bmatrix}.$$

The parameters of the nonlinear observer part (27.13) are as follows: $\alpha_1 = 1.1$, $\alpha_2 = 1.5$, $\alpha_3 = 2$, $\alpha_4 = 3$, $M = 2$. The nonlinear part of the observer is designed as

$$\begin{aligned}
\dot{v}_1 &= w_1 = -3 \cdot 2^{1/4} \,|\, v_1 - y(t) + Cz \,|^{3/4} \operatorname{sign}(v_1 - y(t) + Cz) + v_2, \\
\dot{v}_2 &= w_2 = -2 \cdot 2^{1/3} \,|\, v_2 - w_1 \,|^{2/3} \operatorname{sign}(v_2 - w_1) + v_3, \\
\dot{v}_3 &= w_3 = -1.5 \cdot 2^{1/2} \,|\, v_3 - w_2 \,|^{1/2} \operatorname{sign}(v_3 - w_2) + v_4, \\
\dot{v}_4 &= -1.1 \cdot 2 \operatorname{sign}(v_4 - w_3).
\end{aligned}$$

The observer performance and finite-time convergence for a sample time $\tau = 0.001$ can be seen in Figure 27.5. The transient process is shown in Figure 27.6 for the states x_1 and x_4. It is seen from Figure 27.7 that the system trajectories and their derivatives of any order tend to infinity. Thus, the differentiator could not alone perform the observation. Figure 27.8 shows the effect of discretization in observation with sampling times $\tau = 0.0001$ and $\tau = 0.01$.

27.2.3.2 Unknown-Input Estimation

The state observation and the unknown-input estimation problems are considered here. The input ζ is a bounded function with a Lipschitzian derivative, $k = 1$; both the state x and the disturbance ζ are estimated.

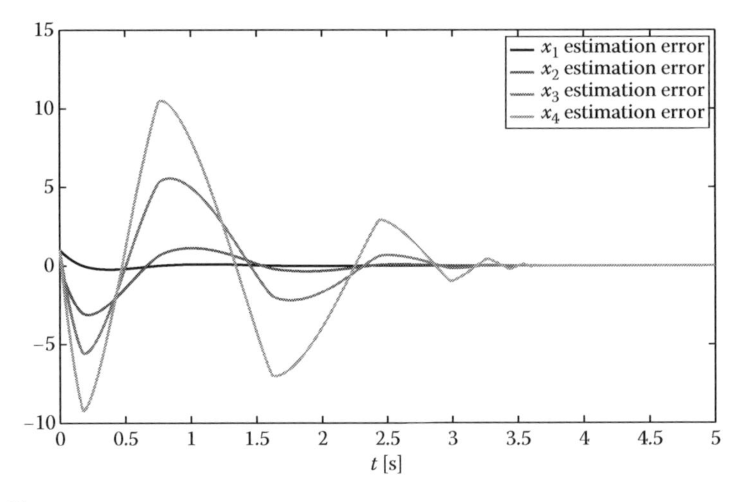

FIGURE 27.5 Observer errors.

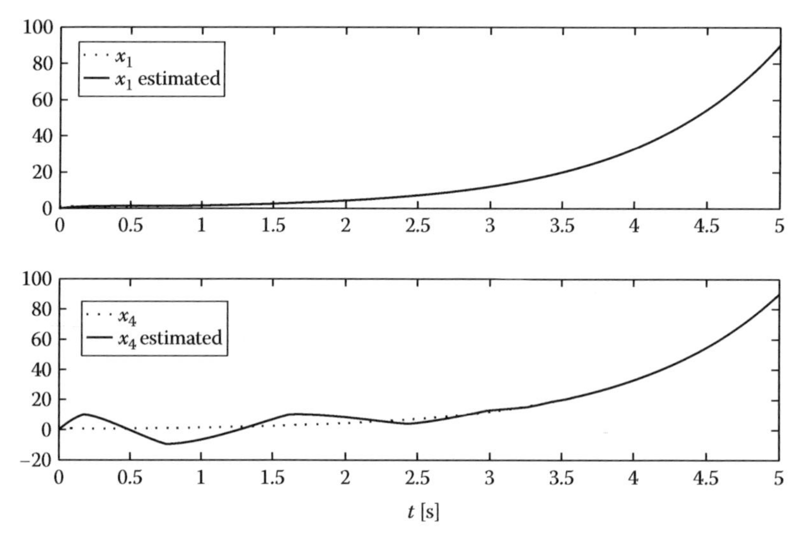

FIGURE 27.6 Convergence of $\hat{x}_1$, $\hat{x}_4$ to x_1, and x_4.

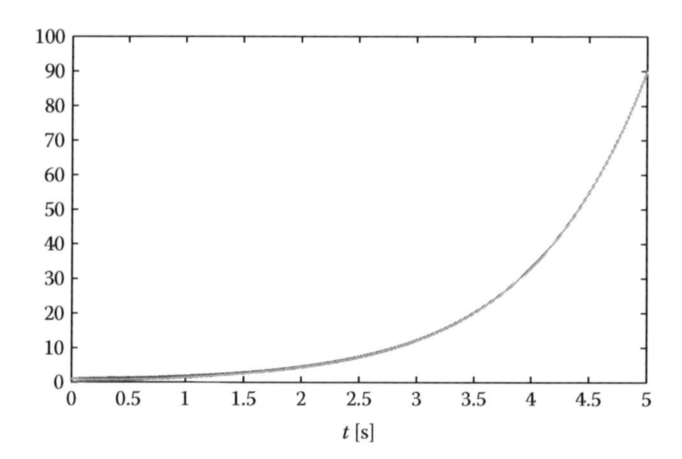

FIGURE 27.7 System coordinates.

The state observer (27.11), (27.12) is designed similarly to the state observation case with $L = \begin{bmatrix} 5 & 5 & 5 & 5 \end{bmatrix}^T$, providing for eigenvalues −1, −2, −3, −4 of matrix $A - LC$. The gain matrix K in (27.12) is also the same. The parameters of (27.17) $\alpha_1 = 1.1$, $\alpha_2 = 1.5$, $\alpha_3 = 2$, $\alpha_4 = 3$, $\alpha_5 = 5$, $\alpha_6 = 8$, $M = 1$ are chosen, such that

$$\begin{aligned}
\dot{v}_1 &= w_1 = -8\,|v_1 - y(t) + Cz|^{5/6}\,\mathrm{sign}(v_1 - y(t) + Cz) + v_2,\\
\dot{v}_2 &= w_2 = -5|\,v_2 - w_1|^{4/5}\,\mathrm{sign}(v_2 - w_1) + v_3,\\
\dot{v}_3 &= w_3 = -3\,|v_3 - w_2|^{3/4}\,\mathrm{sign}(v_3 - w_2) + v_4,\\
\dot{v}_4 &= w_4 = -2\,|v_4 - w_3|^{2/3}\,\mathrm{sign}(v_4 - w_3) + v_5,\\
\dot{v}_5 &= w_5 = -1.5\,|v_5 - w_4|^{1/2}\,\mathrm{sign}(v_5 - w_4) + v_6,\\
\dot{v}_6 &= -1.1\,\mathrm{sign}(v_6 - w_5).
\end{aligned}$$

The finite-time convergence of estimated states to the real states is shown in Figure 27.9 with the sampling interval $\tau = 0.001$. The unknown-input estimation obtained using relation (27.20) is demonstrated

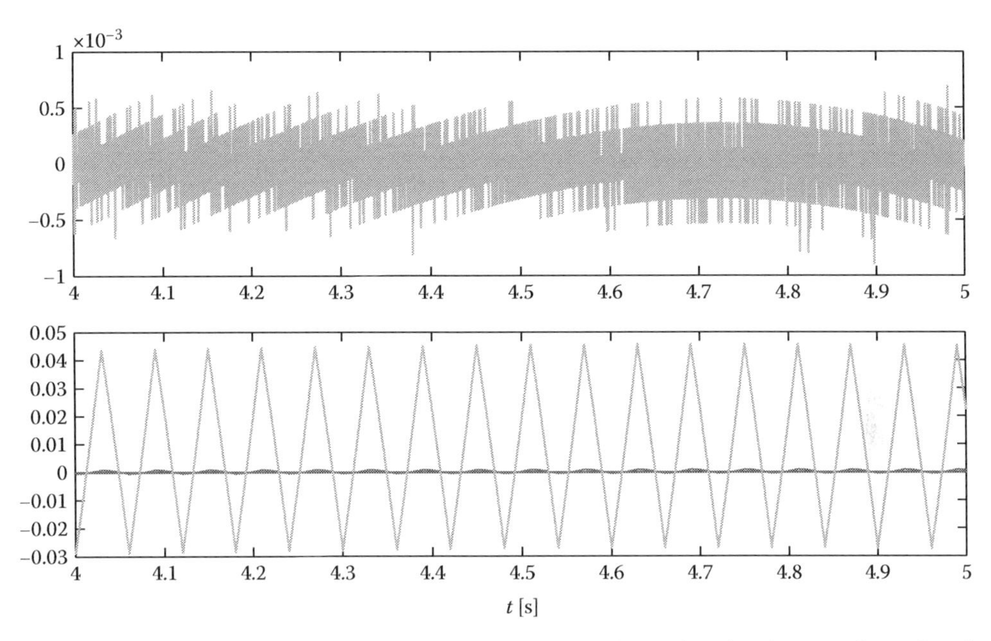

FIGURE 27.8 Observer errors (detail) with sample times $\tau = 0.0001$ (upper figure) and $\tau = 0.01$ (lower figure).

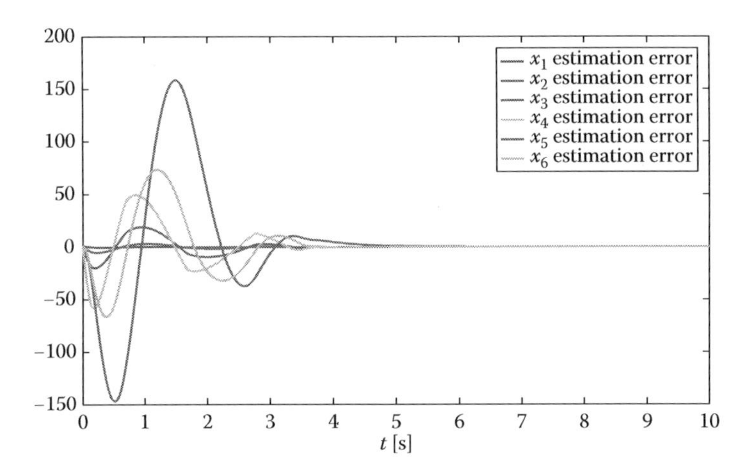

FIGURE 27.9 Observer errors for the unknown input estimation case.

in Figure 27.10. The effects of discretization can be seen in Figure 27.11 for the sampling intervals $\tau = 0.0001$ and $\tau = 0.01$.

27.3 System Identification

In this section, the problem of parameter identification for linear systems is solved using the sliding modes main features such as high accuracy and finite-time exact convergence of the estimations. With respect to the observation analysis given in the past section, the uncertain system is considered as a particular case of a system affected by unknown inputs.

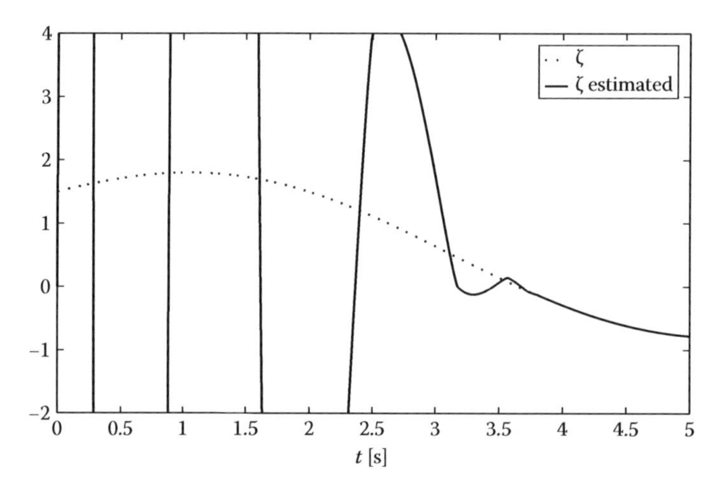

FIGURE 27.10 Unknown input estimation.

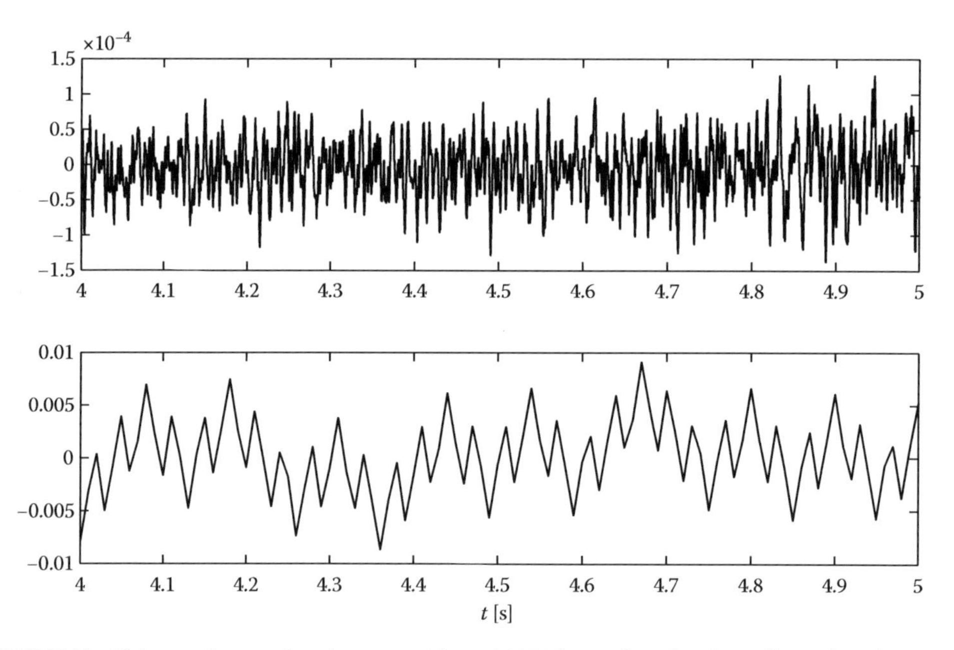

FIGURE 27.11 Unknown input estimation error with $\tau = 0.0001$ (upper figure) and 0.01 (lower figure).

Consider a special case of system (27.10), a linear system with parametric uncertainty given by the following equation:

$$\begin{aligned} \dot{x} &= (A + \Delta A)x + (B + \Delta B)u, \\ y &= Cx, \end{aligned} \tag{27.39}$$

where $x \in \mathbb{R}^n$ is the state vector, $u \in \mathbb{R}^q$ is a vector of known inputs (i.e., control), the output of the system is given by $y \in \mathbb{R}^p$, matrices A, B, C are known and have proper dimensions, and matrices ΔA, ΔB represent parametric uncertainties and are considered unknown.

In order to guarantee the stability of the estimation error, the matrices ($A + \Delta A$) and A should be Hurwitz.

Let us rewrite the system in a manner consistent with the unknown-input observer presented before. Let m be the number of rows of the extended matrix $[\Delta A\ \Delta B]$ in which at least one element different from zero appears. For each row i that contains at least one element different from zero, define k_i as the total number of columns with elements different from zero.

The uncertain part of the system can be written as

$$\Delta Ax + \Delta Bu = D\Theta(x,u),$$

where $D \in \mathbb{R}^{n\times m}$ is a matrix containing only values 1 and 0, and which only has m rows with an element different from zero; the vector of nonlinear elements $\Theta(x, u)$ is given by

$$\Theta(x,u) = \begin{bmatrix} \theta_1\varphi_1(x,u) \\ \vdots \\ \theta_m\varphi_m(x,u) \end{bmatrix},$$

where $\theta_i \in \mathbb{R}^{1\times k_i}$ represent the uncertainties and $\varphi_i(x,u) \in R^{k_i}$ is the corresponding regressor vector [16].

Without loss of generality, system (27.39) can be written as

$$\begin{aligned} \dot{x} &= Ax + Bu + D\Theta(x,u), \\ y &= Cx, \end{aligned} \tag{27.40}$$

Now system (27.40) has the form of a system with unknown inputs.

Let us assume the output vector $y = Cx$ poses a full relative degree vector with respect to the uncertainty. It is, there exist a selection of outputs such that the vector $r = (r_1, \ldots, r_m)$ associated to such outputs satisfies

$$c_i A^k D = 0, \quad k = 0,1,\ldots,r_i - 2, \tag{27.41}$$

$$c_i A^{r_i - 1} D \neq 0, \quad i = 1,2,\ldots,m, \tag{27.42}$$

$$T = \begin{bmatrix} c_1 \\ c_1 A \\ \vdots \\ c_1 A^{r_1-1} \\ \vdots \\ c_m \\ \vdots \\ c_m A^{r_m-1} \end{bmatrix}, \quad \operatorname{rank} T = n. \tag{27.43}$$

Additionally, consider the input u and its first derivative $\dot{u}$ to be bounded, i.e., there exist constants u_0^+, u_1^+ such that $\| u \| < u_0^+, \| \dot{u} \| < u_1^+$.

The high-order sliding mode observer can be applied for state observation. For this particular case, the observer takes the form

$$\begin{aligned}\dot{z} &= Az + Bu,\\ y_z &= \bar{C}z,\\ \hat{x} &= z + T^{-1}\bar{v}(y - y_z),\end{aligned} \tag{27.44}$$

where T is defined according to the selection of the vector of relative degrees r.

Vector $\bar{v}(y - y_z)$ is given by a proper selection of the rows of $v(y - y_z)$ which is composed by the solutions of the following vector differential equation computed for each component of the outputs y and y_z:

$$\begin{aligned}\dot{v}_{i,1} &= w_{i,1} = -\alpha_{r_i+1}M_i^{1/(r_i+1)}\,|v_{i,1} - y_i + y_{z_i}|^{(r_i)/(r_i+1)}\,\mathrm{sign}(v_{i,1} - y_i + y_{z_i}) + v_{i,2},\\ \dot{v}_{i,2} &= w_{i,2} = -\alpha_{r_i}M_i^{1/(r_i)}\,|v_{i,2} - w_{i,1}|^{(r_i-1)/(r_i)}\,\mathrm{sign}(v_{i,2} - w_{i,1}) + v_{i,3},\\ &\vdots\\ \dot{v}_{i,r_i} &= w_{i,r_i} = -\alpha_2 M_i^{1/2}\,|v_{i,r_i} - w_{i,r_i-1}|^{1/2}\,\mathrm{sign}(v_{i,r_i} - w_{i,r_i-1}) + v_{i,r_i+1},\\ \dot{v}_{i,r_i+1} &= -\alpha_1 M_i\,\mathrm{sign}(v_{i,r_i+1} - w_{i,r_i}),\end{aligned} \tag{27.45}$$

for each component of the output is computed

$$v_i^T = [v_{i,1}^T \quad v_{i,2}^T \quad \cdots \quad v_{i,r_i}^T].$$

It is remarkable that the last component of the vector differential equation ($v_{i,ri+1}$) is not considered to form part of the vector v_i. Vector $v(y - y_z)$ is given by

$$v(y - y_z)^T = \begin{bmatrix} v_1^T & v_2^T & \cdots & v_m^T \end{bmatrix}. \tag{27.46}$$

Vector $\bar{v}(y - y_z)$ is composed by the rows of vector v corresponding to the outputs selected to form the vector of full relative degree.

Notice that the observer (27.44) is a particular case of (27.25), (27.31), (27.35) with $L = 0$ and a particular selection of matrix T.

Summary: Let matrices A and $A + \Delta A$ of system (27.40) be Hurwitz and the input u be bounded and Lipschitz. With a proper gain selection α_j and M_i, the state x of system (27.39) is exactly observed after a finite-time transient by the observer (27.43) through (27.46).

The application of the high-order sliding mode observer guarantees that the state vector x is completely known and as a consequence the next vector equality is satisfied:

$$\begin{bmatrix} v_{1,r_1+1}\\ \vdots\\ v_{p,r_p+1}\end{bmatrix} = \begin{bmatrix} c_1A^{r_1}\\ \vdots\\ c_pA^{r_p}\end{bmatrix}\tilde{e} + \begin{bmatrix} c_1A^{r_1-1}D\\ \vdots\\ c_pA^{r_p-1}D\end{bmatrix}\Theta(x,u)$$

The variable $\tilde{e}$ can be estimated as

$$\begin{bmatrix} c_1 A^{r_1} \\ \vdots \\ c_p A^{r_p} \end{bmatrix} \tilde{e} = \begin{bmatrix} c_1 A^{r_1} \\ \vdots \\ c_p A^{r_p} \end{bmatrix} T^{-1} \bar{v}(y - y_z). \tag{27.47}$$

Define

$$z_{eq}(x,z,y) = \begin{bmatrix} v_{1,r_1} \\ \vdots \\ v_{p,r_p} \end{bmatrix} - \begin{bmatrix} c_1 A^{r_1} \\ \vdots \\ c_p A^{r_p} \end{bmatrix} T^{-1} \bar{v}(y - y_z). \tag{27.48}$$

The convergence in finite time of the observer allows us to guarantee that the equalities $\hat{x} = x$, $e = 0$ are satisfied for all $t \geq t_0$, where $t_0 > 0$ is the observer convergence time. According to the representation of matrix D, vector z_{eq} satisfies the following vector equality:

$$z_{eq} = Q\Theta(x,u) = \begin{bmatrix} q_1 \theta_i \varphi_1(x,u) \\ \vdots \\ q_p \theta_p \varphi_p(x,u) \end{bmatrix},$$

where

$$Q = \begin{bmatrix} q_1 \\ \vdots \\ q_p \end{bmatrix} = \begin{bmatrix} c_1 A^{r_1 - 1} D \\ \vdots \\ c_p A^{r_p - 1} D \end{bmatrix}.$$

Each row of the equivalent output injection satisfies the following equality:

$$z_{eq_i}(\hat{x},z,y) = q_i \theta_i \varphi_i(\hat{x},u), \tag{27.49}$$

where θ represents the uncertainties term, and the functions $z_{eq_i}(\hat{x},z,y)$ and $\varphi_i(\hat{x},u)$ are continuous and completely known.

By following (27.49), it is possible to write down the following equality:

$$\frac{1}{t}\int_{t_0}^{t} z_{eq_i}(\hat{x},z,y)\varphi_i(\hat{x},u)^T \, d\tau = q_i \frac{1}{t}\int_{t_0}^{t} \theta_i \varphi_i(\hat{x},u)\varphi_i(\hat{x},u)^T \, d\tau.$$

Define the following variable

$$\Gamma_i = \left(\int_{t_0}^{t} \varphi_i(\hat{x},u)\varphi_i(\hat{x},u)^T \, d\tau \right)^{-1}.$$

The parametric uncertainty θ_i can be written in terms of (27.49) and Γ as

$$\hat{\theta}_i = \frac{1}{q_i} \left(\int_{t_0}^{t} z_{eq_i}(\hat{x},z,y)\varphi_i(\hat{x},u)^T \, d\tau \right) \Gamma_i \tag{27.50}$$

Variable θ_i can be reconstructed by application of Equation 27.50 if the following condition is satisfied:

$$sup \, \| t\Gamma_i \| < \infty. \tag{27.51}$$

This last condition is known as persistent excitation condition (see, for example, [15,16]).

Summary: Let inequality (27.51) be satisfied. The application of the algorithm (27.48) through (27.50) guarantees the reconstruction of the parameter θ_i.

27.3.1 Example

Consider the following uncertain system:

$$\dot{x} = \begin{bmatrix} 0 & 1 & 0 & 0 & 0 \\ 0 & 0 & 1 & 0 & 0 \\ -1 & -2 & -3 & 0 & 0 \\ 1 & 0 & 0 & 0 & 1 \\ -1 & 0 & -2 & -1 & -1 \end{bmatrix} x + \begin{bmatrix} 0 & 0 \\ 0 & 1 \\ 2 & 0 \\ 0 & 1 \\ 0 & 3 \end{bmatrix} u,$$

$$y = \begin{bmatrix} 1 & 0 & 0 & 0 & 0 \\ 0 & 0 & 0 & 1 & 0 \end{bmatrix} x.$$

The initial conditions for the system are $x_0 = \begin{bmatrix} 1 & 1 & 1 & 1 & 1 \end{bmatrix}^T$. The nominal values of the matrices are given by

$$A = \begin{bmatrix} 0 & 1 & 0 & 0 & 0 \\ 0 & 0 & 1 & 0 & 0 \\ -1 & -13/2 & -3 & 0 & 0 \\ 1 & 0 & 0 & 0 & 1 \\ -1 & 0 & -2 & -4 & -1 \end{bmatrix}, \quad B = \begin{bmatrix} 0 & 0 \\ 0 & 1 \\ 2 & 0 \\ 0 & 1 \\ 0 & 1 \end{bmatrix}, \quad C = \begin{bmatrix} 1 & 0 & 0 & 0 & 0 \\ 0 & 0 & 0 & 1 & 0 \end{bmatrix}.$$

Hence, the uncertainty is given by

$$\Delta A = \begin{bmatrix} 0 & 0 & 0 & 0 & 0 \\ 0 & 0 & 0 & 0 & 0 \\ 0 & 9/2 & 0 & 0 & 0 \\ 0 & 0 & 0 & 0 & 0 \\ 0 & 0 & 0 & 3 & 0 \end{bmatrix}, \quad \Delta B = \begin{bmatrix} 0 & 0 \\ 0 & 0 \\ 0 & 0 \\ 0 & 0 \\ 0 & 2 \end{bmatrix}.$$

The eigenvalues of $(A + \Delta A)$ are -2.3247, $-0.3376 \pm 0.5623i$, $0.5000 \pm 0.8660i$, and the eigenvalues of A are -0.1658, $-0.5 \pm 1.9365i$, $-1.4171 \pm 0.0055i$. The output of the system $y = Cx$ has a full relative degree vector with respect to the unknown input.

The input of the system is given by

$$u = \begin{bmatrix} 0.2\,\text{sen}(9t) + 0.3\,\text{sen}(5t) + 0.3\,\text{sen}\left(2t + \dfrac{\pi}{4}\right) - 0.3 \\ 0.2\,\text{sen}(9t) + 0.3\,\text{sen}\left(7.5t + \dfrac{\pi}{8}\right) + 0.3\,\text{sen}\left(6t + \dfrac{\pi}{4}\right) \end{bmatrix}.$$

It is clear that u and its first derivative are bounded. The high-order sliding mode observer takes on the form

$$\dot{z} = \begin{bmatrix} 0 & 1 & 0 & 0 & 0 \\ 0 & 0 & 1 & 0 & 0 \\ -1 & -13/2 & -3 & 0 & 0 \\ 1 & 0 & 0 & 0 & 1 \\ -1 & 0 & -2 & -4 & -1 \end{bmatrix} z + \begin{bmatrix} 0 & 0 \\ 0 & 1 \\ 2 & 0 \\ 0 & 1 \\ 0 & 1 \end{bmatrix} u,$$

$$y_z = \begin{bmatrix} 1 & 0 & 0 & 0 & 0 \\ 0 & 0 & 0 & 1 & 0 \end{bmatrix} z.$$

Matrix T is given by

$$T = \begin{bmatrix} 1 & 0 & 0 & 0 & 0 \\ 0 & 1 & 0 & 0 & 0 \\ 0 & 0 & 1 & 0 & 0 \\ 0 & 0 & 0 & 1 & 0 \\ 1 & 0 & 0 & 0 & 1 \end{bmatrix}.$$

Vector $v(y - y_z)$ is defined as

$$v(y - y_z)^T = \begin{bmatrix} v_{1,1} & v_{1,2} & v_{1,3} & v_{1,4} & v_{2,1} & v_{2,2} & v_{2,3} \end{bmatrix}^T,$$

where each element is defined as part of the solution of the vector differential equation:

$$\begin{aligned}
\dot{v}_{1,1} &= w_{1,1} = -\alpha_4 M_1^{1/4} \left| v_{1,1} - y_1 + y_{z_1} \right|^{(3)/(4)} \operatorname{sign}(v_{1,1} - y_1 + y_{z_1}) + v_{1,2}, \\
\dot{v}_{1,2} &= w_{2,2} = -\alpha_3 M_1^{1/3} \left| v_{1,2} - w_{1,1} \right|^{2/3} \operatorname{sign}(v_{1,2} - w_{1,1}) + v_{1,3}, \\
\dot{v}_{1,3} &= w_{1,3} = -\alpha_2 M_1^{1/2} \left| v_{1,3} - w_{1,2} \right|^{1/2} \operatorname{sign}(v_{1,3} - w_{1,2}) + v_{1,4}, \\
\dot{v}_{1,4} &= -\alpha_1 M_1 \operatorname{sign}(v_{1,4} - w_{1,3}), \\
\dot{v}_{2,1} &= w_{2,1} = -\alpha_3 M_2^{1/3} \left| v_{2,1} - y_2 + y_{z_2} \right|^{2/3} \operatorname{sign}(v_{2,1} - y_2 + y_{z_2}) + v_{2,2}, \\
\dot{v}_{2,2} &= w_{2,2} = -\alpha_2 M_2^{1/2} \left| v_{2,2} - w_{2,1} \right|^{1/2} \operatorname{sign}(v_{2,2} - w_{2,1}) + v_{2,3}, \\
\dot{v}_{2,3} &= -\alpha_1 M_2 \operatorname{sign}(v_{2,3} - w_{2,2}).
\end{aligned}$$

The estimation error convergence for the observation process $e = x - \hat{x}$ is shown in Figure 27.12. Notice that after a finite-time transient, the observer ensures the convergence to zero of the error e. The parametric uncertainty can be written in the form (27.48) as

$$\begin{aligned}
\theta_1 &= \theta_{1,1}, \\
\varphi_1(x,u) &= x_2, \\
\theta_2 &= \begin{bmatrix} \theta_{2,1} & \theta_{2,2} \end{bmatrix}, \\
\varphi_2(x,u) &= \begin{bmatrix} x_4 & u_2 \end{bmatrix}^T.
\end{aligned}$$

The parameters to be identified are θ_1 and θ_2, given by

$$\begin{aligned}
\theta_1 &= \begin{bmatrix} 4 & 5 \end{bmatrix}, \\
\theta_2 &= \begin{bmatrix} 3 & 2 \end{bmatrix}.
\end{aligned}$$

Apply the identification algorithm with $t_0 = 4$. The identification of the parameter $\theta_{1,1}$ is shown in Figure 27.13. The identification of parameters, $\theta_{2,1}$ and $\theta_{2,2}$, is shown in Figure 27.14.

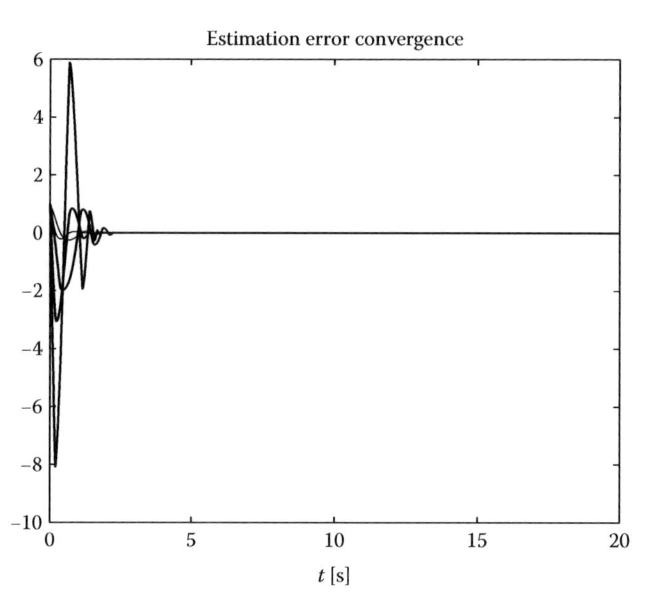

FIGURE 27.12 Observation error.

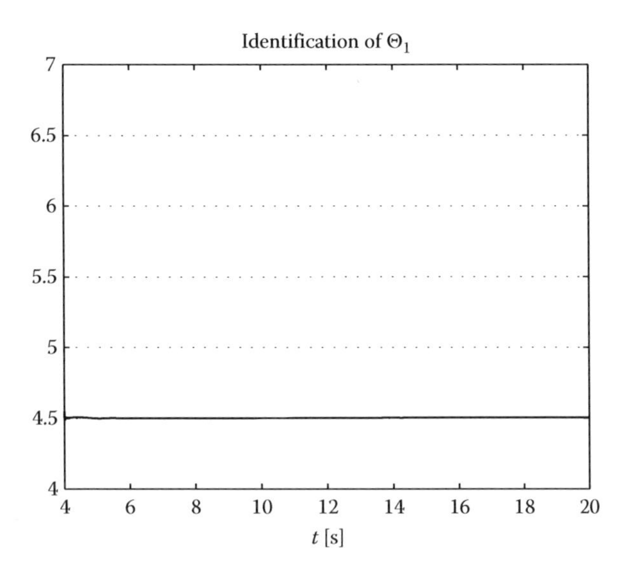

FIGURE 27.13 Identification of parameter $\theta_{1,1}$.

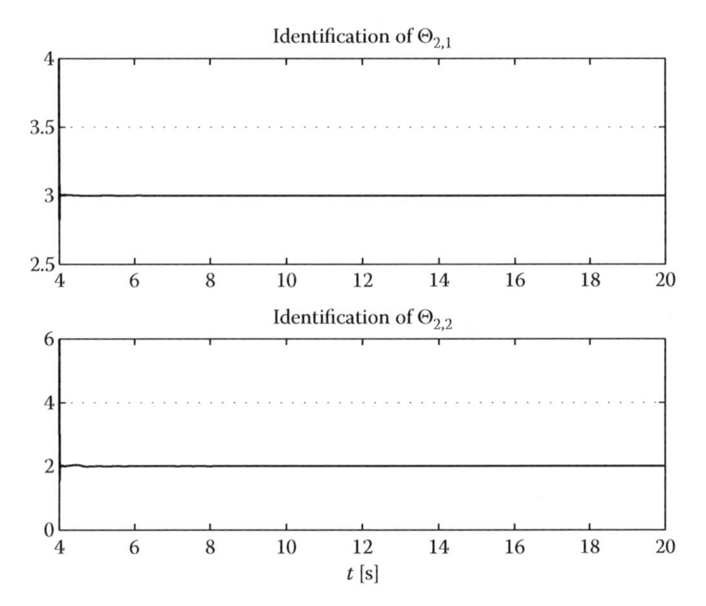

FIGURE 27.14 Identification of parameters $\theta_{2,1}$ and $\theta_{2,2}$.

27.4 Bibliography Review

The techniques for state observation can be extended to systems whose relative degree is not well defined and also to nonlinear systems.

The problem of state observation in the presence of unknown inputs for systems without a well-defined relative degree has been studied in [2,6].

The nonlinear case is considered in [4,7] providing finite-time convergence.

References

1. A.N. Atassi and H.K. Khalil. Separation results for the stabilization of nonlinear systems using different high-gain observer design. *Systems and Control Letters*, 39:183–191, 2000.
2. F.J. Bejarano, L. Fridman, and A. Poznyak. Exact state estimation for linear systems with unknown inputs based on hierarchical super-twisting algorithm. *International Journal of Robust and Nonlinear Control*, 17(18):1734–1753, 2007.
3. C.T. Chen. *Linear Systems: Theory and Design*. Oxford University Press, New York, 1999.
4. J. Davila, L. Fridman, A. Pisano, and A. Usai. Finite-time state observation for nonlinear uncertain systems via higher order sliding modes. *International Journal of Control*, 82(8):1564–1574, 2009.
5. A.F. Filippov. *Differential Equations with Discontinuous Right-Hand Sides*. Kluwer Academic Publishers, Dordrecht, the Netherlands, 1988.
6. L. Fridman, A. Levant, and J. Davila. Observation of linear systems with unknown inputs via high-order sliding-modes. *International Journal of System Science*, 38(10):773–791, 2007.
7. L. Fridman, Y. Shtessel, C. Edwards, and X. Yan. Higher-order sliding-mode observer for state estimation and input reconstruction in nonlinear systems. *International Journal of Robust Nonlinear Control*, 18(4–5):399–413, 2008.
8. M.L. Hautus. Strong detectability and observers. *Linear Algebra and Its Applications*, 50:353–368, 1983.
9. A. Isidori. *Nonlinear Control Systems*. Springer-Verlag, London, U.K., 1996.
10. A.N. Kolmogoroff. On inequalities between upper bounds of consecutive derivatives of an arbitrary function defined on an infinite interval. *American Mathematical Society Translations* 2:233–242, 1962.
11. A. Levant. Robust exact differentiation via sliding mode technique. *Automatica*, 34(3):379–384, 1998.
12. A. Levant. High-order sliding modes: Differentiation and output-feedback control. *International Journal of Control*, 76(9–10):924–941, 2003.
13. A. Levant. Homogeneity approach to high-order sliding mode design. *Automatica*, 41(5):823–830, 2005.
14. A. Levant. Quasi-continuous high-order sliding-mode controllers. *IEEE Transactions on Automatic Control*, 50(11):1812–1816, 2005.
15. L. Ljung. *System Identification. Information and System Sciences Series*, 2nd edn. Prentice Hall PTR, Englewood Cliffs, NJ, 1999.
16. T. Soderstrom and P. Stoica. *System Identification*. Prentice Hall International, Cambridge, U.K., 1989.

IV

Modeling and Control

28

Modeling for System Control

A. John Boye
University of Nebraska

and

Neurintel, LLC

28.1 Introduction

A mathematical model of a physical system is a representation of the system by a set of mathematical equations. As stated by others, modeling is a "technique of system analysis and design using mathematical or physical idealizations of all or a portion of the system. Completeness and reality of the model are dependent on the questions to be answered, the state of knowledge of the system, and its environment" [BM00].

Mathematical models are constructed for a wide range of systems and are discussed in most basic references and textbooks [B91,CRD00,DB08,DHS03,FPE10,KG09,O10,VE97]. They are used for physical engineering systems such as electrical, mechanical, hydraulic, pneumatic, and thermal, as well as for systems in the areas of applied sciences, life sciences and medicine, social sciences, economics, and business and management. Models are used when it is neither possible nor practical (due to complexity, expense, or safety) to build the actual system, then test and modify it until desired performance is achieved. A model is a substitute for the actual system when designing, testing, and modifying various components of the system. The model is used to examine the system without having to do any physical experiments. Obviously it would be easier and safer to study and modify the operation of a new controller for the space shuttle using a mathematical model than it would be to actually use the space shuttle itself.

Obtaining a good mathematical model of a system is perhaps the most important task in mathematical modeling. No mathematical model is exact and certain basic assumptions are made. In this case, it is crucial that the model includes all of the necessary critical factors in order to be adequate for the particular needs of the particular problem, but it should not be overly complex. Completely describing the performance and operation of a system will require a large number of equations that need to be simplified. At the same time, however, the model must not be overly simplified, so that important properties are omitted. A good model is one that is both accurate and simple.

The knowledge of the particular problem, the skills, and the practical experience of the engineer play an important role in creating a good model. Eighty to ninety percent of the effort in control system analysis and design should be devoted to constructing an accurate model [PH99]. The importance of obtaining the correct system model cannot be stressed enough. Many problems that arise in applying control theory to real applications are directly related to the use of an incorrect or inappropriate plant or measurement model. Developing an accurate model is perhaps the most important and difficult aspect of control engineering. Ultimately, how well the model meets the two possibly conflicting criteria of simplicity and accuracy will be decided by the performance of the actual physical system.

There are two basic approaches to modeling: analytical modeling and empirical or experimental modeling [B91]. Both of them will be discussed below, with the emphasis on analytical modeling.

28.2 Analytical Modeling

There are four basic steps in the process of analytical modeling.

1. *Determine the model structure*. The purpose and intended use of the model needs to be defined. The limitations and boundaries of the problem and assumptions, such as linearity and stationarity, need to be determined.
2. *Determine the system components*. The fundamental physical laws affecting the system need to be used to identify the inputs, outputs, and important system parameters, including constants and variables. Complex systems should be divided into subsystems whose behaviors and interactions are known and defined, which is greatly facilitated by block diagrams or signal flow graphs of the system.
3. *Determine the system equations*. The necessary equations describing the system now need to be written using the fundamental physical relationships and the laws of conservation and compatibility.
4. *Determine the model accuracy*. The model needs to be evaluated and its performance needs to be validated, which can be done analytically or by computer simulation using a more complete (or more realistic) model. Such verification by simulation will not catch errors common to both models, and the ultimate verification must be based on experiments. If the model is not accurate enough for the particular application, it needs to be corrected by returning to one of the above steps and modifying accordingly.

These four steps will be discussed further in the following sections.

28.2.1 Determine the Model Structure

The first step in constructing a mathematical model is to determine how the system can be described according to its intended purposes. As mentioned earlier, a number of assumptions or simplifications should be made on the signals and the system for the purposes of creating the model. A system can be defined in a number of different ways.

28.2.1.1 Continuous Time or Discrete Time

A *continuous-time* system is one that is represented by differential equations, while a *discrete-time* (or *sampled data*) system is one that is represented by difference equations. Although most systems are actually continuous, they are often modeled as discrete systems, so that a digital computer may be used for design and/or control. Also, systems may be affected by continuous changing events, sometimes called *change oriented*, or by discrete changing events, called *discrete event oriented*.

28.2.1.2 Deterministic or Stochastic

A *deterministic* system is a system without random parameters or inputs. In other words, the system is known exactly. On the other hand, a *stochastic* system is one in which at least one parameter or input is affected by random disturbances or noise. The external signals that influence a system also have to be modeled. They are also either deterministic or stochastic. These random disturbances affecting the system parameters or inputs could be known and measurable, known and nonmeasurable, or unknown. Smoothing, filtering, and estimation techniques are used to get an accurate response for a stochastic system.

28.2.1.3 Distributed Parameter or Lumped Parameter

A *distributed parameter* system is one that is described by partial differential equations. Systems described in this way are functions of both time and space. Examples include such things as the voltage on a transmission line or a displacement on a flexible structure. A *lumped parameter* system is one described by ordinary differential or difference equations. In control system applications, a distributed parameter system is usually approximated as a lumped parameter system.

28.2.1.4 Dynamic or Static

A *dynamic* system is one that is described by differential or difference equations, while a *static* system is one that is described only by algebraic equations.

28.2.1.5 Linear or Nonlinear

A *linear* system is one that is represented by linear equations, while a *nonlinear* system is one represented by nonlinear equations. In reality every real-world system is nonlinear; however, these are often approximated as linear ones in the operating range of interest. This is often done by using the first two terms of the Taylor series, neglecting the higher order terms. For many purposes, the approximate linear model is sufficient for control system design and gives acceptable performance.

28.2.1.6 Stationary or Nonstationary

Finally, a *stationary* (or *time-invariant*) system is one in which the system parameters do not vary with time, while a *nonstationary* (or *time-variant*) system is one in which the system parameters vary with time. Systems are often approximated as stationary or time invariant, which is especially true if the parameters only change slowly with time.

28.2.2 Determine the System Components

Systems can be described in terms of through and across variables, inductive and capacitive storage elements, and energy dissipaters. Similarities between diverse systems allow systems to be described by analogous circuits [C67,DB08]. A comparison among the various types of engineering systems is shown in Table 28.1. Of course, the actual values of the system parameters need to be identified. This may be a simply educated estimation of the parameter values or it may entail the implementation of some kind of system identification technique using experimental data.

Block diagrams and signal flow graphs are often used here to better represent and simplify the system. Such diagrams help show the interactions within the system itself.

28.2.3 Determine the System Equations

Once the proper form of the elements is selected, these analogs, along with the law of conservation or continuity (relating the *through* variables) and the law of compatibility (relating the *across* variables),

TABLE 28.1 Comparison of Engineering Systems

	Electrical	Translational Mechanical	Rotational Mechanical	Fluid	Thermal
Through variable	i, current	f, force	τ, torque	q, flow rate	q, flow rate
Across variable	v, voltage	v, velocity	ω, angular velocity	P, pressure	T, temperature
Storage (inductive)	L, inductor $i=\frac{1}{L}\int v\,dt$	K, spring constant $f=K\int v\,dt$	K, spring constant $\tau=K\int \omega\,dt$	I, fluid inertia $q=\frac{1}{I}\int P\,dt$	
Storage (capacitive)	C, constant $i=C\frac{dv}{dt}$	M, capacitor $f=M\frac{dv}{dt}$	J, moment of inertia $\tau=J\frac{d\omega}{dt}$	C, fluid capacitance $q=C\frac{dP}{dt}$	C, thermal capacitance $q=C\frac{dT}{dt}$
Dissipator	R, resistor $i=\frac{1}{R}v$	B, damping, viscous friction $f=Bv$	B, damping, viscous friction $\tau=B\omega$	R, fluid resistance $q=\frac{1}{R}P$	R, thermal resistance $q=\frac{1}{R}T$
Notes	$i=\dot{q}$ (charge)	$v=\dot{x}$ (position)	$\omega=\dot{\theta}$ (angle)	$q=\dot{v}$ (volume)	$q=\dot{h}$ (heat)

are used to write the differential or difference equations. Linear equations are sufficient for many system problems. Symbolic algebra computer programs such as MATLAB®, Simulink®, Maple™, and LabVIEW™ are becoming increasingly popular in manipulating these equations.

There are two approaches to represent these equations. The first approach represents the system as one nth-order differential or difference equation. As an example, a continuous-time system with a single input u, a single output y, and all initial conditions equal to zero would be represented by

$$\frac{d^n y}{dt^n}+a_{n-1}\frac{d^{n-1}y}{dt^{n-1}}+\cdots+a_1\frac{dy}{dt}+a_0 y=b_0 u+b_1\frac{du}{dt}+\cdots+b_m\frac{d^m u}{dt^m}. \tag{28.1}$$

Taking the Laplace transform of Equation 28.1 and rearranging, the transfer function can be obtained as

$$\frac{Y(s)}{U(s)}=H(s)=\frac{b_m s^m+b_{m-1}s^{m-1}+\cdots+b_1 s+b_0}{s^n+a_{n-1}s^{n-1}+\cdots+a_1 s+a_0}. \tag{28.2}$$

The other approach to these equations is to use the state-space model. Here state variables and matrices are used to represent the system by a total of n first-order differential or difference equations. Using this approach, Equation 28.1, with $m = n$, can be written as the following state equation [B91]:

$$\dot{\mathbf{x}}(t)=\begin{bmatrix}-a_{n-1} & 1 & 0 & 0 & \cdots & 0\\ -a_{n-2} & 0 & 1 & 0 & \cdots & 0\\ \vdots & \vdots & \vdots & \vdots & \vdots & \vdots\\ -a_1 & 0 & 0 & 0 & \cdots & 1\\ -a_0 & 0 & 0 & 0 & \cdots & 0\end{bmatrix}\mathbf{x}(t)+\begin{bmatrix}b_{n-1}-a_{n-1}b_n\\ b_{n-2}-a_{n-2}b_n\\ \vdots\\ b_1-a_1 b_n\\ b_0-a_0 b_n\end{bmatrix}u(t), \tag{28.3}$$

and output equation

$$\mathbf{y}(t)=\begin{bmatrix}1 & 0 & 0 & \cdots & 0\end{bmatrix}\mathbf{x}(t)+b_n u(t). \tag{28.4}$$

Equations 28.3 and 28.4 represent just one of many valid forms of the state and output equations, but (for linear systems) they all can be represented by the compact matrix form as $\dot{\mathbf{x}}=\mathbf{A}\mathbf{x}+\mathbf{B}\mathbf{u}$ and $\mathbf{y}=\mathbf{C}\mathbf{x}+\mathbf{D}u$.

Direct digital control and digital computers require the system to be described in a digital form by difference equations. These discrete system difference equations can be obtained in two different ways. The first way is to rewrite the system difference equations in discrete form directly from the original system. The second way is to approximate the continuous equations as discrete ones using any of a number of methods, such as forward or backward differences. The conversion to the transfer function uses the z-transform instead of the Laplace transform. As an example, a discrete-time system with a single input u, a single output y, and all initial conditions equal to zero would be represented by

$$\begin{aligned} &y(k+1)+a_0y(k)+a_1y(k-1)+\cdots+a_ny(k-n) \\ &\quad = b_0u(k+1)+b_1u(k)+b_2u(k-1)+\cdots+b_mu(k+1-m). \end{aligned} \tag{28.5}$$

Taking the z-transform and rearranging, the transfer function can be obtained as

$$\frac{Y(z)}{U(z)} = H(z) = \frac{b_0+b_1z^{-1}+\cdots+b_mz^{-m}}{1+a_0z^{-1}+\cdots+a_nz^{-(n+1)}}. \tag{28.6}$$

These equations can also be written in state variable form similar to what was done for the continuous case. These discrete-time state equations can be obtained by either approximating the original differential equations (e.g., Equation 28.1) or the transfer functions (e.g., Equation 28.2) and then selecting the states or by approximating the continuous-time state equations (e.g., Equations 28.3 and 28.4).

For a nonlinear system, the state and output equations will be given by $\dot{\mathbf{x}} = f(\mathbf{x}, \mathbf{u}, t)$ and $\mathbf{y} = h(\mathbf{x}, \mathbf{u}, t)$. The transfer function is of no value in the nonlinear case.

28.2.4 Determine the Model Accuracy

The final and perhaps the most important step in the process of analytical modeling is to verify the model's accuracy. There is always a limited valid range for any model, and it is necessary to be certain that the model is realistic for the specific problem and purpose. If the model is not acceptable, a return to one or more of the previous steps is necessary to redefine or reformulate the problem.

The verification can be done analytically or by computer simulation and it is most easily done by simulating the system using a computer. One of the main advantages of computer-based simulation is that equations do not need to be in closed form. However, when using a computer for model simulation, it is necessary that the model be in digital form. A continuous-system problem must be converted to a discrete one. This conversion will either have to be done by the engineer or, more typically, by the particular simulation software. In either case, numerical methods, which introduce their own unique approximations, need to be used. Consequently, it is important to realize that even if the simulation shows that the particular model is sufficient, the model still may have differences with the actual physical system.

Computer simulations may be done using a higher level language such as C or C++, using spreadsheets, or using any one of the many software packages such as MATLAB®, Simulink®, Maple™, and LabVIEW™. Besides simulating the model of the system, these software packages are very useful for designing the control system and verifying its performance as well. Most modern software packages can also handle a variety of system types and have good graphic capabilities.

28.3 Empirical or Experimental Modeling

In addition to analytical modeling, the second basic approach to modeling is empirical or experimental modeling. Empirical or experimental modeling uses experimental measurements to create a model that best fits the data. If frequency data is available, for example, a model may be created with the use of a

Bode diagram. Sometimes a model structure is assumed and then experimental measurements are used to estimate the parameter values so that the model best fits the data.

Usually, time series models are used for empirical modeling. These models are classified as autoregressive (AR), moving average (MA), and the combination of the two known as ARMA. For these time series models, the difference equations that relate the inputs to the outputs are of the form given in Equation 28.5. The z-transform transfer function is given in Equation 28.6. For the MA (all zero) model, all $a_i = 0$, for the AR (all pole) model all $b_i = 0$, except b_0. The ARMA model has both poles and zeros.

Empirical or experimental modeling can often be done either off-line or online. Adaptive systems, for example, carry out the estimations online. Neural networks, fuzzy systems, and expert systems are other experimental techniques used for creating models.

28.4 Examples

28.4.1 Example 1

A model of an electromechanical system consisting of a separately excited direct current (DC) motor is developed here. For this example, assumptions are made to model the DC motor as a linear, lumped parameter, deterministic, continuous-time, and stationary system. A schematic of one is shown in Figure 28.1. Separately excited DC motors may be connected either as field-controlled or armature-controlled. For field-controlled, the armature current i_a is held constant and the motor is controlled by varying the field current i_f. For armature-controlled, the field current i_f (and therefore the flux ϕ_f) is held constant and the motor is controlled by varying the armature current i_a. Armature-controlled motors are more stable and therefore used more often.

Referring to Figure 28.1, the torque $T(t)$ produced by the armature is proportional to the armature current, or

$$T = K_t i_a, \tag{28.7}$$

where K_t is the torque constant. Writing Kirchhoff's voltage law at the armature,

$$v_s = R_a i_a + L_a \frac{di_a}{dt} + v_b, \tag{28.8}$$

where v_b is the back electromotive force (emf) and

$$v_b = K_b \omega, \tag{28.9}$$

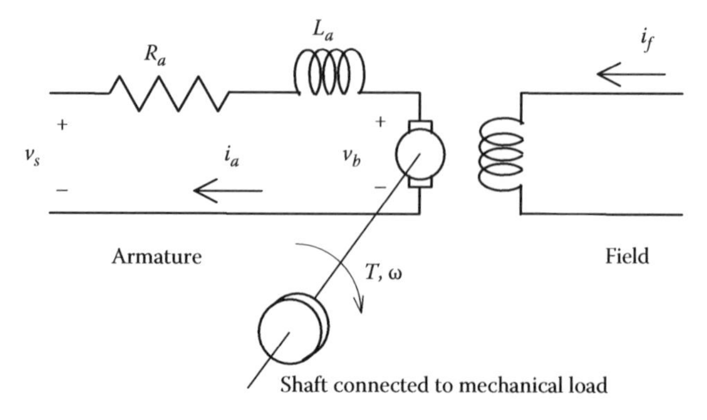

FIGURE 28.1 Separately excited DC motor.

where K_b is the back emf constant and ω is the motor's speed. Since the position of the armature θ is related to the speed by $\omega = (d\theta/dt)$, Equation 28.9 can be rewritten as

$$v_b = K_b \frac{d\theta}{dt}. \tag{28.10}$$

Substituting Equation 28.10 into Equation 28.8

$$v_s = R_a i_a + L_a \frac{di_a}{dt} + K_b \frac{d\theta}{dt}. \tag{28.11}$$

With a mechanical load on the motor, the motor torque is used to accelerate the total inertia of the motor and load J and in overcoming the friction, which is assumed proportional to the angular velocity $F\omega$ or

$$T = J\frac{d\omega}{dt} + F\omega = J\frac{d^2\theta}{dt^2} + F\frac{d\theta}{dt}. \tag{28.12}$$

Combining Equation 28.7 with Equation 28.12

$$i_a = \frac{J}{K_t}\frac{d^2\theta}{dt^2} + \frac{F}{K_t}\frac{d\theta}{dt}. \tag{28.13}$$

Taking the Laplace transform, with all initial conditions equal to zero, Equation 28.11 becomes

$$V_s = R_a I_a + L_a s I_a + K_b s\Theta, \tag{28.14}$$

and Equation 28.13 becomes

$$I_a = \frac{J}{K_t} s^2\Theta + \frac{F}{K_t} s\Theta. \tag{28.15}$$

Therefore, the transfer function, obtained by combining Equations 28.14 and 28.15, is

$$\frac{\Theta}{V_s} = \frac{K_t}{JL_a s^3 + (JR_a + FL_a)s^2 (FR_a + K_t K_b)s}.$$

The state-space model of this system can also be written. One such configuration assigns the states as $x_1 = \theta$, $x_2 = (d\theta/dt)$, and $x_3 = (d^2\theta/dt^2)$. This leads to the state equation

$$\dot{\mathbf{x}} = \begin{bmatrix} 0 & 1 & 0 \\ 0 & 0 & 1 \\ 0 & a_{32} & a_{33} \end{bmatrix} \mathbf{x} + \begin{bmatrix} 0 \\ 0 \\ b_3 \end{bmatrix} v_s,$$

where

$$a_{32} = -\frac{FR_a + K_t K_b}{JL_a}, \quad a_{33} = -\frac{JR_a + FL_a}{JL_a}, \quad \text{and} \quad b_3 = \frac{K_t}{JL_a}.$$

If the output is the armature velocity, $\omega = x_2$, the output equation is

$$y = \begin{bmatrix} 0 & 1 & 0 \end{bmatrix} \mathbf{x} + \begin{bmatrix} 0 \end{bmatrix} v_s.$$

It should be emphasized that this, as in all models, is only an approximation to the true system transfer function and state-space representation since a number of approximations are performed.

28.4.2 Example 2

A classic example often used for demonstration of control techniques is the inverted pendulum. This problem can be described as follows. A pendulum is hinged onto a cart and is free to rotate around a pivot point in one direction, as shown in Figure 28.2. For this system the following parameters are defined:

g = gravitational acceleration (m/s^2)
J = moment of inertia about the center of gravity of the pendulum (kg-m^2)
L = length of the pendulum (m)
m = mass of the pendulum (kg)
M = mass of the cart (kg)
μ_w = coefficient of friction for the cart wheels (N-s/m)
s = displacement of the cart (m)
θ = angular rotation of the pendulum (rad)

When the pendulum begins to fall, a force u is applied to the cart by a motor that will push or pull the cart to prevent the pendulum from falling.

28.4.2.1 System Equations

A free body diagram of the pendulum is shown in Figure 28.3. The horizontal displacement of the center of gravity of the pendulum is $s + (L/2)\sin\theta$ and the vertical displacement is $(L/2)\cos\theta$. The forces exerted on the pendulum are the force due to gravity mg, a horizontal reaction force at the pivot point due to the cart F_x, and a vertical reaction force at the pivot point due to the cart F_y.

Summing the forces on the pendulum gives

$$F_x = m\frac{d^2}{dt^2}\left(s + \frac{L}{2}\sin\theta\right), \tag{28.16}$$

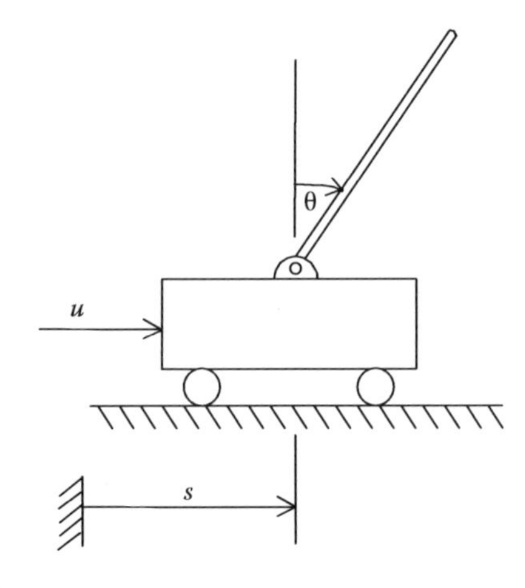

FIGURE 28.2 Inverted pendulum.

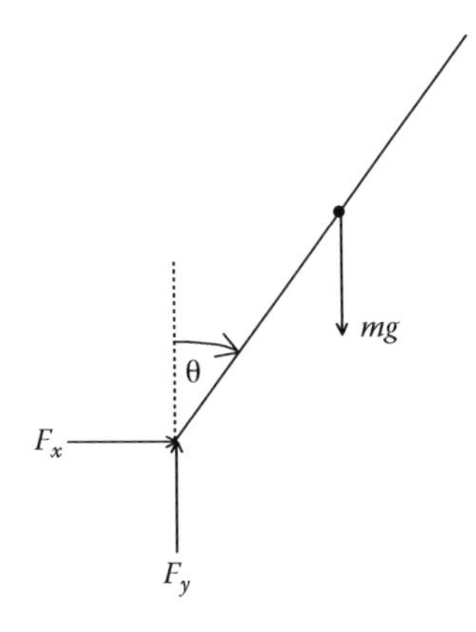

FIGURE 28.3 Free body diagram—pendulum.

and

$$F_y - mg = m\frac{d^2}{dt^2}\left(\frac{L}{2}\cos\theta\right). \quad (28.17)$$

Summing the moments of inertia with respect to the center of gravity of the pendulum gives

$$J\frac{d^2\theta}{dt^2} = F_y\frac{L}{2}\sin\theta - F_x\frac{L}{2}\cos\theta. \quad (28.18)$$

A free body diagram of the cart is shown in Figure 28.4. The horizontal forces exerted on the cart are the external applied force u, a horizontal reaction force due to the pendulum F_x, and the force due to friction at the wheels of the cart $\mu_w v$.

Summing the horizontal forces on the cart gives

$$M\frac{d^2 s}{dt^2} = u - F_x - \mu_w\frac{ds}{dt}, \quad (28.19)$$

Combining the differentiation Equations 28.16 through 28.19 become

$$F_x = m\ddot{s} + m\frac{L}{2}\ddot{\theta}\cos\theta - m\frac{L}{2}\dot{\theta}^2\sin\theta, \quad (28.20)$$

$$F_y = mg - m\frac{L}{2}\ddot{\theta}\sin\theta - m\frac{L}{2}\dot{\theta}^2\cos\theta, \quad (28.21)$$

$$J\ddot{\theta} = F_y\frac{L}{2}\sin\theta - F_x\frac{L}{2}\cos\theta, \quad (28.22)$$

and

$$M\ddot{s} = u - F_x - \mu_w\dot{s}. \quad (28.23)$$

Substituting Equations 28.20 and 28.21 into Equations 28.22 and 28.23 gives the nonlinear system equations

$$J\ddot{\theta} = mg\frac{L}{2}\sin\theta - m\left(\frac{L}{2}\right)^2\ddot{\theta} - m\frac{L}{2}\ddot{s}\cos\theta, \quad (28.24)$$

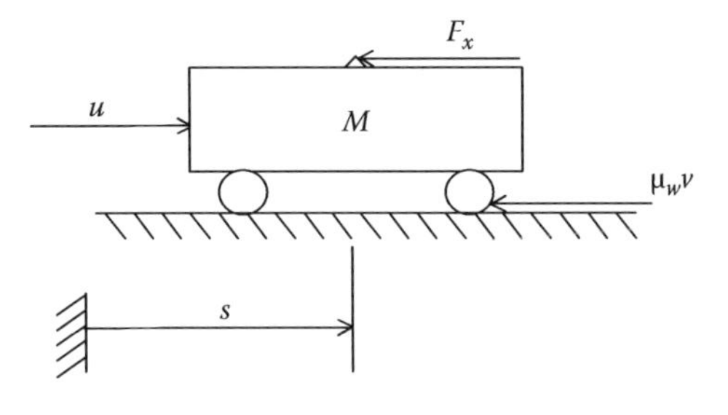

FIGURE 28.4 Free body diagram—Cart.

and

$$M\ddot{s} = u - m\ddot{s} - m\frac{L}{2}\ddot{\theta}\cos\theta + m\frac{L}{2}\dot{\theta}^2\sin\theta - \mu_w\dot{s}. \tag{28.25}$$

28.4.2.2 Linearization

Using the first two terms of the Taylor series, Equations 28.24 and 28.25 are linearized to give

$$J\ddot{\theta} = mg\frac{L}{2}\theta - m\left(\frac{L}{2}\right)^2\ddot{\theta} - m\frac{L}{2}\ddot{s}, \tag{28.26}$$

and

$$M\ddot{s} = u - m\ddot{s} - m\frac{L}{2}\ddot{\theta} - \mu_w\dot{s}. \tag{28.27}$$

Rearranging, Equation 28.26 becomes

$$\left[J + m\left(\frac{L}{2}\right)^2\right]\ddot{\theta} = mg\frac{L}{2}\theta - m\frac{L}{2}\ddot{s}. \tag{28.28}$$

Defining $a = [J + m(L/2)^2]/(mL/2)$, Equation 28.28 becomes

$$a\ddot{\theta} + \ddot{s} = g\theta. \tag{28.29}$$

Finally, substituting Equation 28.29 into Equation 28.27

$$\left(M + m - \frac{mL}{2a}\right)\ddot{s} = u - \frac{mL}{2}\frac{g}{a}\theta - \mu_w\dot{s}, \tag{28.30}$$

and defining $b = M + m - (mL/2a)$ and $c = ((mL/2)\,(g/a))/ab$, Equation 28.30 becomes

$$b\ddot{s} = u - abc\theta - \mu_w\dot{s}. \tag{28.31}$$

28.4.2.3 State Assignment

The two second-order equations, Equations 28.29 and 28.31, require four states. These can be defined as

$$\begin{aligned} x_1 &= s, \\ x_2 &= \dot{s}, \\ x_3 &= s + a\theta, \\ x_4 &= \dot{s} + a\dot{\theta}. \end{aligned} \tag{28.32}$$

Using Equations 28.29, 28.31, and 28.32, the following state equations are found:

$$\dot{x}_1 = x_2$$

$$\dot{x}_2 = -cx_1 - \frac{\mu_w}{b}x_2 + cx_3 + \frac{1}{b}u$$

$$\dot{x}_3 = x_4$$

$$\dot{x}_4 = -\frac{g}{a}x_1 + \frac{g}{a}x_3$$

These equations can be written in standard state variable matrix form as

$$\dot{\mathbf{x}} = \begin{bmatrix} 0 & 1 & 0 & 0 \\ -c & -\frac{\mu_w}{b} & c & 0 \\ 0 & 0 & 0 & 1 \\ -\frac{g}{u} & 0 & \frac{g}{a} & 0 \end{bmatrix}\mathbf{x} + \begin{bmatrix} 0 \\ \frac{1}{b} \\ 0 \\ 0 \end{bmatrix}u. \tag{28.33}$$

The following values were used in this example:

$g = 9.81\ \text{m/s}^2$
$L = 1.0\ \text{m}$
$m = 1.0\ \text{kg}$
$M = 10\ \text{kg}$
$\mu_w = 1.0\ \text{N-s/m}$

Using these values and assuming the pendulum has a uniform mass distribution

$$J = \frac{mL^2}{12} = \frac{1}{12} = 0.0833\ \text{kg-m}^2.$$

In addition, with the above values, the following are obtained:

$$a = \frac{[J + m(L/2)^2]}{mL/2} = \frac{2}{3} = 0.6667\ \text{m},$$

$$b = M + m - \frac{mL}{2a} = 10.25\ \text{kg},$$

and

$$c = \frac{(mL/2)(g/a)}{ab} = 1.0767\ \text{s}^{-2}.$$

With these values, Equation 28.33 becomes

$$\dot{\mathbf{x}} = \begin{bmatrix} 0 & 1 & 0 & 0 \\ -1.0767 & -0.0976 & 1.0767 & 0 \\ 0 & 0 & 0 & 1 \\ -14.715 & 0 & 14.715 & 0 \end{bmatrix} \mathbf{x} + \begin{bmatrix} 0 \\ 0.0976 \\ 0 \\ 0 \end{bmatrix} u.$$

The output can be chosen in a variety of ways. Here, we chose the output to be the cart position s. Therefore the output equation becomes

$$\mathbf{y} = \begin{bmatrix} 1 & 0 & 0 & 0 \end{bmatrix} \mathbf{x} + \begin{bmatrix} 0 \end{bmatrix} u.$$

28.5 Conclusion

Designing a control for an engineering system usually requires the use of a mathematical model of the system. This mathematical model is obtained either analytically or empirically. Analytical models are used when the basic structure of the system is known, while empirical or experimental models are used when the structure of the system is not known and experiments need to be performed to obtain it.

Most engineering systems, where noise is not a major problem, are most commonly modeled as deterministic, linear, lumped parameter systems in either continuous or discrete time. These are represented in one of two ways: (1) as differential or difference equations and transfer functions, or (2) as a state-space model. For most higher order systems, the state-space representation is most common.

In any case, an accurate mathematical model of the physical system is essential for successful control system design.

References

[B91] Brogan, W. L., *Modern Control Theory* (3rd edn.), Prentice Hall, Englewood Cliffs, NJ, 1991.

[BM00] Breitfelder, K. and Messina, D., editors, *IEEE 100: The Authoritative Dictionary of IEEE Standard Terms* (7th edn.), The Institute of Electrical and Electronics Engineers, Inc., New York, 2000.

[C67] Cannon, R. H., *Dynamics of Physical Systems*, McGraw-Hill, Inc., New York, 1967.

[CRD00] Cha, P. D., Rosenberg, J. J., and Dym, C. L., *Fundamentals of Modeling and Analyzing Engineering Systems*, Cambridge University Press, Cambridge, U.K., 2000.

[DB08] Dorf, R. C. and Bishop, R. H., *Modern Control Systems* (11th edn.), Pearson Prentice Hall, Upper Saddle River, NJ, 2008.

[DHS03] D'Azzo, J. J., Houpis, C. H., and Sheldon, S. N., *Linear Control System Analysis and Design* (5th edn.), Marcel Dekker, Inc., New York, 2003.

[FPE10] Franklin, G. F., Powell, J. D., and Emami-Naeini, A., *Feedback Control of Dynamic Systems* (6th edn.), Pearson, Upper Saddle Ridge, NJ, 2010.

[KG09] Kuo, B. C. and Golnaraghi, F., *Automatic Control Systems* (9th edn.), John Wiley & Sons, New York, 2009.

[O10] Ogata, K., *Modern Control Engineering* (5th edn.), Prentice Hall, Upper Saddle River, NJ, 2010.

[PH99] Phillips, C. L. and Harbor, R. D., *Feedback Control Systems* (4th edn.), Prentice Hall, Englewood Cliffs, NJ, 1999.

[VE97] Vu, H. V. and Esfandiari, R. S., *Dynamic Systems Modeling and Analysis*, McGraw-Hill, Inc., New York, 1997.

29

Intelligent Mechatronics and Robotics

Satoshi Suzuki
Tokyo Denki University

Fumio Harashima
Tokyo Metropolitan University

29.1 Introduction

Intelligent mechatronics is a machine system that has its own entity and equips human-like or creature-like smart abilities. The concept of "intelligent" is, however, ambiguous; hence, there is no clear definition because the meanings and the nuance are modified according to individual intention of a designer who deals with the concept. A robot embedded with complex artificial intelligent functions based on a human cognitive model is a good example of the intelligent mechatronics. Meanwhile, in the industrial world, a product with self-diagnostic functions, or a mechatronics device that has higher function than the former version, are sometimes called "intelligent mechatronics." As another example, so-called wet-mechatronics fusing the mechatronics technology into the biological organic body appears to be a kind of intelligent mechatronics. As seen above, there is no clear-cut boundary about intelligent mechatronics. In this chapter, however, the authors would like to omit such path-breaking types of intelligent mechatronics, and the representative intelligent mechatronics will be explained.

29.2 System Structure of Intelligent Mechatronics

Intelligent mechatronics basically consists of the electronics-mechanism unit and the information processing unit for control of the motion. Hence, a virtual agent by computer graphics without the physical body is not accepted as an intelligent mechatronics. The intelligent mechatronics differs from ordinary automatic machines since the intelligent mechatronics is designed with consideration for dynamic change of environment, human, and events. To cope with these exogenous factors, usual intelligent mechatronics has the processors of perception, cognition, and motor, and the information storage like *model human processor* (MHP) shown in Figure 29.1. MHP was proposed by Card et al. as a human cognitive architecture. MHP consists of 3 processors, 4 memories, 19 parameters, and 10 principles of

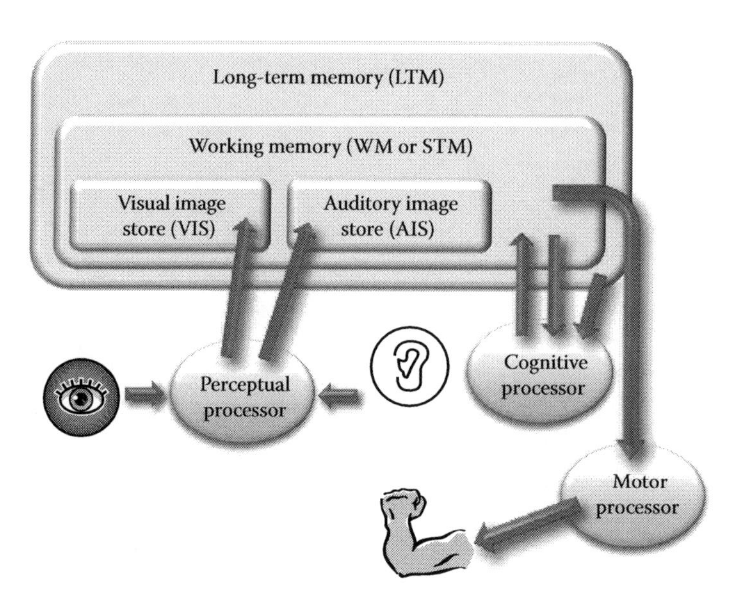

FIGURE 29.1 Model human processor (MHP).

operation [CMN83]. Although no one proves whether human beings actually work like MHP, the model is widely accepted and is utilized for the design of the intelligent mechatronics.

The basic strategy of the adaptation is to modify the control law and algorithms according to the state of affairs around the machine. Kawato-Ito model [DKK01] is often utilized to build logical structures of the adaptation ability in many types of intelligent mechatronics. This model was originally proposed to explain human brain function for the voluntary motion control. In the scheme of the model, the voluntary motion is generated by a feedforward control using the inner dynamics models that have been already learned, and the inner models are modified by feedback-error learning when the output from the inner model deviates from the reference motion [KG92]. Further, Wolpert extended the feedback-error learning into module selection and identification control (MOSAIC), which consists of multiple modules having local characteristics. Based on the concept of MOSAIC, human sequential actions are generated by switching the modules according to the state of affairs [WK98]. The idea of MOSAIC resembles the NASRAM that will be explained later in Section 29.4, and these module-based architectures are the fundamental method for the intelligent mechatronics in order to adapt to dynamically changeable factors.

As mentioned above, basic control structure of the intelligent mechatronics is frequently designed based on human models. Moreover, informatics, cognitive psychology, brain science, sociology, and biology are also required to build up a sophisticated intelligent mechatronics that can adapt to various circumstances. From recent trend on the mechatronics research, one of categories about intelligent mechatronics is illustrated by Figure 29.2. Representative types of the intelligent mechatronics will be introduced below, and their key technologies, theories, and examples are explained.

29.3 Network Intelligent Mechatronics

The purpose of this type of intelligent mechatronics is realization of high performance by integrating information. The difference between other information integration systems and this intelligent mechatronics is that sensor nodes of the intelligent mechatronics are movable by the mobile platforms like wheeled mobile robots. The network intelligent mechatronics is often called *networked robot*. The intelligent abilities are created by cooperation with other factors via network. Therefore, not only robots but also the environment of the work area including the infrastructure should be designed adequately. One of the illustrative cases is *intelligent space* [SH07] that makes whole space of the room intelligent by using

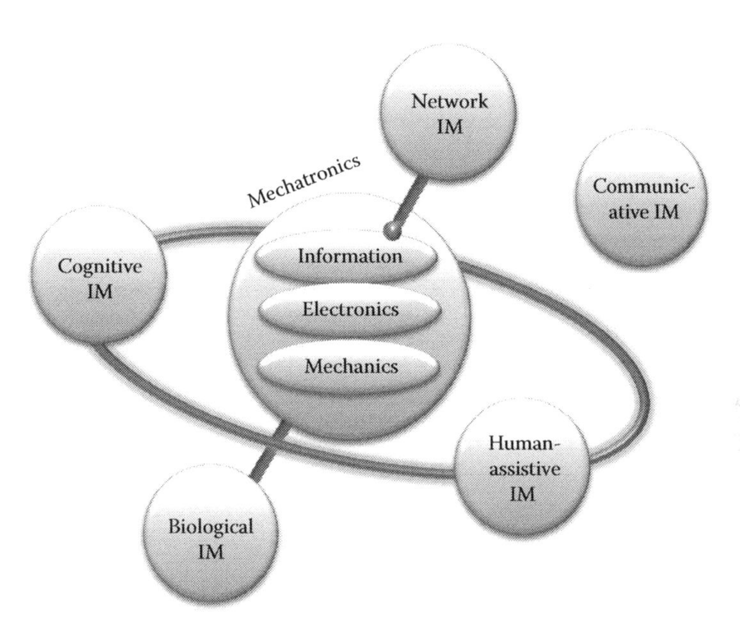

FIGURE 29.2 Categories of intelligent mechatronics (IM).

FIGURE 29.3 Network intelligent mechatronics: *Intelligent space*. (Courtesy of H. Hashimoto).

sensors distributed to the room and the mobile robots (Figure 29.3). *Intelligent space* resembles *intelligent room* [B97] studied by the former AI laboratory at MIT. *Intelligent room* enabled control of equipments in the room by the gesture and voice recognition. *DreamSpace* (IBM) and *Oxygen Project* (MIT) [Oweb] are similar researches; however, these approaches except *intelligent space* do not consider physical interaction between robot agents and humans. In Japan, three types of network robots were proposed: the virtual type, the unconscious type, and the visible type [MIA03]. The virtual type is an agent or an icon robot in virtual space in order to communicate with a human. The unconscious type is embedded in the environment and detects both human action and circumstance by various kinds of sensing technology. The visible type is a conventional robot that works in actual space. In other countries, e.g., in Korea, the *ubiquitos robot companion* is proposed and that has similar categorization to the Japanese version. In Euro-America, *networked robots*, which consists of the unconscious type and the visible type but does not include the virtual type, have been studied. The field tests, for instance the URUS project (ubiquitos networking robotics in urban settings project) in EU [SA06], are proceeding all over the world.

For the network intelligent mechatronics, the mission-critical technology is position estimation of each mobile sensor node. The orthodox method is the dead reckoning with compensation by using landmarks of which position are known. When there is no available landmark or the environment is unknown, the technique of simultaneous localization and mapping (SLAM) [BD06] is utilized. SLAM is the process by which a mobile robot can build a map of the environment around the robot and, at the same time, utilizes the map to compute its location. Mainly, two approaches of the SLAM are used: the probabilistic localization algorithms known as Monte Carlo localization [TFB01] and the scan matching method [CM91]. The example of application is the surveillance system COMETS for the management of environmental and industrial disaster [CMP07]. In the system, SLAM is executed by using unmanned air vehicles and mobile robots to investigate the supervised area. Since this type of intelligent mechatronics is already equipped with the network for communication with other mobile nodes, extension of the local network to the Internet enables us to realize a new type of network robot termed "network intelligence" or "web intelligence" [MIH06]. The mobile robots access the Internet and extract adequate information by using the Semantic Web technology for environment recognition and further processing. The key technology of Semantic Web is "ontology," which classifies the synonymous keywords and the associated concepts [HYK08]. Such data-mining approach using the Internet appears to evolve the network intelligent mechatronics for the future.

29.4 Cognitive Intelligent Mechatronics

In order to understand the circumstance and to determine own behavior, some artificial cognitive function is implemented in this type of intelligent mechatronics. Mutual interaction by the machine's embodiment to a human and the environment is significant. This intelligent mechatronics differs obviously from an existing environmental discrimination based on only passive observation or a classical artificial intelligence that is built only by the computer database. The embedded algorithms are designed mainly by referring various human cognitive models. Norman's "Seven Stages of Action" [N88] and Rasmussen's "Three Levels of Skilled Performance" [RPG94] provide us with useful information. Model of "Seven Stages of Action" is made up of perception, interpretation, evaluation, goal, intention, planning, and execution, as shown in Figure 29.4. Norman emphasized the repetitive cycle that the result

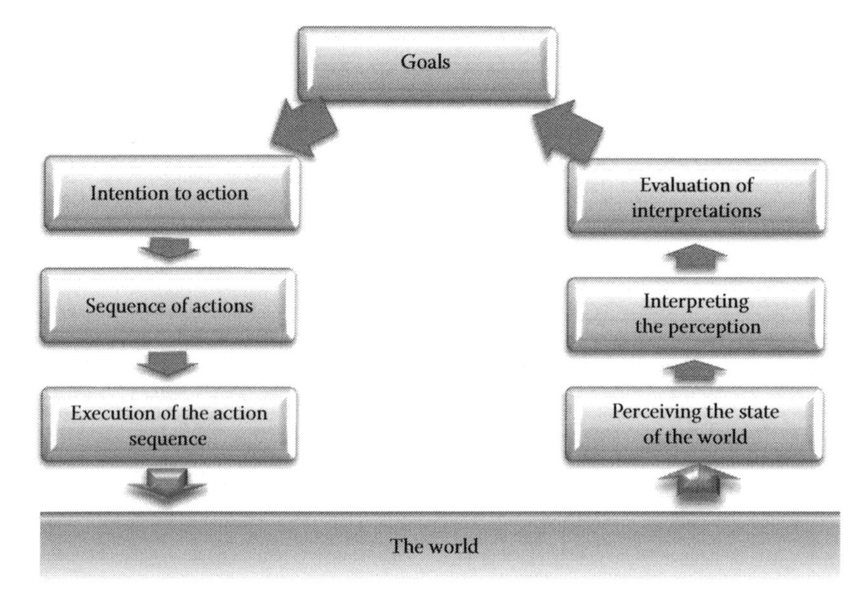

FIGURE 29.4 Seven stages of action.

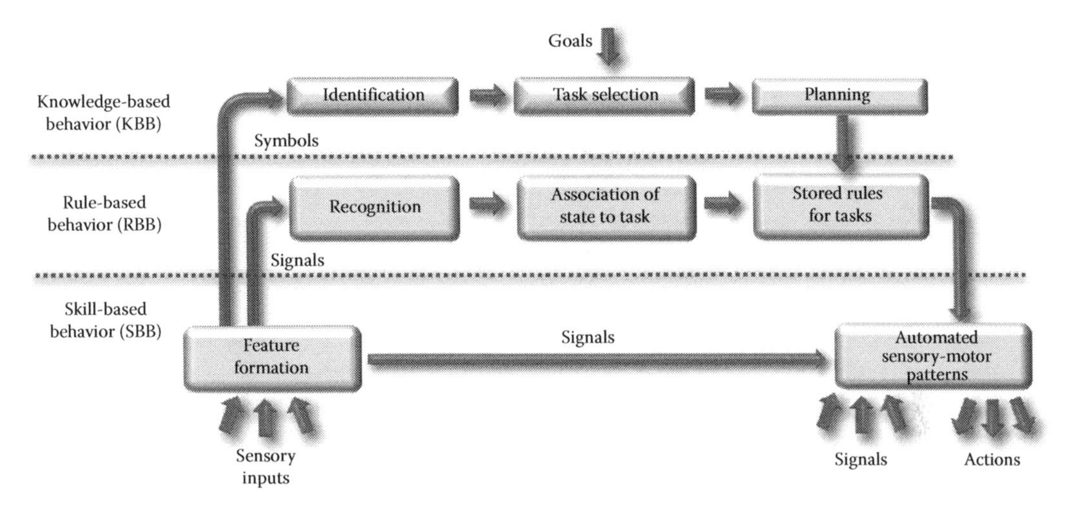

FIGURE 29.5 Three levels of skilled performance.

of execution of action was perceived again. In this model, human activity is considered as circulative mental actions for achievement of the goal. On the other hand, Rasmussen's model defines three qualitatively different levels of the skill-based behavior (SBB), the rule-based behavior (RBB), and the knowledge-based behavior (KBB), as shown in Figure 29.5. Sensory input from the real world is processed in any of these levels, and the outcomes computed from each level are integrated to emerge as an action. GOMS [GJA93] for explanation of the human cognitive process is also used to equip the machine with the function of the action selection. GOMS is a generic name of the modeling groups that include four basic elements: goal, operator, method, and selection rule. "Learning by exploration" model [PL90] is also utilized to express trials-and-errors process of human behavior. This model classifies actions of the user by using four phases: the goal-setting, search, selection, and evaluation.

For the control algorithm and its implementation, the representative approaches are as follows with relation to the above-mentioned models:

- Sequential processing of perception–decision–action (Norman's model–like approach)
- Top-down hierarchical control mechanism (Rasmussen's model–like approach)
- Bottom-up hierarchical control mechanism (subsumption architecture, SSA)

The first approach is so-called classical artificial intelligence, and this is often used in practice because of the simplicity of the algorithm. As shown in Norman's model, first, some physical state is measured by sensor from the world (perception phase). Next, adequate behavior is determined by computing based on the collected information (decision phase), and the motor control of the mechanism is executed to achieve the behavior (action phase). The program algorithm is basically a repetition processing of sensing, computing, and controlling. There is, however, limitation known as *frame problem* that an exception processing has to be programmed adequately beforehand against the dynamical domain. This approach tends to be difficult to give a quick response to dynamic change of the real world.

The second approach has a hierarchical structure with multilayers corresponding to the complexity of the knowledge and intelligence that are required to perform actual task. Processing of this intelligent control is done by switching the numerical computation and the symbolic computation according to difference of the task quality. Pursuant to the Rasmussen's model, the signal, the sign, and the symbol are assigned to the skill-based behavior (SBB), the rule-based behavior (RBB), and the knowledge-based behavior (KBB), respectively. In the hierarchical control, each layer is driven by the different type of information. Since the whole structure and the knowledge have to be given by the system designer, this system mechanism is called a top-down approach. NASREM (NASA/NBS standard reference model for telerobot control system architecture [FLA87]) by Albus and the *Multiresolutional Control Architecture*

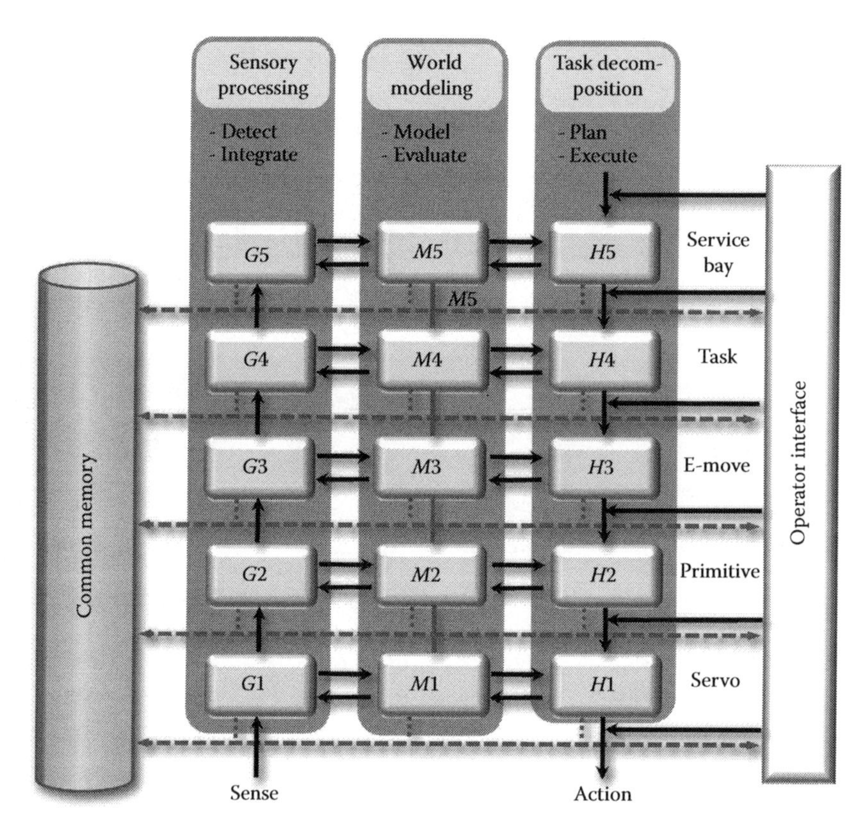

FIGURE 29.6 NASA/NBS standard reference model for telerobot control system architecture (NASREM).

by Meystel are representative cases [FE00]. NASREM consists of three modules: the task decomposition, the world modeling, and the sensory processing. Each module has hierarchical network structure, as shown in Figure 29.6. In NASREM, *common memory* stores information of the model of the external world and provides the information to each module. The task decomposition module is a scheduler that splits up the task goal into subtasks so as to execute subtasks in the lower level module. This module assigns the task, makes the plan, and executes the task at each layer. The world modeling module builds and maintains the database of models of the internal and external world and responds to the inquiries from the other modules. This module plays a role of the so-called internal model, and the predicted value is computed there and is passed to the other modules. The sensory processing module measures the status by sensors, performs pattern recognition, detects events, and controls of filtering of the sensor information. This module affects the control system by comparing the predicted values with the measured values and stores information of the newly detected events, objects, and relations into the common memory via the world modeling module. Thus, each module and each layer affect each other.

The third approach, SSA, was proposed by Brooks. Brooks advocated that it was quite difficult to encode the real world sufficiently and the adjustability of actions generated by the symbolic computation were low. SSA is a parallel distributed processing using the excitatory and inhibitory mechanism of multisensory-motor systems. Action of the machine is changed reflexively by interaction to the environment without an explicit model of the environment. Since this mechanism is driven directly by the sensory information, SSA is called bottom-up approach. SSA is instantaneous against the environment; however, the behavioral pattern tends to be simplistic, and the learning ability is primitive. Hence, SSA is not adequate to complex tasks that require ratiocination or communication since SSA does not have model of the inner and outer worlds. Despite these, this mechanism was impactful compared to orthodox AI methods since SSA has counterbalancing ability to the frame problem at least in principle.

Recently, the "affordances" theory attracts attention, and the behavior design of robots based on the affordances concept becomes an active area of research. There are two kinds of definitions about "affordance" in a precise sense, and Norman's affordance is frequently cited in the context of human–machine interaction [N88]. Definition of the affordance is the action possibilities and opportunity given by an environment to the creature that lives there. For instance, a human changes his or her hand shape according to difference of the handle type of the door involuntarily and changes direction of the hand motion. An intelligent mechatronics designed using the affordance theory extracts information that is implicitly embedded in the environment and selects the final action from several action possibilities based on the information [K02]. The selection is done by the probabilistic decision with the perceived action possibilities induced by environmental factors. The perception of the environment and decision of the action often corresponds to discretization from the continuous-time sensor signal to discrete value of status; hence, a dynamic clustering method to convert from analog signal to the discrete status is required. Soft computing, such as neural network (NN), genetic algorithm (GA) [SSZ97], Fuzzy theory, and self-organizing map (SOM) [K90] are common maneuvers [SFT97]. Hidden Markov model (HMM), which can classify the time-series data into discrete states by the expression of the data generation process with probability parameters, is also well utilized. For action acquisition, the evolutionary computation [KMK01] and the reinforcement learning are utilized since these techniques enable modification of own parameters of the control model through interaction to the environment. As an advanced cognitive intelligent mechatronics, studies of self-development to obtain framework of general behavior by positive interaction to the environment have been continued. The pioneering research appears to be the COG project [BS93]. In recent study using the humanoid robot QRIO, several actions of the children's toy-play were autonomously realized depending on the situation by the recurrent neural network and the self-organizing map (RNN-SOM) [F07].

29.5 Communicative Intelligent Mechatronics

This intelligent mechatronics is one of the human–machine interfaces, but this is quite different from mere information terminal since the entity of the utterance exists with the body of the mechatronics. The benefit is supported by the fact that communication using a visible robot is superior to a virtual agent on the computer display [KB04]. There are two main types of communication: the verbal communication and the nonverbal one. For the verbal communication, the voice recognition and the response system are required, and the various kind of the algorithms have been studied for a number of years. The popular method is to extract the voice feature as typified by the missing feature theory (MFT) [BCG01] and the maximum likelihood linear regression (MLLR) [LW95]. A mechanism to understand dynamics of communication is also necessitated, and the applied examples using the recurrent neural network and the interactive evolutionary computing [SIN05] are known.

On the other hand about the nonverbal communication, the face direction and the pointing behavior [SYH07], the joint attention [S00], the conjugate gaze [KKB07], the nod [OW01], and the facial expression [HO106] have been studied. Any of these approaches tries to communicate intuitively and instantaneously. In fact, the eye contact using the robotic head enhances natural communication to transmit the machine's intention to a human [IOI03]. An essence of these nonverbal communication methods is that human's interaction with the machine is induced by anthropomorphic machine that imitates human-like behavior. In order to equip the machine with this kind of intelligence, not only the sophisticated motion control of the machine body but also the mathematical "mental model" is required (for instance, [PK06]). As the nonverbal communication can induce strong *mental interaction* from a human, socially intelligent mechatronics, so-called therapy robots have been developed. For instance, the seal robot Paro [S04] was created in order to attempt to improve physiological and psychological health in elderly patients (Figure 29.7). The social/sociable robots such as Kismet and Leonardo had been developed to study cognitive-affective architecture based on the embodied cognition theories such as the "theory of mind," the mind reading, and the social commonsense [B02].

FIGURE 29.7 Seal-type mental commit robot: Paro. (Courtesy of T. Shibata, AIST, Japan).

Physical interaction using reaction force on the cooperative task appears to be nonverbal communication that is peculiar for the mechatronics although this is not strictly nonverbal communication in the customary sense. As a study of adaptive intelligent mechatronics to read the human intention from the force information, partner ballroom dance robot (PBDR) shown in Figure 29.8 is known. PBDR estimates the intention of the human dance partner from the force-sensor information, selects adequate dance step, and generates own motion [KTH07]. Such technologies of the collaborative control based on

FIGURE 29.8 Partner ballroom dance robot. (Courtesy of K. Kosuge).

the mechanical interaction and the adaptive motion planning involving estimation of human intention are needed for the intelligent mechatronics to create everyday space where human and robot coexist.

29.6 Biological Intelligent Mechatronics

This type of intelligent mechatronics form variable colony that is assembled by more than one similar unit. The main concern is how to create living creature's adaptation capability such as the self-organizing and the differentiated function. One example is the "modular robot" that imitates self-assembly of the cell membrane and tissues [K07,FU98]. Although each individual unit moves by relatively simple control logic, we feel some intellectual intention from the whole of population behavior of the colony. From this aspect, this approach is expected to play an important role in revealing the adaptation of the creature. The biological intelligent mechatronics has attracted attention as new orientation of the mechatronics.

29.7 Human-Assistive Intelligent Mechatronics

The origin of the intelligent mechatronics is to mimic human sophisticated abilities by using technologies of mechatronics. From this historical background, dynamic adaptation to an environment had been studied from the early stages since the adaptation was considered as the most significant human ability. The other adaptation to a human was, however, not considered sufficiently in the olden days. For this issue, positive enhancement of the human abilities by the machine is taken into account considerably in recent studies, and many international research projects are continued. In German MORPHA project, the communication and interaction with intelligent robot assistants were treated. Coexistence of the human and the robot for the housekeeping and the manufacturing were studied in the project [Mweb]. In the European COGNIRON project, robot companions to exhibit cognitive capacities have been studied for human-centered environments [Cweb]. In the Japanese HAM (human adaptive mechatronics) project, the analyses of human skill and the establishment of assist methods for human operators are main concerns [HS08,H05,F03]. HAM is aimed at assisting the human according to his or her skill level by changing the own function. The surgery support system (scrub nurse robot system) [MMS05] as shown in Figure 29.9, the haptic assist device [SKF06], and

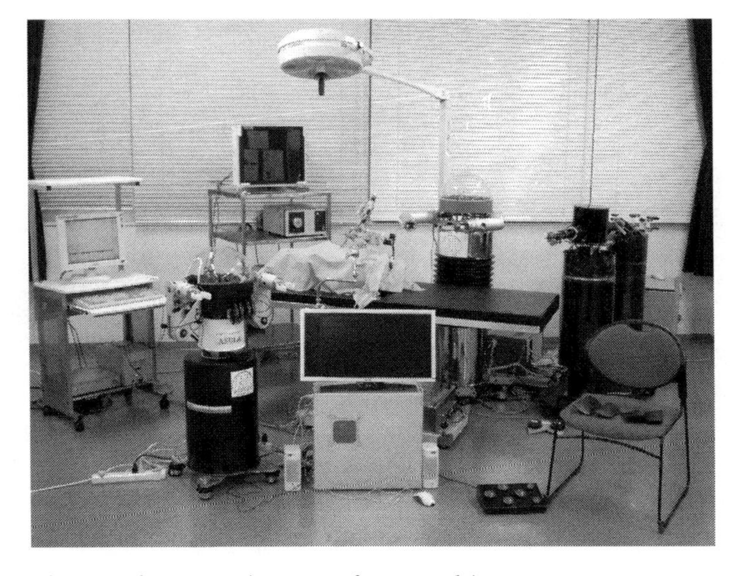

FIGURE 29.9 Scrub nurse robot system. (Courtesy of F. Miyawaki).

the teleoperated legged robot [ITS05] have been studied as the application of HAM. Skill analyses by monitoring brain with the near-infrared spectroscopy in the mirror drawing test [SK08] and the virtual pendulum stabilization task [SPH05] have been continued. The safe manual control [AIF06] and the rate saturation control of actuators [SHF10] were proposed as theories for HAM. As shown here, studies about the intelligent mechatronics to support the human are nurtured by many multidisciplinary international research projects.

29.8 Conclusion and Epilogue

In the above sections, categories of the intelligent mechatronics, the related theories, and the application were introduced. It is, however, hard to guarantee that the authors' subjective opinion does not remain there. There must be other opinions about adequate classification and better definition for the intelligent mechatronics. Conversely, this may express the diversity of the intelligent mechatronics.

Although various studies entitled with a keyword of intelligent mechatronics are active, the completeness of the quality is still low compared with the human intelligence. The reason is that knowledge of human psychology cannot be written sufficiently by computer algorithms; hence, the present intelligent mechatronics cannot understand the human perfectly. If abstract expression could be permitted here, the role of the intelligent mechatronics is to enhance human intelligence by mechatronics systems.

Acknowledgments

The HAM project and several topics mentioned in this chapter were supported by a Grant-in-Aid for 21st Century COE (Center of Excellence) Program and a Grant-in-Aid for Scientific Research (A) in the Ministry of Education, Culture, Sports, Science and Technology, Japan. The therapy robots Paro, the dance robot PBDR, and *intelligent space* were part of the Interaction and Intelligence project of Precursory Research for Embryonic Science and Technology (PRESTO) supported by the Japan Science and Technology Agency. The authors wish to express their gratitude to the Ministry and the Agency, Prof. H. Hashimoto, Prof. K. Kosuge, Dr. T. Shibata, Prof. F. Miyawaki, and their colleagues for their cooperation.

References

[AIF06] K.J. Aström, M. Iwase, K. Furuta, and J. Akesson, Safe manual control of pendulums—A human adaptive mechatronics perspective, *International Journal of Assistive Robotics and Mechatronics*, 7(1), 3–11, 2006.

[B02] C. Breazeal, *Designing Sociable Robots*, MIT Press, Cambridge, MA, 2002.

[B97] R. Brooks et al. The intelligent room project, *International Cognitive Technology Conference*, Aizu, Japan, pp. 271–278, 1997.

[BCG01] J. Barker, M. Cooke, and P. Green, Robust ASR based on clean speech models: An evaluation of missing data techniques for connected digit recognition in noise, *European Conference on Speech Communication Technology (Eurospeech-2001)*, Aalborg, Denmark, vol. 1, pp. 213–216, 2001.

[BD06] T. Bailey and H.D. Whyte, Simultaneous localization and mapping (SLAM): Part II, *IEEE Robotics and Automation Magazine*, 13(3), 108–117, 2006.

[BS93] R.A. Brooks and L.A. Stein, Building brain for bodies, *MIT AI Memo*, Cambridge, MA, no. 1439, pp. 1–15, 1993.

[CM91] Y. Chen and G. Medioni, Object modeling by registration of multiple range images, *IEEE International Conference on Robotics and Automation*, 3, 2724–2729, 1991.

[CMN83] S.K. Card, T. Moran, and A. Newell, *The Psychology of Human-Computer Interaction*, Hillsdale, NJ, Lawrence Erlbaum Associates, 1983.

[CMP07] D. Cruz, J. McClintock, B. Perteet, O.A.A. Orqueda, Y. Cao, and R. Fierro, Decentralized cooperative control, *IEEE Control Systems Magazine*, 27(3), 58–78, 2007.

[Cweb] COGNIRON Official website. (available at: http://www.cogniron.org/). August 5, 2010.

[DKK01] K. Doya, H. Kimura, and M. Kawato, Neural mechanisms of learning and control, *IEEE Control Systems Magazine*, 21(4), 42–54, 2001.

[F03] K. Furuta, Control of pendulum: From super mechano-system to human adaptive mechatronics, *The 42nd IEEE Conference on Decision and Control*, plenary, Maui, HI, pp. 1498–1507, 2003.

[F07] M. Fujita, How to make an autonomous robot as a partner: Design approach vs Emergent approach, *Philosophical Transactions of the Royal Society A: Mathematical, Physical and Engineering Sciences 365(1850)*, 15, 21–47, 2007.

[FE00] T. Fukuda et al., *Intelligent System—Adaptation, Learning, Evolutional System, and Computational Intelligence* (in Japanese), Sho-Ko-Dou, Tokyo, Japan, 2000.

[FIA05] K. Furuta, M. Iwase, and K.J. Aström, Control of pendulum from human adaptive mechatronics, *Chinese Conference in Decision and Control*, Harbin, China, plenary, pp. 15–21, 2005.

[FLA87] J.C. Fiala, R. Lumia, and J.S. Albus, Servo level algorithms for the NASREM telerobot control system architecture, *SPIE 1987 Cambridge Symposium of Advances Intelligent Robotic Systems*, Cambridge, MA, vol. 851, no. 78, pp. 103–108, 1987.

[FU98] T. Fukuda and T. Ueyama, Cellular robotics and micro robotic systems, *World Scientific Series in Robotics and Automated Systems*, vol. 10, World Scientific Publishing Co., Singapore, 1994.

[GJA93] W.D. Gray, B.E. John, and M.E. Atwood, Project Ernestine: Validating a GOMS analysis for predicting and explaining real-world tasks performance, *Human-Computer Interaction*, 8, 237–309, 1993.

[H05] F. Harashima, Human adaptive mechatronics, keynote speech, *The IEEE Industrial Electronics Society*, Raleigh, USA, in CD-ROM, 2005.

[HO106] K. Hayashi, Y. Onishi, K. Itoh, H. Miwa, and A. Takanishi, Development and evaluation of face robot to express various face shape, *The 2006 IEEE International Conference on Robotics and Automation*, Orlando, FL, pp. 481–486, 2006.

[HS08] F. Harashima and S. Suzuki, Intelligent mechatronics and robotics, keynote speech, *The 2008 IEEE International Conference on Emerging Technologies and Factory Automation (ETFA2008)*, Hamburg, Germany, in CD-ROM, 2008.

[HYK08] F. Harashima (editorial supervisor), T. Yamaguchi, N. Kubota, and Y. Takama, *Introduction to Intelligent Network Systems*, (in Japanese) CORONA Publishing Co., Ltd., Tokyo, Japan, 2008.

[IO103] M. Imai, T. Ono, and H. Ishiguro, Physical relation and expression: Joint attention for human-robot interaction, *IEEE Transaction on Industrial Electronics*, 50(4), 636–643, 2003.

[ITS05] H. Igarashi, A. Takeya, S. Shirasaka, S. Suzuki, and M. Kakikura, Adaptive teleoperation system with HAM, *IEEE International Workshop on Robots and Human Interactive Communication*, Nashville, TN, pp. 484–489, 2005.

[K02] N. Kubota, Perception-based robotics based on perceiving-acting cycle with modular neural networks, *The 2002 IEEE World Congress on Computational Intelligence*, Honolulu, HI, in CD-ROM, 2002.

[K07] E. Klavins, Programmable self-assembly, *IEEE Control Systems Magazine*, 27(4), 3–56, 2007.

[K90] T. Kohonen, The self-organizing map, *Proceeding of the IEEE*, 78(9), 1464–1480, 1990.

[KB04] C. Kidd and C. Breazeal, Effect of a robot on user perceptions, *IEEE/RSJ International Conference on Intelligent Robots and Systems (IROS2004)*, Sendai, Japan, in CD-ROM, 2004.

[KG92] M. Kawato and H. Gomi, A computational model of 4 regions of the cerebellum based on feedback-error learning, *Biological Cybernetics*, 68(2), 95–103, 1992.

[KKB07] T. Kooijmans, T. Kanda, C. Bartneck, H. Ishiguro, and N. Hagita, Accelerating robot development through integral analysis of human-robot interaction, *IEEE Transactions on Robotics*, 23(5), 1001–1012, 2007.

[KMK01] N. Kubota, M. Mihara, and F. Kojima, Evolutionary robotics for quasi-ecosystem, *The 2001 Congress on Evolutionary Computation*, Seoul, Korea, pp. 115–120, 2001.

[KTH07] K. Kosuge, T. Takeda, Y. Hirata, M. Endo, M. Nomura, K. Sakai, M. Koizumi, and T. Oconogi, Partner ballroom dance robot (PBDR), *SICE Journal of Control, Measurement, and System Integration*, 1(1), 74–80, 2007.

[LW95] C.J. Leggetter and P.C. Woodland, Maximum likelihood linear regression for speaker adaptation of continuous density hidden Markov models, *Computer Speech and Language*, 9, 171–185, 1995.

[MIA03] Ministry of Internal Affairs and Communications, Reports from working group concerning network robot technologies (in Japanese), 2003.

[MIH06] Y. Muto, Y. Iwase, S. Hattori, K. Hirota, and Y. Takama, Web intelligence approach for human robot communication under TV watching environment, *SCIS & ISIS2006*, Tokyo, Japan, pp. 426–429, 2006.

[MMS05] F. Miyawaki, K. Masamune, S. Suzuki, K. Yoshimitsu, and J. Vain, Scrub nurse robot system—intraoperative motion analysis of a scrub nurse and timed-automata-based model for surgery, *IEEE Transactions on Industrial Electronics*, 52(5), 1227–1235, 2005.

[Mweb] Morpha Official website. (available at: http://www.morpha.de/php_d/index.php3). August 5, 2010.

[N88] D.A. Norman, *The Design of Everyday Things*, Basic Books, New York, 1988.

[OW01] H. Ogawa and T. Watanabe, InterRobot: Speech-driven embodied interaction robot, *Advanced Robotics*, 15(3), 371–377, 2001.

[Oweb] Oxygen Project website. (available at: http://oxygen.lcs.mit.edu/overview.html), August 5, 2010.

[PK06] K.-S. Park and D.-S. Kwon, A cognitive modeling for mental model of an intelligent robot, *International Journal of Assistive Robotics and Mechatronics*, 7(3), 16–24, 2006.

[PL90] P.G. Polson and C. Lewis, Theory-based design for easily learned interfaces, *Human-Computer Interaction*, 5(2–3), 191–220, 1990.

[RPG94] J. Rasmussen, A.M. Pejtersen, and L.P. Goodstein, *Cognitive System Engineering*, John Wiley & Sons, Inc., New York, 1994.

[S00] B. Scassellati, Foundations for a theory of mind for a humanoid robot, PhD thesis, Massachusetts Institute of Technology at Cambridge, Cambridge, MA, June 2001.

[S04] T. Shibata, An overview of human interactive robot for psychological enrichment, *Proceedings of the IEEE*, 92(11), 1749–1758, 2004.

[SA06] A. Sanfeliu and J. Andrade-Cetto, Ubiquitous networking robotics in urban settings, *IEEE/RSJ IROS Workshop on Network Robot Systems*, Beijing, China, pp. 14–18, 2006.

[SFT97] T. Shibata, T. Fukuda, and K. Tanie, Synthesis of fuzzy, artificial intelligence, neural networks, and genetic algorithm for hierarchical intelligent control, Chapter 108, in *The Industrial Electronics Handbook*, 1st edn., Ed. J.D. Irwin, CRC/IEEE Press, Boca Raton, FL, pp. 1364–1368, 1997.

[SH07] T. Sasaki and H. Hashimoto, Intelligent space as a platform for human observation, Chapter 17, *Human Robot Interaction*, Ed. Nilanjan Sarkar, I-Tech Education and Publishing, Vienna, Austria, pp. 309–324, 2007.

[SHF10] S. Suzuki, F. Harashima, and K. Furuta, Human control law and brain activity of voluntary motion by utilizing a blancing task with an inverted pendulum, *Advances in Human-Computer Interaction*, vol. 2010, pp. 16, 2010. (available at: http://www.hindawi.com/journals/ahci/2010/215825.html)

[SIN05] Y. Suga, Y. Ikuma, D. Nagao, T. Ogata, and S. Sugano, Interaction evolution of human-robot communication in real world, *International Conference on Intelligent Robots and Systems (IROS2005)*, Edmonton, Alberta, Canada, pp. 1482–1487, 2005.

[SK08] S. Suzuki and H. Kobayashi, Brain monitoring analysis of voluntary motion skills, *International Journal of Assistive Robotics and Mechatronics*, 9(2), 20–30, 2008.

[SKF06] S. Suzuki, K. Kurihara, K. Furuta, and F. Harashima, Assistance control on haptic systems for human adaptive mechatronics, *Journal of Advanced Robotics*, 20(3), 323–348, 2006.

[SSZ97] E. Sanchez, T. Shibata, and L.A. Zadeh, *Genetic Algorithms and Fuzzy Logic Systems: Soft Computing Perspectives*, World Scientific, River Edge, NJ, 1997.

[SYH07] E. Sato, T. Yamaguchi, and F. Harashima, Natural interface using pointing behavior for human-robot gestural interaction, *IEEE Transactions on Industrial Electronics*, 54(2), 1105–1112, 2007.
[TFB01] S. Thrun, D. Fox, W. Burgard, and F. Dellaert, Robust Monte Carlo localization for mobile robots, *Artificial Intelligence*, 128, 99–141, 2001.
[WK98] D.M. Wolpert and M. Kawato, Multiple paired forward and inverse models for motor control, *Neural Networks*, 11, 1317–1329, 1998.

30

State-Space Approach to Simulating Dynamic Systems in SPICE

Joel David Hewlett
Auburn University

Bogdan M. Wilamowski
Auburn University

30.1 Introduction

Since the advent of digital computing, the modeling and simulation of dynamic systems has been an integral part of the engineering process. Every increase in computing power has led to the simulation of evermore detailed and accurate models. Naturally, this has created a demand for the necessary set of tools. As a result, numerous software suites and packages have been developed for this very purpose, many of which also offer integrated environments for processing and analyzing the generated data. Some of the more familiar names such as MATLAB®, SIMULINK, LabVIEW, or VisSim may come to mind (though dozens of others have also seen significant use).

There is, however, another option, namely, SPICE. Though rarely considered, it too represents a viable and, in some cases, even preferable solution. This may come as a surprise to some, and understandably so. After all, while SPICE has long been the predominant choice for simulating and analyzing electronic circuits, it is hardly considered as a serious solution for the simulation of more general dynamic systems. In truth, however, this need not be the case.

In this chapter, a simple and straightforward method is shown, which makes it possible to simulate nearly any linear or nonlinear dynamic system in SPICE; the same approach used in [1] and [2] to successfully simulate a nonlinear 25th-order oil pumpjack system.

30.2 Building Blocks

To begin constructing and simulating models, the proper set of tools must be assembled first. Although many of the necessary building blocks cannot be implemented in SPICE directly, they can still be implemented indirectly with little effort. The implementations of some of the most commonly used blocks are presented in this section. While a comprehensive knowledge of SPICE should not be necessary, at least some familiarity is assumed. However, there are a number of excellent SPICE references available if needed [3,4].

30.2.1 Integrator

The integrator is the fundamental building block of the state-variable model. While most versions of SPICE do not support this operation directly, it can be indirectly implemented using the circuit in Figure 30.1b. The derivation is straightforward. The current across a capacitor can be expressed as

$$i = C\frac{dv}{dt}. \tag{30.1}$$

Solving (30.1) for v yields

$$v(t) = \frac{1}{C}\int_{-\infty}^{t} i(\tau)d\tau = v(t_0) + \frac{1}{C}\int_{t_0}^{t} i(\tau)d\tau. \tag{30.2}$$

Next, assuming the initial condition $v(t_0) = 0$ and letting $C = 1$ F, (30.2) reduces to

$$v(t) = \int_{t_0}^{t} i(\tau)d\tau. \tag{30.3}$$

When this result is applied to the circuit in Figure 30.1b, it is clear that the current across the capacitor is directly proportional to the input voltage v_{in}. Therefore, by (30.3), the voltage across the capacitor is

$$v_{out}(t) = \int_{t_0}^{t} v_{in}(\tau)d\tau. \tag{30.4}$$

It should be noted that not all versions of SPICE can handle the circuit in Figure 30.1b. While some versions will simulate this circuit without incident, most (including the ever popular PSpice) will refuse to cooperate. This is because for this particular configuration of G and C, most versions of SPICE will consider node 1 to be floating. Thus, having no reference for node 1, the simulation cannot proceed. Fortunately, there is a simple way to work around this. By placing a resistor in parallel with the capacitor, the problem is eliminated. Of course, the resulting integrator is no longer ideal; however, for very large resistor values, say $R \geq 1\,\text{G}\Omega$, the resulting loss is negligible.

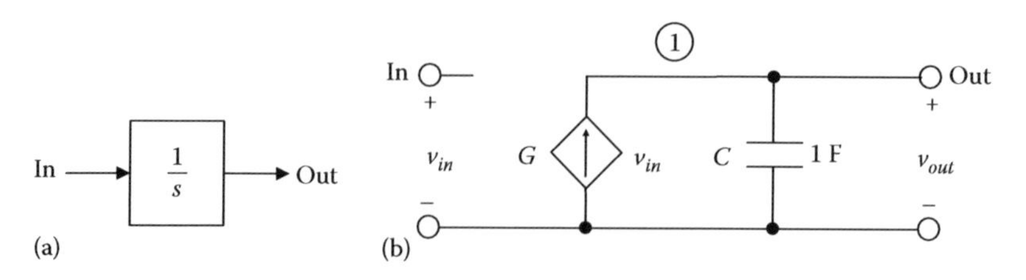

FIGURE 30.1 (a) Integrator block and (b) SPICE equivalent.

The integrator circuit in Figure 30.1b is implemented in SPICE using the following syntax:

```
1: C1 0 1 1
2: R1 0 1 1GIG
3: G1 0 1 VALUE = {V(in)}
```

On the first line, the integrating capacitor C1 is defined. The resistor on line 2 is used to circumvent the floating-node issue mentioned previously. Finally, the dependent current source G1 is used to define the function that is to be integrated. Later, when this circuit is used for constructing state-variable models, each integrator will be associated with a unique state variable. Thus, the element names and node numbers should match their corresponding state variables. For instance, a state variable labeled as x_3 would be integrated using

```
1: C3 0 3 1
2: R3 0 3 1GIG
3: G3 0 3 VALUE = {V(in)}
```

30.2.2 Differentiator

Like integration, differentiation is another useful operation, which can be indirectly implemented in SPICE with very little effort. In fact, the two circuits are quite similar. Here, the capacitor in Figure 30.1b has been replaced by the inductor in Figure 30.2b, in order to take advantage of the differential relationship between voltage and current. Naturally, the derivation of this circuit is also similar to that of the integrator and is therefore omitted. It suffices to say that the end result is

$$v_{out}(t) = \frac{\delta}{\delta t} v_{in}(t). \tag{30.5}$$

The syntax for this circuit is similar as well

```
1: L1 0 1 1
2: G1 0 1 VALUE = {V(in)}
```

30.2.3 User-Defined Expressions

One of SPICE's most powerful features is the ability to implement arbitrary mathematical expressions using dependent sources. One way of doing this is to define the desired function using the .FUNC statement. For instance, suppose a Gaussian curve is to be implemented in the form

$$f(x) = a\exp\left(-\frac{(x-b)^2}{2c^2}\right), \tag{30.6}$$

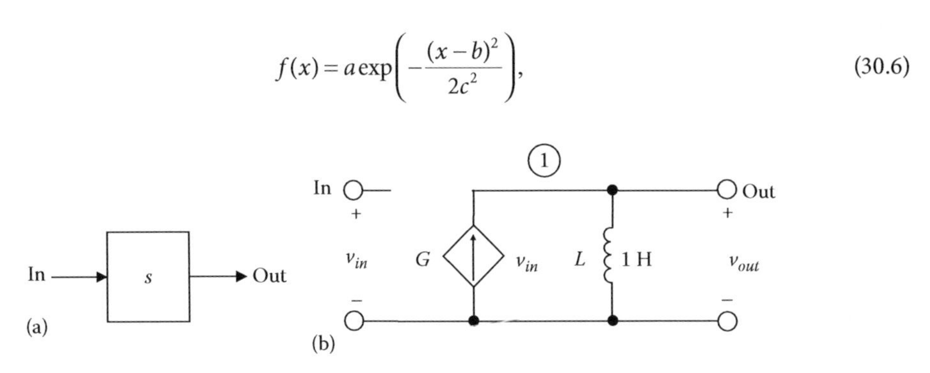

FIGURE 30.2 (a) Differentiator block and (b) SPICE equivalent.

TABLE 30.1 Functions Supported in SPICE

SPICE Syntax	Function
`ABS(x)`	Absolute value
`SQRT(x)`	Square root
`EXP(x)`	e^x—exponential function
`LOG(x)`	In(x)—natural logarithm
`LOG10(x)`	Logarithm with base 10
`PWR(x,y)`	$\|x\|^y$—power
`PWRS(x,y)`	x^y—signed power
`SIN(x)`	sin(x)—sine (x in radians)
`COS(x)`	cos(x)—cosine (x in radians)
`TAN(x)`	tan(x)—tangent (x in radians)
`ASIN(x)`	$\sin^{-1}(x)$—arc sine (result in radians)
`ACOS(x)`	$\cos^{-1}(x)$—arc cosine (result in radians)
`ATAN(x)`	$\tan^{-1}(x)$—arc tangent (result in radians)
	min if $x < min$
`LIMIT(x,min,max)`	*max* if $x > max$
	Limit results to (*min,max*) range; otherwise x
`TABLE(x, y1, x1,...)`	Lookup table and interpolate points between listed values

where $a = 1$, $b = 2$, and $c = 1$. Using the `.FUNC` statement, this function could be implemented in the following way:

```
.FUNC f(x) = {exp(-(x-2)^2/2}
```

Now, a dependant source can be used to obtain a voltage V(out) related V(in) by (30.6). That is,

```
E1 out 0 VALUE = {f(V(in))}
```

Of course, the same result could have been achieved using `VALUE = exp(-(V(in)-2)^/2`, without the `.FUNC` statement. However, using the `.FUNC` statement often makes the listing more readable, especially when longer and more complex expressions are involved.

A number of functions are supported in SPICE. A list of some of these built-in functions and their corresponding syntax are presented in Table 30.1.

30.2.4 Parameter Definitions

It is often necessary to define certain parameters that characterize the various attributes of a system, such as mass or dimension. In SPICE, this can be done by defining any necessary parameter values via the `.PARAM` statement. This can alleviate the need to search out all instances of a given parameter in order to modify its value.

In the case of the Gaussian curve from the previous section, the `.PARAM` statement could be utilized in the following way:

```
.PARAM a = 1, b = 2, c = 1
.FUNC f(x) = {a/exp( (x-b)^2/(2*c^2))}
```

Instead of directly placing the values of a, b, and c in the function definition, they are initialized as parameters using the `.PARAM` statement.

TABLE 30.2 Examples of Common Input Signals

SPICE Syntax	Signal Type
`E 1 0 VALUE = {stp(time)}`	Unit step
`E 1 0 VALUE = {stp(1-time)}`	Unit pulse
`E 1 0 VALUE = {time}`	Ramp
`E 1 0 VALUE = {sin(2*pi*(0 + 5/2*time)*time)}`	Linear chirp

30.2.5 Stimuli

SPICE is also well equipped for handling external stimuli and is capable of supplying nearly any sort of input signal needed. While only a handful of built-in signal types are offered, it is possible to generate nearly any time-dependent signal imaginable using dependent sources. This is because SPICE supports the use of the keyword *time* as a variable in user-defined expressions. This provides for a nearly limitless number of possibilities. A few examples of some common input types are listed in Table 30.2.

30.2.6 Initial Conditions

Another important consideration is the handling of initial conditions. Fortunately, SPICE has its own built-in method for this in the form of the `.IC` statement. The `.IC` statement allows for the initialization of any node voltages used in transient analysis. Representing the system's state variables as voltages allows full control over the initial state of the system. The syntax for this is straightforward. For example, an initial state of $x = [1, 2, 3]$ would be specified by

```
.IC V(1) = 1 V(2) = 2 V(3) = 3
```

30.2.7 Simulation

The final step in generating the system netlist is specifying the desired simulation parameters. This is done using the transient analysis command `.TRANS`. The `.TRANS` command takes up to five arguments, which are used to specify the start time (T_{start}), stop time (T_{stop}), print step (T_{print}), maximum time step (T_{step}), and the bias point calculation switch (`UIC`). The syntax for this is

`.TRANS` T_{print} T_{stop} T_{start} T_{step} `UIC`

For example, suppose a system is to be simulated for a period of 20 s with a maximum simulation step of 10 ms, and the resulting data is to be plotted over the interval from 1 to 20 s, with a resolution of one point per ms. To generate the desired data, the following syntax would be used,

```
.TRANS 1ms 20s 1s 10ms
```

If a set of initial conditions were also desired, the above statement would be appended with the `UIC` switch.

30.3 Examples

30.3.1 Mass–Spring–Damper System

For the first example, a simple linear second-order mass–spring–damper system is examined. As the name implies, this system consists of a mass attached to a fixed surface by a damped spring, as shown in Figure 30.3. Ignoring friction, the mass m is under the influence of three distinct forces: the spring force

$$F_s = -kx, \tag{30.7}$$

the damping force

$$F_d = -bv = b\frac{dx}{dt} = b\dot{x}, \tag{30.8}$$

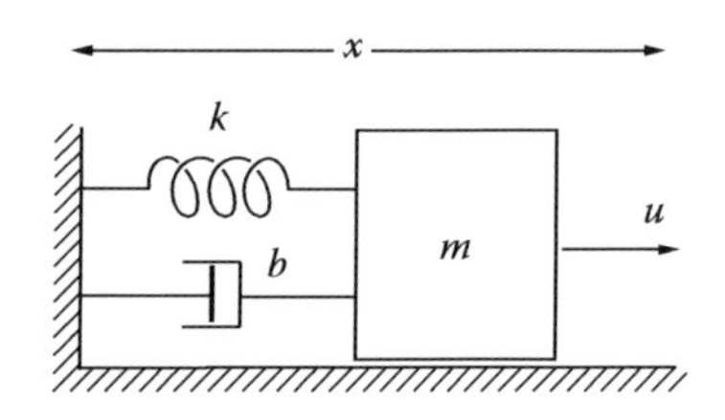

FIGURE 30.3 Second-order mass–spring–damper system.

and the external force u. Combining these forces yields the total force acting on the mass, i.e.,

$$F_{tot} = m\ddot{x} = -b\dot{x} - kx + u. \tag{30.9}$$

Next, an equivalent state-space representation of the system must be found. Since the system is second order, two state variables are required, which correspond to the displacement and velocity of the mass. Thus, the following substitutions are made:

$$z_1 = x, \quad \text{and} \quad z_2 = \dot{x} = z_1. \tag{30.10}$$

Now, (30.9) is rewritten in terms of the assigned state variables, resulting in the following state-space representation of the system:

$$\begin{aligned} z_1 &= z_2 s, \\ z_2 &= -\frac{b}{m} z_2 - \frac{k}{m} z_1 + \frac{1}{m} u. \end{aligned} \tag{30.11}$$

With the system in state-variable form, the task of generating the listing shown in Figure 30.4 is trivial. The process can be handled in only a few simple steps:

1. (Lines 1–2) Define any necessary constants using a `.PARAM` statement. In this case, let $m = 1$, $k = 50$, and $b = 20$. Note also that the first line of a SPICE netlist should always be a comment.
2. (Lines 3–5) Assign a 1F capacitor for each state variable to be integrated. For this particular system, there are two state variables, which require two capacitors.
3. (Lines 6–8) If necessary, add 1 GΩ resistors in parallel with each capacitor in order to avoid floating-node errors.
4. (Lines 9–11) Implement each state equation using a separate dependent current source.
5. (Lines 12–13) Define any external stimuli. For this example, a unit step function is used for the input u.
6. (Lines 14–15) Specify any nonzero initial conditions using the `.IC` command. Here, an initial state $z = 0$ is assumed, meaning no initial velocity and no initial displacement.
7. (Lines 16–19) Finally, define the parameters for simulation using the `.TRAN` command.

Figure 30.5 shows the SPICE simulation results for the completed netlist.

```
1 :  * Source Mass-Spring-Damper
2 :  .PARAM m=1, k=50, b=3
3 :  * Assign capacitors for integration
4 :  C1 0 1 1
5 :  C2 0 2 1
6 :  * Eliminate floating nodes
7 :  R1 0 1 1GIG
8 :  R2 0 2 1GIG
9 :  * Define state equations
10:  G1 0 1 VALUE={V(2)}
11:  G2 0 2 VALUE={-b/m*V(2)-k/m*V(1)+1/m*V(u)}
12:  * Add any stimuli to the system
13:  Eu u 0 VALUE={stp(time-1)}
14:  * Specify initial conditions
15:  .IC V(1)=0, V(2)=0
16:  * Setup simulation
17:  .TRAN 10ms 4s 0ms 1ms UIC
18:  .PROBE
19:  .END
```

TWO VARIABLE HEADER (lines 3–8)

V(1) = z_1, displacement
V(2) = z_2, velocity

FIGURE 30.4 SPICE netlist for the mass–spring–damper system.

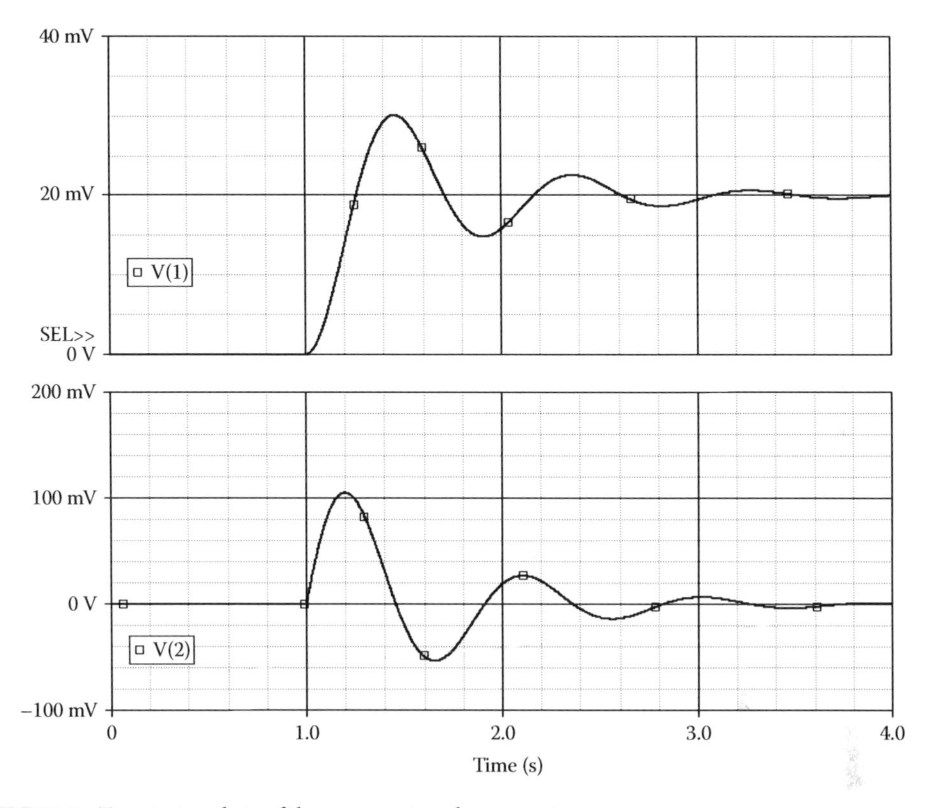

FIGURE 30.5 Transient analysis of the mass–spring–damper system.

Because the syntax for the integrating elements is independent of all but the order of the system, they can be viewed as a sort of standardized header, as indicated by the shaded box in Figure 30.4. Therefore, this portion of the listing is omitted in the remaining examples, due to its trivial nature.

30.3.2 Lorenz Attractor

Now we turn our attention to a rather famous example from the field of chaos theory known as the Lorenz attractor. While this system is deceivingly simple in appearance, in reality, its behavior is surprisingly complex. The term chaos, when applied to dynamic systems, implies an abnormally high level of sensitivity to initial conditions. For these systems, even minute changes in the initial state can result in wildly different trajectories. Of all such systems, the Lorenz attractor is perhaps the most widely documented and has become an icon of sorts in the field. Its popularity is due in large part to its simplicity, as well as the aesthetic nature of its solution. In this example, we will use SPICE to demonstrate its unique behavior.

As always, we begin by representing the system in state-variable form. Here, we have a relatively simple third-order nonlinear system described by the following equations, commonly known as the Lorenz equations:

$$\begin{aligned} x_1 &= \sigma(x_2 - x_1), \\ x_2 &= x_1(\rho - x_3) - x_2, \\ x_3 &= x_1 x_2 - \beta x_3. \end{aligned} \tag{30.12}$$

The parameters σ, ρ, and β are known, respectively, as the Prandtl number, the Rayleigh number, and the physical proportion. The classical examples of this system use values of $\sigma = 10$, $\rho = 28$, and $\beta = 8/3$. Also, because there is no external stimulus, the system must be excited using a set of nonzero initial conditions. Therefore, since the goal is to observe and verify its chaotic nature, the results are simulated and compared for both $\mathbf{x}_0 = [-8, 8, 27]$ and $\mathbf{x}_0 = [-7.99, 8, 27]$. The corresponding netlist is shown in Figure 30.6.

The results for both sets of initial conditions are shown in Figure 30.7. Notice how quickly the two trajectories deviate. For the first 6 s, the two solutions appear quite similar. However, by the 8 s mark, they bear almost no resemblance whatsoever. A phase portrait for the first solution is shown in Figure 30.8. Also notice how the state trajectory seems to oscillate randomly between two separate basins of attraction. This iconic image truly captures the seemingly paradoxical combination of elegance and complexity that this system embodies.

```
1 :  * Source Lorenz Attractor
2 :  .PARAM sigma=10, rho=28, beta={8/3}
  :  THREE VARIABLE HEADER
11:  G1 0 1 VALUE={sigma*(V(2)-V(1))}
12:  G2 0 2 VALUE={V(1)*(rho-V(3))-V(2)}
13:  G3 0 3 VALUE={V(1)*V(2)-beta*V(3)}
14:  .IC V(1)=-8, V(2)=8, V(3)=27
15:  .TRAN .01ms 20s 0 .1ms UIC
16:  .PROBE
17:  .END

V(1) = x1
V(2) = x2
V(3) = x3
```

FIGURE 30. 6 SPICE netlist for the Lorenz attractor.

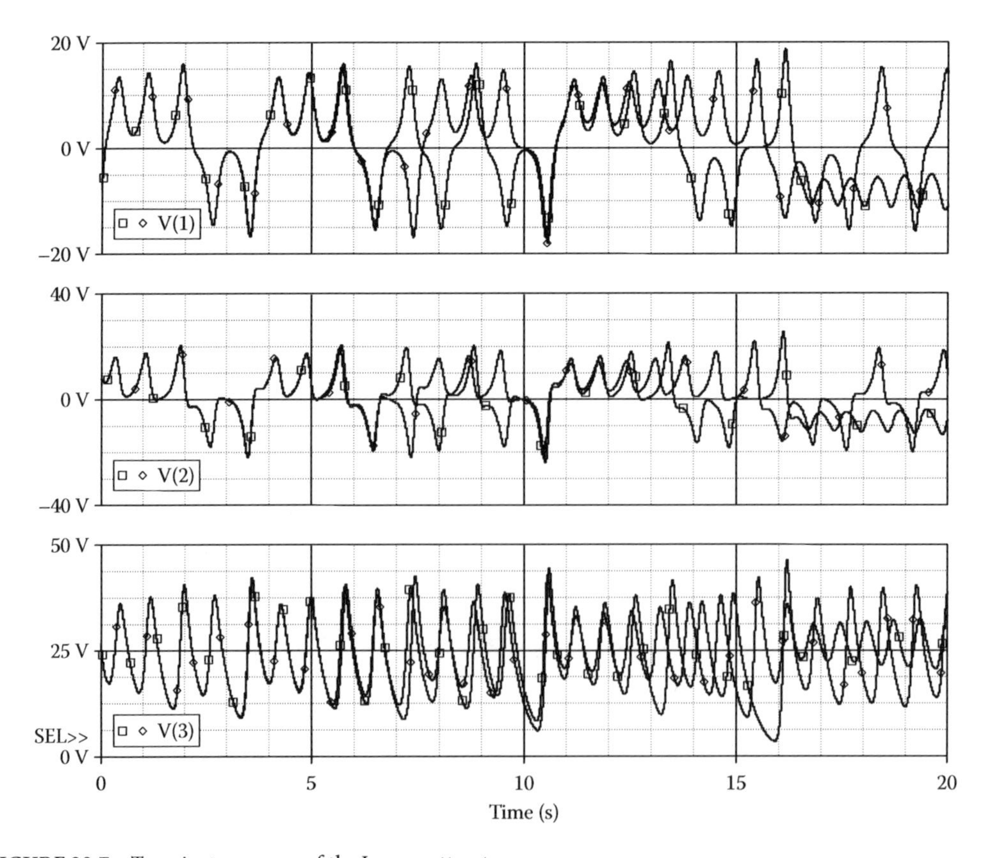

FIGURE 30.7 Transient response of the Lorenz attractor.

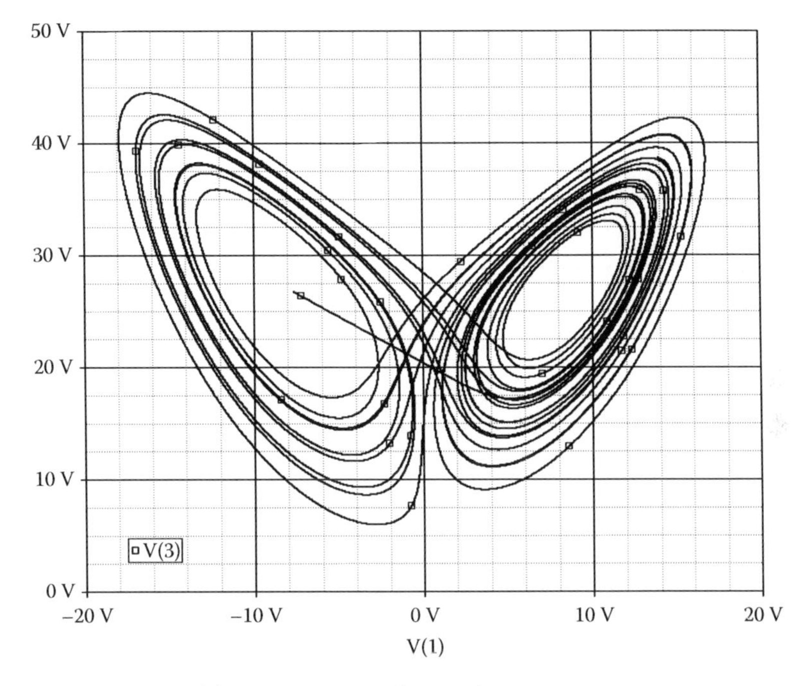

FIGURE 30.8 State trajectory of the Lorenz attractor for x_1 and x_3.

30.3.3 Inverted Pendulum

Now, consider a classic system from nonlinear dynamics known as the inverted pendulum [5]. The system, shown in Figure 30.9, is comprised of a small mass attached to the end of a shaft, which pivots atop a wheeled cart. Despite its relative simplicity, the inverted pendulum is often used as a benchmark in the field of control theory due to the fact that it is both nonlinear and inherently unstable.

In order to simulate this system, a nonlinear state-variable model must be found of the form

$$\dot{z} = f(z,u,t),$$
$$y = Cz. \tag{30.13}$$

Thus, a set of differential equations describing the system dynamics need to be derived.

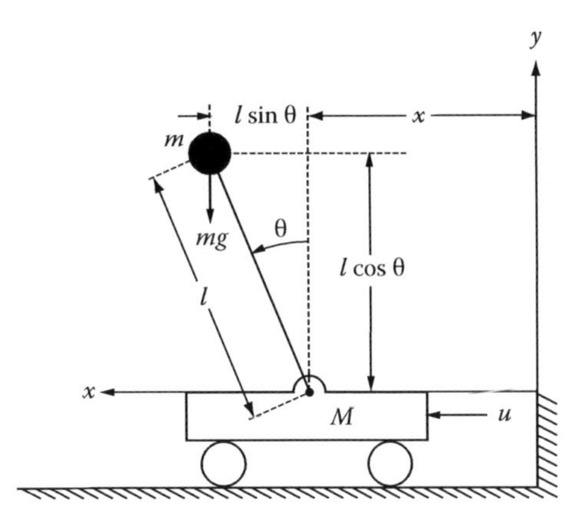

FIGURE 30.9 Inverted pendulum with cart.

```
1 :  * Source InvertedPendulum
2 :  .PARAM Mc=2, mp=0.1, l=0.5, g=9.81
  :  FOUR VARIABLE HEADER
13:  * Define plant state equations
14:  G1 0 1 VALUE={V(2)}
15:  G2 0 2 VALUE={(V(u)*cos(V(1))-(Mc+mp)*g*sin(V(1))+
16:  + mp*l*(cos(V(1))*sin(V(1)))*V(2)**2)
17:  + /(mp*l*cos(V(1))**2-(Mc+mp)*l)}
18:  G3 0 3 VALUE={V(4)}
19:  G4 0 4 VALUE={(V(u)+mp*l*sin(V(1))*V(2)**2-
20:  + mp*g*cos(V(1))*sin(V(1)))/
21:  + (Mc+mp-mp*cos(V(1))**2)}
22:  * Step response for t=1
23:  Eu u 0 VALUE={stp(time-1)}
24:  * Specify initial conditions
25:  .IC V(1)=0, V(2)=0, V(3)=0, V(4)=0
26:  * Setup simulation
27:  .TRAN .1e-3 10S 0s 1e-3 UIC
28:  .PROBE
29:  .END
```

V(1) = z_1, pendulum angle
V(2) = z_2, angular velocity
V(3) = z_3, cart position
V(4) = z_4, cart velocity

FIGURE 30.10 SPICE listing for the inverted pendulum.

The system has 2 degrees of freedom, which correspond to the cart position (x) and the angle of the pendulum with respect to the vertical axis (θ). Therefore, the system dynamics can be described using the following pair of second-order differential equations:

$$\ddot{x} = \frac{u + ml\sin(\theta)\dot{\theta}^2 - mg\cos(\theta)\sin(\theta)}{M + m - m\cos^2(\theta)}, s$$

$$\ddot{\theta} = \frac{u\cos(\theta) - (M+m)g\sin(\theta) + ml\cos(\theta)\sin(\theta)\dot{\theta}^2}{ml\cos^2(\theta) - (M+m)l}. \tag{30.14}$$

Assigning the state variables $z_1 = \theta$, $z_2 = \dot{\theta}$, $z_3 = x$, $z_4 = \dot{x}$, $y_1 = \theta$, and $y_2 = x$ yields an equivalent set of state equations:

$$\begin{aligned}
z_1 &= z_2, \\
z_2 &= \frac{u\cos(z_1) - (M+m)g\sin(z_1) + ml\cos(z_1)\sin(z_1)z_2^2}{ml\cos^2(z_1) - (M+m)l}, \\
z_3 &= z_4, \\
z_4 &= \frac{u + ml\sin(z_1)z_2^2 - mg\cos(z_1)\sin(z_1)}{M + m - m\cos^2(z_1)}, \\
y_1 &= z_1, \\
y_2 &= z_3.
\end{aligned} \tag{30.15}$$

Finally, using (30.15), the netlist can be composed in the usual fashion. The completed listing is shown in Figure 30.10. Note that in SPICE, a statement can be made to span multiple lines by placing a "+" symbol in front of each subsequent line. This is especially useful when dealing with long lists of parameters or equations like the ones shown here.

The step response for this system is shown in Figure 30.11. The nonlinearity and instability of this system both can be seen clearly.

30.3.4 State-Feedback Control

In this final example, a state-feedback controller is used to stabilize the inverted pendulum model from the previous example. One of the most attractive features of state-feedback is the ability to arbitrarily

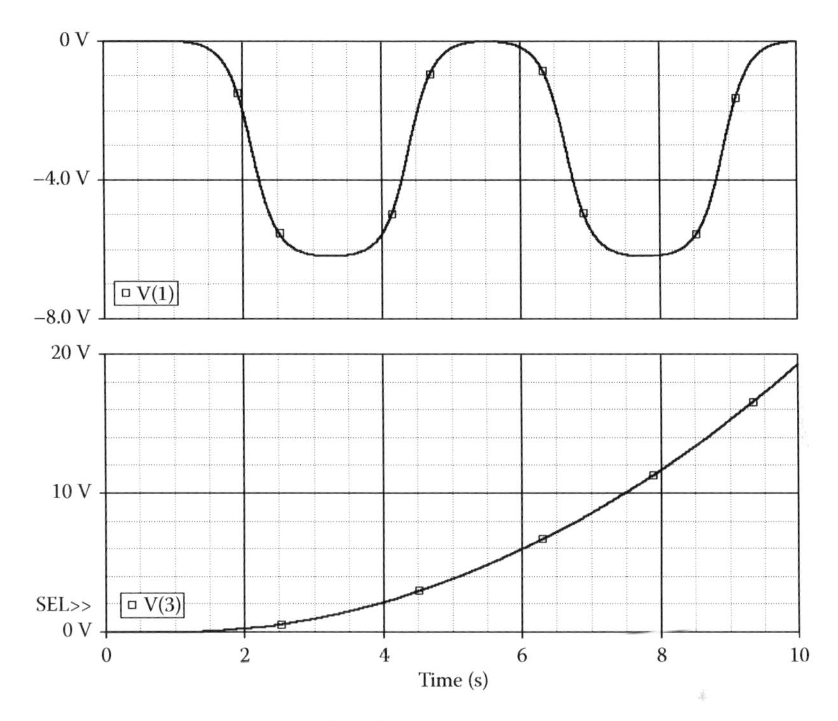

FIGURE 30.11 Step response for the inverted pendulum.

place the poles of the closed-loop system. To do this, however, two major issues must be addressed. First, the desired method of pole placement is limited to linear state-variable models, which (30.15) is clearly not, and second, the use of state feedback requires that all of the system's state variables are accessible for measurement. Looking at (30.15), it is clear that this is also not the case. Fortunately, there are ways of overcoming both obstacles.

With regard to the linearity requirement, it can be shown that within a sufficiently small neighborhood of equilibrium, the stability of a nonlinear system is equivalent to that of its first-order Taylor approximation [6]. Therefore, if the poles of the linearized model are placed in such a way that under normal operating conditions, the dynamics of the closed-loop system are restricted to the necessary region, then the resulting set of state-feedback gains should be sufficient for the stability of the nonlinear plant as well. While this can sometimes cause problems in cases where the "typical operating conditions" are unknown or ill defined, it is not a major issue for the inverted pendulum. This is because at any significant distance from equilibrium, the energy necessary to drive the system back would be prohibitively expensive anyway.

Evaluating the first Taylor expansion of (30.15) about $z, u = 0$, the following linearized model is obtained:

$$\dot{z} = \overbrace{\begin{bmatrix} 0 & 1 & 0 & 0 \\ 20.6 & 0 & 0 & 0 \\ 0 & 0 & 0 & 1 \\ -0.49 & 0 & 0 & 0 \end{bmatrix}}^{A} z + \overbrace{\begin{bmatrix} 0 \\ -1 \\ 0 \\ 0.5 \end{bmatrix}}^{B} u, \qquad (30.16)$$

$$y = \underbrace{\begin{bmatrix} 1 & 0 & 0 & 0 \\ 0 & 0 & 1 & 0 \end{bmatrix}}_{C} z.$$

Now, using (30.16), the system's closed-loop poles (λ) can be arbitrarily placed by solving

$$\left|\lambda I - A + BK\right| = 0 \tag{30.17}$$

for the proper state-feedback matrix K. For λ = [–5,–5.1,–5.2,–5.3], this yields a gain matrix K = [–215.5,–48.4,–71.6,–55.7].

Now, the second obstacle arises. That is, in order to implement the designed state feedback, access to all four state variables is required. However, only two of the system's four state variables are directly available for measurement. To work around this, we introduce what is known as a full state observer [7], which is used to estimate the remaining states based on the form of the linearized model (30.16).

The observer is a fourth-order dynamic system of the form

$$\dot{\hat{z}} = [A - LC]\hat{z} + Ly + Bu, \tag{30.18}$$

where

$\hat{z}$ is the estimated state

u and y contain the inputs and outputs of the plant

The matrix L is selected by the same process as K, and is used to place the closed-loop poles of the observer. Choosing the same pole locations used for the controller design, λ = [–5,–5.1,–5.2,–5.3], yields

$$\dot{\hat{z}} = \overbrace{\begin{bmatrix} -10.28 & 1 & -0.10 & 0 \\ -26.40 & 0 & -0.52 & 0 \\ -0.09 & 0 & -10.32 & 1 \\ -0.48 & 0 & -26.63 & 0 \end{bmatrix}}^{A-LC} \hat{z} + \overbrace{\begin{bmatrix} 10.28 & 0.10 \\ 47.01 & 0.52 \\ 0.09 & 10.32 \\ -0.01 & 26.63 \end{bmatrix}}^{L} y + \overbrace{\begin{bmatrix} 0 \\ -1 \\ 0 \\ 0.5 \end{bmatrix}}^{B} u. \tag{30.19}$$

Now, the controller is implemented using the estimated state $\hat{z}$ in place of z.

The completed listing is shown in Figure 30.12. The observer, on lines 23–26, is implemented in the usual fashion. Also, notice that the previous system input u has been replaced by the reference r. Now, the plant input is the difference between the reference and the state feedback, defined on line 28.

```
1 :  * Source InvertedPendulum
2 :  .PARAM Mc=2, mp=0.1, l=0.5, g=9.81
3 :  + k1=-215.5, k2=-48.4, k3=-71.6, k4=-55.7]
  :  EIGHT VARIABLE HEADER
22:  * Define the full state observer
23:  G14 0 14 VALUE=-.48*V(11)-26.63*V(13)-.01*V(1)+26.63*V(3)+.5*V(u)
24:  G13 0 13 VALUE=-0.09*V(11)-10.32*V(13)+V(14)+0.09*V(1)+10.32*V(3)
25:  G12 0 12 VALUE=-26.40*V(11)-0.52*V(13)+47.01*V(1)+0.52*V(3)-V(u)
26:  G11 0 11 VALUE=-10.28*V(11)+V(12)-0.10*V(13)+10.28*V(1)+0.10*V(3)
27:  * State feedback controller
28:  Eu u 0 VALUE={V(r)-k1*V(11)-k2*V(12)-k3*V(13)-k4*V(14)}
  :  NONLINEAR PLANT MODEL*
38:  * Step response for t=1
39:  Er r 0 VALUE={stp(time-1)}
40:  * Specify initial conditions
50:  .IC V(1)=0, V(2)=0, V(3)=0, V(4)=0,
51:  + V(11)=0, V(12)=0, V(13)=0, V(14)=0
52:  * Setup simulation
53:  .TRAN .1e-3 10S 0s 1e-3 UIC
54:  .PROBE
55:  .END
```

V(11) = $\hat{z}_1$, pendulum angle estimate
V(12) = $\hat{z}_2$, angular velocity estimate
V(13) = $\hat{z}_3$, cart position estimate
V(14) = $\hat{z}_4$, cart velocity estimate

*Fig. 1.10, lines 13-21

FIGURE 30.12 Listing for the state-feedback controller and full state observer.

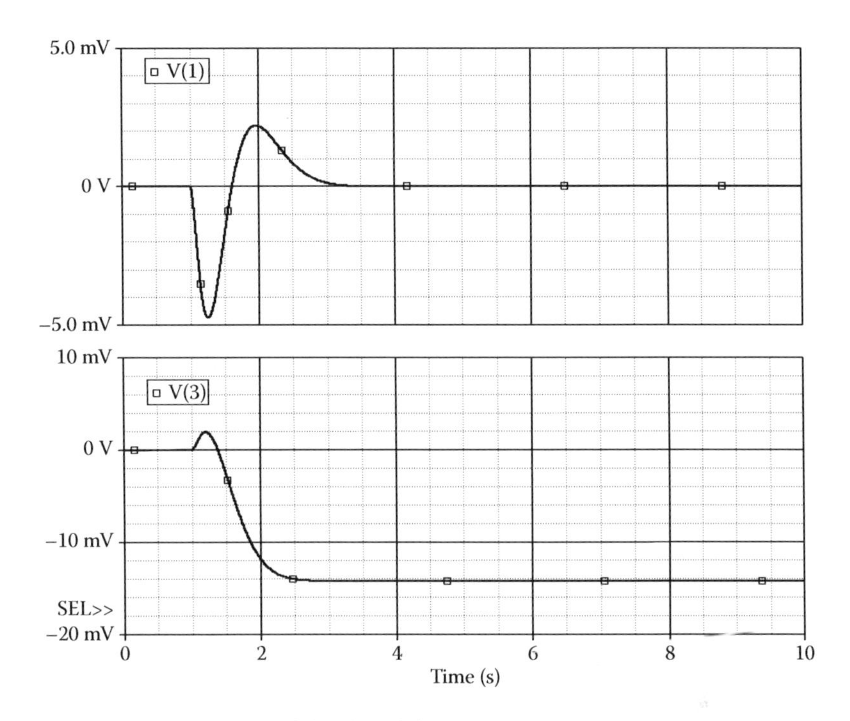

FIGURE 30.13 Step response for the stabilized pendulum.

The step response for the composite system is shown in Figure 30.13. It can clearly be seen that both the pendulum angle and the cart position are now stable.

30.4 Conclusion

In this chapter, a method was presented for simulating dynamic systems in SPICE. It was shown that while SPICE is largely known for its handling of electronics, it is also quite capable of handling more general systems. This offers many advantages in the field of electrical engineering where one often encounters models that combine both electrical and mechanical systems. Furthermore, it was shown that the method used to implement these models is straightforward and requires minimal effort from the user.

References

1. B.M. Wilamowski and O. Kaynak, Oil well diagnosis by sensing terminal characteristics of the induction motor, *IEEE Transactions on Industrial Electronics*, 47(5), 1100–1107, October 2000.
2. B.M. Wilamowski, Faults identification of oil wells using neural networks, *Recent Advances in Mechatronics*, Springer-Verlag, Berlin, Germany, 1999, pp. 459–467.
3. B.M. Wilamowski and R.C. Jaeger, *Computerized Circuit Analysis Using SPICE Programs*, The McGraw-Hill Companies, Inc., New York, 1997.
4. A. Vladimirescu, *The Spice Book*, 2nd edn., John Wiley & Sons, Inc., New York, 2010.
5. H.K. Khalil, *Nonlinear Systems*, Prentice-Hall, Inc., Englewood Cliffs, NJ, 2002.
6. W.L. Brogan, *Modern Control Theory*, Prentice-Hall, Inc., Englewood Cliffs, NJ, 1991.
7. D.G. Luenberger, An introduction to observers, *IEEE Transactions on Automatic Control*, AC-16(6), 596–602, December 1971.

31

Iterative Learning Control for Torque Ripple Minimization of Switched Reluctance Motor Drive

Sanjib Kumar Sahoo
National University of Singapore

Sanjib Kumar Panda
National University of Singapore

Jian-Xin Xu
National University of Singapore

31.1 Introduction

Many industrial automation applications need variable-speed operation for improving energy efficiency or product quality. Electric drives are preferred in variable-speed applications for their ease of control, clean operating environment, and easy access of electricity at the point of use. DC motors are easiest to control and are commonly used in such applications. However, brush-commutator of DC motors adds to cost, complexity, and needs frequent maintenance. AC motors are more robust than DC motors but generally difficult from control point of view. With advancement in power electronics and microprocessor technology, AC motor control performance has been improved substantially. Due to this reason, more and more AC motors are used in variable-speed applications. In the field of electric drives, research is directed toward adopting more robust motors in variable-speed drive applications.

Switched reluctance motors (SRMs) have the simplest and most robust construction among all electric motors [1]. Both stator and rotor are stacks of laminated sheets, with only stator having concentric coils. There are no permanent magnets or rotor bars on the rotor. These are also quite economical for mass manufacturing. However, due to the double-salient construction and magnetic saturation, torque production is highly nonlinear. In conventional operation of SRM, the phase windings are switched on sequentially, one at a time. This mode of operation and the nonlinear torque production lead to large amount of torque ripples. Torque ripples can cause speed ripples, particularly at low-speed operation.

Such excitation also produces radial force on the rotor, leading to substantial vibration and acoustic noise. Due to these reasons, SRM could not be used for high-performance industrial applications.

Over the last few decades, researchers have suggested different techniques [2–12] for mitigating this problem. This is still an open research problem and currently there is a lot of interest in it from the drives research community. This chapter shows the use of iterative learning control (ILC) to improve the torque control performance of SRM. With improved torque control, SR drives can be used for high-performance motion control applications.

31.2 Operating Principle of SRM

SRM works on reluctance torque principle [1]. It is equivalent to an electromagnet pulling a piece of soft iron toward it, so that the reluctance of the associated magnetic circuit is minimized. In SRMs, both stator and rotor have salient poles, of different numbers, as shown in Figure 31.1 for an SRM with eight stator poles and six rotor poles. Each stator pole has a concentric coil wound around it. The windings on two stator poles, which are at exactly 180° with each other (1–1′, 2–2′, 3–3′, and 4–4′), are connected in series to constitute the phase winding (phase 1 winding is shown in Figure 31.1). The magnetic path for one phase is shown, consisting of the stator core, stator poles, air gap, rotor poles, and rotor core. When two diametrically opposite rotor poles are aligned with the stator poles in a phase (the corresponding rotor position is called "aligned" position), the reluctance of the magnetic path is at its minimum. The corresponding phase inductance would then be maximum. When the rotor poles are away from the stator poles, the reluctance increases and is maximum when the rotor poles are right at the middle of two consecutive stator poles (the corresponding position is called "unaligned" position). The phase inductance at unaligned position would be the minimum.

When any of stator phase windings is energized, the nearest rotor poles will experience a pull so as to align with the energized stator poles. Once the rotor poles fully align with the stator poles, the pulling torque becomes zero. The fully aligned phase is then switched off, and the next phase is switched on. Such sequential switching of phases produces a continuously rotating motion.

Figure 31.2 shows the Ψ-i curve (flux-linkage vs phase current) when rotor position is fixed at θ. In this case, the electrical energy input to the winding is stored as magnetic energy. The stored magnetic energy W_f (horizontally shaded area OAB) can be calculated as

$$W_f = \int_0^t vi\,dt = \int_0^t \frac{d\psi}{dt} i\,dt = \int_0^\psi i\,d\psi \tag{31.1}$$

where v is the voltage across the phase winding. The vertically shaded area (OAC) under the ψ vs. phase current i curve represents a fictitious quantity called *co-energy*. This quantity does not have any physical

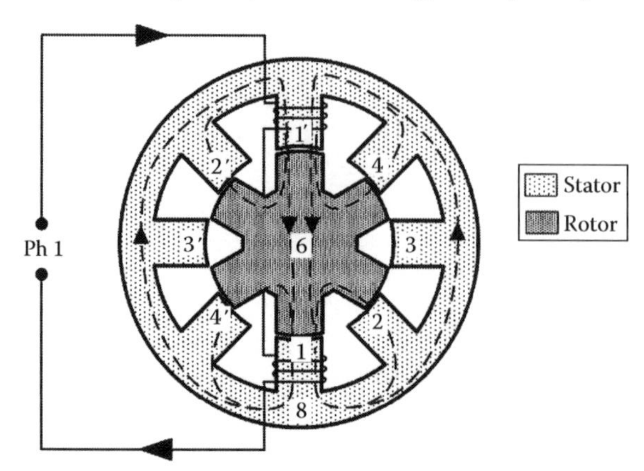

FIGURE 31.1 Cross-sectional view of SRM showing the stator, rotor, one phase winding, and magnetic-flux path.

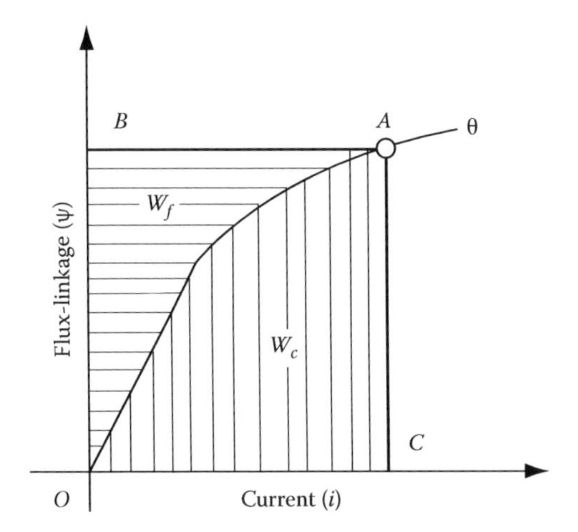

FIGURE 31.2 Field-energy (W_f) and co-energy (W_c) in SRM.

meaning, but change in co-energy is the same as the mechanical work done in electromagnetic system. Co-energy in Figure 31.2 can be calculated as

$$W_c = \int_0^i \psi\, di. \tag{31.2}$$

Figure 31.3 shows the change in co-energy when phase current is maintained constant and rotor is allowed to move from position θ_1 to position θ_2. When motor operates in linear magnetic region (as can be seen in Figure 31.3a), change in the co-energy is given by the triangular area (OA_1A_2) with vertical shading. The change in co-energy (ΔW_c), which is the amount of mechanical work done during the rotor movement, will be exactly one-half of the electrical energy input (ΔW_e) to the phase, given by the area ($B_1A_1A_2B_2$):

$$\Delta W_e = \int ei\,dt = \int \frac{d\psi}{dt} i\,dt = i\left(\psi^2 - \psi^1\right) = \left(L^2 - L^1\right)i^2, \tag{31.3}$$

where

ψ^2 and ψ^1 are flux-linkages at rotor positions θ_2 and θ_1, respectively
L^2 and L^1 are the phase inductance values at these rotor positions

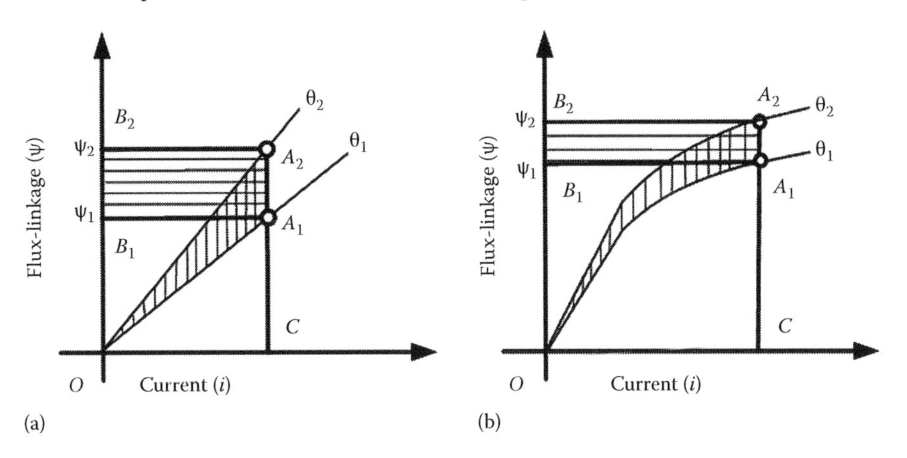

FIGURE 31.3 (a) Change in co-energy under linear magnetization. (b) Change in co-energy under saturated magnetization.

The change in co-energy is equal to one-half of the electrical energy ΔW_e:

$$\Delta W_c = \frac{1}{2}\left(L^2 - L^1\right)i^2. \tag{31.4}$$

The average reluctance torque produced when rotor moves from θ_1 to θ_2, due to phase current i, is given by

$$T_{av} = \frac{\Delta W_c}{\Delta\theta} = \frac{1}{2}i^2\frac{\Delta L}{\Delta\theta} = \frac{1}{2}i^2\frac{dL}{d\theta}, \tag{31.5}$$

where $\Delta\theta$ is the displacement. Figure 31.3b describes change in co-energy when the system enters into magnetic saturation region. Phase inductance depends on both rotor position as well as phase current. This complicates the calculation of average torque. However, it can be observed that the change in co-energy (mechanical work done) is more than one-half of the input electrical energy. This results in better conversion ratio and hence, SRM is usually operated in deep magnetic saturation.

31.2.1 Trapezoidal Phase Inductance Profile

Assuming SRM to operate in linear magnetic region and that magnetic flux crosses the air gap only at 90°, phase inductance can be idealized to be directly proportional to the overlap angle between stator and rotor poles. When stator and rotor poles are unaligned, this idealized phase inductance will be at the minimum (L_u). It will remain at this value as rotor pole approaches the stator pole, until the rotor pole tip meets the stator pole tip. Thereafter, it will rise at a constant rate as overlap angle increases and attain the maximum value (L_a) when there is maximum overlap. As per design practice, stator pole arc length (θ_s) is less than the rotor pole arc length (θ_r):

$$\theta_s < \theta_r, \tag{31.6}$$

$$\left(\theta_s + \theta_r\right) < \frac{2\pi}{N_r}, \tag{31.7}$$

where N_r is the number of rotor poles. Due to this, the pole overlap remains constant for a period when the idealized phase inductance remains at the maximum value, L_a. As the rotor moves away from the stator pole, it will fall at a constant rate, until overlap becomes zero and it becomes L_u again. Therefore, the idealized phase inductance profile would have a trapezoidal shape as shown in Figure 31.4:

$$\begin{aligned} L(\theta) &= L_u \quad \text{for } 0 \le \theta \le \theta_1 \\ &= L_u + K(\theta - \theta_1) \quad \text{for } \theta_1 \le \theta \le (\theta_1 + \theta_s) \\ &= L_a \quad \text{for } (\theta_1 + \theta_s) \le \theta \le (\theta_1 + \theta_r), \end{aligned} \tag{31.8}$$

$$K = \frac{L_a - L_u}{\theta_s - \theta_r},$$

where K is the position rate of change of phase inductance.

The dotted line in Figure 31.4 belongs to a stator phase adjacent to the phase discussed above. Hence, the phase inductance of all the phases can be obtained by suitably shifting the phase inductance profile of any one phase along the rotor position axis.

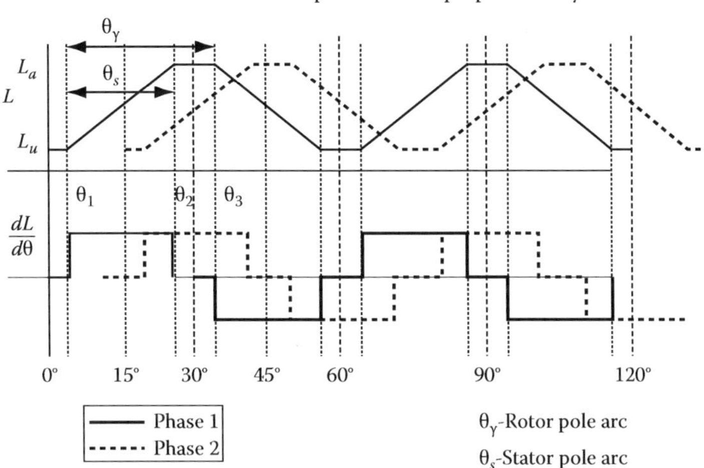

FIGURE 31.4 Trapezoidal profile for SRM phase inductance.

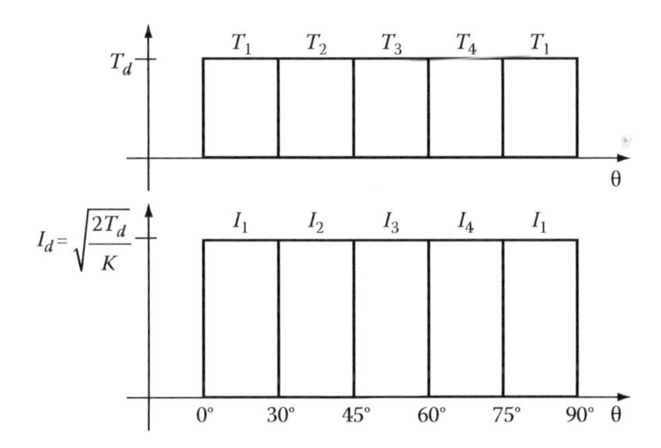

FIGURE 31.5 Phase torque shares and phase current references assuming linear magnetization.

For this idealized phase inductance profile, instantaneous torque (T_{inst}) will be same as the average torque as in (31.5), i.e.,

$$T_{inst} = \frac{1}{2}Ki^2, \tag{31.9}$$

where $K = (dL/d\theta)$ is the position rate of change of phase inductance. It is like the torque constant in DC series motor. In this case, SRM can generate constant torque, if the phase windings are injected with rectangular current pulses as shown in Figure 31.5. As torque direction is independent of phase current direction, both motoring and braking torque production is possible with unidirectional current; by only placing the current pulse in the region of positive or negative K value.

31.3 Electronic Phase Commutation

The stator phase windings are switched in sequence to produce a rotating motion. It was shown that, at any time, only one phase conducts current and produces the total motor torque. The process of torque transfer from one phase to another is called phase commutation. Due to the nonzero phase inductance, and finite DC-link voltage available, it is impossible for the current to rise or fall instantaneously. Hence, in this operating mode where only one phase is excited at a time, there will be a large torque dip during the phase commutation. The peak-to-peak torque ripples due to nonideal phase commutation can be even more than 100% of the average motor torque.

However, it can be seen in the Figure 31.4 that there is some overlapping in the torque-producing regions of two adjoining phases. At any position, two phases can produce torque in same direction. During phase commutation, two nearby phases can share the total torque demand so as to avoid the torque dips during phase commutation. This can be put mathematically as

$$T_d = T_1 + T_2,$$
$$T_1 = f(\theta) * T_d, \tag{31.10}$$
$$T_2 = (1 - f(\theta)) * T_d,$$

where

T_d is the demanded motor torque
T_1 and T_2 are the torque produced by the two conducting phases
$f(\theta)$ is the position-dependent torque-sharing function (TSF)

Choice of $f(\theta)$ cannot be arbitrary. Suitable TSF needs to be chosen as per various constraints.

31.3.1 Nonlinearity of SRM Magnetization Characteristics

There are some deviations in the phase inductance profile toward both the aligned and unaligned positions. Usually, SRM air gap is smaller compared to other motors. When the rotor poles approach stator poles, there will be some flux-fringing effect. This causes phase inductance to start rising even before the rotor pole tip reaches the stator pole. This is one cause of the actual phase inductance profile being different from the trapezoidal profile.

As seen in Figure 31.3a, when SRM operates in linear magnetic region, only half of the electrical energy input to the phase winding gets converted to mechanical work. When SRM is operated in deep magnetic saturation region, more than half of the electrical energy input will be converted into mechanical work as shown in Figure 31.3b. This reduces the amount of volt-amperes handled by the converter, for the same amount of work done by the motor. Hence, SRM is usually designed to work with deep magnetic saturation.

At the start of overlap between stator and rotor poles, only the pole tips carry the total magnetic flux. Hence, they get saturated at a very small current. As the overlap increases, saturation starts at a large current. With saturation, the effective phase inductance falls. The combined effect of flux-fringing and magnetic saturation makes the phase inductance a nonlinear function of both rotor position and current. These effects result in a highly nonlinear and coupled relationship between phase torque, current, and rotor position. Hence, a rectangular current pulse in the increasing inductance region does not produce a rectangular torque pulse. To generate rectangular phase torque, it is essential to produce a position-dependent profile of the phase current. These nonlinearities demand suitable nonlinear controller for good torque control accuracy. Nonlinear control laws are usually computation intensive and require high-speed digital controller.

31.4 Direct Torque Control of SRM

Direct torque control (DTC) scheme treats phase torque as plant output and generates the desired phase voltage directly, as shown in Figure 31.6. The electromagnetic system dynamics is given in the following equation:

$$\frac{d\psi}{dt} = v - iR, \tag{31.11}$$

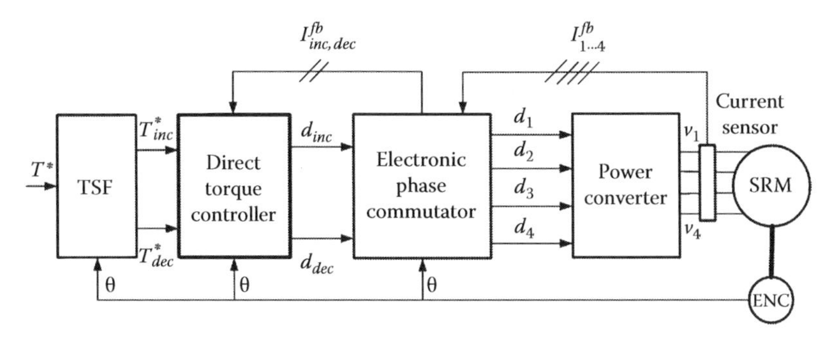

FIGURE 31.6 Direct torque controller for SRM.

where
- ψ is the flux-linkage
- v is the phase voltage
- i is the phase current
- R is the phase winding resistance

By expanding (31.11), we get

$$\frac{d\psi}{dt} = \frac{\partial\psi}{\partial i}\frac{di}{dt} + \frac{\partial\psi}{\partial\theta}\frac{d\theta}{dt} \tag{31.12}$$

and

$$\frac{di}{dt} = \left(\frac{\partial\psi}{\partial i}\right)^{-1}\left(v - iR - \frac{\partial\psi}{\partial\theta}\frac{d\theta}{dt}\right). \tag{31.13}$$

The state equation of interest for the DTC scheme is

$$\frac{dT}{dt} = \frac{\partial T}{\partial i}\frac{di}{dt} + \frac{\partial T}{\partial\theta}\frac{d\theta}{dt}. \tag{31.14}$$

By substituting *di/dt* from (31.13), we get

$$\frac{dT}{dt} = \left(\frac{\partial T}{\partial i}\right)\left(\frac{\partial\psi}{\partial i}\right)^{-1}\left(-iR - \frac{\partial\psi}{\partial\theta}\frac{d\theta}{dt}\right) + \frac{\partial T}{\partial\theta}\frac{d\theta}{dt} + \left(\frac{\partial T}{\partial i}\right)\left(\frac{\partial\psi}{\partial i}\right)^{-1} v. \tag{31.15}$$

A spatial ILC-based DTC scheme has been presented in Figure 31.7. For torque feedback, various torque estimation schemes are available [13]. The most common approach is to measure the motor torque data for different phase current and rotor position. Then, a look-up table containing this torque data is implemented online. The look-up table takes in the rotor position feedback and the measured phase current and returns the estimated phase torque. This approach requires some online memory but involves very little computation.

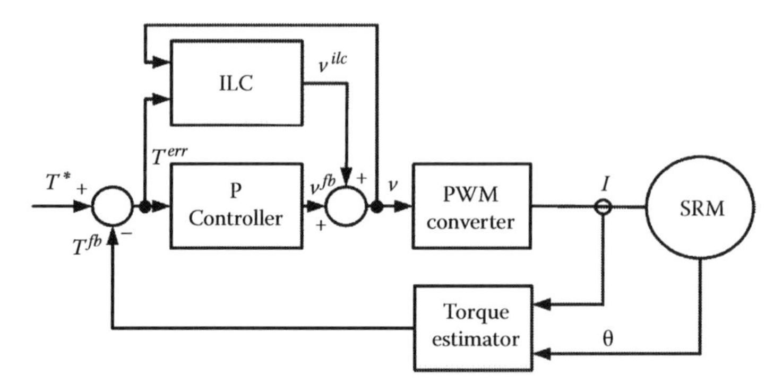

FIGURE 31.7 ILC-based DTC scheme for SRM.

31.5 Proposed Spatial ILC-Based Controller

31.5.1 Iterative Learning Controller

Consider a control task that requires perfect tracking in a finite interval when both the control reference and disturbance are repeatable. Most control methods, including adaptive or robust control, may not be suitable for such a class of tasks for two reasons. First, these control methods are characterized by asymptotic convergence, thus it is difficult to guarantee a perfect tracking even if the initial discrepancy is zero. Second and more importantly, those control methods are not able to learn from the previous task execution that may succeed or fail. Without learning, a control system can only produce the same performance without improvement even if the task is repeatedly executed. ILC was proposed to meet this kind of control requirement. The idea of ILC is straight forward: using control information of preceding execution to improve the present execution. This is realized through memory-based learning [11]. ILC-based torque control scheme had been used for other motor drives such as PMSM [11,12]. ILC control design has to satisfy initial resetting conditions, i.e., the actual phase torque matches the desired phase torque at the start (time $t = 0$) for every iteration.

31.5.2 Phase Torque Periodic in Rotor Position

Each phase torque reference is decided by the motor torque demand and a TSF. As shown in Figure 31.8 for TSF, each phase is active for 30°, producing the increasing share (T^*_{inc}) for first 15° and the decreasing share (T^*_{dec}) for the remaining 15°. For a constant motor torque demand, phase torque reference will be periodic in *rotor position*, but not necessarily in time. Hence, the ILC for the proposed DTC scheme has to be designed and implemented in terms of rotor position. As phase windings are inactive for 30°, there is enough time for actual phase torque to be zero at the beginning of each period. Hence, the resetting condition $T^*(m,0) = T(m+1,0) = 0$, where m is the iteration number, is satisfied for the phase torque tracking problem. As a result, ILC approach can be applied, and each period of phase torque is regarded as an iteration.

31.5.3 Implementation of the Spatial ILC Scheme

Usually, digital controller samples output and calculates control input at fixed time intervals. Due to variation of motor speed, the distance between two time samples $\theta(t_n)$ and $\theta(t_{n+1})$ in Figure 31.9 may vary.

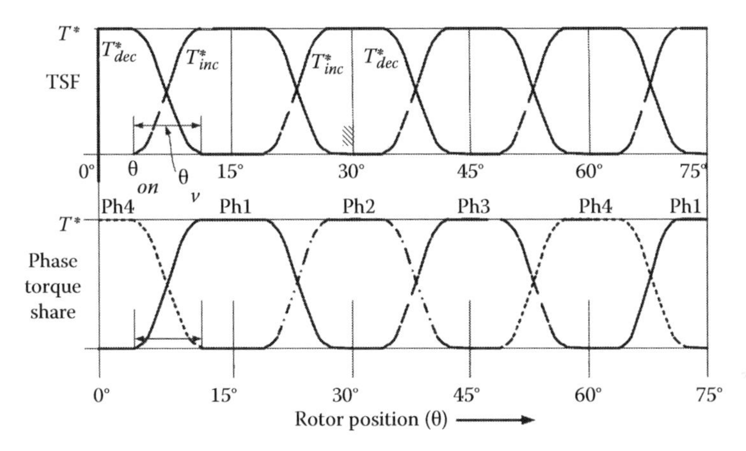

FIGURE 31.8 Cubic torque-sharing function with the phase torque shares.

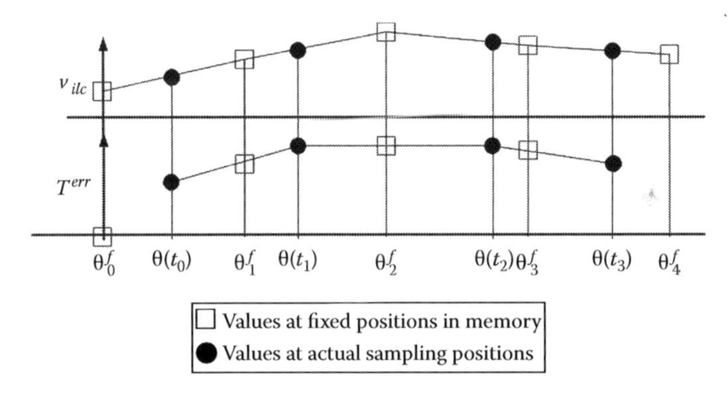

FIGURE 31.9 Description of position-based ILC: θ^f_n, fixed rotor positions in memory'; $\theta(t_n)$, rotor positions at sampling instants during an iteration; v_{ilc}, the ILC compensation voltage; and T^{err}, torque tracking error.

Second, as the number of samples in one rotation (360 electrical degrees) may not be an integer number, rotor positions at sampling instants will not be identical in all iterations. Therefore, the proposed ILC scheme is based on rotor position, i.e., ILC learning is done at fixed equidistant rotor positions. The control voltages and torque errors at these fixed rotor positions (θ^f_n) are mapped to the sampled values at $\theta(t_n)$ through interpolation, as shown in Figure 31.9. Linear interpolation is used for simplicity:

$$T^{err}\left(m,\theta^f_n\right) = T^{err}(m,\theta(t_b)) + \frac{T^{err}(m,\theta(t_a)) - T^{err}(m,\theta(t_b))}{\theta(t_a)-\theta(t_b)}\left(\theta^f_n - \theta(t_b)\right), \tag{31.16}$$

where

$T^{err} = T^{ref} - T$ is the tracking error
m is the mth iteration number
θ^f_n is the nth fixed position
$\theta(t_b)$ is the sampled position just before θ^f_n
$\theta(t_a)$ is the sampled position just after θ^f_n

ILC compensation voltage is updated using the ILC updating law

$$v_{ilc}\left(m,\theta_n^f\right)=v_{ilc}\left(m-1,\theta_n^f\right)+G^*T^{err}\left(m-1,\theta_{n+1}^f\right), \tag{31.17}$$

where

- $v_{ilc}(m,\theta_n^f)$ is the ILC compensation voltage at the nth fixed position in memory during the mth iteration
- G is the learning gain
- $\theta(t)$ is the rotor position at sampling instant t
- θ_b^f is the fixed position in memory just before $\theta(t)$
- θ_a^f is the fixed position in memory just after $\theta(t)$

Finally, the ILC compensation voltage for the sampling instant t, is obtained from the compensation voltages stored in memory as given by

$$v_{ilc}\left(m,\theta(t)\right)=v_{ilc}\left(m,\theta_b^f\right)+\frac{v_{ilc}\left(m,\theta_a^f\right)-v_{ilc}\left(m,\theta_b^f\right)}{\theta_b^f-\theta_a^f}\left(\theta(t)-\theta_b^f\right). \tag{31.18}$$

31.5.4 ILC Convergence

In each iteration, ILC updates the control input by a quantity proportional to the error. This should result in reduction of tracking error from iteration to iteration. Error convergence is defined as reduction of tracking error to within a tolerable limit, after a number of iterations. The learning gain is designed so as to ensure convergence. The discrete-time representation of the DTC system (31.15) can be written as

$$T(n+1)=FT(n)+Bu(n), \tag{31.19}$$

where $T(n)$ is the value of the state variable, namely, the torque, at the nth sampling instant. The ILC convergence criteria for such a system is given in references [14,15]:

$$|1-GB|_\infty<1, \tag{31.20}$$

where G is the learning gain. The rectangular method for converting the continuous-time system into discrete-time system can be used, i.e.,

$$\dot{T}=\frac{T(n+1)-T(n)}{T_s}\Rightarrow T(n+1)=T(n)+T_s\dot{T}. \tag{31.21}$$

For spatial ILC, the learning occurs at fixed rotor positions, and hence the time-sampled data have to be mapped to values corresponding to the rotor positions. Let two consecutive rotor positions be θ_k^f and θ_{k+1}^f. The equivalent sample time between the two consecutive rotor positions is

$$T_s=\frac{\theta_{k+1}^f-\theta_k^f}{\frac{d\theta}{dt}}. \tag{31.22}$$

This equivalent sampling time should be used for discretization of the torque dynamics. The discrete form of the torque dynamics can be obtained from (31.21) by replacing T_s as given in (31.22) and $\dot{T}$ as given in (31.15) as

$$T\left(\theta_{k+1}^{f}\right)=T\left(\theta_{k}^{f}\right)+\frac{\partial T}{\partial \theta}\left(\theta_{k+1}^{f}-\theta_{k}^{f}\right)+\frac{\partial T}{\partial i}\left(\frac{\partial \psi}{\partial i}\right)^{-1}\left(-iR-\frac{\partial \psi}{\partial \theta}\omega\right)T_s$$
$$+\frac{\partial T}{\partial i}\left(\frac{\partial \psi}{\partial i}\right)^{-1}T_s v(k). \tag{31.23}$$

Thus, for the discrete-time torque dynamics,

$$B=\frac{\partial T}{\partial i}\left(\frac{\partial \psi}{\partial i}\right)^{-1}T_s. \tag{31.24}$$

The convergence criteria for the proposed spatial ILC is obtained as

$$0\leq G\leq \frac{2}{\dfrac{\partial T}{\partial i}\left(\dfrac{\partial \psi}{\partial i}\right)^{-1}T_s}. \tag{31.25}$$

Accurate knowledge of torque and flux-linkage model is not necessary for accurate tracking of phase torque. Approximate values can be used to obtain the leaning gain. By ignoring the flux saturation effect, the instantaneous torque relationship (31.9) for the SRM can be used for this purpose.

The range for the learning gain can be obtained as follows:

$$\begin{aligned} T&=\frac{1}{2}Ki^2,\\ \frac{\partial T}{\partial i}&=Ki,\\ \frac{\partial \psi}{\partial i}&=L(\theta), \end{aligned} \tag{31.26}$$

$$0\leq G\leq \frac{2L_u}{Ki_{rated}}\frac{\omega}{\Delta\theta}, \tag{31.27}$$

where $\Delta\theta$ is the sampling interval in the rotor position domain.

31.5.5 Performance of the Controller at High Speed

The rotor positions for ILC updating are fixed from period to period. The values of torque tracking error and phase voltages at these rotor positions are obtained using interpolation of actual sampled data at neighboring positions. The desired phase voltage and phase torque vary with the rotor position in highly nonlinear manner and hence, linear interpolation is accurate when the data samples are spaced closely; else, the error will increase in proportion to the interpolation interval. Error in estimation of phase torque and phase voltage at the learning points will lead to error in updated ILC voltage. As motor speed increases, the number of data samples per learning period will be less. For instance, if the speed increases by 10 times, the available number of samples per learning period will be one-tenth. Hence, the phase torque and voltage values at the learning points will be updated based on sample values at farther rotor positions and therefore will be less accurate. Thus, torque tracking would deteriorate as motor speed increases.

31.6 Experimental Results

The ILC-based DTC for SRM has been experimentally verified on our laboratory set-up with a four-phase, 8/6 pole, 1 hp SRM. The motor has a full-load torque of 1.8 N m and rated current of 10 A. A dSPACE DS1104 controller board was used for implementing the control algorithm. The program execution takes 150 μs. This decides sampling frequency for the torque control loop as 6.6 kHz. The maximum value of learning gain G as in (31.27) for the system is calculated with $L_u = 10\,\text{mH}$, $K = 0.1$, $I_{rated} = 10$ A, $\omega = 1$ rad/s, and $\Delta\theta = \theta^f_{k+1} - \theta^f_k = 0.1°$. The upper bound of G is 11.45 according to (31.27). The ILC learning gain G is set at 10 for the experimental results. All the experimental results are for motor speed below 200 rpm.

The controller is tested at 100% motor rated torque. First, the feedback controller is tuned without any ILC compensation. A high-gain proportional feedback controller is applied with feedback gain set to 150 as any further increase makes the torque output oscillatory near the unaligned rotor position. The results of proportional feedback controller are shown in Figure 31.10, where CH1 denotes the motor demanded torque, CH2 denotes the estimated motor torque, and CH3 denotes the motor torque error. Thus, the average torque error is about 0.12 N m and the torque ripples are about 0.3 N m peak-to-peak or about 33% of the demanded motor torque. In Figure 31.10, CH4 denotes the measured phasel current, which is controlled automatically within the rated current limit.

The results for both the P-type feedback controller and ILC compensator together for the DTC scheme are shown in Figure 31.11. It shows the low torque ripples (CH3 is less than 0.2 N m peak-to-peak) and a

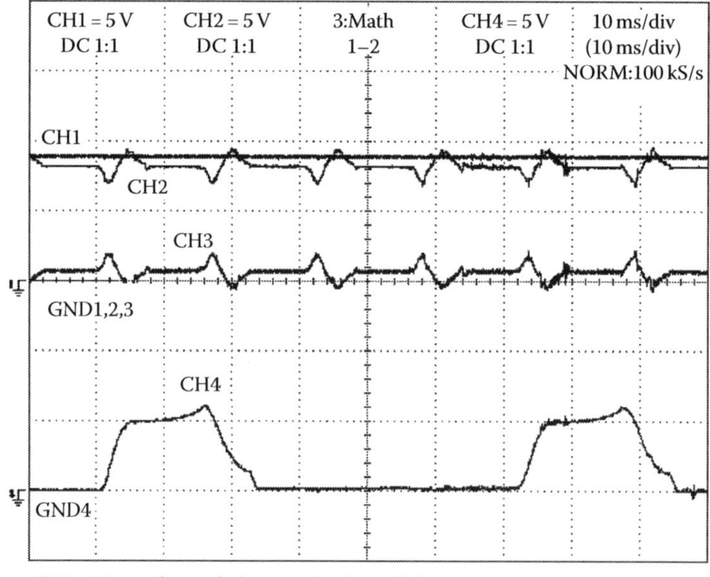

CH1—Motor demanded torque (1.0 N m/div)
CH2—Estimated motor torque (1.0 N m/div)
CH3—Motor torque error (1.0 N m/div)
CH4—Phase 1 measured current (5 A/div)

FIGURE 31.10 Performance of proportional feedback-type torque controller.

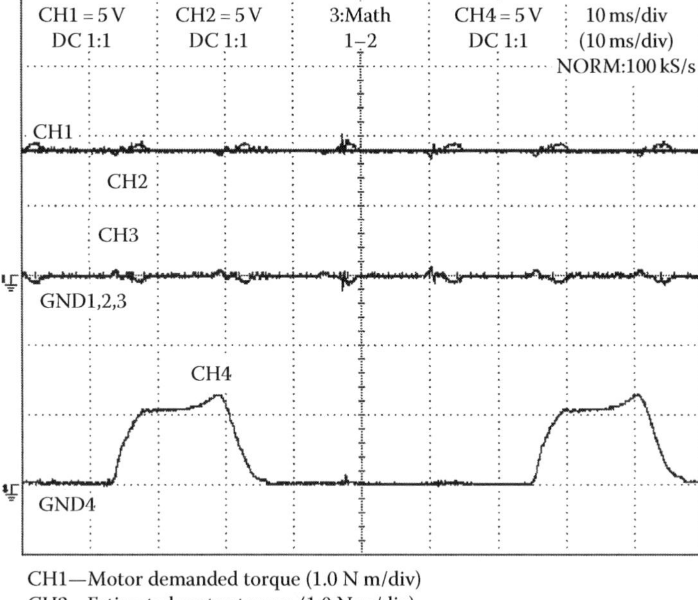

CH1—Motor demanded torque (1.0 N m/div)
CH2—Estimated motor torque (1.0 N m/div)
CH3—Motor torque error (1.0 N m/div)
CH4—Phase 1 measured current (5 A/div)

FIGURE 31.11 Performance of ILC compensation to proportional feedback-type torque controller.

good average torque control (CH1 and CH2 matching well) are possible at full-load torque. CH4 denotes that the phase current stays within rated current limit when the DTC scheme ensures that phase torque tracks the desired phase torque accurately.

These results demonstrate that proposed DTC scheme can provide accurate torque control for a constant motor torque reference, for any torque level. The torque ripples are reduced to approximately 5% of average torque.

31.7 Conclusions

It has been shown that ILC can be used for torque ripples minimization of SRM. A spatial ILC scheme was used to realize each phase torque for constant motor torque demand, as each phase torque reference is periodic in rotor position. The proposed ILC-based DTC scheme uses the knowledge of a linearized phase inductance variation for obtaining the learning gain. Experimental results demonstrate the accuracy of the proposed ILC-based DTC scheme on a prototype SRM. This method is suitable for applications requiring ripple-free constant torque at low speeds.

References

1. R. Krishnan, *Switched Reluctance Motor Drives: Modeling, Simulation, Analysis, Design, and Applications*, CRC Press, Boca Raton, FL, 2001.
2. F. Blaschke, The principle of field-orientation as applied to the transvector closed-loop control system for rotating-field machines, *Siemens Rev.*, 34, 217–220, 1972.
3. I. Takahashi and T. Noguchi, A new quick-response and high efficiency control strategy of an induction machine, *IEEE Trans. Ind. Appl.*, IA-22, 820–827, September/October 1986.
4. M. Depenbrock, Direct self control of inverter-fed induction machines, *IEEE Trans. Power Electron.*, 3, 420–429, October 1988.

5. G.S. Buja and M.P. Kazmierkowski, Direct torque control of PWM inverter-fed AC motors—A survey, *IEEE Trans. Ind. Electron.*, 51(4), 744–757, August 2004.
6. I. Husain, Minimization of torque ripple in SRM drives, *IEEE Trans. Ind. Electron.*, 49(1), 28–39, February 2002.
7. S.K. Sahoo, S.K. Panda, and J.X. Xu, Indirect torque control of switched reluctance motors using iterative learning control, *IEEE Trans. Power Electron.*, 20(1), 200–208, January 2005.
8. R.B. Inderka and R.W. De Doncker, DITC-direct instantaneous torque control of switched reluctance drives, *IEEE IAS Annual Meeting*, Pittsburgh, PA, vol. 3, pp. 1605–1609, 2002.
9. C.R. Neuhaus, N.H. Fuengwarodsakul, and R.W. De Doncker, Predictive PWM-based direct instantaneous torque control of switched reluctance drives, *IEEE Power Electronics Specialists Conference*, Jeju, Korea, pp. 3240–3246, June 18–22, 2006.
10. G. Baoming, W. Xiangheng, S. Pengsheng, and J. Jingping, Nonlinear internal-model control for switched reluctance drives, *IEEE Trans. Power Electron.*, 17(3), 379–388, May 2002.
11. J.-X. Xu, S.K. Panda, and T.H. Lee, *Real-Time Iterative Learning Control: Design and Applications* (Advanced Industrial Control Series), Springer-Verlag, Berlin, Germany, 2009.
12. Q. Weizhe, S.K. Panda, and J.X. Xu, Speed ripple minimization in PM synchronous motor using iterative learning control, *IEEE Trans. Energy Conversion*, 20(1), 53–61, March 2005.
13. S.K. Sahoo, Q. Zheng, S.K. Panda, and J.X. Xu, Model-based torque estimator for switched reluctance motors, *Conference Proceedings PEDS 2003*, Singapore, vol. 2, pp. 958–963, November 2003.
14. D.H. Hwang, Z. Bien, and S.R. Oh, Iterative learning control method for discrete-time dynamic systems, *IEE Proc. Part-D*, 138(2), 138–144, March 1991.
15. J.-X. Xu, Analysis of iterative learning control for class of nonlinear discrete-time systems, *Automatica*, 33(10), 1905–1907, 1997.

32

Precise Position Control of Piezo Actuator

Jian-Xin Xu
National University of Singapore

Sanjib Kumar Panda
National University of Singapore

32.1 Piezo Actuator

The history of piezoelectric materials dates back to 1880, when Pierre and Jacques Curie published the first experimental demonstration of piezoelectricity in various materials such as rochelle salt, quartz, and tourmaline [1]. When these crystalline materials are subjected to tensile or compressive forces, they become electrically polarized. Conversely, the crystals deform when exposed to an electric field. Together, these two effects are known as piezoelectric effect. These two aspects are distinguished as positive and inverse piezoelectric effects.

Piezoelectric motors are a kind of solid-state motors that generate gross mechanical motion through the amplification and repetition of micro-deformations of active materials like piezoceramics. Numerous types of solid-state motors have been developed, which can be classified into quasistatic and ultrasonic/resonant types according to their working principle, as well as linear and rotary motors according to the motion pattern [2,3]. These motors work on inverse piezoelectric effect. There is an orbital motion of the stator at the rotor contact point, which can be achieved through proper control of the active material. Second, the frictional interface between the rotor and stator rectifies the micro-motion to produce macro-motion of the rotor.

Piezoelectric motors made of solid states are a new kind of motors suitable for low-speed, high-precision positioning applications. The rise in the trend to use piezoelectric motors for low-speed high-precision applications is mainly due to the absence of electromagnetic field, which leads to simple construction, compact size, and high torque at low speed. The advent of high-speed switching power devices and material with less frictional wear and tear have made the use of piezoelectric motor feasible in industrial applications that demand precise positioning. In this chapter, a linear ultrasonic piezo motor (LUSM) is used as an illustrative example, and its structure, working principle, and speed characteristics are discussed.

32.1.1 Structure and Working Principle of LUSM

The stator of the linear ultrasonic motor consists of a rectangular piezoceramic plate and two electrodes. Each electrode on the front face of the plate consists of two parts (*A* and *A′*, *B* and *B′*) placed diagonally as shown in Figure 32.1, namely in the checker board pattern. A common electrode, which fully covers the plate, is bonded on the opposite side of the plate and grounded through a tuning inductor. The movement of the ceramic plate is constrained by a pair of fixed support and two high stiffness springs supporting along the long edge. These supports are designed so as to allow the ceramic plate to slide in *x*-direction, and bend along the *y*-direction. A rigid ceramic spacer is attached at the center of the short edge and undergoes an elliptical motion due to vibration of the ceramic layer. With the help of preload force applied, the elliptical motion of the spacer will be transferred to the moving stage through friction. The motor uses both the bending and longitudinal modes of vibration simultaneously to generate the driving force. It uses bending mode in the *y*-direction and longitudinal mode in *x*-direction resulting in the elliptical motion in *x*–*y* plane. When the electrodes *A* and *A′* are excited with sinusoidal voltage, the stage moves to the positive *y*-direction. The stage moves to negative *y*-direction when electrodes *B* and *B′* are excited.

The amplitude of the displacement is maximum when the frequency of the exciting voltage is equal to the natural resonant frequency of the ceramic plate. The inductor between the common plate C and the ground tunes the electrical resonance frequency to its mechanical resonance frequency.

32.1.2 Control Voltage–Velocity Characteristics of the Motor

Figure 32.2 illustrates the motor control voltage versus velocity characteristic. The motor moves in positive or negative direction depending upon the polarity of the applied control voltage V_c. When a positive control voltage is applied, electrodes *A* and *A′* will be excited and the motor moves to the positive direction. Similarly for a negative control voltage V_c, electrodes *B* and *B′* will be excited and the motor moves in negative direction. The applied voltage amplitude determines the oscillation amplitude and, hence, the range of speed and force that the motor can produce.

Due to static friction between the spacer and moving stage, the behavior of piezo motor is characterized by a big deadzone, namely, the motor does not move till the control voltage exceeds the offset value.

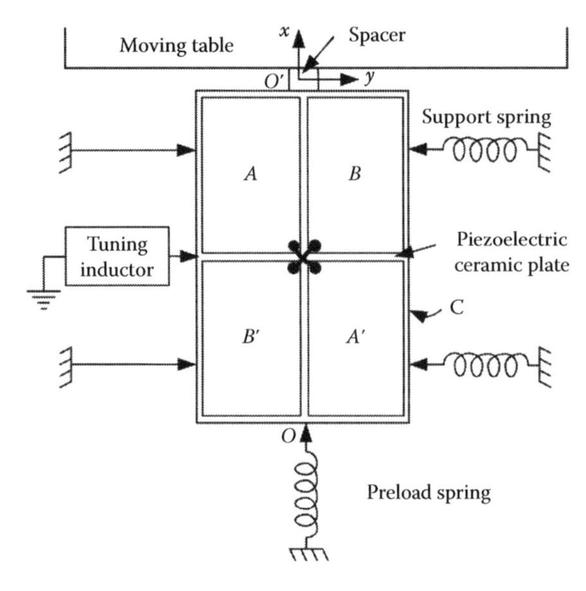

FIGURE 32.1 Linear ultrasonic motor structure.

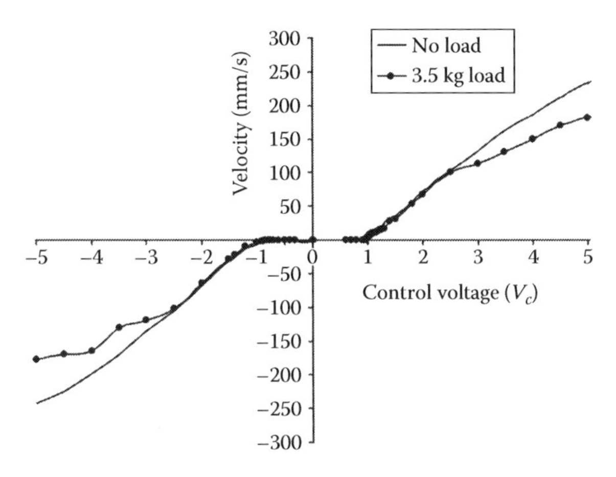

FIGURE 32.2 LUSM control voltage–velocity characteristics, which presents a deadzone due to static friction.

From Figure 32.2, the deadzone is almost invariant with the load on the stage since the load on the stage is acting vertically downward while the driving force is acting tangentially sideways on the stage. A special type of roller bearing is used with negligible friction. The deadzone is not symmetric, with 0.9 V for the positive direction and −0.8 V for the negative direction.

Though the static friction gives position holding feature to the ultrasonic motor, it introduces nonlinearity in motor characteristics. Compensation for the deadzone is imperative at the control level to generate high-performance motion profiles.

A simplified model of the LUSM is

$$F = M\frac{d^2x}{dt^2} + B\frac{dx}{dt} + F_d, \tag{32.1}$$

where

- F is the force generated by the piezo vibration
- x is the position of the moving stage
- M is the net mass of the moving stage
- B is the coefficient of viscous friction
- F_d is the disturbance force due to variations in motor parameters

The force generated by the piezo vibration is the function of the control voltage V_c and is given by $F = K_f V_c$, where K_f is known as force constant.

32.2 Deadzone Compensation and Proportional-Integral Control

Various deadzone compensation schemes have been proposed [4–6]. Since the magnitudes of the static friction is almost invariant, a feedforward compensation scheme can be implemented using the velocity reference and position error signals. The schematic of the deadzone compensation is shown in Figure 32.3.

With the feedforward deadzone compensation scheme shown in Figure 32.3, the motor characteristics becomes linear as shown in Figure 32.4, which is highly desirable, as the input nonlinearity is removed and any standard controller can be designed and applied directly without distortion.

To find the accurate model for the piezoelectric motor is difficult as the parameters of the motor are time varying. Therefore, a model-free controller is desirable. With the deadzone compensation, a linear proportional-integral (PI) controller can be used for the precise position control of the piezo motor. The PI controller in discrete form is

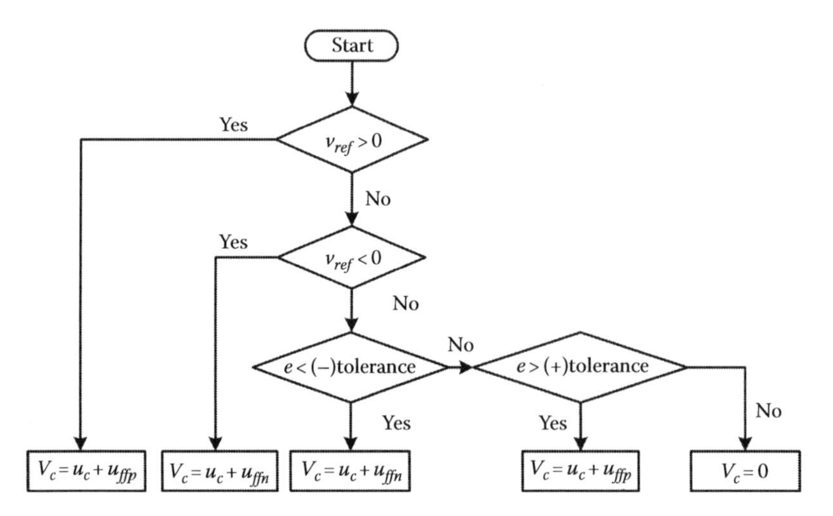

FIGURE 32.3 Feedforward deadzone compensation scheme. This scheme works for both the time-varying trajectory and setpoint tracking. v_{ref} is the reference velocity. u_{ffp} and u_{ffn} are compensation voltages for positive and negative reference velocities, respectively. u_c is a signal generated by an appropriate controller. V_c is the total control signal. For continuously changing reference velocity, the compensation is determined according to the reference velocity. When the reference velocity is zero, such as in setpoint control, the compensation is determined according to the sign of the position error. When the position error is within the tolerable limit, no control signal is generated. The tolerance for the position error can be set proportional to the resolution of the encoder.

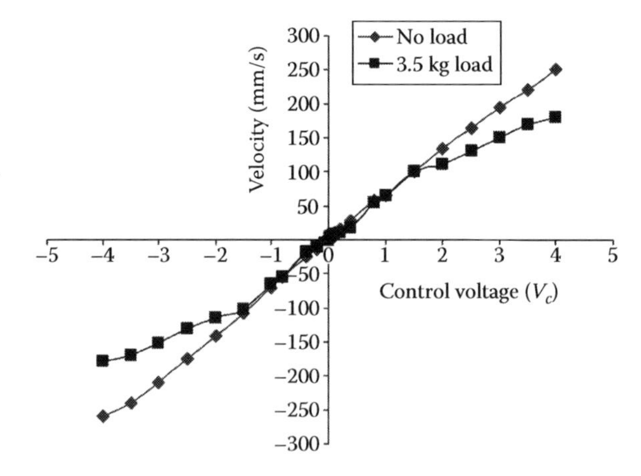

FIGURE 32.4 The voltage–velocity characteristics with the deadzone compensation, which is linear, and thus the deadzone effect is eliminated. The velocity is lesser under the loaded condition. This is due to the added inertia of the load to the stage.

$$u_c(k) = K_P e(k) + K_I T_s \sum_{i=0}^{k} e(i), \tag{32.2}$$

where

$e(k) = x_d(k) - x(k)$ is the tracking error
$x_d(k)$ is the reference position
$x(k)$ is the position of the moving stage
T_s is the sampling interval
K_P and K_I are the proportional and integral gains of the PI controller, respectively

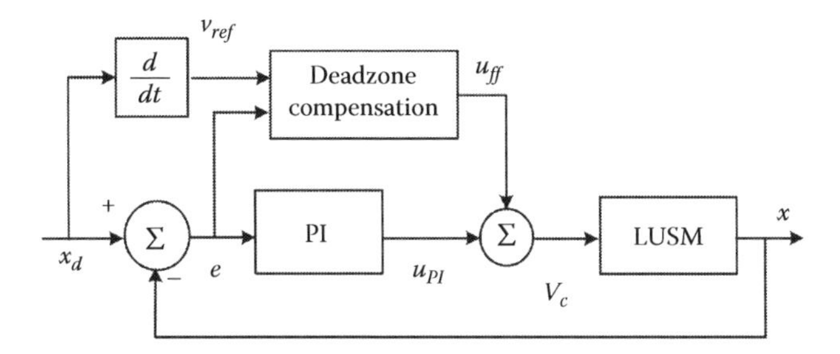

FIGURE 32.5 PI position control scheme with deadzone compensation, where u_{PI} is the output of the PI controller.

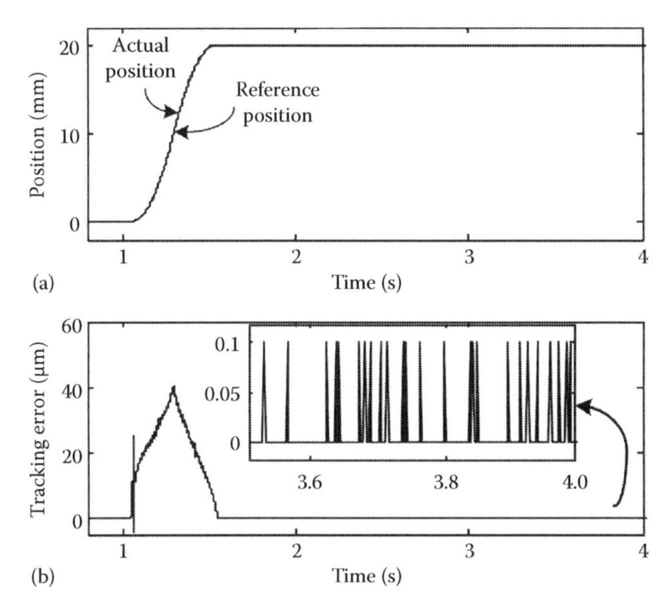

FIGURE 32.6 Experimental results on tracking performance with PI controller: (a) Slider reference and actual position, (b) tracking error. The PI gains are chosen to be $K_P = 35$ and $K_I = 10.5$, respectively. Sampling interval $T_s = 50\,\mu s$. The reference trajectory consists of two segments. The first segment, from 1 to 1.6 s, is a sinusoidal curve. The second segment, starting from 1.6 s, is a constant (20 mm). The steady-state error, zoomed in (b), is 0.1 μm, which is the resolution of the encoder used. In contrast, PI cannot perform precise tracking during the first segment when the reference is time varying.

The closed-loop PI position control scheme is shown in Figure 32.5.

PI controller can achieve precise positioning at the steady state, as shown in Figure 32.6. For comparison purpose, a periodic sinusoidal position reference tracking with PI controller is shown in Figure 32.7. The tracking errors are slightly below 20 μm under no-load condition and slightly above 20 μm with 3 kg load.

It is known in control theory that an integral controller can eliminate steady-state error with respect to a setpoint reference but is unable to achieve precise tracking for time-varying references. By increasing the proportional controller gain, it is possible to reduce the tracking error, but also incur large control signals when the tracking error is large. A desired controller should be able to provide a varying gain, which is low or high when the tracking error is large or small, respectively. Consider a switching controller

$$u_c = \beta\,\mathrm{sgn}(e), \tag{32.3}$$

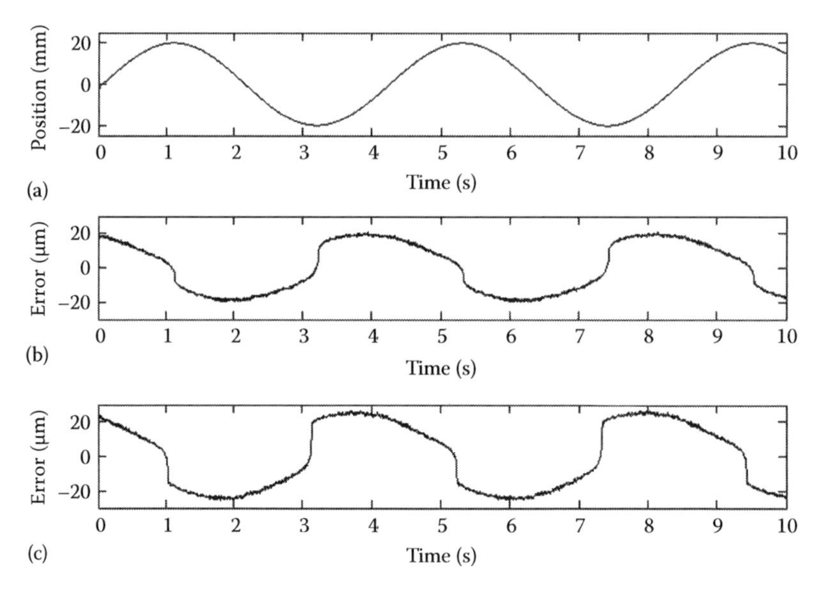

FIGURE 32.7 Periodic sinusoidal position tracking with PI controller: (a) reference position, (b) no-load tracking error, and (c) tracking error under 3 kg load.

where

$$\operatorname{sgn}(e) = \begin{cases} 1 & \text{if } e > 0 \\ -1 & \text{if } e < 0 \end{cases}. \tag{32.4}$$

The controller possesses an infinite gain at $e = 0$, and a varying proportional gain K when $e \neq 0$

$$K = \frac{u_c}{e} = \frac{\beta}{|e|},$$

which is reciprocal to the tracking error. Such a switching controller is in essence a sliding mode controller. The value of β can be chosen according to the hardware constraint, such that the controller works within the constraint, for instance, the maximum voltage of the power supply.

32.3 Sliding Mode Controller

Sliding mode control (SMC) has been proven to be a powerful control method for the complex nonlinear systems [7,8]. The underlying idea of SMC is to design a variable structure control law such that the system state is forced to reach and stay on a predefined surface, called sliding surface.

The first step in SMC design is to choose a switching surface. For the second-order piezo motor (32.1), the P-D type switching surface is commonly selected

$$\sigma = \lambda e + \frac{de}{dt}, \tag{32.5}$$

where λ is a positive constant. The aim of the SMC is to force the system state to reach the sliding surface $\sigma = 0$, and thereafter maintain it.

A typical SMC has the form

$$u_{\text{smc}} = u_{eq} + \beta \operatorname{sgn}(\sigma), \tag{32.6}$$

where u_{eq} is a model-based compensation term derived from the condition $\dot{\sigma} = 0$, known as the equivalent control, which keeps the state at $\sigma = 0$. From (32.1), neglecting the disturbance F_d the equivalent control is

$$u_{eq} = \frac{M(\lambda \dot{e} + \ddot{x}_d) + B\dot{x}}{K_f}. \tag{32.7}$$

The switching term $\beta \operatorname{sgn}(\sigma)$ in (32.6) is designed to handle the unknown disturbance or any parametric uncertainties. β is the switching gain whose value depends upon the maximum system disturbance or bounds of parametric uncertainties.

A major drawback of SMC is the chattering in the neighborhood of the sliding surface caused by switching delays. One way to limit this oscillation is to introduce a narrow boundary layer near the sliding surface to smoothen the control action [9]. The control input will be

$$u_{\text{smc}} = u_{eq} + \beta \operatorname{sat}\left(\frac{\sigma}{\phi}\right), \tag{32.8}$$

where the factor ϕ defines the thickness of the boundary layer, and sat(·) is a saturation function defined as

$$\operatorname{sat}\left(\frac{\sigma}{\phi}\right) = \begin{cases} \dfrac{\sigma}{\phi} & \text{for } \left|\dfrac{\sigma}{\phi}\right| < 1 \\ \operatorname{sgn}\left(\dfrac{\sigma}{\phi}\right) & \text{for } \left|\dfrac{\sigma}{\phi}\right| \geq 1 \end{cases}. \tag{32.9}$$

The performance of the controller can be further improved by augmenting a proportional part to the control law (32.8). The additional term $\alpha\sigma$ forces the motor to approach the switching surface faster when σ is large, where α is chosen much smaller than β [10]. The final control law is

$$u_{\text{smc}} = u_{eq} + \alpha\sigma + \beta \operatorname{sat}\left(\frac{\sigma}{\phi}\right), \tag{32.10}$$

which can reduce the chattering greatly and reach the sliding surface fast.

The parameters of the SMC used in experiment are $\lambda = 10{,}000$, $\beta = 0.4$, $\alpha = 0.02$, and $\phi = 100$. Figure 32.8 shows the setpoint tracking results. The steady-state error is 0.1 μm without using an integrator. Though the chattering is reduced by using the sat function, it still exists at a smaller scale. The results for tracking the same sinusoidal reference trajectory as in Figure 32.7 with SMC are shown in Figure 32.9. The peak error is approximately 10 μm, which is half of that with PI controller. For the loaded condition, the peak error is still of the same magnitude. As the sliding mode controller is robust to disturbances, the tracking error does not deteriorate with load variations.

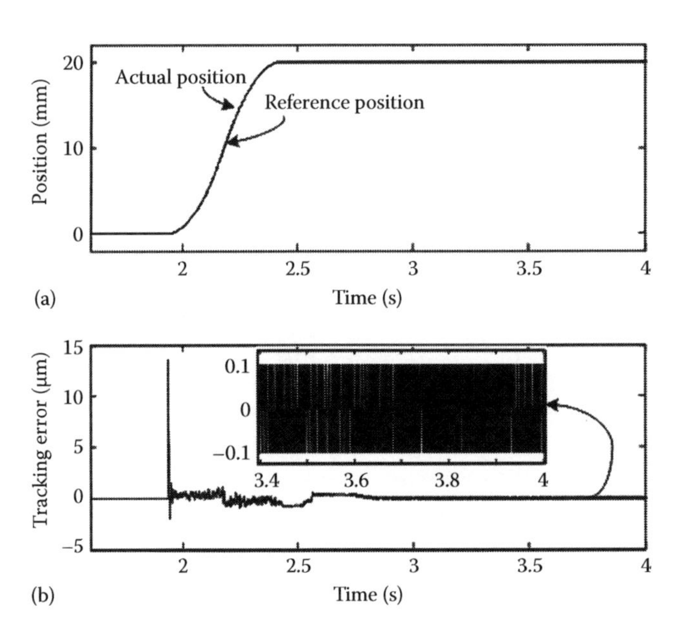

FIGURE 32.8 Setpoint position control profiles with SMC. (a) Stage position and (b) tracking error.

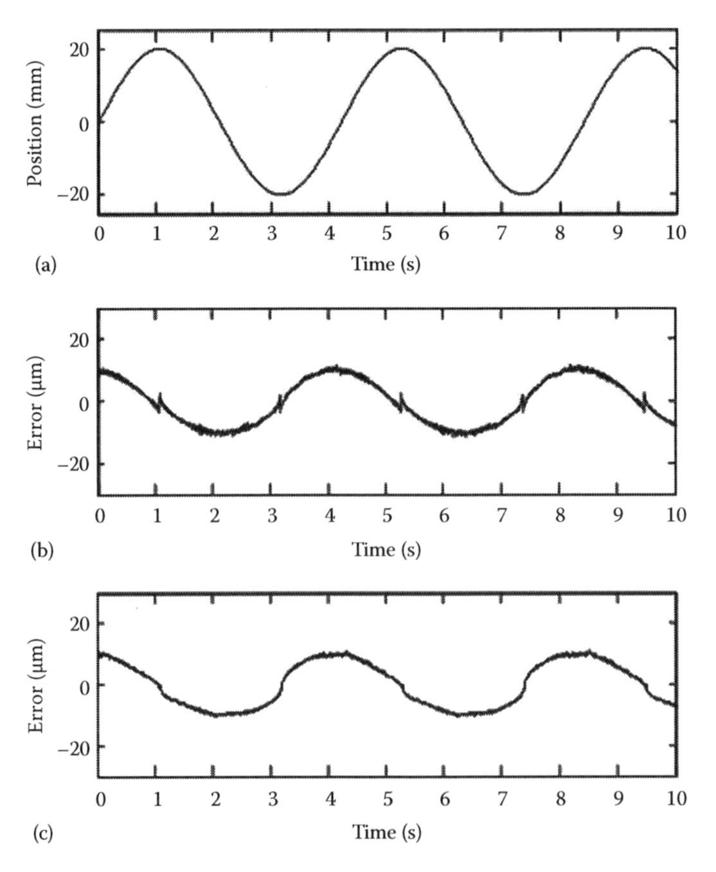

FIGURE 32.9 Periodic sinusoidal position tracking with SMC. (a) Reference and actual position, (b) tracking error under no-load, and (c) tracking error under 3 kg load.

32.4 Repetitive Learning Controller

In many high-performance applications, the tracking error of the SMC, which is around 10 μm, is still considered to be substantial. SMC is a high-gain feedback control designed for the systems with bounded disturbances. To further improve the tracking performance, feedforward control should be used. Repetitive learning control (RLC) offers a unique feedforward scheme, which is based on the repetitiveness of the task. For the periodic reference or disturbance, RLC can improve the tracking performance from cycle to cycle.

The key principle of RLC is to use the previous cycle tracking errors and control signals to update the current cycle control signals directly, hence gradually reduce the tracking errors when the control task repeats [11]. When the reference trajectory is periodic, the tracking errors are also periodic, as shown in Figures 32.7 and 32.9. To reduce the periodic tracking error an RLC is added to the exiting sliding mode controller as shown in Figure 32.10. The output of repetitive learning controller, $u_{rlc}(k)$, is

$$u_i(k) = (1-\gamma)u_{i-1}(k) + \kappa e_{i-1}(k+L), \tag{32.11}$$

where

γ is a forgetting factor
κ is a learning gain
$u_{i-1}(k)$ is the previous cycle output of the RLC
$e_{i-1}(k + L)$ is the previous cycle error at $(k + L)$th instant
L is used to adjust the phase delay of the error

The present error and present control output from the learning controller are stored in the memory and are used to calculate the compensation required in the next cycle in order to eliminate the error cycle by cycle. The RLC designs, are discussed in [12].

The experimental results obtained with SMC plus RLC for the same reference trajectory as in Figure 32.7 are shown in Figures 32.11 and 32.12. The parameters of the SMC are same as before and that of RLC are $\gamma = 0.05$ and $\kappa = 10$. Figure 32.11 shows the results for no-load condition. The experimental results for loaded condition are presented in Figure 32.12. From these results, we conclude that the performance of the scheme is equally good both under no-load and loaded conditions.

In both figures it is seen that once the RLC is put into action, the error reduces from cycle to cycle till it converges to a small magnitude. RLC output control voltage keeps adjusting so as to reduce the error. The error converges from 20 μm by pure SMC to approximately 1 μm after 15 learning cycles. It is clear from the results that the RLC mechanism, once learned to produce the required cyclic control profiles, completely takes over the control action, meanwhile the SMC control profile gradually disappears. Namely, the control action changes from feedback-based SMC into feedforward-based RLC.

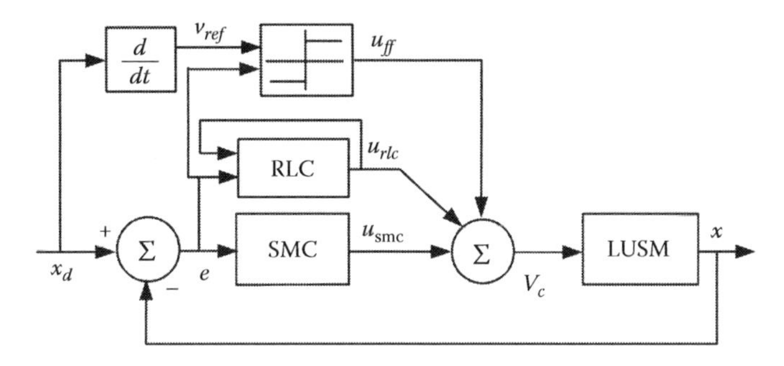

FIGURE 32.10 Configuration for closed-loop position control of LUSM with RLC and SMC.

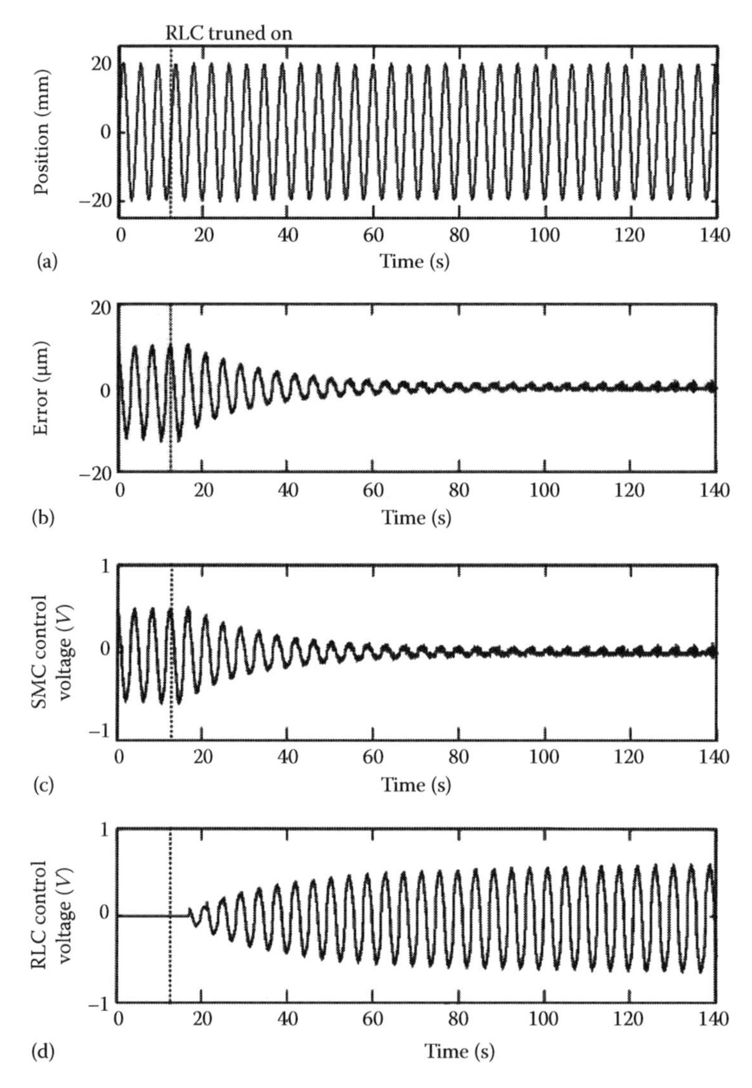

FIGURE 32.11 Position control of LUSM with SMC and RLC under no-load. (a) Actual and reference position, (b) tracking error, (c) control voltage from SMC, and (d) control voltage from RLC.

32.5 Conclusions

The piezoelectric motors are a new type of motors having wide applications in high-precision positioning. The inherent deadzone in the motor characteristics, which leads to nonlinearity in the system, can be compensated. PI control can achieve precise position control with respect to setpoint reference at the steady state. SMC can be used to achieve more precise tracking control when the reference is time varying. For periodic references, the tracking error can be further reduced by combining SMC with RLC. The experimental results obtained show that PI plus deadzone compensation can achieve a precision of 0.1 μm at the steady state, which reaches the same level of encoder resolution. SMC can achieve a prevision about twice as high as PI, and SMC plus RLC can achieve a precision about 10 times as high as SMC alone.

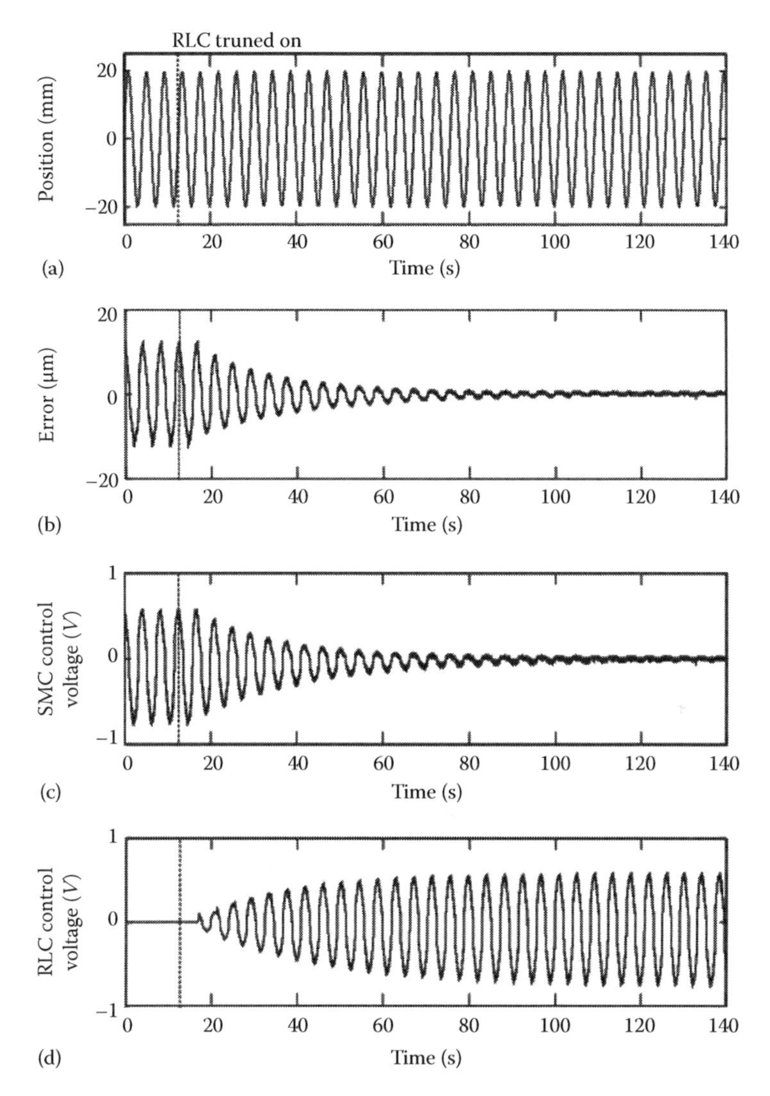

FIGURE 32.12 Position control of LUSM with SMC and RLC under 3 kg load. (a) Actual and reference position, (b) tracking error, (c) control voltage from SMC, and (d) control voltage from RLC.

References

1. J. P. Shields, *Basic Piezoelectricity*. W. Foulsham, Slough, Bucks, U.K., 1966.
2. T. Sashida and T. Kenjo, *An Introduction to Ultrasonic Motors*, Clarendon Press, Oxford, New York, 1993.
3. N. W. Hagood IV and A. J. McFarland, Modeling of a piezoelectric rotary ultrasonic motor, *IEEE Transactions on Ultrasonics, Ferroelectrics, and Frequency Control*, 42(2), 210–224, 1995.
4. T. Senjyu, T. Yoshida, K. Uezato, and T. Funabashi, Position control of ultrasonic motors using adaptive backstepping control and dead-zone compensation with fuzzy inference, *IEEE International Conference on Industrial Technology*, 1, 560–565, December 2002.
5. T. Senjyu, S. Yokoda, Y. Gushiken, and K. Uezato, Position control of ultrasonic motors with adaptive dead-zone compensation, *IEEE Industry Applications Conference: Thirty-Third IAS Annual Meeting*, vol. 1, St. Louis, Misouri, pp. 506–512, October 1998.

6. T. Senjyu, T. Kashiwagi, and K. Uezato, Position control of ultrasonic motors using marc and dead-zone compensation with fuzzy inference, *IEEE Transactions on Power Electronics*, 17, 265–272, March 2002.
7. V. Utkin, J. Guldner, and J. Shi, *Sliding Mode Control in Electromechanical Systems*, Taylor and Francis, Philadelphia, PA, 1999.
8. X. H. Yu and J.-X. Xu, *Variable Structure Systems: Towards the 21st Century, Lecture Notes in Control and Information Sciences*, vol. 274, Springer Verlag, Berlin, Germany, 2002, ISBN 3-540-42965-4.
9. K. David Young, V. I. Utkin, and Ü. Özgüner, A control engineer's guide to sliding mode control, *IEEE Transactions on Control Systems Technology*, 7(3), 328–342, May 1999.
10. D. Q. Zhang and S. K. Panda, Chattering-free and fast response sliding mode controller, *IEE Proceedings Control Theory and Applications*, 146(2), 171–177, March 1999.
11. R. W. Longman, Iterative learning control and repetitive control for engineering practice, *International Journal of Control*, 73(10), 930–954, 2000.
12. J.-X. Xu and R. Yan, On repetitive learning control for periodic tracking tasks, *IEEE Transactions on Automatic Control*, 51(11), 1842–1848, 2006.

33

Hardware-in-the-Loop Simulation

Alain Bouscayrol
University of Lille 1

33.1 Introduction

Electric drives are being used more and more frequently in industrial applications. In order to obtain the required performances of the drive and its control, software simulation is becoming an essential preliminary step. This involves replacing the power system with models in order to define and tune the control algorithm. Hardware-in-the-loop (HIL) simulation is used for validation tests of real-time embedded systems before implementation on actual processes. Contrary to software simulation, HIL simulation uses one or several actual devices instead of their simulation models. The other parts of the process are simulated into a controller board or in parallel computers [M97]. Even though many HIL simulations are dedicated to assessing controller boards, drive validations are nowadays more and more developed using this methodology. HIL simulation thus enables us to check availability and reliability of drives (machines, power electronics, and control) before their insertion into an entire system. Indeed, implementation constraints are taken into account, such as sensor accuracy, the sampling period, the modulation frequency, active limitations, and so on. Moreover, fault operations can easily be tested in various cases. From another point of view, it could be valuable to insert a device, which is difficult to model (an internal combustion engine, for example), in the loop. In this way, the influence of the model uncertainties could be tested using HIL simulation.

For many years, HIL simulation has been intensively used for controller assessment. The aerospace industry has used this technique, since flight control systems are critical in ensuring safety [M97]. This methodology yields exhaustive testing of a control system to prevent costly and damaging failures. Moreover, HIL simulations reduce development time and can enable more tests to be performed than on the actual system.

TABLE 33.1 Pictograms Used

Pictogram	Meaning	Pictogram	Meaning
→ (open arrow)	Signal variables	Dotted box	Power subsystems
→ (filled arrow)	Power variables	Empty box	Subsystems model or control
Dotted ellipse	Power sources	Hatched box	Electronic Control Unit (ECU)

From the 1990s, many groups in the automotive industry have employed HIL simulation for testing embedded electronic control units (ECUs) [H96,RABB02]. Indeed, this methodology avoids intense and complex integration tests on the actual vehicle. Thus, the time development can be reduced and a high quality can be ensured. HIL simulation is becoming a standard for ECU development in the automotive industry [H96].

HIL simulation is currently being used more and more to develop new components and actuators in many fields. We can cite, for example, vehicle component evaluation [ZRW01], assessment of drive controls [AAC99,CDFS04], power electronics and electric grids [WPJWD05,LDGSR07,LWFM07], servo control and robotics [PXX04], railway traction systems [TKS99,ABVDCE10], and education applications [BGDL09]. More recently, electrical generators of wind energy conversion systems have been tested using HIL simulation [KCB04,LSWSZ06,BGDL09]. In this case, sometimes reduced-scale power systems are firstly used to validate control algorithms before implementation on a full-scale power system. Power propulsion systems for electric vehicles (EVs) and hybrid electric vehicles (HEVs) [AD04,O05] are also examples of new applications for HIL simulation. In these cases, actual drives can be tested before integration on the vehicle chassis. More applications of HIL simulation are presented in [B08].

The main concepts of HIL are presented in this chapter. A simple example is provided to illustrate the different types of HIL simulation. All pictograms used throughout the chapter are presented in Table 33.1.

33.2 Software Simulation and Hardware-in-the-Loop Simulation

In this section, the difference between pure software simulation and HIL simulation is presented.

33.2.1 Control Design of an Energy Conversion System

33.2.1.1 Architecture of an Energy Conversion System

An energy conversion system converts energy from a generator to a load. Only unidirectional energy flow conversions are considered in this section. However, all concepts can be extended to bidirectional energy flow systems.

The system can be decomposed into several subsystems (Figure 33.1). All power parts are connected by action and reaction variables (for instance voltage and current), whose product is the exchanged power.

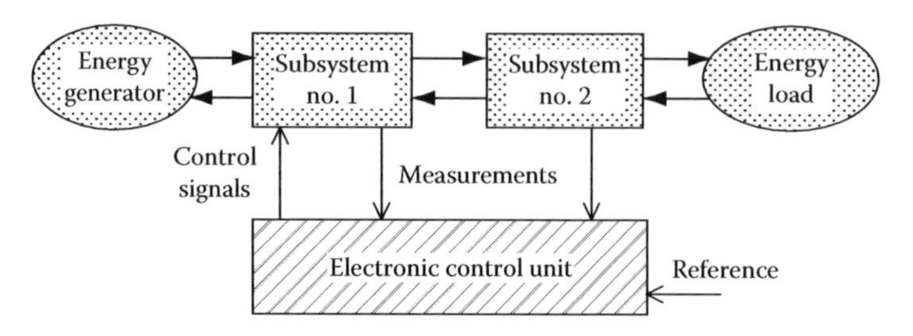

FIGURE 33.1 Control of an energy conversion system.

The ECU defines control signals to manage the energy conversion. Some measurements are required for the control algorithm. All variables managed by the ECU are signal variables.

33.2.1.2 Different Steps in the Control Development

Several steps are required to develop the control in order to achieve the best possible energy conversion performance. First, a model is derived from the actual process. Second, a control scheme is proposed and the control parameters are tuned. Most of the time, a simulation of the entire system and of its control can be obtained. This intermediary step enables various tests to be performed, avoiding damage on the actual process during the development of the control. The control algorithm is then implemented in an ECU taking into account new constraints, such as the accuracy of the sensors or the effect of the sampling period. HIL simulation can be used before the implementation of the control on the real system in order to provide intermediary validation tests.

33.2.2 Software Simulation

33.2.2.1 Architecture of Software Simulation

Simulation software is used to check performance of the control. The system model and the control are both studied in the same simulation environment (Figure 33.2). The actual process is then replaced by a process model, which can also be decomposed in subsystems. All variables between subsystems are signal variables (virtual power variables).

33.2.2.2 Limitations of Software Simulation

Generally a simple model is elaborated according to well-defined assumptions in order to ensure fast simulation. Moreover, the model parameters are obtained by identification methods that lead to certain model errors. Finally, some parameters are nonstationary, such as the machine resistance, which varies with the temperature. Thus, model uncertainties are difficult to take into account, even if robustness tests are generally provided in simulation.

The technology to implement the control is not taken into account at this stage. Indeed, the real-time development is often carried out later, including the choice of the controller board. The effects of quantification, the sensor accuracy, and the discrete-time operation (sampling period) are often neglected in software simulation. Moreover, nondesirable effects such as noise or electromagnetic compatibility are also difficult to be presented solely by simulation. Thus, HIL simulation could be a useful intermediary step before the control implementation on the real process.

33.2.3 Hardware-in-the-Loop Simulation

33.2.3.1 Architecture of HIL Simulation

In HIL simulation, one part of the closed-loop system is replaced by a real part (Figure 33.3). This actual part corresponds to the subsystems under test before their implementation in the whole

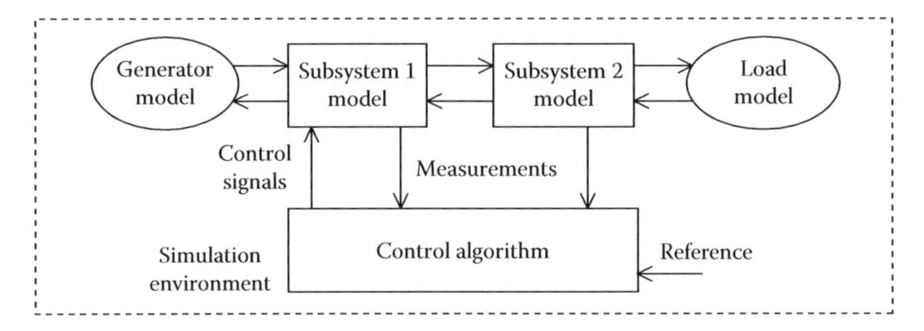

FIGURE 33.2 Software simulation of an energy conversion system.

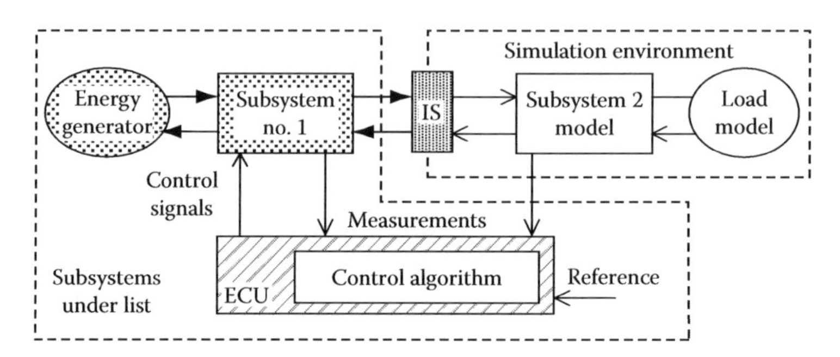

FIGURE 33.3 HIL simulation of an energy conversion system.

system. This actual part, however, has to communicate with the simulation part. Thus an interface system (IS in Figure 33.3) has to be developed to ensure the connection between the hardware and the simulation parts.

33.2.3.2 Constraints of HIL Simulation

Since an actual part is used and an IS is added, the cost of HIL simulation is obviously greater than the cost of software simulation. The IS has to generate different kinds of variables in order to connect both parts. Moreover, this interface has to induce no time delay to yield the real behavior of the closed-loop system.

Given that the actual system will work in real time, the simulation model also has to be computed in real time. A more powerful computer or controller board is thus required to enable real-time operation.

33.3 Signal and Power HIL Simulations

Two different HIL simulations can be defined depending on different objectives.

33.3.1 Signal HIL Simulation

33.3.1.1 Objective of Signal HIL Simulation

Signal HIL simulation is used to test ECUs before their implementation on the real processes. The control algorithm developed in software simulation is thus coded in the chosen ECU. This kind of HIL simulation is used in industry for checking the ECUs in different operation modes, especially in fault operations. This method enables various tests, which can be repeated many times, to be performed without impact on the real power conversion system.

33.3.1.2 Principle of Signal HIL Simulation

The entire model of the power conversion system is simulated in a real-time simulation environment (Figure 33.4). The IS only manages signal variables. This explains why this HIL method is called signal HIL simulation. The ECU (subsystem under test) is connected to the IS. Its input and output signals must be the same as those provided by the actual process.

33.3.1.3 Requirements of Signal HIL Simulation

The ECU must think that it is connected to the real process. Generally, a second ECU, called the emulation ECU, is used to simulate the power conversion system in real time (Figure 33.5). The interface of the emulation ECU is used as the IS.

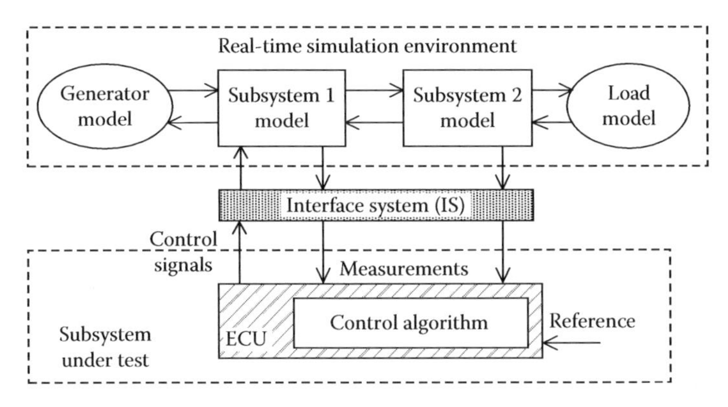

FIGURE 33.4 Signal HIL simulation of an energy conversion system.

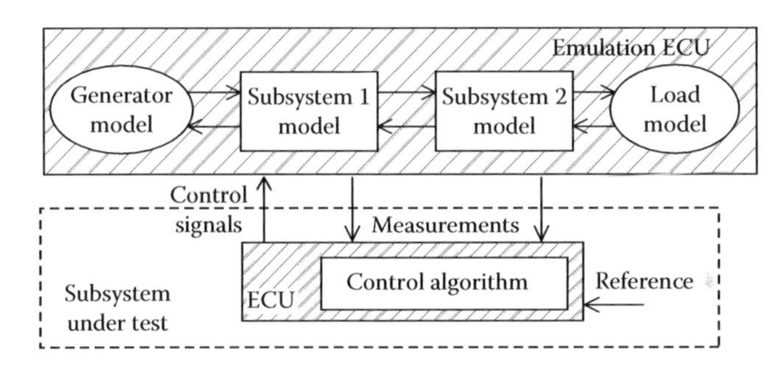

FIGURE 33.5 Practical implementation of signal HIL simulation of an energy conversion system.

The measurements must be provided in the same way as for the real process. For this reason, the sampling period of the emulation ECU must be adapted to take into account the smallest time constant of the system. This induces an important constraint when simulating systems with power electronics, which have small time constants due to modulation.

The sensor effects must also be reproduced. For analog measurements, the emulation ECU must provide variables, which seem analog for the tested ECU. The quantification output of the emulation ECU requires the smallest step possible. Moreover, the sampling period of the emulation ECU must be lower than the sampling period associated with the measurements in the tested ECU. The synchronization of both sampling periods can be used to avoid undesirable effects.

33.3.2 Power HIL Simulation

33.3.2.1 Objective of Power HIL Simulation

Power HIL simulation is used to test ECUs and also one part of the energy conversion system before their implementation on real processes. This kind of HIL simulation is useful to test new subsystems (including their control) that will be inserted in the system. Power HIL simulation is also used as an intermediary step before implementation on the whole system for safety and/or flexibility reasons. A new traction system, for example, is generally tested in a static experimental set-up before implementation in the moving vehicle.

33.3.2.2 Principle of Power HIL Simulation

In power HIL simulation, the energy conversion system is split into two parts: the power part to be tested and the power part to be simulated (Figure 33.6). The IS exchanges power variables with the power

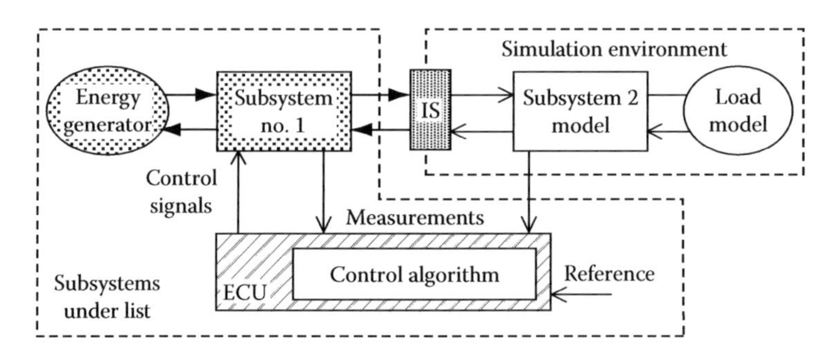

FIGURE 33.6 Power HIL simulation of an energy conversion system.

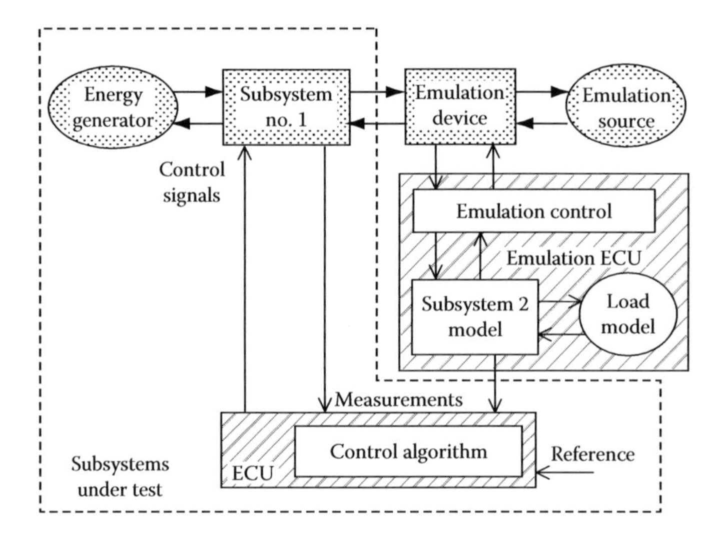

FIGURE 33.7 Practical implementation of power HIL simulation of an energy conversion system.

part to be tested, and signal variables to the power part to be simulated. The virtual power part is simulated in real time using subsystem models. The control algorithm is managed by the ECU.

33.3.2.3 Requirements of Power HIL Simulation

Due to its connection to the power part to be tested, the IS requires an emulation device (Figure 33.7) that delivers compatible variables to the real subsystem (see example in Section 33.4). Moreover, because energy must to be converted by this emulation device, an emulation source is also required to receive this energy.

The emulation device has to impose the same behavior as the simulated part. An emulation control is thus developed, and it communicates with the subsystem models. All these functions have to be carried out in real time. An emulation ECU can be used. The IS is thus composed of an emulation device, an emulation load, and the emulation control (Figure 33.8).

The same technical requirements as those for the signal HIL simulation are necessary. However, other conditions are required. The closed-loop control of the emulation device must be faster than the response time of the simulated subsystem. In this way, the dynamics of the simulated subsystem can be reproduced.

The power of the emulation device must be greater than the power of the simulated subsystem. In this way, the power conversion of the system can be simulated on the whole power range. Moreover, the

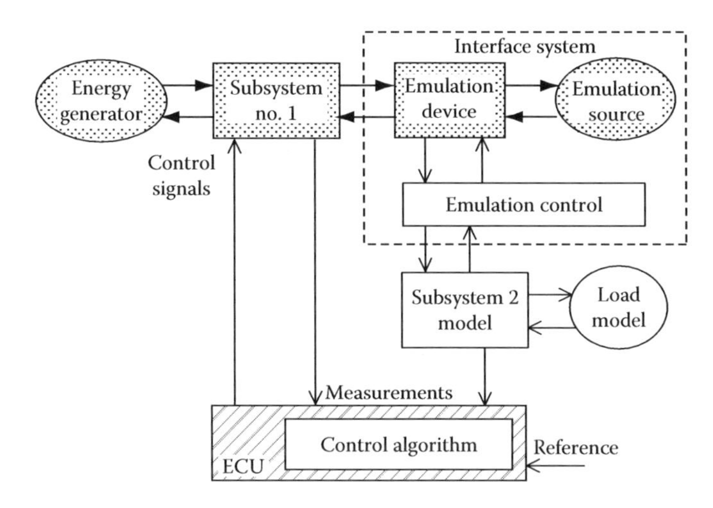

FIGURE 33.8 IS of power HIL simulation.

maximum values of the variables of the emulation device must be greater than those of the simulated subsystem. If this is not the case, the validity range of the HIL simulation will be reduced in case of limitation operation.

33.3.3 Reduced-Scale Power HIL Simulation

33.3.3.1 Objective of Reduced-Scale Power HIL Simulation

For high-power systems, the development of a full-scale power HIL simulation may require a supplementary intermediary step involving a reduced-scale HIL simulation. A reduced-scale power HIL simulation is used to check the operation principles of new subsystems using a reduced-scale experimental set-up. In this way, initial tests can be carried out, avoiding the risk of damage to the full-power subsystem to be tested.

33.3.3.2 Principle of Reduced-Scale Power HIL Simulation

The reduced-scale power HIL simulation is derived from the full-scale version (Figure 33.9). In this case, however, the tested power part is replaced by an equivalent subsystem with reduced power. The power of the emulation device is also reduced.

In order to maintain the characteristics of the simulated subsystem, a power adaptation (PA in Figure 33.9) is inserted between the emulation control and the subsystem model. This power adaptation involves a linear power amplification. In this way, nonlinear effects can be properly taken into account. Moreover, another power adaptation is needed between the controls of the tested and simulated parts. The adaptation blocks define the limits between the reduced-scale and full-scale parts (Figure 33.10).

33.3.3.3 Technical Requirements of Reduced-Scale Power HIL Simulation

The same technical requirements are needed as for the full-scale power HIL simulation. Moreover, special attention must be paid to the power adaptation. It has to be chosen in order to respect the power ratio between the full-scale and the reduced-scale parts. This power adaptation, however, must also avoid the limitation of the reduced-scale variables before the limitation of the full-scale variables, or nonphysical effects will be obtained when the full-scale is not in limitation [ABVDCE10].

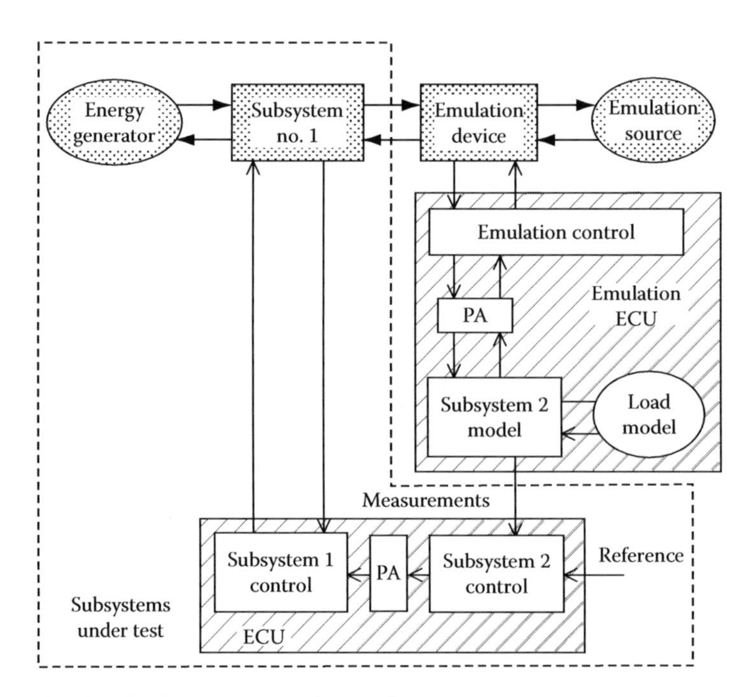

FIGURE 33.9 Reduced-scale of power HIL simulation of an energy conversion system.

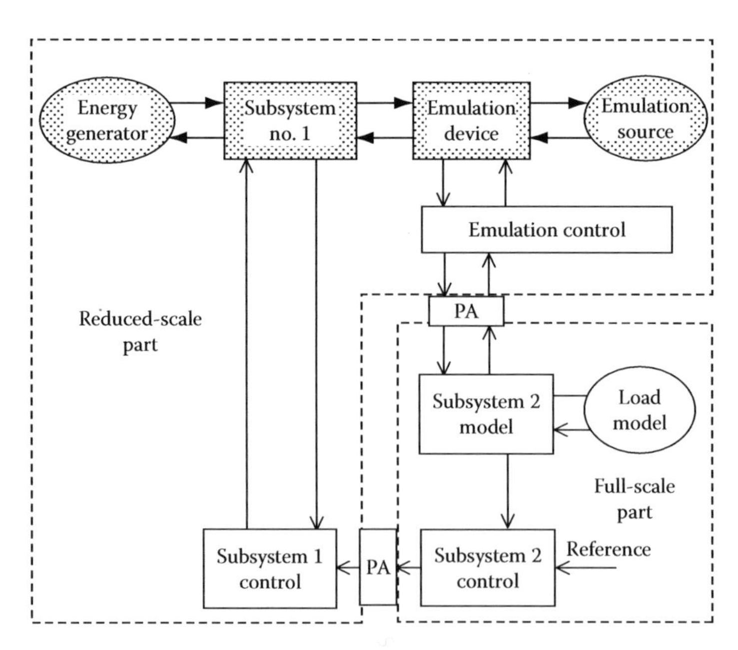

FIGURE 33.10 Reduced-scale of power HIL simulation: Reduced-scale and full-scale parts.

33.4 Example of the Traction of an EV

A simple traction system of an EV is taken as an example to illustrate the different HIL simulation concepts. In this application, a bidirectional energy flow is considered because the regenerative braking enables to recharge the battery.

33.4.1 Studied System

The studied system deals with a simplified traction drive of an EV (Figure 33.11). It is composed of a battery (energy source), a chopper, a permanent magnet DC machine, a gearbox, a mechanical differential, and two driven-wheels. The different relationships of this system are summarized in this section. Since no slips, no curves, and no brake systems are taken into account, an equivalent single-wheel system is considered. More details can be found in [BLDLA06].

33.4.1.1 Model of the Traction System

The chopper yields the voltage u_{chop} from the battery voltage V_{bat}, and the current i_{chop} is obtained from the machine current i_{dcm}:

$$\begin{cases} u_{chop} = m_{chop} V_{bat} \\ i_{chop} = m_{chop} i_{dcm} \end{cases} \tag{33.1}$$

with m_{chop} the modulation ratio of the chopper. The armature winding of the DC machine yields the armature current i_{dcm}, as state variable, from the e.m.f. e_{dcm} and the chopper voltage u_{chop}:

$$L_{arm} \frac{d}{dt} i_{dcm} = u_{chop} - e_{dcm} - R_{arm} i_{dcm}, \tag{33.2}$$

where L_{arm} and R_{arm} are the inductance and resistance of the armature winding. The torque of the DC machine T_{dcm} is obtained from the machine current i_{dcm}, and the e.m.f. e_{dcm} is linked to the rotation speed Ω_{gear}:

$$\begin{cases} T_{dcm} = k_{dcm} i_{dcm} \\ e_{dcm} = k_{dcm} \Omega_{gear} \end{cases}, \tag{33.3}$$

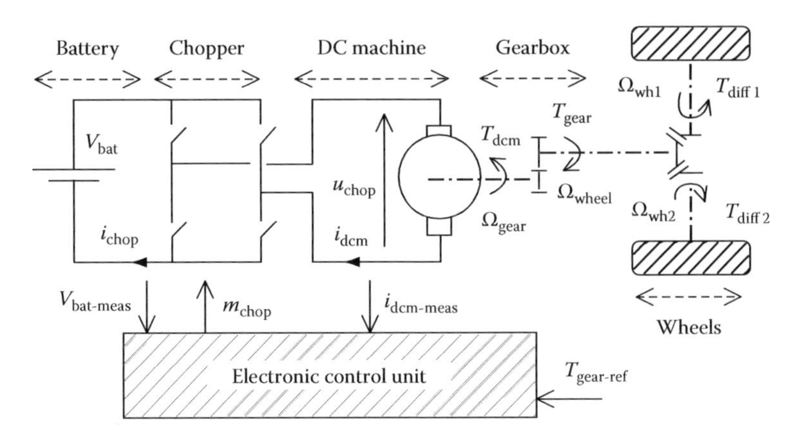

FIGURE 33.11 Scheme of the traction system of the EV.

where k_{dcm} is the torque coefficient. The gearbox yields the gearbox torque T_{gear} and the rotation speed Ω_{gear}, respectively, from the machine torque and the rotation speed of the wheel Ω_{wheel} using the gearbox ratio k_{gear}:

$$\begin{cases} T_{gear} = k_{gear} T_{dcm} \\ \Omega_{gear} = k_{gear} \Omega_{wheel} \end{cases}. \tag{33.4}$$

The wheel converts the rotational motion into a linear motion to obtain the traction force F_{tract}:

$$\begin{cases} F_{tract} = \dfrac{1}{R_{wheel}} T_{gear} \\ \Omega_{wheel} = \dfrac{1}{R_{wheel}} v_{ev} \end{cases}, \tag{33.5}$$

where R_{wheel} is the wheel radius. The EV velocity v_{ev} is obtained using the classical dynamics relationship with the traction and resistant forces, F_{tract} and F_{res}:

$$M \frac{d}{dt} v_{ev} = F_{tract} - F_{res}, \tag{33.6}$$

where M is the equivalent mass of the vehicle, including the rotating masses. The environment yields the resistive force F_{res} from the vehicle velocity and characteristics of environment:

$$F_{res} = F_0 + a v_{ev} + b v_{ev}^2 + Mg \sin\alpha, \tag{33.7}$$

where

F_0 is the initial rolling force
a is the rolling coefficient
b is the drag coefficient
α is the slope rate
g is the gravity

33.4.1.2 Control of the Traction System

The ECU contains the control algorithm, which delivers the modulation ratio m_{chop} to the chopper from the reference torque $T_{gear\text{-}ref}$, deduced from the acceleration required. It is composed of a current control loop (Figure 33.12) [BLDLA06]. A synthetic scheme of the EV traction system is described in accordance with the principles developed in the previous sections (Figure 33.13).

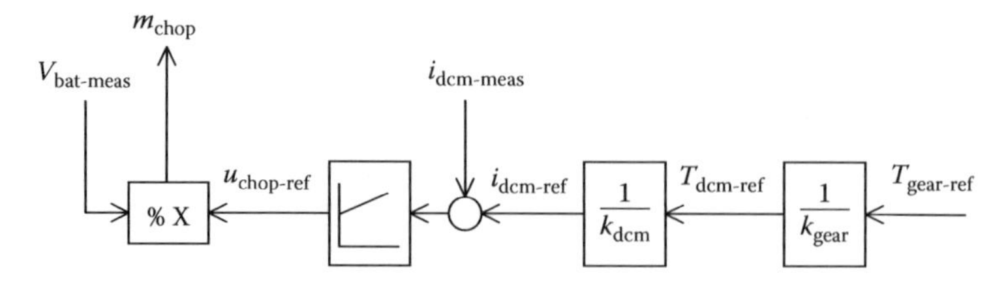

FIGURE 33.12 Control scheme of the EV traction system.

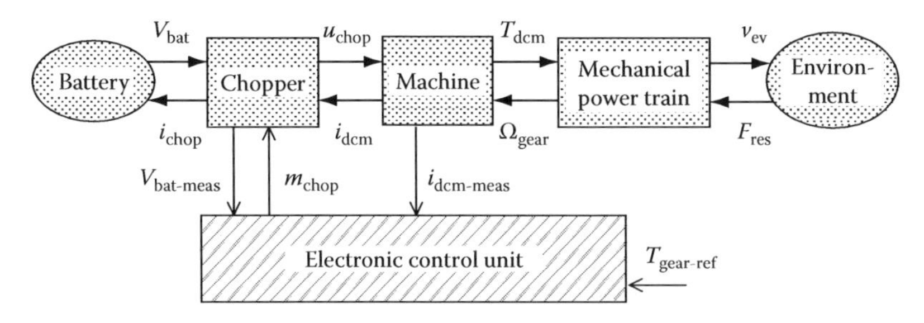

FIGURE 33.13 Synthetic scheme of the EV traction system.

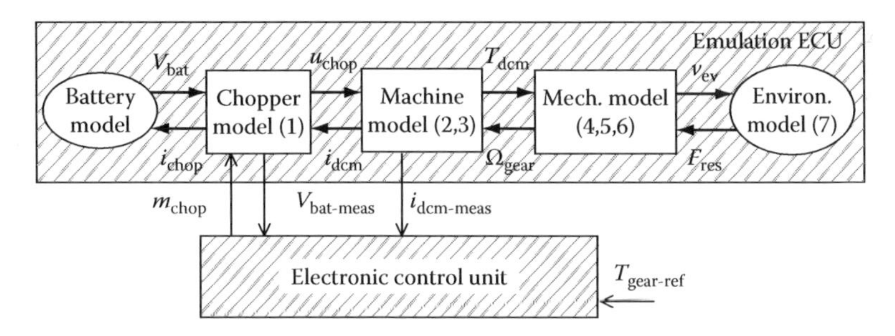

FIGURE 33.14 Hardware configuration of the signal HIL simulation.

33.4.2 Signal HIL Simulation of the Studied System

In this case, the ECU, which contains the control algorithm, is tested. Another ECU (emulation ECU) is used to compute the model of the traction system in real time (Figure 33.14). The relationships within the model are indicated in brackets. Three subsystems are simulated in this case.

33.4.3 Electrical Power HIL Simulation of the Studied System

In this case, the ECU, the battery, and the chopper are tested. The effect of the modulation frequency can thus be taken into account, especially its electromagnetic compatibility influence on the ECU. Moreover, the effect of the battery discharge can also be studied. Indeed, obtaining a precise modeling of the battery is a complex task.

In order to impose the same current as the DC machine in the chopper, a specific interface has to be used. Another controlled chopper is used (Figure 33.15). Moreover, an inductor is inserted to smooth the current, as in the machine. Furthermore, another battery is used to store energy when the vehicle accelerates and to provide energy for regenerative braking.

The emulation control is composed of a current loop, which controls the current in the inductor. The reference current is generated by the machine model. Moreover, the measurement of the chopper voltage is required as input for the machine model. Due to its modulated value, this measurement is delicate. A filter can be used if a mean value of this voltage is sufficient. The IS is thus composed of an inductor, a chopper, a second battery, and the emulation control.

33.4.4 Mechanical Power HIL Simulation of the Studied System

In this case, the ECU, the battery, the chopper, and the DC machine are tested. It may be valuable to study the electric drive on a static bench before its implementation on a moving part (the vehicle). The mechanical power train behavior has to be imposed on the DC machine shaft. Another drive (for instance, an induction machine drive) is used to reproduce this interaction (Figure 33.16).

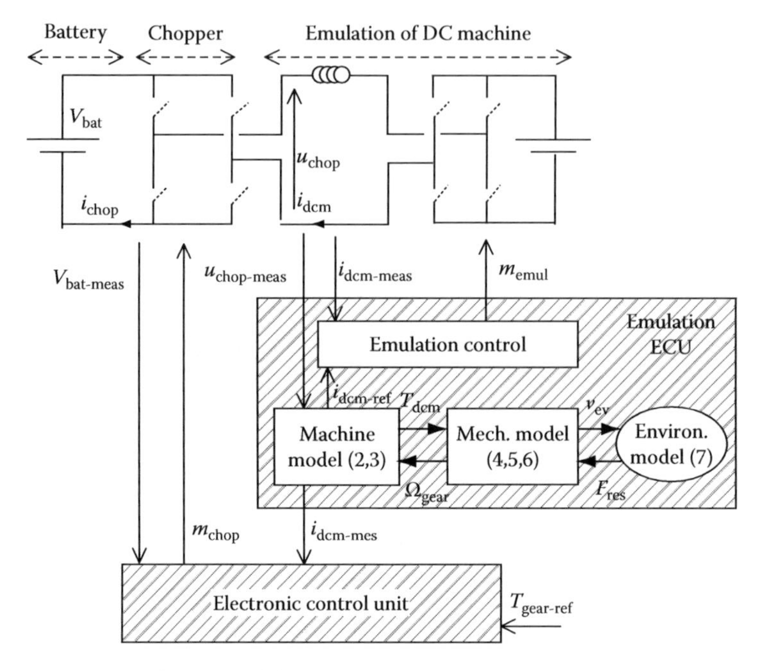

FIGURE 33.15 Hardware configuration of the electrical power HIL simulation.

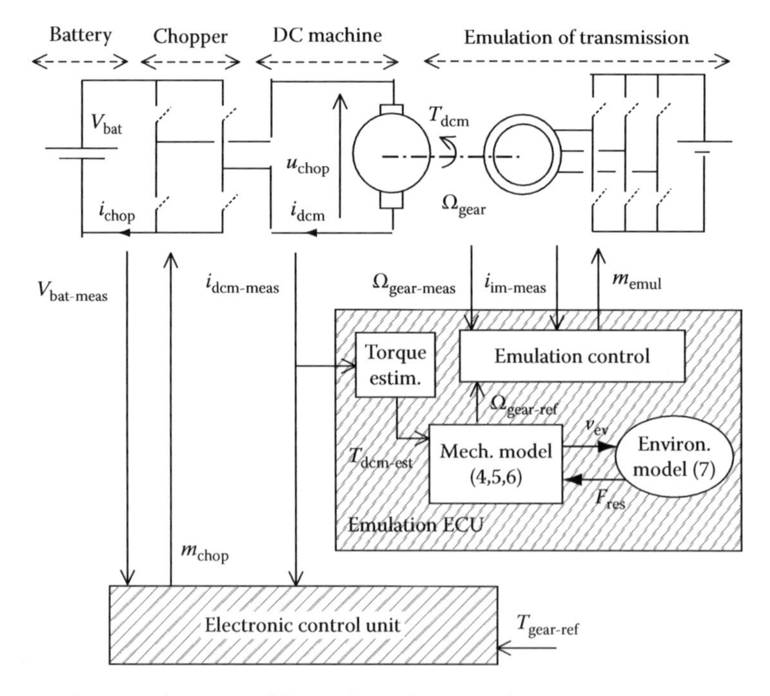

FIGURE 33.16 Hardware configuration of the mechanical HIL simulation.

The emulation control allows us to impose the same rotation speed as expected for the mechanical power train. The gear speed reference is provided by the mechanical model. The measurement of the load drive current is used (for example, to achieve a field-oriented control). Moreover, a torque measurement is required as input for the mechanical model. Instead of this sensitive sensor, a torque estimation can be used from the measurement of the DC machine current. The IS is thus composed of an induction machine, an inverter, a second battery, and the emulation control.

33.4.5 Reduced-Scale Mechanical Power HIL Simulation of the Studied System

A reduced-scale experimental set-up is used instead of the full-scale drive. This reduced-scale mechanical power HIL simulation could be an intermediary step in checking, in real time, the control scheme. The second step involves the full-scale HIL simulation. The HIL scheme (Figure 33.17) is derived from the full-scale mechanical power HIL simulation. The power adaptation (PA) is defined as follows:

$$\begin{cases} \Omega_{\text{gear-ref}} = k_{\text{speed}} \Omega_{\text{gear-mod}} \\ T_{\text{dcm-mod}} = k_{\text{torque}} T_{\text{dcm-est}} \end{cases}. \tag{33.8}$$

In the system control, the power adaptation is derived from the power adaptation in the model:

$$T_{\text{dcm-ref}} = \frac{1}{k_{\text{torque}}} T_{\text{dcm-mod}}. \tag{33.9}$$

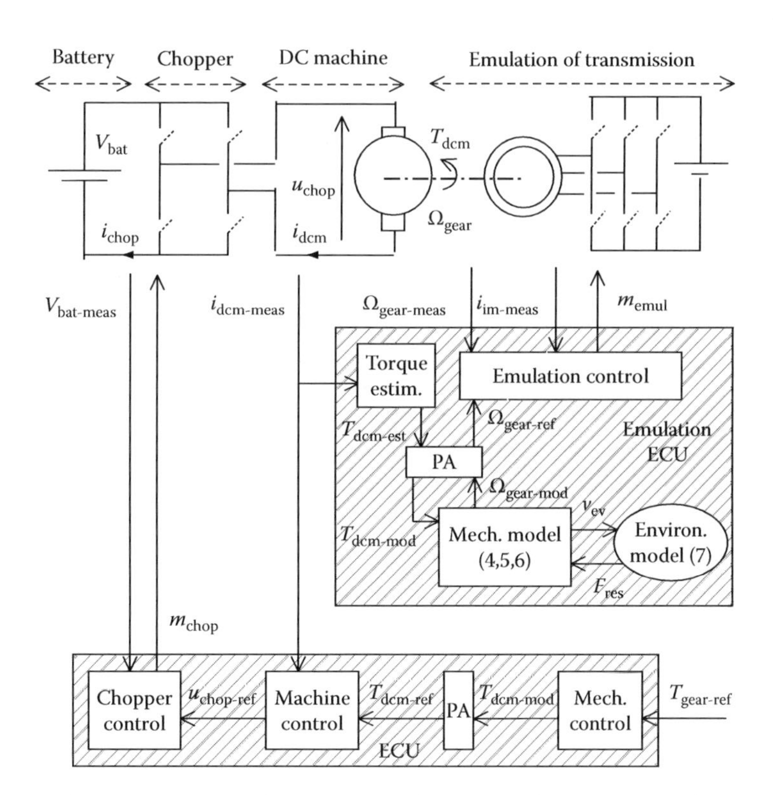

FIGURE 33.17 Hardware configuration of the reduced-scale mechanical power HIL simulation.

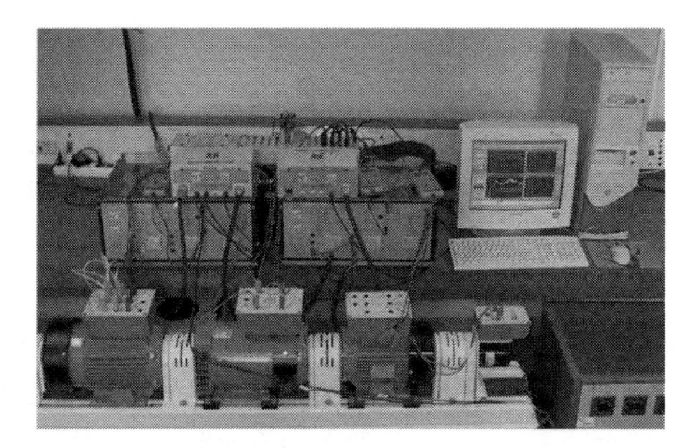

FIGURE 33.18 Experimental set-up of the reduced-scale mechanical HIL simulation.

In [BLDLA06], an example of a reduced-scale mechanical power HIL simulation is provided in the case of a traction system, using an induction machine. The experimental set-up consisted of a 1.5 kW induction machine (Figure 33.18) and the actual vehicle traction system contained a machine of 20 kW. In this HIL simulation, a single ECU was used for the control algorithm and for the emulation part (model and emulation control) because no other ECUs were available.

33.5 Conclusion

HIL simulation is more and more used to study performances of ECUs, electric drives, and other actuators. It is a valuable intermediary step between the simulation of the entire studied system and the implementation on the actual process. Different kinds of HIL simulation can be used in function of the part of the system to test before its implementation. Signal HIL simulations are devoted to test ECUs and power HIL simulations are devoted to test power subsystems including their control.

The first step consists in defining where is the limit of the subsystem to be tested, and what part will be simulated. Then a specific IS has to be defined to reproduce the behavior of the simulated part on the tested subsystem. A special attention must be paid to the IS because it is related to the quality of the HIL simulation.

In some cases, reduced-power HIL simulation can be first achieved to propose progressive tests, specifically for high-power systems. This reduced-power HIL simulation is an intermediary step before developing a full-power HIL simulation. In this case, the full-power model of the simulated subsystem has to be connected with a reduced-scale tested subsystem. A specific attention must be paid in the power adaptation and the different limitations of both connected parts of different rated power.

References

[AAC99] Z. H. Akpolat, G. M. Asher, and J. C. Clare, Experimental dynamometer emulation of non-linear mechanical loads, *IEEE Transactions on Industry Applications*, 35(6), 1367–1373, 1999.

[ABVDCE10] A. Allègre, A. Bouscayrol, J. N. Verhille, P. Delarue, E. Chattot, and S. El Fassi, Reduced-scale power hardware-in-the-loop simulation of an innovative sukway, *IEEE Transactions on Industrial Electronics*, 57(4), 1175–1185, April 2010.

[AD04] K. Athanasas and I. Dear, Validation of complex vehicle systems of prototype vehicle, *IEEE Transactions on Vehicular Technology*, 53(6), 1835–1846, 2004.

[B08] A. Bouscayrol, Different types of hardware-in-the-loop simulation for electric drives, *Proceedings of the IEEE-ISIE'08*, Cambridge, U.K., 2008.

[BGDL09] A. Bouscayrol, X. Guillaud, P. Delarue, and B. Lemaire-Semail, Energetic macroscopic representation and inversion-based control illustrated on a wind energy conversion systems using hardware-in-the-loop simulation, *IEEE Transactions on Industrial Electronics*, 56(12), 4826–4835, December 2009.

[BLDLA06] A. Bouscayrol, W. Lhomme, P. Delarue, B. Lemaire-Semail, and S. Aksas, Hardware-in-the-loop simulation of electric vehicle traction systems using energetic macroscopic representation, *Proceedings of the IEEE-IECON'06*, Paris, France, 2006.

[CDFS04] R. Champagne, L. A. Dessaint, H. Fortin-Blanchette, and G. Sybille, Analysis and validation of a real-time AC drive simulator, *IEEE Transactions on Power Electronics*, 19(2), 336–345, 2004.

[H96] H. Hanselmann, HIL simulation testing and its integration into a CACSD toolset, *Proceedings of the IEEE-CACSD'96*, Dearborn, MI, 1996.

[KCB04] H. M. Kojabadi, L. Chang, and T. Boutot, Development of a novel wind turbine simulator for wind energy conversion systems using an inverter-controlled induction motor, *IEEE Transactions on Energy Conversion*, 19(3), 547–552, 2004.

[LDGSR07] P. Lok-Fu, V. Dinavahi, C. Gary, M. Steurer, and P. F. Ribeiro, Real-time digital time-varying harmonic modelling and simulation techniques IEEE Task Force on harmonics modelling and simulation, *IEEE Transactions on Power Delivery*, 22(2), 218–1227, 2007.

[LSWSZ06] H. Li, M. Steurer, S. Woodruff, K. L. Shi, and D. Zhang, Development of a unified design, test, and research platform for wind energy systems based on hardware-in-the-loop real time simulation, *IEEE Transactions on Industrial Electronics*, 53(4), 1144–1151, 2006.

[LWFM07] B. Lu, X. Wu, H. Figueroa, and A. Monti, A low cost real-time hardware-in-the-loop testing approach of power electronics control, *IEEE Transactions on Industrial Electronics*, 54(2), 919–931, 2007.

[M97] D. Maclay, Simulation gets into the loop, *IEE Review*, 43(3), 109–112, May 1997.

[O05] S. C. Oh, Evaluation of motor characteristics for hybrid electric vehicle using the HIL concept, *IEEE Transactions on Vehicular Technology*, 53(3), 817–824, 2005.

[PXX04] F. Pan, D. Xue, and X. Xu, The research and application of dSPACE based hardware-in-the-loop simulation technique in servo control, *Journal of System Simulation*, 16, 936–939, 2004.

[RABB02] C. A. Rabbath, M. Abdoune, J. Belanger, and K. Butts, Simulating hybrid dynamic systems, *IEEE Robotics and Automation Magazine*, 9(2), 39–47, 2002.

[TKS99] P. Terwiesch, T. Keller, and E. Scheiben, Rail vehicle control system integration testing using digital hardware-in-the-loop simulation, *IEEE Transactions on Control System Technology*, 7(3), 352–362, 1999.

[WPJWD05] Z. Weidong, S. Pekarek, J. Jatskevich, O. Wasynczuk, and D. Delisle, A model-in-the-loop interface to emulate source dynamics in a zonal DC distribution system, *IEEE Transactions on Power Electronics*, 20(2), 438–445, 2005.

[ZRW01] Q. Zhang, J. F. Reid, and D. Wu, Hardware-in-the-loop simulator of an off-road vehicle electrohydraulic steering system, *Transactions of the ASAE*, 43(6), 1323–1330, September 2001.

V

Mechatronics and Robotics

34

Introduction to Mechatronic Systems

Ren C. Luo
National Taiwan University

Chin F. Lin
National Chung Cheng University

34.1 What Is Mechatronics?

Mechatronics is an integrated technology that synthesizes several evolving engineering disciplines and technologies, including precision mechanical engineering, electrical engineering, computer science, control algorithms, and systems thinking in the design of products and manufacturing processes. 3C (computer, communications, and consumer electronics) products, automobiles, and household appliances are all examples of mechatronic systems. Mechatronics deals with the methods of designing and manufacturing intelligent electromechanical products. It covers a wide range of application areas including most of the electronic devices used in, e.g., consumer products, instrumentation, automobile industry, machine tool industry, computer industry, aerospace industry, medical industry, robot industry, and so on.

Mechatronics is generally defined as a field encompassing mechanical components and electronically based decision making (control circuitry). But systems engineers consider that mechatronics is a form of concurrent design and manufacturing engineering practices (Hunt, 1988; Buur, 1990; Keys, 1991). Other definitions excerpted from the literature are as follows:

> Mechatronics is defined as the field of study involving the analysis, design, synthesis, and selection of systems that combine electronic and mechanical components with modern controls and microprocessors (Alciatore and Histand, 2005).

> Mechatronics is the synergistic combination of precision mechanical engineering, electronics, control engineering, and computer science (Craig and Stolfi, 1994).
>
> Mechatronics is the application of complex decision making to the operation of physical systems (Auslander and Kempf, 1996).
>
> Mechatronics is an engineering process that involves the design and manufacture of intelligent products or systems involving hybrid mechanical and electronic functions (Hsu et al., 1995; Hsu, 1998).

People often mistakenly regard mechatronics as synonymous to "electromechanics." Electromechanical devices such as motors are common in electric systems but not in mechatronic systems. A mechatronic system involves intelligent process control related to technologies based on mechanical engineering disciplines like mechanical dynamics, thermodynamics, and fluid dynamics. Most of the mechatronic systems have the ability to sense changes in the real-world environment for intelligent control. For this purpose, various sensors are employed to assist mechatronic systems in achieving the desired actions of a mechanical component or to overcome the technological restrictions encountered in a specific manufacturing process. Often, researchers are required to apply various sensing technologies in addition to electromechanical principles. Therefore, to view a mechatronic system as an "intelligent electromechanical system" is thus appropriate.

Mechatronics is a multidisciplinary engineering science that involves four major disciplines: mechanical engineering, electronic engineering, information engineering, and intelligent control and automation, described as follows:

1. Mechanical engineering (machines, mechanical elements, dynamics, and kinetics)
2. Electronic engineering (microelectronics, power electronics, instrumentation, and signal processing)
3. Information engineering (operation system, control theory, software engineering, and artificial intelligence)
4. Intelligent control and automation (multisensor data fusion, artificial intelligence, fault diagnosis, safety, and fault tolerance)

Mechatronics technology brings about several improvements, e.g., better system performance, higher product quality, lower manufacturing cost, lower power consumption, higher precision, efficient conversion of electrical power into mechanical power and vice versa, and so on (Hewit and King, 1996).

Mechatronics is a key cornerstone of modern industrial technologies, and it is being developed to be more efficient, convenient, intelligent, integrated, and human friendly.

During the design process of mechatronic systems, the interplay between mechanical system design and electronic system design is crucial and requires the employment of simultaneous engineering technologies to achieve the goal of designing an overall integrated system and create synergistic effects. In future, more sophisticated control algorithms will be realized, e.g., for the estimation of nonmeasurable variables, self-tuning and adaptation of controller parameters, reconfiguration, supervision and fault diagnosis, safety, and fault tolerance. Hence, mechatronic systems with adaptive or even learning behavior can also be genuinely called intelligent mechatronic systems (Isermann, 2008).

34.1.1 Historical Development of Mechatronic Systems

Technological history indicates that much effort has been made for improving mechatronic systems. The application fields of primitive mechatronic systems, e.g., automotive mechatronics, manufacturing equipment, industrial robots, household appliances, and entertainment devices, highlight the marvelous ingenuity of their inventors. In the nineteenth century, for purely mechanical systems, also called premechatronic systems, decision making (computation) was done in the same medium through which the major power flows in the system, and the control mechanism operated without electrical, hydraulic, or pneumatic auxiliary energy. There are several oft-quoted examples, e.g., steam engine, typewriter, mechanical adding machine, pneumatic process controller, carburetor, sewing machine, tracer lathe, and cam grinder (Auslander, 1996).

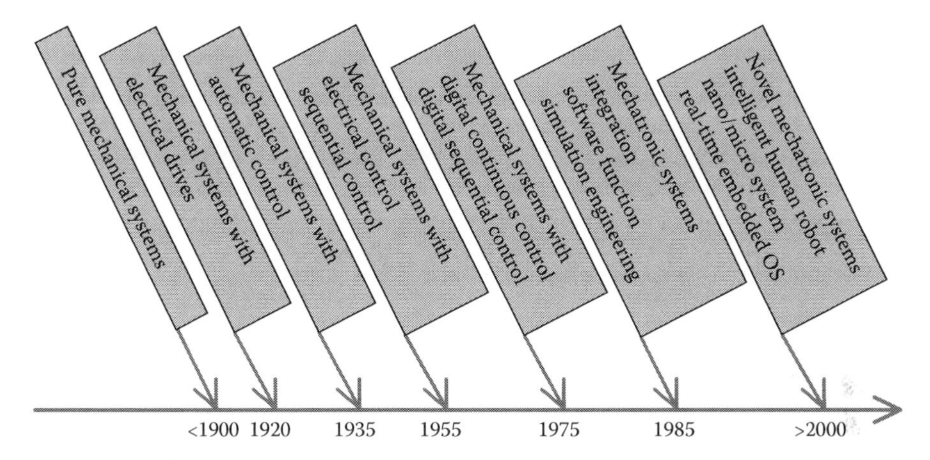

FIGURE 34.1 Historical development of mechanical, electronic, and mechatronic systems. (Adapted from Isermann, R., *Control Eng. Pract.*, 16(1), 14, January 2008.)

Due to the development in electronic techniques year by year, efforts are constantly being made to achieve complexity and significant progress in mechanical systems. The technological progress from purely mechanical systems to mechatronic systems is depicted in Figure 34.1 (Isermann, 2008).

34.1.2 Components and Functions

A mechatronic system organization includes a central processing unit (CPU), an actuator, instrumentation, a controlled target system, an operation interface, and other elements, e.g., a communication unit, a power supply unit, and a tiny operation system. The most common type of CPU in use today is the microcontroller; very small, highly integrated microprocessors are used as decision-making elements in mechatronic systems. The actuator is capable of controlling power delivery to achieve the desired actions of a mechanical component. The instrumentation acquires the variation of the controlled target system, and then sends feedback to the microcontroller. A diagram of a controlled mechatronic system with four central components is shown in Figure 34.2.

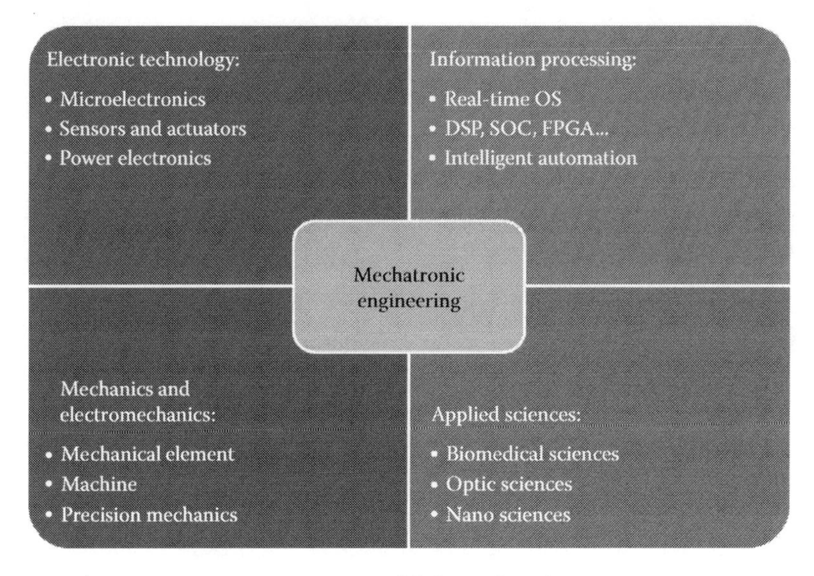

FIGURE 34.2 Mechatronics: synergistic integration of different disciplines.

Basically, the major function of a mechatronic system is to sense changes in the real-world environment, and various sensors are employed to assist mechatronic systems in achieving the desired actions of a mechanical component or to overcome the technological restrictions encountered in a specific manufacturing process. Modern mechatronic systems provide many improved and smart functions, e.g., the estimation of nonmeasurable variables, self-tuning and adaptation of controller parameters, reconfiguration, supervision and fault diagnosis, safety, and fault tolerance. In the future, many autonomic functions, including reconfigurability, self-organization, self-repair, autonomous evolution, self-replication, and so on, will be implemented (Bekey, 2005).

34.2 Interdisciplinary Fields of Mechatronics

34.2.1 Mechanical Engineering

Mechanical engineering is one of the most important fields discovered by human beings. An old adage says "machinery is the mother of industry," and it implies that mechanical engineering is one of the oldest and broadest engineering disciplines. Mechanical engineering involves the analysis, design, manufacturing, and maintenance of various mechanisms and was developed for the application of principles from the core concepts of mechanics, kinematics, thermodynamics, fluid mechanics, materials science, energy, and so on. For instance, the applications of mechanical engineering are in automobiles, refrigerators, air conditioners, and many other appliances, and various types of machines like milling, drilling, and lathe are used. At the core of all the applications of mechanical engineering is mechanical engineering design. During the design process of a mechanism, some fundamental physical subjects of mechanical engineering are considered, and they are described individually in Sections 34.2.1.1 through 34.2.1.3.

34.2.1.1 Primary Mechanism

The design of mechanisms involves many factors. These include their structure, kinematics, dynamics, stress analysis, materials, lubrication, wear, tolerances, production considerations, control and actuation, vibrations, critical speeds, reliability, costs, and environmental considerations. Modern trends in the design of mechanisms emphasize economical design analysis by means of computer-aided design techniques.

A mechanism is a machine or a mechanical appliance that is disposed to perform the desired action or function. Classically, it is a mechanical means for the conversion of motion, the transmission of power, or the control of these parameters. Mechanisms are at the core of the workings of many machines and mechanical devices. In modern usage, mechanisms are not always limited to mechanical means. In addition to mechanical elements, they may include pneumatic, hydraulic, electrical, and electronic elements.

Most mechanisms consist of combinations of a relatively small number of basic components, in which the most important are cams, gears, links, belts, chains, and logical mechanical elements, which include devices such as ratchets, trips, detents, and interlocks. In order to understand how any mechanism works, its degree of freedom, structure, and kinematics must be considered.

The degree of freedom is conveniently illustrated for mechanisms with rigid links as follows:

The general degree-of-freedom equation is

$$F = \lambda(l - j - 1) + \sum f_i$$

where

F is the degree of freedom of the mechanism

l is the number of links of the mechanism

j is the number of joints of the mechanism

f_i is the degree of freedom of relative motion at the ith joint

λ is the mobility number; the most common cases are $\lambda = 3$ for plane mechanisms and $\lambda = 6$ for spatial mechanisms

There are also several basic components and mechanisms, such as belt drive, cam mechanism, chain drive, escapement, gear, linkage (mechanism), and ratchet. For more detailed information about the basic components, it is advised to refer to a textbook on mechanisms.

34.2.1.2 Statics, Dynamics, and Kinematics

Statics is the branch of mechanics that is concerned with load analysis, e.g., force, torque, and moment, on mechanical systems in static equilibrium. Static equilibrium is a state in which the relative positions of any subsystems, components, and structures do not vary over time, or change at a constant velocity. In other words, the system is either still, or its center of mass moves at constant velocity.

According to Newton's first law, the situation of static equilibrium implies that the net force and net torque at every point in a system is zero. From the two constraints of static equilibrium, the distributed quantities of stress and pressure can be derived.

In the field of classical mechanics, there are three branches: "statics" describes the state of equilibrium and its relation to forces, "kinetics" explains the situation of motion and its relation to forces (Wright, 1896), and "kinematics" deals with the implications of observed object motions without considering the causes leading to the motion (Whittaker, 1936). In another approach, statics and kinetics are combined under the name dynamics, which deals with the determination of the motion of bodies resulting from the action of specified forces (MacGregor, 1887). Another approach, which combines kinetics and kinematics under the name rubric dynamics, is explained in most engineering books on mechanics and is still widely used by engineers (Timoshenko and Young, 1956; Lakshmana Rao et al., 2004).

34.2.1.3 Strength of Materials

In material science, strength of materials is one of the branches that studies the relation between applied stress and relative deformation in size and probes into the ability of a material to withstand an applied stress without failure. It provides mechanical engineers with an awareness of various responses exhibited by solid engineering materials when subjected to mechanical and thermal loadings. It is an introduction to the physical mechanisms associated with the design-limiting behavior of engineering materials, especially stiffness, strength, toughness, and durability. It provides an understanding of the basic mechanical properties of engineering materials, testing procedures used to quantify these properties, and ways in which these properties characterize material response. It proposes quantitative skills to deal with materials-limiting problems in engineering design, and suggests a basis for materials selection in mechanical design.

The following are basic definitions and coefficients that are considered during material selection and machinery design:

Stress: One-dimensional stress is defined as the distributed quantities of force per area, which is expressed as

$$\sigma = \frac{F}{A},$$

where F is the force [N] acting on an area A [m^2].

Depending on the direction of the applied force on a particular area, there are three kinds of stresses: compressive stress, tensile stress, and shear stress. *Compressive stress* (or compression) is the stress state caused by an applied load that induces a reduction in the length of a material along the direction of the applied load. A simple case of compressive stress is one-dimensional compression induced by the pushing coaxial forces. The compressive strength of materials is generally higher than that of tensile stress.

Tensile stress is the stress state caused by an applied load that tends to elongate a material in the same direction as of the applied load; in other words, it is the stress caused by pulling the material. *Shear stress* is the stress state caused by a pair of opposing forces acting along parallel lines of action through the material; in other words, it is the stress caused by sliding faces of the material relative to one another. An example is cutting paper with scissors.

Strain: Strain or reduced deformation is a mathematical term to express the trend of the deformation change among the material field. For coaxial loading displacements of a specimen, e.g., a bar element, it is expressed as the quotient of the displacement and the length of the specimen.

Deflection: Deflection is a term to describe the magnitude to which a structural element bends under a load. For example, the deflection of a spring beam depends on its length, its cross-sectional shape, the material to which the deflecting force is applied, and how the beam is supported.

34.2.1.3.1 Coefficients of Materials

Yield strength: The yield strength or yield point of a material is the lowest stress that induces permanent deformation in a material. Prior to the yield point, the material will deform elastically and will return to its original shape when the applied stress is removed. Once the yield point is passed, some fraction of the deformation will be permanent and nonreversible.

Compressive strength: Compressive strength is the capacity of a material to withstand axially directed pushing forces. When the limit of compressive strength is reached, materials are crushed.

Tensile strength: Tensile strength or ultimate tensile strength is indicated by the maxima of a stress–strain curve and, in general, indicates when necking will occur. When the tensile stress is more than the tensile strength, the material will undergo tensile failure in the manner of ductile failure or in the manner of brittle failure. Tensile strength can be given as either true stress or engineering stress.

Fatigue strength: Fatigue strength is a measured strength of a material or a component under cyclic loading and is usually more difficult to assess than the static strength measures. The maximum stress values are less than the ultimate tensile stress limit and may be below the yield stress limit of the material.

Impact strength: Impact strength is the capability of a material to withstand suddenly applied loads in terms of energy. It is often measured with the Izod impact strength test or Charpy impact test, both of which measure the impact energy required to fracture a sample.

34.2.2 Electrical and Electronic Engineering

Electrical and electronic engineering is the field of engineering that deals with the study and application of electricity, electronics, and electromagnetism. The field first became an identifiable occupation in the late nineteenth century after commercialization of the electric telegraph and electrical power supply. It now covers an extensive range of areas including power, electronics, control systems, signal processing, smart sensors, and telecommunications.

The distinction between electrical and electronic engineering is usually confused. Generally, electrical engineering is considered to deal with the equipments associated with large-scale electrical systems, such as for power generation, power transmission, and motor control, whereas electronic engineering deals with the production of small-scale electronic systems including communications, consumer electronics, computers, and integrated circuits (ICs). Alternatively, electrical engineers are usually concerned with using electricity to convert energy, as in heaters, motors, and transformers, while electronic engineers are concerned with using electricity to transmit information, as through microchips and sensors.

During electrical system design, some fundamental subjects in electrical and electronic engineering fields are considered, and they are described individually in Sections 34.2.2.1 through 34.2.2.3.

34.2.2.1 Microelectronic Systems

For over half a century, the technology of microelectronics has been advancing through miniaturization, leading to significant increases in computing power and continuous decreases in manufacturing cost. Microelectronics is a subfield of electronic engineering. Microelectronics is the subject of the study and manufacture of electronic components that are usually small scale and are made from semiconductors.

Twenty years ago, the manufacturing technologies of semiconductors were still in their initial stage, which limited the applications of microelectronic products. In those times, discrete electrical components, e.g., transistors, capacitors, inductors, resistors, and diodes, were typical in electrical systems, which resulted in complicated design processing for a sample function, heavy work, and wastage of time. Digital ICs consist mostly of transistors. Analog circuits commonly contain resistors and capacitors as well. Inductors are used in some high-frequency analog circuits but tend to occupy a large chip area if used at low frequencies; gyrators can replace them in many applications.

In recent years, the dynamic development in the manufacturing technologies of semiconductors has resulted in a continuous scale down of microelectronic components. The new-generation ICs have a higher integration capability, a higher computing efficiency, show higher functional performance, and are more intelligent. As techniques have improved, it has been possible to integrate analog and digital circuits on a chip to form a mixed-signal embedded unit, and this significantly reduces the difficulty during design processing and largely decreases the time wasted, especially in wireless communication systems.

Microelectronic systems always behave as electrical cybernetics systems, which deal with the control engineering of mechatronic systems. Through collaboration, the mechatronic modules perform the production goals and inherit flexible and agile manufacturing properties in the production scheme. Modern production equipment consists of mechatronic modules that are integrated according to a control architecture. The most popular architectures involve hierarchy, multiplicity, diversification, and hybrid structures. The methods for achieving a technical effect are described by control algorithms, which may or may not utilize formal methods in their design.

34.2.2.1.1 Electronic Components

The following text briefly describes the key electronic components that are fundamental building blocks of modern electronic devices and are ubiquitous in modern electronic systems.

Resistors: Resistors are components used in electrical networks and electronic circuits and are ubiquitous in most electronic equipments. A resistor is a two-terminal electronic component that produces a voltage across its terminals that is proportional to the electric current through it in accordance with Ohm's law.

Capacitors: A capacitor is a passive electronic component consisting of a pair of conductors separated by a dielectric. When a voltage potential difference exists between the conductors, an electric field is present in the dielectric. This field stores energy and produces a mechanical force between the plates. The effect is greatest between wide, flat, parallel, narrowly separated conductors.

Inductors: An inductor is also a passive electrical component that can store energy in a magnetic field created by the electric current passing through it. Typically, an inductor is a conducting wire shaped as a coil, the loops helping to create a strong magnetic field inside the coil due to Faraday's law of induction. Inductors are one of the basic electronic components used in electronics where current and voltage change with time due to the ability of inductors to delay and reshape alternating currents.

Diodes: In electronics, a diode is a two-terminal electronic component that is provided with a two-terminal semiconductor P–N junction and conducts electric current asymmetrically, that is, it conducts current more easily in one direction than in the opposite direction. The most common function of a diode is to allow an electric current in one direction (called the forward direction) while blocking current in the opposite direction (the reverse direction). Thus, the diode can be thought of as an electronic version of a check valve. This unidirectional behavior is called rectification, and is used to convert alternating current to direct current, and remove modulation from radio signals in radio receivers.

Transistors: A transistor is a semiconductor device commonly used to amplify or switch electronic signals. A transistor is made up of a solid piece of a semiconductor material, with at least three terminals for connecting to an external circuit. A voltage or current applied to one pair of the transistor's terminals changes the current flowing through another pair of terminals. Because the controlled (output) power can be much more than the controlling (input) power, the transistor provides amplification of a signal. Some transistors are packaged individually, but most are found in ICs.

34.2.2.1.2 Integrated Circuits

34.2.2.1.2.1 Operational Amplifier An operational amplifier (op-amp) is a two-port electronic component, a DC-coupled high-gain electronic voltage amplifier with differential inputs and, usually, a single output. Typically, the output of the op-amp is controlled either by negative feedback, which largely determines the magnitude of its output voltage gain, or by positive feedback, which facilitates regenerative gain and oscillation. High-input impedance at the input terminals (ideally infinite) and low-output impedance (ideally zero) are important typical characteristics.

The operational amplifier, as the name suggests, is the primary device of operational circuits for analog signals, e.g., adder, multiplier, integrator, and differentiator. An implemental circuit of a control system is composed of these operational circuits.

The op-amp is one type of differential amplifier. Other types of differential amplifiers include the fully differential amplifier (similar to the op-amp, but with two outputs), the instrumentation amplifier (usually built from three op-amps), the isolation amplifier (similar to the instrumentation amplifier, but which works fine with common-mode voltages that would destroy an ordinary op-amp), and the negative feedback amplifier (usually built from one or more op-amps and a resistive feedback network). Op-amps are among the most widely used electronic devices today, being used in a vast array of consumer, industrial, and scientific devices.

34.2.2.1.2.2 Signal Converter Typically, an analog-to-digital (A/D) converter is an electronic device that converts an input analog continuous voltage (or current) into a digital number proportional to the magnitude of the voltage or current. The digital output may use different coding schemes, such as binary code, gray code, or two's complement binary code. The A/D converter is a key component to bridge the gap between levels of continuous physical signals acquired from the sensor and discrete digital signals, which is suitable for electrical computing and processing in a CPU.

The reverse operation is performed by a digital-to-analog converter, which is a device for converting a digital (usually binary) code to an analog signal (current, voltage, or electric charge).

34.2.2.1.2.3 Multiplexer and Demultiplexer An electronic multiplexer and demultiplexer makes it possible for several signals to share one device or resource, e.g., one A/D converter or one communication line, instead of having one device per input signal. A multiplexer is a device that selects one of many analog or digital input signals and forwards the selected input into a single line. A multiplexer with the nth power of two inputs has n select lines, which are used to select the input line to be sent to the output. A demultiplexer is a device taking a single input signal and selecting one of many data-output lines, which is connected to the single input. A multiplexer is often used with a complementary demultiplexer on the receiving end.

34.2.2.1.2.4 Logic Gate and Arithmetic Logic Integrated Circuit In electronics, an arithmetic logic circuit, such as adder, is a digital circuit that performs logical operation of numbers in Boolean logic. In modern computers, arithmetic operators reside in the arithmetic logic unit (ALU), where all operations, such as addition, subtraction, multiplication, and division, are performed. Although an arithmetic logic circuit can be constructed for many numerical representations, such as a binary-coded decimal like excess-3, the most common arithmetic logic circuits operate on binary numbers. In the arithmetic logic circuit, in which two's complement or one's complement is used to represent negative numbers, it is trivial to modify an adder into an adder–subtractor.

The arithmetic logic circuit is composed of a group logic gate that performs a Boolean logical operation on a set of meaningful logic inputs and produces a set of corresponding logic outputs. Logic gates are primarily implemented electronically using diodes or transistors, and a logic level is represented by a voltage or current, which depends on the type of electronic logic in use. A truth table is a table that describes the operational rules of a logic gate and lists the value of the output for every possible combination of the inputs. In order to obtain an efficient implementation, a minimization procedure, some optimizing approach, e.g., Karnaugh maps, the Quine–McCluskey algorithm, or a heuristic algorithm, is used to reduce the circuit complexity.

34.2.2.1.3 Programmable Chip

34.2.2.1.3.1 Central Processing Unit The CPU or processor is the main component of a computing or controlling system that carries out the functional program, which is composed of a series of instructions. The CPU consists of five specific units including instruction control unit, ALU, register and cache memory, memory management unit, and data bus I/O (input/output).

34.2.2.1.3.2 Microcontroller A microcontroller is a miniature computer on a single chip consisting of a relatively small CPU combined with support functions such as a crystal oscillator, timers, a watchdog timer, serial and analog I/O, a programmable flash ROM, and a small amount of RAM. Some microcontrollers, which are designed for sample or dedicated applications, may operate at lower clock-rate frequencies, enabling power consumption as low as milliwatts or microwatts. The capability of retaining functionality while waiting for a triggered event such as a button press is important, e.g., the power consumption during the sleeping mode, when the CPU clock and most peripherals turn off, may be just nanowatts, making many of the microcontrollers well suited for long-lasting battery applications. Microcontrollers are applied in automatically controlled products and devices, e.g., automobile engine control systems, office machines, information appliances (IA), and smart toys. In comparison to a design that reduces cost and size and uses separate components (including a microprocessor, external memory, and I/O devices), microcontrollers make it economical to digitally control even more devices and processes. Hybrid signal microcontrollers are popularizing in recent years; these chip-integrated analog and digital components are needed to control non-digital electronic systems.

34.2.2.1.3.3 Embedded System An embedded system is a microprocessor-based system that is built to deal with some desired functions and is not designed to be programmed by the end user. In other words, a user can make choices concerning functionality but cannot change the functionality of the system in any way. An embedded system is designed to perform a particular and dedicated task attached a few of choices or options.

The majority of embedded systems in use today are embedded in machinery such as automobiles, telephones, home appliances, and peripherals for desktop computers. While some embedded systems are very sophisticated, many have minimal requirements for memory and program length, with no operating system, and low software complexity. Typical input and output devices include switches, relays, solenoids, LEDs, small or custom LCD displays, radio frequency devices, and sensors for acquiring data such as temperature, humidity, and light level. Embedded systems usually have no keyboard, screen, disks, printers, or other recognizable I/O devices of a personal computer and may lack human interaction devices of any kind.

A modern car may have over 50 specific functions constructed and controlled by a microcontroller, such as an engine management system, electronic antilock brakes, an electronically controlled gearbox, a safety system with an air bag, motor-driven windows, air conditioning, and so on (Heath, 2003). Another type of programmable microchip may play performance-critical roles, where it may need to act more like a digital signal processor (DSP), with higher clock speeds and power consumption.

34.2.2.2 Power Supply Unit

A power supply unit is a device or system that supplies electrical or any other type of energy to loads, such as lamp, controller, motor, and so on. The term is most commonly applied to electrical energy supplies, less often to mechanical ones, and rarely to others. Power supply is capable of converting one form of electrical power to another desired form and voltage, typically involving the conversion of AC line voltage into a well-regulated lower-voltage DC for electronic devices. Low-voltage, low-power DC supply units are commonly integrated with the devices to which they supply power.

Considering the issues of commonly affected power supplies, how much power they can supply, how long they can supply without recharging, how stable their output voltage or current under varying load conditions, whether they provide continuous power or pulses, a regulated power supply can tightly control the output voltage and/or current to a specific value. A power management module is capable of power saving function during system idle to improve the power efficieny.

Power supply can be broadly divided into linear and switching power supply. The linear power supply is simple to make but is extremely bulky and heavy for high-current devices; voltage regulation in a linear power supply can result in low efficiency. A switched-mode power supply is smaller than a linear power supply and is usually more efficient, but more complex.

An electrical battery is a set of many electrochemical cells, used to convert stored chemical energy into electrical energy. In recent years, due to the popularization of mobile products, such as digital camera, cell phone, and personal media player, the acid- or alkali-based battery has become a common power source for many household and industrial applications. Batteries may be used once and discarded, or recharged for years, as in standby power applications. Primary batteries can produce current immediately on assembly. Disposable batteries are intended to be used once and discarded. Secondary batteries must be charged before use; they are usually assembled with active materials in the discharged state. Rechargeable batteries or secondary cells can be recharged by applying electrical current, which reverses the chemical reactions that occur during their use. A battery's characteristics may vary over load cycle, over charge cycle, and over lifetime due to many factors, including internal chemistry, current drain, and temperature.

34.2.2.3 Sensor and Actuator Technology

The five senses—vision, touch, hearing, smell, and taste—provide real-time information for perceiving the external environment to humans. Through the vision organs, we can quickly recognize how an object looks and what it is. Touch sensing includes the capability to sense the degree of slip, surface compliance, and temperature; supplies important sensory information; and helps us manipulate and recognize objects. The sense of hearing is also indispensable for maintaining our daily lives. In addition to these external five sensors, the muscle spindle is well known for detecting the length of the muscle, and otolith organs perceive the direction in which acceleration is imparted.

In the past 20 years, sensor devices have become highly advanced, especially in MEMS-related fields, and include optical sensors, piezo-based force or acceleration sensors, chemical sensors, and so on. An intelligent mechatronic system is supported by various sensing components. A vision sensor is the most important source of information about surroundings for a mechatronic system. However, the state-of-the-art artificial vision is still far behind human recognition capability. In recent advancements, the artificial tactile sensors cannot detect the tangential force component, which results in unsatisfactory performance. Various internal sensors, e.g., potentiometers, encoders, tachometers, acceleration sensors, and gyrosensors, have been utilized in mechatronic systems such as robots, manufacturing systems, automotive vehicles, aircraft vehicles, and so on.

Yan et al. (2006) presented new mechatronic devices, e.g., a multi-degree-of-freedom electrical actuator, for advanced control applications. Odhner and Asada (2006) presented a method for using thermoelectric devices (TEDs) together with a low-order estimator to achieve feedback control of a shape memory alloy (SMA) actuator array. In the paper by Paynter and Juarez (2000), a novel type of pneumatic actuator has been designed to be used as a spherical drive in various jointed-arm or jointed-leg robots.

In the electrohydraulic actuator proposed in the paper by Habibi and Goldenberg (2000), high torque/mass ratio and modularity are desirable features of the new design. The paper by Maas et al. (2000) presents a model-based control for ultrasonic motors. The paper contributed by Hu and Su (2007) proposes a novel procedure for detecting environmental changes by using a pan-tilt-zoom (PTZ) camera.

Formerly, the unobtainable force output of the ionic polymer metal composite (IPMC) actuator resulted in limited practical applications. Richardson et al. (2003) investigated the control of IPMCs and demonstrated that force and position controllers can be effectively implemented on the polymer actuator. Goldfarb et al. (2003) proposed a power supply and actuation system appropriate for a position- or force-controlled human-scale robot and described the design and control of a liquid monopropellant actuation system. In their paper, Stein et al. (2003) present mathematical models of spherical encoders, which encode spherical motion in the feedback control of spherical motor systems, based on a finite number of binary sensors and a two-color painting of a ball rotating within a housing. In the paper by Aghili et al. (2007), the design and control of a high-performance direct-drive system with an integrated motor was presented.

34.2.3 Information Technology

Information technology is an essential recipe in the design of mechatronic systems. The directly measurable output quantities, such as positions, speeds, accelerations, forces, and currents, are a sequence of analog or digital signals. Sections 34.2.3.1 and 34.3.2.2 present information theory and analysis in terms of their classification, representation, and processing methods.

34.2.3.1 Signal Analysis and Processing

Signal theory, which studies signal analysis and processing, is an essential recipe in designing mechatronic systems. Signals can be classified in groups based on whether they are random or nonrandom, stationary or nonstationary, periodic or nonperiodic, continuous or discrete, and analog or digital.

An appropriate description of a signal is that "a signal is analogous and intrinsically it may have many signals (Mahalik, 2003)." This is a funny definition. The meaning of the first clause of the above definition describes that the strength or intensity of these process variables acquired from sensors is converted into the same equivalent magnitude, called equivalence. The second clause means that the signal actually consists of many signal components, which we are unable to visualize in the time-amplitude space. A signal is thus simply a summation of a set of components, and this set is unique. Each component within the signal is characterized by three parameters, such as amplitude, frequency, and phase. The variation of each parameter influences the shape of the signal.

The classification of signals is defined via various ways, the following four groups are considered imperative with regard to engineering study:

1. **Random and nonrandom**
   One of the classifications of signals is made based on whether they are random or nonrandom. The randomness of a signal, which embraces some degree of uncertainty at the time of its appearance, is determined from the statistics, and the future values of the random signal are predicted based on the observation of past values.
   A nonrandom signal, which does not contain any uncertainty before it occurs, can be written by an explicit mathematical expression. In fact, a nonrandom signal does not exist in the real world, but it is used extensively for assumptions, simplicity, and understanding.
2. **Stationary and nonstationary**
   Signals whose stationary parameters, such as mean and standard deviation, do not change with time are called stationary signals, whereas the others are called nonstationary signals. Stationary signals exist in the real world due to random processes, but a signal that is close to a stationary signal is called a weak stationary signal, and it can be observed in the real world and be assumed to be a stationary signal.

3. **Periodic and nonperiodic**
   A signal whose amplitude is reiterated over a fixed interval is said to be periodic, whereas the others are classified as nonperiodic. Periodic and nonperiodic signals are also called power signals and energy signals, respectively.
4. **Analog, sampled, discrete, and digital**
   The signals that exist in the real world are analog signals, and are illustrated as a continuous function of time. In order to take advantage of the digital data processing technology, we need to process the analog signals in terms of converting them into a discrete form. The process of transferring continuous models and equations into discrete counterparts is called *discretization*. In the discrete domain, the signal is certainly defined only by discrete instants of time. The value of the analog signal is sampled and measured at certain intervals of time, and each result is referred to as a *sample*. The signals that are generated via the above processes are called discrete time signals, which are represented as a sequence of sampled values.

The signals are sampled from the analog domain via the processes of quantization to form discrete signals, and then the coding process encodes the discrete signals to generate digital signals that can be used in digital computers.

What analysis do we need to do? A signal is defined with three parameters: amplitude, frequency, and phase. The answer to this question is that the analysis to be performed is through studying the amplitude, frequency, and phase of the components either at a particular point of time or in an interval. When we understand the signals via analysis, we can control and manipulate them as much as possible. Unless we understand the characteristics, properties, and contents, it is way ahead to manipulate the signal appropriately.

A periodic analog signal is a continuous signal with period T. In other words, the amplitudes of the signal remain same at an interval of T. In the case of a periodic analog signal, a Fourier series analysis can induce a great deal of mathematical treatment. The Fourier series analysis states that "a periodic signal of period T can be expressed as a summation of infinite number of trigonometric functions such as sine and cosine functions as long as the signal has finite number of maxima, minima and discontinuity within T (Mahalia, 2003)." This statement also implies that the periodic signal is absolutely integrable and has finite energy.

Nevertheless, mostly the output signals acquired from many sensors or transducers are nonperiodic, such as temperature sensors and pressure sensors. The frequency components contained in the nonperiodic signal are not visible from the time-domain plot. For this reason, a nonperiodic signal must be transformed from the time domain to the frequency domain via the mathematical tool called *Fourier transform* (FT). The FT of a signal is the summation of products of two terms, the time-domain signal and the exponential $\boldsymbol{e}$. When the exponential term, which consists of a frequency parameter, is multiplied with the signal, it gets amplified only if the signal has the frequency component included in the exponential term, or else the product becomes zero. In other words, the process of multiplication in FT extracts the frequency components present in the signal. The integration does not mean the sum of all products. It actually means the summation of products over all times. The FT of a signal is a complex function, and the integration takes the limit from $-\infty$ to $+\infty$.

While continuous signals are useful, they are not suitable to work in digital processing equipments. Instead, we usually represent continuous signals by a sequence of values and recommend a mathematical tool, called *discrete Fourier transform* (DFT), to engross discrete signals. DFT is a mathematical operation specially used to transform an ordered sequence of data samples from the time domain to the frequency domain, and the spectral information within the discrete signal can be obtained. In fact, the DFT is a method of digital signal processing (DSP) that makes it possible to allow computer implementation for the extraction of amplitude, frequency, and phase of the discrete signal.

As far as the calculation of DFT is concerned, the number of multiplications involved in the calculation of DFT is equal to the square of the number of the data points. It means that the calculation of DFT takes a large amount of computer time. The DFT algorithm has been superseded by the *fast Fourier transform* (FFT) algorithm, which reduces calculation redundancies in terms of taking much less computer time. FFT has the advantage of reducing differential equations to essentially algebraic problems, and is considered as one of the most significant numerical techniques.

There are some signals, such as nonstationary signals, that even if they differ from each other, cannot be distinguished from their frequency-domain plot. The appropriate mathematical tools to be used to solve a problem are either *wavelet transform* (WT) or *short-time Fourier transform* (STFT). WT and STFT, which transform the signal to a "time–frequency" domain, reveal the frequency information as well as the information of the time of occurrence of the frequency components.

Real-world dynamic systems are continuous in nature, and most of the continuous systems are described by differential equations. The characteristics and behavior of the system are understood by solving such differential equations. The mechatronic system design often places great emphasis on studying the dynamic behavior. Irrespective of the involvement between dissimilar systems, many mechatronic systems are described by linear differential equations. Usually, it is required to solve these differential equations in order to

1. Study the system
2. Predict the behavior of the system
3. Improve the performance of the existing system
4. Optimize and develop a better system

The *Laplace transformation* is a technique mostly used for solving linear differential equations with initial conditions applied. Engineering systems are described or modeled by the linear differential equations, also called model equations. The behavior of the system is well understood by analyzing the model equations, and the analysis is carried out in terms of solving these model equations. Laplace transform methods make it easier to solve differential equations.

34.2.3.1.1 Difference Equation and Z-Transformation

In order to take advantage of digital technology, continuous time signals are converted into discrete time signals. In the context of the differential equation in relation to a continuous signal, there exists the *difference equation* in relation to a discrete signal. The *Z-transform* is used to solve difference equations, and it is applied to take discrete time-domain signals into a complex-variable frequency domain. The Z-transform discloses new ways of solving problems and analyzing systems in the discrete domain. In general, the Z-transform can be used to

1. Represent discrete time signals and systems
2. Analyze the stability of discrete time systems
3. Observe the frequency response of discrete time systems
4. Design discrete time applications
5. Solve difference equations

34.2.3.2 Operation System

An operating system is a basic system program that controls overall the computer, including hardware and other software, and when the computer system is turned on, the operating system builds an environment where it is easier to run other applications. An operating system plays the role of an abstraction layer between the application software and the low-level hardware. Windows, Linux, Unix, etc., are all names of the most commonly used operating systems. According to the application domain, operating systems can be classified into three types: general-purpose operating systems, distributed operating systems, and real-time operating systems.

As numerical control is present in most modern mechatronic applications, real-time aspects are vitally important in terms of performance, flexibility, safety, cost, etc. Real-time requirements originate mainly from the dynamic characteristics of the mechanical device or machine that is to be controlled. Obviously, this information is somewhat superfluous in most cases since any feedback control must perform in real time in order to be useful. In recent years, the emphasis on real time in the mechatronic engineering context is important since it has become difficult to guarantee real-time performance considering the ever-increasing complexity of the control applications and the relatively high-bandwidth requirements which are present in mechatronic systems. Also, there is a trend in the development of mechatronic systems toward the decentralization of control functions and the spatial distribution of processing capacity throughout the machine. This distribution requires, apart from robust embedding and integration technologies, specific real-time operating features.

The most important factor to determine how a real-time system works is that an external event must be responded with a predictable and controllable method. For this reason, a real-time operating system must be provided with features such as timeliness, predictability, synchronicity, and reliability.

34.2.4 Intelligent Control and Machine Automation

Intelligent control and machine automation are the most important developments in the mechatronics field. Intelligent control attempts to build up and enhance conventional control methodologies and design to solve new, challenging control problems.

34.2.4.1 Control System Design

A control system is a device that behaves as an electrical cybernetics system, to manage, command, direct, and regulate the behavior of other devices or systems. Typical control systems can be classified as open loop and closed loop, linear and nonlinear, feedforward and feedback, time variant and time invariant, etc.

An open-loop controller is the simplest controller that computes its input into a system using only the current state. Open-loop control is useful for well-defined systems where the relationship between the input and the resultant state can be modeled by a mathematical formula. However, the varied dynamic characteristics of a real controlled system resulting in the feedback of error based on direct measurable input and output signals is essential to avoid the problems of the open-loop controller. Closed-loop controllers have several advantages over open-loop controllers, such as disturbance rejection, guaranteed performance even with model uncertainties, higher stability, reduced sensitivity to parameter variations, and improved reference-tracking performance. A common closed-loop controller architecture is the PID controller, which is appropriate when the process or system is linear and time invariant. It looks at the current value of error, the integral of the error over a recent time interval, and the current derivative of the error to determine not only how much of correction to apply but for how long. In summary, the proportional controller is capable of reducing the rise time, but never of eliminating the steady-state error. The integral action contributes to the effect of eliminating the steady-state error, but it may make the transient response worse. The derivative function induces the effects of increasing the stability of the system, reducing the overshoot, and improving the transient response.

34.2.4.2 Advanced Intelligent Controller

Usually in motion control, the velocity is estimated from the encoder position, which is the most typical sensory method based on the difference of successive encoder counts. The velocity may also be estimated based on the model-based state estimation theory (Kim and Sul, 1996; Song and Sul, 1998; Kwon et al., 2003). Recently, the kinematic Kalman filter (KKF) (Jeon and Tomizuka, 2007), an accelerometer-assisted velocity estimation method, was investigated and confirmed to be superior in reliability and cost-effectiveness. Lau et al. (2007) proposed a new approach to vibration damping and control. Liu and Yao (2006) presented an automated on-board model of cartridge valve flow mapping based on a localized estimation method. Fite and Goldfarb (2006) presented a detailed actuator design with a pair of proportional injector valves and a force controller. Munir and Book (2002) proposed a stable bilateral teleoperation, which employs a wave-based predictor using the Kalman filter and an energy regulator to cope with unexpected time delays. Panzieri et al. (2002) presented an outdoor navigation system using the Kalman filter as the localization algorithm. A sensor fusion problem is discussed by Baeten and De Schutter (2002), who demonstrated how a vision system can effectively complement a force control to detect corners along the contour. The modeling of chemical kinetics is crucial for the modeling of the three-way catalytic converters (TWC), but it is very difficult. Glielmo et al. (2000) discussed the derivation of a dynamic model for a TWC, which is an interesting yet wide area of automotive mechatronics. They proposed a machine learning algorithm for which the reaction kinetics of the system is modeled by a neural network, the parameters of which are tuned via a genetic algorithm (GA).

34.2.4.3 Multisensor Data Fusion

In recent years, multisensor fusion methods play gradually an important role in intelligent applications, such as mobile robot navigation, multiple target tracking, and aircraft navigation. In order to use multiple sensors effectively and intelligently, specially designed methods are required for integrating and fusing the data acquired by these sensors of the systems. Some general multisensor fusion methods include weighted average, Kalman filter, Bayesian estimate, and fuzzy logic (Luo and Kay, 1989). The diagram shown in Figure 34.3 represents an integrated framework as being a composite of these basic functions for multisensor data fusion. A group sensor provides the input information, which must first

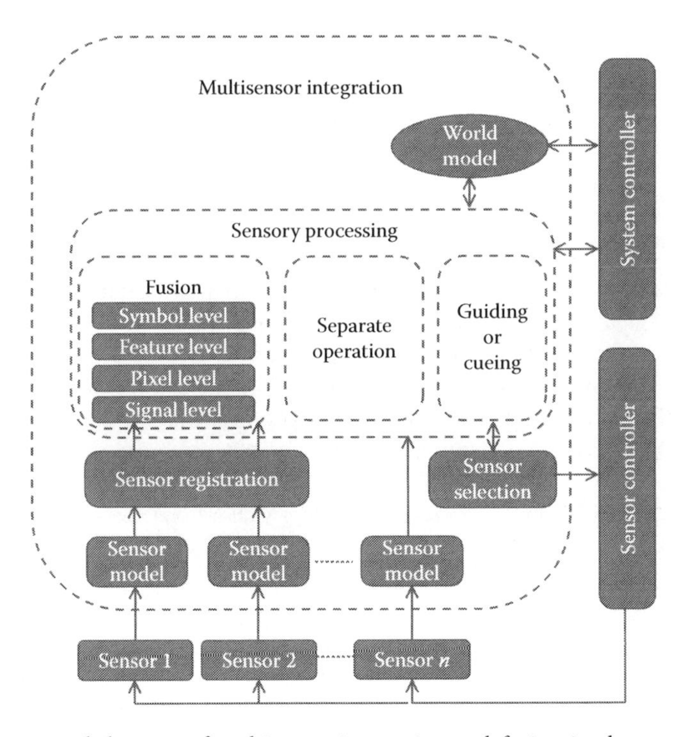

FIGURE 34.3 Functional diagram of multisensor integration and fusion in the operation of a system. (Adapted from Luo, R.C. and M.G. Ray, *IEEE Trans. Sys. Man Cyber.*, 19, 901, September/October 1989).

be effectively modeled to be used for integration. A common assumption is that the uncertainty in the sensor data can be adequately modeled as a Gaussian distribution. The modeled data from each sensor can be integrated into the operation of the system in accord with three different types of sensory processing: fusion, separate operation, and guiding or cueing. The results of the sensory processing functions serve as inputs to the world model. A world model is used to store information concerning any possible state of the environment the system is expected to be operating in.

34.3 Design Procedures

At the core of all the various applications of mechatronics is design. To understand its importance, let us for example consider a digital camera. If a new digital camera is to be manufactured, the manufacturing engineer just cannot manufacture the digital camera haphazardly on his or her own. The most important part before manufacturing the digital camera is designing it. The digital camera has to be designed to give it a particular shape considering the aesthetics, ergonomics, costs, safety, impact-withstanding capacity, etc. It cannot be designed all at once; various iterations are carried out, which means if one option is not suitable, the other is considered. The iterative design and calculations are continued unless an attractive digital camera with all the modern facilities that can be made available at affordable costs to the potential customers has been designed. The manufacture of the digital camera is the last step.

Thus, designing is the most crucial part of any applications of mechatronics. Design procedures can be defined as an iterative decision-making process that formulates the procedure to convert resources optimally into systems, processes, or devices, so as to perform the desired functions and meet the human needs. In brief, in design we first consider the various resources available and then formulate procedures to convert these resources into suitable systems, processes, or devices to meet certain functional and human needs, by making the best use of the resources at the lowest possible cost.

The main aim of Sections 34.3.1 through 34.3.6 is, rather than describing design procedures, to help us apply knowledge that we have gained from our studies in science, engineering, social science, and humanities, to solve practical design problems. The object of these sections is not to present new designing methods; rather, it is to help us utilize our knowledge purposefully and effectively when we design a new application of mechatronics (Khemani, 2008).

34.3.1 Bio-Inspired Concepts for Mechatronic System Design

In the previous decade, the move toward emulating nature and mimicking the wonders uncovered in biology resulted in biologically inspired systems. By observation and inspiration from the biological phenomena, the correlative principles can be derived or understood and used to solve a nonbiological problem. There are some observed aspects of the ways in which animals regulate their internal environments and control their movements in the world. These control and regulation methods were compared and contrasted with those used in engineering control systems and inspired researchers to devise a framework for the design of intelligent controllers. Researchers have proposed excellent results in a wide range of scientific domains, inspired by nature, enabling exploration, communication, and advances that were never dreamed possible just a few years ago.

Biological inspiration can play many different roles in mechatronics, and confusion about these multiple meanings accounts for a wide spectrum of belief about the value of biology for developing better mechatronic systems and achieving improved performance. One of the most important roles is that it can serve as an existence proof of performance, implying that some desirable behavior is possible. On the other hand, it must be understood that a biological metaphor is applicable and relevant to a mechatronic problem, but this does not mean that the corresponding biological phenomena can necessarily be understood in mechatronic terms.

For innovation or creation, a deeper understanding of biological phenomena is required, and there are at least three additional potential roles that have been proposed (Wooley and Lin, 2001):

1. Biology as a source of principles, e.g., several well-known cases are artificial neural network, fuzzy logic, GA, expert system, and so on.
2. Biology as an implementer of mechanisms, e.g., a variety of mobile robots including wheeled robots, legged robots (with two, four, six, and eight legs), flying robots, underwater robots, snake-like robots, climbing robots, jumping robots, and other kinds of robots.
3. Biology as a physical substrate for mechatronics, e.g., some biomechatronic systems, medical mechatronics, auxiliary mechanisms for disabled humans, and so on.

34.3.2 Machinery Design Progress

When we design the elements of a machine or the complete machine, we have to consider several important factors, such as (Khemani, 2008)

1. Cost
2. High output and efficiency
3. Strength
4. Stiffness and rigidity
5. Wear resistance
6. Lubrication
7. Operational safety
8. Ease of assembly
9. Ease and simplicity of disassembly
10. Ease and simplicity of servicing and control
11. Light weight and minimum dimensions
12. Reliability
13. Durability
14. Economy of performance
15. Accessibility
16. Processability
17. Compliance with state standards
18. Economy of repairs and maintenance
19. Use of standard parts
20. Use of easily available materials
21. Appearance of the machine
22. Number of machines to be built

When designing a machine, we cannot apply rigid rules or obey a fixed machine design procedure to get the best design for the machine at the lowest possible cost. When a new product is to be developed, the problems keep on arising at the design stage, and these can be solved only by having a flexible approach and considering various ways; even so, there are some common steps to be followed. Here are some guidelines as to how the machine design engineer can proceed with the design:

1. Prepare a written statement
2. Consider the possible mechanisms
3. Analyze transmitted forces
4. Select the material
5. Find allowable stress
6. Finalize dimensions of the machine elements

7. Consider past experience
8. Prepare drawings

A good machine design engineer possesses some skills that help him or her design the elements of a machine such that they meet all desired requirements and reach high quality at lowest possible costs. These skills help the engineer consider all the relevant parameters in a broad sense, understand their effects on the machine, and find out the best solution to the problem. Some important skills that a good machine design engineer should possess (Khemani, 2008) are

1. Inventiveness
2. Engineering analysis
3. Engineering science
4. Interdisciplinary ability
5. Mathematical skills
6. Decision making
7. Manufacturing processes
8. Communication skills

34.3.3 From Design to Realization

A systemic progress and several modern software design tools are required during the design of mechatronic systems. Mechatronic system design is an iterative and integrated procedure, which involves the knowledge of several traditional domain-specific engineering fields, e.g., mechanical, electrical/electronics, software, information, and automation, as also user interface, and integrates multidisciplinary software and hardware environments for successful design, testing, prototyping, implementation, and validation. Design aspects such as requirements, environment, system objectives, application scenarios, functions, active structure, shape, and behavior are considered. The important design steps include the distribution of interdisciplinary tasks, the use of sensors and actuators, electronic and software architectures, the controller design, and the creation of synergies, resulting in desired functions. The development scheme is represented in the form of a "V"-model, which distinguishes especially between the mechatronic system design and system integration, as shown in Figure 34.4 (Beuth-Verlag GmbH, 2004; Isermann, 2008).

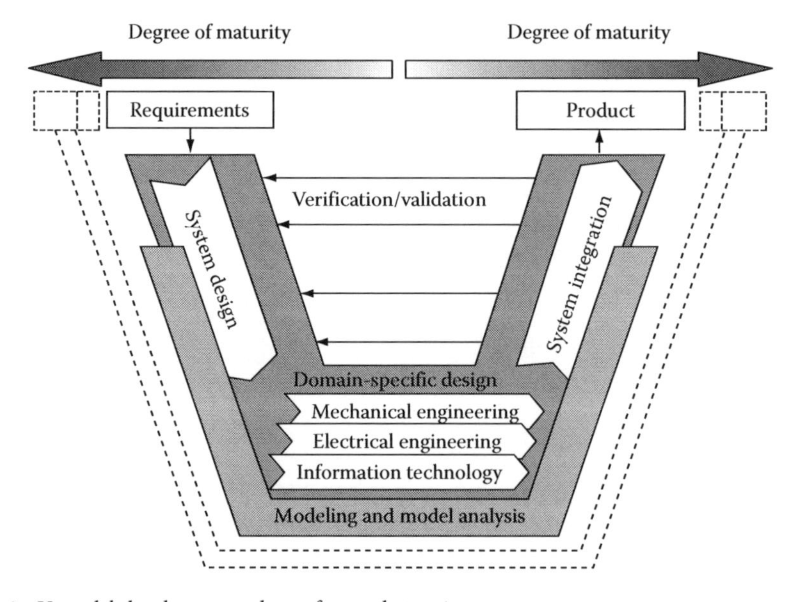

FIGURE 34.4 V-model-development scheme for mechatronic systems.

34.3.4 System Integration Aspects

A mechatronic system refers to a synergistic combination of interdisciplinary tasks, including mechanical, electrical/electronics, control, and precision engineering, as well as information technology and systems thinking, in the design of products and manufacturing processes. The key term of this definition is "synergistic combination," which refers to the system integration approach. A mechatronics product should include features such as functional interaction among the contributing components, spatial integration, intelligence, reconfigurability, and flexibility. Some typical mechatronic products include the 3C products, automobiles, household appliances, railways and vehicles, the antilock braking system (ABS), the autonomous mobile robot, etc. The relative literature on integration aspects are as follows. Burmester and Giese (2005) presented the visual integration of object-oriented modeling techniques. Pil and Asada (1996) described a recursive experimental design method for simultaneously optimizing both the mechanical structure and control. Youcef-Toumi (1996) proposed the necessary steps in mechatronics. Lin and Huang (1996) presented the integration of a compact mechanism design for an NTU five-fingered dexterous robot hand. The paper by Sancho-Pradel and Coodall (2007) described the architecture of a multiprocessor system on a chip that is being developed for adaptive, high-performance, embedded real-time control applications.

34.3.5 CAD Tools for Design

Computer-aided software packages for mechatronic system design include

(a) Mechanism design and analysis in the engineering development stage using CAD tools
(b) Mechanism design and analysis in the engineering development stage using CAE tools
(c) Mathematical model building for obtaining static and dynamic models
(d) Data transformation for system simulation and tooling
(e) Firmware and software programming, and implementation for the final mechatronic application

A broad range of CAD/CAE/CAM tools is available for two- and three-dimensional mechanical designs, such as Pro/E with a total solution for mechanism design, and Protel or PADS for a multilayer printed circuit board (PCB) layout. Object-oriented languages, such as DYMOLA and MOBILE, based on specified ordinary differential equations, algebraic equations, and discontinuities, are employed to model a large combined system. For computer-aided control system design, packages such as ACSL, SIMPACK, MATLAB®/Simulink®, and MATRIX-X have been used broadly. These simulation techniques are valuable tools for design and for variations of design parameters before manufacturing.

34.3.6 Mathematical Models and Systematical Simultaneous

Computer technologies have become an essential tool as far as the design and development of mechatronic systems are concerned. Mathematical models for the static and dynamic behavior of mechanisms are required at various steps of mechatronic system designing, e.g., simulation, control design, and reconstruction of variables. The first way to obtain this model is through theoretical modeling based on physical principles. Because the components are derived from different domains, the issue of theoretical modeling of mechatronic systems shall be interesting. The modeling of electrical circuits, multi-body mechanical systems, or hydraulic systems employs a well-developed domain-specific knowledge and also corresponding software packages. However, it is still not satisfactory to model and simulate the components from different domains via a computer-assisted general methodology.

A theoretical modeling method for a mechatronic system design is lumped-parameter processes described by Isermann (1999). Lumped parameters are a simplification of a mathematical model of a physical system, where variables of spatially distributed fields are represented as single scalars. A good example of a lumped-parameter model is the graphical representation of an electrical network as a circuit diagram in which voltages are assigned to the vertices and currents to the edges of the diagram. The mathematical analysis of such a circuit model is much simpler than solving the Maxwell equations for the actual physical system. It can be a reasonable approach if the system is homogeneous enough.

The general procedure for the theoretical modeling of lumped-parameter processes can be briefly described as follows:

1. Define the flows, e.g., electrical and mechanical.
2. Define the process elements, e.g., sources, sinks, storages, transformers, and converters.
3. Graphically represent the process model, e.g., multi-port diagrams, block diagrams for signal flow, and bond graphs for energy flow.
4. State equations, e.g., balance equations for storages, constitutive equations for process elements, and phenomenological laws for irreversible processes.
5. State interconnection equations, e.g., continuity equations for parallel connections and compatibility equations for serial connections.
6. Calculate the overall process for the model, e.g., establish input and output variables, state space representation, and I/O models.

All process elements except the sources have no definite causality. For this reason, an appropriate approach was proposed to model the system via a software language available with the following properties:

1. *Basic process elements*, described by their physical laws
2. *Connection of process elements* via terminals or ports, like the real system
3. *Inclusion of other models*
4. *Easy user handling* by adding or neglecting elements and through a graphical interface

The object-oriented modeling environment just conforms to these properties realized in the modeling languages SIMULA and DYMOLA (Elmqvist and Mattson, 1989). The definitions of, e.g., objects, classes, encapsulation, and inheritance are used for object-oriented modeling. Based on such a framework, unified object-oriented languages for physical systems modeling in different domains have been developed recently, like MODELICA, OMOLA, and VHDL-AMS.

To verify the theoretical models, the several well-known experimental modeling (identification) approaches with measured input and output variables can be used, e.g., correlation analysis and frequency response measurement, and Fourier and spectral analysis. In other situations, when the parameters are frequently unknown or change with time, parameter estimation methods can be applied for models with continuous time or discrete time, especially if the models are linear in their parameters (Isermann, 1992).

For a nonlinear system with multidimensional characteristics, artificial neural networks are flexible methods to identify and approximate the physical system. These methods can be expanded for nonlinear dynamic processes (Isermann et al., 1997; Isermann, 1999).

In recent years, the real-time simulation is in demand to minimize iterative development cycles and to meet a short time to market. With regard to this requirement, the hardware and the software have to be developed simultaneously.

The *hardware-in-the-loop* (HIL) *simulation* is characterized by operating real components in connection with real-time simulated components. Because actuators and the control hardware very often form one integrated subsystem or because actuators are difficult to model precisely and to simulate in real time, the simulation system usually consists of real actuators and a simulated driving circuit.

34.4 Applications

The rapid development of mechatronic technologies results from ingenious mechanical engineering and marvelous electrical engineering in the past 20 years and is applied to various industry to develop a great deal of applications. According to their construction, traditional classification, and industrial

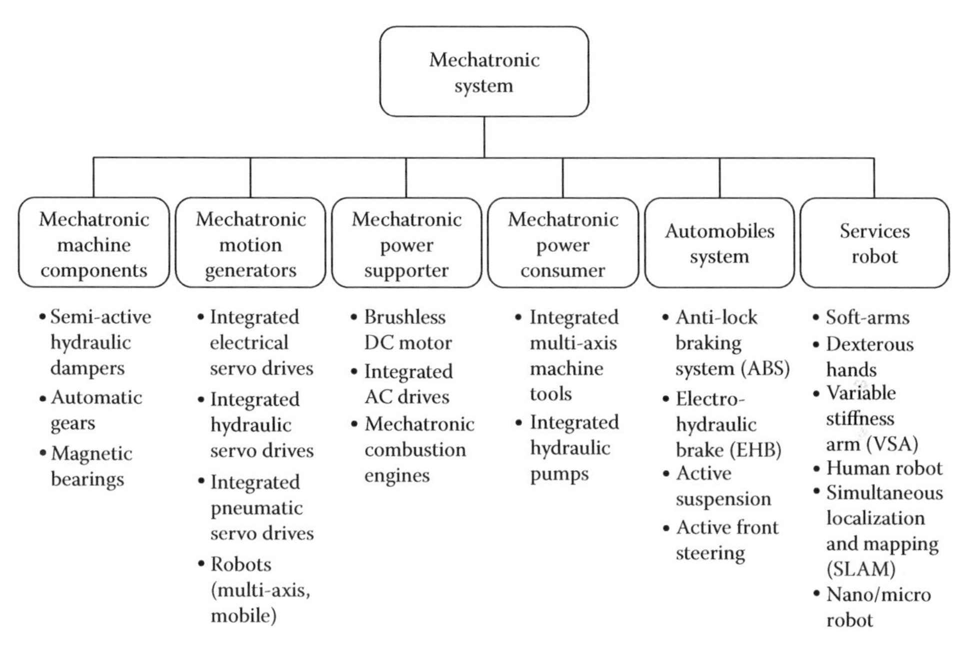

FIGURE 34.5 Examples for mechatronic systems. (Adapted from Isermann, R., *Control Eng. Pract.*, 16(1), 14, January 2008.)

branches, these applications, so-called mechatronic industry, can be subdivided into nanotechnology; automatic control and robotics; biomedical engineering; design, manufacturing, and testing of MEMS; metrology; photonics; mechatronic products; and so on. The development until now is shown in Figure 34.5 and can be found in Isermann (2008).

34.4.1 Automatic Control and Robotics

Along with technical progress, there were some breakthroughs, such as the development of the theory of intelligent and robust control, the rapid progress of the system-on-a-chip digital processor, the advancement of CAE/CAD/CAM for mechanical construction implementation, and so on. In the mechatronics field, motion control was the key technology to achieving the solutions required for a wide variety of problems faced in industry, such as improvement of productivity, reliability, and quality.

A control method to estimate the unknown friction in a dynamical situation was proposed by Iwasaki et al. (1999) and Sakai et al. (1999), to control the position of the mechanical object with the existence of friction. Berns et al. (1999) presented an innovative design of a four-legged walking machine, which is able to simulate both mammal- and reptile-walking modes. A different type of motion technology in the field of walking machines is provided by Muller et al. (2000). They proposed a comprehensive report on nonlinear dynamic models for describing the overall behavior of the large-scale walking machine ALDURO, a hydraulically driven anthropomorphically legged and wheeled robot. Natale et al. (2000) addressed a systematic procedure of designing the structure and tuning the parameters of a force controller for robots endowed with a positional interface. Inoue et al. (2001) described a novel stable motion control strategy for a mobile manipulator with the consideration of external force from the environment. An acceleration controller is used to recover from perturbations due to external force. Soft computing is a popular research topic among mechatronic engineers in recent years. In the papers by Ito et al. (2001), an evolutionary algorithm for the robust motion controller design in mechatronic systems using a GA was presented. The proposed design algorithm requires no complicated procedure unlike the conventional design schemes, thus allowing the compensator parameters tuning to be autonomous. Beck and Turschner (2001) proposed the automated startup procedure of a PI state-controlled rolling-mill motor by using evolutionary algorithms as an auto-tuning tool to optimize the motion controller and to make the motion controller autonomous. Godler et al. (2001) presented a method to compensate the periodic sensing error of a harmonic-drive built-in torque sensor and evaluated its performance under load torque and rotational speeds, and also addressed a practical sensor integration problem in motion control, i.e., torque sensing using strain gages cemented directly on the reduction gear. Carrozza et al. (2002) proposed a novel biomechatronic design and fabrication of prosthetic hands, which consists of integrating multiple-DOFs finger mechanisms, multisensing capabilities, and distributed control in order to obtain a humanlike appearance, simple and direct controllability, and low mass.

A direct-drive robotic manipulator used in semiconductor manufacturing is proposed and demonstrated experimentally by Hosek (2003). Kovecses et al. (2003) made use of differential variational principles of constrained dynamic systems to research the dynamics of parallel robots and mechanisms. A dynamics and control model for a heavy-duty electrohydraulic harvester manipulator is presented by Papadopoulos et al. (2003), and the parameters are experimentally identified and validated, leading to a good tracking behavior. The paper by Caccavale et al. (2003) focused on the dynamic model and control of the Tricept hybrid industrial robot. The macro/micro control problem in the presence of macro flexibilities is studied in the paper by George and Book (2003). Lee et al. (2003) studied the manipulator interactions with a stiff environment and a nonlinear bang-bang impact controller that is proposed independent of the mode of operation. Nho and Meckl (2003) presented a control method, which used the theories of neural network, fuzzy logic, and proportional derivative, to solve the problem of providing accurate parameter estimates for computed-torque control. Doulgeri et al. (2003) presented a feedback grasp controller and stable grasping in rolling manipulations with soft deformable fingertips under two soft-contact motion models. Wang and Xu (2003) described the analyses of the stable full-state tracking and internal dynamics problems of nonholonomic wheeled mobile robots under output-tracking control laws, and the presented formulation offers new insight into the zero dynamics of the mobile robot. Sujan and Dubowsky (2003) presented an algorithm based on a mutual information theoretical metric for the excitation of vehicle dynamics, to efficiently estimate the dynamic parameters of mobile robots. The mutual-information-based metric measures the uncertainty of each parameter's estimate, and is used to optimally select the external excitation required to excite the dynamic system effectively. Ji et al. (2003) developed a bond-graph model to model the system dynamics and perform online tracking, fault detection, fault isolation, and fault identification. They also proposed a hierarchical fault accommodation framework that allows the continued operation of the mobile robot with some performance guarantee in the presence of actuator faults.

In most mobile robot localization schemes, the neglected factors of slip, sinkage, and other wheel-terrain dynamic interactions are the principal cause of the tractive problem and the odometric accuracy loss. Reina et al. (2006) implemented a multimodal sensor-fusion approach and innovated a vision-based algorithm for wheel sinkage estimation to provide a deterministic detection of slip and sinkage, especially on unpaved rough terrains. The paper by Dean-Leon et al. (2006) presented the theoretical contribution and experimental validation in the problem of constrained force control with an uncalibrated camera/robot and unknown contact friction. The paper by Matsuno et al. (2006) presented an excellent solution to the problem—robotic manipulation of deformable objects (such as a rope). Furthermore, they confirmed the effectiveness of their method experimentally in solving a difficult robotic problem commonly encountered in daily life. Huang et al. (2006) considered the automated deployment and maneuver of safety cones used in highway maintenance work zones and proposed a special centralized localization method for such a group of extremely sensor-limited follower robots. A statistical simulation of the localization method is investigated, which analyzes the choice of an appropriate distance between two tracked positions for the minimum allowable radius of curvature for the follower robot's motion. Tang et al. (2006) made use of screw-theoretical analysis tools to provide a systematic framework for the formulation and evaluation of system-level performance and examined cooperative payload transport by robot collectives in an "army of ants" approach. Luo and Su (2007) proposed an intelligent security robot provided with a fire-detecting system, which used an adaptive fusion algorithm to fuse

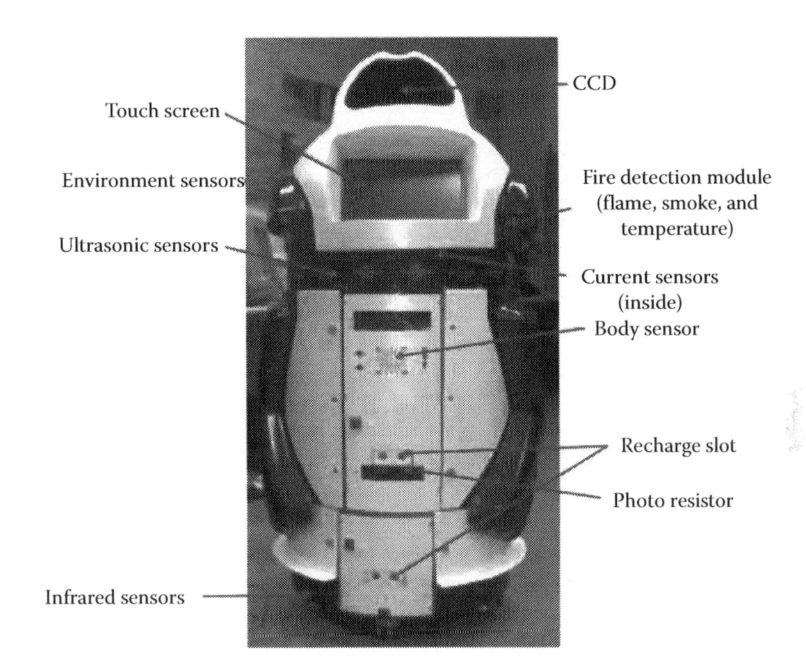

FIGURE 34.6 Overview of intelligent security robot. (Adapted from Luo, R.C. and Su, K.L., *IEEE/ASME Trans. Mechatron.*, 12(3), 274, June 2007.)

the information acquired from the smoke sensor, the flame sensor, and the temperature sensor (Figure 34.6). Hwang and Chang (2007) developed a mixed H2 and H∞ decentralized control for a car-like mobile robot (CLMR) in an intelligent space to obtain a piecewise-line trajectory tracking and obstacle avoidance. Many of the problems, e.g., localization, information about the environment, high computational power, different software, and sensor-based control for each mobile robot, are solved.

Murphy and Sitti (2007) designed and implemented the agile small-scale wall-climbing robot, utilizing dry elastomer adhesives. The quasi-static model, determining the forces acting on the robot, and the adhesive performance model for a climbing robot were developed, which demonstrated the margin of safety and steady-state operating points. Dollar and Howe (2006) described the realization of novel single-piece compliant robotic graspers that mimic the integrated structural, sensing, and actuation functionalities of the human hand. They present a gripper fabricated using a simple prototyping technique that minimizes construction complexity and increases robustness, while preserving the advantages of the passive joint compliance. Agrawal and Fattah (2006) described the unique approach of designing a planar biped, for which the model is nearly linear by a judicious placement of counterweights within the system. Dynamic equations of motion for the swing phase and the impact of the biped with the ground were considered as nearly linear with nonlinear perturbations, which tremendously simplifies the development of nonlinear feedback-linearized controllers. Zoss et al. (2006) proposed the first-of-a-kind untethered energy-autonomous exoskeleton prototype, Berkeley's lower extremity exoskeleton (BLEEX)—a powered leg exoskeleton that shadows the operator's movement to enhance the overall load-carrying capabilities in rough, unstructured, and uncertain terrains. McIntosh et al. (2006) and Banala and Agrawal (2005) discussed the development and experimental validation of a novel mechanism for biaxial rotation, using a single actuator, of an ornithopter wing, which is a compact, lightweight mechanism to achieve such coordinated wing motions.

Climbing robots can perform many tasks inaccessible to traditional vehicles or humans, such as inspection, repair, cleaning, surveillance, and exploration. In the papers by Yano et al. (1997), Shen et al. (2005), and Balaguer et al. (2006), the design, fabrication, and evaluation of several novel bio-inspired climbing robots are presented and discussed. Menon et al. (2004, 2005), Santos et al. (2007),

Jia et al. (2006), and Unvri et al. (2006) investigated a gecko-inspired climbing robot. Several biologically inspired quadruped robots were proposed by Makita et al. (1999), Zhang et al. (2005), Son et al. (2006), Lei et al. (2007), and Duy et al. (2007). Linnemann et al. (1999), Izu et al. (2002), Brown et al. (2007), Yang et al. (2007), and Brunete et al. (2007) investigated the snakelike robot. Dog-inspired quadruped pet robots were showed by Luo et al. (2001, 2005), Suzuki et al. (2007), and Seiji et al. (2004). Some interesting swarm-inspired robots were developed by Gao et al. (2007), Choi et al. (2005), Kingsley et al. (2006), and Arena et al. (2007). Recently, Rifai et al. (2007) introduced flapping-wing micro unmanned air vehicles (UAVs), and a bounded state feedback control torque, based on the theory of cascade, was applied to deal with the attitude stabilization.

Based on current trends, this chapter offers some projections to the future, including various mechatronic applications as well as developments in group machines interaction; human–machine communication and cooperation; and machine automation and intelligent control, such as learning, planning, communicating, and socializing. The development of mechatronic products for many fields (such as entertainment, military applications, aerospace industry, industrial manufacture, household services, care of the elderly and people with disabilities, and so on) has become the primary factor for maintaining high living standards for all industrialized nations in today's world. For instance, the case study of household services presents an intelligent security robot system, NCCU Security Warrior, capable of fire alarm and intruder detection (Luo et al., 2007) (Figures 34.7 and 34.8). When the Security Warrior detects a fire alarm, it immediately rushes to the fire field and tries to extinguish the fire. At the same time, the Security Warrior sounds alert, flashes the light, and sends a warning message by a GSM device to the cell phone of the house owner. In another scenario, the Security Warrior, on detecting an intruder, immediately triggers the alarm, sends a GSM message to the security department, chases the intruder, tracking his face via a CCD camera in the robot, and then sends the image to the security personnel.

In fact, the Security Warrior is a multifunction service robot that can achieve more complex tasks than foregoing functions, such as care of the elderly and people with disabilities and home cleaning. In order to autonomically accomplish a complex task, the developed service robot must possess the fundamental capabilities including object following and tracking, detecting and avoiding obstacles, autonomous navigation, supervising via an electro-network, a remote manipulated vision system, human–robot interaction (HRI), multisensor data fusion, and so on. Various multifunction robots, similar to the Security Warrior, have been developed. The famous R&D groups, MIT, IROBOT in America,

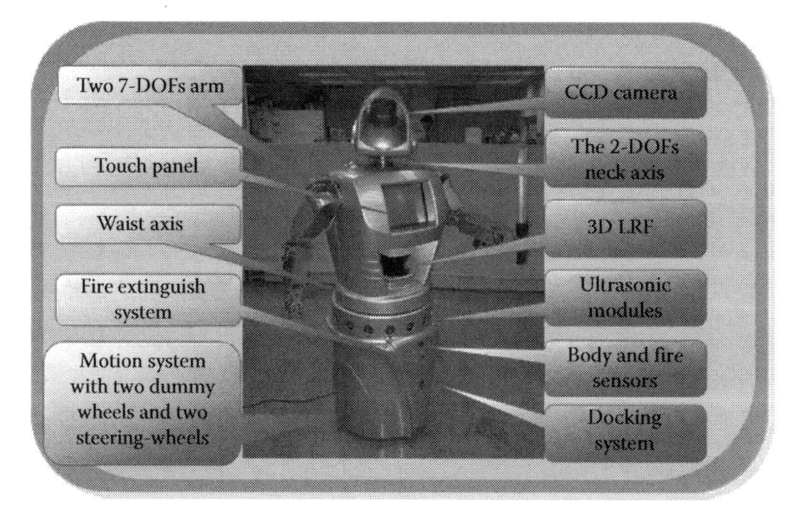

FIGURE 34.7 System architecture of intelligent security robot. (Adapted from Luo, R.C. et al., NCCU security warrior: An intelligent security robot system, in *IEEE International Conference on Industrial Electronics* (*IECON 2007*), Taipei, Taiwan, November 5–8, 2007.)

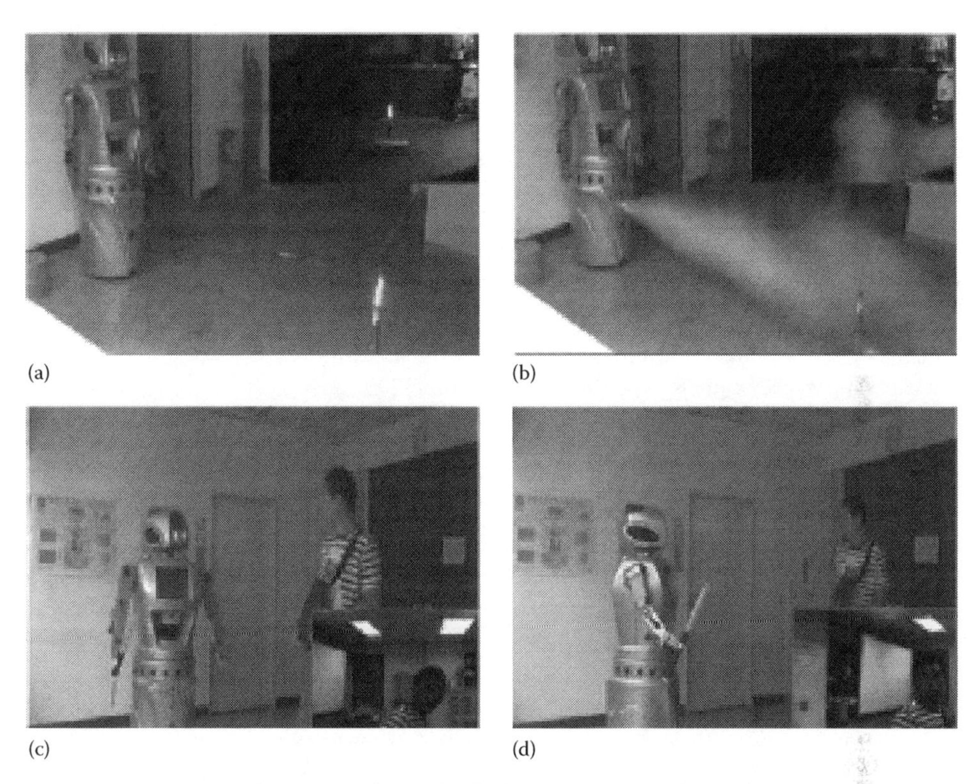

FIGURE 34.8 Fire extinguish system and intruder-frightening system. (Adapted from Luo, R.C. et al., NCCU security warrior: An intelligent security robot system, in *IEEE International Conference on Industrial Electronics (IECON 2007)*, Taipei, Taiwan, November 5–8, 2007.)

and SONY and HONDA in Japan, have invested in advanced robotics, such as entertainment robots, military robots, and service robots. The developments in robotics are continually being published in recent years by more and more research organizations. For instance, ASIMO (Sakagami et al., 2002) is made by HONDA, and the University of Tokyo proposes the human robot HRP (Kaneko et al., 2004). In Japan, the security robot has become more and more popular among people and companies such as ALOSK, FSR, and SECOM.

Shadpey et al. (2009) contributed to the demonstration of robust position and contact force control based on the augmented hybrid impedance control scheme for seven-DOF redundant robot arms. The interaction forces are measured by a six-DOF force/torque sensor and are fed back to the outer-loop controller that implements either a force or an impedance controller in each of the six DOFs of the tool frame. In the force controller, a force set point is arranged, and desired inertia and damping are introduced in the force control loop to improve transient performance. The inner loop consists of a Cartesian-level potential difference controller, a redundancy resolution scheme at the acceleration level, and a joint-space inverse dynamics controller. Experimental results for seven-DOF robot arms are given to illustrate the performance of the force control strategy. A successful application of the proposed scheme to a surface-cleaning task and position- and force-tracking experiments are described.

Křepela and Singule (2007) proposed a mathematical model of a multi-body mass system of the *C* axis, which is controlled with position feedback, and this mathematical model is recommended for the observation of dynamic characteristics in the loading working cycles before machine prototype realization. Changing the dumping of the worm gear has an advantageous influence on the stability. A high ratio of the worm gear reduces the influence of the moment of inertia of the workpiece on the eigenfrequency. The knowledge of the eigenfrequencies for this mechanical system enables accurate regulator optimization of both motors.

Vlachý et al. (2007) dealt with the design of a control unit for a biped robot "Golem 2" with 12 DOFs and accomplished static walking. Zezula et al. (2007) described a humanoid robot construction, which consists of two legs with six DOFs. A computational model was built in MAPLE, a technical computing software, for forward and inverse kinematic simulations. The coordination of robot move was simulated in the environment of MATLAB, Simulink, SimMechanics, and VRML.

Grepl (2007) discussed the extended kinematic model for a quadruped mobile walking robot. The model has been built based on a homogeneous coordinates approach and is comprehended as an open-tree manipulator, and therefore standard algorithms for forward and inverse kinematics can be employed. However, the inverse model, named 12–18, works with the redundant manipulator structure, and the pseudoinverse of the Jacobian matrix should be used. The inverse model 3-3 uses a regular manipulator and allows separate positioning of each leg. The algorithms briefly characterized above have been processed in Maple. Symbolic modifications and a simplification of algebraic expressions have been performed, and the result has been implemented in the MATLAB environment with the SimMechanics applied for verification. The resulting computational models have been successfully tested using VRML visualization in MATLAB.

The rapid growth of the multimedia stream over the Internet, such as digital television, DVD editions, and video transmission, has increased the demand for effective image-compression techniques and the methods of coding/decoding systems evaluation. Ostaszewska and Kłoda (2007) presented the incompatibility of those parameters that were discovered with human perception and gave a proposal of modification in the algorithm, which improves the agreement of parameters with observers' opinion.

The main advantage of mobile robot undercarriages with omnidirectional wheels is mainly their excellent maneuverability. Particularly for this reason, it can be considered as an ideal tool for the verification of various types of algorithms proposed for local navigation, path planning, mapping, and further development with respect to university-indoor environment and robotics classes. Kubela and Pochylý (2007) contributed to the results of work carried out during the proposal and simulation modeling of the mobile robot undercarriage equipped with omnidirectional wheels.

An environment detection and recognition system is essential for a mobile robot. Timofiejczuk et al. (2007) dealt with elaborated methods of acquirement, transmission, processing, analysis, and recognition for visual data. While developing the system, numerous approaches to different methods of lighting and image recognition were tested. Moczulski et al. (2007) proposed a concept and design of a mobile robot capable of inspecting ventilation ducts made of steel sheets. The robot can operate in several modes, including autonomous, manual, and training ones.

Modern mobile robots are usually composed of heterogeneous hardware and software components. Przystałka and Adamczyk (2007) proposed a control architecture that may be used either in the early development phase or in the final robot prototype. This work focuses only on key aspects such as the hardware layout, the low-level real-time operation system, and some software components implemented on the robot.

Panfil et al. (2007) dealt with the implementation of a behavior-based control and a learning controller with neural-network-based coordination methods for autonomous inspection robots. The control architecture is designed to be used in the mobile robot (Amigo) for the visual inspection of ventilation systems.

Racek et al. (2007b) contributed a design of a combined snake robot with its mechanical construction and simulation. This construction consists of independent segments design. Each of designed segments can realize not only linear movement but curving movements too. Racek et al. (2007a) dealt with the verification of the mechanical construction by the simulation of the combined snake robot design.

Bodnicki and Sęklewski (2007) designed and constructed a robot prototype that generates two types of gaits, walking gait and swimming gait, and can move, like an amphibian, e.g., a salamander. This robot is constructed with fully interchangeable modules, which gives a recombinant ability and is suitable for various tasks.

Robot localization and path planning are the actual problems encountered in robotics. Marada (2007) proposed the design of a small autonomous robot for practically verifying artificial intelligence methods concretely on a Micro-mouse task. The physical model was designed for simple construction, unpretentious production, and relatively little cost, but sufficient capability for performing different experiments.

Věchet et al. (2007) provided the identification of the wheel robot position and orientation when dealing with the global localization problem. They used a method called PCSM (precomputed scan matching), which is based on the fusion of precomputed match data and the analysis of the history of robot motion, for solving the problem for autonomous robot in a known environment.

Ondroušek et al. (2007) contributed to the automatic generation of a four-legged robot's walking gait using heuristic search state space algorithms, which improves the efficiency of the classical A* algorithm via branch and bound methods. Simulation verification shows a reduction in the number of state space nodes generated during the search.

34.4.2 Biomedical Engineering

An emerging variant of mechatronics is biomechatronics that integrates the fields of biological science and mechatronic technology. The paper by Mori et al. (2006) presented a "real-life" exoskeleton for healthcare. They described a standing-style transfer system for people with disabilities, which simulated and overcame serious instability problems of conventional powered exoskeleton systems. On the same line, the paper by Kong and Jeon (2006) proposed the tendon-driven exoskeletal assistive device, EXPOS, which consists of a wearable exoskeleton and caster walker that overcome the drawbacks of the exoskeleton, for the elderly and the patients to move, walk, sit down, and stand up fairly well. Menciassi et al. (2003) proposed an innovative prototype miniature robotic instrument consisting of a microfabricated microgripper, instrumented with semiconductor strain gages as force sensors, and capable of characterizing the mechanical properties of tiny biological tissues useful in medical diagnosis. Liu et al. (2007) proposed a neurosurgical robot system that they have developed for clinical trials and discussed how to improve the positioning accuracy calibration by using a revised Denavit–Hartenberg kinematic model and compensation for joint transmission errors using a back-propagation (BP) neural network. Rehabilitation medicine is also the main application in the biomechatronics field. Masia et al. (2007) investigated a mechatronic device, a single-DOF mechanism with a novel statorless configuration, for rehabilitation of grasping functions. Perry et al. (2007) proposed a seven-DOF anthropomorphic powered exoskeleton for the upper limb. The cable-actuated dexterous exoskeleton offers remarkable opportunities as a versatile human–machine interface and as a new generation of instrumentation for assistive technology. Zollo et al. (2007) proposed a biomechatronic approach to the design of an anthropomorphic artificial hand with self-adaptive grasp that mimics the natural motion of human fingers. Tanaka et al. (2007) investigated a compact tactile sensor system based on the polyvinylidene fluoride (PVDF) film and intended to enhance tactile capabilities of artificial hands for Braille automatic reading. Tung et al. (2007) presented a mechatronic device for minimally invasive and teleoperated surgery. The proposed actuator is made from laser-machined SMA tubes, and provides actuation locally to the desired point of manipulation, thus, greatly improving the physician's ability to intervene in diseases. Mitsuishi et al. (2007) proposed a remote operating system for laparoscopic minimally invasive teleoperated surgery that provides a force feedback to the operator.

In recent years, the hasty lifestyle brings about a degradation of the human organism and, subsequently, considerable health problems. The spine and big joints are degraded most often. Clinical solutions are often based on the application of a fixator, and eventually on complete endoprosthesis. Březina et al. (2007) contributed the designed device for biomechanical components testing. The movements of the device are designed to be as large as real physiological movements, and its motion and effector action forces are reached by a pair of robots with parallel kinematics.

34.4.3 Mechatronic Products

The word "haptic" was derived from the Greek "haptikos," meaning "to touch," which also may be defined as proceeding from the sense of touch. The sensory information of haptic is divided into tactile and kinesthetic. Tactile sensing gives a perception of surface textures and geometry, such as temperature, skin curvature and stretch, vibration, slip, pressure, and contact force. Kinesthetic information caused by physical forces applied to the body is sensing and awareness of the body position and orientation, so-called proprioception.

Haptic interface is a device with the ability to transform the sensory information in its original form to the operator, e.g., surgical simulation where students can practice surgical procedures on virtual models, and human–machine interfacing where operators can feel the graphical user interface. To experience the actual conditions, the supplier provides a haptic interface for customers to feel and touch a product in a remote location before making a purchase. On the other hand, the applications that allow for the remote control of robots, so-called telerobotics, are those based on real haptics that is sensed and then regenerated by the haptic devices. It enables the possibility to manipulate and feel objects remotely.

In the past decade, the extensive advances in information science, nano/microelectromechanical systems (NEMS/MEMS), networking technology, and semiconductor industry have had effects on the field of haptics. Not only are such components now reachable, but the commercial products are only limited by the imagination.

A novel haptic mouse was developed by Kyung et al. (2007), which offers the ability of displaying properties such as patterns, gratings, and roughness. Gao and Book (2006) examined the use of passive electromagnetic brakes and presented a novel multimodal haptic user interface with enhanced motion redirection and force control. Human hands exhibit highly asymmetric input–output bandwidths and sensing sensitively to the high-frequency accelerations. The capability of discerning material properties, such as hardness and texture, depends on high-frequency feedback. Yet, in order to keep the system stable, the high-frequency information in the feedback band is always neglected entirely during teleoperating a remote robot. Tanner and Niemayer (2006) presented the control architecture that both incorporates this important information and provides the natural ability to scale and shape the high-frequency content independent of the low-frequency force feedback.

Janschek et al. (2007) studied the performance of a camera stabilization, consisting of a real-time camera vibration detector and a piezoelectric actuator as a high-precision focal plane adjuster. They said that the visual servoing was used to minimize the size of the optical device and to improve the sensitivity to attitude disturbances.

Stefanopoulou and Suh (2007) provided a comprehensive overview on the fuel cell technology, and proposed the technical integration in the mechatronics field, including chemical, fluid, mechanical, thermal, electrical, and electronic subsystems. The paper by Horowitz et al. (2007) discussed the most important advancements in recent years in magnetic hard disk drive servo systems, which may have to be deployed in the near future to sustain the continuing 60% annual increase in the storage density of these devices. They proposed two mechatronic innovations: First, in order to improve the precision and track-following capability of the read/write head-positioning control system, the use of high-bandwidth dual-stage actuator servo systems was innovated. Second, the instrumentation of disk drive suspensions with vibration sensing strain gages was employed to enhance airflow-induced suspension vibration suppression in hard disk drives.

Bzymek et al. (2007) provided the results of the applications of these two types of image processing. They described the methods of image processing applied in a vision system, which consists of two CCD and one infrared camera, for assessing the quality of welded joints. The base of elaborated approaches to joint assessment is a set of images taken in infrared and visible light during the welding process. Images taken in visible light are sources of information about outer conditions of the joint, while thermograms help us obtain information concerning the joint interior. Fidali (2007) presented the results of the application of selected methods of the thermogram analysis. Thermo-vision finds more often its application in machinery and apparatus diagnostics. A thermographic camera provides the noncontact simultaneous temperature measurements in many points of an object and records them in the form of a thermographic image, which can be a source of diagnostic information.

Measuring breaking moments precisely in the mNm range is difficult. Horváth and Nagy (2007) proposed a new procedure based on purely electrical measurements, which, instead of traditional methods, uses brake and additional mechanical elements to determine torque versus angular velocity characteristics of micromotors.

An optimization of the actuator design is considered by Piskur and Tarnowski (2007). The overall criteria are the maximal energy efficiency ratio and the minimal mass and volume of the actuator for the required kinetic energy of the core. A mathematical model for the two-criteria optimization is elaborated, and a poly-optimization is completed.

Kudła (2007) discussed the leading technologies for machining of precise components or patterned surfaces in the micrometer scale that are miniaturized versions of tools used in a conventional range of dimensions or specially developed tools for cutting. The designs, materials, and fabrication processes for such tools are diverse, but some common and specific problems become noticeable.

Ultra capacitors and super capacitors, also called the next-generation battery, are an emerging technology that promises to play an important role in meeting the demands of electronic devices and systems. It as an independent energy source capable of powering everything from power tools to power trains. Miecielica and Demianiuk (2007) presented the internal structure of these components, their advantages compared with batteries and conventional capacitors, and the most interesting applications of ultra capacitors in industry applications.

Oiwa (2007) described a precise machine concept based on compensating for the hexapod-type parallel kinematics mechanism (six DOFs) motion errors between the tool and the workpiece. The six-DOF motions are measured regardless of temperature fluctuation and external forces because the mechanism has a compensation system for the elastic and thermal errors of the joints and the links. Hence, the tool position and orientation are compensated by using the measured six-DOF errors.

Zelenika et al. (2007) contributed in the effort to optimize both mechanical and electronic/control components of ultraprecision devices. The considered mechanics are conformed to overcome the nonlinearities of conventional devices. The design guidelines for hinge optimization are given, and a preliminary consideration of the scaling effects is performed. The developed control system is based on a rapid controller prototyping platform consisting of a Compact-PCI system running under the Linux RTAI real-time extension.

34.4.4 Micro-Nano Mechatronic Systems

The micromechatronic system, which encompasses the same concepts as mechatronics in the "macro" world, is the synergistic integration of mechanical, electrical, information, communication, and nano/microlevel technologies in the design, manufacture, and operation of industrial products and processes. A micromechatronic system emphasizes the key factors in integration, intelligence, communication functions, which includes actuators/sensors, control of mechatronic systems, NEMS/MEMS, micromechatronic devices, human–machine interface/haptics, embedded computing and software engineering, networked and embedded micromechatronics, design/integration methodologies for micromechatronic systems, etc. Micromechatronics plays key roles on specific engineering systems, such as automotive systems, mobile robots, precision motion control systems, consumer electronics, telecommunications, space, medical devices, biomechatronics, nano/micromanipulation, and nanorobotics. In this section, we briefly discuss the state of the art of intelligent micromechatronics in the macro and nano world and present recent advances made in research results.

Ferreira and Fontaine (2003) presented a new generation of compliant multi-DOF microconveyors that use direct-drive standing-wave ultrasonic actuators, which are inspired from high-force actuator concepts, and proposed a dynamic model consisting of the piezoelectric arrayed resonator dynamics and the frictional force model. The paper by Shen et al. (2006) considered the use of a closed-loop optimally controlled piezoelectric microforce sensor for micromanipulation and microassembly applications. The developed microsensor is based on a cantilevered composite beam structure with embedded piezoelectric

PVDF actuating and sensing layers. When an external load causes a deformation in the sensing layer, immediately a signal is sent via the optimal controller to the actuating layer, which generates a force to counteract the external load, and then remains in the equilibrium position. This sensing determines the applied force device through the measurement of the balance force and shows enhanced stiffness and dynamic range with good sensitivity and high manipulability. Hélin et al. (1998) presented the theory, simulation, and experimental results of a critical component of micromechatronic systems, that is, ultrasonic micromotors, using Lamb and Rayleigh waves. They proposed a model with a quantitative description of motion in friction conditions for calculating key parameters to improve the design of ultrasonic micromotors. Dario et al. (1998) addressed a microrobot with a new type of electromagnetic micromotor, and described in detail the principle, design, fabrication, and performance of the wobble-type micromotor. The microrobot with a volume of 4 mm is applied in new instruments for minimally invasive therapy. Arai et al. (1998) illustrated a micromechatronic approach for the design of an important subsystem (the microgripper) of a more complex system for micromanipulation. They proposed a method to deal with this problem of adhesive forces in micromanipulation. Reynaerts et al. (1998) described the implementation of concurrent engineering on the microsystem design and pointed out that this MEMS technology facilitates the way for concurrent engineering of real 3D micromechanical systems integrating sensors, actuators, and processing electronics. Tendick et al. (1998) presented micromechatronic design methods and applications of a micromechatronic millirobotic system for the minimally invasive surgery (MIS) developed. Itoh (2000) investigated the biomechatronics problem regarding the possibility of treating protozoa as living micromanipulators. The rapid turning controller and the experimental results were presented. Arai et al. (2003) reported a pinpoint injection method of microtools (MTs) at the desired location of a microchamber and made use of the enhanced laser micromanipulation technology effective in the discovery of the target microbe without damaging it. Sitti (2003) proposed a new principle of a piezoelectrically actuated motion amplification mechanism for micromechanical flying insects. A four-bar mechanism with two flexible links is applied in a micromechanical flying insect thorax design for stroke amplification. El Rifai and Youcef-Toumi (2007) proposed a robust adaptive controller applied in an atomic force microscope, which is capable of auto-tuning gains for different cantilever-sample combinations, and compensating for the uncertainties resulting from the choice of scan parameters. Fatikow et al. (2007) contributed to the development of an automated nanohandling station in a scanning electron microscope (SEM). This station enables handling tasks in the micrometer and submicrometer ranges, like the handling of TEM lamella. Hung et al. (2007) constructed a modulation/demodulation scheme to design a novel dual-stage piezoelectric nanopositioner that successfully extended the measurement range beyond the limit of the wavelength of the optical fiber Fabry–Perot interferometer. The paper by Yuan and Yang (2007) presented a novel approach of automated planning for multirobot-based nanoassembly that provides an improved self-organizing map to fit all the nanoassembly tasks into a seamless process and makes it possible for handling environmental uncertainty and generating optimized motion paths at run time with a modified shunting neural network.

Savia and Koivo (2009) reviewed contact micromanipulation strategies, especially from the adhesion point of view. At microscale, the most important adhesive forces are van der Waals force, electrostatic force, and capillary force; the scaling-effect-induced adhesion forces severely complicate micromanipulation when compared to conventional macromanipulation, and further, adhesive forces are also the reason behind the fairly low level of automation in microassembly systems. Adhesion forces bring unpredictability to manipulation operations, especially in the release phase. Lack of repeatability in the operations and different deviations require a strong involvement of a human operator. Adhesion forces should be taken into consideration already in the design phase of a micromanipulation system or a microassembly line. In other words, the manipulation strategies and the end-effectors should be designed such that the adhesion forces can be mastered. A proper design reduces the requirements set on the (higher-level) control system. Computationally light-adhesion force models for arbitrary object shapes are an essential tool when designing the manipulation operations and the end-effectors.

Shirinzadeh and Liaw (2009) presented a motion-tracking control methodology based on a neural network for piezo-actuated flexure-based micro/nanomanipulation mechanisms. In particular, the radial basis function neural networks are adopted for function approximations, and a lumped-parameter dynamic model that combines the piezoelectric actuator and the micro/nanomechanism is established for the formulation of the proposed approach. The motion-tracking control methodology has to overcome the presence of unknown system parameters, nonlinearities including the hysteresis effect, and external disturbances. The important improvement of this motion-tracking control methodology is that no prior knowledge is required for the system parameters, as well as for the thresholds and weights of the neural networks in the physical realization of the control system.

The images of standard calibration samples can be used to model and correct positioning errors caused by dynamics effects in scanning probe microscopes (SPMs), rather than using external sensors. Clayton and Devasia (2009) contributed to the development of conditions, on the calibration sample and the scan trajectory, that allow for the image-based identification of SPM nanopositioner dynamics, and improve the positioning speed. In particular, there are two relations developed between (1) the minimum spacing between features and the orientation of the calibration sample and (2) the amplitude and frequency content of the SPM probe's position trajectory. A trade-off between calibration sample and scan trajectory properties was discussed in the context of a scanning tunneling microscope (STM) example. These conditions were applied to identify a dynamics model of an STM using images of a highly oriented pyrolytic graphite (HOPG) calibration sample.

Chibum and Salapaka (2009) proposed a novel design method of two-DOF robust optimal control for achieving multiple objectives of resolution, bandwidth, and robustness to modeling uncertainties in nanopositioning systems. The main theoretical contribution of this paper was the formulation of a multiobjective two-DOF optimal control problem in terms of the linear matrix inequality (LMI). The controller is obtained by standard convex optimization tools. The main distinguishing feature of this approach is demonstrated as the flexibility this method provides in formulating and solving the optimization problems, which results in achieving a larger set of performance specifications. The experimental results demonstrate over a 200% improvement in the bandwidth over optimal one-DOF designs and achieve specifications that are impossible in a one-DOF framework.

Mestrom et al. (2009) contributed a phase feedback approach for an oscillator circuit containing a nonlinear, electrostatically actuated, clamped-clamped beam microelectromechanical (MEM) resonator. The principle of the proposed approach is illustrated for a nonlinear Duffing resonator, which is the representative of many types of MEM oscillators. Phase feedback allows for operation of the resonator in its nonlinear regime. The closed-loop technique enables control of both the frequency of oscillation and the output power of the signal. Phase feedback allows for operation of the resonator in its nonlinear regime. In order to select optimal operation points for the phase feedback in oscillator circuits incorporating a nonlinear resonator, both the frequency of oscillation and the output power of the signal are used. The application of phase feedback results in more robustness with respect to the dynamic pull-in instability than the open-loop case. However, the drawback is a deteriorated phase noise response.

Elbuken et al. (2009) contributed a novel microrobot technology using magnetic levitation for wireless micromanipulation tasks. The wireless microrobot, fabricated based on MEMS technology and weighing less than 1 g, can be positioned in three dimensions by external magnetic fields. A photothermal polymer microgripper, which can be actuated remotely by laser focusing, is equipped for the microrobots to perform micromanipulation operations. It makes the microrobot free of any wiring and leads to an increased motion range with more functionality, in addition to dust-free motion and the ability to work in closed environments. The operation of the microrobot is such that a closed chamber opens the path for the micromanipulation of biological samples and hazardous items. The 3D motion capability of the microrobots is verified experimentally to operate in a working space of $4 \times 4 \times 5\,cm^3$. Micromanipulation experiments such as pick and place, push, and pull were demonstrated using objects with $100\,\mu m$ and 1 mm diameter.

Sangkyung et al. (2009) investigated a novel approach based on a phase-domain analysis for the mode-matched control of a MEMS vibratory gyroscope. By tuning the electrostatic stiffness to adjust the resonant frequency of the sensing mode to that of the drive mode, the gyroscopic sensitivity to the Coriolis force is improved. In the mode-matched control loop design, the resonant characteristics of the driving axis are used as the reference mode. Then, the phase difference between sense and drive modes at the resonant frequency of the drive mode is used to generate a control signal for phase error regulation. Through the simulation using practical MEMS gyroscope parameters and the results of experiments in a manufactured MEMS gyroscope prototype, the desired performance and robustness can be achieved when there are phase errors.

Jingang et al. (2009) presented a novel hysteresis compensation method for piezoelectric actuators. The hysteresis nonlinearity of the actuator is considered as a disturbance over a linear system. A disturbance observer (DOB) is then used to estimate and compensate for the hysteresis nonlinearity. As a benefit, the exact knowledge of the hysteresis is not required, which makes the design easy to implement in practical applications. Experimental validation of the proposed hysteresis compensation is performed on the lead magnesium niobate–lead titanate ($Pb\,Mg\,Nb\,Tio_3$) (PMN-PT) cantilever piezoelectric actuator. The experimental results demonstrate the effectiveness and efficiency of the approach.

The microtribology has a significant effect on MEMS, particularly in the design and modeling of bearings in micromotors. Zhang and Meng (2009) apply a new slip model and a modified Reynolds equation based on the kinetic theory of gases for the gas-lubricated slider bearings beneath the bushings of a micromotor in MEMS. The theoretical analyses show that gas bearings can benefit from the scaling down of their geometric parameters, mainly resulting from the rise in the pressure within the bearings and the decrease of the weight of the micromotors.

The numerical solutions demonstrated that the present new slip model shows better accuracy in predicting the bearing properties than the first-order and second-order slip models and the modified gas lubrication equation (MMGL) model is utilized to analyze the step-shaped bushings, and is a good approximation to variable hard sphere (VHS), and variable soft sphere (VSS) models. The solutions provided by the present new slip model match the solution obtained from the linearized Boltzmann equation very well.

34.4.5 Automotive Mechatronics

The burgeoning automotive industry impacts the social, environmental, economic, and technological aspects of our daily life. More and more mechatronic systems were investigated for automotive applications, including ABS, supplemental restraint systems (air bags), cruise control, and traction control. Mechatronic systems play a more important role in improving automotive functionality, safety, economy, and comfort, and the designed functions assist the driver to prevent unstable or unpredictable behavior and to stabilize the motion of the automobile. The complex capabilities of fault detection and diagnosis are achieved by the integration of several burgeoning technologies in the field of actuators, sensors, data processing, and so on.

In the traditional valve-controlled hydraulic elevator, when the vehicle moves downward, the entire potential energy of the vehicle is wasted and converted into fluid heat by throttling. To reduce the energy consumption and power installation requirements, Yang et al. (2007) designed a novel microcontroller-based energy-regenerating hydraulic elevator using an accumulator as the energy-restoring component, which would be reduced to a level at par with that of an electric elevator. Fischer et al. (2007) proposed the concepts of model-based fault detection and diagnosis along with sensor fault tolerance for an automotive system and realized a vehicle lateral dynamics system and an active suspension system. Arimitsu et al. (2007) developed a new safety driving system that employs seat-belt vibration to stimulate and awake drivers. In the field of combustion engine control, the applications of hybrid modeling and receding horizon optimal control techniques are proposed by Giorgetti et al. (2006). Hattori et al. (2006) overcame the problem of obstacle avoidance in automotive motion.

They proposed an optimum trajectory based on the framework of the so-called H-VDIM algorithm for vehicle dynamics control. Deur et al. (2006) presented a survey paper of the authors' work in the field of modeling, analysis, and validation results of automotive power train systems and their components. They gave a functional description of an automotive power train system, followed by the modeling of an electronic throttle mechanism, manifold air temperature and pressure measurement in engine, torque-converter fluid dynamics modeling, wet clutch fluid dynamics, automatic transmission, and tires. Isermann (2008) contributed a review paper describing the increasing mechatronic design, systematic development chains, from modeling and design to implementation and testing in the automotive industry.

34.5 Conclusions

This chapter describes the development of mechatronic engineering in the past and the present and discusses its future trend. A multitude of commercial products were realized by the integrated technologies of mechatronic engineering. From the observation of the direction of progress in the mechatronic domain, more and more powerful, intelligent, and instamatic (instant plus automatic) functions are required to be realized. The integrated technologies have not been limited to the mechanical and electronic domains. More scientific disciplines are involved in improving the capability of mechatronic systems. In recent years, the development of tools is more and more consummate, e.g., RP/RT/CAE/CAD/CAM for mechanism designs and the integrated development environment (IDE) platform for electronic function design and implementation. The most important improvements are modular and hierarchical design concepts, which help the designer escape from a mire of moderating a complicated, interdisciplinary problem, and be absorbed in the development of individual functions, especially in intelligent control, autonomous motion, and so on. This chapter summarizes the ongoing developments in mechatronic engineering, reviews technical literatures, and proposes the viewpoints of future development.

References

Aghili, F., J.M. Hollerbach, and M. Buehler, A modular and high-precision motion control system with an integrated motor, *IEEE/ASME Transactions on Mechatronics*, 12(3): 317–329, June 2007.

Agrawal, S.K. and A. Fattah, Motion control of a novel planar biped with nearly linear dynamics, *IEEE/ASME Transactions on Mechatronics*, 11(2): 162–168, April 2006.

Alciatore, D. and M. Histand, *Introduction to Mechatronics and Measurement Systems*, 3rd edn, New York: McGraw-Hill, p. 544, ISBN: 978-0-07-296305-2, 2005.

Arai, F., D. Andou, Y. Nonoda, T. Fukuda, H. Iwata, and K. Itoigawa, Integrated microendeffector for micromanipulation, *IEEE/ASME Transactions on Mechatronics*, 3(1): 17–23, March 1998.

Arai, F., H. Maruyama, T. Sakami, A. Ichikawa, and T. Fukuda, Pinpoint injection of microtools for minimally invasive micromanipulation of microbe by laser trap, *IEEE/ASME Transactions on Mechatronics*, 8(1): 3–9, March 2003.

Arena, P., L. Fortuna, M. Frasca, L. Patane, and M. Pavone, Realization of a CNN-driven cockroach-inspired robot, *Circuits and Systems, 2006. ISCAS 2006. Proceedings. 2006 IEEE International Symposium on*, Kos, Greece, pp. 4 and 21–24, May 2006.

Arimitsu, S., K. Sasaki, H. Hosaka, M. Itoh, K. Ishida, and A. Ito, Seat belt vibration as a stimulating device for awakening drivers, *IEEE/ASME Transactions on Mechatronics*, 12(5): 511–518, October 2007.

Auslander, D.M., What is mechatronics? *IEEE/ASME Transactions on Mechatronics*, 1(1): 5–9, March 1996.

Auslander, D. and C.J. Kempf, *Mechatronics, Mechanical System Interfacing*, Englewood Cliffs, NJ: Prentice-Hall, 1996.

Baeten, J. and J. De Schutter, Hybrid vision/force control at corners in planar robotic-contour following, *IEEE/ASME Transactions on Mechatronics*, 7(2): 143–151, June 2002.

Balaguer, C., A. Gimenez, A.J. Huete, A.M. Sabatini, M. Topping, and Bolmsjo, The MATS robot: Service climbing robot for personal assistance, *IEEE Robotics and Automation Magazine*, 13(1): 51–58, March 2006.

Banala, S. and S.K. Agrawal, Design and optimization of a mechanism for out of plane insect wing like motion with twist, *ASME Transactions Journal of Mechanical Design*, 127: 841–844, 2005.

Beck, H.-P. and D. Turschner, Commissioning of a state-controlled high-powered electrical drive using evolutionary algorithms, *IEEE/ASME Transactions on Mechatronics*, 6(2): 149–154, June 2001.

Bekey, G.A., *Autonomous Robots: From Biological Inspiration to Implementation and Control*, Cambridge, MA: MIT Press, 2005.

Berns, K., W. Ilg, M. Deck, J. Albiez, and R. Dillmann, Mechanical construction and computer architecture of the four-legged walking machine BISAM, *IEEE/ASME Transactions on Mechatronics*, 3(1): 32–38, March 1999.

Beuth-Verlag GmbH, *VDI 2206 Design Methodology for Mechatronic Systems*, Berlin, Germany: Beuth-Verlag GmbH, 2004.

Bodnicki, M. and M. Sęklewski, Design of small-outline robot—Simulator of gait of an amphibian, *Recent Advances in Mechatronics*, pp. 77–81, Heidelberg, Germany: Springer, ISBN 978-3-540-73955-5, 2007.

Breazeal, C.L., *Designing Sociable Robots*, 1st edn, Cambridge, MA: The MIT Press, ISBN: 978-0262025102, p. 281, May 4, 2002.

Březina, T., M.Z. Florian, and A.A. Caballero, The design of the device for cord implants tuning, *Recent Advances in Mechatronics*, pp. 195–199, Heidelberg, Germany: Springer, ISBN 978-3-540-73955-5, 2007.

Brown, H.B., M. Schwerin, E. Shammas, and H. Choset, Design and control of a second-generation hyper-redundant mechanism, *IEEE/RSJ International Conference on Intelligent Robots and Systems, 2007* (*IROS 2007*), Rome, Italy, pp. 2603–2608, October 29, 2007–November 2, 2007.

Brunete, A., J.E. Torres, M. Hernando, and E. Gambao, Multi-drive control for in-pipe snakelike heterogeneous modular micro-robots, *IEEE International Conference on Robotics and Biomimetics, 2007* (*ROBIO 2007*), Shenyang, China, pp. 490–495, December 15–18, 2007.

Burmester, S. and H. Giese, Visual integration of UML 2.0 and block diagrams for flexible reconfiguration in mechatronic UML, *2005 IEEE Symposium on Visual Languages and Human-Centric Computing*, vols. 20–24, Dallas, TX, pp. 109–116, September 2005.

Buur, J., Mechatronics in Japan—Strategies and practice in product development, *Mechatronics: Designing Intelligent Machines*, in *Proc. Inst. Mech. Eng.*, 131–136, 1990.

Bzymek, A., M. Fidali, and A. Timofiejczuk, Methods of image processing in vision system for assessing welded joints quality, *Recent Advances in Mechatronics*, pp. 258–262, Heidelberg, Germany: Springer, ISBN 978-3-540-73955-5, 2007.

Caccavale, F., B. Siciliano, and L. Villani, The tricept robot: Dynamics and impedance control, *IEEE/ASME Transactions on Mechatronics*, 8(2): 263–268, June 2003.

Carrozza, M.C., B. Massa, S. Micera, R. Lazzarini, M. Zecca, and P. Dario, The development of a novel prosthetic hand-ongoing research and preliminary results, *IEEE/ASME Transactions on Mechatronics*, 7(2): 108–114, June 2002.

Choi, J., B.L. Rutter, D.A. Kingsley, R.E. Ritzmann, and R.D. Quinn, A robot with cockroach inspired actuation and control, *Proceedings of the 2005 IEEE/ASME International Conference on Advanced Intelligent Mechatronics*, Monterey, CA, pp. 1569–1574, July 24–28, 2005.

Clayton, G.M. and S. Devasia, Conditions for image-based identification of SPM-nanopositioner dynamics, *IEEE/ASME Transactions on Mechatronics*, 14(4): 405–413, June 2009.

Craig, K. and F. Stolfi, Introduction to mechatronic system design with applications, Prod. No. HV0542-1 (Notes), IEEE and ASME, San Jose, CA, 1994.

Dario, P., M.C. Carrozza, C. Stefanini, and S. D'Attanasio, A mobile microrobot actuated by a new electromagnetic wobble micromotor, *IEEE/ASME Transactions on Mechatronics*, 3(1): 9–16, March 1998.

Dean-Leon, E.C., V. Parra-Vega, and A. Espinosa-Romero, Visual servoing for constrained planar robots subject to complex friction, *IEEE/ASME Transactions on Mechatronics*, 11(4): 389–400, August 2006.

De Doncker, R.W., Modern electrical drives: Design and future trends, *CES/IEEE Fifth International Power Electronics and Motion Control Conference, IPEMC '06*, vol. 1, Shanghai, China, pp. 1–8, August 2006.

Deur, J., J. Petric, J. Asgari, and D. Hrovat, Recent advances in control-oriented modeling of automotive power train dynamics, *IEEE/ASME Transactions on Mechatronics*, 11(5): 513–523, October 2006.

Dollar, A.M. and R.D. Howe, A robust compliant grasper via shape deposition manufacturing, *IEEE/ASME Transactions on Mechatronics*, 11(2): 154–161, April 2006.

Doulgeri, Z. and J. Fasoulas, Grasping control of rolling manipulations with deformable fingertips, *IEEE/ASME Transactions on Mechatronics*, 8(2): 283–286, June 2003.

Elbuken, C., M.B. Khamesee, and M. Yavuz, Design and implementation of a micromanipulation system using a magnetically levitated MEMS robot, *IEEE/ASME Transactions on Mechatronics*, 14(4): 434–445, June 2009.

Elmqvist, H. and S.E. Mattson, Simulator for dynamical systems using graphics and equations for modeling, *IEEE Control Systems Magazine* 9: 53–58, 1989.

El Rifai, O.M. and K. Youcef-Toumi, On automating atomic force microscopes: An adaptive control approach, *Control Engineering Practice*, 15(3): 349–361, March 2007.

Fatikow, S., T. Wich, H. Hulsen, T. Sievers, and M. Jahnisch, Microrobot system for automatic nanohandling inside a scanning electron microscope, *IEEE/ASME Transactions on Mechatronics*, 12(3): 244–252, June 2007.

Ferreira, A. and J.G. Fontaine, Dynamic modeling and control of a conveyance microrobotic system using active friction drive, *IEEE/ASME Transactions on Mechatronics*, 8(2): 188–202, June 2003.

Fidali, M., Application of analysis of thermographic images to machine state assessment, *Recent Advances in Mechatronics*, pp. 263–267, Heidelberg, Germany: Springer, ISBN 978-3-540-73955-5, 2007.

Fite, K.B. and M. Goldfarb, Design and energetic characterization of a proportional-injector monopropellant-powered actuator, *IEEE/ASME Transactions on Mechatronics*, 11(2): 196–204, April 2006.

Fischer, D., M. Börner, J. Schmitt, and R. Isermann, Fault detection for lateral and vertical vehicle dynamics, *Control Engineering Practice*, 15(3): 315–324, March 2007.

Gao, D. and W.J. Book, Steerability for planar dissipative passive haptic interfaces, *IEEE/ASME Transactions on Mechatronics*, 11(2): 179–184, April 2006.

Gao, Y., W. Chen, Z. Lu, and H. Dang, Kinematic analysis and simulation of a cockroach robot, *2nd IEEE Conference on Industrial Electronics and Applications, 2007* (*ICIEA 2007*), Harbin, China, pp. 1208–1213, 23–25 May 2007.

George, L.E. and W.J. Book, Inertial vibration damping control of a flexible base manipulator, *IEEE/ASME Transactions on Mechatronics*, 8(2): 268–271, June 2003.

Giorgetti, N., G. Ripaccioli, A. Bemporad, I.V. Kolmanovsky, and D. Hrovat, Hybrid model predictive control of direct injection stratified charge engines, *IEEE/ASME Transactions on Mechatronics*, 11(5): 499–506, October 2006.

Glielmo, L., M. Milano, and S. Santini, A machine learning approach to modeling and identification of automotive three-way catalytic converters, *IEEE/ASME Transactions on Mechatronics*, 5(2): 132–141, June 2000.

Godler, I., M. Hashimoto, M. Horiuchi, and T. Ninomiya, Performance of gain-tuned harmonic drive torque sensor under load and speed conditions, *IEEE/ASME Transactions on Mechatronics*, 6(2): 155–160, June 2001.

Goldfarb, M., E.J. Barth, M.A. Gogola, and J.A. Webrmeyer, Design and energetic characterization of a liquid-propellant-powered actuator for self-powered robots, *IEEE/ASME Transactions on Mechatronics*, 8(2): 254–262, June 2003.

Goldfarb, M., E.J. Barth, M.A. Gogola, and J.A. Webrmeyer, Design and energetic charaterization of a liquid-propellant-powered actuator for self-powered robots, *IEEE/ASME. Transactions on Mechatronics*, 8(2): 254–262, June 2003.

Grepl, R., Extended kinematics for control of quadruped robot, *Recent Advances in Mechatronics*, pp. 126–130, Heidelberg, Germany: Springer, ISBN 978-3-540-73955-5, 2007.

Habibi, S. and A. Goldenberg, Design of a new high-performance electrohydraulic actuator, *IEEE/ASME Transactions on Mechatronics*, 5(2): 158–164, June 2000.

Hattori, Y., E. Ono, and S. Hosoe, Optimum vehicle trajectory control for obstacle avoidance problem, *IEEE/ASME Transactions on Mechatronics*, 11(5): 507–512, October 2006.

Heath, S., *Embedded Systems Design (EDN Series for Design Engineers)*, 2nd edn, Burlington, MA: Newnes, pp. 11–12, 2003.

Hélin, P., V. Sadaune, C. Druon, and J.-B. Tritsch, A mechanical model for energy transfer in linear ultrasonic micromotors using Lamb and Rayleigh waves, *IEEE/ASME Transactions on Mechatronics*, 3(1): 3–8, March 1998.

Hewit, J.R. and T.G. King, Mechatronics design for product enhancement, *IEEE/ASME Transactions on Mechatronics*, 1(2): 111–119, June 1996.

Horváth, P. and A. Nagy, Determination of DC micro-motor characteristics by electrical measurements, *Recent Advances in Mechatronics*, pp. 278–282, Heidelberg, Germany: Springer, ISBN 9783-540-73955-5, 2007.

Horowitz, R., Y. Li, K. Oldham, S. Kon, and X. Huang, Dual-stage servo systems and vibration compensation in computer hard disk drives, *Control Engineering Practice*, 15(3): 291–305, March 2007.

Hosek, M., Observer-corrector control design for robots with destabilizing unmodeled dynamics, *IEEE/ASME Transactions on Mechatronics*, 8(2): 151–164, June 2003.

Hsu, T.R., Mechatronics in manufacturing, a contributing part of *CRC Mechanical Engineering Handbook*, ed. F. Kreith. Boca Raton, FL: CRC Press, February 1998.

Hsu, T.R., J.C. Wang, F. Barez, B. Furman, P. Hsu, and P. Reischl, An undergraduate curriculum in mechatronic systems engineering, *Proceedings of the International Mechanical Engineering Congress and Exposition*, San Francisco, CA, November 12–17, 1995.

Hu, J.-S. and T.-M. Su, Robust environmental change detection using PTZ camera via spatial-temporal probabilistic modeling, *IEEE/ASME Transactions on Mechatronics*, 12(3): 339–344, June 2007.

Huang, J., S.M. Farritor, A. Qadi, and S. Goddard, Localization and follow-the-leader control of a heterogeneous group of mobile robots, *IEEE/ASME Transactions on Mechatronics*, 11(2): 205–215, April 2006.

Hung, S.-K., E.-T. Hwu, M.-Y. Chen, and L.-C. Fu, Dual-stage piezoelectric nano-positioner utilizing a range-extended optical fiber Fabry–Perot Interferometer, *IEEE/ASME Transactions on Mechatronics*, 12(3): 291–298, June 2007.

Hunt, V.D., Thinking mechatronically, *Management Automation*, 44–45, February 1988.

Hwang, C.-L. and L.-J. Chang, Trajectory tracking and obstacle avoidance of car-like mobile robots in an intelligent space using mixed H_2/H_∞ decentralized control, *IEEE/ASME Transactions on Mechatronics*, 12(3): 345–352, June 2007.

Inoue, F., T. Muralami, and K. Ihnishi, A motion control of mobile manipulator with external force, *IEEE/ASME Transactions on Mechatronics*, 6(2): 137–142, June 2001.

Isermann, R., *Identifikation Dynamischer Systeme*, vols. 1 and 2, 2nd edn, Berlin, Germnay: Springer, 1992.

Isermann, R., *Mechatronische Systeme*, Berlin, Germnay: Springer, 1999a.

Isermann, R., Modeling, identification and simulation of mechatronic systems, *IFACCongress*, Beijing, China/Amsterdam, the Netherlands: Elsevier, 1999b.

Isermann, R., Mechatronic systems—Innovative products with embedded control, *Control Engineering Practice*, 16(1): 14–29, January 2008.

Isermann, R., S. Ernst, and O. Nelles, Identification with dynamic neural networks—Architectures, comparisons, applications, *Plenary, IFAC Symposium on System Identification* (*SYSID*), Fukuoka, Japan, 1997.

Itoh, A., Motion control of protozoa for bio-MEMS, *IEEE/ASME Transactions on Mechatronics*, 5(2): 181–188, June 2000.

Ito, K., M. Iwasaki, and N. Matsui, GA-based practical compensator design for a motion control system, *IEEE/ASME Transactions on Mechatronics*, 6(2): 143–148, June 2001.

Iwasaki, M., T. Shibata, and N. Matsui, Disturbance-observer-based nonlinear friction compensation in table drive system, *IEEE/ASME Transactions on Mechatronics*, 3(1): 3–8, March 1999.

Izu, H., H.W. Date, K. Shigeta, T. Yamanaka, S. Nakaura, and M. Sampei, Locomotion and coiling motion control of snakelike robot using pneumatic actuators, *SICE 2002. Proceedings of the 41st SICE Annual Conference*, vol. 3, Tokyo, Japan, pp. 1476–1480, August 5–7, 2002.

Janschek, K., V. Tchernykh, and S. Dyblenko, Performance analysis of opto-mechatronic image stabilization for a compact space camera, *Control Engineering Practice*, 15(3): 333–347, March 2007.

Jeon, S. and M. Tomizuka, Benefits of acceleration measurement in velocity estimation and motion control, *Control Engineering Practice*, 15(3): 325–332, March 2007.

Ji, M., Z. Zhang, G. Biswas, and N. Sarkar, Hybrid fault adaptive control of a wheeled mobile robot, *IEEE/ASME Transactions on Mechatronics*, 8(2): 226–233, June 2003.

Jia, Q., J. Liao, H.-X. Sun, J.-Z. Song, and T. Cheng, Research on theoretical models of synthetic geckos' adhesion technology, *2006 IEEE Conference on Robotics, Automation and Mechatronics*, Bangkok, Thailand, pp. 1–5, December 2006.

Kaneko, K., F. Kanehiro, S. Kajita, H. Hirukawa, T. Kawasaki, M. Hirata, K. Akachi, and T. Isozumi, Humanoid robot HRP-2, *Proceedings. 2004 IEEE International Conference on Robotics and Automation* (*ICRA'04*), vol. 2, pp. 1083–1090, New Orleans, LA, April 26–May 1, 2004. http://www.robotdiy.com/; http://pc.watch.impress.co.jp/docs/2005/0624/alsok.htm

Keys, L.K., Mechatronics, systems, elements, and technology: A perspective, *IEEE Transactions on Components Hybrids and Manufacturing Technology*, 14: 457–461, September 1991.

Khemani, H., What is engineering design? On-line available at http://www.brighthub.com/engineering/mechanical/articles/9932.aspx#ixzz0VVxfkhS5, published October 6, 2008.

Kim, H.-W. and S.-K. Sul, A new motor speed estimator using Kalman filter in low-speed range, *IEEE Transactions on Industrial Electronics*, 43(4): 498–504, August 1996.

Kingsley, D.A., R.D. Quinn, and R.E. Ritzmann, A cockroach inspired robot with artificial muscles, *2006 IEEE/RSJ International Conference on Intelligent Robots and Systems*, Beijing, China, pp. 1837–1842, October 2006.

Křepela, J. and V. Singule, Dynamical behaviors of the C axis multibody mass system with the worm gear, *Recent Advances in Mechatronics*, pp. 1–5, Heidelberg, Germany: Springer, ISBN 978-3-540-73955-5, 2007.

Kong, K. and D. Jeon, Design and control of an exoskeleton for the elderly and patients, *IEEE/ASME Transactions on Mechatronics*, 11(4): 428–432, August 2006.

Kovecses, J., J.-C. Piedboeuf, and C. Lange, Dynamics modeling and simulation of constrained robotic systems, *IEEE/ASME Transactions on Mechatronics*, 8(2): 165–177, June 2003.

Kubela, T. and A. Pochylý, Simulation modeling and control of a mobile robot with omnidirectional wheels, *Recent Advances in Mechatronics*, pp. 22–26, Heidelberg, Germany: Springer, ISBN 978-3-540-73955-5, 2007.

Kudła, L., Design and fabrication of tools for micro-cutting processes, *Recent Advances in Mechatronics*, pp. 303–307, Heidelberg, Germany: Springer, ISBN 978-3-540-73955-5, 2007.

Kwon, S.J., W.K. Chung, and Y. Youm, A combined observer for robust state estimation and Kalman filtering, *Proceedings of the 2003 American Control Conference*, vol. 3, pp. 4–6, Denver, CO, pp. 2459–2464, June 2003.

Kyung, K.-U., S.-C. Kim, and D.-S. Kwon, Texture display mouse: Vibrotactile pattern and roughness display, *IEEE/ASME Transactions on Mechatronics*, 12(3): 356–360, June 2007.

Lakshmana Rao, C., J. Lakshminarasimhan, R. Sethuraman, and S.M. Sivakumar, *Engineering Mechanics*, New Delhi, India: PHI Learning Pvt. Ltd., p. vi. ISBN 8120321898, 2004.

Lau, K., D.E. Quevedo, B.J.G. Vautier, G.C. Goodwin, and S.O.R. Moheimani, Design of modulated and demodulated controllers for flexible structures, *Control Engineering Practice*, 15(3): 377–388, March 2007.

Lee, C. and S.M. Salapaka, Fast robust nanopositioning—A linear-matrix-inequalities-based optimal control approach, *IEEE/ASME Transactions on Mechatronics*, 14(4): 414–422, June 2009.

Lee, E., J. Park, K.A. Loparo, C.B. Schrader, and P.H. Chang, Bang-bang impact control using hybrid impedance/time-delay control, *IEEE/ASME Transactions on Mechatronics*, 8(2): 272–277, June 2003.

Lin, L.R. and H.P. Huang, Integrating fuzzy control of the dexterous National Taiwan University (NTU) Hand, *IEEE/ASME Transactions on Mechatronics*, 1(3): 216–229, September 1996.

Linnemann, R., K.L. Paap, B. Klaassen, and J. Vollmer, Motion control of a snakelike robot, *1999 Third European Workshop on Advanced Mobile Robots* (*Eurobot '99*), Monterey, CA, pp. 1–8, September 6–8, 1999.

Liu, S. and B. Yao, Automated onboard modeling of cartridge valve flow mapping, *IEEE/ASME Transactions on Mechatronics*, 11(4): 381–388, August 2006.

Liu, J., Y. Zhang, and Z. Li, Improving the positioning accuracy of a neurosurgical robot system, *IEEE/ASME Transactions on Mechatronics*, 12(5): 527–533, October 2007a.

Liu, W., X. Qu, and Y. Yan, Design and motion analysis of wheel-legged climbing robot, *Chinese Control Conference*, CCC 2007, pp. 149–153, June 26–31, 2007b.

Luo, R.C. and M.G. Kay, Multisensor integration and fusion in intelligent system, *IEEE Transactions on Systems, Man, and Cybernetics*, 19: 901–931, September/October 1989.

Luo, R.C. and S.-C. Hsieh, The development of multisensor integrated quadruped pet robot, *IEEE International Conference on Mechatronics, 2005. ICM '05*, Taipei, Taiwan, pp. 203–207, July 10–12, 2005.

Luo, R.C. and K.L. Su, Autonomous fire-detection system using adaptive sensory fusion for intelligent security robot, *IEEE/ASME Transactions on Mechatronics*, 12(3): 274–281, June 2007.

Luo, R.C., K.L. Su, J.H. Tzou, and S.H.H. Phang, Multisensor based control of pet robot through the Internet, *The 27th Annual Conference of the IEEE Industrial Electronics Society, 2001. IECON '01*, vol. 1, Denver, CO, pp. 416–421, November 29–December 2, 2001.

Luo, R.C., Y.T. Chou, C.T. Liao, C.C. Lai, and A.C. Tsai, NCCU security warrior: An intelligent security robot system, *IEEE International Conference on Industrial Electronics* (*IECON 2007*), Taipei, Taiwan, November 5–8, 2007.

Maas, J., T. Schulte, and N. Frohleke, Model-based control for ultrasonic motors, *IEEE/ASME Transactions on Mechatronics*, 5(2): 165–180, June 2000.

MacGregor, J.G., *An Elementary Treatise on Kinematics and Dynamics*, London, U.K.: Macmillan, p. v, 1887.

Mahalik, N.P., *Mechatronics Principles, Concepts and Applications*, New Delhi, India: Tata McGraw-Hill, pp. 30, ISBN 0070483744, 2003.

Makita, S., N. Murakami, M. Sakaguchi, and J. Furusho, Development of horse-type quadruped robot, *1999 IEEE International Conference on Systems, Man, and Cybernetics, 1999. IEEE SMC '99 Conference Proceedings*, vol. 6, Tokyo, Japan, pp. 930–935, October 12–15, 1999.

Marada, T., The robot for practical verifying of artificial intelligence methods: Micro-mouse task, *Recent Advances in Mechatronics*, pp. 102–106, Heidelberg, Germany: Springer, ISBN 978-3-540-73955-5, 2007.

Masia, L., H.I. Krebs, P. Cappa, and N. Hogan, Design and characterization of hand module for whole-arm rehabilitation following stroke, *IEEE/ASME Transactions on Mechatronics*, 12(4): 399–407, August 2007.

Matsuno, T., D. Tamaki, F. Arai, and T. Fukuda, Manipulation of deformable linear objects using knot invariants to classify the object condition based on image sensor information, *IEEE/ASME Transactions on Mechatronics*, 11(4): 401–408, August 2006.

McIntosh, S.H., S.K. Agrawal, and Z. Khan, Design of a mechanism for biaxial rotation of a wing for a hovering vehicle, *IEEE/ASME Transactions on Mechatronics*, 11(2): 145–153, April 2006.

Menciassi, A., A. Eisinberg, M.C. Carrozza, and P. Dario, Force sensing microinstrument for measuring tissue properties and pulse in microsurgery, *IEEE/ASME Transactions on Mechatronics*, 8(1): 10–17, March 2003.

Menon, C. and M. Sitti, Biologically inspired adhesion based surface climbing robots, *Proceedings of the 2005 IEEE International Conference on Robotics and Automation, 2005. ICRA 2005*, Seattle, WA, pp. 2715–2720, April 18–22, 2005.

Menon, C., M. Murphy, and M. Sitti, Gecko inspired surface climbing robots, *IEEE International Conference on Robotics and Biomimetics, 2004. ROBIO 2004*, 22–26, Shenyang, China, pp. 431–436, August 2004.

Mestrom, R.M.C., R.H.B. Fey, and H. Nijmeijer, Phase feedback for nonlinear MEM resonators in oscillator circuits, *IEEE/ASME Transactions on Mechatronics*, 14(4): 423–433, June 2009.

Miecielica, M. and M. Demianiuk, Ultra capacitors—New source of power, *Recent Advances in Mechatronics*, pp. 308–312, Heidelberg, Germany: Springer, ISBN 978-3-540-73955-5, 2007.

Mitsuishi, M., N. Sugita, and P. Pitakwatchara, Force-feedback augmentation modes in the laparoscopic minimally invasive telesurgical system, *IEEE/ASME Transactions on Mechatronics*, 12(4): 447–454, August 2007.

Moczulski, W., M. Adamczyk, P. Przystałka, and A. Timofiejczuk, Mobile robot for inspecting ventilation ducts, *Recent Advances in Mechatronics*, pp. 47–51, Heidelberg, Germany: Springer, ISBN 978-3-540-73955-5, 2007.

Mori, Y., J. Okada, and K. Takayama, Development of a standing style transfer system "ABLE" for Disabled Lower Limbs, *IEEE/ASME Transactions on Mechatronics*, 11(4): 372–380, August 2006.

Muller, J., M. Schneider, and M. Hiller, Modeling, simulation, and model-based control of the walking machine ALDURO, *IEEE/ASME Transactions on Mechatronics*, 5(2): 142–152, June 2000.

Munir, S. and W.J. Book, Internet-based teleoperation using wave variables with prediction, *IEEE/ASME Transactions on Mechatronics*, 7(2): 124–133, June 2002.

Murata, S. and H. Kurokawa, Self-reconfigurable robots, *IEEE Robotics and Automation Magazine*, 14(1): 71–78, March 2007.

Murphy, M.P. and M. Sitti, Waalbot: An agile small-scale wall-climbing robot utilizing dry elastomer adhesives, *IEEE/ASME Transactions on Mechatronics*, 12(3): 330–338, June 2007.

Natale, C., R. Koeppe, and G. Hirzinger, A systematic design procedure of force controllers for industrial robots, *IEEE/ASME Transactions on Mechatronics*, 5(2): 122–131, June 2000.

Neugebauer, R., B. Denkena, and K. Wegener, Mechatronic systems for machine tools, *CIRP Annals—Manufacturing Technology*, 56(2): 657–686, 2007.

Nho, H.C. and P. Meckl, Intelligent feedforward control and payload estimation for a two-link robotic manipulator, *IEEE/ASME Transactions on Mechatronics*, 8(2): 277–282, June 2003.

Odhner, L.U. and H.H. Asada, Sensorless temperature estimation and control of shape memory alloy actuators using thermoelectric devices, *IEEE/ASME Transactions on Mechatronics*, 11(2): 139–144, April 2006.

Oiwa, T., Ultra-precision machine feedback-controlled using hexapod-type measurement device for six degree-of-freedom relative motions between tool and workpiece, *Recent Advances in Mechatronics*, pp. 330–334, Heidelberg, Germany: Springer, ISBN 978-3-540-73955-5, 2007.

Ondroušek, V., S. Věchet, J. Krejsa, and P. Houška, Verification of the walking gait generation algorithms using branch and bound methods, *Recent Advances in Mechatronics*, pp. 151–155, Heidelberg, Germany: Springer, ISBN 978-3-540-73955-5, 2007.

Ostaszewska, A. and R. Kłoda, Quantifying the amount of spatial and temporal information in video test sequences, *Recent Advances in Mechatronics*, pp. 11–15, Heidelberg, Germany: Springer, ISBN 978-3-540-73955-5, 2007.

Panfil, W., P. Przystałka, and M. Adamczyk, Behavior-based control system of a mobile robot for the visual inspection of ventilation ducts, *Recent Advances in Mechatronics*, pp. 62–66, Heidelberg, Germany: Springer, ISBN 978-3-540-73955-5, 2007.

Panzieri, S., F. Pascucci, and G. Ulivi, An outdoor navigation system using GPS and inertial platform, *IEEE/ASME Transactions on Mechatronics*, 7(2): 134–142, June 2002.

Papadopoulos, E., B. Mu, and R. Frenette, On modeling, identification, and control of a heavy-duty electrohydraulic harvester manipulator, *IEEE/ASME Transactions on Mechatronics*, 8(2): 178–187, June 2003.

Paynter, H.M. and J.M. Juarez, Jr. Thermodynamic analysis of a mechatronic pneumatically driven spherical joint, *IEEE/ASME Transactions on Mechatronics*, 5(2): 153–157, June 2000.

Perry, J.C., J. Rosen, and S. Burns, Upper-limb powered exoskeleton design, *IEEE/ASME Transactions on Mechatronics*, 12(4): 408–417, August 2007.

Pil, C. and H. H. Asada, Integrated structure and control design of mechatronics systems using a recursive experimental optimization method, *IEEE/ASME Transactions on Mechatronics*, 1(3): 191–203, September 1996.

Piskur, P. and W. Tarnowski, Poly-optimization of coil in electromagnetic linear actuator, *Recent Advances in Mechatronics*, pp. 283–287, Heidelberg, Germany: Springer, ISBN 978-3-540-73955-5, 2007.

Przystałka, P. and M. Adamczyk, EmAmigo framework for developing behavior based control systems of inspection robots, *Recent Advances in Mechatronics*, pp. 37–41, Heidelberg, Germany: Springer, ISBN 978-3-540-73955-5, 2007.

Racek, V., J. Sitar, and D. Maga, Simulation and realization of combined snake robot, *Recent Advances in Mechatronics*, pp. 67–71, Heidelberg, Germany: Springer, ISBN 978-3-540-73955-5, 2007a.

Racek, V., J. Sitar, and D. Maga, Design of combined snake robot, *Recent Advances in Mechatronics*, pp. 72–76, Heidelberg, Germany: Springer, ISBN 978-3-540-73955-5, 2007b.

Reina, G., L. Ojeda, A. Milella, and J. Borenstein, Wheel slippage and sinkage detection for planetary rovers, *IEEE/ASME Transactions on Mechatronics*, 11(2): 185–195, April 2006.

Reynaerts, D., J. Peirs, and H. Van Brussel, A mechatronic approach to microsystem design, *IEEE/ASME Transactions on Mechatronics*, 3(1): 24–33, March 1998.

Richardson, R.C., M.C. Levesley, M.D. Brown, J.A. Hawkes, K. Watterson, and P.G. Walker, Control of ionic polymer metal composites, *IEEE/ASME Transactions on Mechatronics*, 8(2): 245–253, June 2003.

Rifai, H., N. Marchand, and G. Poulin, Bounded attitude control of a biomimetic flapping robot, *IEEE International Conference on Robotics and Biomimetics, 2007. ROBIO 2007*, Bangkok, Thailand, December 15–18, 2007, pp. 1–6.

Sakagami, Y., R. Watanabe, C. Aoyama, S. Matsunaga, N. Higaki, and K. Fujimura, The intelligent ASIMO: System overview and integration, *International Conference on Intelligent Robots and System, 2002. IEEE/RSJ*, vol. 3, 30, pp. 2478–2483, Tokyo, Japan, September–October 5, 2002.

Sakai, S., H. Sado, and Y. Hori, Motion control in an electric vehicle with four independently driven in-wheel motors, *IEEE/ASME Transactions on Mechatronics*, 3(1): 9–16, March 1999.

Sancho-Pradel, D.L. and R.M. Goodall, Targeted processing for real-time embedded mechatronic systems, *Control Engineering Practice*, 15(3): 363–375, March 2007.

Santos, D., S. Kim, M. Spenko, A. Parness, and M. Cutkosky, Directional adhesive structures for controlled climbing on smooth vertical surfaces, *2007 IEEE International Conference on Robotics and Automation*, 10–14, Roma, Italy, pp. 1262–1267, April 2007.

Savia, M. and H.N. Koivo, Contact micromanipulation—Survey of strategies, *IEEE/ASME Transactions on Mechatronics*, 14(4): 504–514, June 2009.

Schafer, W. and H. Wehrheim, The challenges of building advanced mechatronic systems, *Future of Software Engineering, FOSE'07*, vols. 23–25, pp. 72–84, Washington, DC, May 2007.

Seiji, Y. and Y. Tomohiro, Training AIBO like a dog—Preliminary results, *13th IEEE International Workshop on Robot and Human Interactive Communication, 2004. ROMAN 2004*, Kurashiki, Japan, pp. 431–436, September 20–22, 2004.

Shadpey, F., H.A. Talebi, J. Jayender, R.V. Patel, A robust position and force control strategy for 7-DOF redundant manipulators, *IEEE/ASME Transactions on Mechatronics*, 14(5): 575–589, June 2009.

Shen, W., J. Gu, and Y. Shen, Proposed wall climbing robot with permanent magnetic tracks for inspecting oil tanks, *Proceedings of the 2005 IEEE International Conference on Mechatronics and Automation*, vol. 4, Ontario, Canada, pp. 2072–2077, July 29–August 1, 2005.

Shen, Y., E. Winder, N. Xi, C.A. Pomeroy, and U.C. Wejinya, Closed-loop optimal control-enabled piezoelectric microforce sensors, *IEEE/ASME Transactions on Mechatronics*, 11(4): 420–427, August 2006.

Shirinzadeh, B. and H.C. Liaw, Neural network motion tracking control of piezo-actuated flexure-based mechanisms for micro-/nanomanipulation, *IEEE/ASME Transactions on Mechatronics*, 14(5): 517–527, June 2009.

Sitti, M., Piezoelectrically actuated four-bar mechanism with two flexible links for micromechanical flying insect thorax, *IEEE/ASME Transactions on Mechatronics*, 8(1): 26–36, March 2003.

Son, H.-M., J.-B. Gu, S.-H. Park, Y.-J. Lee, and T.-H. Nam, Design of new quadruped robot with SMA actuators for dynamic walking, *International Joint Conference SICE-ICASE, 2006*, Busan, South Korea, pp. 344–348, October 2006.

Song, S.-H. and S.-K. Sul, An instantaneous speed observer for low speed control of AC machine, *Thirteenth Annual Applied Power Electronics Conference and Exposition, 1998. APEC '98. Conference Proceedings 1998*, vol. 2, 15–19, Guilin, China, pp. 581–586, February 1998.

Stefanopoulou, A.G. and K.-W. Suh, Mechatronics in fuel cell systems, *Control Engineering Practice*, 15(3): 277–289, March 2007.

Stein, D., E.R. Scheinerman, and G.S. Chirikjian, Mathematical models of binary spherical-motion encoders, *IEEE/ASME Transactions on Mechatronics*, 8(2): 234–244, June 2003.

Sujan, V.A. and S. Dubowsky, An optimal information method for mobile manipulator dynamic parameter identification, *IEEE/ASME Transactions on Mechatronics*, 8(2): 215–225, June 2003.

Sun, L., M.Q.-H. Meng, W. Chen, H. Liang, and T. Mei, Design of quadruped robot based CPG and fuzzy neural network, *2007 IEEE International Conference on Automation and Logistics*, 18–21, Jinan, China, pp. 2403–2408, August 2007.

Sung, S., W.-T. Sung, C. Kim, S. Yun, and Y.J. Lee, On the mode-matched control of MEMS vibratory gyroscope via phase-domain analysis and design, *IEEE/ASME Transactions on Mechatronics*, 14(4): 446–455, June 2009.

Suzuki, H., A. Aburadani, H. Nishi, and S. Inoue, Animal gait generation for quadrupedal robot, *Second International Conference on Innovative Computing, Information and Control, 2007. ICICIC'07*, Tokyo, Japan, pp. 20–20, September 5–7, 2007.

Tanaka, M., K. Miyata, and C. Seiji, A wearable braille sensor system with a post processing, *IEEE/ASME Transactions on Mechatronics*, 12(4): 430–438, August 2007.

Tang, C.P., R.M. Bhatt, M. Abou-Samah, and V. Krovi, Screw-theoretic analysis framework for cooperative payload transport by mobile manipulator collectives, *IEEE/ASME Transactions on Mechatronics*, 11(2): 169–178, April 2006.

Tanner, N.A. and G. Niemeyer, High-frequency acceleration feedback in wave variable telerobotics, *IEEE/ASME Transactions on Mechatronics*, 11(2): 119–127, April 2006.

Tendick, F., S.S. Sastry, R.S. Fearing, and M. Cohn, Applications of micromechatronics in minimally invasive surgery, *IEEE/ASME Transactions on Mechatronics*, 3(1): 34–42, March 1998.

Timoshenko, S. and D.H. Young, Engineering mechanics. New York: McGraw Hill, 1956.

Timofiejczuk, A., M. Adamczyk, A. Bzymek, and P. Przystałka, Environment detection and recognition system of a mobile robot for inspecting ventilation ducts, *Recent Advances in Mechatronics*, pp. 27–31, Heidelberg, Germany: Springer, ISBN 978-3-540-73955-5, 2007.

Tung, A.T., B.-H. Park, G. Niemeyer, and D.H. Liang, Laser-machined shape memory alloy actuators for active catheters, *IEEE/ASME Transactions on Mechatronics*, 12(4): 439–446, August 2007.

Uneri, A., A. Aydemir, and M. Sitti, Geckobot: A gecko inspired climbing robot using elastomer adhesives, *Proceedings of the 2006 IEEE International Conference on Robotics and Automation (ICRA 2006)*, Orlando, FL, pp. 2329–2335, May 15–19, 2006.

Věchet, S., J. Krejsa, and P. Houška, The enhancement of PCSM method by motion history analysis, *Recent Advances in Mechatronics*, pp. 107–110, Heidelberg, Germany: Springer, ISBN 978-3-540-73955-5, 2007.

Vlachý, D., P. Zezula, and R. Grepl, Control unit architecture for biped robot, *Recent Advances in Mechatronics*, pp. 6–10, Heidelberg, Germany: Springer, ISBN 978-3-540-73955-5, 2007.

Vo, H.D., T.P. Nguyen, M.Y. Suk, K.K. Hak, and B.K. Sang, A new approach for designing quadruped robot, *4th IEEE International Conference on Mechatronics, ICM2007*, 8–10, Kumamoto, Japan, pp. 1–5, May 2007.

Wang, D. and G. Xu, Full-state tracking and internal dynamics of nonholonomic wheeled mobile robots, *IEEE/ASME Transactions on Mechatronics*, 8(2): 203–214, June 2003.

Whittaker, E.T., *A Treatise on the Analytical Dynamics of Particles and Rigid Bodies: With an Introduction to the Problem of Three Bodies*, 4th edn, ed. Sir William McCrea, Cambridge, U.K.: Cambridge University Press, Chapter 1, p. 1, ISBN 0521358833, 1936.

Wooley, J.C. and H.S. Lin, *Catalyzing Inquire at the Interface of Computing and Biology*, Washington, DC: The National Academies Press, 2001.

Wright, T.W., *Elements of Mechanics Including Kinematics, Kinetics and Statics: With Applications*, eds. E. London and F.N. Spon, New York: D. Van Nostrand Company, p. 85, 1896.

Youcef-Toumi, K., Modeling, design, and control integration: A necessary step in mechatronics, *IEEE/ASME Transactions on Mechatronics*, 1(1): 29–38, March 1996.

Yan, L., I.-M. Chen, G. Yang, and K.-M. Lee, Analytical and experimental investigation on the magnetic field and torque of a permanent magnet spherical actuator, *IEEE/ASME Transactions on Mechatronics*, 11(4): 409–419, August 2006.

Yang, B., Y.H. Lv, and W.-H. Yang, Analysis of crosstalk between snakelike circuits in proximity, *International Symposium on Electromagnetic Compatibility, 2007. EMC 2007*, Munich, Germany, pp. 565–567, October 23–26, 2007a.

Yang, H., W. Sun, and B. Xu, New investigation in energy regeneration of hydraulic elevators, *IEEE/ASME Transactions on Mechatronics*, 12(5): 519–526, October 2007b.

Yano, T., T. Suwa, M. Murakami, and T. Yamamoto, Development of a semi self-contained wall climbing robot with scanning type suction cups, *Proceedings of the 1997 IEEE/RSJ International Conference on Intelligent Robots and Systems, 1997. IROS'97*, vol., 2, 7–11, Victoria, BC, pp. 900–905, September 1997.

Yi, J., S. Chang, and Y. Shen, Disturbance-observer-based hysteresis compensation for piezoelectric actuators, *IEEE/ASME Transactions on Mechatronics*, 14(4): 456–464, June 2009.

Yuan, X. and S.X. Yang, Multirobot-based nanoassembly planning with automated path generation, *IEEE/ASME Transactions on Mechatronics*, 12(3): 352–356, June 2007.

Zelenika, S., S. Balemi, and B. Roncevic, An integrated mechatronics approach to ultra-precision devices for applications in micro and nanotechnology, *Recent Advances in Mechatronics*, pp. 355–359, Heidelberg, Germany: Springer, ISBN 978-3-540-73955-5, 2007.

Zezula, P., D. Vlachý, and R. Grepl, Simulation modeling, optimalization and stabilisation of biped robot, *Recent Advances in Mechatronics*, pp. 120–125, Heidelberg, Germany: Springer, ISBN 978-3-540-73955-5, 2007.

Zhang, W.-M. and G. Meng, Property analysis of the rough slider bearings in micromotors for MEMS applications, *IEEE/ASME Transactions on Mechatronics*, 14(4): 465–473, June 2009.

Zhang, X., H. Zheng, Z. Cheng, Z. Cheng, and L. Zhao, A biological inspired quadruped robot: Structure and control, *2005 IEEE International Conference on Robotics and Biomimetics (ROBIO)*, Hong Kong, China, pp. 387–392, 2005.

Zollo, L., S. Roccella, E. Guglielmelli, M.C. Carrozza, and P. Dario, Biomechatronic design and control of an anthropomorphic artificial hand for prosthetic and robotic applications, *IEEE/ASME Transactions on Mechatronics*, 12(4): 418–429, August 2007.

Zoss, A.B., H. Kazerooni, and A. Chu, Biomechanical design of the Berkeley lower extremity exoskeleton (BLEEX), *IEEE/ASME Transactions on Mechatronics*, 11(2): 128–138, April 2006.

35

Actuators in Robotics and Automation Systems

Choon-Seng Yee
National University of Singapore

Marcelo H. Ang Jr.
National University of Singapore

35.1 Overview

Actuators are devices that produce actions, which typically are forms of motion or forces/torques exerted. The input to the actuator is some form of energy (usually electrical) and the output is mechanical motion or force exertion. Actuators are therefore key components in an electromechanical system such as a robotic manipulator. The input to the manipulator is a controlled amount of electrical energy and the result is the manipulator motion corresponding to the robotic task.

A robotic manipulator is a multi-degree-of-freedom mechanism consisting of a series of linkages that is connected in some fashion to provide dexterity in the manipulator end-effector (or "hand" which effects the manipulator task). Each link moves relative to each other through joints. Some joints may be passive like pin joints, whereas other joints are coupled to actuators which cause relative motion between the links connected by these joints. The actuators therefore serve as the muscles of the manipulators.

There is a wide range of actuators in use in industry. They vary depending on the loads they actuate. In robotic manipulators, actuators can be in the form of motors that move the robot links. Actuators, such as solenoids, are also used to actuate small devices such as to turn switches on or off. Actuators can be classified into electric, hydraulic, and pneumatic, depending on the type of input energy they accept.

Motors are electric actuators that accept electrical energy and the output is mechanical rotation of the motor shaft (rotor). The resulting motion is effected through electromagnetic means where the electrical input and the magnetic fields inside the electric actuator cause the rotor to rotate. Solenoids are also electric actuators because the current flowing through the coil causes motion of a ferromagnetic cylindrical rod longitudinally located at the center of the coil. The ferromagnetic material inside electric actuators saturates at certain levels of magnetic flux density, thus limiting their torque or force capabilities [1]. The load to weight ratios of the electric actuators are therefore limited compared to hydraulic actuators.

Hydraulic actuators use very high-pressurized liquid to effect motion or force. The liquids are usually oils because they are noncompressible, and the hydraulic pressures are in the order of a few thousand psi. Their load capabilities are at least an order of magnitude larger than electric actuators. Hydraulic actuators therefore have the best torque to weight ratios. Another important feature of hydraulic actuators is the high stiffness (viewed from the load side) provided due to the noncompressibility of oil as compared to a lower stiffness provided by an electromagnetic medium.

Pneumatic actuators use pressurized air as the actuating mechanism. Their advantage is that they offer a simple and low cost method for linear motion. Because air is compressible, they are not stiff and their responses are slow. Pneumatic actuators are therefore well-suited for lighter load applications and lower performance servo-control applications.

Electric, pneumatic, and hydraulic actuators all find their use in robots. Hydraulic actuators are used in large, high-performance robots with large payloads, while pneumatic actuators are used more in smaller, lower performance (and cost) robots for motions. Pneumatic actuators are ideal for applications requiring the robot end-effector (or hand) to move to fixed positions. They usually require no feedback of position, and mechanical stops are typically employed to define the fixed positions. If precise positioning is required, hydraulic or electric actuators with servo control are used. The advantage of hydraulic and pneumatic actuators are their ruggedness and safety. Because they are sealed, they can operate in harsh (dirty, wet, etc.) environments. They can also be operated in explosive environments because unlike electric motors, there are no dangers of brush arcing causing sparks. Their disadvantages compared to electric motors are their higher cost and maintenance requirements, higher noise levels, lower efficiencies, the need for fluid transport systems, and flammability of oils in hydraulic systems [2]. For applications that require medium loads and accurate positioning control, electric actuators provide the best price/performance solution. Most robots today employ electric actuators to achieve accurate positioning control in light to medium load applications.

A special category of actuators are those that rely on special materials and their properties. These "exotic" actuators are not commonly used in robotic manipulators because of inadequate loading capabilities and slow response times [2]. They are also very costly. Piezoelectric materials can be used as actuators because they change shape when subjected to an electric field. This change in shape can therefore effect motion. Magnetostrictive materials are similar to piezoelectric ones except that an applied magnetic field causes a change in shape. Thermal actuators rely upon expansion or contraction when subjected to temperature changes in a resistance type heating element inside the actuator. Another interesting actuator is the shape memory alloy which is a metal which adopts a "memorized" shaped when it achieves a certain transition temperature. Other actuators in this category include electrostatic actuators consisting of two conducting plates that are either electrostatically attracted to or repelled against each other depending on the DC voltages applied to the plates [2].

Transmissions are also important in robots. Transmissions serve two purposes. The first purpose is to transmit the power output of the actuator to another location that is remote from the actuator. In a robot, for example, the actuators can be located at the base of the robot thus freeing the need for the robot links to carry the actuators. The second purpose is to transform (i.e., amplify) the load capability of the actuator at the expense of speed. Transmissions are gear systems, belts and pulleys, and sprockets and chains.

Actuators need peripheral circuitry to "drive" them. These circuits are called drivers or amplifiers. The drivers serve as an intermediate stage between the actuator inputs and the energy sources. They also serve as interfacing circuits to allow input from a computer or other input device. In this section, we also explain the different drivers typically used with the actuators. We concentrate upon electric actuators (i.e., motors) because they are the most commonly used actuators in robotic manipulators. For completeness, we include discussions on the drives associated with the electric motors, and also transmission devices.

35.2 Direct Current Motors

35.2.1 Principle of Operation

DC motors in general operate according to Ampere's law of magnetic force, which explains and quantifies the force acting on a current carrying body (motor coil) in the presence of a magnetic field. According to Ampere's law of magnetic force, the force, $\boldsymbol{F}$, acting on the body is the cross product of the current, $\boldsymbol{I}$, flowing across the body and the magnetic flux density, $\boldsymbol{B}$, of which the body is in, as shown in Figure 35.1. It is expressed in Equation 35.1 and the vector representation is shown on the right of Figure 35.1.

$$\boldsymbol{F} = \boldsymbol{I} \times \boldsymbol{B} \tag{35.1}$$

Current $\boldsymbol{I}$ flowing through a conductor induces a force $\boldsymbol{F}$ that is normal to the current and the magnetic flux density $\boldsymbol{B}$. The current is the electrical input to the DC motor; the induced force is coupled to the output shaft of the motor to produce the output torque. The clever arrangement of the flow of current, flux, and induced force results in a smooth torque delivery of the motor.

35.2.1.1 DC Brushed Motor

A simple model of the DC motor can be seen in Figure 35.2. It consists of a pair of permanent magnets as the *stator* (which serves as the fixed housing of the motor) and a motor coil as the *rotor* (or rotating shaft) connected to a commutator. The commutator, which forms the heart of DC brushed motors, basically consists of two (or more) commutator bars (the half cylinders in Figure 35.2) and two brushes, both made of conducting materials. The brushes are spring-loaded to maintain contact with the cylinder formed by the commutator bars. Each of the two ends of the motor coil is connected electrically to an independent commutator bar. When the rotor coil rotates, the commutator cylinder (formed by all the commutator bars, insulated from one another) rotates together and the current flow changes according to the contact made by the brushes on the commutator bars, thus creating an alternating current flow in the coil, depending on the position of the coil.

In the example shown in Figure 35.2, there will be a time when the brushes are in contact with both commutator bars, that is, when the motor coil is perpendicular to the magnetic field lines. This will cause the power to short circuit, but the angular momentum of the rotor will bring the coil to move

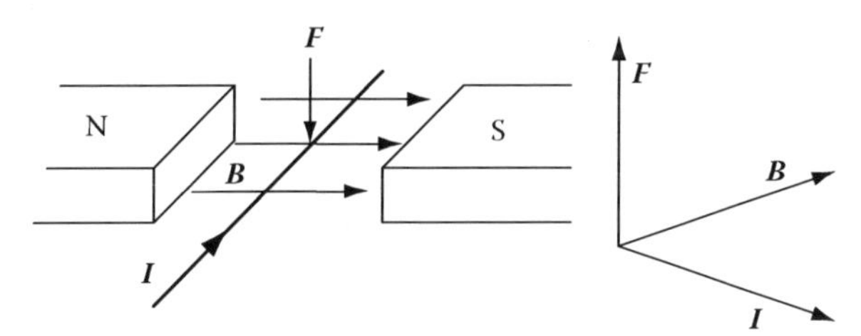

FIGURE 35.1 Ampere's law of magnetic force.

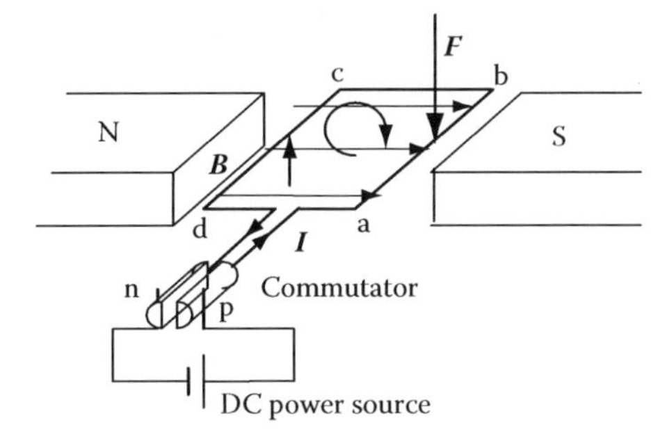

FIGURE 35.2 DC motor with commutator.

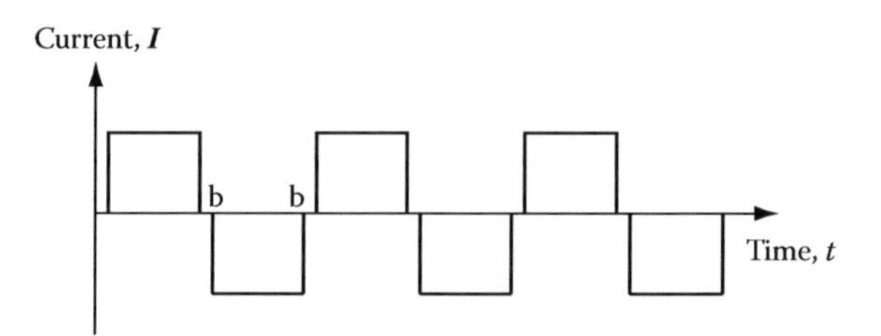

FIGURE 35.3 Current flow in motor coil.

through this position, minimizing the short circuit time. Alternatively, more than one set of coils can be used and the number of commutator bars increased accordingly to eliminate the short circuit problem, which most motors nowadays implement.

With the commutator in place, the effective voltage across the motor coil, which is also equivalent to the current flowing across the motor coil with fixed resistance, will have an alternating effect as shown in Figure 35.3 for the segment "ab" of the motor coil. The small sections denoted by "b" shows the time when short circuit occurred.

35.2.1.2 DC Brushless Motors

DC Brushless motors do not make use of commutators to regulate the power or current flowing into the coils [3]. Instead, brushless motors regulate the current flow through semiconductor switches with position feedback of the motor shaft. An inherent characteristic of brushless motors is the requirement of sensors to sense the absolute angular position of the motor shaft. The construction of the brushless motor also differs from that of brushed motors. In brushed motors, permanent magnets are used as stators while the motor coils are attached to the rotor. In brushless motors, the stator normally consists of the motor coil windings while a permanent magnet takes over as the rotor.

Brushless motors have the following advantages over brushed motors [3]:

1. Brushless motors have higher maximum speed and greater capacity because of the construction of the motor. Rotor shaft friction is reduced because there is no need for commutator brushes. Furthermore, heat dissipation of the stator coils is more effective through the stationary motor housing or casing.
2. They work in less favorable surroundings. The size of brushless motors are comparatively smaller than brushed motors, making them suitable for compact applications, such as in robotic arms. Brushless motors provide much better torque to weight ratios, that is, for the same weight of the motors, brushless motors provide much higher torque. The absence of a commutator also eliminates the possibilities of sparks or arcing, making it safe for applications in locations with flammable gases such as in the petrochemical industries.
3. Since no commutator is used, brushless motors practically requires no maintenance. They also produce less noise compared to brushed motors.

Brushless motors are also similar in construction to induction motors or AC motors, but the characteristics of the two are different. Brushless motors have the characteristics of a DC brushed motor, namely having a linear speed–torque relationship when the power fed into the motor system is fixed or constant.

35.2.2 Brushed Motor Drives

In general, there are two popular types of driver circuits being used for the brushed DC motors: linear drive and the widely used pulse width modulation (PWM) drive [4].

35.2.2.1 Linear Drive

As the name implies, a linear drive provides a continuous flow of current to the motor that is linearly proportional to the torque or speed required of the motor. Linear drive controls the amplitude of the

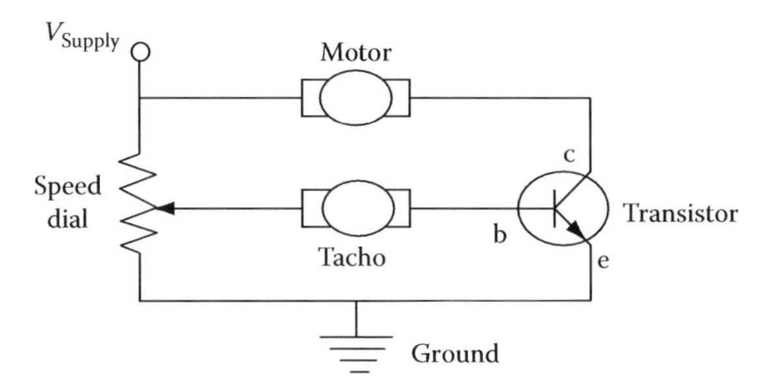

FIGURE 35.4 Simple linear drive for DC motor.

voltage or current being sent to the motor. The simplest way of implementing a linear drive in a closed loop system is with a single power transistor, as shown in the sample in Figure 35.4.

The current to the base of the transistor is controlled, thus controlling the current flowing through the motor. The input to the base of the transistor may come from the Digital to Analog (D/A) card or chip output of a microcontroller (or computer), or as shown in Figure 35.4, from the output of the tachometer in a speed control system. Linear drives are ideal for high performance speed control systems. By adjusting the speed dial, the speed of the motor can be controlled by limiting the current to the base of the transistor, which is generated from the tachogenerator feedback of the system. The example above applies only for unidirectional motor speed. To perform reversible speed control, an additional reversible switch can be added to the circuit to switch the connections between the motor and the driver. The power rating of the transistor is appropriately chosen to match the current capacities or requirements of the motor.

To interface the linear drive to a digital controller such as a personal computer, the controller must have some form of analog input and output facilities such as the analog to digital and digital to analog converter (ADDA) card for the PC. The tacho feedback is read through the A/D converter into the PC for processing before an output through the D/A converter is generated. The D/A output is then fed to the base pin of the transistor to trigger the motor motion.

Motor manufacturers also provide standard drivers for the different makes of motors. The input signals required are typically the bipolar analog command voltage and a transistor–transistor logic (TTL) enable signal. The linear driver could also provide current feedback, that is, the tachogenerator feedback signals at the controller's disposal. Known motor manufacturers who provide motors with drivers are Maxon Motors from Switzerland and Baldor Motion Products of the United States.

35.2.2.2 PWM Drive

The *pulsed width modulated* (PWM) drive is more popular among the motor driver circuits used. A PWM driver is driven by a single DC source, with an internal amplifier switching the power on and off at a fixed frequency and at a variable "firing angle," so that the average power (in terms of voltage and current) is controlled. The frequency of the output from the PWM system is determined by an external resistor–capacitor (RC) circuit as expressed in Equation 35.2 and the "firing angle" is controlled by an analog input typically between the range of 0 and 3 V. A simple PWM driver circuit is shown in Figure 35.5 using the Motorola/TI TL.494 PWM Driver chip.

$$f_{osc} = 1.1 \div (R_T \times C_T) \tag{35.2}$$

The generation of output pulse from the PWM is shown in the timing diagram in Figure 35.6, with three different stages of input signal, two being the minimum and the maximum duty cycle and another at varying dead-time control signal.

The frequency of the PWM is normally set to the nonaudible range to keep the system quiet. A higher operating frequency would also ensure a more even distribution of power and smoother motion. In the example shown in Figure 35.5, the frequency used is about 1.1 kHz.

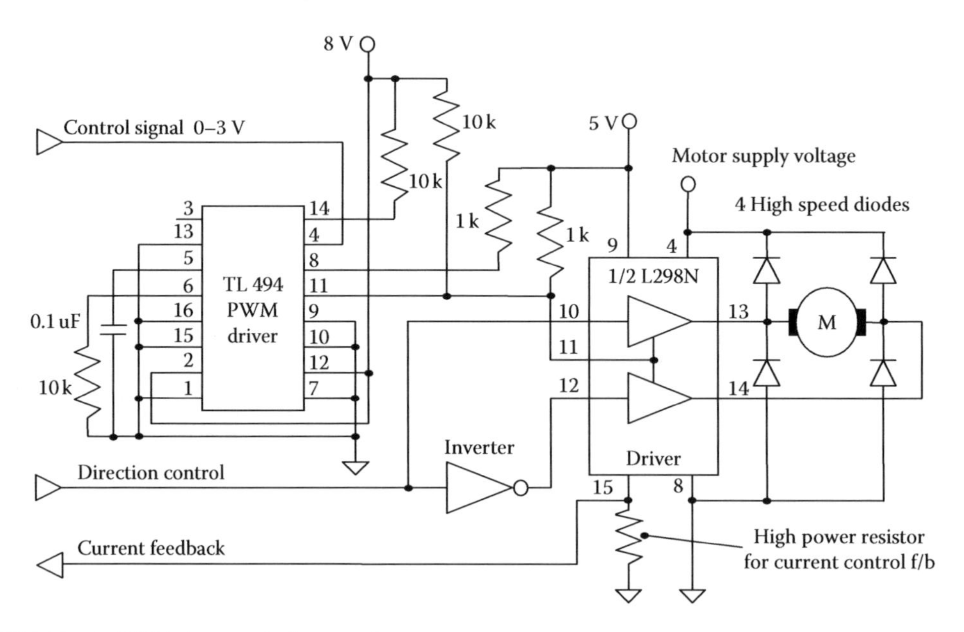

FIGURE 35.5 PWM circuit with driver and motor.

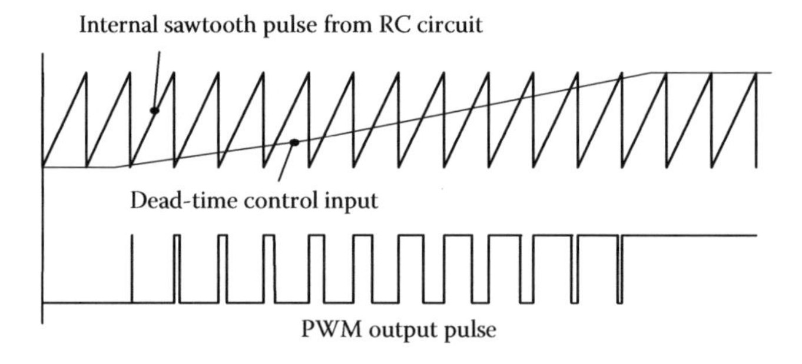

FIGURE 35.6 PWM output signal.

From Figure 35.6, the general operation of the PWM can be seen clearly. As the dead-time control input varies between 0 and 3 V, in the case of the TL494 being used here, the output PWM pulses changes from 0% duty cycle to 100% duty cycle linearly. The output PWM pulses generated then go through a dual full-bridge driver, the SGS L298N, which will translate the TTL pulse train into a DC pulse train with an amplitude of the motor rated voltage.

With the implementing of this PWM driver, the controller need only to provide the analog voltage for the dead-time control signal and a TTL direction signal. From the circuit shown, the controller could also read in the analog signal of the current drawn by the motor during the operation. Other forms of feedback can also be added, such as tachometer or digital encoder, which will only affect the controller but not the PWM driver directly. Motor manufacturers also produce PWM drivers for the different makes of motors. Typical input signals are the same as for the described PWM driver circuit.

35.2.3 Brushless DC Motor Drive

The driver for a DC brushless motor normally consists of a sine wave generator, a PWM driver array followed by a transistor array and feedback devices attached to the brushless motor. The circuit has to read the position and (or) velocity feedback from the brushless motor and interpret the signals in order

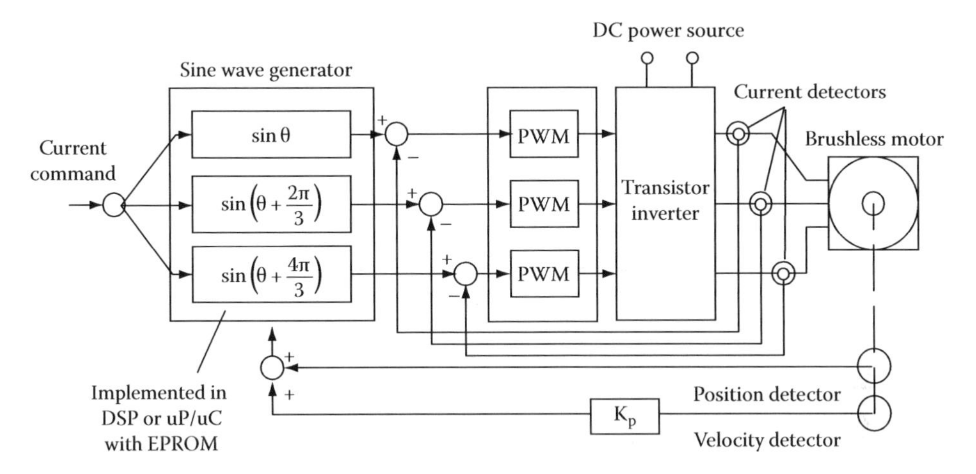

FIGURE 35.7 Control block diagram for brushless DC motor.

to produce the output to the few (normally 3 or 4) stator coils. The overall control block diagram for a three-phase brushless motor driver is shown in Figure 35.7.

Unlike an AC motor or induction motor, where the input to the actuator is an AC supply, the brushless motor uses a DC power source to drive the motor. A sine wave generator generates the alternating (or sinusoidal) waveform for each phase of the motor after processing the feedback information on the position of the rotor shaft. In view of the processing power needed, either a microcontroller with an erasable programmable read-only memory (EPROM) or a digital signal processor (DSP) is used for this task. In order to maintain the DC motor characteristics, the sinusoidal waveform for each phase coil is fed through a PWM before sending the power into the motor.

There are also some other chips available in the market to drive brushless DC motors, such as the SGS L6230, which is a bi-directional three-phase brushless DC motor driver [5], which can be implemented on its own or integrated with a microprocessor/controller.

35.2.4 DC Motor Performance and Characteristics

As the magnetic flux of the permanent magnets for typical DC permanent magnet motors, be it brushless or brushed, does not vary much, the speed–torque and the current–torque relationships are linear over the extended range of operation.

The maximum speed shown in Figure 35.8 is the no-load speed of the motor at a fixed voltage. The maximum torque on the *y*-axis where the speed is zero is the stall torque. The current flowing through the motor coil is proportional to the torque or load applied onto the motor rotor. This simplifies the control system significantly in many applications.

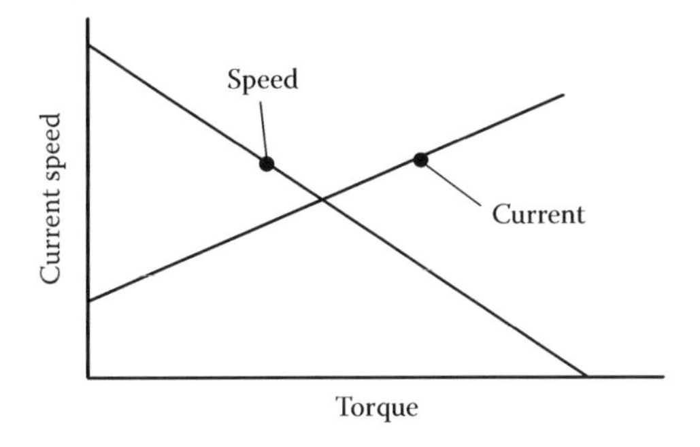

FIGURE 35.8 Typical DC motor characteristics.

35.3 Stepper Motors

35.3.1 Principle of Operation

Stepper motors are slightly different from DC motors in construction and application, although both categories make use of electromagnetic flux in coils for their operations. In a stepper motor, the construction is similar to that of a DC brushless motor; the coils are wound around mild steel cores with teeth-like surfaces facing the rotor, attached to the motor housing as stator. The rotor is made up of either a mild steel multi-toothed core, sometimes with a permanent magnet at the center of the core. The total number of teeth for the stator is normally one pair more than the number of teeth at the rotor. A simple example can be seen in Figure 35.9.

From Figure 35.9, when coil number I is energized, the magnetic flux generated pulls the rotor teeth pair A toward teeth I, as shown in Figure 35.9a. This is the first step in the stepper motor sequence. With coil I still activated, coil II is now energized as well, creating another flow of magnetic flux, which finds the shortest route through to teeth B of the rotor. As the path for the magnetic flux of coil II is much larger, teeth B is pulled toward teeth II, causing the rotor to rotate in the counter clockwise direction until both A-I and B-II strike a balance, moving the rotor 15° to the position as shown in Figure 35.9b. After this position is reached, coil I is de-energized so that only the magnetic path B-II is present, moving the rotor another 15° to the position shown in Figure 35.9c, so that B is aligned to teeth II. This is a full step of the stepper motor and the steps repeat itself with the third (or even fourth) coil.

Stepper motors are normally used in open loop control systems. As shown in the example above, the stepper motor rotates at fixed steps for every change in the driver signal, that is, from step (a) to step (b) in Figure 35.9. Unless the load applied to the stepper motor exceeds the rating of the motor, which causes it to slip, the performance of the motor should be very accurate. The typical step size of stepper motors is in the range of 0.9° and 1.8°. By using a half step driving, the step size could be halved, making the resolution of the stepper motor even finer.

There are a few types of stepper motors available in the market. A bipolar permanent magnet stepper motor has only four wires coming out from the motor, the AB pair and the CD pair as shown in Figure 35.10a. The unipolar permanent magnet stepper motor, on the other hand has at least six wires coming out of the motor, of which two of them are ground wires normally in black and white. Coils A and B share one ground line and coils C and D share the other ground line as shown in Figure 35.10b. The color scheme for the wires varies widely for different manufacturers. One should always consult the manufacturer or the distributor for the specifications before connecting the wires to the driver circuit. The driver circuit provides the current output to each of the coils of the stepper motor.

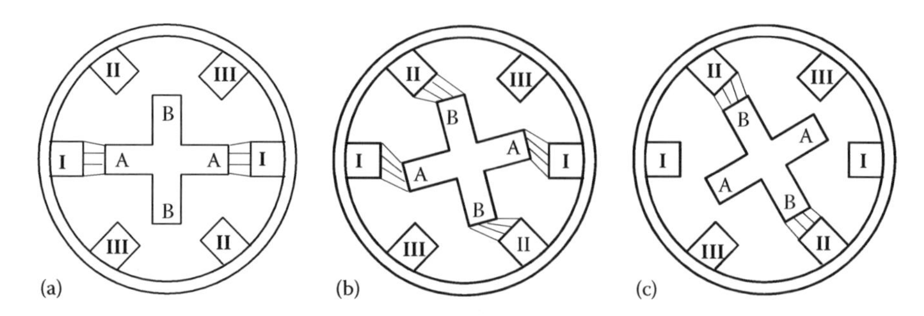

FIGURE 35.9 Simple stepper motor construction and operations.

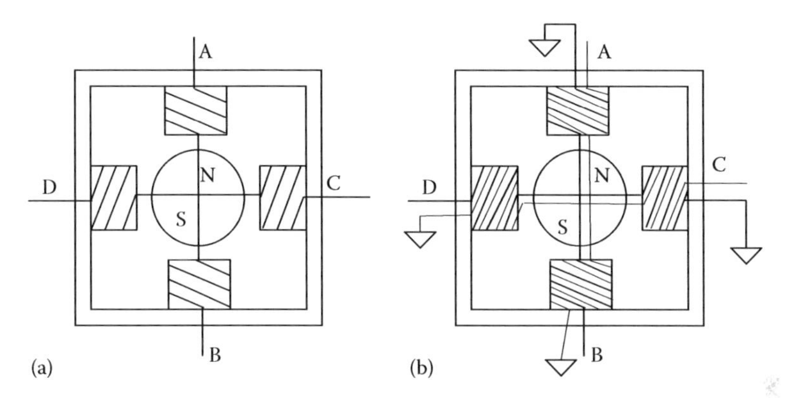

FIGURE 35.10 Bipolar and unipolar permanent magnet stepper motors: (a) bipolar PM stepper and (b) unipolar PM stepper.

35.3.2 Stepper Motor Driver

The stepper motor in general requires four logical signals to activate the motion. The four signals are identified as A, B, C, and D channels. The signals sent to each of the channels are logical high (for ON state at the motor rated voltage) and logical low (for OFF state at 0 V) signals. These signals have to be sent in the particular sequence as described in the previous section.

One popular decoder chip being used for the stepper motor driver is the SGS L297, coupled with the dual full-bridge driver chip L298N, also from SGS. The standard connection for the stepper motor driver circuit is shown in Figure 35.11 [5].

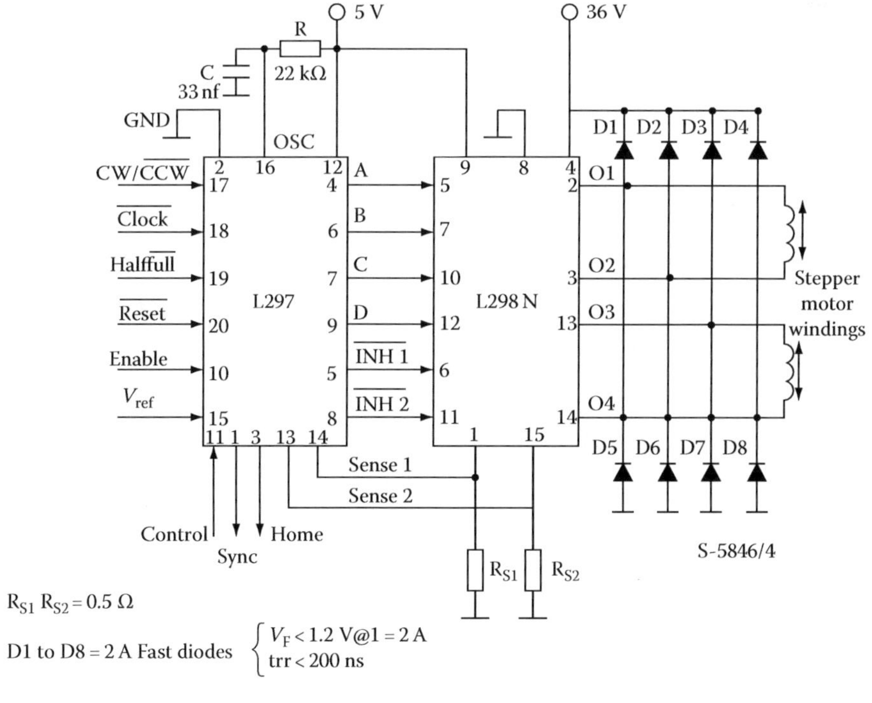

FIGURE 35.11 Two-phase bipolar stepper motor control circuit. (From SGS, *SGS Motion Control Application Manual*, SGS, Englewood Cliffs, NJ, January 1987. With permission.)

The SGS L297 is a stepper control that generates the required sequence of current flow through the coils. The sequence output of the L297 needs to be amplified at levels required by the motor, and this is accomplished by the L298 N driver chip.

The implementation of this driver circuit allows easy interface with other digital controllers such as a personal computer or a microcontroller. The driver requires essentially only three signals from the controller:

1. The CLOCK signal, which progresses the output stage by one step on the rising edge of the pulse received. This clock signal input to the driver is a pulse train whose frequency is linearly related to the angular velocity of the motor. For each pulse received, the motor rotates one step.
2. The clockwise/counterclockwise (CW/CCW) signal, which decides whether to change the state of the output in the forward or backward direction, that is, A-B-C-D for forward and D-C-B-A backwards.
3. The ENABLE signal, which allows the driver chip to accept input signals and produce output pulses. A TTL high (+5 V) signal means get to work! (This signal could be permanently tied to high if it is a continuous system which does not require much setting up.)

Another optional signal is the HALF/FULL signal, which determines whether to carry out full step driving or half step driving, depending on the resolution requirement of the system.

For interfacing with a personal computer, a digital input/output (I/O) card such as the 8255 card needs to be installed to provide the digital signals to the driver chip. For position control, one of the 8255 ports can be used to provide the clock pulses at fixed interrupts with a counter keeping track on how much the motor has moved. Another 8255 port can be used separately to provide fixed signals to both the ENABLE and direction lines.

For speed control, the 8253 counters of the 8255 card can be configured to give a pulse train at fixed frequency until the configuration is changed.* By using it, the host PC can be free to perform other tasks. In addition, one 8255 port is required to provide the direction and to enable signals to the driver.

The same signals can also be used for stepper driver circuits supplied by the motor manufacturers and the prices are quite reasonable. One added feature might be the option for two different pulse trains for the clockwise and counter clockwise directions.

35.3.3 Stepper Motor Performance and Characteristics

Stepper motors are excellent open-loop control actuators. They are widely used in PC industries such as printers and floppy disk drives. On the heavy industry side, most XY tables are driven by powerful stepper motors. They provide excellent interface capabilities with digital controllers such as microcontrollers and PCs.

On the other hand, there are some limitations to the implementation of stepper motors in certain systems. Stepper motors use fixed holding torque at every step and excessive load causes slippage in the motor positioning. In the same manner, the acceleration rate and the maximum velocity of the motor shaft is also limited by the load of the system.

One other characteristic is the resonance frequency of the stepper motor. For every stepper motor, there is a particular range of frequency where the motor does not function properly with full load. A typical speed–torque curve for the stepper motor is shown in Figure 35.12.

* Digital I/O cards typically contain at least one 8255 or equivalent chip that provides digital I/O (one 8255 provides 3 8-bit ports or 24 I/O lines), and at least one 8253 (or equivalent) timer (one 8253 chip provides 3 16 bit counters that can be used as square wave generators and others.). One card with two 8255's and one 8253 costs less than US $40.00.

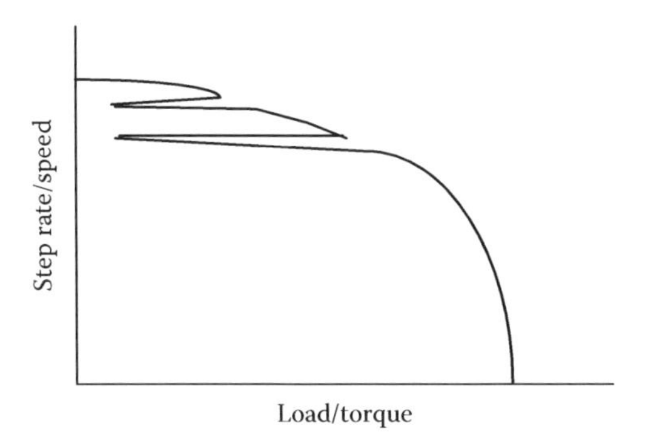

FIGURE 35.12 Stepper motor characteristics. (From Kenjo, T., *Stepping Motors and Their Microprocessor Controls*, Oxford Science Publications, Oxford, U.K., 1984. With permission.)

35.4 RC Servo Motors

An radio-controlled (RC) Servo Motor is a compact device integrating its own electronics, position feedback, and motor. This motor is normally referred to as a "servo." Its output shaft can be positioned to specific angular positions by sending a special PWM signal to the servo motor. As long as the special PWM signal exists on the signal line, the servo will maintain the angular position of the output shaft. As the special signal changes, the angular position of the shaft changes accordingly. In practice, servos are used in radio controlled devices such as RC racing cars (where the servo is used to actuate the steering wheels and set different speeds) and RC airplanes (to control the position of surfaces like the elevators and rudders.) They are also used in RC puppets, and of course, robots.

Servos are extremely useful in robotics. The motors are relatively small, with built in control circuitry, and are extremely powerful for their size. A standard servo such as the Futaba S-148 can produce 42 oz-inches of torque, which is quite strong for its size. It also draws power proportional to the mechanical load. A lightly loaded servo, therefore, does not consume much energy. A dismantled servo motor is shown in the Figure 35.13, with the control circuitry, the motor, the potentiometer (which serves as an angular position sensor), a set of gears, and the case. All servos are controlled through a three-wire cable. The wire for power (+3.5 to 6V) is normally Red, the ground wire in generally Black, and the White (sometimes Yellow) wire is for the control signal (i.e., the PWM signal).

The servo motor's main component is a miniature DC brushed motor, with its output shaft connected to a series of three to four stages of reduction gears. The gear train converts the high-speed-low-torque output of the miniature motor to a high-torque-low-speed output at the output shaft of

FIGURE 35.13 Dismantled servo motor.

the servo. It takes slightly more than a second for the output shaft of a servo to travel 180°. The output shaft of the servo is connected to a potentiometer which provides feedback of the output angle to the servo's control circuit. The control circuit performs a close loop position control of the output shaft of the servo according to the special PWM signal. The power for the servo motor is drawn directly from the power lines of the supply cable, namely the Red and Black wires. Servo motors are normally designed to operate within the range of 180°, but some manufacturers have models which boast of extended ranges of up to 270°. A normal servo is mechanically incapable of turning any farther due to a mechanical stop on the main output gear, although some servos have been modified to provide continuous rotations.

The control signal which has to be supplied to the White or Yellow wire is called a Pulse Coded Modulation, which is a variation of Pulse Width Modulation. A typical control circuit of the servo expects to see a pulse with a minimum high period of 1.25 ms every 20 ms (0.02 s). The frequency of the signal is 1/20 ms or 50 Hz. The length or period of the pulse determines the position of the motor output shaft. For a general servo motor with a 180° range, a 1.5 ms pulse makes the servo motor turn to the 90° position, normally referred to as the neutral position. If the pulse is lower than 1.5 ms, then the servo motor turns the shaft toward 0° position. If the period of the pulse is set higher than 1.5 ms, the shaft turns toward 180° position, as shown in Figure 35.14.

A simple controller for a servo motor can be patched using two 555 timer chips, as shown in Figure 35.15. The first 555 timer on the left sets the trigger for the second 555 timer at 50 Hz, which means a period of 20 ms for the servo motor. The second 555 timer will generate the Pulse Coded Modulation signal for the servo motor at the desired duty cycle, depending on the 5 kΩ variable resistor setting. In the example shown, the duty cycle will range from 0.5 to 2.5 ms. This signal can also be generated easily using micro-controllers for more complex applications.

Figure 35.16 shows a simple humanoid robot constructed out of nine low-cost servo motors, designed by a group of robotics hobbyists from Singapore. The robot can be controlled from a PC through an RS232 cable connected to a PIC based servo controller. Details of the robot can be found at http://www.mini-Robotics.com.

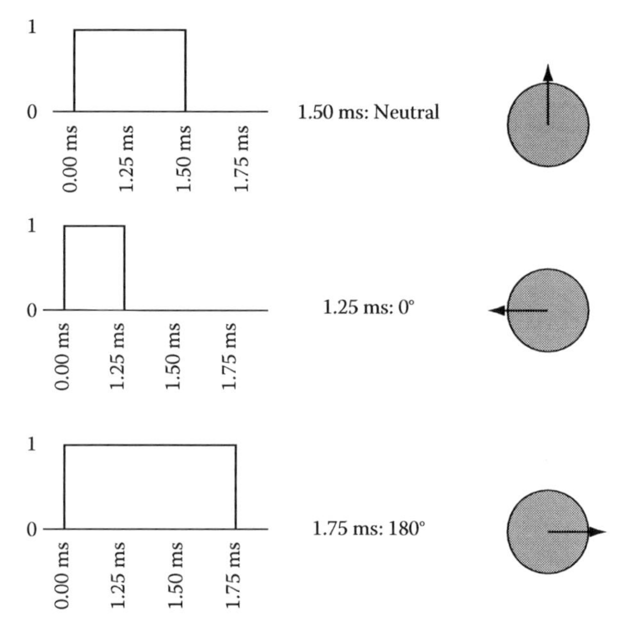

FIGURE 35.14 Pulse coded modulation and its matching servo output.

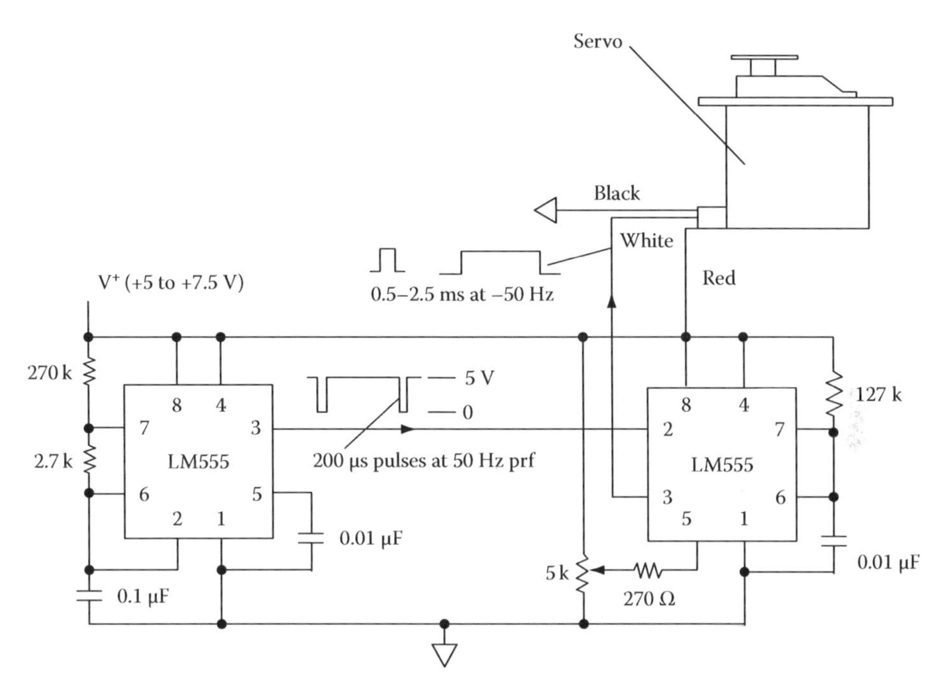

FIGURE 35.15 Simple servo control circuit.

FIGURE 35.16 Simple humanoid robot with nine servo motors.

35.5 Intelligent Motors

The intelligent motor or smart motor is a new range of motors which emerged in the twenty-first century due to greater demands in many applications, especially in edutainment robotics. One major market for these intelligent motors are in humanoid robots for hobbyists. Since the turn of the century, many researchers have poured in a lot of effort in the study of humanoids. A number of international robot competitions focusing on humanoid robots also emerged to boost the interest, such as the two the main players, the RoboCup from Japan and FIRA of South Korea. For medium size humanoid robots, standing around 500 mm off the ground, intelligent/smart motors play an important role in the design and performance of the robot.

35.5.1 Principles of Operation

Intelligent motors look just like RC servo motors but maybe slightly bulkier. The construction of an intelligent motor is similar to that of the servo motor, as shown in Figure 35.17 of a disassembled AI Motor produced by a Korean company, Megarobotics Ltd [7]. The middle portion comprises mainly the miniature DC motor on the right and the potentiometer on the left. The top part mainly comprises the three to five stages of gears to achieve high working torque. The bottom enclosure houses the microcontroller which performs the motor control and communication with the supervisory controller. The addition of communications functions to the RC servo motor allows many of them to be daisy chained to a supervisory computer.

At the front and back of the motor, there are two 4-pin (specific to the AI Motors) connectors to facilitate TTL level serial communication (RS-232) between the motor and its supervisory controller, which can be either a PC or a microcontroller. Two of the four wires also provide the power required by the motor during operation. The advantage of this configuration is that motors can be connected within a single chain of wires, as each motor is identified with a unique ID, provided the power supply is sufficient to drive all motors within the chain. In other words, a single control cable can be used to control a series of motors within a mechanical system instead of one set of wires for each actuator. For the AI Motor, the ID range as well as the maximum number of devices within a single channel is 31 (ID 0–30), while the Dynamixel motors by Robotis can take up to 254 devices (with ID 0 × 00 to 0 × FD) within a single channel.

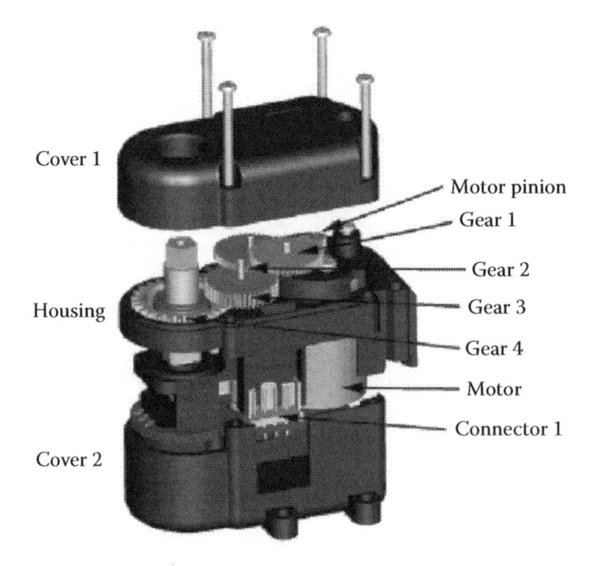

FIGURE 35.17 Disassembled AI motor showing internal construction. (Modified from Megarobotics Ltd., AI Motor-1001 Manual Version 1.02, Megarobotics Ltd., Seoul, Korea.)

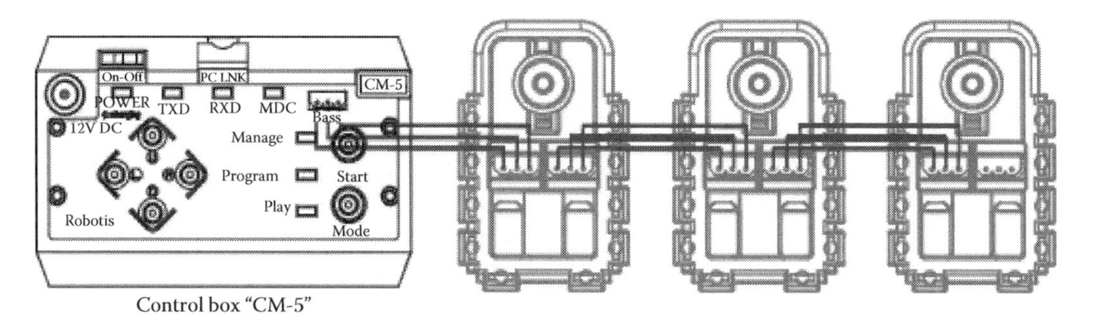

FIGURE 35.18 Connection between controller and Dynamixel smart motors. (From Robotis, *Dynamixel AX-12 User's Manual*, Robotis, Seoul, Korea, 2006. With permission.)

The disadvantage of such a configuration is that there would be inherent delays in the propagation of motor commands, thus the overall system cannot perform synchronous actuation of all actuators. An example of this link is shown in Figure 35.18, extracted from the user manual of another series of intelligent motors, the Dynamixel series of smart motors from Robotis [8], which is also based in South Korea.

Apart from reducing the messy wiring, intelligent or smart motors also allow the user/designer to reconfigure individual motors for specific functions, as the performance of each motor is controlled by individual microcontrollers. In general, the motors can be configured to operate in position control mode just like a standard servo motor or in 360° continuous rotation mode just like any DC motor with speed specified. In addition, each motor within the series may be interrogated to provide feedback information to the supervisory controller through the communication channel. For the Dynamixel motors by Robotis, a three-wire cable is used, two for power and one for data. For this purpose, the half duplex universal asynchronous receiver/transmitter (UART) is employed in both the controller and motor to enable bidirectional communication on the single data line, as shown in Figure 35.19.

The communication protocol for smart motors is relatively simple, beginning with one or two bytes header and ending off with a checksum byte. The data section generally comprises the device ID and specific motor commands with their accompanying parameters, if any. No conflict will arise so long as

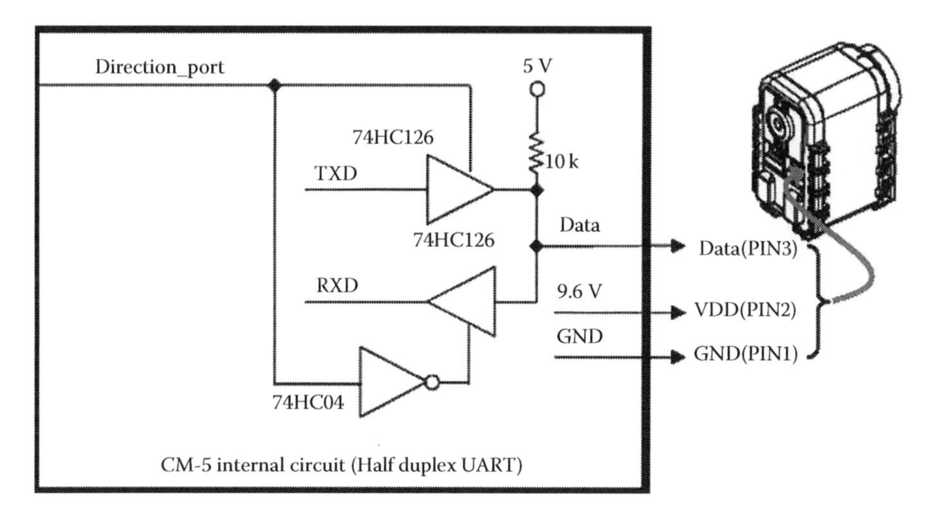

FIGURE 35.19 Communication between the CM-5 controller and smart motor. (From Robotis, *Dynamixel AX-12 User's Manual*, Robotis, Seoul, Korea, 2006. With permission.)

FIGURE 35.20 Some robots built with the Dynamixel smart motors. (From Robotis, *Dynamixel AX-12 User's Manual*, Robotis, Seoul, Korea, 2006. With permission.)

the user/designer ensures no conflicting ID number has been issued to more than one device. The communication bandwidth can also be pushed to the limit depending on the number of devices used within the system and the environment the machine is being used in. Theoretically, the highest achievable communication rate is 1 Mbps.

With the mentioned characteristics, robotic systems as shown in Figure 35.20 can be designed and controlled without much difficulties. These robots would be able to perform complicated movements within their stable static movements, but dynamic movements may require a breakthrough in actuator technologies and power.

35.6 Electroactive Polymer Actuators

Electro active polymers (EAPs) are materials that produce a displacement (strain) when subjected to a voltage [9]. Electrodes are typically placed on opposite surfaces of the EAP. The electrodes need to deform following the deformation of EAPs; therefore, electrodes made of graphite powder or conducting pastes (e.g., MG Chemicals Carbon Conductive Grease 846) are typically chosen. Figure 35.21 shows an EAP with grease applied on the top and the bottom surfaces of the EAP. The intersecting electrodes on the EAP surface form the active region for actuation. When relatively high voltage (0–6 kV) is applied to the electrodes, the active region compresses the EAP in thickness and expands it in the planar direction. The area of active region expanded with increasing voltage applied (Figure 35.22).

EAPs can therefore be used as actuators that can potentially produce large forces. The way the EAP is used determines the actuation configuration. One possibility is the development of robotic fingers using "roll actuators." The rolls are basically made up of a compressed spring around which is wrapped the electro active material (3M VHB4910), as shown in Figure 35.23.

The active part of the roll EAP film is wrapped around a compressed spring. This film has exhibited up to 380% strain in area expansion at 5–6 kV when it is highly prestrained. The spring is initially in

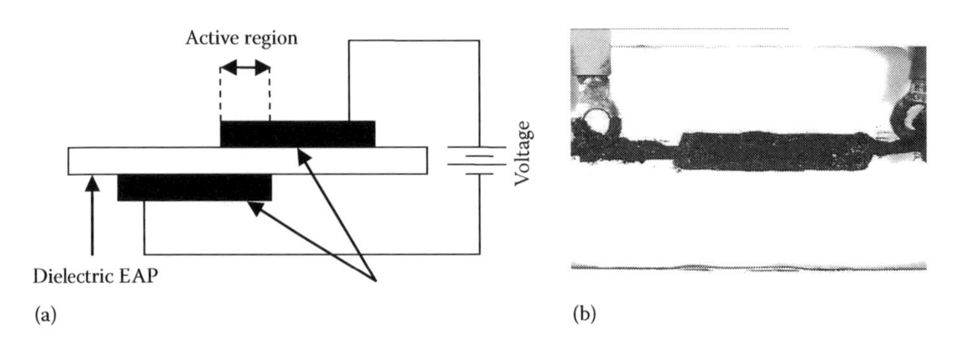

FIGURE 35.21 (a) Side view of active region on EAP. (b) Top view of active region on EAP.

FIGURE 35.22 Length and width expansion of active region.

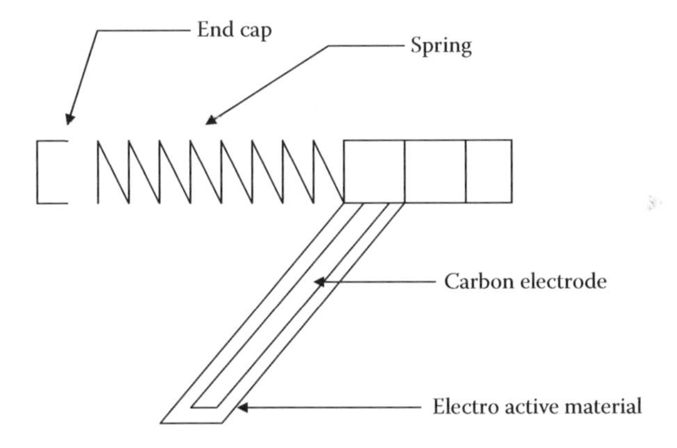

FIGURE 35.23 Construction of a robotic finger using EAP.

a compressed state and the wrapping is made. The spring tends to extend thus keeping the tape in a strained state. Similarly, the tape tends to keep the spring in a compressed state. When a voltage is applied across the tape, it expands thus releasing the spring that can produce a force. It is to be noted that by patterning electrodes on two or more spans of the roll and independently actuating them a bending motion is produced. Such roll actuators can be used as robotic fingers as shown in Figure 35.24.

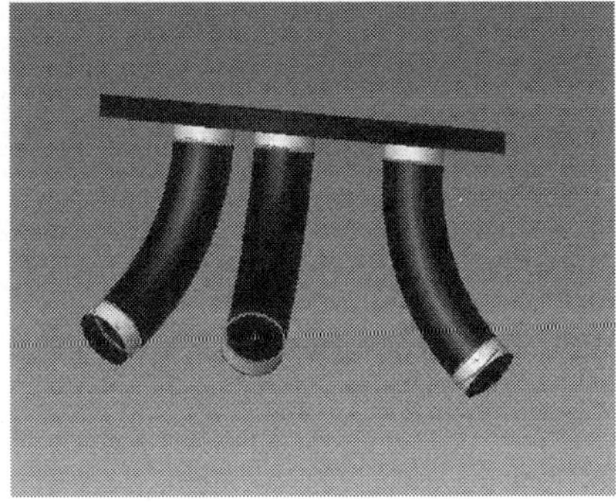

FIGURE 35.24 Robotic fingers based on EAP roll actuators.

35.7 Transmissions

As mentioned in the introduction, transmissions amplify the load capabilities of actuators and/or allow the actuators to be remotely located to the devices being actuated. The following are the transmissions used in machinery: gear systems, belts and pulleys, and sprockets and chains, with gear systems and cables and pulleys being more commonly used in robotic manipulators.

35.7.1 Belts, Cables, and Chains

Flat belts are common for large crowned pulleys to allow remote location of actuators but are not appropriate for power transmission. They are not applicable for robots. V belts have V-shape cross sections that provide wedging action in the pulley, thus giving better power transmitting capabilities. However, belts can slip and this can be an advantage in robots because of the resulting overload protection [2]. Timing belts have ridges that run in grooves on the pulleys ensuring no slip occurs. Cables are similar to belts but have the advantage of being used in more than one plane. Some belts have studs on them to allow for timing and/or no-slip operations. Figure 35.25 shows a possible arrangement of a belt or cable to drive a robot link.

Although rarely used for power transmission, straps consisting of flat metal cables can be used as tendons to hold joints together as shown in Figure 35.26.

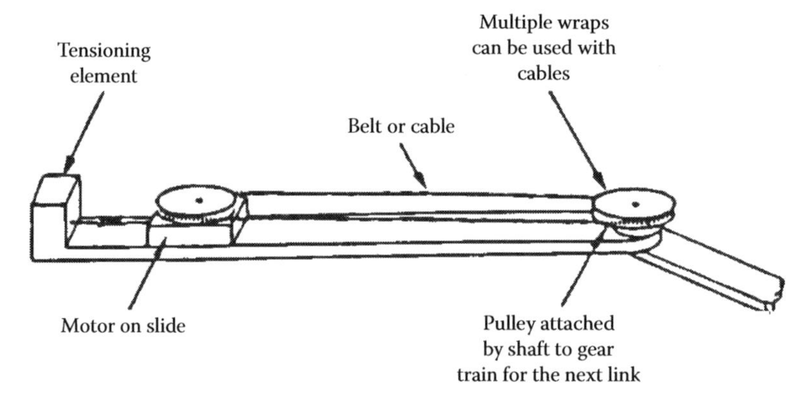

FIGURE 35.25 Belt or cable drive for a robot link. (From Andeen, G.B. (ed.), *Robot Design Handbook*, McGraw-Hill, New York, 1988. With permission.)

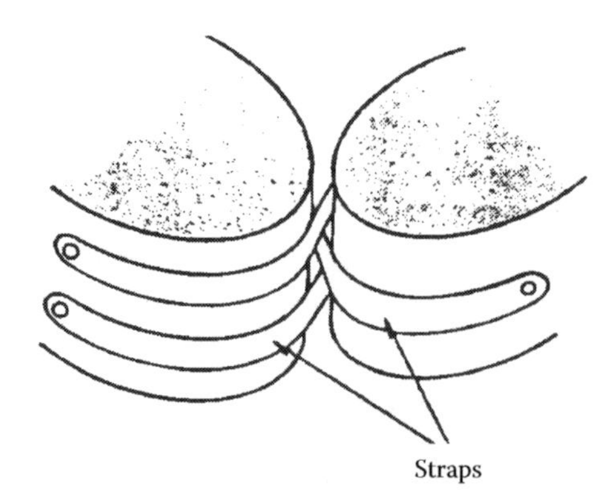

FIGURE 35.26 Flat metal cables used as tendons holding the joints together. (From Andeen, G.B. (ed.), *Robot Design Handbook*, McGraw-Hill, New York, 1988. With permission.)

35.7.2 Gear Trains

Gear trains can be classified according to the relative geometric configurations of the gear shafts.

35.7.2.1 Parallel Shafts

The spur, helical, and herringbone gear arrangements have the gear shafts in parallel as shown in Figure 35.27. The spur gear is most common with teeth parallel to the centerline. The mating teeth makes instantaneous full width contact when meshing, thus resulting in possible vibrations and shakings. The helical gear, on the other hand, allows gradual contact of the engaging teeth and results in smoother load transmission. The helical gear, however, provides an axial force as a side effect, thus requiring thrust bearings to withstand this load. The herringbone gear have left- and right-hand helices cut into a single gear thus balancing the axial load in the gear itself.

Another possible configuration for parallel gear shafts is the internal gear arrangement as shown in Figure 35.28. The advantage of this design is the compactness which makes it attractive for robots. This arrangement is commonly used for reduction gear and planetary gear arrangements.

35.7.2.2 Nonparallel Shafts

For gear shafts that are intersecting (i.e., shafts lie in a plane), bevel gears are used as shown in Figure 35.29. The teeth of bevel gears can be spiral too to acquire similar advantages of helical gears over spur gears.

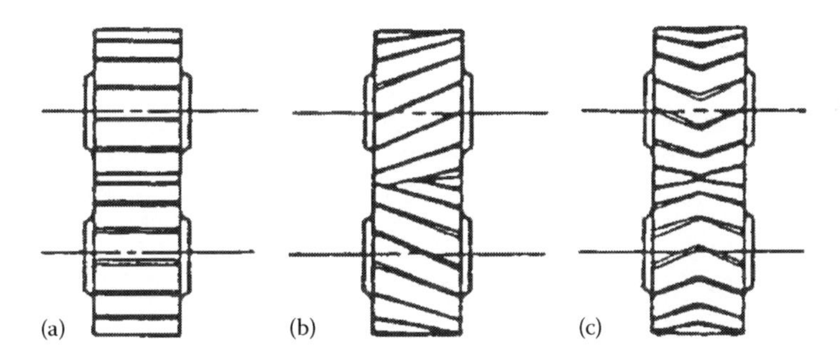

FIGURE 35.27 Spur (a), helical (b), and herringbone (c) gear couplings. (From Andeen, G.B. (ed.), *Robot Design Handbook*, McGraw-Hill, New York, 1988. With permission.)

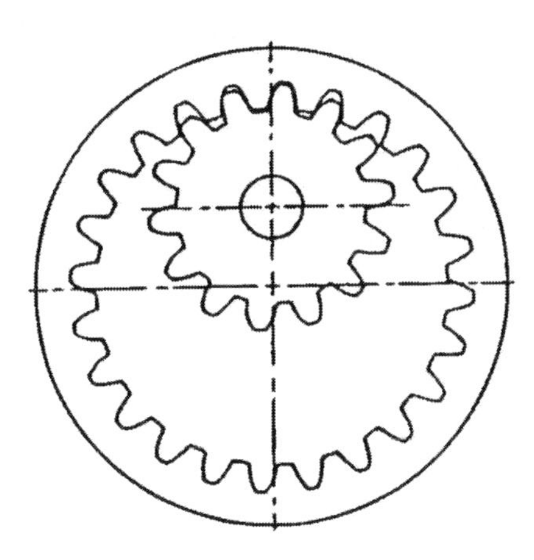

FIGURE 35.28 Internal gear arrangement. (From Andeen, G.B. (ed.), *Robot Design Handbook*, McGraw-Hill, New York, 1988. With permission.)

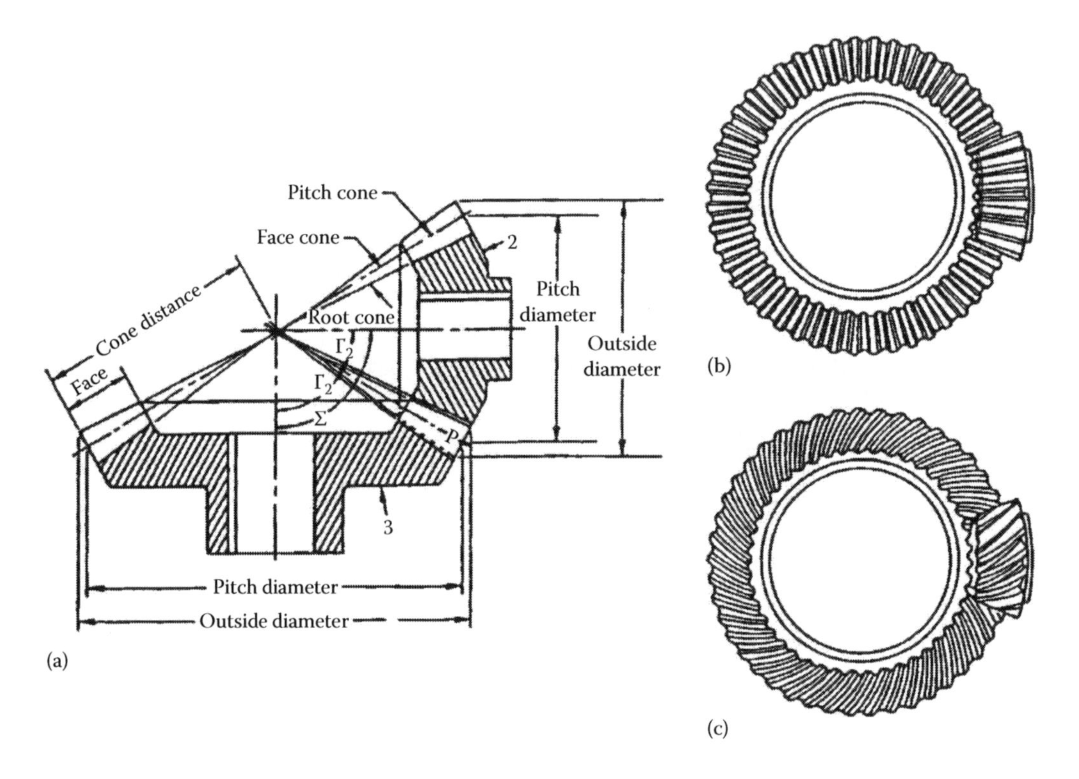

FIGURE 35.29 Bevel gear arrangement (a) geometry, (b) straight, and (c) spiral bevel gears. (From Andeen, G.B. (ed.), *Robot Design Handbook*, McGraw-Hill, New York, 1988. With permission.)

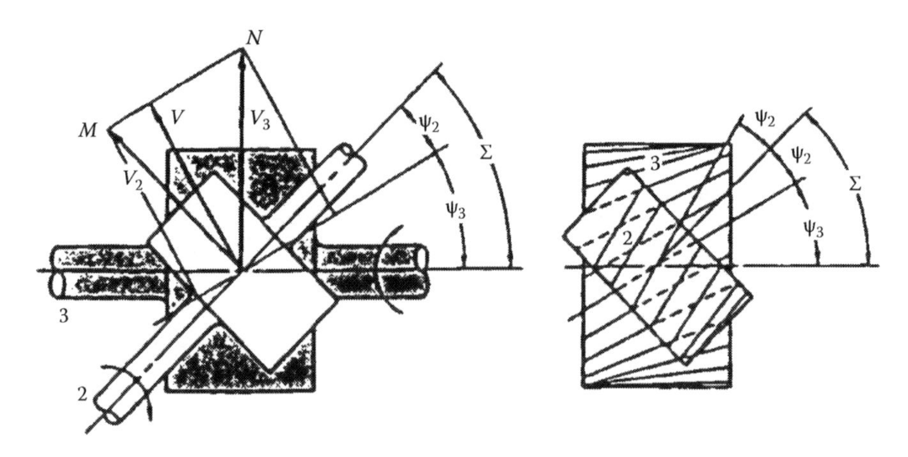

FIGURE 35.30 Crossed-axis helical gears. (From Andeen, G.B. (ed.), *Robot Design Handbook*, McGraw-Hill, New York, 1988. With permission.)

Crossed-axis helical gears (Figure 35.30) are used when the gear shafts are nonparallel and nonintersecting. They are used solely for power transmissions in light load applications.

When the nonintersecting shafts are at right angles to each other, worm gears are used as shown in Figure 35.31, where the spur gear drives a worm gear.

Hypoid gears are similar to spiral bevel gears except the gear shafts are nonintersecting as shown in Figure 35.32.

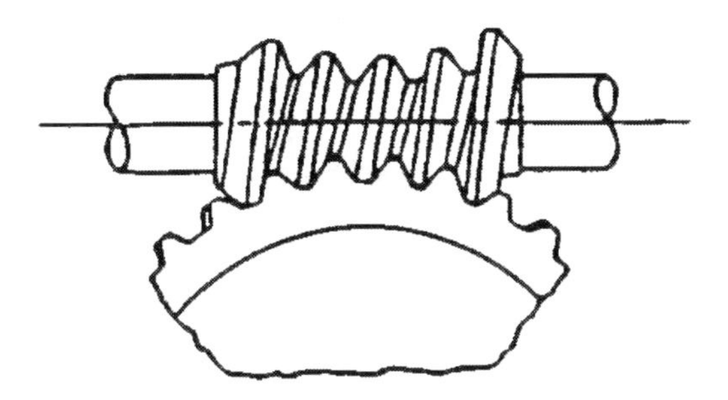

FIGURE 35.31 Worm gears used in non-intersecting shafts at right angles. (From Andeen, G.B. (ed.), *Robot Design Handbook*, McGraw-Hill, New York, 1988. With permission.)

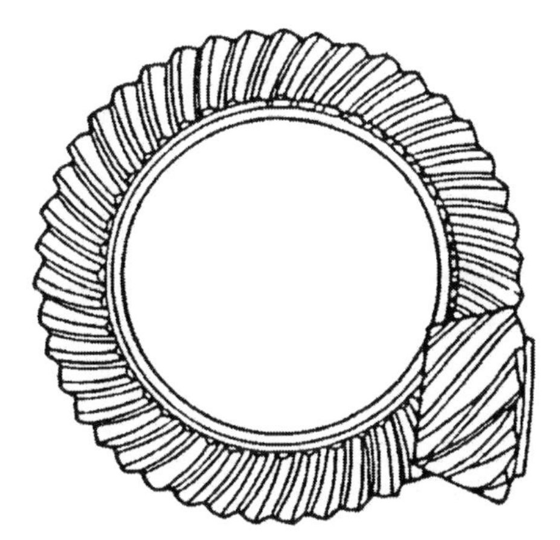

FIGURE 35.32 Hypoid gears. (From Andeen, G.B. (ed.), *Robot Design Handbook*, McGraw-Hill, New York, 1988. With permission.)

35.7.2.3 Others

The rack and pinion type gearing arrangement is useful for transforming rotational motion to a linear one. Teeth on a gear mates with teeth on a plane or rack as shown in Figure 35.33.

Another interesting transmission mechanism is the ball screw. Ball screws employ ball bearings incorporated between the mating surfaces resulting in friction-free transmissions.

There are other means of effecting motion direction transformations such as the one shown in Figure 35.34, wherein a linear motion from a hydraulic cylinder or ball screw causes rotation of the upper link about the hinged joint.

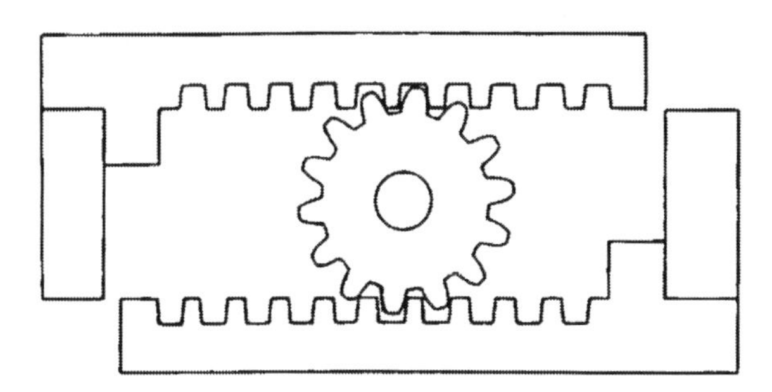

FIGURE 35.33 Gear and rack arrangement. (From Andeen, G.B. (ed.), *Robot Design Handbook*, McGraw-Hill, New York, 1988. With permission.)

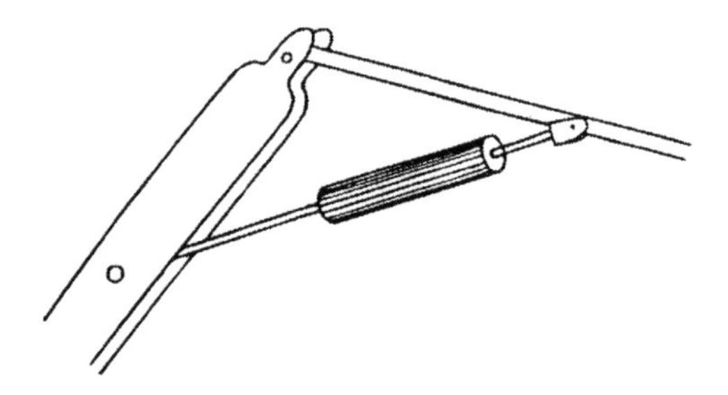

FIGURE 35.34 Linear to rotational motion using a hydraulic cylinder. (From Andeen, G.B. (ed.), *Robot Design Handbook*, McGraw-Hill, New York, 1988. With permission.)

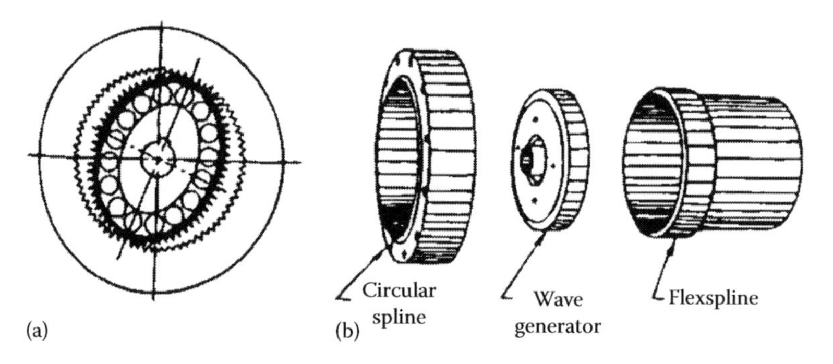

FIGURE 35.35 Harmonic drive assembled system (a) and its components (b). (From Andeen, G.B. (ed.), *Robot Design Handbook*, McGraw-Hill, New York, 1988. With permission.)

For large load capacities in a compact space, harmonic drives are very useful in robots [1,2]. Harmonic drives provide very high speed reductions (e.g., off-the-shelf units available for 64:1 to 320:1 [2]) and can therefore provide very high torque. The principle of operation of a harmonic drives is shown in Figure 35.35. The harmonic drive consists of a rigid spline (circular spline) with internal teeth that forms the housing of the drive, an annular spline ("flexspline") that has external teeth that meshes with the internal teeth of the housing, and a wave generator that is driven by the actuator. The external radius of the annular spline is slightly smaller than the internal radius of the circular spline (housing). The annular spline is also called a flexspline because it undergoes elastic deformation during the meshing process. The wave generator is the input side of the drive and is coupled to the actuator; the flexspline or the circular spline are the output side of the drive and is coupled to the load. When the wave generator is rotated by an actuator, the flexspline is deformed thus engaging teeth in diametrically opposite points coincident with the major axis of the elliptical wave generator and disengaging points at the minor axis. The wave generator has a double-ended cam in place that achieves the rotation of the major and minor axes of the ellipse. If the circular spline is fixed and the load is coupled to the flexspline, rotation of the wave generator would produce a reverse rotation of the flexspline; and if the rigid spline has 202 teeth and flexspline 200 teeth, then a single revolution of the wave generator will precess the flexspline backward two teeth resulting in a velocity ratio of 100:1. If the flexspline is fixed and the load is coupled to the circular spline, then the input and output shafts rotate in the same direction. The relationship between input and output angular velocities is [2]:

$$\frac{\omega_{in}}{\omega_{out}} = \frac{N_o}{N_c - N_f}$$

where

N_o is the number of teeth on output member (flexspline or circular spline)
N_c is the number of teeth on circular spline
N_f is the number of teeth on flexspline

35.7.2.4 Backlash

An inherent problem in gear transmissions is *gear backlash* which arises from the clearance between the tooth and tooth space. Backlash is the amount by which the width of a tooth space is wider than the thickness of the engaging tooth as references to the pitch circles [2]. Gear backlash has limited the positioning accuracy achievable by robots. With gear backlash, fine motion may not be achieved because small motions of the input side may not cause motion on the output side of the drive. Furthermore, when the teeth finally mesh, extreme stress due to impact occur and the inertia of the mechanisms may cause positioning errors. There are many ways to alleviate this problem. One way is to have gear preloading, that is, to use an auxiliary torque motor so that the gear tends to favor one side of the gear tooth; or one can have spring loaded gears for light duty applications.

References

1. C.W. de Silva, *Control Sensors and Actuators*, Prentice Hall, Upper Saddle River, NJ, 1989.
2. G.B. Andeen (ed.), *Robot Design Handbook*, McGraw-Hill, New York, 1988.
3. Y. Dote and S. Kinoshita, *Brushless Servomotors—Fundamentals and Applications*, Oxford Science Publications, Oxford, U.K., 1990.
4. Electro-Craft Corp., *DC Motors, Speed Controls, Servo Systems—An Engineering Handbook*, Electro-Craft Corp., Hopkins, MN, July 1980.
5. SGS, *SGS Motion Control Application Manual*, SGS, Englewood Cliffs, NJ, January 1987.
6. T. Kenjo, *Stepping Motors and Their Microprocessor Controls*, Oxford Science Publications, Oxford, U.K., 1984.
7. Megarobotics Ltd., *AI Motor-1001 Manual Version 1.02*, Megarobotics Ltd., Seoul, Korea, 2004.
8. Robotis, *Dynamixel AX-12 User's Manual*, Robotis, Seoul, Korea, 2006.
9. Y. Bar-Cohen, *Electroactive Polymer (EAP) Actuators as Artificial Muscles: Reality, Potential, and Challenges*, SPIE Press, Bellingham, WA, 2004.

36

Robot Qualities

Raymond Jarvis
Monash University

The fascination many people have for robots is often associated with their supposedly human-like qualities; a comparison with human capabilities is a good starting point in describing what robots are generally about.

There are essentially two kinds of robotic devices: those that are somewhat like the human arm (robotic manipulator) and are generally used in industry at a fixed location, and those that are mobile and are required to navigate through their defined workplaces efficiently and without collision.

In industrial settings, robotic manipulators are currently used for sorting, welding, de-burring, painting, and manufacture; mobile robots, often referred to as automated guided vehicles (AGVs), are used for the carriage of goods and components between workstations in a factory or warehouse.

Mechanically speaking, robots can be powerful, tireless, are capable of repeating actions accurately, and can survive hostile environments that are either uncomfortable or hazardous for humans. Many robot manufacturers are now able to design and produce units that are able to work for thousands of hours at high precision between maintenance sessions.

Robotic manipulators have a number of configurations according to how their links operate and the types of space they sweep out. Polar robotic manipulators have a rotation joint at the waist and an arm that can extend or contract in length and tilt up and down in a vertical plane. Cartesian robots are able to move their end-effectors (hands) independently in three orthogonal directions. Anthropomorphic robots (human arm like) have three main revolute joints, the waist, shoulder and elbow, each with one degree of freedom. Scara robotic manipulators are very rigid in one plane to suit particular operations. In addition to being able to place the end of their arms in various positions in their work spaces, robotic manipulators also have degrees of freedom related to the orientation of the end-effector attached to the wrist. For example, a standard six degree of freedom anthropomorphic revolute jointed robotic manipulator would in addition to the waist, shoulder, and elbow rotations, also have an end-effector attached to a wrist with another three degrees of freedom associated with yaw, pitch, and twist rotations. Also, the opening and closing of the end-effector would add other degrees of freedom (at least one).

The energy required to operate robotic manipulators is usually one of three kinds: electrical, hydraulic, or pneumatic. Since the efficiency, accuracy, and speed of a robotic manipulator is directly related to how a rigid set of links can be moved without carrying unnecessary loads, considerable thought at the design stage goes into how to keep the mass associated with the energy sources away from moving links, particularly the ones furthest away from the waist, and at the same time not introduce sources of imprecision such as gear-trains, wires, and long-drive shafts. Hydraulic energy systems are often preferred over electric ones where very powerful but fast manipulators are required. However, through the use of direct drive motors (no gears), very high precision and fast electric energy driven robotic manipulators have become very attractive since such a system is highly controllable and has a few moving parts. For smaller robotic manipulators, the use of shape memory alloy for tendon-like actuation is being carefully researched. For very small robotic manipulator devices (micro-robots) processes similar to those used for very-large-scale integration (VLSI) circuitry in the microelectronics industry are being used to manufacture electrical motors and linear actuators. Friction wear is one of the difficulties to be overcome for such devices and magnetic bearing technology promises one solution to this problem.

The high accuracy and repeatability (not the same) enjoyed by many robotic manipulators derives from the precision of manufacture of their components, the linear and shaft encoder resolution of their joint/link position monitoring elements and the quality of the control system used. Many very sophisticated control theoretic systems are now being applied in robotics; these are capable of better control than the classical proportional, integrative and differential (PID) controllers that have been used for decades. The complexity of the dynamic distribution of load forces among the joints, links, and pay load while a robotic manipulator is moving along an optimal trajectory at high speed represents a considerable challenge to the control system.

AGVs have their distinct attributes while sharing some (particularly those relating to motion control) with robot manipulators. An important distinction derives from the different ways in which, on the one hand, the configuration of the links of a robotic manipulator and, on the other, the position and orientation of an AGV, are determined. Encoders on the joints of a robotic manipulator are used to monitor its configuration at all times; converting outputs from encoders to end-effector positions and orientations is just a matter of calculation. Unless an AGV is just a cart moving on toothed wheels along a rack, its exact position and orientation cannot be determined using the type of encoders used for robotic manipulators. The use of encoders on plain wheels (odometry) does not achieve the same result since noncircularity under load, slippage, and ground irregularity all contribute to accumulative error, so that eventually the position and orientation of the robotic vehicle becomes uncertain. Constraining the vehicle to a track specified by marks on the ground or signal wires buried beneath it does limit localization (position and orientation) error, but at the cost of severe path limitation and/or considerable site preparation. Realizing a free-ranging AGV requires the use of localization instruments such as beacons or natural landmarks.

Outside uses in traditional factory environments, mobile robots can be used to operate in rugged terrain, on and under water, in mines, space, and eventually perhaps, inside the human body. Various means of locomotion include wheels, tracks, legs, hover, jets, and propellers. Power sources include combustion engine, electrical, and combinations of these, and hydraulics.

The survivability of robotic manipulators and vehicles in hostile environments is a matter concerning the variety of materials that can be used in their construction, their rigidity and power, the choice of energy sources available, and the many ways by which the manipulator can be sealed from dangerous atmospheres and liquids they may be immersed in. For example, a water-sealed manipulator with a noncorrosive outer layer could work at a great depth in the ocean with electrical energy being provided from the surface, as part of a seabed exploration mission on a tele-operated vehicle, which could stay submerged for long periods. The use of robotic manipulators in handling radioactive materials or very high temperature components as well in the vacuum of deep space are other examples of hostile environment application.

An interesting complimentary activity to that of surviving hostile environments is the use of robots in clean rooms where human-centered contaminants cannot be tolerated, the robot being sterilized by methods not applicable to humans before being introduced to that environment. Also of interest is the potential use of micro-robots for exploratory and surgical tasks within the human body, where small size and sterilization are critical so as to minimize the evasiveness of such operations.

Thus, overall, from the mechanical perspective, robots have certain advantages over humans. These include speed, power, endurance, accuracy, smallness, and capability of operating in hostile environments and not contaminating others. Cost effectiveness must be gauged with respect to specific activities, but, in general, tasks that are repetitive and require precision and speed are those where the cost effectiveness of using robots is often very great.

As working environments become more complex, less structured, and less controllable, more intelligence in humans as well as machines is needed to cope with the situation. For the type of intelligence associated with vast amounts of arithmetic calculations, computers are clearly superior to humans. There are other types of reasoning processes that, through developments in artificial intelligence, have been shown to be computationally feasible to a degree that does now or will soon be superior to human capabilities in the same area. However, in the area of perception, including speech recognition, vision and tactile, force and olfactory sensing computational feasibility and capability have been slow in development. It is in these domains where nearly every human is expert by comparison.

Perception governs the way in which we deal with our environment in the sense of manipulating objects in it, or navigating through it. To the extent that perceptive mechanisms must be particularly active during learning about an initially unknown environment and continuously so when coping with an ever-changing one, so must robots acquire the mechanical/electronic analogs of these human skills to accommodate to the types of unstructured environments that humans have mastered and then to also apply these skills in domain humans find hostile. This is the realm of intelligent robotics, which is the sensory and reasoning skill enhancement of robotic manipulators and mobile robots to enable them to carry out useful tasks in unstructured and variable environments.

Intelligent robotics can be defined as the melding of perception, reasoning, and action into an integrated mechanical/electronic device that is capable of operating on and/or in its environment to carry out useful tasks. The sensory aspects of this domain are described in sections to follow.

The reasoning aspects have to a large extent been covered in earlier sections. The most specific and generally low-level reasoning essential to both robotic manipulators and mobile robotics is path planning. Numerous examples of how to generate optimal trajectories abound in the literature. This problem is usually defined in terms of determining an optimal collision-free path from a given start point to a given goal point and a number of variations on this requirement to include search and exploratory modes. Optimality can be in terms of minimal length, minimal time, minimal energy, or combinations of these, perhaps with a modifying reliability factor based on safety tolerances. In initially unknown and/or time-varying environments, sensor data acquisition and analysis are essential to allow piecewise optimal trajectories to be determined.

Communication between the subsystems of an intelligent robotic system and between separate systems, perhaps required to operate in cooperation, is an important requirement. A communication system, whether wired or wireless, ought to provide adequate bandwidth, high reliability, and consistency of protocol to maximize the efficiency and reliability of the whole system. For mobile robots, the appropriate distribution of sensors and computational support between on-board use and attached to stationary control stations is critically dependent on the communications requirements in relation to the flow of command and sensory data in both directions. When multiple robots need to operate cooperatively, the means by which they communicate with each other, directly or via a central command station, is crucial to the success of such operations. In well-established industrial environments, the need for using internationally standardized communication protocols throughout a manufacturing enterprise is now almost universally recognized as essential.

Thus, from the sensory/intelligence perspective, humans are still far in the lead by comparison with robots, but the inspiration that has come from the study of intelligent and perceptive biological systems has led to a growing research effort in improving artificial perception and intelligence for use on robots to enable them to efficiently carry out a range of tasks that until now have been the exclusive domain of humans. The recent rapid improvement in affordable sensory and computational technology promises to deliver a variety of intelligent robot systems for use in unstructured environments in the manufacturing, mining, health care, catering, and service industries and in the home to the market place within the next decade.

In recent times, three application domains have gained new significance, essentially through global circumstances of historic significance and the maturity of the robotics technologies in being able to support these domains with sufficient reliability and cost effectiveness. The first is the military/security domain for mine detection, patrolling functions, aggressive action, and terrorist detection and neutralization. In many of these applications, sensor-supported teleportation is sufficient, but varying degrees of autonomous operation is helpful. The second is in the assistive technology domain to support an aging population; linked to this is the use of robotics for physical prosthesis. The third domain is that of space exploration, exemplified by recent Martian exploration successes. Human exploration is very expensive and risky, and thus robotic substitutes are much in demand.

Cutting across a large scope of robotic application domains is the recent fascination with humanoid robots that are able to take their place alongside humans and are better able to interact with humans due to similarities of form, function, and communications abilities. In addition to the physical challenges of bipedal gait locomotion, dexterous manipulation ability, and compact and efficient energy requirements, the need for human/machine communication via speech, gesture, and tactile cues has become a major research endeavor.

Also, a relatively recent development is the idea of building multiple (from a few to swarms) mobile robots that cooperate to carry out tasks no one robot can achieve well on its own and to provide the scope of applying graceful degradation strategies that allow for substitution of duties to cover tasks relinquished by a failed unit. Such developments are often inspired by the behaviors of social insects and their ability to communicate reliably with very low bandwidth channels, sometimes using small modifications of the environment to relay messages (e.g., odor trails).

Somewhat related to the multiple robot scenario is the notion of building modular, self-configurable robots that can alter their structure to meet the requirements of their mission. The provision of energy, communication, and intelligence among the modules is a complex challenge.

Overall, the variety and application scope of robots have expanded over recent years, and the ever-improving fields of sensor technology and affordable computational resources, together with developments of both material and battery technologies, have provided the basic tools to permit both old and new ideas to be implemented in ways not possible until now.

37

Robot Vision

Raymond Jarvis
Monash University

Seeing, in sighted humans, accounts for in excess of 85% of all sensory input data, the analysis of which supports our physical interactions with our everyday environment and our safe navigation through it as well as our appreciation of its richness as a quality-of-life component.

Not surprisingly, therefore, robot vision in its full realization would expand the scope of applicability of robotic systems, both manipulators and mobile vehicles, from the relative structure of classical industrial environments to the unstructuredness of natural and human-centered environments to embrace applications in mining, space exploration, undersea, service industries, healthcare, warehousing, mineral exploration, search and rescue, fire fighting, security, agriculture, forestry, recreation, catering, and domestic operations. Considerable research and development effort has gone into the theoretical as well as practical issues of robot vision (as a component of artificial intelligence). This effort continues today with the powerful support of rapidly improving and affordable sensor and computational technologies. This support promises an explosion of commercially viable vision-based robotic systems within the next decade or two.

Computer vision is usually distinguished from computer image processing by the explicit consideration of the three-dimensionality of the world of objects in the former case, even when the extraction of the depth dimension may be by the analysis of one or more two-dimensional images. However, this distinction is not always acknowledged and many image-processing procedures are claimed to be exercises in computer vision analysis.

Digital images are essentially 2D arrays of values, each representing intensity. The dimensionality of the array specifies the spatial resolution of the image, while the number of data bits per cell specifies the intensity resolution. Each cell is referred to as a pixel (picture element). Three arrays of cells, one representing the red intensity, the second green, and the third blue, can collectively be regarded as a digital color image. Other color coding schemes specifying hue, saturation, and intensity in various ways are also used. A typical digital color image may consist of three 512×512 element arrays with 8 bits per cell per array representing red, green, and blue values, respectively. The whole data structure is represented as $512 \times 512 \times 3$ bytes (8 bits) of information. In recent times, multi-megapixel digital cameras have become easily affordable so the resolution trend is upward. A color pixel is the set of three values at a particular spatial location in the arrays.

Panoramic camera systems using "fish-eye" lenses or parabolic (and variants) mirrors have recently become popular as means of providing surrounding visual information for surveillance and robotic navigation support. Unwrapping such imagery is a relatively easy task often carried out before processing the images in conventional ways.

When range data (whether extracted from images or laser range finders) are used or derived with respect to a single viewpoint, it is usually also represented in a rectangular array of cells, each being referred to as a rangel (range element). This type of representation is not strictly 3D, as it represents distances from the viewpoint to surfaces visible from that location. The surfaces of objects not visible because of self-obscuration or obstruction by other objects cannot be ranged to from that viewpoint and are thus not represented in the range array. Since, for a typical 3D scene, approximately half the surface points cannot be ranged to from a single specific viewpoint, this representation is often referred to as 2 1/2D. A full 3D representation might take the form of a 3D volumetric array with solid or surface occupancy indicated. Such a representation is composed of volume elements (voxels). Simple occupancy can be supplemented by values indicating color components or surface normal directionality or other property values of interest.

A full 3D volumetric representation including surface properties such as color, normal directionality, and perhaps other attributes can take up a considerable number of memory locations in a computer. For example, a $512 \times 512 \times 512$ volume array with eight 8 bit property elements per cell would occupy 2^{30} bits ($\approx$1 billion bits of memory).

Robot vision can be interpreted as that part of computer vision that is ultimately intended to be used to guide the actions of robots. The quality of a robot vision system is judged in terms of it providing timely, reliable, and accurate information, extracted from visual or range data (or both), which can be used to direct correct action of a robot in carrying out a specified task and the extent to which its structure permits its use over a wider range of possible tasks. A superior system should also be affordable, physically robust, safe, and energy efficient.

While vision strictly refers to 3D perception derived from images, the analysis of directly acquired range data is usually accepted as part of this domain. The use of laser rangefinders has proliferated in recent times with the appearance of new products in the marketplace accelerating. Relatively high-speed and dense ranging systems are now readily available. While the price of such devices is high in comparison with digital video cameras, volume production to meet demand is likely to bring the cost down.

In the context of intelligent robotics (a cooperative interplay of perception, reasoning, and action [Figure 37.1]), robot vision can be thought of as a powerful perceptual component. Since a robotic task can only be completed by also including reasoning (perhaps path planning and event sequencing) and action (perhaps following a planned path, avoiding obstacles, gripping an object, etc.), it is best to see robot vision as part of a whole system. In this way, its quality can be judged in relation to its contribution to the whole task.

Two robotic task scenarios are used to illustrate the functions of useful vision systems. While these examples are somewhat simplistic and traditional, they place vision unambiguously within a robot perception/reasoning/action loop to which humans can easily relate, thus improving the tutorial value of this exercise.

The first example is a robot/camera "hand–eye" coordination task in which a robotic manipulator is used to pick up identified objects from a table or shelf and carrying out some manipulation task on them, whether sorting, rearranging or retrieving them to assist a disabled person. As an example (Figure 37.2), a color camera and a rangefinder collect color intensity and range data in registered 2D arrays, which collectively result in a 2 1/2D representation of the 3D scene of the pile of objects to be dealt with. A database describing the types of objects likely to be encountered in terms of structure, geometry, and

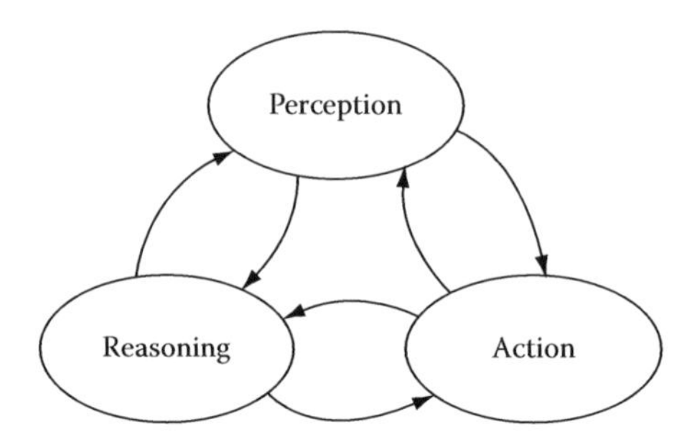

FIGURE 37.1 Intelligent robotics.

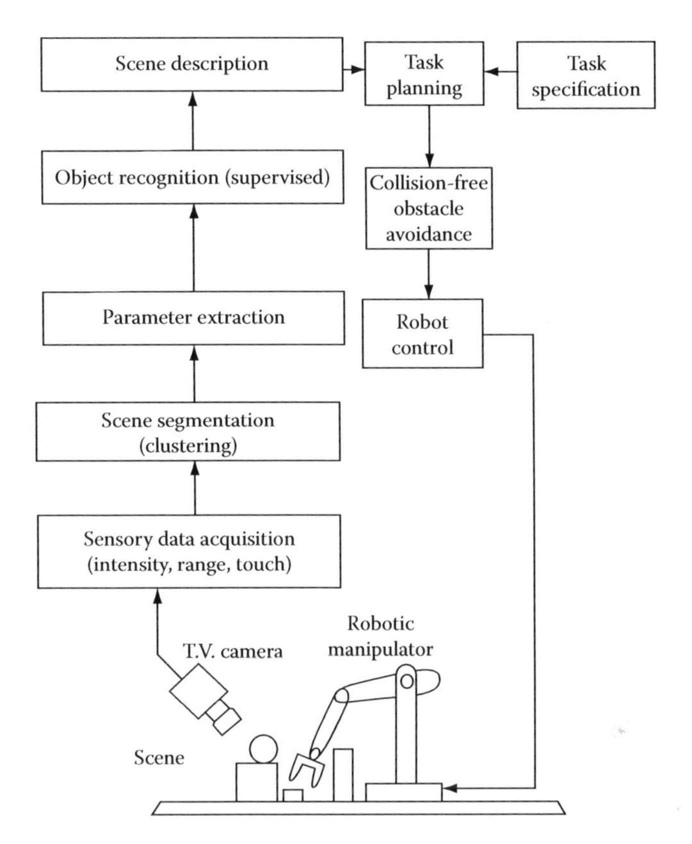

FIGURE 37.2 Hand–eye coordination cycle.

surface appearance is available. It is also possible that specific objects be identified once their detailed characteristics have been extracted. This latter task can be achieved using, among other methodologies, a scale-invariant feature transform (SIFT) algorithm that extracts a number of very localized features for marking purposes, where each set of such features essentially captures "a signature" of a specific object. While the SIFT approach can match specific details between two images taken from different viewpoints or between a viewed object image and the image of the same (or identical) object collected in a database of many images of various objects, thus being able to identify it, generic objects (e.g., cups or apples) can not be directly recognized in this way because shape information has not been processed.

Since, for reasonably high-resolution color intensity/range, large amounts of data have to be processed, the usual first task of the vision system is to segment the scene into its major components without referring to the database of object descriptions. Some type of "semantic-free" clustering procedure can be used to group pixels/ranges together in clumps, which are homogeneous or continuous in some way. These initial clumps can be further refined, perhaps with some reference to broad distinguishing characteristics derived from the database. Once individual objects are more or less isolated, a search for a match with an object represented in the database can take place for each object in the scene. This is essentially a 2 1/2D to 3D matching process and can be very difficult to carry out reliably, particularly for free-form objects. Some objects may be oriented, so that they cannot be unambiguously matched, even if they are not severely obscured by other objects. Severely obscured objects may not be able to be identified using only one viewpoint or before the obscuring objects are removed.

Even objects not yet identified have to be taken account of in a volumetric occupancy sense when determining possible grip sites with respect to identified objects and planning approach, grip, and withdraw trajectory components for the manipulator, all of which should preferably be collision-free and optimal in some sense (e.g., minimum path length, time, or energy).

It may be that, after one or more objects are removed, analysis of what remains becomes easier and a new cycle of analysis commenced.

An interesting variation of the above cycle is to discover grip sites for object removal and to carry out trajectory planning and actuation prior to identifying the object. In this way, the object might be more easily identified once it is physically separated from the pile, as not only has segmentation been carried out physically but also the robot can present the object in various orientations to the vision system so that unambiguous identification can be carried out before completing the sorting action. If all objects must be handled (as is the case for this example), this is a good approach; however, in a situation where only a particular object in a pile must be retrieved, this post-handling recognition approach will be cumbersome.

The type of vision system described above (both the pre- and post-handling recognition modes) is referred to as a model-reference system since an explicit database describing the various objects to be encountered is used for identifying scene objects by using matching search procedures. Most industrial robot vision systems are of this type since model parameters can usually be expected to be available from computer-aided design (CAD) data. Of course, there are many variations to this robotic eye–hand coordination theme. For example, the sensory instrumentation may be fixed in the environment or mounted on the robot, whether on its body or hand. Being able to deliberately change the viewpoint of sensory data gathering can greatly assist in removing ambiguity of some interpretation.

The second example concerns an autonomous mobile robot (Figure 37.3.) equipped with beacon localization and range sensors, which is required to navigate through an initially unknown and slowly time-varying obstacle-strewn space from a specified start position to a specified goal position via a quasi-minimal length collision-free path. Since a complete map of the environment is not initially available and any partial map generated incrementally using sensor data is subject to change, absolute global

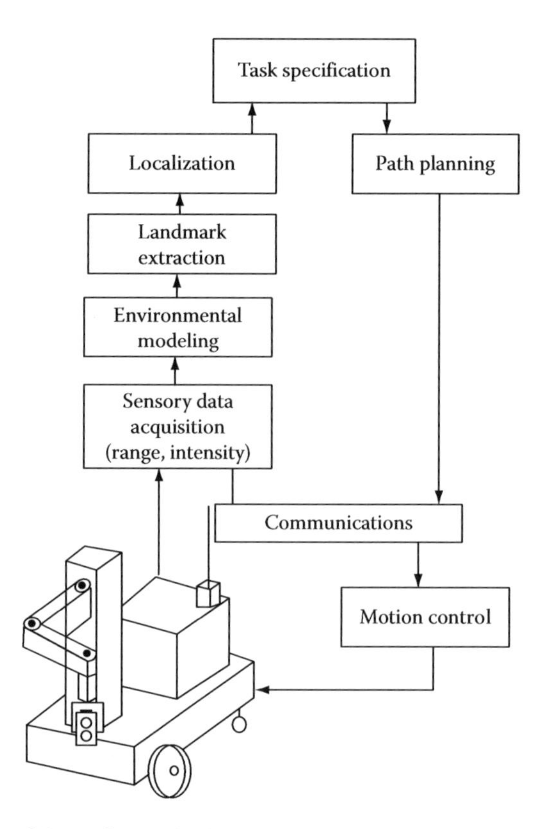

FIGURE 37.3 Sensory data driven robot navigation.

optimality of the path actually taken cannot be guaranteed; all that can be hoped is that what is known at any particular moment is taken fully into consideration at each stage of forward path planning.

At the start, the position and orientation of the robot are known; at any stage of its path toward its goal, position, and orientation data (localization) can be determined using a bar code localizer, which determines the identity and angular placement of bar codes at known positions around the room (beacons).

The on-board rangefinder gathers position data from all surfaces visible all around the robot up to some maximum range measurable by the rangefinder. Obstacle identification is not required for the avoidance strategy although it may be necessary to identify a navigation target (if a simple geometric location has not been specified). The floor projection of this range data is expanded in all directions by a distance equal to the radius of the robot (assumed circularly symmetric) plus a small collision tolerance amount that must accommodate the maximum range error. A high-resolution floor grid map of occupied and free space is built using the data available so far. Space whose occupancy status has not yet been determined is presumed empty until shown otherwise (the optimistic strategy).

A minimal length collision-free path is determined and the robot follows this path for a specified distance. Localization data collected during the move can be used to adjust the actual path to closely follow the planned path.

Range collection and incremental environmental modeling followed by path planning and partial actuation continues until the goal is reached. Change of occupancy status due to movement of obstacles must also be accommodated in the incremental map-building process.

A more ambitious task would be to carry out the navigation task without the use of beacon localization, relying only on natural landmarks for localization. One way of doing this is to match range data against the map so far derived without actually identifying particular landmarks. The continuity of the robot's movement can simplify this process by constraining the search space of the match task. A more elegant approach would be to identify appropriate landmarks and to recognize these same landmarks as the robot moves; this clearly requires more of the vision system. Note that the recognition of a landmark for localization purposes does not mean that the landmark must be recognized in the model-reference sense as was the situation for the first example.

In recent times, many robotics researchers have defined and solved, to various degrees, a robot navigation problem known as the SLAM (simultaneous localization and mapping) problem, which concerns the construction of relatively accurate maps of a previously unknown environment by collecting sensory data, as the robot explores that environment, determining its location by matching current sensor readings with the map so far constructed and correcting the geometry of the map when a particular location is reached after a wandering phase. The confirmation of this event is known as the "closure problem" and remains a difficult requirement if it is to be carried out with high confidence, as an error in this aspect can be very disorienting. Sensors including visual, radar, ultrasonic, iterative navigation, and others have been used for such experiments.

From the above two examples, though incomplete and not universal, it is easy to see the important role that vision can play in supporting robotic hand–eye coordination and autonomous robot navigation. In both cases, it was the 3D nature of the problem domain that made vision a particularly appropriate perceptive mechanism to apply, especially since no-contact analysis of the scenes involved was an important aspect of the tasks involved. Note also that vision may be applied both for volumetric modeling as well as shape recognition or both according to what is needed to solve the problem.

While instrumenting a system to collect or derive range data is only one part of robot vision, clearly the quality of such range data in terms of spatial density, accuracy, reliability, and timeliness is critical to the whole analysis process if it is to lead to robust plans for robots to follow. It is of interest to consider the variety of ways by which range might be extracted from a 3D scene.

Humans use a wide variety of depth cues, usually in selective combinations, to disambiguate many competing hypotheses about how images acquired by the retinas of two eyes might be explained in terms of 3D structures in the scene. Human depth cues include texture gradient, out-of-focus blur, aerial perspective, stereo disparity, binocular vergence, size constancy, and relative obscurance. Many robot

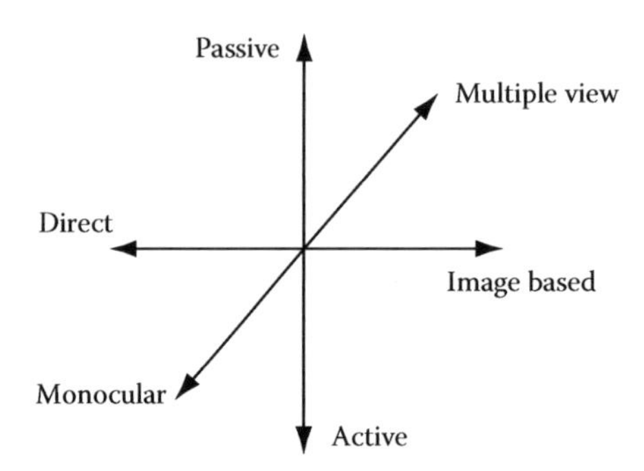

FIGURE 37.4 Rangefinder classification dichotomies.

vision ranging schemes have been inspired by human (and other animals) range extraction mechanisms. Some of these are better suited to robot vision than others and combinations can often be used to produce better results than can be obtained by using one approach alone. However, there are many ranging systems that are based on visual mechanisms not enjoyed by humans; these are also worth considering.

Three dichotomies define eight rangefinding domains, not all of which are populated by feasible realizations. These dichotomies are (see Figure 37.4)

1. Active or passive
2. Image based or direct
3. Single (monocular) or multiple viewpoint

A passive method uses only ambient lighting sources, while an active system probes the scene with an energy beam or a structured light pattern. Image-based methods analyze pixel arrays to derive range, while direct methods measure time-of-flight of known velocity light or ultrasonic energy beams to calculate range. Single-viewpoint methods use one fixed camera or one time-of-flight system, while multiple-viewpoint methods are essentially based on triangulation. The more popular ranging modes within this categorizing structure will be briefly described below.

In indoor industrial settings, where the intrusion of contrived energy sources is not a severe disadvantage and where the contrived light source can be discriminated from the ambient light, the two most popular ranging methods are laser time-of-flight and striped lighting systems. Ultrasonic time-of-flight systems are also of interest but are not covered here.

Laser time-of-flight range instruments either measure the time it takes for a short pulse of laser light to beam to and be reflected back from a point on the surface of an object or the phase shift encountered by a modulated continuous laser beam during the round trip. Since light travels at approximately 30 cm/ns (10^{-9} s), sub-centimeter accuracy range measurement require time to be measured with approximately ±10 ps (10^{-12} s) accuracy using the first approach. With both approaches, averaging over a number of cycles can be used to reduce range error, but at the cost of increased measurement time.

A number of laser rangefinders are now readily available in the marketplace at modest cost ($2000–$7000). One supplier is Erwin Sick, another is Laser Optronix.

Figure 37.5a through c shows a rotating Erwin Sick Rangefinder, and two examples of its 3D scan result. This configuration provides a "foveal" attention bias toward the forward looking direction that could be of distinct advantage for head-on obstacle detection and avoidance.

Of less modest cost (US$75,000) but one that may reduce in time is a device manufactured by Velodyne Pty Ltd. This device is designed to be mounted on the roof of a vehicle and is weather proof. It rotates 64 laser range finders in a vertical distribution at up to 15 Hz and retrieves over 1 million range values (up to 120 m, depending on the reflective properties of the target) per second. Collimated

FIGURE 37.5 (a) Rotating Sick laser range finder. (b) Outdoor 3D laser scan result. (c) Indoor 3D laser scan result.

intensity information is also gathered. An image of a Velodyne instrument is shown in Figure 37.6a and a scan from such a device shown in Figure 37.6b.

Even more expensive (approximately $200,000) and accurate (but slower) range scanner equipment is manufactured by Riegl. Such devices can collect range data up to over 800 m away, but the data collection rate is in the vicinity of 12,000 samples per second. Figure 37.7a shows such a scanner and Figure 37.7b an indoor scan result with color image data fused with the range data. These large-scale laser range

(a)

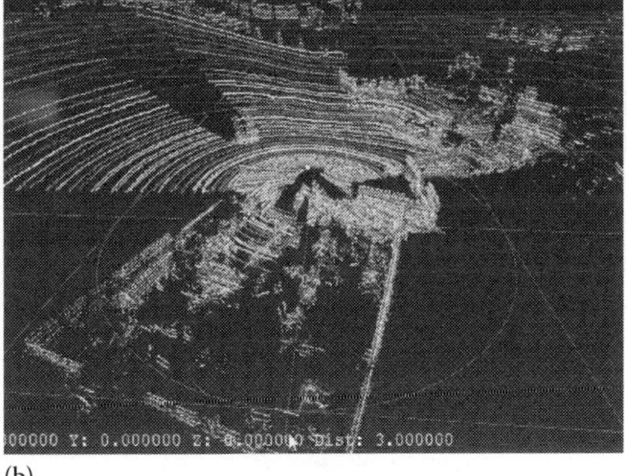

(b)

FIGURE 37.6 (a) Velodyne rotating multi-laser range finder. (b) Velodyne indoor/outdoor range scan.

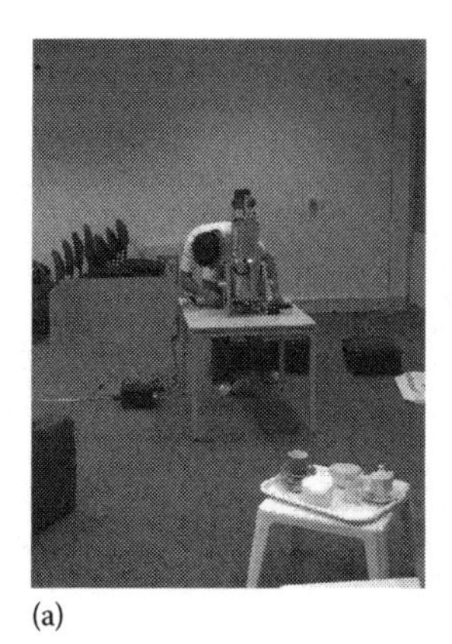
(a)

(b)

FIGURE 37.7 (a) Riegl laser range scanner. (b) Indoor Riegl range scanner result (with color imagery fused).

finders are excellent for use in outdoor robot navigation for both localization and path planning (as well as target recognition, if necessary).

Active methodology can have the disadvantages of detection (say in a military context) and cross interference (if many systems are present in the same environment). Whether this will prevent their domination over passive methods is yet to be seen.

Striped lighting ranging methods are triangulation based and rely on analyzing the distortion in the image of a stripe of light projected on the scene obtained from a camera displaced from the contrived light source in a direction perpendicular to the stripe light plane. Using many stripes simultaneously can lead to ambiguity in identifying the individual lines in the image, while using only one line at a time and sweeping this across the scene can be slow for high-density ranging, particularly if a standard frame rate video camera is used. An elegant compromise consists of collecting a set of images for a stripe lighting pattern set, which allows each stripe to be binary coded among the images. For n stripes, only $(\log_2 n) + 1$ images need to be collected; doubling the number of stripes requires only one extra image.

In terms of the categorization scheme introduced earlier, the laser time-of-flight ranging method would be regarded as in the direct/active/single-view class, while the striped lighting ranging method would be regarded as being in the image-based/active/multiple-view class. Other members of the active class are range from brightness (with contrived light sources) and range from attenuation (again with contrived light sources). Both time-of-flight and striped lighting systems have been used in industrial robotics for some years now.

Passive range methods have been actively researched over the last three decades, partly because their biological source of inspiration brings with it the confidence that these approaches have been tested through Darwinian natural selection over countless generations and partly because they promise a very wide application scope and are intrinsically safe. Computational and microelectronic advances in recent times have brought fast and affordable general-purpose passive rangefinding to the marketplace for practical use in many industries and in the home with or without robotic implications.

Figure 37.8 shows a disparity map of a scene using a Digiclops stereo (actually three cameras are used) system available from Point Grey Research Inc. The larger disparities (separation of matching components in separate camera view) are shown as brighter. Disparity is inverse range related.

Passive stereopsis is clearly the most popular approach to passive rangefinding. There are two stereopsis modes. Lateral stereopsis (Figure 37.9) refers to methods that use at least two cameras in a known baseline configuration in a plane perpendicular to the depth coordinate to derive range, while temporal stereopsis refers to methods where time sequences of images from a camera moving through the 3D scene are used to derive range. In either case, there are two essential process components that must be

FIGURE 37.8 Digiclops passive stereo disparity map.

dealt with since both methods rely on matching elements among images, using the shifts of location (disparity) as inverse range measures (scaled appropriately):

1. The extraction or definition of the elements to be matched among images
2. The search for correspondence between those selected elements among the images

The image elements chosen for matching can be edge points, lines, corners, area patches, regions, object segments, or entire objects. More preparatory processing is required by choosing more complex elements, and, in the case of entire objects, the segmentation of these objects may be simplified by the

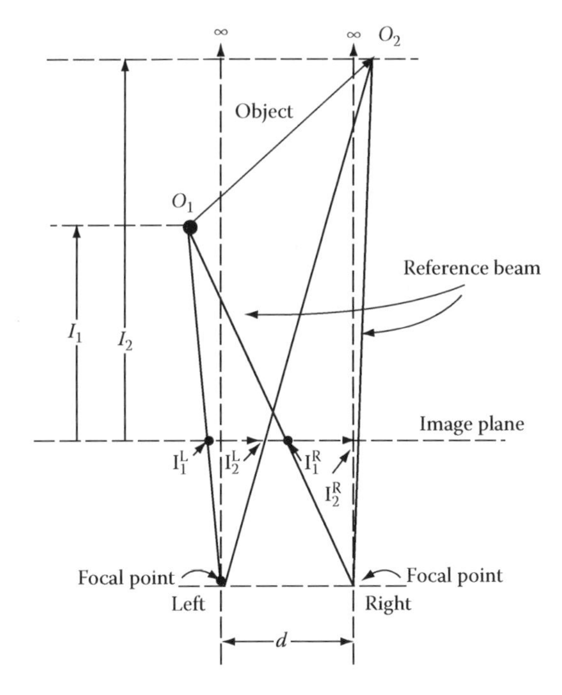

FIGURE 37.9 Geometry of passive stereopsis.

use of range data, which is to be the result of the analysis, thus creating a potential deadlock situation. Thus, there is some advantage in matching the most primitive elements that can be discriminated clearly and matched unambiguously.

In recent times, mask correlation has been favored as a correspondence matching approach since the high computational complexity, this entails now falls within the capacity of off-the-shelf computational resources.

For lateral stereopsis, the use of more than the minimum number of two cameras can improve both accuracy and reliability. Special hardware for very fast extraction of multiple camera stereopsis range data is under active research and development.

A variant of temporal stereopsis is known as optical flow; in this case, the vectors indicating the short-term movement of pixels are calculated from a video sequence with a single camera without explicitly solving the correspondence problem. Depth can be calculated from this vector flow field as can the egomotion of the camera through the scene. Analysis of moving objects in the scene can also be carried out using optical flow.

Once dense, accurate, and reliable range data have been extracted, the resulting 2 1/2D representation of the 3D scene must be further processed. Full 3D scene representation can be in the form of an ordered or unordered set of 3D Cartesian coordinates with each point having a location in space as well as attributes such as surface color and normal vector data. Alternatively, a voxel-based representation may be preferred. Other data structures have also been proposed. The appropriateness of a data structure used to represent 3D scenes is measured in terms of compactness, lack of ambiguity, easy access, modifiability, and its ability to support the application in mind.

The results of a complete scene analysis carried out by a robot vision system may take various forms but would generally provide information on the identity, pose, and placement of individual items and their functional interrelationship (e.g., support, adjacency, occlusion, linkage, containment, etc.). In particular applications, perhaps a subset of this information would suffice; building vision systems to provide more than required is hardly a sensible thing to do for a specific task, but stretching vision methodologies toward generalization of applicability is also a worthwhile endeavor.

While extracting reliable 3D information has been emphasized here as a crucial aspect of robot vision, whether for hand–eye coordination or navigation, clearly the remaining processing steps required to generate a complete scene analysis and subsequent robotic interactions can be complex and varied. The very rich and plentiful literature on these methodologies cannot be properly dealt with here.

In summary, robot vision attempts to emulate human vision to the extent of correctly interpreting the makeup of 3D scenes, with the purpose of guiding robotic manipulators and autonomous mobile robots in fulfillment of useful tasks; many methods of range extraction support these ambitious goals, and rapidly improving sensory device technologies and computational systems are likely to lead to an explosion of commercially available systems within the next decade.

38 Robot Path Planning

Raymond Jarvis
Monash University

Both robotic manipulator arm trajectory and mobile robot path planning can be cast as point trajectory planning in a Euclidean space of as many dimensions as the degrees of freedom of the robotic device. Such a construction is called "configuration space." For example, a mobile robot moving in a two-dimensional plane can be represented by a point in a three-dimensional Euclidean space specifying two location coordinates, x and y, and one orientation coordinate, θ. All obstacles in this space are grown to accommodate the extent of the robot.

For a six-degree-of-freedom robotic manipulator arm, the configuration space is six dimensional, each dimension representing 1 degree of freedom, whether of a revolute or prismatic movement. Growing the obstacles, including itself, to represent the forbidden regions in configuration space can be very complex, the mapping being possibly one to many, since a simple obstacle can fragment into many forbidden hypervolumes in configuration space.

There are a great many path-planning approaches presented in the literature and there is scope to cover only a few here. The ones chosen should give some small taste of the whole field.

Here, it will be assumed we are working in configuration space where path planning is a matter of finding collision-free trajectories (continuous paths) in this space.

The three path-planning approaches to be described and compared here are A* path planning [1,2], the distance transform (DT) approach [3–5], and rapidly exploring random trees (RRT) approach [6]. These are quite distinct from each other and yet combinations among them can certainly be implemented.

Each of the above methods will be described in turn and then compared in terms of complexity, flexibility, and application scope. A discussion section follows.

38.1 A* Search Path Planning

The A* Search strategy is a general tree search procedure with a number of interesting properties. It appears in Nilsson's text of 1971 [2]. Lozano Perez and Wesley [1] showed how to apply A* Search to robot path planning in convex polygonal obstacle fields to extract the minimal length collision-free path from a nominated start point to a nominated goal point, both being assumed to be in free space (space not occupied by obstacles). All such paths in two dimensions are made up of straight lines chosen among the start and the goal point and the set of polygonal obstacle vertices. Each straight-line component has

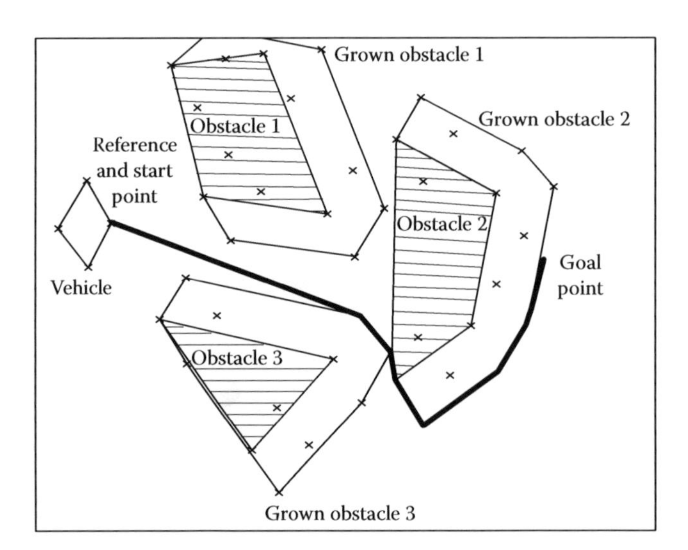

FIGURE 38.1 A* path planning example with grown obstacles for a translation-only robotic vehicle.

end points that are mutually visible, since the path cannot penetrate an obstacle. This search procedure is efficient in the sense of opening the least number of tree nodes as the search proceeds and also guarantees the optimal least length path solution if such a path exists.

Jarvis [7] showed how to exactly accommodate convex polygonal shaped vehicles moving through the obstacle field by pure translation and how to approximately accommodate a combined translating and rotating polygonal (convex) robot vehicle. Figure 38.1 shows an example of A* optimal path planning for a translation-only robot vehicle, where the exact obstacle growth for configuration space mapping is shown.

The A* algorithm is a strategy for exploring paths through nodes (vertices) associated with the least estimated cost (path length in this case) of reaching the goal. This cost estimate is made up of two parts, the known cost of reaching a node currently under examination from the start point and the estimated cost of reaching the goal from the node. The direct Euclidean distance from the node to the goal is used in the Lozano Wesley algorithm [1]; this meets the requirement of optimal search that such an estimate should not be greater than the actual collision-free distance to the goal.

The algorithm is as follows. The set of nodes considered is made up of the start and goal points and all obstacle vertices:

1. Place the start node on a list designated as OPEN along with its direct Euclidean distance from the goal.
2. From the OPEN list take the node with the least cost (distance along path so far to start plus estimated distance to goal, such an estimate being the direct Euclidean distance) and place it on a list designated CLOSED. Attach a backward pointer to its parent node.
3. Find the visible successors of the last node placed on the CLOSED list. If any of these are the goal node, tracing back to the start node establishes the optimal path and the search is terminated. Otherwise, place all visible successors on the OPEN list together with their exact backward (to start) and estimated forward (to goal) costs.
4. Return to 2 and repeat.

The admissibility and optimality properties of the A* search and methodology are given in Nilsson [2].

If N = VUSUG, the set of nodes under consideration, V standing for vertices, S for start, and G for goal, then N is of size two greater than that of V. The size of N is denoted $|N|$.

Since the optimal path must link a subset of nodes with straight lines there is a limit to the number of such paths, this being

$$\sum_{k=1,|N|-1} \frac{(|N|-2)!}{(|N|-k-1)!} \tag{38.1}$$

The derivation of this formula is given in Appendix B of Jarvis [7].

The number of such paths grows rapidly with $|N|$. For example, while for $|N| = 4$, the number of such paths is 5, when $|N| = 10$ the number of paths is 129, 101. It does not take much imagination to realize that if the obstacle field is very complex the computational burden is very onerous. Nevertheless, the number of paths remains finite in the two-dimensional case. However, for three dimensions (and above) there are an infinite number of possible paths since, although any path that would possibly be optimal must be made up of straight-line components, these may go via the edges of the obstacles and not just their vertices. Pseudo-vertices can be marked along edges to obtain an approximate solution, but, given the computationally complex growth potential, this is not a very satisfactory solution.

38.2 Rapidly Exploring Random Trees

This approach abandons the requirement for path optimality for the acceptance of feasibility for the sake of efficiency, particularly in searching high dimensional spaces. The basic algorithm is given in [6] and a number of extensions and developments can be found in [8]. The approach is very simple. Again the start, goal, and obstacle field are all known. The obstacle field can be specified simply by a binary test of whether a point in the search space is inside or outside an obstacle, and thus these are not restricted to convex polygonal shapes.

The basic algorithm is as follows, ignoring any dynamic or turning circle constraints:

1. Choose a uniformly distributed random point in the configuration space. From the start point, extend along the straight line to the chosen point an increment, say, δ, the whole length of which is in free space. This new point joined to the start point constitutes the first branch of the rapidly exploring random tree that will ultimately explore free space and connect with the goal point.
2. Keep choosing random points and extending the nearest node of the growing tree toward such points (one at a time) an increment of δ in free space until one such node of the tree is close enough to the goal so that the last link can be made without further calculation.
3. Tracing back from the goal to the start via tree branches defines the found, feasible, collision-free path from the start to the goal.

Since the random points are independent of the structure of the obstacle distributions, and while the exploration is fast and simple, the finding of the nearest-neighbor node of the growing tree becomes more and more tedious. Also, narrow gaps are hard to grow through and the tree can become very cluttered in the process. However, relatively uncluttered obstacle spaces can be explored very rapidly even in high dimensions. Starting at both ends (start and goal) is an obvious improvement. A number of other extensions have also been developed [9]. It is very hard to compute the complexity of such an approach since it depends very much on the shapes of the obstacles and their distribution. Clearly, the complexity rises steeply with the complexity of the obstacle structure since, as the tree grows, extending its growth becomes more and more expensive unless clever nearest-neighbor methods such as kd-trees [10–12] are used. It can also be questioned whether the exact nearest neighbor is essential or whether a mere near neighbor is good enough.

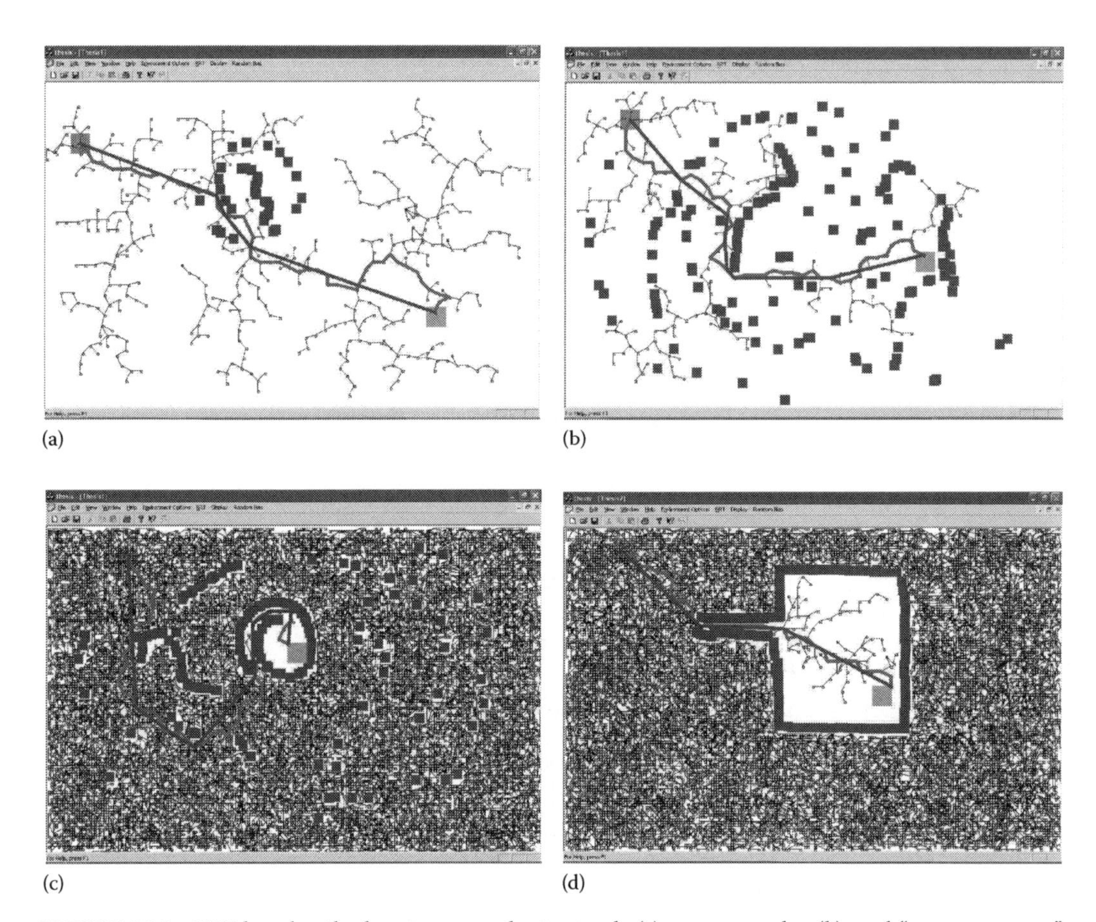

FIGURE 38.2 RRT-based path planning examples in simple (a), more complex (b), and "narrow passage" environments (c,d).

Figure 38.2 shows a number of examples of RRT path planning. The results include some path straightening [9] extensions.

38.3 Distance Transform Path Planning

Distance transforms are propagation algorithms that designate distance through tessellated space with respect to simple or multiple cells nominated as zero.

This type of algorithm appeared in Rosenfeld and Pfaltz [13] as a way of describing the shape of blobs, typically photomicrograph images. If the cells on the outline of a shape are designated zero distance values, growing the distance transform (DT) will generate wavefronts of increasing distance into the shape. The local maxima of such a propagation can be used as skeleton to regenerate the whole DT field exactly.

For the purposes of collision-free path planning, Jarvis [3] turned this procedure inside out to propagate distances out from goal points to fill all of free space, flowing around the obstacles. The raster-scan method of doing this requires multiple passes to guarantee growth into complex-free space shapes. The basic algorithm in two-dimensional configuration space is as follows:

Let cell [x, y], x = 0, 1 .... to xmax + 1
y = 0, 1 ... to ymax + 1

be a rectangularly tessellated space representing the configuration workspace of a robot operating in two dimensions. The robot position is represented by a simple cell, the obstacles having been grown to accommodate the robot extent. The outer edges of cell [x, y] are regarded as obstacle cells that enclose the remainder of the workspace. Let "goal," "start," and "blocked" be three Boolean arrays of the same dimension as cell. If c[x, y] is a goal point, then goal [x, y] is "true" but "false" otherwise. Similarly, start [x, y] is "true" if cell [x, y] is a start point and "false" otherwise; blocked [x, y] is "true" if cell [x, y] is occupied (part of an obstacle) and "false" otherwise. In Pascal-like pseudocode

```
(* border initializations*)

for y: = 0 to ymax+1 do
  for x: = 0 to xmax+1 do
      if goal[x,y] then cell[x,y]: = 0;
                         else cell[x,y]: = xmax*ymax; (*a large number*)

(*Calculate distance transform*)

repeat
for y: = 2 to ymax do
  for x: = 2 to xmax do
     if not blocked[x,y] then
     cell[x,y]:=min(cell[x-1,y]+1,cell[x-1,y-1]+1,cell[x,y-1]+1,
          cell[x+1,y-1]+1,cell[x,y,]);
for y: = ymax = 1 downto 1 do
    for x: = xmax-1 downto 1 do
     if not blocked[x,y] then
      cell[x,y]: = min(cell[x+1,y]+1,cell[x+1,y+1]+1,
        cell[x,y+1]+1, cell[x-1,y+1]+1,cell[x,y]);
until no change;
```

The number of iterations of the repeat cycle depends on the topography of the obstacle field. Blocked and goal cells remain unaltered; all other cells contain integers specifying exactly how many steps are required to reach the goal from that point. The optimal trajectory from any point in free space is that of steepest descent to the goal. Thus, only one DT evaluation for trajectories from any free cell needs to be developed. Multiple goal points and start points can be introduced without any change to the algorithm; each steepest-descent trajectory from a start point will find the nearest goal. Figure 38.3 shows some examples of DT-based path planning.

It turns out that, with perfect prediction concerning the movement or appearance or disappearance of obstacles, a temporal dimension can be coped with very simply [4]. It is also possible to weight the costs of visiting cells along a path to reflect tractability, as long as the costs are positive. For example, a whole suite of variations collectively termed "covert path planning" [14–17] can be generated by considering visibility variations among the cells. Also, any number of dimensions can be handled by simple variations of the two-dimensional algorithm.

The complexity of the DT computation depends only on the obstacle topology (which determines the number of iterations of the repeat cycle needed) and the resolution of the work space and is otherwise independent of the number of obstacles. The other major cost as the resolution becomes very fine or the space represented very large is memory. For example, a 10×10 km work space with 1×1 m cells will require 100 million cells. However, the computation structure itself is very regular and simple.

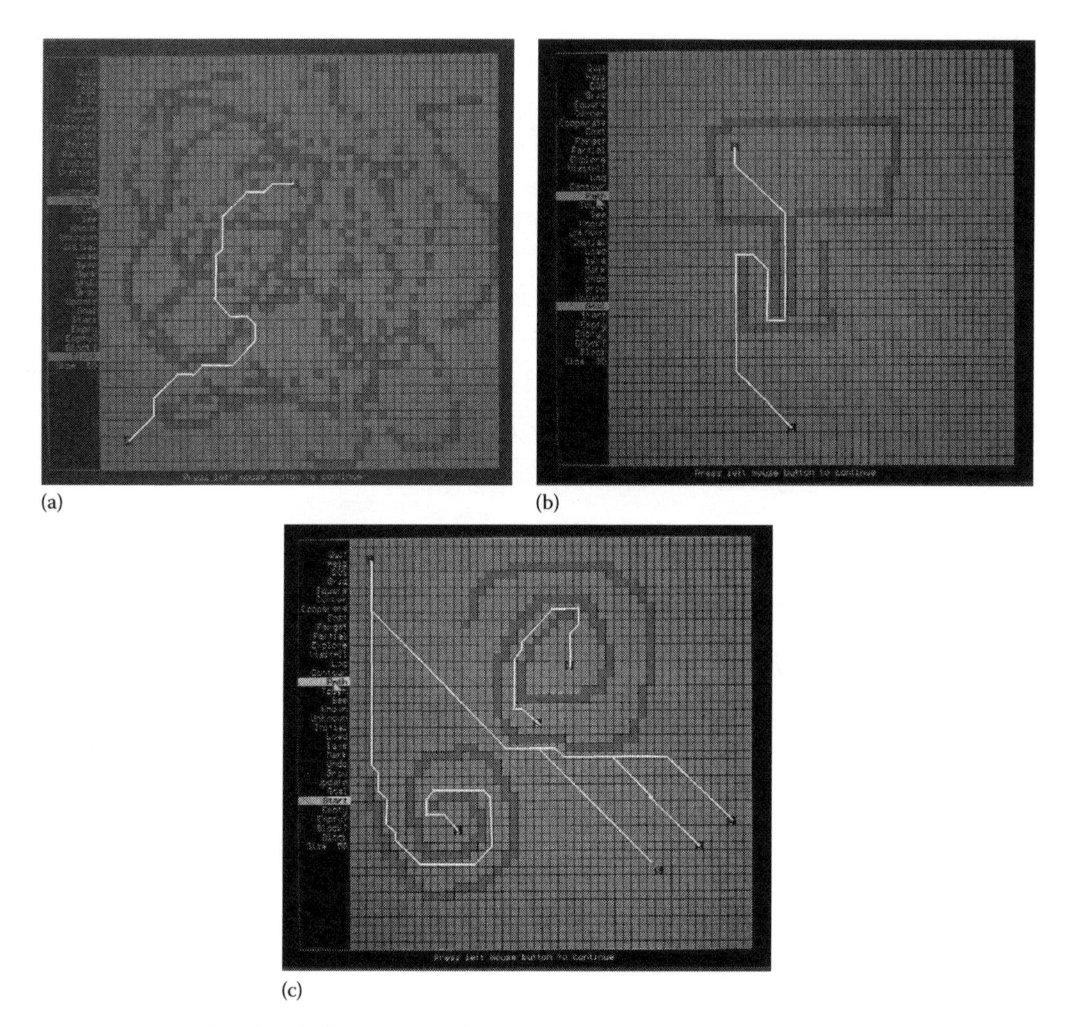

FIGURE 38.3 DT-based path planning examples (a,b,c), including "narrow passage" (b) cases and a multiple start/multiple goal variant (c).

38.4 Comparisons: Complexity, Flexibility, and Application Scope

It should be noted at the outset, before embarking on comparisons between the three path planners described above, that the A* search methodology and the RRT approach are both normally presented as operating in Euclidean spaces where the dimensions are in real numbers, while the DT is intrinsically linked with tessellated spaces. Thus, the A* and RRT approaches are not directly effected by the extent of a robot working environment whether it be over meters or kilometers for a typical robot of dimensions from fractions to multiple meters, though both are sensitive to the complexity and topology of the obstacle field.

On the other hand, the DT approach incurs a computational complexity directly related to the spatial resolution but is not severely sensitive to the complexity or topology of the obstacle field. Of course, both the A* and the RRT approaches can be approximately mapped into a type of tessellated space by replacing real number measures by integers. The A* and DT formulations provide optimal path solutions but the RRT only feasible ones.

The A* and RRT approaches normally provide what are called "single query" solutions that produce point-to-point paths. However, multiple goals can easily be specified for both where the attainment of a single goal of the set is sought. The DT, of its nature, propagates distances out from single, multiple, or collective goals over all reachable free space. A steepest-descent path from any cell within this free space provides an optimal solution.

Interestingly, the computational cost of the RRT in real space increases incrementally as the tree grows since nearest-neighbor calculations grow with the number of free nodes (linearly or otherwise according to the method used). If the RRT is posed in tessellated space, a spiral search out from each new random point can discover the nearest neighbor, and this becomes less and less costly as the number of nodes grow. In fact, one could think of some way of switching nearest-neighbor search procedures at some point of growth of the exploratory tree.

In addition to the considerations of optimality or feasibility, there is also the question of safety in the face of imprecision or the inclusion of notions of tolerance in following the prescribed path. Unfortunately, optimal paths tend to "graze" obstacles to minimize path lengths. The outer borders of obstacles can be expanded beyond merely allowing for the physical extent of the robot in the process of mapping the environment to configuration space (as mentioned earlier) to also permit tolerance layers. All three methods described here can use such a technique. However, the DT permits, in addition to simple obstacle growth, a softer way of encouraging safe paths with some (hopefully small) additional cost with respect to path length. This approach is quite elegant in a double use of the DT propagation algorithm. First, ignoring goal points, a DT is generated in all free space out from all obstacles, collectively marked as zero distances from themselves. The resultant DT specifies the least number of steps from anywhere in free space to the nearest obstacle cell; the larger the number the safer it is for the robot to include this cell on its path. Subtracting all free cell values from the maximum provides a cost field that reflects risk of collision due to imprecision of both the position of obstacles and that of the robot. Adding these costs to distance costs by minor adjustment of the basic algorithm provides steepest-descent paths that are safe rather than distance optimal. The tolerance risk and distance costs can be weighted according to preference of emphasis on tolerance or shortness of paths.

It is also of interest to consider how various weightings can be introduced in the path evaluation to reflect the nature of the work space or other contributing factors. For example, how would one consider varying tractability of terrain types or whether positions along the path are more or less visible from particular vantage points or more generally [17].

For the A* algorithm, both the backward (to start point) exact cost and the forward (to goal) estimate can each include values that are not simply path-length related. For the RRT, since only feasible paths are sought, weightings related to traversal of various parts of free space have no influence on the algorithm. The DT approach is most easily adjusted to reflect varying cost weights. When deciding what value to write in each cell visited, one need simply choose the least value of neighbor's value plus the cost of entering that cell from that neighbor. Thus, entry costs can vary according to entry direction, even allowing for nonsymmetric costs as might be associated with one way only paths. Costs associated with visibility can easily be accommodated as in covert path planning [17].

Another consideration of comparison relates to the possibility that the obstacles in the working environment may be initially unknown and can only be discovered and mapped using sensors on-board the robot as it navigates through the environment. The need to continually update path planning as new knowledge is acquired must be met by a consistent process that continues to provide paths of some global viability. One assumption that can be adopted for all three approaches is that space not yet mapped be considered empty (until proven otherwise). Replanning must be carried out whenever new knowledge may be critical for proper path planning.

For the A* approach, all obstacles, when discovered, are restricted to be represented as convex polyhedra. Complex shapes must be segmented into convex polyhedral components. Enclosing edges not yet observed must be closed to complete convex polyhedra. Thus, while it is possible to introduce new obstacles into the procedure, it is by no means particularly simple to incrementally accommodate these additions. Also, if new obstacles are discovered along the best path so far computed, replanning must consider such a point as the new start point for the navigation to continue.

For the RRT approach, accommodating newly discovered obstacles is possible but clumsy since only feasible paths are sought in the first place and various paths of a grown tree must be pruned when obstacle intrusions are discovered. Thus, the RRT approach does not make much sense for initially unknown environments.

In the DT case, newly discovered obstacle cells are simply put into the environment map and the DT recomputed. Since the DT is starting-point nonspecific, one can simply continue a steepest-descent path from the current position. Recomputing the DT need only be done when newly discovered obstacles might threaten the optimality of the current planning path. Thus, only when new obstacle cells are found "down hill" from the current robot location relative to the goal need the DT be recomputed.

The RRT and DT can accommodate any dimensional search spaces with trivial adjustments, but the A* method is critically changed when the dimension is greater than two since the finiteness of the number of paths, one of which might be the optimal one, is no longer assured.

The DT can very simply accommodate time as one search dimension where perfect predictability is assured [4]. The irreversibility of time actually simplifies the computation to a single pass.

The flexibility of the A* algorithm is severely restricted by the convex polyhedral requirement of its obstacles and the difficulty of going beyond two dimensions of configuration space. The RRT is difficult to apply in initially unknown obstacle fields and inefficient where narrow passages must be navigated through but is otherwise unencumbered by obstacle shape, though somewhat impeded by obstacle distribution density. The DT is sensitive to resolution but not obstacle shape or density of distribution except when topological complexities require high iteration cycles. However, the topologies requiring a very high number of cycle iterations are very rare in realistic situations.

Both the A* and DT methods can accommodate traversability cost variation, the DT with considerable simplicity, but such considerations make little sense for RRT application.

The comparison made in the previous section forms the basis for deciding, for a particular application, just which of the three path-planning methods described is best suited to the task. The DT suits most cases where some tolerance is acceptable with regard to strict optimality (since the tessellation structure introduces some distortion to Euclidian distance) and where very high resolution is not required. The RRT is very powerful in high-dimensional spaces without narrow passages where the feasibility of the path rather than optimality is acceptable. The A* method is a source of inspiration, but its restrictions regarding obstacle representation, dimensionality, and steep complexity growth with the number of obstacle vertices does not recommend it for practical implementation.

Some considerations that might render the DT and RRT methods even more capable can be touched on here. First, the RRT may produce acceptable results when nearest-neighbor requirements are relaxed to mere near neighbor. The usually jagged path can be smoothed [9]. Also, probability distribution adjustments for random point selection may be able to take advantage of known obstacle field properties, particularly to reduce the narrow neck problems. DT algorithms can be defined over nonuniformly tessellated spaces including pyramid structures [18]. Such extensions may partially overcome the resolution-related complexity costs. Fine tessellations may only be required in congested spaces and not in open, obstacle-sparse spaces.

There are even ways of mixing RRT and DT methods, but these have not yet been thoroughly investigated. For example, tessellated strips (possibly quite broad) along a feasible path discovered by an RRT algorithm could then be treated as a restricted region for DT generation to obtain the best path within that strip.

References

1. Lozano-Perez, T. and Wesley, M.A. An algorithm for planning collision-free paths among polyhedral obstacles, *Commun. ACM*, 22(10) (1979) 560–570.
2. Nilsson, N.J. *Problem Solving Methods in Artificial Intelligence*, McGraw-Hill, New York (1971).

3. Jarvis, R.A. Collision-free trajectory planning using distance transforms, *Proceedings of the National Conference and Exhibition on Robotics*, Melbourne, August 20–24, 1984, also in *Mech. Eng. Trans. J. Inst. Eng.*, ME10(3) (1985), 187–191.
4. Jarvis, R.A. On distance transform based collision-free path planning for robot navigation in known, unknown and time-varying environments, invited chapter for a book entitled *Advanced Mobile Robots*, ed. Y.F. Zang, World Scientific Publishing Co. Pty. Ltd., Singapore (1994), pp. 3–31.
5. Jarvis, R.A. Collision-free path planning in time-varying obstacle fields without perfect prediction, *Proceedings of the '93 ICAR, International Conference on Advanced Robotics*, Tokyo, Japan, (1993), pp. 701–706.
6. LaValle, S.M. Rapidly-exploring random trees: A new tool for path planning, Technical Report No. 98–11, Department of Computer Science, Iowa State University, Ames, IA (1998).
7. Jarvis, R.A. Polyhedra obstacle growing for collision-free path planning, Department of Computer Science, A.N.U., Technical Report TR-CS-81-16. Also *Proceedings of the Australian Computer Science Conference*, Perth, Australia (1982), pp. 203–213.
8. LaValle, S.M. and Kuffner, J.J. Randomized kinodynamic planning. In *Proceedings of the IEEE International Conference on Robotics and Automation*, Orlando, FL (1999).
9. Deak, Z. and Jarvis, R.A. Robotic path planning using rapidly exploring random trees, *Proceedings of the Australian Conference on Robotics and Automation 2003*, Brisbane, Australia, December 1–4, (2003).
10. Arya, S., Mount, D., Netanyahu, R., Silverman, R., and Wu, A. An optimal algorithm for approximate nearest neighbor searching. *J. ACM*, 45 (1998), 891–923.
11. Dickerson, M., Duncan, C., and Goodrich, M. K-D trees are better when cut on the longest side. In *Proceedings of the Eighth Annual European Symposium on Algorithms*, Saarbrücken, Germany (2000), pp. 179–190.
12. Duncan, C., Goodrich, M., and Kobourov, S. Balanced aspect ratio trees: Combining the advantages of k-d trees and octrees. In *Proceedings of the 10th ACM-SIAM Symposium Discrete Algorithms*, Baltimore, MD (1999), pp. 300–309.
13. Rosenfeld, A. and Pfaltz, J.L. Sequential operations in digital image processing, *J. ACM*, 13(4) (1966) 471–494.
14. Marzouqi, M.S. and Jarvis, R.A. Covert robotics: Hiding in known environments. In *Proceedings of the IEEE Conference on Robotics, Automation and Mechatronics (RAM)*, Traders Hotel, Singapore (December 1–3, 2004), pp. 804–809.
15. Marzouqi, M.S. and Jarvis, R.A. Covert path planning in unknown environments with known or suspected sentries locations, submitted to the *IEEE International Conference on Robotics and Automation (ICRA)*, Barcelona, Spain (2005).
16. Marzouqui, M. and Jarvis, R. Covert robotics: Covert path planning in unknown environments. In *Proceedings of Australasian Conference on Robotics and Automation*, Canberra, Australia (December 6–8, 2004), p. 8.
17. Marzouqi, M.S. and Jarvis, R.A. Covert path planning for autonomous robot navigation in known environments, *Proceedings of the Australasian Conference on Robotics and Automation*, Brisbane, Australia (2003).
18. Pavlidis, T. *Structural Pattern Recognition*, Springer-Verlag, New York (1977).

39

Mobile Robots

Miguel A. Salichs
University Carlos III

Ramón Barber
University Carlos III

María Malfaz
University Carlos III

39.1 Introduction

The objective of this chapter is to provide a brief overview of mobile robots, presenting the main issues corresponding to these robots. A mobile robot can be dedicated to many different tasks (e.g., transport, manipulation, and exploration). The chapter focuses on the topics associated with mobility, such as locomotion and navigation, which are common to all mobile robots. Other topics (e.g., object handling and human–robot interaction) that can be only of interest for some specific tasks or some specific types of mobile robots have not been included.

39.2 Locomotion

A mobile robot needs locomotion mechanisms that enable it to move unbounded throughout its environment. A large variety of possible ways to move exist in nature—walk, jump, run, slide, etc.— and these have inspired most of the locomotion mechanisms of mobile robots. Nevertheless, the wheel is a human invention that achieves extremely high efficiency on flat ground. Two main kinds of mobile robots can be distinguished based on their locomotion mechanisms:

1. Legged robots. Legged locomotion is characterized by a series of point contacts between the robot and the ground. The key advantage is its adaptability and maneuverability in rough terrain. On the other hand, the main disadvantages of legged locomotion include power consumption and mechanical complexity.
2. Wheeled robots. The wheel has been by far the most popular locomotion mechanism in mobile robotics. It can achieve very good efficiencies and does so with a relatively simple mechanical implementation.

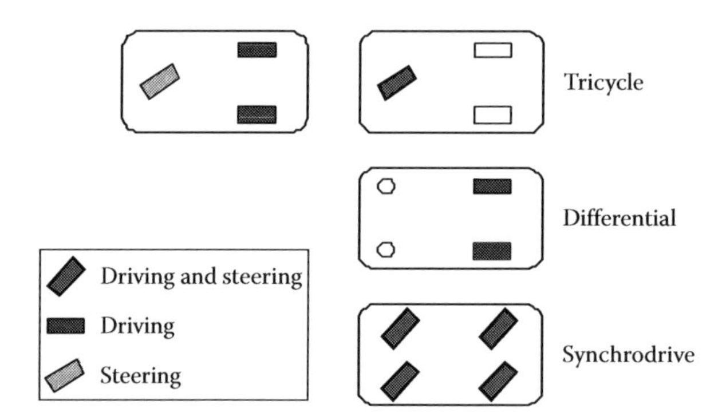

FIGURE 39.1 Steering configuration for WMRs.

Mobile robots generally locomote either using wheeled mechanisms or using a small number of articulated legs. In general, legged locomotion requires higher degrees of freedom and, therefore, greater mechanical complexity than wheeled locomotion [21]. Since the analysis of legged robots would imply a full chapter, in the current chapter, we are going to focus our attention on wheeled mobile robots (WMRs). An in-depth review of legged robots can be found in [11].

WMRs are increasingly present in industrial and service robotics, particularly when flexible motion capabilities are required on reasonably smooth grounds and surfaces [17]. Balance is not usually a research problem in WMR designs because WMR are almost always designed, so that all wheels are in contact with the ground at all times. Therefore, three wheels are sufficient to guarantee stable balance, although there are also two-wheeled robots that can be also dynamically stable [21].

Several mobility configurations can be found in applications depending on the number and type of wheels, their location, and actuation. There are several classifications of wheels, and one of them can be the following: driving, steering, driving/steering, and passive.

The most common drive configurations for single-body robots are differential drive, synchro drive, and tricycle or car-like drive [17] (see Figure 39.1):

- Differential drive: Differentially steered vehicles have two drive wheels that are responsible for driving and steering. The steering action is accomplished by having each wheel to rotate at different speeds. This type of configuration provides some additional advantages like forward and backward movements that can be performed at the same speed. In addition, the vehicle requires a smaller area to maneuver.
- Synchro drive: This technique, also known as all-wheel steering, is based on at least three driving/steering wheels. This configuration allows the vehicle to move in any direction.
- Tricycle drive: The driving action can be through the front wheel and/or the rear wheels with the steering action being performed by the front wheel. The minimum radius of curvature depends on the distance between the front and the rear axes and the maximum steering angle of the front wheel. From the motion point of view, four-wheeled mobile robots with car-like configuration have similar capabilities.

39.2.1 Kinematics

Kinematics is the most basic study of how mechanical systems behave. In mobile robotics, we need to understand the mechanical behavior of the robot both in order to design the appropriate mobile robot for each task, as well as to understand how to create control software for a defined mobile robot hardware [21].

Since the WMRs are the most widely used in general, in this section we are going to describe the kinematics of this kind of mobile robots. To control a WMR, it is important to know the relationships

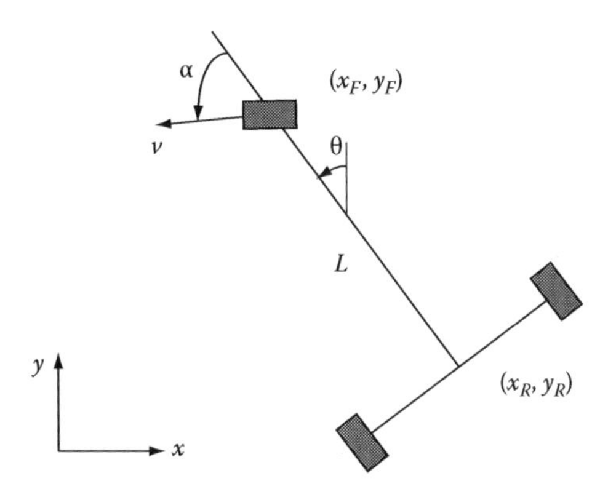

FIGURE 39.2 Tricycle steering configuration.

between the actions on the actuators and the movements of the robot. These relationships are used for two purposes. First, to calculate the actions necessary to move the robot from one position to another; second, to evaluate the displacement of the robot from the movements of the wheels (i.e., odometry).

In the following section, we provide the kinematic equations that correspond to each of the drive configurations previously described.

Tricycle steering configuration. The driving signals are the steering wheel angle α and the linear speed of the front wheel v (see Figure 39.2). The kinematic equations of a point placed in the front wheel are

$$\begin{aligned} \dot{x}_F &= -v\sin(\theta+\alpha) \\ \dot{y}_F &= v\cos(\theta+\alpha) \\ \dot{\theta} &= \frac{v}{L}\sin(\alpha) \end{aligned} \tag{39.1}$$

The kinematic equations of a point placed in the middle of the rear axis are

$$\begin{aligned} \dot{x}_R &= -v\cos(\alpha)\sin(\theta) \\ \dot{y}_R &= v\cos(\alpha)\cos(\theta) \\ \dot{\theta} &= \frac{v}{L}\sin(\alpha) \end{aligned} \tag{39.2}$$

Differential steering configuration. The driving signals are the linear speed of the two driving wheels: v_1, v_2 (Figure 39.3). The kinematic equations of a point placed in the middle of the driving wheels are

$$\begin{aligned} \dot{x}_C &= -\frac{v_1+v_2}{2}\sin(\theta) \\ \dot{y}_C &= \frac{v_1+v_2}{2}\cos(\theta) \\ \dot{\theta} &= \frac{v_1-v_2}{D} \end{aligned} \tag{39.3}$$

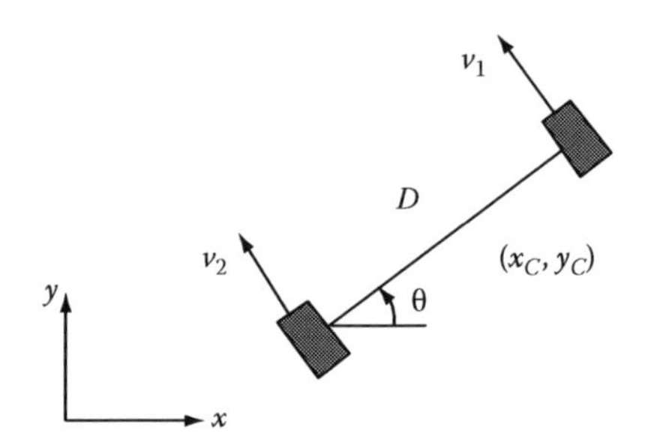

FIGURE 39.3 Differential steering configuration.

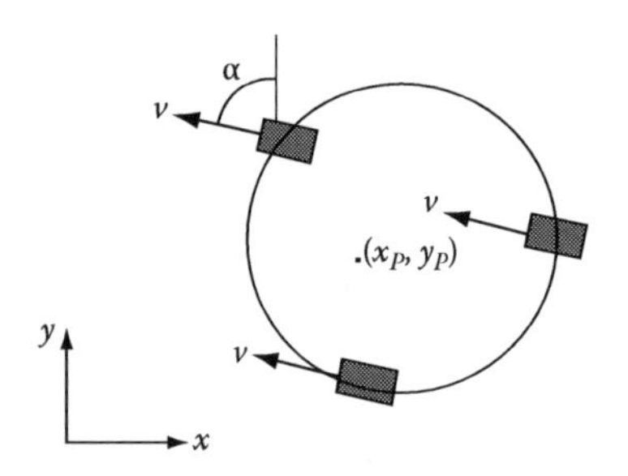

FIGURE 39.4 Synchro drive steering configuration.

Synchro drive steering configuration. The driving signals are the linear speed v and the steering angle of the wheels α (Figure 39.4). The kinematic equations of any point in the driving structure are

$$\begin{aligned}\dot{x}_P &= -v\sin(\alpha)\\ \dot{y}_P &= v\cos(\alpha)\end{aligned} \tag{39.4}$$

With this configuration, the robot orientation is independent of the actions on the driving wheels.

39.3 Perception

Perception involves a set of algorithms that transform the data obtained by the sensors into a representation that could be used by the components of the system responsible for decision making. Perception is one of the key elements that enables a mobile robot to fulfill its intrusted missions.

Regardless of the applications task or how it is performed, a mobile robot has to be displaced from one point to another within an environment to accomplish a task. Regarding its environment, the mobile robot asks itself the following questions: "where I am?" and "what surrounds me?" questions that any perceptual system must help to answer.

The characteristics that a designer should analyze when setting up the sensorial system are size, power consumption, simplicity, redundancy, capacity of operating in real time, ability to detect all sorts of objects in the environment, resolution, accuracy, maximal and minimal effective distance, and angle of view. As can be imagined, the ideal sensor with optimum parameters for these characteristics does not exist. Therefore, a compromise between virtues and defects has to be made in order to choose the sensor depending upon the specific application for which the mobile robot is designed.

Sensors can measure internal information of the robot including the movements of its own elements (proprioceptive sensors) or external information (exteroceptive sensors). As far as the types of sensors that can be used, the most common exteroceptive sensors are rangefinders and vision systems.

39.3.1 Rangefinder Sensors

Rangefinder sensors are usually based on ultrasound or laser signals. The operation of ultrasound sensors consists of emitting a pulse of ultrasound and measuring of the time (designated time of flight) elapsed until the echo is received. The advantages of ultrasound sensors are operational simplicity and price. Therefore, one can assemble a set of sensors to cover a large surface at a relatively low price. The usual physical layout of sensors is to form a ring around the robot. The factors to consider when designing a perceptual system based on sensors of this type are the maximal and the minimal distance that covers the sensor and the dispersion angle of the wave. These kinds of sensors have several disadvantages such as multiple reflection paths, variations of sound speed, echo intensity highly decreases with the distance to the object producing the echo, low resolution due to the dispersion angle, a possible interference among the sensors of one robot and the sensors from the other robots, the dependency on the geometry of the surface of the objects producing the echo, the dependency on the acoustic characteristics and size of the reflected object, and the incident angle.

Laser sensors also provide distances to objects in front of the robot. These systems are much more accurate than ultrasound systems. Depending upon the way in which the wave is transmitted, laser sensors are classified as pulse signal or continuous signal. In the former, the distance is obtained from the time elapsed between the emission and the receipt of a pulse. In the latter, the distance depends on the difference of the phase between the emitted and the received wave.

39.3.2 Vision Sensors

Another sort of perception is one based on computer vision. Vision is usually applied in order to extract different features from the environment for localizing and mapping. Compared with rangefinder sensors, the cost of cameras has decreased, but more computational cost algorithms are required to process the information. Currently, one camera is enough to extract information from the environment. But to obtain three-dimensional information, a setup of two cameras with a stereo vision system is necessary. An alternative to obtain 3D information is to use several images taken with one camera from different positions (e.g., optical flow).

Vision systems can be used for the detection of the movement of objects too. This is done by obtaining the correspondence among some characteristics (e.g., corners, edges, and textures) that correspond to the projection of the same point in several images.

The main problem of vision-based systems is that the computation time can be too long for applications in real time as is the case of the mobile robots.

39.4 Control Architectures

The design of software for the control systems of advanced robots is usually made using a modular approach. This makes the system more scalable and, moreover, facilitates the participation of several persons on a robot design project, each of them being in charge of developing part of the software. However, there is not a single way to organize a modular structure. In the last few decades, different trends in software architectures of robots have appeared.

Early robots, in the 1960s and 1970s, used planning-based architectures. Any movement of these robots had to be planned in advance. Planners needed models to predict the results of each action. The main goal of these robots was motion, and therefore the models were maps of the environment and the planners were motion planners. At that time, the perception system of the robots was very imprecise, making it very difficult for the robots to directly construct maps. In fact, in most cases, maps were built by human operators.

The applicability of this approach, based on maps, was limited to static environments, but many environments are not static and change frequently. This is particularity true when the robots move in areas with people. Of course, the position of each person cannot be established in a static map because people move. Another important source of inaccuracies was the localization system. Localization systems of those robots were quite imprecise, and this imprecision produced a very bad performance of the system. This was because, in most occasions, motion commands were established by the planner assuming that the robot was in a certain position different from the real one.

Poor results of planning-based architectures obliged to search for other alternatives. In the mid-1980s, the reactive architectures began to be developed [4]. In reactive architectures, the use of planners and models was minimized. In fact, in most robots with reactive architectures, there were neither maps nor planners. Decisions were based on real-time information from sensors, making the creation of maps with that information unnecessary. This approach produced very good results in comparison with planning-based architectures. For instance, robots were able to move quite fast in dynamic environments, avoiding obstacles. Somehow, reactive architectures meant the rediscovering of the closed-loop control system by a scientific community that in many cases was far from the control engineering approaches.

Robots programmed with reactive architectures are able to show different behaviors as a function of the control algorithms used by the robot. In robots with reactive architectures, it is common to have several modules, each of them producing a different behavior. Simple behaviors can be also fused to produce more complex behaviors. Behaviors are a consubstantial element with reactive architectures. In some occasions, the term "behavior-based" is used as a synonym of "reactive," when referring to robot software architectures.

Reactive architectures meant a significant advance in the development of robots, although not everything was positive in reactive architectures. They also have some drawbacks. Behaviors of robots with reactive architectures usually do not include the consecution of an explicit goal as in the planning-based architectures. For instance, a classical task of robots with planning-based architectures was moving the robot from one position to another by minimizing some function cost, as distance or time. Nevertheless, this type of tasks could not be usually solved with reactive architectures. The information from the environment in pure reactive systems is limited to the direct information coming from sensors, and the robot does not use information that is not currently available to sensors (e.g., maps of another room that is not the room where the robot is currently located). This lack of global information makes the consecution of global goals difficult and produces undesirable effects as, for instance, cyclic behaviors or blocking situations caused by local minima.

In sum, both approaches, reactive and planning-based, offer some advantages, but they also show some drawbacks. Trying to get the best of both, in the mid-1990s, hybrid architectures began to appear. These architectures usually adopt a reactive approach at low level, the modules closer to sensors and actuators, and a planning-based approach at high level. That means that motion-control loops are closed at low level producing different behaviors, and at the same time it is possible to reach planned decisions based on models. Reactive modules make short-term decisions in local areas (e.g., immediate movements in the area close to the robot) and planning modules make mid- and long-term decisions in global areas (e.g., future movements to distant areas). Reactive modules permit effective motions of the robot, even in complex and dynamic environment, and planning modules permit the consecution of explicit goals.

The main difference between hybrid architectures and planning-based architectures is the autonomy of low-level modules to make its own decisions. In planning-based architectures, low-level modules work in open loop with respect to the environment. The control loop is closed through the planner. In hybrid architectures, low-level modules are quite autonomous to make decisions. This change also modifies the functions of the planners in hybrid systems compared to planning-based systems. In full planning systems, a planner might supply very detailed orders, but in a hybrid system, the orders provided by a planner are higher level commands and they do not include many details because the details are fulfilled by the reactive modules. For instance, in a hybrid system to have motion plans that only include the route that the robot must follow is enough, making it unnecessary to include the exact position of the robot at each time. In this case, the reactive modules are in charge of selecting the best trajectory, avoiding possible obstacles that could find the robot in the planned route.

Many hybrid architectures have three tiers [7]. The lower tier is composed of modular skills able to process information coming from sensors and control the robot in a reactive mode, providing a repertoire of behaviors. At the higher tier lie the planners—in most cases there is only one planner—and in the middle tier there is a sequencer, which is in charge of activating skills according to the plan produced by the planner. Other hybrid architectures have two tiers [3]. In this case, the upper tier includes planners, sequencers, and other modules that process information on models and knowledge databases.

39.5 Localization and Mapping

To be able to reach goals in the environment, a model of this environment is needed in order to know where the robot is localized at that moment and where the goal is. There are different options to represent the environment that will condition the way the robot localization is carried out and how the planning to reach one point from another is performed. The model of the environment can be made a priori or while the robot is moving, though the information from the sensors is always needed to validate the robot's position.

Next, different ways to represent the environment and to localize positions within it are going to be enumerated, focusing on the SLAM system as the current trend for the map construction and localization while the robot moves.

39.5.1 Uncertainty

The construction of maps and the estimation of the localization of the robot is affected by the errors from the sensors. For that reason, the algorithms generally used are based on methods for the estimation of parameters with uncertainties that use probability functions. One way to estimate the map and localization is to use the Kalman filter [12]. In the context of mobile robotics, where the movement model and the observation model are typically nonlinear, the optimality assumptions of the Kalman filter are not met. On the other hand, the Kalman Filter assumes that the probability densities for the errors of both models are Gaussian, which is not normally met either. For these reasons, the extended Kalman filter (EKF) [10] is used, which allows to take into account the nonlinearities of the processes when they are linearized around their operation point. In spite of this, this filter still has some drawbacks, such as the Gaussian assumptions for the models and its computational cost.

Another method for the estimation of maps and the localization of the robot is the particles filter, a method to estimate the state of a system that is changing with the time. It was proposed in 1993 by Gordon et al. [8] as a bootstrap filter to implement recursive Bayesian filters. Basically, the particles filter consists of a set of samples (the particles) and values (weights) associated with each of these samples. The particles are possible states of the process that can be represented as points in the state space of this process. In relation to localization, the starting point is an initial set of particles in the environment. Each particle represents a certain probability for the robot to be localized in that position. Through a process of observation, movement, and re-sampling, the probabilities for the robot to be localized in this position are updated. In the re-sampling process, the particles that increase their probability are divided, and those with low probability are eliminated, generating new particles around the localizations of maximum probability.

In case of topological maps, the most frequent way to manage uncertainty are algorithms based on partially observable Markov decision processes [13].

39.5.2 Environment Representations

A global map is defined as the one representing the whole area of possible movements of the robot. This is a data structure that provides information about the environment, localizing static obstacles and allowing an overall planning for long distances.

The maps can be divided into two main types: topological representations and metric representations, as shown in Figure 39.5.

The topological representations consist of logical connections between representative elements of the environment, while the metric ones localize the objects in specific positions within the map, which usually has a scale and a coordinate system associated to measure distances and angles.

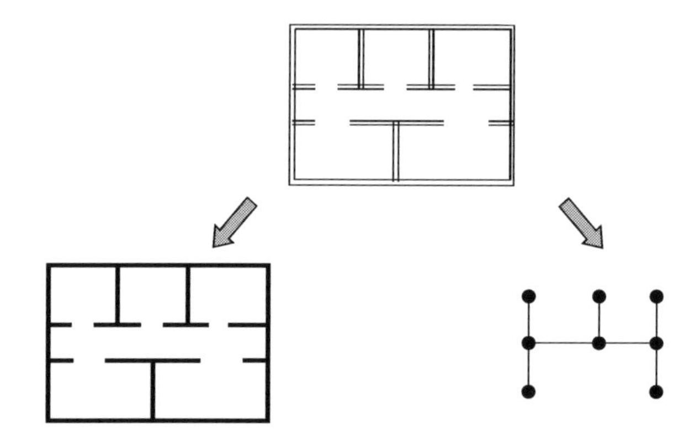

FIGURE 39.5 Metric (left) and topological (right) representation of an environment (center).

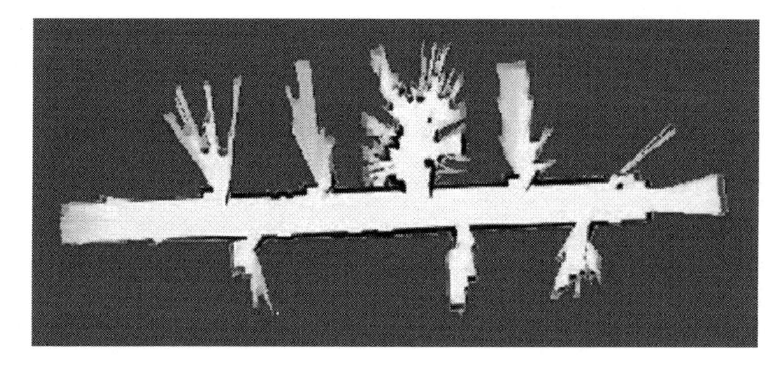

FIGURE 39.6 Example of an occupation grid from the measurements of a laser.

39.5.2.1 Geometric Representation

The geometric representations model the environment in which the robot moves through the coordinates of the geometric elements making the environment up.

The most frequently used techniques are based on occupation grids and on those maps that shape the environment by geometric representation of elements, such as lines, polygons, landmarks, and the like. Both maps can be constructed from the information of the robot sensors. Figure 39.6 shows an occupation grid obtained by integrating the measurements from a laser rangefinder, and Figure 39.7 shows an example in which the environment is modeled by using lines from these measurements [1].

39.5.2.2 Topological Representation

The topological representation allows to build paths in a simple and natural way, similarly to the indications that humans give each other to reach a specific localization [2]. It is based on the information related to the different elements localized along the path and the relative movements around them. Therefore, it is a useful representation for a robot with a navigation system based on behaviors.

The topological representations consist of the alternation of nodes and arcs. Several ways of topological representations exist, which can be grouped into region maps and motion maps. In region maps, each node is associated with a region of the environment, and the arcs are paths connecting one region to another. Figure 39.8 shows a region map extracted from an occupancy map [22].

In motion maps, the world is represented in an explicit way through a graph in which the nodes are related to significant information from the environment (for example, sensory events) and the arcs are associated with motion behaviors that lead from one node to another. One of the first implementations of this representation is the one by Kuipers and Byun [14], shown in Figure 39.9.

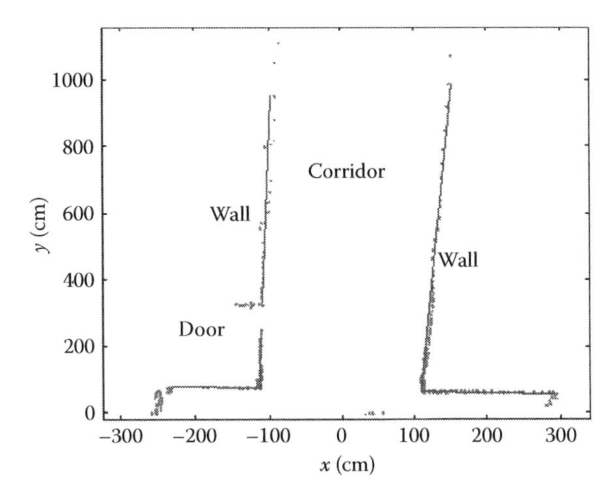

FIGURE 39.7 Model by lines from the measurements of a laser.

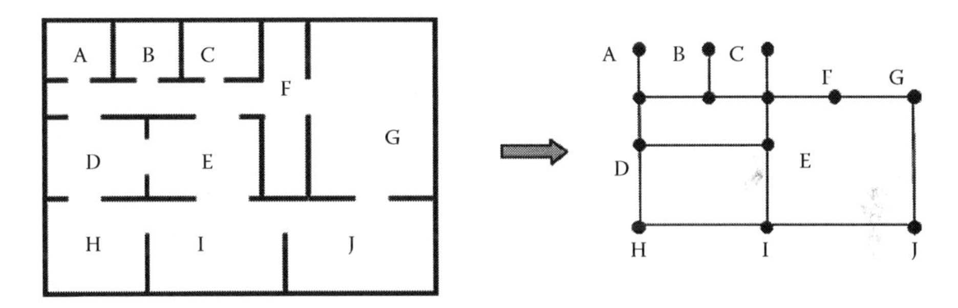

FIGURE 39.8 Region map extracted from an occupancy map.

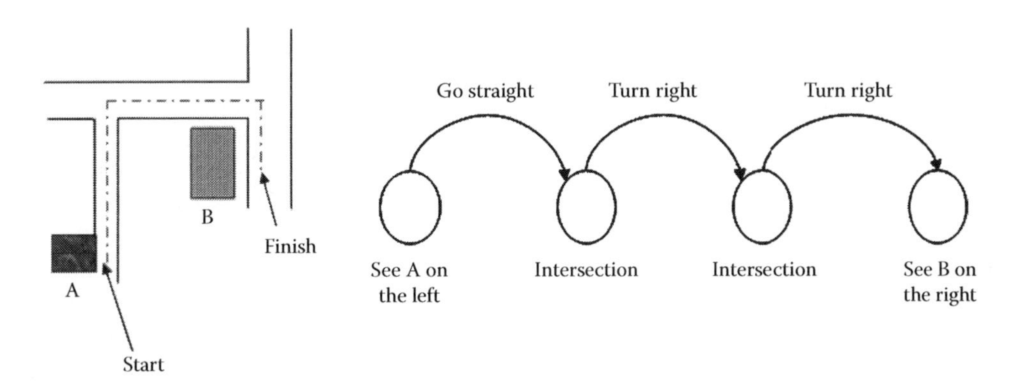

FIGURE 39.9 Topological representation based on motion maps.

39.5.3 Localization

The localization problem consists of identifying the place and pose in which the robot is localized. In order for a robot to perform tasks such as path following or map generation, it must be able to determine its localization. Geometric localization means to establish position and orientation with respect to an absolute reference system. Topological localization means to establish the situation of the robot in the graph associated to the topological map.

39.5.3.1 Localization without Information from the Environment

One way to know the position of the robot in the map consists of estimating its new position, from another one that is already known, through the use of the information coming from proprioceptive sensors. This technique is called odometry. It uses incremental information, which have accumulative errors. This is the reason why it is not valid for long distances. Errors are the consequence of inaccuracies in the robot's locomotion system (different wheel diameters, misalignment of the axes) or can be produced by the elements of the environment (slippery floor, lost contact with the terrain, and the like).

39.5.3.2 Localization with Information from the Environment

When there is a priori information available from the environment, either from an occupancy map or from the position of landmarks, the robot's position can be calculated from this information.

In the case of positional information coming from beacons, whether natural or artificial, the most frequently used technique is the triangulation, which estimates the position of the robot from angles and distances from beacons. In the case the robot moves in an external environment, the triangulation can be performed with the support of the satellite system known as GPS.

If there are no beacons available, other techniques consisting on comparing the measurements from the sensors with a map of the environment are applied. The robot takes some measurements with sensors, such as laser rangefinders or sonar, and, after translation and rotation movements, it estimates its position within the environment. Scan matching is a very common technique used to compare occupancy maps with the measurements obtained by rangefinder sensors.

39.5.4 Simultaneous Localization and Mapping

Although the localization and mapping functions for the robot have been analyzed separately, it must be remarked that they require a parallel solution, since the robot needs a map to know where it is, but it also needs to know where it is moving around to create the map. Therefore, if there is no a priori information, both tasks must be carried out at the same time to find a correct solution. This is the purpose of the SLAM systems: simultaneous localization and mapping for robots navigation [16]. Most common SLAM algorithms are based on two techniques: the EKF and the particles filter (e.g., FastSLAM).

The classical approach to SLAM is based on the EKF. In this case, the information included in the maps consists of the position of a set of environment landmarks. The state vector estimated by EKF includes not only the localization of the robot but also the position of the landmarks of the environment (i.e., the map). The drawback of this method is the Gaussian assumptions for the observation of the environment and robot movement models.

FastSLAM [18] uses a particles filter to solve the SLAM problem. In this case, each particle has associated not only a hypothesis of the robot localization but also the information of a possible map.

An important problem of SLAM systems is the problem of data association. If, by mistake, the measurements obtained from a landmark are associated to another landmark, then the system provides inaccurate estimations.

In SLAM, the uncertainty in the localization of the robot and the landmarks is reduced by the repetition of measurements in the same places where the robot was localized previously. It implies that the robot must go again through the same places in order to build a correct map.

39.6 Trajectory Planning

The path-planning methods are strongly linked to the representation of the environment used. There are different path-planning techniques, such as graph-based planners, and those based on occupancy maps.

39.6.1 Planners Based on Graphs

These planners can be applied to both topological and geometric representations from which a graph can be extracted. They are based on graph-search algorithms. The algorithms analyze the nodes to choose the best path according to a specific criteria. In case of graphs with strong connections (completely connected), as the case of regular grids, the computational cost is very high. Therefore, it is more appropriate to use algorithms based on pruning and branching. The Dijkstra algorithm [5] is the most used for parameterized topological graphs and the A* algorithm [6] is one of the most common methods used to find optimum paths in robotics. Its advantages are the possibility to apply it to any configuration space representation (e.g., Cspace) transformable to a graph and its remarkable usefulness in a very connected regular grid, where the computational cost is highly reduced.

39.6.2 Planners Based on Occupancy Maps

One of the most used techniques on this type of representation is the one based on potential fields [15]. With this technique, the robot moves toward an object that exerts an attractive force to it simulating a gravitational pull. The further the goal is from the robot, the greater this force is. However, this is not the only force acting on the robot. Since the ultimate goal of this technique is to avoid the obstacles that are close to the robot, they generate a repulsive force to it, which is higher when the robot is closer to the obstacle. So, the motion of the robot is governed by the vector sum of all the forces acting on it, as shown in Figure 39.10. This technique ensures the movement away from obstacles and toward the goal once they have been dodged.

In theory, the robot is considered as a particle affected by the force fields. There is also a problem of local minima, since when a force field is generated, it is likely that the sum of any of the magnitudes is zero. This can cause the robot to stop.

Other types of planners used on grids are the planners based on wavefront propagation [20]. The basic idea is to consider the representation space as a conductive material, so that a stimulus spreads from the starting node toward the nearby cells as a waveform, until the goal is reached.

It is important to remark that these types of planners have an interesting side effect since not only is the path from the starting node to the ending node obtained but also the optimal path from all points to the goal cell.

The best-known algorithm for path planning based on waves propagation is the so called Trulla, developed by Ken Hughes [9]. It is based mainly on the assumption that the resulting map is similar to a potential field (Figure 39.11) where the optimum path simulates the value of the sensors.

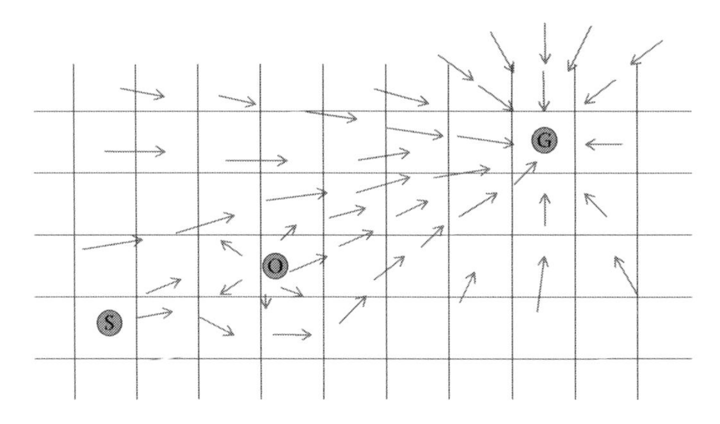

FIGURE 39.10 Example of a trajectory based on potential fields.

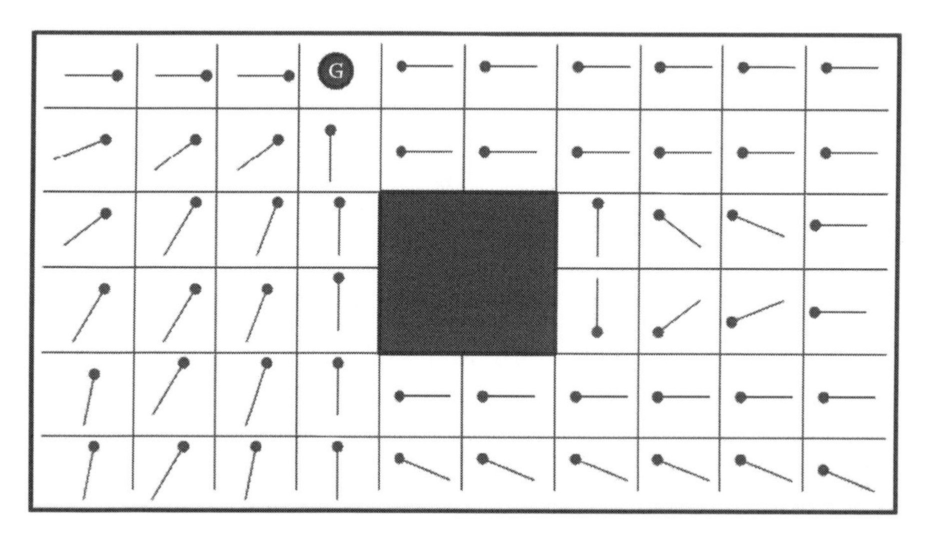

FIGURE 39.11 Example of path planning using the Trulla algorithms.

An interesting advantage of these planners is that they allow to take into account undesirable terrains, modeled as cells with low conductivity (assuming that the inaccessible places have a zero conductivity). Therefore, it is possible that the optimal path contains undesirable conditions versus a longer path with more desirable ones.

39.7 Conclusions

Nowadays, mobile robots technology is in constant evolution. In this chapter, the main concepts and tendencies in robotics have been presented. As locomotion systems mostly used, legged and wheeled robots have been introduced, describing the kinematics of wheeled robots in a more detailed way. The perception of the internal information and the surrounding world is an important issue in robotics. Perception involves a set of algorithms that transform the data from the sensors into a representation that could be used for decision making. Range finder and vision sensors have also been commented.

The control architectures appear as a way of organizing the performance of the robot. The design of software for the control of advanced robots is usually made using a modular approach. There exist several approaches and trends for this purpose.

To be able to reach goals in the environment, a model of this environment is needed in order to know where the robot is localized, existing different options to represent the environment. The most important ones are the geometrical and topological representations. In the construction of maps and in the estimation of the localization of the robot, the uncertainty of the measures must be taken into account, since they are affected by the errors from the sensors, especially in the localization of the robot. The localization problem consists of identifying the place and pose in which the robot is localized using odometry or information from the environment. The SLAM appears as a solution to solve the problem of localization and mapping in parallel.

Finally, in order to establish the path between the goals of the robots, path planning methods are needed. These methods are strongly linked to the representation of the environment used. Planners based on graphs and occupancy maps are the most used.

References

1. R. Barber, M. Mata, J. M. Armingol, and M. A. Salichs. A perception system based on laser information for mobile robot topologic navigation. *Proceedings of the IEEE 28th Conference on Industrial Electronics, Control and Instrumentation*, 38:2779–2784, 2002.
2. R. Barber and M. A. Salichs. Movile robot navigation based on event maps. *FSR2001 Proceedings of the Third International Conference on Field and Service Robotics*, Espoo, Finland, pp. 61–66, 2001.
3. R. Barber and M. A. Salichs. A new human based architecture for intelligent autonomous robots. In *The Fourth IFAC Symposium on Intelligent Autonomous Vehicles, IAV 01*, pp. 85–90, 2001.
4. R. A. Brooks. A robust layered control system for a mobile robot. *IEEE Journal of Robotics and Automation*, 2(1):14–23, 1986.
5. A. Cara, G. Taylorb, and C. Brunsdonc. An analysis of the performance of a hierarchical wayfinding computational model using synthetic graphs. *Computers, Environment and Urban Systems*, 25:69–88, 2001.
6. R. Dechter and J. Pearl. Generalized best-first search strategies and the optimality of a*. *Journal of the ACM*, 32(3):505–536, 1985.
7. E. Gat. *Artificial Intelligence and Mobile Robots: Case Studies of Successful Robot Systems*, chapter Three-layer architectures. MIT Press, Cambridge, MA, 1998.
8. N. Gordon, D. Salmond, and A. F. M. Smith. Novel approach to nonlinear/non-gaussian bayesian state estimation. *IEE Proceedings F: Radar and Signal Processing*, 140(2):107–113, 1993.
9. K. Hughes, A. Tokuta, and N. Ranganathan. Trulla: An algorithm for path planning among weighted regions by localized propagations. *Intelligent Robots and Systems*, 1:469–476, 1992.
10. S. J. Julier and J. K. Uhlmann. A new extension of the kalman filter to nonlinear systems. In *International Symposium Aerospace/Defense Sensing, Simulation and Controls*, Orlando, FL, pp. 182–193, 1997.
11. S. Kajita and B. Espiau. Legged robots, *Handbook of Robotics*. Springer, Berlin, Germany, pp. 362–387, 2008.
12. R. E. Kalman. A new approach to linear filtering and prediction problems. *Journal of Basic Engineering*, 82(1):35–45, 1960.
13. S. Koenig and R. Simmons. A robot navigation architecture based on partially observable markov decision process models, *Artificial Intelligence Based Mobile Robotics: Case Studies of Successful Robot Systems*. MIT Press, Cambridge, MA, pp. 91–122, 1998.
14. B. J. Kuipers and Y. T. Byun. A robot exploration and mapping strategy based on a semantic hierarchy of spatial representations. *Journal of Robotics and Autonomous Systems*, 8:47–63, 1991.
15. S. M. LaValle. *Planning Algorithms*. The MIT Press, Cambridge, MA, 2006.
16. J. J. Leonard and H. F. Durrant White. Simultaneous map building and localization for an autonomous mobile robot. In *IEEE/RSJ International Workshop on Intelligent Robots and Systems IROS'91*, Osaka, Japan, pp. 1442–1447, November 1991.
17. A. De Luca, G. Oriolo, and M. Vendittelli. *Control of Wheeled Mobile Robots: An Experimental Overview*, pp. 181–226. 2001.
18. M. Montemerlo, S. Thrun, D. Koller, and B. Wegbreit. FastSLAM: A factored solution to the simultaneous localization and mapping problem. In *Proceedings of the AAAI National Conference on Artificial Intelligence*. AAAI, Edmonton, Canada, 2002.
19. R. R. Murphy. *Introduction to AI Robotics*. The MIT Press, Cambridge, MA, 2000.
20. J. A. Sethian. *Theory, Algorithms, and Applications of Level Set Methods for Propagating Interfaces*. The MIT Press, Cambridge, MA, 1996.
21. R. Siegwart and I. R. Nourbakhsh. *Introduction to Autonomous Mobile Robots*. The MIT Press, Cambridge, MA, 2004.
22. S. Thrun. Learning metric-topological maps for indoor mobile robot navigation. *Artificial Intelligence*, 99:21–71, 1998.

Index

A

B

C

E

I

J

K

L

M

N

O

P

Q

R

S

T

U

V

W

Y

Z